Period	Dates (approximate)	…res and Ideas
		Michelangelo. "Last Judgment" (1541). *Benvenuto Cellini.* "Perseus" (1547).
Modern Age of Science (Classical period from 1600 to 1900)	1600	*Galileo.* Law of falling bodies (1602). *Harvey.* "Circulation of blood" (1618). *Shakespeare.* "Hamlet" (1601). *Kepler.* Laws of planetary motion (1609–15). *Gilbert.* "On Magnetism" (1602). *Bacon.* "Novum Organum" (1612).
	1650	*Descartes.* "Analytic Geometry" (1637) marks the beginning of the middle period of mathematics. *Boyle.* Important work on gases and chemistry. *Pascal and Fermat.* Foundations of probability theory. *Hobbes.* "Leviathan" (1651), a work in political theory. *Newton.* "The Mathematical Principles of Natural Philosophy" (1687), often regarded as the greatest scientific work of all times. Leibniz, a contemporary, is given equal credit for invention of the calculus.
The Enlightenment (18th Century)	1750	*Locke, Montesquieu, Voltaire, Rousseau, and Adam Smith* extend and develop the idea of natural law implicit in Newton's system, and apply it to government. *Bach.* "Well Tempered Clavichord" (1722). *Watt.* Steam engine patented (1769).
	1800	*Jefferson* and contemporaries apply natural law doctrine. Declaration of Independence (1776). *Gauss (prince of mathematicians)* opens era of modern mathematical analysis in his work of 1802. *Dalton.* Atomic theory (1808). *Avogadro.* Number of molecules in a given volume. *Volta and Galvani.* First experiments in electricity.
	1850	*Joule and Mayor.* Law of conservation of energy. *Darwin.* "Origin of Species" (1859). *Marx.* "Communist Manifesto" (1848). *Riemann.* Non-Euclidean geometry. *Faraday.* Experiments on electricity. *Maxwell.* "Treatise on Electricity and Magnetism" (1873). *Gibbs.* Work on statistical mechanics.
The "New Era" in science; first principles are questioned. Social and political philosophy also challenged. The new style in arts.	1900	*Einstein.* Special theory of relativity (1905). *Planck.* Quantum Theory (1903). *Roentgen.* X-rays (1895). *Becquerel, Curies, Rutherford.* Radioactivity. *Bohr.* Atomic structure (1913). *Heisenberg, Dirac, and Schroedinger.* Quantum mechanics. *Fermi and others.* Nuclear physics. *Urey, Gamow, Hoyle, and others.* New astronomy. *Freud.* Theory of the unconscious. *Cantor, Russell, Whitehead, Hilbert, and Gödel.* Abstract mathematics. *Joyce, Bartok, and Picasso* experiment with new art forms.

12-5-69

second edition

The elements and structure of the physical sciences

second edition

The elements and structure of the physical sciences

J. A. Ripley, Jr.

Associate Professor of Physical Sciences
Stanford University

R. C. Whitten

Research Scientist
Ames Research Center
National Aeronautics and Space Administration

John Wiley & Sons, Inc.
New York London Sydney Toronto

Library of Congress Catalog Card Number: 72-82979
SBN 471 72322 3
Printed in the United States of America

10 9 8 7 6 5 4 3 2 1

Preface to first edition

This book is the outgrowth of material prepared for a course in the physical sciences required of liberal arts students at Wilkes College. It is primarily an exposition of the major concepts and theories of physical science, covering both their logical structure and the empirical evidence validating that structure. It is, additionally, an attempt to set these ideas into a broad historical and philosophical context. The understanding of science and its relation to our culture requires much more than the mastery of scattered principles, a few theories, an assortment of facts, and a handful of formulas into which the facts are fitted. The reader of this book is invited to study the genesis of theories and their logical relationship to one another and to the observational evidence. Then, and only then, will he be in a position to appreciate the scope and limitations of science and the significance it has, for good or ill, on the philosophy that underlies our culture. Over a century ago Shelley wrote, in his *Defense of Poetry,* the often quoted words "The artist's task is to absorb the new knowledge of the scientists and assimilate it to human needs, color it . . . and transform it into the blood and bone of human nature." In Shelley's time, "natural philosophy"—as science was then called—was essentially closer to everyday concepts and principles that are familiar to all of us. This is no longer true. As science has evolved (particularly in the last 50 years) into a more organic unity, it has also become more complex, abstract, and indirect. This, inevitably, was one aspect of growth;

as it proceeded toward greater generality, sophistication, and power, its articulation removed scientific ideas farther from a simple reference to immediately apprehended data. The sleek machines—television sets, automatic mechanisms, and computers—can be operated with little knowledge of the principles upon which they work. Nevertheless, there still remain those large ideas of science, ideas which all thoughtful persons should understand and analyze because they have such an impact on our philosophy.

This book, like many others addressed primarily to nonscientists, concentrates upon a relatively few selected topics. These are treated with considerable thoroughness, using elementary algebra and geometry as necessary mathematical tools. With but one or two exceptions, no principle or theory is discussed without a careful statement of the evidence and logic supporting it. (I believe that no concept of science is really understood until its logical and experimental validity is appreciated.) Where proofs and derivations are required, they are given. To help those whose algebra may have become somewhat rusty or those who have neglected mathematics out of distaste or fancied inability, a section reviewing the operations of algebra is included. In addition, there is a whole chapter on the nature of mathematics and its relationship to science which, I hope, will prove illuminating to those who have been exposed to mathematics merely as a routine tool of calculation.

Throughout, great emphasis is placed on the symbolic structure of science. It is, after all, this symbolic structure that provides the correspondence to and interpretation of the external world—a map, so to speak—that enables man through science to operate more effectively in the universe. Much has been written about the "revolution in science" that has taken place since the turn of the century: relativity theory, quantum theory, atomic and nuclear physics, and modern cosmology, all of which are presented in this book. Whether the emergence of these theories represents revolution or evolution, I shall not argue. Certainly the end of the nineteenth century marked the culmination of a long development begun over three centuries earlier by Copernicus, Galileo, and Newton and extended by myriads of their contemporaries and followers. Often stressed by those discussing the changes in science that have taken place in our century are the philosophical consequences that have arisen out of one or another particular theory, such as the relativity theory, or the uncertainty principle. These are important, no doubt, and will be discussed in this book. However, in my opinion, the more profound change does not depend upon particular discoveries or theories, no matter how important these may be for technology, but is reflected instead in a more sophisticated interpretation of science by modern thinkers, both scientists and philosophers. Perhaps the shift in the climate of opinion may best be described as one that has moved away from a certain "literal mindedness" of a previous generation toward science. (There were many exceptions, of course, such as Poincaré, Mach, Duhem, and others.) Today, there exists a deeper realization that the scientific approach is only one among many approaches to "truth" and "reality." Hence, there will be found in this study continuous reference to the analogies and metaphors, which abound as much in the symbolic structure of science as they do in art, literature, religion, or philosophy. The symbolic structure of science is distinguishable from other symbolic structures that guide men's actions, primarily because it is validated differently. Ultimately, *meaning*—as guaranteed on the one hand by internal logic, ideally expressed in mathematical relationships, and on the other by external reference, ideally expressed in well-defined operations of measurement and observation—takes precedence in science over all other criteria. This emphasis on meaning explains the contrast between science as a symbolic structure and poetry, for example, whose rhythm, imagery, sound, and meaning all play their role, but whose meaning is secondary to the total response evoked. ("A poem should not mean, but be" is the way one modern poet phrases it.) The elements of simplicity, elegance, and imagery are all certainly present in the sciences—particularly as motivating forces in the minds of the creative workers, as is illustrated again and again in this book; but these play a secondary role in the ultimate acceptance or rejection of a scientific theory. "I had such a beautiful theory," Pauli, the noted physicist, is reported to have said to Niels Bohr, "but alas, it did not accord with observation." Bohr responded, "It could not have been a beautiful theory if it did not account for the facts."

Although modern physical science is characterized by its deductive approach, and although this approach is the simplest and most effective one to use in the training of professional scientists, I have chosen to present the material of science in its historical development for several reasons.

First, by starting with a discussion of astronomy, the earliest science, a student is led gradually to build up a competence to handle more and more abstract concepts. The motions of the stars and planets can be observed by everyone, and the arguments culminating in the modern theory of the solar system can be followed directly and simply in terms of what can be seen. Furthermore, the nature of indirect measurement and the power of mathematics as a logical tool can be effectively illustrated in astronomy.

Second, by reviewing the genesis of theories, we develop a keener realization that scientific principles, laws, and theories are creative constructs used to organize experience and are not merely systems of "facts" discovered by carefully pursuing a single "scientific method." Even such a simple generalization as the "law of the lever," which is so frequently studied in elementary schools, is an abstraction (friction is assumed to be absent); how much more of an abstraction, full of imagination and bold conjecture, is the kinetic theory of dancing and spinning atoms and electrons which cannot be seen.

Finally, the slow, cumulative, but always tentative, evolution of scientific ideas should be studied in historical perspective. A great deal can be learned from the struggle that once raged over the Copernican theory, both regarding the nature of scientific theories and their effect on mankind's way of looking at the universe.

Although the historical development of scientific theories structures this book, I have also tried to present each theory, even the modern ones, with adequate thoroughness on the basis of evidence and modern criticism. For example, the ideas of the quantum theory and the relativity theory have been carried, I believe, about as far as they can be, without a much more formidable use of mathematics.

I am aware that many critical problems in the philosophy of science have been touched upon only lightly. For example, for the sake of clarity and simple exposition, I have drawn too sharp a distinction between the internal logical definition of a symbol—given by the mathematics in which it is imbedded—and the "operational definition"—given by references to what are somewhat glibly termed the "pointer readings" of a laboratory instrument. P. Bridgman, the American physicist, who originated the term and pushed its use as a critical instrument, recognized that many unconscious assumptions and unsettled problems surround the concept of an operational definition. A vast complex of theory is always involved in modern measuring instruments. Those who wish to pursue this and allied problems in the philosophy of science should refer to the many books now available on the subject, including, in particular, the excellent work of Ernst Nagel, *The Structure of Science,* recently published by Harcourt Brace & World, Inc., New York.

Some of the sections in Chapter 8, especially the one Normative Theories As They Are Related to Scientific Theories, will probably be sharply criticized because I have stepped well beyond the scope of science proper. In spite of the anticipated criticism, I leave the material as it stands, with the hope of encouraging readers to think through their own point of view. My goal has been to try to place science within the unity of knowledge, experience, and insight but, ultimately, this task must be undertaken by each individual for himself. I wish to encourage this process by some tentative, but certainly not authoritative, suggestions. More than all else today, we need to develop what Stephen Spender calls the "connecting imagination."

At the end of each chapter there will be found a series of exercises and problems. As a general rule, these were selected to provoke thought and discussion rather than merely to provide material for a drill. Hence, there are relatively few problems that can be answered simply by substituting values in equations. It is felt that any teacher using this book as a text could better provide such drill material himself, appropriate to his own aims. In several of the exercises the questions asked are such that no definitive answer can be given. Again, I have thought that including such questions would give emphasis to the open-endedness of science. Many questions upon which science has a bearing will, however, never receive a final and definitive answer.

Throughout this book the mks (meter-kilogram-second) system of units has been given preference. This system makes the proper handling of equations simpler, on the whole, than any other system of units. By converting solutions to English units the student soon becomes familiar with the mks units. I have emphasized the use of approximations, not only because calculations are made easier, but also because it is important for a student of science to realize that, since all measurements contain elements of imprecision, results based upon measurement are never completely "exact" in science. Very often, just to discover the order of magnitude of a given quantity is itself a great scientific achievement.

JULIEN A. RIPLEY, JR.

Wilkes-Barre, Pennsylvania, 1964

Preface to second edition

The purpose and general tone of the first edition of this book have been maintained. In this edition the opportunity has been taken for the expansion, revision, and modification of previous topics. The section on chemistry, for instance, has been extended to include a more thorough discussion of chemical bonding, chemical kinetics, and some of the basic concepts of organic chemistry. Because of contemporary interest in "space science" and in the spatial environment of man, two new chapters on atmospheric physics and the physics of space have been added. (The material referred to is the contribution of Dr. Robert Whitten, co-author of this new edition.)

New sections have been added on lasers, the tides, quasars and pulsars, earthquake waves, and fundamental particles. In general where new developments and discoveries of significance have been made, these have either been referred to or incorporated in the text.

The number of illustrations has been increased. Questions and problems have been added to. Numerical answers to the majority of the problems are now given in the Appendix. Hopefully these changes will make the text more usable for the general student.

What to include and what to omit in a text of this nature is always a puzzling problem. For example, more space might have been accorded to that important branch of physical sciences—geology. Such an extension would have made the text unwieldy, so the additions have been restricted to the chapters on atmospheric and space physics, except for the brief sections on tides and seismic waves.

A new section on fundamental particles, which may be omitted without loss of continuity, has been introduced. Even though this section is abstract and difficult, it deals with basic ideas about the ultimate structure of our universe. The student should be given the opportunity to become aware of the "shape of" modern scientific thought even if he does not completely master the contemporary formal structure erected by scientists to "explain" the universe.

The authors, of course, remain convinced that the proper social control of technology, which becomes daily more important, requires increasingly greater understanding of the methods and the philosophy upon which science is based. Just as it is the responsibility of the scientist to weigh and to measure the consequences of his research and make them as clear as possible to the public, so it is the responsibility of the citizen living in this dominantly technological age to understand the nature of science and its relationship to other human activities and beliefs.

Many are the suggestions and criticisms received by the authors. They cannot all be acknowledged. Dr. Whitten wishes to express especial thanks to Dr. R. C. Robbins, Elmer Robinson, and Ilia Poppoff for their helpful criticisms. Dr. Ripley extends his appreciation to the interested colleagues and students at Stanford for their criticisms. To Dr. William Perkins special gratitude is due for his suggestions relative to the chapter on cosmology; to Jeffrey Johnson for his contributions on seismic waves. We wish also to thank Dr. Alvarez-Tostado for trenchant criticisms arising from many conversations.

J. A. Ripley, jr.
R. C. Whitten

Stanford, California
Summer, 1969

Contents

chapter one

Introduction

It was a scant four hundred years ago that modern science began to emerge as a separate discipline. Newton's major work, published in 1687, still carried in its title the term "natural philosophy." A century later Voltaire, Benjamin Franklin, and Goethe, all better known in other fields of endeavor, were nevertheless vitally interested in the science of their day and actually carried out scientific research. With each passing century, however, the growth of organized scientific knowledge has become more specialized and the pace of that growth is becoming more and more rapid. Science now with its institutional support and drastic technological consequences has become the dominant influence in western civilization.

Until the mid-nineteenth century, craftsmanship and technology consisted of many more or less independent skills and recipes, garnered from fairly direct experience and passed on from one generation to the next. Basic science was pursued primarily out of curiosity in an attempt to understand and discover the laws of nature. The motivating force behind scientific work was not aimed primarily at improving technology, although the existing technology did present, as it does now, much of the raw data for scientists to analyze.

By the twentieth century basic science was nourishing technology instead of vice versa. Today research and development is carried out mostly by large organizations and institutions in the hope of finding practical technical applications rather than in the hope of understanding our universe. (It is also true, of course, that the modern scientist depends heavily upon technical advances for the instruments and tools of research which he uses.)

The dominance of science in our society is therefore twofold. In the first place, technology has changed our institutions, as well as the face of the earth, in a most dramatic fashion. Secondly, the way we see, think, and act is profoundly influenced by scientific knowledge.

This giant, technology, seems to have taken on a life of its own, almost out of control of man's will and guidance, presenting promises, it is true, but also and more importantly, posing great dangers. The incredible development of weaponry, the large-scale degradation of the environment—land, sea, and air—are proceeding inexorably and at an accelerated pace. "Reverence for nature as well as for man" is losing ground rapidly in spite of such quiet voices as those of Einstein and Rachel Carson. There is no escape for modern man into a natural environment untouched by technology. For this reason alone, it is necessary for all men to examine the warnings of those who are concerned with the larger implications. Inevitably both "basic science" and technology are being funded and supported to a greater and greater degree by institutions that have *particular* goals in mind. Too often the experts within a special field of applied science are tempted to regard their work in terms of the special goals for which support is being given. It is incumbent, therefore, upon the larger public to acquire enough understanding so that priorities can be judged and caution applied. The lessons of modern physics reinforce what biologists have long known—the processes of nature are intricately connected with one another. Disturb them *here,* and the effects *there* may be vastly greater than we can foresee in our impatience to exploit some technological advance promising immediate but possibly temporary results.

But it is not merely the control and the guidance of technology which require a better and more widespread understanding of science. Complementary and closely related to this reason for incorporating science into a liberal arts education is the need to understand "basic science" because it has come to influence the way we see and interpret things—our very attitudes and actions towards the world about us. Religion, philosophy, political theories, governmental decisions, even the arts and the humanities are all affected by the science of our age.

We are living in a new revolution of scientific thought as profound as was the Copernican revolution of the sixteenth and seventeenth centuries. The impact of new discoveries and new theories is still being absorbed. Relativity, quantum theory, nuclear energy, discoveries in astronomy, molecular biology, electronics, cybernetics, and space science lead not only to vast technical changes, but to new, more imaginative ways of looking at the universe. The discoveries and new conceptions have been so surprising that our critical guards have been all but ablated. Speculations—wild, woolie, and weird—are encouraged and defended in the name of science, which currently admits to so many "unorthodox" or "unclassical" possibilities. For example, interest in astrology is growing (more words are printed giving astrological forecasts in daily newspapers than are printed on science); flying saucers with "little green men from Mars" are seriously reported; magic rituals and prophesies are encouraged; one semi-mystic group refers to itself by the name, "Scientologists"; over a million copies have been sold of *The Dawn of Magic* by Pauwels and Bergier, which suggests that the alchemists had some profound and secret insight into the true nature of the universe. Not that these hypotheses and speculations need be met always with scorn or contempt. Indeed, they may contain suggestive or psychologically helpful ideas! But they are not to be accepted as scientific, precisely because they are not subject to test or to disproof. (How can one possibly *disprove* the possibility that a flying saucer from some other planet has at one time or another landed on the earth?) If we are to discriminate between what science is and what it is not, and determine the degree of assent, belief, or skepticism we attach to each, the nature of science must be understood.

If the lack of understanding science leads to a certain gullibility on the part of the public, it also leads to an inability to grasp the power and the beauty of scientific generalizations as *one* route to comprehend natural processes and events. Like any artist, the scientist seeks to find and to express connections between phenomena which seem to be totally unrelated. More than this, the scientist seeks to find the largest possible body of coherent generalizations under which all phenomena can be explained and understood. He creates and uses mathematical structures and models to express the connections, even when the mathematical structures or models only faintly resemble the immediately experienced phenomena. For example, the mathematical equation of a wave form is found to apply to sound, to light, to radiant heat, to earthquakes, to water waves, to electrons, and to atmospheric disturbances. It is true that the equations seem forbiddingly abstract, but their generality and their fruitfulness in prediction is carried in these abstract symbolic expressions. A scientist can hardly proceed without using models or mathematical symbols of one kind or another. There is, of course, some danger in that he will take his models too literally. Even greater is the danger that the public will take them too literally. (Have we not all seen films of the Bohr atom with its tight central nucleus and spherical electrons flying about in their planetary orbits—a useful heuristic model, but very misleading.) It is much easier to be sophisticated about Picasso's abstract representation of the human figure than the physicist's model of the atom which is never directly observed.

Nature is vastly more complicated for the modern scientist than it was for his 19th century predecessor. There are some ways in which we can accomplish miraculous feats of engineering, such as the Apollo launchings. Great strides have been made in connecting one branch of science with another. But the more knowledge we achieve, the greater becomes our recognition of how little we know, and how tentative are our theories about the fundamental processes of the universe. The scientist, in moments of contemplation about his subject, is led to "a reverence towards nature," much as the artist, the poet, or the philosopher.

William Blake's often-reproduced portrait of Newton, (see Fig. 1.1) showing him with his back to the universe while he concentrates on a trivial diagram, was a satirical comment on the uses of science in the 19th century. Alas, science is being as much misused today. Modern technology spawns as much brutality, cruelty, and ugliness as in Blake's lifetime at the height of the industrial revolution—with an added possibility that humanity itself may be exterminated.

Nonetheless, science at its best remains one of the noblest free creations of

Fig. 1.1. William Blake's ironic portrait of Sir Isaac Newton. Courtesy of the Tate Gallery, London.

the human spirit—an activity which affects all our other values. The patient pursuit of truth, tolerance towards dissent, free inquiry into any subject, the creation and testing of alternative concepts, willingness to question authority, and a basic optimism—all of these are prescribed by the scientific community, even if they are too often ignored.

In this book we will trace the development of the physical sciences. The major concepts, principles, and methods will be discussed with attention to the role of science in the context of man's intellectual history. Obviously, selection from the vast supply of material on the physical sciences is necessary. This is not an introductory text for scientists, but all of the more important concepts and laws will be examined. With few exceptions, no theories will be discussed without attention to the evidence which validates them. Sometimes there will be no way of avoiding a minimum of mathematics as the only economical method of showing the logical structure of a theory. But only the simplest algebra will be used, and the operating rules for using this algebra will be reviewed in Chapter 3. Many students complain that they have a "blind spot" where mathematics is concerned. This is largely imaginary. Counting, measuring, numbering, and performing arithmetic operations of multiplication and division are themselves highly abstract operations, as we shall see, but few adults are incapable of such operations. (In contrast, citizens of Rome or Athens, with only Roman numerals at their command, would have found tremendous difficulties in multiplying or dividing one three-digit number by another.) The simple rules of algebra which we shall use, including the handling of ratios and exponents, together with a few geometric relationships, are basically no more abstract nor complicated than the rules of multiplication and division that the student has already mastered.

1. THE NATURE OF SCIENCE AND THE PROCESS OF VERIFICATION

The subject matter of science and its basic concepts are in a state of constant revision. Even the methods pursued by different scientists vary greatly according to subject matter, according to the stage of the science, and according to the background and attitude of the scientists involved. Any explicit definition of science or scientific method is therefore necessarily inadequate and incomplete. Only by seeing what scientists have done and are doing, and analyzing the results can an adequate appreciation of the nature of science be gained.

But it may be helpful at the start of this book to discuss briefly some of the aims and general procedures of science, to show how they resemble those of other disciplines and activities and in what ways they differ.

Presumably the aim of all intellectual activity is to attain understanding and foresight. We seek to find out how the present resembles the past, how the future will resemble the present, or in other words, what recognizable patterns persist in the flux of events. We pursue this attempt both as a satisfying end in itself and as a guide to action and further understanding. In this, science has much the same end as all the humanities, philosophy, and religion. But while seeking the same general goal of understanding and foresight, which scientists usually term prediction and explanation, scientists use more specific criteria of judgment and of testing than do those pursuing other disciplines.

In the first place, a scientific theory must be capable of refutation. Let us return to the discussion of why astrology is not regarded today as a science whereas astronomy is. The astrologer claims that the positions of the stars at a person's date of birth and his destiny are intimately connected. Predictions can even be made as to the favorable and unfavorable days for undertaking specific ventures. Possibly some such connection does exist. Certainly many natural philosophers and even such eminent astronomers such as Tycho Brahe and Johannes Kepler cast horoscopes. (Kepler did so, however, with considerable skepticism and reluctance.) But astrology is not generally regarded as a science today precisely because its theories are not subject to any test which could falsify them. Either the horoscope is so vague and ambiguous that the predictions cannot ever be clearly refuted, or, if they are specific, the astrologers always have an "out." For example, astrologers have even predicted that a horse will win a certain race. If he does not win, it is invariably explained that it is because the starting time was a few minutes late or early. "But the stars, themselves," say astrologers, "do not lie." Thus the conclusions of astrology can never be clearly disproved. In fact theories of astrology are not even drastically modified. Most of the systems used today go back to the one constructed by Ptolemy, without any significant change.

Perhaps this seems like an extreme example, but it is a useful one to bear in mind. Many religious, philosophic, or poetic statements may very well be convincing and appeal to men as "true." But they cannot be regarded as "scientific" unless they can be so stated as to be subject to disproof. In contrast, a theory such as the phlogiston theory was a good scientific theory, precisely because it could be and was eventually shown to be false. The most fruitful theories in science are those which make predictions which can be tested. The best scientific theories rouse many scientists to the activity of trying to find its refutation and its limits. A false theory often proves more useful in stimulating a chain of investigation, than one which commands general assent but is so vague as to lead to no clear test.

Of course, in the early and "natural history" stages of a science, (for example the biology of a previous generation, when the science consisted largely of classifications) the science may not abound in theories which can be put to such a test. But, in general, the theories of the physical sciences have progressed to this point. The physical sciences, embracing astronomy, physics, chemistry, and geology, are sometimes referred to as the "exact sciences." This is very misleading. Their preeminence is only in small measure due to the relative precision and reliability of prediction in these fields. More important is their clear logical structure, and the possibility of verification which entails their susceptibility to disproof.[1]

Perhaps we have labored too much this point of "disproof" when general opinion about science seems to emphasize "proof." Partly this is out of impatience with such commonly expressed statements as "Science proves that . . ." or "It can be scientifically shown that . . ." This is nonsense. Empirical science can prove nothing. Some theories are "well-confirmed" or "verified," but they are not *proved.* In the physical sciences, for example, most of which have reached a more or less advanced state of development, there are no theories, generalizations, or "laws" formerly regarded by many as certainties, which have not been drastically modified in this century. In fact, it might not be too much of an exaggeration to state that the purpose of a theory, at least within pure science, is to lead to investigations which will eventually bring about its demise. The progress of science is marked by a continuous cycle: A theory or generalization suggests new observations or experiments which disprove or modify the original theory, thus leading to a new theory or generalization, which again must be tested, etc. This does *not* imply, however, that an older theory when replaced is completely discarded. More often the newer theories, of greater generality, subsume many of the older and well-established laws. Einstein's theory does not invalidate the application of Newtonian physics to engineering practices. Even portions of the older Ptolemaic earth-centered system which seems to have been totally replaced by the sun-centered Copernican system are retained. One simple example may be cited: In the Ptolemaic system, the morning and evening stars are identified as the single planet, Venus—an identification by no means simple to make without a well-developed model of the solar systems. Copernicus, of course accepted this identification. The development of the sciences are cumulative or progressive in a way which distinguish the sciences from other human endeavors.

Let us now look a little more closely into these dual processes of disproof and verification as they apply to the physical sciences. You are asked to keep these comments in mind as you study the later chapters.

Note first of all that the physical sciences aim at statements of the widest pos-

[1]See K. Popper, *The Logic of Scientific Discovery,* Harper & Row, New York, 1959, for a careful discussion of the criterion of the falsifiability of scientific theories.

sible generalization about the universe. (Presumably if some discovery were made, let us say, in biology which seemed to contravene an established "law" of physics, not only would biology be affected, but physical theories themselves would have to be recast.) Now any generalization is a universal statement. It goes beyond the facts presented by immediate observation. It is therefore hypothetical, a guess—you might almost say—as to how the world is best to be interpreted. Such statements of the widest generality are rarely verified by direct observation but rather by the consequences that can be deduced from them. For example, Newton's first law of motion states that any body in a state of rest or in a state of motion will continue in that state of rest or in a state of rectilinear uniform motion forever unless acted on by some outside force. If we look about us, observing the heavens above and the earth beneath, no object is ever observed which is either in a state of rest or of uniform motion. Every object in the universe is indeed acted upon by outside forces. This "law of motion" is not therefore established by observation, nor can it be so verified. Rather it is a hypothesis, which along with other hypotheses, leads to certain *consequences* which can be put to experimental test. Furthermore, it is not often a *single* generalization or law, but a whole nexus of interlocking and coherent hypotheses and interpretations which are tested through the predictions and consequences to which they lead.

Perhaps the logical structure and the validation of statements in the physical sciences can best be expressed by considering the similarity to a mathematical system. Those laws or hypotheses of the broadest generality play a role in science similar to the role of the axioms or postulates of mathematics. From a few postulates, various conclusions or theorems are drawn by logical implication. Thus in the most advanced sciences we have a set of basic laws or principles A, A′, A″, etc., which entail consequences B, B′, B″, etc. which are of an infinite number. The verification of A, A′, A″, etc., nearly always consists of testing a large variety of conclusions B, B′, B″, etc. through observation or experiment. It is an obvious logical fallacy (the fallacy of affirming the consequence) to say that we "prove" the validity of A, A′, A″ from the factual validity of B, B′, B″, etc. On the other hand, if observation or experiment entails the falsity of any one of the conclusions B, B′, B″, etc., we must abandon or modify one or more of our original laws or hypotheses from which the conclusions are drawn.

In summary we can say that the great generalizations of science, no matter how well verified by present knowledge, cannot be thought of as representing absolute "truth." For, although science may be said to be in constant pursuit of elusive "truth," final certainty is always a giant step away. With every new "discovery," with every new law or generalization, new problems are generated. The most widely accepted theories of yesterday have often faded into what we now regard as superstitions. And every theory, every law of science is subject to modification and reinterpretation. This was the meaning that Einstein intended to convey in his often repeated statement, "Although no amount of experiments, no amount of observation will ever prove my theory to be true, one experiment, one observation may prove it false."

Notable also is the great difference between pure mathematics and a physical science, however close they may be in structural appearance. For, in mathematics, as we shall see in Chapter 3, we are never concerned with whether the postulates from which we start are true, nor whether the theorems that are deduced fit the facts or have any relationship to experience. In contrast, in the physical sciences, the validity of every statement is directly or indirectly subject to the touchstone of experienced data.

The above remarks will become clearer when they are illustrated by the examples that we shall analyze. Here, it is well to emphasize two more aspects that we shall study in detail. One of these involves the use of definitions; the other has to do with the nature of experiments.

The nonscientists, undertaking the study of physics, may become somewhat impatient at the emphasis laid upon the necessity of precise and complete definitions of all the terms that are used in science, particularly in cases where he feels he understands the meaning of a word that is a part of his everyday vocabulary and also is used in a highly technical sense in physics. (For example: Why distinguish between "speed" and "velocity?") But precise definitions are absolutely essential in physics for two reasons: (*1*) the logical structure of the science requires it; and (*2*) the possibility of an unambiguous appeal to the facts also demands it.

It will be helpful to keep in mind two types of definition used in the sciences: the so-called "operational definition," which means that a concept is defined in terms of some very clearly stated directions about observing or measuring (for example, the second of time may be defined in terms of the successive crossing of the meridian by a star); and the formal definition—one of logic—which means that a concept is defined in terms of other concepts. These, in turn, are either "operationally defined" or are defined in terms that *ultimately* will be operationally defined. Speed, for example, is logically defined in terms of distance and time, where both distance and time are each operationally defined in terms of directions regarding their individual measurement.

Thus science, on account of the nature and precision of its definitions, is on the one hand distinguished from philosophy, in which there always appears to be great argument regarding the meaning of terms; and on the other hand, by the ultimate necessity of having to refer to experience, it is distinguished from mathematics, in which the terms are formally, not operationally defined.

With respect to the experimental aspect of the physical sciences, we need emphasize only one or two general points now, postponing a more rigorous discussion until later when we analyze examples. Experiments are always guided by theory. They are a way of asking nature a question. In order to ask a question, some problem must have occurred that can be stated with some degree of clarity. The experiment usually answers a question that falls into one of three types. (*1*) Is a hypothesis valid in a certain particular situation? Can a previously held hypothesis or law be extended beyond the range previously tested? (*2*) Are two or more quantities related and, if so, what is the law of their relationship? (*3*) Which of two or more hypotheses has the greater validity? In all these cases, more or less elaborate theoretical considerations are almost inevitably present.

2. SCIENCE AND THE HISTORY OF IDEAS

In science, as much as in other fields, each generation is apt to smile at the "errors" and crude absurdities of previous generations. Even in matters that we might think could be settled by the simplest type of observation, mankind is constantly plagued by errors or strongly held conceptions and opinions. A striking example of this can be found in some of the early drawings of microscopists which show the human sperm cell in the form of a homunculus or a little man. Another example occurs in Leonardo da Vinci's drawings. Leonardo ordinarily was extremely precise in his observations and in his anatomical sketches. But Sarton[2] reports that he was so much under Galen's influence that, in some of his sketches, he drew certain details which just were not present at all. Galen's explanation of the functioning of the heart called for small holes in the wall separating the auricle from the ventricle. These were required in Galen's theory to allow the blood to pass from one chamber to the other. Leonardo, in Sarton's words, was "so obfuscated by Galen's theory" that he proceeded to "observe" and sketch these "invisible holes" in his anatomical drawings.

Science is not therefore a smooth process of arriving at certain conclusions, based on "objective" observations. Rather its development is marked by legions of wrong guesses, faulty observations, tenaciously held theories, and even errors in logic. Usually it is extraordinarily difficult to confirm or disconfirm a particular theory. It took a century to settle the controversy over the nature of light, that is, to reach a decision between Newton's theory that light is composed of particles and Huyghen's alternative theory that it is a wave phenomenon. Again, Lavoisier, the eminent French scientist who formulated the first modern table of chemical elements, included in his table "caloric" or heat among the elements. Long after it was felt that the caloric theory involved serious difficulties and possible contradictions it maintained its sway because no verifiable alternative theory had been proposed.

[2]George Sarton, *Six Wings: Men of Science in the Renaissance,* Meridian, New York, 1966.

It is therefore useful to study the historical development of scientific theories.

For example, we shall consider in some detail the "Copernican Revolution," which represented a new view, displacing a great cosmological theory integrated with religious, political, and social beliefs. Such a study not only makes clearer the nature and structure of strictly scientific theories, but may also help the nonexpert, who is interested in other fields of knowledge, by leading him to consider continuously the nature, structure, and validity of other theories and beliefs such as democracy, Marxism, and Freudianism.

Furthermore, theories and beliefs, whether religious, political, ethical, social, or aesthetic, do not each arise in a vacuum and go their separate ways but are closely related and mutually dependent. "The relationship between scientific findings and man's general views, are indeed deep, intimate and subtle," as Oppenheimer has expressed it.

As an aid in making one's own connections between the various facets of the development of ideas, and in order to keep aware of these possibilities, a table of chronology has been provided at the front of the book, which lists many of the more important scientific ideas in juxtaposition with developments in other fields of thought. It is suggested that this chronology be referred to as a map of the history of ideas. References will also be given to recommended reading, which should both intensify and extend, beyond the limits of this book, the pursuit of a study in special fields of interest.

3. QUESTIONS

1. What is the difference between an experiment and an observation?

2. What is meant by the term "a scientific explanation?"

3. List as many "philosophical assumptions" as you can think of which underlie scientific investigations.

4. A dictionary definition is usually circular, i.e., each definition is defined by other words in the dictionary. How is such circularity escaped? How do you define a straight line? Relate your discussion to the concept of an operational definition as discussed in this chapter.

5. How is the language of science distinguished from ordinary language?

6. What is meant by "reality?" Is an atom "real?" Is a conceptual entity that cannot be directly referred to experiment, observation, or to measurement scientifically valid? Think of various examples, such as the "center of the earth," "absolute time," a completely "isolated" physical system, the exact "time of death" of an individual.

7. Look up the term "Occam's razor" if you are not already familiar with it. What is the justification of applying this principle to scientific investigations?

8. Contrast a photograph, a painting, a map, and a scientific model. What is the function of each?

9. Where do the rules of arithmetic come from?

10. Why is it "forbidden" to divide a number by zero?

11. Can you, at this point, cite any specific evidence or argument for believing the earth travels about the sun rather than vice versa? If not, what are the grounds for your belief? Are they better than the grounds for the Medieval conviction that the sun traveled about the earth?

12. Can you think of any reason other than your own consciousness for distinguishing the "direction" of time? How is "past" different from "future?"

13. Ask a group of persons to draw a diagram of a gibbous (that is, three-quarters) moon and note how many of the group illustrate it incorrectly.

14. The terms "cause" and "effect" are slippery ones. In formal discourse, they are not often used by physical scientists. Criticize this statement: Since the first frost *invariably* comes *after* the leaves have commenced to change color, the color changes in leaves cause winter.

15. A certain tribe in Asia has believed for centuries that an eclipse is caused by a dragon swallowing the sun. When the natives make much noise by banging on cymbals and bells, the dragon gets frightened and invariably regurgitates the sun. Criticize this belief.

16. Is a "fact" a proposition (that is, a sentence connecting concepts in an unambiguous manner and purporting to be true or false), or is it an immediately sensed experience? Give several examples of facts and analyze them from the point of view of their validity, their universal acceptability, and their interpretation. Can "scientific facts" ever change?

17. The term "frame of reference" is one that has arisen from science, in particular out of the theory of relativity. What does it mean? Consider what you mean by "up" and "down." Is your meaning the same to all observers on the earth? Would it be acceptable to an observer on Mars, or to an observer on a planet revolving around a star other than the sun? How would you propose to make the meaning clear to *all* observers? Relate what is meant by a frame of reference to the discussion of "operational definitions."

18. What is logic? Look up the definitions of induction and deduction. If proposition A (or set of propositions A) implies proposition B (or set of propositions B) does it follow that B implies A? Under what circumstances does A imply B *and* B imply A? Think of several examples.

19. Assume A implies B. If B is false, what does this "prove logically" about A? If B is true, what does this "prove" about A? If A is false, what does this prove about B? Is a scientific theory ever proved? What is the difference between proof and verification?

20. There has been a considerable amount of attention, discussion, and controversy regarding the apparent change in the "size" of the full moon as it is first perceived on the horizon and then as it is perceived high in the sky. As an introduction to making simple measurements, devise and carry out a measurement of the angular size of the full moon as it moves up in the sky.

21. One plus one, mathematically, is equal to two. Add one cloud to another cloud and you may have but one. Does this mean that mathematical statements are sometimes false?

Recommended Reading

Karl Popper, *The Logic of Scientific Discovery,* Harper, New York, 1959. See chapters II–IV. Chapter IV discusses criterion of falsifiability.

Morris Cohen, *A Preface to Logic,* Henry Holt, New York, 1944. Chapter I includes a good elementary discussion of logical formalism in science and mathematics.

Bertrand Russell, *The Impact of Science on Society,* Simon & Schuster, New York, 1953. A light and entertaining book emphasizing empiricism.

S. F. Mason, *Main Currents of Scientific Thought,* Henry Schuman, New York, 1953.

H. Butterfield, *The Origins of Modern Science,* Macmillan, New York, 1951.

E. A. Burt, *The Metaphysical Foundations of Modern Science,* Anchor Paperback, Doubleday, New York, 1524.

John Kemeny, *A Philosopher Looks at Science,* Van Nostrand, New York, 1959.

chapter two

On size, shape, and measurement

We are so accustomed to expressing knowledge in terms of numbers that we often overlook how complicated the process of measurement may be, how many hidden assumptions may have been brought into this process, and how important it is to distinguish between very different types of measurements that are used in our daily lives and thinking.

At the same time we may overlook the effectiveness with which measurements may be used to solve what appear to be primarily qualitative problems. For example, the problem of the function of the heart and the manner in which blood flows in our bodies was once considered to be an anatomical problem whose solution could be found if enough qualitative observations on anatomical structures were carried out. Some measurements made by Harvey, a contemporary of Galileo, led to the correct solution.

Harvey estimated the amount of blood ejected during the systole at two fluid ounces. If the heart beats 72 times per minute, this means that the quantity of blood ejected by the organ in the course of an hour is 8640 ounces. That is three times the average weight of the adult body.[1]

[1] Henry E. Sigerist, *The Great Doctors,* Doubleday, New York, 1958, p. 123.

Continuing this analysis, Harvey examines the various hypotheses as to where this much blood comes from: Does it come from food? Air? He rejects every hypothesis except the one that the blood circulates. Upon testing this hypothesis he found supporting evidence, and published his new theory. The new theory had qualitative implications, as in the case of Galileo's new theories of motion, which caused bitter controversies. Obviously Harvey had to make a number of assumptions for this argument to be valid. In this case the assumptions were reliable enough to lead to a reliable conclusion. Here, then, is an example of a crude but effective use of measurement, leading to a qualitative result.

Important as measurement is for the sciences, however, the measuring process by itself does not always lead directly to significant knowledge. There is an oft-quoted dictum expressed by Lord Kelvin, one of the greatest nineteenth century scientists: "When you can measure what you are speaking about, and express it in numbers, you know something about it; but when you cannot measure it, when you cannot express it in numbers, your knowledge is of a meagre and unsatisfactory kind." Few moderns, either scientists or non-scientists, would agree with Kelvin. Logic embraces more than number. Topology, a branch of mathematics, does not use numbers. And surely, few would try to assign a definitive measurement to love or beauty, although love and beauty are known in other than a "meagre and unsatisfactory" manner.

Furthermore, assigning numbers is a very tricky business. Not only does one have to make a great many assumptions, but one also has to be careful of how the results are used. Does an IQ of 150 imply that such a person is one and one-half times as intelligent as another person with an IQ of 100? Can we operate with a price index as we do with quantities in physics? Can we add densities? Or temperatures? If not, does mercury really have a density 13.6 times that of water? Is water at 100°C "twice as hot" as water at 50°C?

1. MEASUREMENT AS AN ANALOGY

Every measurement can be regarded as an assertion of an analogy between physical properties and a number system. When one says "this table is five feet long," one is asserting that the whole length is to a standard length called a foot as the number five is to the number one. There is, in other words, an assertion of a formal similarity between two lengths and two given numbers. Several different lengths, when measured, have a relationship to one another which is declared to have the same relationship as numbers do to one another. In every case there is a reference to an abstract number system.

Even at this point, a problem arises. Some measurements establish an analogy between the measurable property and *ordinal* numbers. (We shall inquire into the nature of number systems in the next chapter. Here, it need only be said that the important property of ordinal numbers is, as the name implies: they are ordered by the relationship of greater or less or equal, and the rules of arithmetic such as addition and multiplication do not apply as they do to cardinal numbers. Cardinal numbers, not only are ordered, but they also obey all the rules of arithmetic.) Where we can ascribe an order to a succession of quantities, but cannot attribute to that order the structure of arithmetic, we have established only the *ordinal* properties of the quantities involved and no more. Unfortunately, often we fail to distinguish between the two because the same number symbols apply to both. A typical example of a measurement that is only *ordinal* in nature is the measurement of hardness. The degree of hardness is established in terms of which substance will scratch which other substance. Since diamond scratches glass, it has a higher number than glass. This is an ordinal relationship. These numbers do not imply multiplicative or additive relationships, as they would if the numbers to which the analogy were made were cardinal numbers. If a substance A has a hardness of four, it is not "twice as hard" as substance B with a hardness of two. Other examples that might be cited are the magnitude of a star, or the IQ as a measure of intelligence.

In contrast to ordinally related properties, the more fundamental measurements of the physical sciences are those that establish an analogy between measurable properties and the *cardinal* numbers. (The measurable properties of a body may be thought of rather roughly as "those which are changed by the combinations of similar bodies."[2]) For example, length is a fundamental measurement expressed in cardinal numbers; two lengths can be added together; a length may be divided in half; two lengths can be compared with one another because they are multiples of a standard unit.

Another characteristic of any fundamentally measurable property is that the property either remains constant or we have knowledge of the kinds of variation to which it is subject and can allow for these variations. Thus, our standard of *length* must be such that it will not change from day to day or as it is moved from one position to another. If we use a metal rod as a standard of length, we must know, for example, what corrections to make for changes in temperature. Therefore, the process of accurate measurement may become quite complicated if this type of invariance is to be assured. (Today, in place of a standard platinum rod, which formerly was used to define the unit of length—when held at a specified temperature—the standard that is becoming more widely accepted is the wavelength of a specifically designed type of light. This is considered to be subject to less variation than the platinum rod originally set up as a standard.)

The second fundamentally measurable quantity used in the physical sciences

[2]Norman Campbell, *What is Science?*, Dover Publications, New York, 1952.

is *time.* Units of duration have all the necessary properties that allow us to operate upon them by the rules of arithmetic: they can be added, divided, and multiplied. The problem of establishing a unit of time involves also the problem of invariance. The unit of time is taken as the second, which is defined as 1/86,400 part of a day. But the day itself, defined as the period of rotation of the earth between two successive transits of the sun across the meridian, has been found to vary. Therefore, instead of the variable day, the astronomical *mean* solar day was used as a standard for a long time, it being assumed that the period of revolution of the earth about the sun was invariant. (Since even the mean solar day varies, the *ephemeris* second, a fraction of the tropical year 1900 has been in use since 1956.) As in the case of length, there is a tendency to substitute a new standard for the old. In this case the time of vibration of atoms under specified conditions is now frequently used as a standard in place of astronomical time, on the hypothesis that the former is less variable.

The third unit that *might* have been chosen as fundamental is weight. Weight appears to have all the necessary characteristics for a fundamentally measurable property. However, as we know so well in this age of satellites, weight is not an invariant property of matter. Even on the earth weight varies appreciably from place to place. Therefore, weight is not used as a fundamentally measurable unit of the physical sciences. Instead we use *mass,* a property very closely connected with weight (since, on earth, weight is exactly proportional to mass.) We shall postpone an exact definition of mass until we reach the discussion of Newtonian physics, but we may state as a preliminary definition that the mass of an object is that quantity of matter which can be measured in terms of its weight at a specific point on earth. The international standard is a piece of platinum-iridium, maintained in Paris, which is defined as a kilogram. It has now been proposed that the mass of an atomic particle should be substituted for the arbitrarily chosen mass at Paris.

The three measurable properties we have discussed, namely, *length, mass,* and *time,* are used as the most fundamental measurements in physics. The units we shall use, employed throughout most scientific literature, are those of the metric system. We shall use the *meter* as the standard unit of length, the *kilogram* as the standard unit of mass, and the *second* as the standard unit of time. As the unit of weight or force, which is so often closely associated with the unit of mass, we shall use the unit called the *newton.* For those who have not had any practice in the use of the metric system, the following relationships may make the units more familiar in terms of common experience:

A meter is a little longer than a yard.
A second is the same in the metric as in the English system.

A kilogram has about the same mass as a 2.2-lb weight.
A newton is approximately the same as the gravitational force (weight) of ¼ lb.

(For a complete listing of units and conversions, see Appendix I.)

2. THE LARGE AND THE SMALL; THE QUICK AND THE SLOW; THE MASSIVE AND THE MINISCULE

One by one, let us examine what we know of the basic measurements of the "scale" of the universe. From our day-to-day experience we have an immediate sense of what we mean by a yard or its approximate equivalent, a meter. These measurements are within the scale of our own size. We are able to grasp easily, in a directly experienced way, the meaning of one hundredth of a meter or one hundred meters. When we go much beyond this, things become somewhat vague as far as immediate experience goes. We can distinguish between one thousandth of a meter and two thousandths of a meter by eye; but when we begin to talk, as we must in modern science, of millionths, billionths, or fractions smaller than this, our conceptual imagery breaks down. Similarly, in this day of the jet plane, we can give meaning to distances, such as from Tokyo to London, in terms of thousands of miles (or thousands of kilometers, which implies millions of meters), but the vast distances which we must measure in astronomy cannot be readily translated into experientially meaningful numbers. We are almost forced to use such analogical devices as light-years as the standard. It is expected that we can more easily conceive of a distance in terms of the time it would take light to travel in one year at 186,000 miles per second, than in terms of a million billion miles.

To understand the range of magnitudes to which we must refer, we will use the notation of the scientists—which is as close to everyday experience as any other notation we can devise. We shall attempt to give a "picture" of the various scales of sizes in the universe, in terms of numbers, condensed into a symbolic notation, namely, in powers of ten.

We define 10 as 10^1, 100 as 10^2, 1000 as 10^3, and so on; 10^{-1} as $1/10$, 10^{-2} as $1/100$, etc. In addition, 10^0 is defined as equal to 1.

In this system we have a convenient way of denoting very large and very small numbers. For example, one million is equal to 10^6, one billion is 10^9, and 1000 times one billion is 10^{12}. Similarly, one millionth is denoted as 10^{-6}, one billionth as 10^{-9}, and one thousandth of one billionth as 10^{-12}. (See Chapter 3 for a more complete description of this method of defining numbers.)

Let us glance at the scale of *sizes* which we recognize as marking the characteristics of our universe in Chart 1.

CHART 1. THE APPROXIMATE ORDER OF MAGNITUDE OF SELECTED DISTANCES (IN METERS)

10^{25}	Distance of some of the intermediate galaxies (since a light-year is about 10^{16} meters, this distance is about a billion light-years)
10^{20}	Radius of the Milky Way
10^{16}	Distance to the nearest star (4×10^{16} meters or 4 light-years)
10^{10}	Distance of earth to sun
10^{6}	Diameter of the moon
10^{3}	One kilometer (5/8 of a mile)
10^{1}	Ten meters (32.8 ft)
10^{0}	One meter (39.37 in.)
10^{-5}	Diameter of an intermediate-size organic cell
10^{-7}	Wave length of light (about 5×10^{-7} meters)
10^{-15}	Diameter of nucleus of an atom

Similarly, our directly experienced conception of time is strictly limited to a certain range. Our pulsebeat is not very far from a second; the pendulumlike motion of our arms and legs in walking is in this same general range. Musical time (duration of notes) also does not depart from this general range (from about 1/64 sec to about 1½ sec). We have some directly experienced appreciation of durations as large as several years. But the range both beyond and below this can only be understood conceptually. It is difficult to comprehend either time durations as long as a billion years or as short as a billionth of one second. Yet it is beyond this range by considerable amounts that phenomena in the universe must be measured. We again set up a chart (Chart 2), in terms of powers of ten, which may give some appreciation of the range which must be covered.

CHART 2. THE APPROXIMATE ORDER OF MAGNITUDE OF SELECTED TIMES (IN SECONDS)

10^{17}	Age of the earth (about 4 billion years)
10^{15}	Time since age of dinosaurs
10^{10}	Time elapsed since age of Copernicus
10^{7}	One year (about 3.15×10^{7} sec)
10^{5}	One day (8.64×10^{4} sec)
10^{0}	One second (approximately one heartbeat)
10^{-4}	Time for one vibration of a high-pitched note
10^{-10}	Time for light to travel through lens of the eye
10^{-20}	Time for innermost electron to "revolve around nucleus" of heaviest atom

Finally, let us look at Chart 3, which summarizes the range of *masses* within which we express the laws of modern physics. Again our experience starts with the human scale. Our bodies have a mass of around 60 to 70 kilograms, the smallest differential in mass that we can distinguish by direct feeling is about one gram (one-thousandth of a kilogram), and the largest masses that we see moving about in everyday experience are in the range of several thousands of kilograms (freight trains or ocean liners). The extension of the scale of masses beyond and below these points can be seen in the chart.

CHART 3. THE APPROXIMATE ORDER OF MAGNITUDE OF SELECTED MASSES (IN KILOGRAMS)

10^{30}	Mass of the sun (2×10^{30})
10^{25}	Mass of the earth (6×10^{24})
10^{20}	Mass of the moon (7.4×10^{22})
10^{10}	Mass of the Empire State building
10^{0}	One kilogram (weighs about 2.2 lb)
10^{-10}	A droplet of fog
10^{-25}	Mass of a heavy atom
10^{-30}	Mass of an electron

These charts emphasize the restricted range, relative to the dimensions of the "universe," within which we can make *directly* meaningful observations. How difficult it is for the human being to find "meaning" in the concept of time greater than three generations, or less than one-tenth of a heart-beat, or of a spatial dimension greater than the distance of earth to sun or less than one-millionth of a grain of sand. Most certainly, the meaning that an atom has a dimension of approximately 10^{-10} meter differs greatly from the assertion that the length of a seconds pendulum is close to 0.98 meter. In the latter case we have a fairly direct method of measuring by a comparison with an immediately experienced scale. In the former case all kinds of indirect and highly theoretical assumptions must be made. (For this reason it is not surprising that some magnitudes are changed by large amounts within a comparatively short period in the history of physics. For example, in the last 15 years, the estimate of the age of the solar system has been extended from about 4 billion years to more than 8 billion years.)

3. DERIVATIVE MEASUREMENTS

In addition to the measurements and standard units discussed here, namely those of length, time, and mass, many important measurements and units are used in the physical sciences, which are defined in terms of the fundamental ones. These are measurements or units that are expressed as ratios or multiples of the fundamental units in combination. Thus, for example, velocity is length divided by time (meters per second), density is mass divided by volume (kilograms per cubic meter), and pressure is force divided by area (newtons per square meter).

Areas and volumes have a peculiar place because they may be defined either as derived units (in terms of lengths) or as directly defined units (in terms of unit areas or unit volumes). Although volumes and areas have all the additive properties of the fundamental units, other derived units often lack them. For example, the combining of a quantity of mercury with a quantity of water does not yield a density of the combination equal to the sum of the densities. Even in the case of velocities we must devise certain rules of combination. The velocity of a person rowing north at 6 miles per hour in a stream flowing east at 6 miles per hour does *not* produce a total velocity of 12 miles per hour. The special rules that apply to these combinations of the derived quantities must be considered separately.

One other point needs to be made with respect to derived quantities. They all express some numerical relationship between two quantities that is either asserted as an assumption or as a result of experience. For example, if we define density as weight (or mass) per unit volume, we assert that weight is proportional to volume (for instance, doubling the volume of water will double the weight). Conceivably, it could be that weight increased faster than volume.

A systematic study of various important shapes, figures, and volumes led the Egyptians through their studies of mensuration to conclude that certain relationships invariably existed between areas (and volumes) and the bounding lengths that enclosed these areas (and volumes). The Greeks, not being content with observed and empirical generalizations, proved that on the basis of a limited set of assumptions, many areas or volumes can be expressed in simple terms. Among the more important are the following.

Area of a "rectangle": ab where a and b are the sides.
Area of a circle: πr^2, where r is the radius and π is a constant.
Area of a surface of a sphere: $4\pi r^2$.
Area of a triangle: $\frac{1}{2}ab$, where a is the altitude and b is the base.
Volume of a parallelepiped: abc.
Volume of a sphere: $\frac{4}{3}\pi r^3$.

From the above relationships we can already guess that any area must be expressed in terms of *squared* lengths, any volume must be expressed in terms of *cubed* lengths, and that we may also require some other constant such as π or $\frac{1}{2}$.

4. THE MEASURING PROCESS IS "OPERATIONAL" AND NOT MATHEMATICAL

Already the reader may have gathered that assumptions of a nonmathematical nature must be made as soon as one undertakes a measurement. These assumptions (often unrecognized) are in the nature of "directions" as to how one goes about careful measurement. They become part of the "operational definitions" of length, time, or other quantities.

For example, in measuring the area of a "flat" surface in the shape of a rectangle, all kinds of operational assumptions must be made. The sides must be "straight." How is a straight line to be judged? It does no good to say a straight line is the shortest distance between two points; this merely removes the problem one step, the determination of the shortest distance. Is a stretched rope the "shortest distance?" This involves assumptions regarding tensions, quality of rope, etc. Is (as we most commonly assume) light to be regarded as moving in a straight line, enabling us to use "line of sight" as our criterion of straightness? This again is assuming physical principles that have nothing to do with mathematics. We know that light "bends" under certain circumstances, so we must specify the operational conditions under which we are working. Furthermore, how do we know that the angles of what appears to be a rectangle are actually "right angles," or that the surface is really "flat." (The surface of a pond or a lake is not, we know, really a plane surface but only approximately so—although we would not be far off in measuring the surface area of a lake using Euclid's geometry of planes, yet we would find that a gross error would result if we tried to calculate the surface area of an ocean, such as the Pacific, using the geometry of plane surfaces.)

All these, and many other assumptions must be examined before we apply our mathematics to actual measurements. Mensuration is, in consequence, not strictly a part of mathematics. The separation into two aspects—the deductive, formal portion, which is subsumed under mathematics, and the operational reference, or definition of terms by reference to specific physical operations—becomes an invaluable aid in the analysis of the validity of any physical science. Examples of this recur throughout our studies.

Another point that also becomes clearer as we think about this distinction between a mathematical and a scientific proposition is that, although in the former we often run into exact numerical statements, in the latter any numerical statement is subject to an estimate of the range of accuracy. The use of the term "exact science" is a misnomer; by its nature science cannot be *exact.* Every numerical magnitude in science (except those established by definition, e.g., 1 meter equals 100 centimeters), is subject to an estimate of accuracy. For an idea of the accuracy within which some measurements have been made the reader is referred to Appendix I, in which certain constants are listed, together with a ± indicating the range of precision.

5. INDIRECT MEASUREMENT

If the scales of sizes already shown in the charts of length, mass, and time (Charts 1–3) are considered, some questions arise. How can we extend our measurements into dimensions which are so far from the realm of our experience? How can we measure the size of an atom, which is so small that it is well beyond the reach of our most powerful microscopes? Or, to take a different example from modern astronomy, there is a group of stars, called "white dwarfs," which are so dense that one pint of the substance of these stars has a density of about 15 tons to the pint. How can we arrive at such conclusions about a star so far distant that it takes light many years to reach the earth?

The answer is, of course, that we make such measurements in an indirect way, and that, depending on the type of measurement, we must make many theoretical assumptions to arrive at a figure.

As our first illustration, let us consider the ancient problem of the size and shape of the earth. Every person is taught at an early age that the earth is roughly in the shape of a sphere, somewhat flattened at the poles and having a diameter of about 8,000 miles. Where does such knowledge come from? Certainly no one has ever made a direct measurement. Indeed, the approximate diameter of the earth was calculated two centuries before Christ, at a time when no one had ever circumnavigated the world.

Without reviewing all the reasons why the Greeks were aware that the earth is spherical, in spite of its apparent flatness, we shall refer only to those mentioned by Aristotle: (*1*) the shape of the shadow of the earth cast upon the moon during eclipse; (*2*) the fact that new constellations become visible as one moves north or south over the face of the earth, and that the altitude of any one star increases as one proceeds north; and (*3*) on a calm day at sea the lower portion of a distant ship is not visible while the upper portion can be seen.

But, assuming that the earth is a sphere, how can its diameter be measured? The earliest recorded attempt to give fairly accurate results (within about 5%) was made by Eratosthenes. He observed that at one point in Egypt (near modern Aswan), on the longest day of the year, the sun at its highest point penetrated to the full depth of a deep well. This established the vertical position of the sun with respect to the horizon. On the same day at Alexandria, which is 500 miles due north of Aswan, an associate measured the angle that the sun makes with the vertical when it is at its highest point.

This angle, as shown in Fig. 2.1, was found to be 7½°.

The mathematical analysis is now quite simple. Let x be the circumference of the earth. Then

$$\frac{x}{500} = \frac{360}{7\frac{1}{2}}$$

which yields 24,000 miles as the circumference.

(See if you can state all *mathematical* rules underlying the above calculation.)

Now let us examine this example critically to see what nonmathematical assumptions must be made.

(*1*) Light travels in a straight line.

(*2*) Both the well and the plumb line establishing the "vertical" direction to the center of the earth or, in other words, both are at right angles to the horizon.

(*3*) The earth is spherical.

(*4*) The sun is so far away that its rays may be considered parallel. (Can you suggest an improvement on the method used, which would avoid some of the error involved in assuming the sun to be so far away as to render its rays parallel?)

(*5*) All measurements are accurate to the required degree. (What per cent of error will result if the measurement of the distance is in error by 50 miles, if the 7½° angle is in error by ¼ of 1°?)

(*6*) Alexandria is due north of Aswan. (What difference does this make?)

(*7*) Alexandria and Aswan are both at sea level.

Notice that all these conditions have to do with the physical conditions of the measuring process and are independent of the mathematics. Even if the mathematics is completely correct, great errors can creep into the conclusions if there are errors either in the operation of measurement or in the physical assumptions made that connect the mathematical analysis to the physical reality. Further-

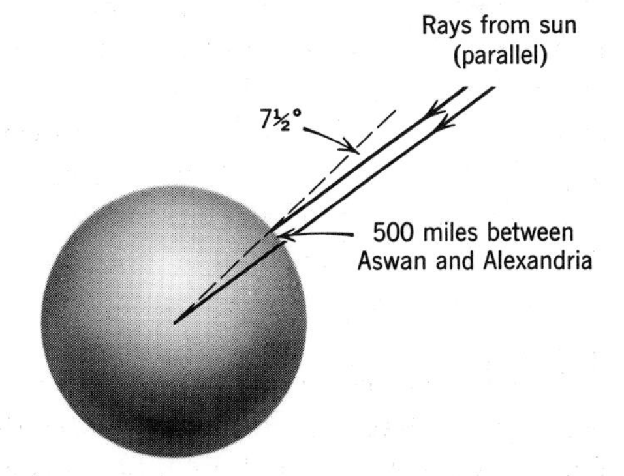

Fig. 2.1. Greek method of determining the size of the earth. Two observations are made of the sun from points 500 miles apart.

more, no matter how carefully the procedures are carried out, the result will not be "exact" in a mathematical sense.

These observations apply to all physical measurements, and become more and more important as the complexity of the measuring procedures increases. Indeed it often arises in physics that a long period of time elapses before necessary refinements are available, and we must be content with estimates that may vary by more than 100%. (It was considered a great advance when the order of magnitude of certain masses, such as the mass of a neutron, became known—where by order of magnitude we mean accurate within a power of ten.)

The second example of indirect measurement is chosen both for its own sake and in order that we may review some elementary mathematical relationships. It consists of the measurement of the distance of the moon, which also was undertaken by a Greek astronomer, Hipparchus.

Essentially the method is that of triangulation. Unconsciously we use an instinctive variation of this method, every day, in our judgment of distances with our eyes. We learn to judge the muscular sensation of changing the angle of our eyes as we look first at an object that is close to us and, then, at an object that is far away.

The basic reason why the method of triangulation works is due to three properties of a triangle. (*1*) If two angles of a triangle and an included side are determined, all the other properties of the triangle are determined, such as the third angle, the other two sides, the altitude, the area, the perimeter, etc. (Technically, we say that two triangles are congruent [identically equal] if either two angles and one side are equal or if two sides and the included angle are equal.) (*2*) Two triangles are said to be similar (that is, the lengths of the sides are respectively proportional) if their angles are equal. (*3*) The Pythagorean theorem, which states that in a right triangle the square of the hypotenuse (side opposite the right angle) is equal to the sum of the squares of the other two sides.

Without considering the origin of these three geometric rules, they can be shown to follow from the assumptions or postulates of Euclidean geometry. If, then, the physical space of the universe is Euclidean, we can with assurance apply these rules to physical problems as needed. (Actually, as we shall see later, there is a high probability that our universe is *not* Euclidean. Over tremendously large distances the sum of the angles of a triangle may not be 180°, and our universe may have to be described in terms of non-Euclidean geometry. We shall discuss this point more completely in subsequent chapters. Mention is made here to again emphasize that we must be very careful in all applications of mathematical rules, however "self-evident" they may appear.)

Take the simplest type of problem. Suppose we wished to measure the height of a flagpole that could not be measured directly. Most of us are probably already familiar with the method used, but it is wise to review the process to see exactly what is involved. Suppose that our only equipment is a protractor for measuring angles, a small ruler, and a tape measure by which we can measure horizontal distances. (It is assumed that the flagpole is vertical and that the ground is horizontal, so that the angle of the flagpole with the ground is actually a right angle.)

After measuring the distance from the flagpole base to some convenient spot at a distance from the base, we then sight from this point to the top of the flagpole to determine the angle. These are the *physical* acts of measurement, which must be done with all suitable precautions.

The determination of the height of the flagpole is now simply accomplished, using the property of similarity of triangles. We can draw a small right triangle of any convenient size, with the base angle set equal to the angle of elevation at which we have sighted the top of the flagpole. Then, as shown in Fig. 2.2, the ratios of the corresponding sides being equal, we can calculate the height of the flagpole from three measurements: the measurement of the base distance from the point of sight to the flagpole, and the measurement of the two sides of the similar triangle which has been drawn.

Look carefully at the simple but general ideas contained in this procedure which underlie that branch of mathematics called trigonometry. The single basic idea is that of similarity of triangles. Trigonometry insofar as it is used for measuring purposes is simple. It merely rings the changes on this one idea, combining it with a few elementary rules from algebra and geometry. As stated before, if two angles and one side of a triangle, or two sides and one angle, are known, the other sides and angles can be found.

It is easiest to start with the right triangle, as we did with the flagpole problem. An extension of the method to other kinds of triangles readily follows because any other kind of triangle can always be analyzed in terms of two right triangles.

Instead of *drawing* a small similar triangle and making measurements, it is

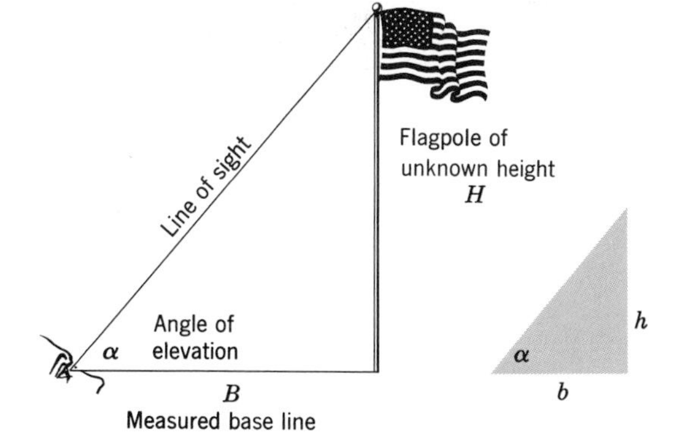

Fig. 2.2. Left, actual triangle. Right, similar triangle (as drawn, any size). The triangles being similar, $H/B = h/b$ or $H = B \times h/b$.

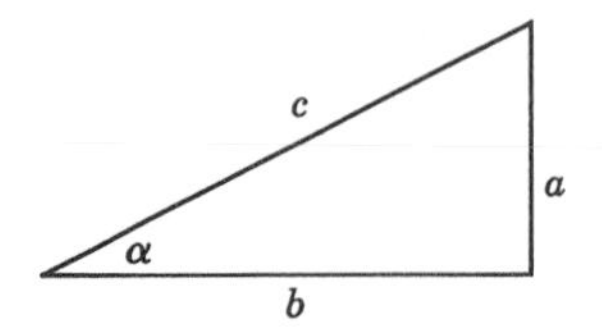

Fig. 2.3 A geometric definition of trigonometric functions: $\sin \alpha = a/c$; $\cos \alpha = b/c$; $\tan \alpha = a/b$

more accurate and more convenient to use ratios which already have been established with the necessary precision for all angles. These are given in trigonometric tables. Three ratios are convenient to work with and the names of these ratios should be memorized. In a right triangle, for any angle α the ratio of the *side opposite* the angle to the *hypotenuse* is called the sine of the angle ($\sin \alpha$); the ratio of the *adjacent side* of the angle to the *hypotenuse* is called the cosine of the angle ($\cos \alpha$); the ratio of the *opposite side* to the *adjacent side* is called the tangent of the angle ($\tan \alpha$). See Fig. 2.3.

(Before actually using any tables, if the reader has not had any practice in trigonometry, he would be wise to draw a number of right triangles with a protractor and make the appropriate measurements to find the sine, cosine, and tangent for a number of angles, such as 30°, 45°, 60°, and 80°. In this way a "feeling for" the meaning of these trigonometric ratios, as well as for the range of their values will be experienced.)

To extend this method to other than right triangles, consider Fig. 2.4. The height of a balloon is to be determined when all that is measured are the two angles of sight to the balloon from the extremities of a measured base line. Two right triangles are constructed from the figure as shown; letting X be the unknown

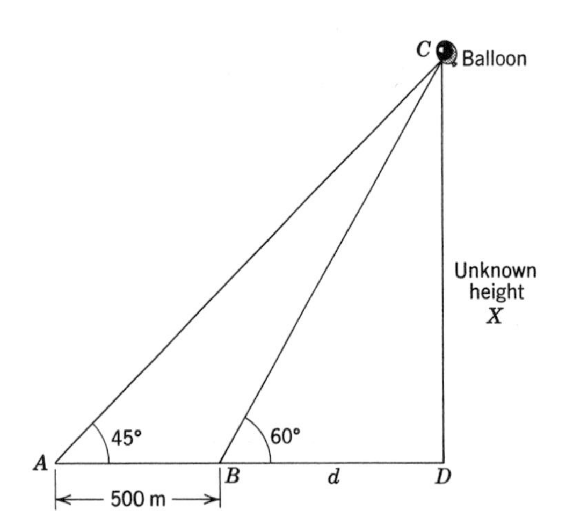

Fig. 2.4. An example of indirect measurement by triangulation.

height and d the unknown distance as shown. There are now two equations: $\tan 45° = \frac{X}{500 + d}$ and $\tan 60° = \frac{X}{d}$ and these are quickly solved.

Can we use this mathematical tool to solve any problems of finding distances, particularly astronomical distances? The answer is a theoretical yes. But as a practical matter, the necessary precision in physical measurements may be lacking. In astronomical distances, the triangles are often so elongated that the necessary precision of angular measurement is not obtainable using a base line on the earth. However, the distance to the moon can be found by using a triangle with a base line on the earth. The general method is illustrated in Fig. 2.5. Note that the determination of the base line requires the previous determination of the earth's circumference as already discussed.

Because the angles used in astronomical calculations are often very small owing to the distances involved, it is useful to introduce a small-angle formula for trigonometric functions. For angles less than 5° the approximation equation:

$$\tan \theta = \sin \theta = \frac{2\pi}{360} \theta$$

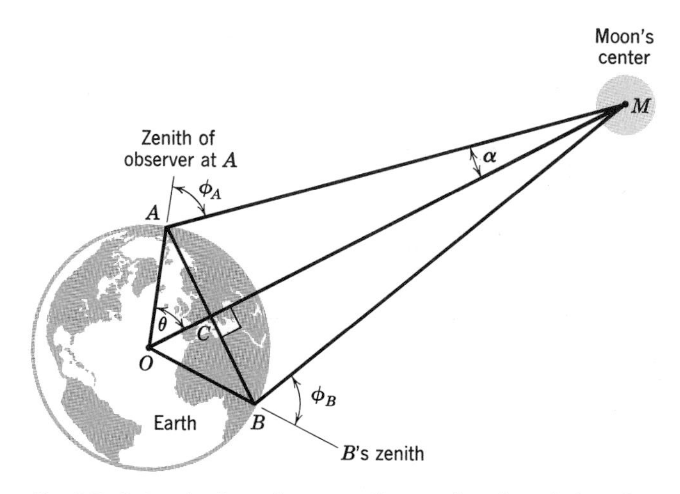

Fig. 2.5. Determination of moon's distance by triangulation. For simplicity in this schematic diagram, zenith angle of moon ϕ_A as observed at A is taken equal to ϕ_B as observed at B. Knowing distance A to B and earth's radius, the central angle and hence θ is known. In above diagram $\alpha = \phi_A - \theta$. Then $\tan \alpha = AC/CM$. $AC = R \sin \theta$ where R is radius of earth. CM (or OM, which is very nearly the same) can now be found.

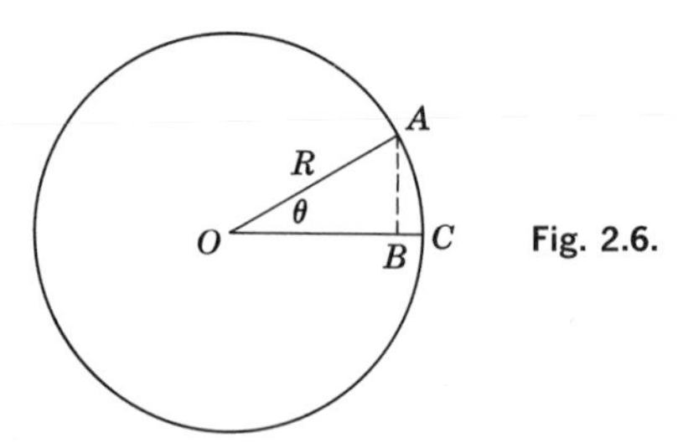

Fig. 2.6.

is accurate within a considerably smaller percentage error than the percentage error in measuring the angle. This equation can be readily derived from Fig. 2.6 and the following assumptions:

Assumptions:

1. $\frac{\theta}{360} = \frac{\text{arc AC}}{2\pi \text{R}}$ (from definition of a degree)

2. $\sin \theta = \frac{\text{AB}}{\text{R}}$ (definition of $\sin \theta$)

3. arc AC = AB (this is a *false* assumption but may be used for *small* angles without introducing any significant error)

It is left for the reader to derive $\sin \theta = \frac{2\pi\theta}{360}$ by algebra from the three assumptions listed. (Also show that for very small angles $\sin \theta = \tan \theta$.)

Should objection be raised to using an approximation formula in an "exact" science like astronomy, it should be pointed out that it is impossible to avoid approximations as soon as mathematics is applied to physical phenomena. Measurements are always approximate. Furthermore, except for some specific angles, the sine, cosine, and tangents, as well as π, are numbers with an endless row of decimals. Rarely does a scientist use for a value of π more than four decimal places. The percentage error in stopping at three or four decimal places usually is negligible compared with the errors of measurement.

The general method should now be clear. If a base line is given, from which two lines can be drawn to a given object, and if the angles of those lines with the base line can be determined, then the altitude of the object above the base line can be determined. The subject of trigonometry as the science of measuring triangles is developed out of the few basic ideas we have outlined. (There is another aspect of trigonometry connected with the idea of periodicity, which is much more subtle, and whose significance is fully as important as triangulation.)

Using the basic ideas developed, can we measure the distance of the moon, the sun, and any star? Surely all we require is a base line between two observation points and the angles that the light from the object in question makes with this base line.

Essentially this is the method pursued in calculating the distance of the moon. (We shall see why the method poses great difficulties in the case of the sun and almost insuperable problems in the case of stars.) Two given points at a known distance apart are used to sight the angle the moon makes with the horizon at these points. Using these data the distance of the moon is calculated.

We shall not examine the many assumptions that must be used in arriving at the determination which follows from Fig. 2.5. However, it is important to realize the extent of the errors that may result from a very small error in measuring angles. Suppose, for example, that it is impossible for the two observers to distinguish any angular difference of less than 0.1°. This will be the case if the base line is too short. Other methods of finding distances to celestial objects must then be found.

So it is that triangulation with a base line on the earth cannot yield an adequate method of measuring the distance of the sun. Yet the Greek astronomers, using logic and mathematics, were able to come to some approximation of this distance. Let us look at a final example of the indirect method of measurement; a way of determining the distance of the sun from the earth by finding the ratio of this distance to the distance of the moon from the earth (which is already determined by methods shown above).

Figure 2.7 shows the essential points in this analysis. The moon is observed

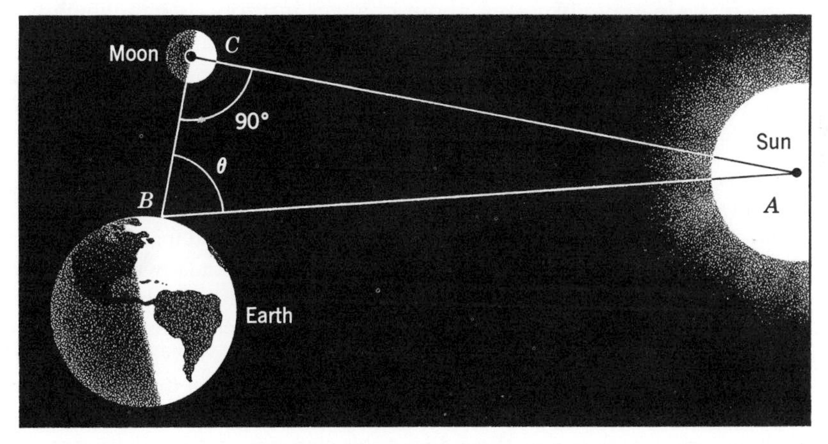

Fig. 2.7. Determination of relative distance of the moon and the sun from the earth. The angle θ is measured. From this, the angle at the sun (at A) can be determined and hence the relative distances, from $\sin A = 2\pi A/360 = BC/BA$.

just before sunset or just after sunrise when it is at its first or last quarter, and when one half of its surface is lighted by the sun. At this point, careful observations are made of the angular distance between the sun and the dividing line between the dark and the light half of the moon. This angle is shown as angle *ABC*. Since the moon is assumed, at this point, to be half lighted by the rays of the sun, the angle *BCA* is considered a right angle. The problem now resolves itself into a determination of the ratio of *AB* to *BC*. If the distance *BC* is known, then the distance to the sun can be calculated.

This method was used by Aristarchus who found the angle *ABC* to be 87°, and who thus concluded that the sun was 19 times as far from the earth as the moon. The angle in question is difficult to determine with any accuracy, and is actually about 89° 50′, which yields a result of almost 400 times the distance of the moon from the earth.

6. ON THE SIZE OF THINGS

Before leaving the topic of measurements, one further comment should be made about magnitudes. Changes in scale of magnitude may bring about radical modification in those laws of physics that we use in dealing with bodies that we can directly observe and handle (macroscopic laws). We cannot necessarily assume that these macroscopic laws will inevitably apply to submicroscopic particles such as electrons and other elementary particles. When we come to the study of nuclear physics (Chapter 21), we shall see how many modifications are required.

Changes in scale may also bring about changes in properties in which we are interested. Galileo pointed out in one of his dialogues that animals cannot be greater than a certain size because, although their weight is proportional to their volume, the strength of their bones is proportional to the cross-sectional area of the bones.

> (From these considerations) you can plainly see the impossibility of increasing the size of structures to vast dimensions either in art or in nature; . . . it would be impossible to build up the bony structures of men, horses, or other animals so as to hold together and perform their normal functions if these animals were to be increased enormously in height; for this increase in height can be accomplished only by employing a material which is harder and stronger than usual, or by enlarging the size of the bones, thus changing their shape until the form and appearance of the animals suggests a monstrosity . . . . Whereas if the size of a body be diminished, the strength of that body is not diminished in the same proportion; indeed the smaller the body the greater its relative strength. Thus a small dog could probably carry on its back two or three dogs of his own size; but I believe that a horse could not carry even one of his own size.[3]

[3]Galilei Galileo, *Dialogues Concerning Two New Sciences,* translated by Henry Crew and Alfonso de Salvio, Macmillan, New York, 1914, reprinted by Dover Publications, New York, pp. 130–131.

Similarly there is a lower limit to the size of mammals, as has often been pointed out. Since a mammal loses heat in proportion to its surface *area,* although its weight is proportional to its *volume,* the proportion of food required to support the necessary metabolism is not proportional to weight. (A housewife need not be surprised at how much her dog or cat consumes in proportion to its weight.) A very small animal will have a large surface area in proportion to its weight as compared to a large animal and, therefore, requires a relatively greater consumption of food. A guinea pig, for example, which weighs a little over one pound, requires in 24 hours about 100 calories per pound to support its metabolic rate; on the other hand, a horse weighing about 1300 pounds requires only about 10 calories per pound of its weight. The smallest mammal, the shrew, requires so much food that it is in danger of starvation if it does not eat every few hours.[4]

Similarly an engineer cannot build two bridges of identical structure but of different size and expect them to have the same strength; the problem of doubling the size of an artificial satellite does not resolve itself into a question of merely doubling the fuel load and dimensions.

7. QUESTIONS

1. Review the argument that for small angles (less than 5°), the sine of the angle and the tangent are almost equal and are proportional to the angle itself. Draw a graph of sin θ vs θ from 0° to 5° to show that $\sin\theta = k\theta$. What is the value of the proportionality constant k?

2. What is meant by the angular diameter of the moon? Explain how you could obtain this diameter by measurements you could make yourself. (*Hint.* Hold up a coin in such a position that it just "covers" the moon. Take the necessary measurements and find the angular diameter. Use small angle relationship $\sin\theta = 2\pi\theta/360$.)

3. When Aristarchus determined the size of the earth, the two points of observation had to be exactly north and south. Explain why. (Consider difference between a great circle and any other circle drawn on a sphere.)

4. If angles could be accurately determined within 0.001°, could the sun's distance be found by observing it from two points on earth?

5. Is the sine of an angle proportional to the angle?

6. Consider a total eclipse of the sun. What is implied about relative distances of moon and sun in relation to their respective diameters? If the distance of the sun from the earth is 400 times greater than the distance of the moon, how do their volumes compare?

[4]A full and interesting discussion of these points is given in D'Arcy Thompson's article "On Magnitude," and also in a more popular essay by J. B. S. Haldane, the biologist, in his article, "On Being the Right Size," in *The World of Mathematics,* edited by James Newman, Simon & Schuster, New York, 1956, Vol. 2.

7. How would you determine the width of a river without crossing it? Illustrate your method by a diagram with specific figures and angles.

8. One method, known since ancient times (and still used by carpenters in staking out lines for foundations of small buildings), of constructing a right angle is to construct a triangle whose sides have lengths in the ratio of 3:4:5. Why must such a triangle be a right triangle? Using your ruler and protractor construct such a triangle and determine the angles.

9. A guide showing visitors a limestone cavern stated that the cavern was 15 million and three years and two months old. When questioned, the guide explained that three years and two months earlier a geologist had assured him that the cave was 15 million years old. You will readily detect the fallacy here. In problems that you are solving, do not attempt to give answers with greater accuracy than is justified by the data. Use approximations whenever they yield results of an appropriate degree of precision. Here is an example: determine the approximate distance that a six-foot person can see out over the ocean on a clear day. (*Hint.* The line of vision is tangent to the circumference of the earth; hence a radius drawn to the point sighted on the horizon is perpendicular to the line of sight. Use the Pythagorean theorem and assume the radius of the earth to be 4000 miles. Note that there is an approximation that you can properly use to save a great deal of useless computation.)

10. Approximately how wide an angle of vision can you see out of one eye?

11. As an exercise in using powers of 10, and also to familiarize yourself with magnitudes, compare the following sizes: a) man to atom; b) earth to man; c) solar system to earth. What is the general range of magnitude a man can measure directly using a ruler?

Problems

1. What is the speed due to the rotation of the earth of a point on the equator in miles per hour? In kilometers per second? What is the speed of a point at 45° latitude? What is the speed of the earth due to its yearly journey about the sun, according to the Copernican model? What would the corresponding answer be for the speed of the sun according to the Ptolemaic system?

2. Find the value of sin 30°, cos 30°, tan 30°, sin 60°, cos 60°, tan 60°, using the fact that in a 30°–60°–90° triangle the shortest side is equal to one-half of the hypotenuse.

3. By definition the "natural" unit for an angle is the radian. It is a central angle in a circle which subtends an arc of length equal to the radius of the circle. Show that one degree $= 2\pi/360$ radians; that one radian equals a little less than 60°.

4. Show that for small angles $\sin \theta$ is nearly equal to θ if θ is measured in radians. Referring to trigonometric tables, determine how great a percentage error will be made in taking $\sin \theta = \theta$, if θ is 5°. Sin 5° is given $= 0.08716$.

5. The distance of an object may be intuitively judged by the converging angle of one's eyes. Estimate what the change of angle is if one views object A at 3 meters and then object B at 6 meters. (*Hint.* Use distance between pupils of your eyes.)

6. How far away would a disk, the size of a quarter, have to be held so that it would subtend one second of arc as seen from your eye? See Fig. 2.8.

Eye

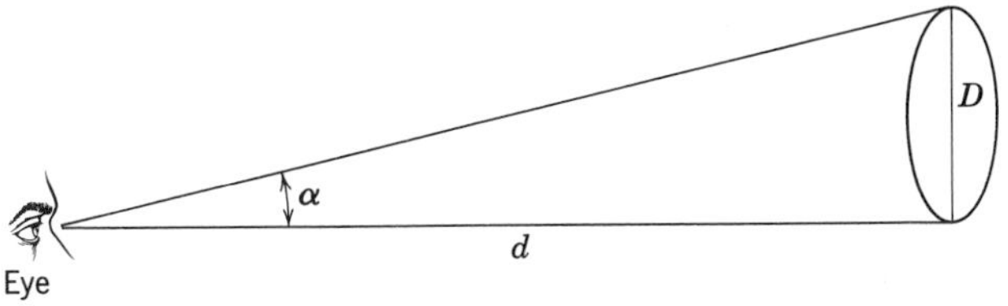

Fig. 2.8. The angular diameter of a circular object (or a sphere whose cross section is observed as a circle) is the angle subtended by the diameter D as observed by the eye. Then $\tan \alpha = D/d$. For small angles (less than 5°) $\tan \alpha = 2\alpha/360$ is a good approximation. If the angular diameter of the moon is $\frac{1}{2}$°, show that d is approximately 115 moon diameters. Observed angular distances between celestial objects is a generalization of the idea of angular diameters. Students of astronomy should learn to gauge such angular distances by eye.

7. No trigonometric tables are supplied in this book. You are expected to find the sine, cosine, and tangent of any angle greater than 10° by using a protractor and ruler. In this way you will become more familiar with the meaning of the trigonometric functions. Use your protractor and ruler to find the sin 30°, sin 60°, cos 30°, cos 60°, tan 30°, and tan 60°. Compare the values thus obtained with the values calculated on the basis of the Pythagorean theorem and the knowledge that in a 30°–60°–90° triangle the hypotenuse is double the shortest leg. Determine sin 45°, tan 45°, cos 15°, and cos 80°.

8. With a protractor and ruler determine the angle (*1*) whose sine is 0.8, (*2*) whose cosine is 0.2, and (*3*) whose tangent is 1.6.

9. It will be convenient, in studying the next two chapters, to learn how to judge angles. Estimate the angular distance between the two stars in the big dipper which are farthest apart.

Recommended Reading

Morris Cohen and Ernest Nagel, *Introductions to Logic and Scientific Method,* Harcourt Brace, New York, 1934, pp. 289–301.

Norman Campbell, *What is Science?,* Dover Publications, New York, 1952, Chapter VI. *The World of Mathematics,* edited by James Newman, Simon & Schuster, New York, 1956. Volume 2, the articles cited above.

Isaac Asimov, *The Realm of Measure,* Riverside Press, Cambridge, 1960. A very elementary book on measurements but one which is entertaining reading. It contains a wealth of material which most readers could review with advantage.

chapter three

The nature and significance of mathematics; some rules of algebra

In almost every discipline relating to human endeavor—philosophy, art, music, the social sciences, technology (including warfare), or the natural sciences—mathematics has played a significant and often a crucial role. Our primary interest will continue in mathematics as "the servant of science," but it would be unfortunate if our interest in the subject as a tool prevented us from paying proper respect to it as "queen."[1] For mathematics may be regarded in Whitehead's words "as the most original creation of the human spirit." In its freedom, its rigor, its beauty of form, generality, and conciseness of expression it remains unmatched. Although we should primarily concern ourselves with learning how mathematics is applied, we should also come to appreciate its nature and beauty.

Any real understanding of the structure of the physical sciences is quite impossible without some use of mathe-

[1] Portions, if not all, of Eric Temple Bell's stimulating book, *Mathematics, Queen and Servant of Science,* Macmillan, New York, 1951, make good reading.

matics. We shall employ in this text, only elementary mathematics, and we shall develop a good many of the general rules in this chapter.

1. THE FORMAL ABSTRACT NATURE OF MATHEMATICS

We have already seen how mathematics enters into science in an intimate fashion through quantitative measurements. We use measurements so often and so frequently that the structural reference behind those measurements tends to escape us. "Any measuring device, however simple and natural it may appear to us, implies the whole apparatus of the arithmetic of real numbers; behind any scientific instrument there is the master instrument, arithmetic [the authority of real numbers] without which the special device can neither be used nor even conceived" is the way that Dantzig[2] sums up the inevitably close relationship between measurement and the mathematics of numbers. Recognizing this, we should not overlook the fact that this is not the only way in which mathematics operates as a tool of science. Mathematics is much broader than a study of the quantitative and the measurable. It is because of its abstract or logical nature that mathematics operates so successfully over such a wide range of subjects. To this formal aspect, we shall now turn our attention.

To repeat the essential point for emphasis: mathematics is only concerned with the relationship between propositions (meaningful statements) or between collections of propositions of the nature A implies B, and has nothing to do with the truth or falsity (agreement with fact) of the propositions concerned. In mathematics we distill the logical relationships without worrying at any step in the procedure whether what we are saying is true or false (that is, whether it "agrees" with facts as known by experience or observation). In science we are, on the other hand, much concerned with agreement or disagreement with "facts," and set up the most elaborate procedures for judging such agreements. In science, every term must be precisely defined either directly or indirectly in terms of a physical operation; every conclusion must be subject to the test of observation, and all connections between statements must be logical. In mathematics only the last condition applies. This is of great advantage to the scientist using mathematics, for it enables him to handle rapidly and simply the logical connections between statements without worrying about the validity of the logical steps taken.

If we agree that the essence of mathematics is the process of deductively deriving conclusions from a few assumptions (called postulates) whose truth or falsity is not examined, we are led to question the origin of the postulates that lead to the theorems. In Euclid's day, and indeed until quite recently, it was consid-

[2]Tobias Dantzig, *Number, the Language of Science,* Macmillan, New York, 1930, p. 243.

ered that the basic assumptions or postulates from which the rest of mathematics was deduced arose from "self-evident truths." The modern point of view, largely resulting from the attention that has had to be paid to non-Euclidean geometry, is that the assumptions are arbitrary as long as they do not lead to inconsistency. The choice of postulates, from this point of view, is determined by the richness or interest of the theorems that can be derived. It is true, of course, that either theorems or postulates are often *suggested* by experience. They are not, however, in any logical sense *dependent* upon experience and may, in fact, appear to be contrary to experience and, yet, be mathematically perfectly proper and valid propositions.[3]

Let us examine the matter somewhat more closely, using the famous Pythagorean theorem. This theorem was probably known by the Egyptians as an observed relationship between the magnitude of the legs of a right triangle and the hypotenuse. The problem which appealed to the Greeks was whether this rule could be reduced to simpler and more fundamental rules, smaller in number, from which a great number of such relationships as that of Pythagoras could be derived. It was one of the notable triumphs of the Greeks that in their system of Euclidean geometry a small set of such postulates was found.

Now when we look at the Pythagorean theorem closely and examine the form of it, not in the configuration of particular numbers, such as $4^2 + 3^2 = 5^2$, but in the configuration of a more general and abstract pattern such as $a^2 + b^2 = c^2$ in which the letters a, b, and c stand not for particular numbers but for any numbers, we are already on the road of generalization and abstract form. Two immediate questions are suggested. (*1*) Can we extend the same law to more than two dimensions? (*2*) Can we conceive of any geometry in which the form $a^2 + b^2 = c^2$ will be modified to some other form such as $a^2 + ab + b^2 = c^2$?

With respect to the first question, it is an easy matter to extend the Pythagorean theorem to three dimensions, and this was of course done within Euclidean geometry. For example, the diagonal from one corner of a rectangular solid of sides, a, b, and c to the opposite corner has a length given by $a^2 + b^2 + c^2 = d^2$.

But, if mathematics does not have to be defined in terms of a model that is experienced, is there any reason why the theorem should not be extended

[3]It may be asked how consistency is established. In 1931, K. Goedel proved that the consistency of a set of postulates cannot itself be proved mathematically. It is inappropriate in this text to enter into all the fascinating implications of this discovery, but the ultimate dependence of mathematics upon some kind of reference outside itself (even if only in terms of a "metamathematical language") appears to be established. However, since the basic concepts of mathematics, called the undefined terms or the logical primitives such as "point," "line," etc., *are defined internally* within the system and not by an operational definition as they must be in the sciences, we find it convenient to preserve the distinction between *mathematics,* regarding it as a strictly formal science, and *physical science,* regarding it as being empirical in reference as well as formal in structure. Those interested might well consult E. Nagel and J. R. Newman, *Goedel's Proof,* New York University Press, 1958.

beyond three dimensions? Is there any inconsistency in conceiving a four-dimensional rectangular solid whose diagonal has the relationship to the sides given by the equation $a^2 + b^2 + c^2 + d^2 = e^2$? We cannot visualize or even draw such a figure but, provided there is no logical inconsistency, since mathematics is considered to be an abstract system, such a relationship is perfectly proper, provided a set of postulates can be established from which a Pythagorean theorem in four dimensions can be derived. (See Fig. 3.1.) This, of course, has been done by modern mathematicians; indeed it has been done for five, six—in fact, an infinite number—of dimensions. It does not actually concern mathematicians whether the extension of Euclidean geometry has any practical applications. It does turn out that four-dimensional geometry has received some useful applications, even though no visualization is possible. Geometry has become abstract in a way that was not considered possible as long as the postulates had to be regarded as "self-evident truths."

Let us look at the second question. Is it possible to have a geometry in which the Pythagorean theorem (that the square of the hypotenuse is equal to the sum of the squares of the other two sides) is replaced by some other theorem? We shall see that this also can be accomplished. To do so, we must take a look at where the theorem comes from. We shall not go through the whole derivation of the theorem (it would, however, be advantageous for the reader to review the derivation), but we shall consider only those aspects of the derivation that are of concern here. To prove the theorem, we use the proposition that the sum of the angles of a triangle is equal to 180°. In turn, this proposition requires the use of the fifth postulate of Euclid, which states that through a point outside a given line one and only one line can be drawn that is parallel to the given line.

This particular postulate has a somewhat different appearance than the other postulates of Euclid, in that it does not appeal to us as quite so "self-evident." For many centuries, mathematicians attempted to show that his particular pos-

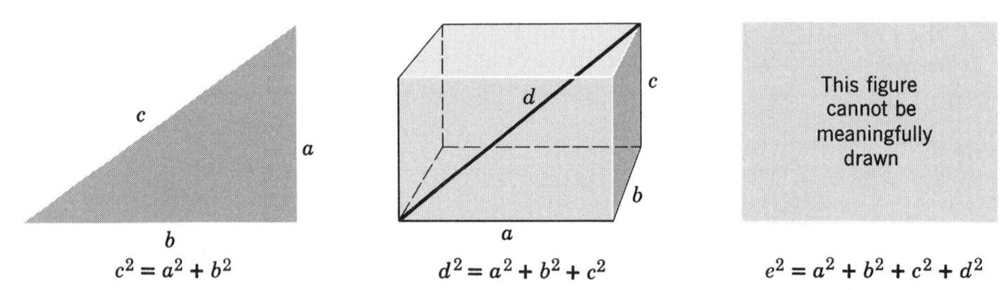

Fig. 3.1. Pythagorean theorem in two, three, and four dimensions. A four-dimensional rectangular parallelepiped cannot be "visualized." However, the mathematical conception is possible and often useful. Its formal properties, such as the application of the Pythagorean theorem to its diagonal, can be studied.

tulate could be derived from the other postulates; if so, it could be eliminated as a postulate, and would then become merely another theorem. All such attempts failed. In 1854 the young German mathematician Riemann tried another approach. He asked the question: what happens if we replace the postulate by one which states that *no* parallel line can be drawn through the given point? By strict logical deduction he established a group of theorems and found that there were no logical inconsistencies. Here then is another system of geometry, which as abstract mathematics is just as good as Euclid's geometry. In this geometry, however, it is found that the sum of the angles of a triangle is greater than 180° and that the Pythagorean theorem has to be modified.

If we stay within two dimensions, that is, on a surface, we can, as a matter of fact, find a model in which the Riemannian geometry will work and can be used. Consider the surface of the earth. If a person starts from a point and travels along the equator due west for 1000 miles, then travels due north until he reaches the North Pole and, finally completes his triangular voyage by going due south to his starting point, he will have described a triangle in which the sum of the angles is certainly greater than 180°. (When he turned from the direction west to north he described a 90° angle; also, since he traveled due south from the north pole, as he approached his starting point his line of travel was again at a 90° angle with the equator. And, since his course toward the North Pole and his course away from it were certainly not identical, he must have turned through some angle at the North Pole.) A little consideration will make it apparent that the Pythagorean theorem no longer applies (the triangle traversed is an isosceles triangle with two right angles). See Fig. 3.2.

If the objection is raised "but the paths taken by the traveler were not really straight lines; they were curves," the reply is that the paths were the "shortest distance" that could be taken on the surface of the globe. If objection is still made that the surface is a "curved" surface, we ask: How can this be established

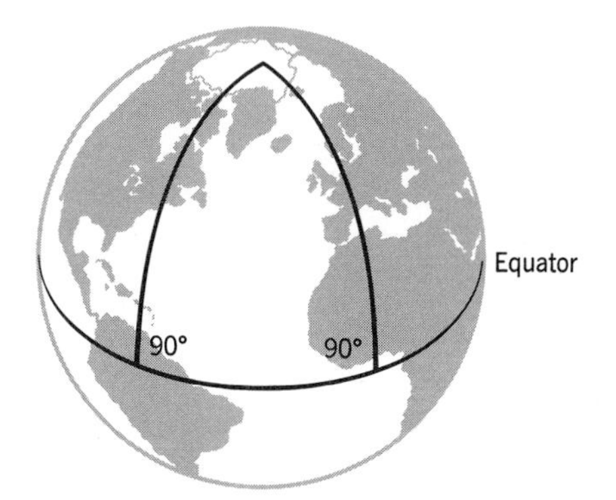

Fig. 3.2. The sum of the angles of a triangle on the earth's surface will be greater than 180°. A plane flying from north pole due south to the equator then turning due east for 1000 miles, then turning due north again to the north pole will describe a triangle the sum of whose angles is more than 180°.

mathematically? One is not allowed to use light rays because that would be using the *physical* assumption that light travels in straight lines. It turns out that there is no other mathematical or metrical method of determining whether one is on a curved surface or a plane surface other than seeing whether the fifth postulate, or some theorem such as the Pythagorean theorem derived from the fifth postulate, remains valid. The net result of all our analysis is that a two-dimensional surface is considered a plane surface if it is Euclidean and if such theorems as the Pythagorean theorem remain valid, but that it is a curved surface if it is non-Euclidean, meaning that the fifth postulate is modified and that all consequent theorems such as the Pythagorean are also modified.

What about the geometry of more than two dimensions? Exactly the same argument applies but, in this case, it must be completely abstract because we cannot visualize any model. Conceptually, there is no objection to considering a three-dimensional space that is non-Euclidean. Such a space will then be curved by definition. The problem of whether the space we live in, the space of the universe, is curved is not a problem of pure mathematics but must be decided by physical observations and experiments. In the twentieth century, Einstein found reason to believe that the space of our universe is actually curved in this sense. He found that the proper expression of laws of physics required a non-Euclidean geometry such as the one that had been developed by Riemann fifty years before as an abstract exercise in pure mathematics. (Once again, as has happened very often in history, abstract mathematics, of no apparent usefulness or applicability when it was developed, was ready and waiting for later application.)

Aside from arousing an interest in what is meant by the term "curvature of space," it is hoped that the example leads to a clearer appreciation of the abstract and formal nature of mathematics. The same theme reappears in the next section where we consider the way in which the idea of number has been extended and generalized.

2. GENERALIZATION; THE GENESIS OF NUMBERS; SYSTEMATIC FOUNDATION OF ALGEBRA

Counting and arithmetic become almost second nature to us at such an early age that we hardly ever, even in maturity, stop to question what lies behind the automatic process. Like a calculating machine, we can turn out the answers; the process is no more questioned by us than it is by the machine.

Numbering is a highly abstract process. It is an assertion that there is something similar between classes or groups of objects that may have no other detectable similarity. We assign the designation "three," for example, to a certain group or collection of objects—which may be fence posts, tadpoles, planets, colors, or mythical gods—signifying that in some way all these objects, real or conceptual have at least one property in common: the property being given the same number.[4]

The postulates (or set of primitive propositions) by which we establish the necessary relations between numbers can be set up in several different ways. We shall not attempt a complete statement of all the postulates needed for operating with all numbers, but we shall discuss only some of the properties and operations involved.

Let us start by assuming that we are familiar with the natural numbers 1, 2, 3, etc. The following two major characteristics are required if we are to use these numbers in our arithmetic.

(1) The cardinal *property of numbers.* This we may call the property of correspondence. It is the property that enables us to match two groups by saying they have the same number. In matching two groups we are not concerned with counting because we can match any two collections of objects without knowing the number of objects in the group; we can, for example, say that in a theater, in which all seats are taken, that the number of seats and the number of people sitting in them are the same regardless of either order or number.

(2) The ordinal *property of numbers.* Ordinality requires that the numbers be arranged in a unique order and that each number has a successor. In addition to these two more general properties, it is necessary to provide for rules of combination of numbers.

Can we give a systematic and consistent account of all the various rules of arithmetic that were drilled into us in school? What is the basis for these rules? Such are mature mathematical problems that are still under discussion, since both the problem of assuring consistency and the problem of reducing the number of rules to a minimum and to the simplest possible type present authentic difficulties.

These problems can only be dealt with if we speak in general terms, that is, if all the rules about which we speak are discussed not in reference to particular numbers or to a particular number system, but in reference to *any* number. Herein lies the power of algebra, in which we let letters stand for *any* number. By symbolizing numbers in this way we can at once begin to study *general* rules, which will hold regardless of the particular numbers involved. A complete set of the most general rules will yield instructions as to how any particular calculation shall be carried out.

[4]Russell has given the following definition of number: "the number of a class is the class of all those classes which are similar to that class." This is, for most purposes, an adequate definition when properly understood, but it does lead to certain difficulties into which we need not enter. It is sufficient for our purposes to leave the concept of number undefined except in so far as it becomes defined by the postulates that we set up for both algebra and arithmetic.

Algebra of Natural Numbers. In the algebra of natural numbers (1, 2, 3, 4, 5, etc.) we require the idea of equality (which is a statement that when $A = B$ in any proposition we may replace A by B or vice versa) and the idea of order (which implies that if A and B are not equal then either A is greater than B or A is less than B and that both of these cannot be true at the same time). In addition to these two basic ideas, which we leave otherwise undefined, there are the following postulates upon which all of the arithmetic of natural numbers can be constructed.

Postulate of equality.

(*1*) If $a = b$ and $b = c$ then $a = c$.

Postulates of addition.

(*2*) If a and b are numbers then $a + b$ is a number that is unique.
(*3*) $a + b = b + a$ (the commutative law).
(*4*) $(a + b) + c = a + (b + c)$ (the associative law).
(*5*) If $a + b = a + c$ then $b = c$ (sometimes called the law of cancellation).

Postulates of multiplication.

(*8*) If a and b are numbers, then $a \times b$ is a number and is unique.
(*9*) $a \times b = b \times a$ (commutative law).
(*10*) $a \times (b \times c) = (a \times b) \times c$ (associative law).
(*11*) If $a \times b = a \times c$ then $b = c$.
(*12*) If $a = b$ then $a \times c = b \times c$.
(*13*) There is a unique number u, such that $u \times a = a$ for any number a (this number u is called 1).

Postulate connecting addition and multiplication.

(*15*) $a \times (b + c) = a \times b + a \times c$ (distributive law).

(Postulates 6, 7, and 14 will be added later. Although we have stated the postulates above using $\times$ as a sign of multiplication, the replacing of this sign by a dot or by juxtaposition of the two symbols being multiplied will often be used in this book hereafter. Thus $a \times b = a \cdot b = ab$).

Provided we know the names (that is, the symbols) and order of the natural numbers (that is, if we can count) the above rules are the basic ones from which all usual arithmetic stems. Theoretically, no more would be required for the arithmetic of addition and multiplication of natural numbers.

Where do these postulates come from? It is too easy an answer to say they are obviously true and self-evident. We are very used to the fact that, of course, $2 + 3$ is the same as $3 + 2$, since both equal 5. Would it be possible to have an algebra in which the commutative law $a + b = b + a$ does not hold? Could such an algebra be consistent? Both these answers are in the affirmative, and such algebras are used today, just as non-Euclidean geometries are used. Furthermore, we have a model to which we can refer so that the abstract idea of a noncommutative algebra can be made clearer by example. (Even without any concrete model or any application, a noncommutative algebra would still remain *sound mathematics*.)

Suppose one takes a rectangular figure such as a book, and designates the three axes of the book as indicated in Fig. 3.3. Consider the operation of rotation through a 90° angle counterclockwise about axis X as operation a; consider the operation of counterclockwise rotation through 90° about the axis Y as b. The problem is: Is the sum $a + b$ the same as $b + a$?

Examination of Fig. 3.3 shows that in the case of rotations about more than one axis, the process of addition is noncommutative. Other examples can be found of systems in which the postulate of the commutative properties of either addition or multiplication is eliminated. Thus, we see that by changing the postulates we can develop totally different types of algebra. Many such algebras have been developed and used. (They do not, of course, apply to ordinary numbers, which are defined basically in terms of the postulates we have listed, plus those additional ones required if numbers are to include fractions and negative numbers as well as positive integers.)

Before we examine the manner in which the natural numbers are extended, it will be helpful to review the *positional* nature of our number system, which allows us to count up to any number using only ten different symbols, plus a rule of position. The discovery of a positional scheme for writing numbers arose in India when the use of 0 as a symbol to designate an empty column on the

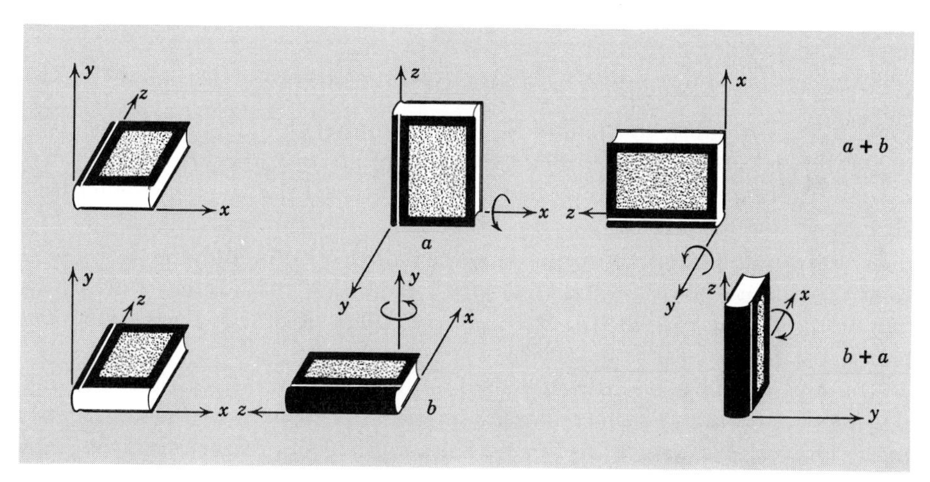

Fig. 3.3. A noncommutative type addition $a + b \neq b + a$. The operation a is a counterclockwise rotation about the x axis (facing the origin); b is the operation of counterclockwise rotation about the y axis. Operation a followed by b (top row) yields a different result than b followed by a (bottom row).

abacus was introduced. On the abacus (as on the desk-calculating machines found today in offices) there was room for 9 elements in the column farthest to the right. If a count went one above 9, then the column farthest to the right was left empty. In the next column a disc was placed, symbolizing the count of ten (or X in Roman numerals). As the count proceeded upward through another count of 9, the right-hand column again was filled, and an additional count was shown by emptying the right-hand column again and placing another element or disc in the second column. Although this had been symbolized by the Roman numerals XX, meaning two tens, the Hindu invention of *a symbol for the empty column* allows us to express the same number twenty as 20, the *position* of the 2 showing that it is to be multiplied by ten. Similarly 200 shows, by the position of the 2, that we mean 2 times one hundred. This process is extended indefinitely, making any number expressible by the nine symbols for integers plus the symbol for zero or the empty column. (In contrast, in the Roman numerals, for every new power of ten a new symbol had to be introduced: 200 became CC, 2000 became MM, and a number such as 3431 had to be expressed in a much more complicated way, namely MMMCCCCXXXI.)

Remember that we already have extended the system of natural numbers 1, 2, 3, 4, 5 . . . by the addition of the symbol 0. We therefore need an additional postulate, which will enable the unique properties of 0 to be included in our system of numbers. This postulate is added to our other postulates for addition.

(*6*) There is a unique element or number z such that $a + z = a$. This element or number z is called 0 by definition.

Extension of Algebra to Include Negative Numbers. The inverse operation of addition is subtraction. Within the system of positive integers this operation can always be performed, provided that we subtract the smaller integer from the larger. We can describe the process in terms of addition as follows.

If $a + x = b$, the process of finding x is called subtraction, and will always work if b is equal to or greater than a. We then have the number x, which we can define as $x = b - a$. (If a is equal to b, then, of course, by the postulate above, $x = 0$.)

This inverse addition was well known to the Greek mathematicians. Subtraction, however, was ruled out as a mathematical operation unless in an equation of the form: $x + a = b$, b was greater than a. As algebra developed, the equation $x + 2 = 3$ was recognized as a perfectly good equation, which yielded a unique solution, although the equation $x + 3 = 2$ was excluded for many centuries from mathematics. (For very good reasons, even today, some possible expressions are excluded from mathematics. We cannot express within modern algebra an equation of the form $x/0 = 4$; this is not allowed as a mathematical expression only because it would lead to inconsistencies.)

Probably, the need of commerce for expression of debts first forced the use of negative numbers upon mathematicians. A debt, or an overdraft, is most conveniently represented by a negative number. Furthermore, from an abstract point of view, there seems to be no formal reason to exclude the equation $x + 3 = 2$, provided a new type of number, the *negative* number, is allowed to take its place among the numbers of our system, and provided they can be so included among the postulates of our system without leading to any inconsistency.

This extension is easily accomplished by the addition of the following postulate.

(*7*) For every element (number) a, there corresponds an element a' such that $a + a' = 0$. This number a' is called the negative of a, and is symbolized as $-a$.

In addition to this postulate, we also need these conventions:

(*1*) A negative times a positive number yields a negative number.
(*2*) A negative number times a negative number yields a positive number.[5]

Extension to Include All Rational Numbers (Fractions as Well as Integers). Just as integers were extended to include negative numbers by considering an operation that is the inverse of addition, so they may be extended to include fractions by considering the inverse of multiplication. In this case the extension was accomplished early in the history of mathematics. The Greeks were completely familiar with fractional numbers. In contrast, it is interesting to note that even in the sixteenth century negative numbers were still regarded as "fictitious or invented numbers," which were less real than fractions or even irrational numbers. How early fractions were introduced is not certain, but we can see how loathe people were to use them by noting that in the Old Testament of the Bible, the ratio of the circumference of a circle to its diameter is considered to be 3.

If we keep to integers, the division or the inverse of multiplication is possible only for some numbers. The equation $2x = 10$ is perfectly permissible, but $2x = 11$ is excluded.

As with negative numbers, we can amend our postulate system to allow for such a possibility. We add the next postulate.

(*14*) For every element (or number) a there is an element (or number) a' such that $a \times a' = u$, where u is the unit element 1, *provided a is not zero.* By definition, this element a' is called the reciprocal of a, and is designated as $1/a$, a^{-1}, or $1 \div a$.

[5]For a good discussion of the need for these conventions, see *Elementary Mathematics from an Advanced Standpoint,* by Felix Klein, Dover Publications, New York, 1945, pp. 25–26.

The number system is now generalized to include fractions. Some of the rules governing fractions which we shall require are reviewed in Section 3 of this chapter.

Extension to Include "Irrational Numbers." Two more ways of generalizing the concept of number are possible. We shall not, in these cases, list the additional postulates required, but shall indicate the general properties of the numbers and some of the rules that are needed.

Let us again look at some equations.

If $x^2 = 4$, we can by trial and error discover that x must be equal to plus or minus 2. This process is called evolution, or the process of extracting a root, and is the inverse of "raising a number to a power." Similarly, if $x^3 = 27$, we can find the value of x, which makes this equation true, namely $+3$.

The natural question arises: What about such equations as $x^2 = 2$?

Does x have any value within our number system, which we have been discussing? (The query arises quite naturally from the problem of trying to find the length of the hypotenuse of a right triangle with legs of length one unit each. Pythagoras made the momentous discovery that the length of the hypotenuse could not be expressed by any fraction. Two legends have been repeated: one is that he celebrated this discovery with a great feast among the initiates, and the other is that he put to death many who knew this secret so that it could be preserved.)

Within the rational number system (that is, the system of numbers that contains all integers, positive and negative, and all ratios or fractions), the equation $x^2 = 2$ cannot be solved.

A system of mathematics unable to use numbers that solve such an equation would be much restricted, both in its formal nature and in its applications. Therefore, it is natural to extend again the number system to include numbers such as the square root of 2, the cube root of 7, etc.

The postulates of such a number system have been worked out. We shall not go into detail here.[6] But a few points need consideration.

First, by the term "irrational" number, one should avoid jumping to the conclusion that we mean "unreasonable." The term merely denotes that a number such as the square root of 2 cannot be expressed as any possible fraction or ratio. (A very neat proof of this was given by Euclid. It may be found in many of the books recommended for reading at the end of this chapter.)

Second, although any fraction can be expressed as a simple or repeating decimal, an irrational number cannot be so expressed. For example, $\sqrt{3} = 1.7305$ ... the numbers after the decimals never repeating themselves in any specific

[6] The full set of postulates is given by J. W. A. Young in *Monographs on Modern Mathematics,* Dover Publications, New York, 1955.

grouping. By real numbers we mean all numbers including integers, fractions, and irrationals. The only numbers excluded are the imaginaries, that is, the square roots of negative numbers, treated in the next section.

Third, a very important correspondence can be shown between the points on a line, as described by Euclidean geometry, and real numbers. *For every point on a line there is a corresponding number within the realm of real numbers.* This means essentially that we can state any theorem having to do with points on a line in terms of real numbers. It is the profound beginning of the ability to express geometric relationships in terms of algebra, or vice versa. (We define such a one-to-one correspondence between two systems as an isomorphic relationship if, in addition to the one-to-one correspondence, the results of addition and multiplication are "preserved.") In 1637, Descartes took the highly imaginative but logical step of showing that all the points on a plane could be expressed in terms of a pair of numbers, one of the pair being identified with points on a horizontal line, the other on a vertical line. This resulted in one of the most fruitful developments in the history of mathematics. (E. T. Bell, historian of mathematics, places the beginning of the middle period of mathematics in 1637, the date of publication of Descartes' geometry, and considers it to have extended until the developments of analysis by Gauss in 1801.) Geometry and algebra were joined together. Any theorem in one field could be expressed in the other, sometimes much more conveniently and succinctly.

Extension to the So-Called "Imaginary Numbers." An interesting study could be made of the way new concepts are created in the history of ideas—scientific, artistic, or otherwise. In some ways a new idea in mathematics is more restricted (logical contradictions *must* be avoided) than in the other imaginative arts. However, there is possible for mathematics a certain inventiveness and freedom, which cannot be achieved by other artistic expressions. To illustrate a "centaur" of mythology, the "ether of space" of nineteenth century physics, the "id" of Freud, all are possible concepts. The first is unrestricted, as long as it can be thought of as a combination of man and beast and can find an artistic or literary representation. The second was invented as a scientific concept to "explain" observed wavelike properties of electromagnetism. Its demise came only when no operational properties corresponding to the concept could be found. The third is a concept that may or may not, in the long run, prove useful to psychologists, but its basic justification will be judged almost entirely by its pragmatic value in hooking together various psychological phenomena into some kind of systematic order. Contrast these kinds of constructs with those of mathematics. On the one hand, mathematical concepts are rigorously limited by the necessity of consistency (for example, division by zero is excluded) and, on the other hand, they are completely free from the necessity of being visualized, or of having any practical or operational application or measurement.

Again, let us consider some possible equations.

First, $x^2 + 10 = 20$. In this case, the value of x can be found within the realm of real numbers.

Second, $x^2 + 20 = 10$. Here, we have an equation that formally is almost identical with the first equation. But, when we inquire into its solution, that is, the value for x for which the equality will be valid, we run into immediate difficulties. Since, by our rules, any negative number multiplied by itself, or any positive number multiplied by itself must, in each case, yield a positive number, the question arises: How can it be possible to have a number such that when multiplied by itself it will yield a negative number? It cannot be done, as long as we restrict ourselves to the postulates that we have established for the real number system. In other words, we again need to extend or generalize our concept of number. This is done by introducing the so-called "imaginary numbers"—which are no more imaginary or unreal than any other numbers but which are only somewhat more difficult to picture.

Rather than list the several postulates needed to make this extension possible, we give the most important one below.

There is an element i such that $(i) \times (i) = -1$, and such that $(-i) \times (-i) = -1$. This element or number is often referred to as the square root of minus one, $\sqrt{-1}$.

All possible numbers are now expressible in the realm that is known as complex numbers, a complex number being composed of the sum of a real and an imaginary number defined by the additional postulates for complex numbers (the most important of which is given above).

This is the widest possible extent to which numbers can be generalized without either running into contradictions or establishing another isomorphic (structurally identical) system.

Unlike the centaur, the ether, or the id, the complex number system depends neither on the possibility of its being visualized nor on its practical applications. The fact that it is a very useful system for handling many problems of physics and engineering is an accidental bonus which has little relevance to mathematics as such. Complex numbers and their properties were well worked out long before their immense practical value became evident in such fields as electrical engineering, acoustics, nuclear engineering, or almost any branch in modern physics.

In this study we shall not need to use complex numbers, but the understanding of how numbers have been generalized required this review.

In the next section we shall not follow the postulational method at all—this is a matter that comes under mathematics—but we shall summarize the more important rules of algebra that we shall be using. They are not given in any logical order, nor are they all independent of one another (that is, some of the rules given can be derived from some of the others; we sacrifice economy for ease of understanding—a sinful procedure for a mathematician but sometimes required for heuristic reasons).

3. REVIEW OF SOME ELEMENTARY MATHEMATICAL RULES

Definitions of Some of the Symbols of Algebra.

MULTIPLICATION

In place of the times sign $\times$ it is convenient to use other symbols in algebra. Thus

$$a \times b = a \cdot b \text{ or } ab \text{ (}ab\text{ is used most often).}$$

DIVISION

$$a \div b = \frac{a}{b} \text{ or } a/b \text{ (}a/b\text{ is more easily printed).}$$

$\neq$ means not equal; $\approx$ means approximately equal.
$\propto$ means proportional to.
The reciprocal of a is $1/a$ or a^{-1}.
$aa = a^2$, $aaa = a^3$, $aaaa \ldots$ (n times) $= a^n$.
$a^{1/2} = \sqrt{a}$, $a^{1/3} = \sqrt[3]{a}$, etc.
$a^{-1} = 1/a$, $a^{-2} = 1/a^2$, etc.

Conversion of Units. Often it is necessary to convert from one system of units to another. The following example may be used to study the manner in which this may be done without confusion.

Convert 30 miles per hour into the equivalent feet/sec.

$$30 \text{ miles/hr} = (30)\left(\frac{\text{miles}}{\text{hr}}\right) =$$

$$(30)\left(\frac{\cancel{\text{miles}}}{\cancel{\text{hr}}}\right)\left(\frac{5280 \text{ ft}}{\cancel{\text{miles}}}\right)\left(\frac{\cancel{\text{hr}}}{3600 \text{ sec}}\right) = 44 \text{ ft/sec}$$

Some Rules for Fractions. (*Note.* Test out numerical examples, substituting numbers for the symbols!)

Addition: $a/b + c/d = \dfrac{ad + bc}{bd}$

Multiplication: $(a/b) \times (c/d) = ac/bd$
Division: $a/b \div c/d = ad/bc$

Cancellation: $ab/cb = a/c$

$$\left(\text{Note that } \frac{a+b}{c+b} \neq \frac{a}{c}\right)$$

Powers and Roots.

$$(a^n)(a^m) = a^{n+m} \qquad (ab)^n = a^n b^n$$
$$a^n/a^m = a^{n-m} \qquad a^0 = 1$$

An Equation Remains Unchanged When the Same or an Equivalent Operation Is Performed on Both Sides of the Equation.

$$\begin{array}{rr} \text{If } x - 3 = & 5 \\ \text{add } \quad 3 = & 3 \\ \hline \text{then } x = & 8 \end{array}$$

$$\begin{array}{rr} \text{If } x + 6 = & 10 \\ \text{subtract } \quad 6 = & 6 \\ \hline \text{then } x = & 4 \end{array}$$

$$\text{If } x^2 = 9$$
$$\text{then } x = \pm\sqrt{9} = \pm 3$$

$$\text{If } \frac{x}{3} = 10$$
$$\text{then } \left(\frac{x}{3}\right)(3) = (10)\cdot(3) \text{ or } x = 30$$

$$\text{If } 3x = 15$$
$$\text{then } \quad x = 5$$

It Is Usual to Express Large or Small Numbers in Powers of 10.

186,000 miles/sec = 1.86×10^5 miles/sec (speed of light)

93,000,000 miles = 9.3×10^7 miles (distance earth to sun)

This makes computations simpler. For example, the approximate time for light to reach earth from the sun:

$$9.3 \times 10^7 \text{ miles divided by } 1.86 \times 10^5 \text{ miles/sec}$$

$$\frac{9.3 \times 10^7}{1.86 \times 10^5}\,\frac{(\cancel{\text{miles}})(\text{sec})}{\cancel{\text{miles}}} = \frac{9.3 \times 10^2}{1.86}\text{ sec} = 5 \times 10^2 \text{ sec}$$

Approximation. In the physical sciences, as opposed to mathematics, we are seldom interested in figures more exact than one per cent. (The result of combining several measurements, each in error by one per cent, will certainly be in error by more than one per cent. If, therefore, we are measuring only with an accuracy of one per cent, it is actually misleading to express results to a supposedly greater accuracy.)

For our purposes, therefore, in carrying out calculations, unless directions are otherwise given, carry out no calculation to an accuracy of more than one per cent. In taking square roots the process of trial and error is satisfactory.

4. VARIATION; RATIO AND PROPORTION; FUNCTIONS AND GRAPHS

The majority of the relationships between measured quantities in the physical sciences can be expressed in *equations* (that is, functional form); these constitute mathematical expressions of the laws. It will be helpful to review some of the more typical forms of such expressions that we shall encounter.

First, let us make clear the nature of a function. If any two quantities are so related that, whenever a definite value is given to one quantity a value is then determined for the other, they are said to be functionally related. Most often, in the cases in which we are interested, the functional relationship is definitely known and can be expressed in a formula or equation; but cases may arise in which we only know that such a relationship exists, but cannot express it in an equation form, there being no known regularity that can be formally expressed (for example, any person's weight varies slightly from hour to hour, but to write an equation showing the functional relationship between weight and hour of day would be extremely difficult).

One of the most universal of functional relationships is proportionality of which there are several types. (*1*) If one quantity y is directly proportional to another quantity x, then the equation that expresses this relationship is $y = kx$, where we refer to the k as the proportionality constant, by which we mean that y increases as x increases and does so at a constant rate k. Thus, if one is traveling at 30 miles/hr, the distance traveled is proportional to the time, and the expression $y = 30x$ implies that for $x = 1$, y will be 30; for $x = 2$, y will be 60, etc. The 30 in this case is a constant of proportionality. (*2*) If one quantity y is inversely proportional to another quantity x, the equation becomes $y = k/x$. (As an example, if one travels a certain distance, k, which is fixed, the time it takes to travel that fixed distance is inversely proportional to the speed [in this case variable] with which one travels. For instance: if one travels 100 miles, or in the above equation $k = 100$ miles then the time, y, is related to velocity, x, by the equation $y = 100/x$.) (*3*) Various combinations of the above proportionalities are possible. We give some examples.

The area of a circle is proportional to the square of its radius:

$$A = kR^2 \text{ where } k = \pi \text{ is the constant of proportionality}$$

The surface area of a sphere is proportional to the square of its radius:

$$S = kR^2 \text{ where in this case } k \text{ is a constant equal to } 4\pi$$

The volume of a sphere is proportional to the cube of its radius:

$$V = kR^3 \text{ where } k \text{ now is equal to } \tfrac{4}{3}\pi$$

The weight of a sphere is proportional both to its density D and to the cube of its radius:

$W = kDR^3$ where k now depends upon both $\frac{4}{3}\pi$ and the units in which density is expressed

The brightness of a light is inversely proportional to the square of the distance from the source:

$B = \frac{k}{d^2}$, k being the constant, B the brightness, and d the distance

It is often useful to see a geometric representation of some of the types of proportionality given in the above examples. Such a representation is called a graph. We let the two variable quantities be represented on distances along the y and x axes as shown in Fig. 3.4. If there is a simple direct proportionality, then we may represent all possible pairs of values of x and y as points that will fall along a straight line through the origin, the slope of which will be equal to the constant of proportionality k.

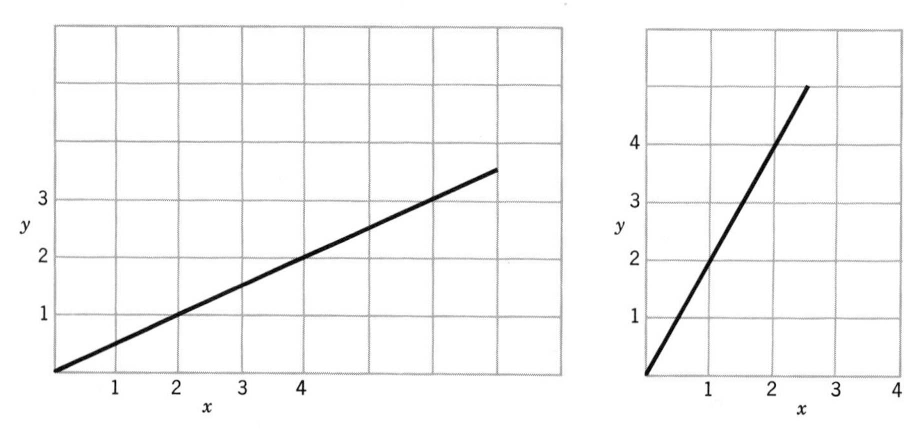

Fig. 3.4. Left: $y = \frac{1}{2}x$; the slope, k, is $\frac{1}{2}$. Right: $y = 2x$; the slope, k, is 2.

We shall consider other graphs of other functions later on, and learn how the interpretation of them, drawn on the basis of observations, often suggests the form of laws or equations that exist between two varying quantities.

5. EXAMPLES OF USE OF PROPORTIONS

As a good example of the use of proportions, let us consider in more detail a previously mentioned topic, that is, the question of how some properties of objects may vary with their dimensions. The following example was selected not only to illustrate how proportionalities can be used, but also because we shall refer to a similar problem later when we consider how the amount of charge on an electron is determined.

Here is the problem.

It is a matter of common observation that although drops of rain fall rather rapidly, droplets of mist fall gently, and the very small drops of fog "settle" with extreme slowness (provided the wind does not prevent their settling at all). We wish to determine why these various sized drops, all in the shape of a sphere, fall at different speeds and how the velocity of their fall depends upon their size or radius.

To answer this question we must get a little ahead of our story, and assume certain laws whose validity we shall later examine. We anticipate only in order to give a good example of the use of the mathematics of proportionality.

The first law, which we use, is that when an object has fallen a long enough time through a resistant medium such as air, it will reach a terminal velocity that is constant. (A man without a parachute must fall from a plane over 1000 ft before he reaches such a steady terminal velocity, but a drop of water need fall a much shorter distance.) We shall see later that the condition for reaching this constant "terminal velocity" is that the weight of the object (the downward force of gravity) is exactly equal to the frictional resistance of the air or resisting medium.

We can therefore set up the condition for a steady terminal velocity (which we assume, on the basis of observation, holds for falling drops) in the form of an equation:

$$W = F \tag{1}$$

where $W =$ weight of the object
$F =$ frictional or upward resistant force

Let us now consider what the weight W depends on, and what the frictional force F depends on.

The weight is proportional to both volume and to the density. But since the density is constant (we are only considering drops of water in each case), we can set weight proportional to volume. However, the volume, in turn, is proportional to the cube of the radius, therefore the weight is proportional to the cube of the radius:

$$W = kR^3 \tag{2}$$

On the other hand, the frictional resistance F is proportional to two quantities combined. Let us assume that for small velocities it is proportional to the velocity of the falling object, and that it is also proportional to the surface area.

As we have seen, the surface area is proportional to the square of the radius. Therefore:

$$F = KVR^2 \qquad (3)$$

where K is now *another* constant of proportionality different from the k above, V is the terminal velocity.

If now we equate the two values of W and F, from the second and third equation, which can be done because of the first equation, we finally get:

$$kR^3 = KVR^2$$

and, using our algebraic rules, this gives us:

$$V = \frac{kR}{K}$$

Since k/K is a ratio of constants, it is constant, and we can set it equal to some new constant, for example, K'. Therefore

$$V = K'R$$

which tells us that the velocity will be proportional to the radius of the drops. It does not tell us what the particular constant K' of proportionality is, but expresses the general way in which (within limits) the terminal velocity of rain drops will depend upon the size of those drops. (For very small drops, the assumption made about frictional resistance is not warranted. In this case the resistance is proportional to the radius, not to its square. How does this affect the result?)

6. QUESTIONS

1. Discuss the meaning and significance of the term *postulate*. How does the modern view of a postulate system differ from the classical view of a system of axioms?

2. Review what is meant by similar figures. If two triangles are similar, what properties are the same between the triangles? What properties differ? In what sense are two circles similar? What ratios remain the same? What properties differ? Compare the general term "analogy" with the mathematical term "similarity."

3. Some geometric relations will be used recurringly throughout this book. Among them are (*1*) given a line tangent to a circle, a radius drawn to the point of tangency will be perpendicular to the tangent line; (*2*) the exterior-interior angles of a line intersecting two parallel lines will be equal; and (*3*) the Pythagorean theorem. While not essential, it would be helpful to learn the proofs of these geometric theorems. An elegant, simple, and most instructive proof of the Pythagorean theorem can be found in *Induction and Analogy in Mathematics* by G. Polya, Princeton University Press, Princeton, 1954, pp. 15–17.

4. Extend the Pythagorean theorem to three dimensions. That is, prove that the diagonal d, of a rectangular parallelepiped (a rectangular box) is related to the dimensions of the sides a, b, and c by the equation $d^2 = a^2 + b^2 + c^2$.

5. If you double the radius of a sphere, how many times do you increase the volume? How many times do you increase the surface area?

6. Is the sine of an angle proportional to the angle itself? Explain why or why not. Is it proportional under certain conditions and not under others?

7. In what way does an equation resemble a sentence? In what way does an equation resemble an analytic balance?

8. Compare the volume of a sphere of diameter a, with the volume of a cube with length of side a. Can you suggest how these relative volumes have a significance in geometric shapes found in nature?

9. Pi (π) may be expressed as the sum of an infinite series. For example, $\pi = 4\,(1 - \frac{1}{3} + \frac{1}{5} - \frac{1}{7} + \frac{1}{9} - \frac{1}{11} \ldots)$. The ratio $\frac{1}{3}$ can also be expressed as an infinite series. (How?) Nevertheless π is *not* a rational number. Can you see the distinction?

10. Using household items, e.g., a length of string and a cylindrical can, determine the value of π experimentally. What is the percentage error?

11. Now try dropping randomly 100 toothpicks from a height of 4 or 5 feet onto a large area marked with parallel lines evenly spaced one toothpick length apart. Count the number of toothpicks which touch or cross any line. It is predicted by statistical theory that the ratio of the number of toothpicks crossing a line to the total number dropped will be in the ratio of $2/\pi$. Determine the percentage error of π thus evaluated and compare with answer of problem 9. (See discussion and proof of the ratio by G. Gamow, *One, Two, Three, Infinity*, Viking Press. New York, 1948, Chapter VIII, Section 3.)

12. One further comment about π. Wherever periodic motions occur, the equation for the period always involves π. For example, the period of a simple pendulum is $P = 2\pi\sqrt{l/g}$ where l is the length and g the constant of acceleration. Can you suggest any reason why π is so intimately associated with periodicity?

Problems

1. In taking square roots and cube roots, sufficient accuracy is reached quickly in most problems by use of the method of trial and error. However, proper handling of powers of ten is also required. Find $\sqrt{12{,}320}$; $\sqrt[3]{6437}$; $\sqrt{0.0081}$; and $\sqrt{0.00072}$. Which is larger, $\sqrt{0.4}$ or $\sqrt{0.2}$; $\sqrt{\frac{1}{4}}$ or $\sqrt{\frac{1}{6}}$?

2. Study methods of solving simple simultaneous equations. (A method that will always work in the simpler equations is first to solve for an unknown using one equation, and then to substitute this value in a second equation.) Solve for x and y in the equations $3/x + 4/y = 16/xy$ and $8x + 2y = 20$.

3. What is the purpose of a graph? A simple proportionality is expressible by a straight line graph through the origin. Think of an example of simple proportionality and illustrate it by a graph. Then consider other relationships such as $y = kx^2$, $y = k/x$, and $y = k/x^2$, and draw their graphs with specific values of k. Graph the equations $y = \sin\theta$, $y = \cos\theta$, and $y = \tan\theta$ between $\theta = 0°$ and $\theta = 90°$.

4. How many molecules of water are there in a glass of water that has a mass of 0.2 kg, if 1 molecule has a mass of 3×10^{-27} kg? (*Use powers of 10 in your calculations.*)

5. On the basis of your general knowledge about oceans, how many glasses of water would you estimate to be contained in all the oceans of the world? (Begin by estimating the surface area of the world, which is covered with water, then estimate the average depth, and arrive at a rough estimate of the cubic contents.) Compare this estimate with the answer of problem 4.

6. How many miles is it to the nearest star, which is about 4 light-years distant? (*Use powers of 10 in your calculations.*)

7. Assume that frictional resistance of air is proportional to the radius of a very small falling droplet instead of being proportional to its square as assumed in the discussion above. What will the terminal velocity now be proportional to?

8. As a review exercise in handling equations, derive the solution of the general quadratic equation $ax^2 + bx + c = 0$. (The solution, namely, $x = \dfrac{-b \pm \sqrt{b^2 - 4ac}}{2a}$ is of such frequent application that it is worth memorizing.)

9. The Greeks found a certain geometric ratio, sometimes called the Golden Mean, to be aesthetically satisfying. It is defined as follows: given any magnitude, if it is divided into the parts x and y, such that $x/y = y/(x + y)$, then y constitutes the Golden Mean. Show that the ratio of x/y is $(-1 \pm \sqrt{5})/2$. A rectangle with sides in this ratio was considered by the Greeks to have the most aesthetically satisfying shape. Compare this ratio to the ratio of the sides of the American flag and that of other common rectangles.

10. Suppose a chain were stretched snugly about the earth at the equator (25,000-mile stretch). If 6 feet were added to the chain, how far would the chain be from the surface of the earth; assume the chain remains concentric with the earth? See whether you can generalize the result. Note how you might apply the result to a quick approximation of the speed of a satellite circling the earth in 90 minutes at a height of 100 miles.

11. Can you show that in the set of all integral numbers the number of even numbers is equal to the number of integral numbers? Can you show that on a line segment AB, with midpoint M, the number of points between A and M equals the number of points between A and B? (Considerations such as these lead to a definition of an infinite class.)

Recommended reading

There is so large a variety of good books on mathematics that a selection from them is difficult. The ones below are listed in the order of increasing complexity.

James Newman, *The World of Mathematics,* 4 volumes, Simon & Schuster, New York, 1956. This is a collection of essays, most of them written by famous mathematicians, for the nonprofessional embracing most of the important aspects of mathematics and its applications.

Morris Kline, *Mathematics in Western Culture,* Oxford Press, New York, 1953. An excellent review of the historical impact of mathematics upon the ideas of our civilization. Very simply written, but authoritative.

Lancelot Hogben, *Mathematics for the Millions,* W. W. Norton, New York, 1937. This very popular book has a mass of practical illustrations and problems. The emphasis is on applications rather than the abstract nature of mathematics, and should be complemented by other references such as the next one by E. T. Bell.

E. T. Bell, *Mathematics, Queen and Servant of Science,* Macmillan, New York, 1951. An excellent book by a scholar in the history of mathematics who is able to present even the most abstract ideas in a straightforward, simple, and entertaining fashion.

W. W. Sawyer, *Prelude to Mathematics,* Penguin Press, Baltimore, 1955. This book and its companion, *Mathematician's Delight,* by the same author and in the same series of Pelican paperbacks gives the nonprofessional a feeling for what mathematics is about. Both are entertainingly written and are easily understood, particularly the latter.

E. C. Titchmarsh, *Mathematics for the General Reader,* Doubleday Anchor Book, New York, 1959. An elementary but systematic account of the major concepts of mathematics from arithmetic to calculus, written distinctly for the general reader as indicated. Ideas are clearly stated and well illustrated.

F. Waismann, *Introduction to Mathematical Thinking,* Harper Torchbook, New York, 1959. This is a careful, rigorous statement of the logical foundations of modern number theory. Somewhat limited in scope and rather specialized for the general reader unless he wishes to fully understand the logical foundation of our number system.

John L. Kelley, *Modern Algebra,* Van Nostrand, New York, 1959. This book was used for Continental Classroom and covers the systematic postulational development of algebra from the modern approach. For those who are unacquainted with the theory of sets, this makes a fine introduction.

J. W. A. Young, *Monographs on Modern Mathematics,* Dover Publications, New York, 1955. An excellent collection of essays on the foundations of mathematics by outstanding mathematicians. Some essays may be found difficult by the general reader unless he is willing to devote considerable time and effort to understanding what is being said.

J. G. Kemeny, H. Mirkil, J. L. Snell, and G. L. Thompson, *Finite Mathematical Structures,* Prentice-Hall, Englewood Cliffs, N. J., 1959. A thorough, somewhat abstract presentation of the foundations of algebra and probability from a set-theory approach.

G. H. Hardy, *A Mathematician's Apology,* rev. ed. Cambridge University Press, N. Y. 1967. An eminent modern mathematician discusses the motives and excitement of a pure mathematician pursuing his art.

chapter four

From Ptolemy to Copernicus

We have seen that a satisfactory scientific theory must possess at least two minimal characteristics: (*1*) the propositions or statements of which it is composed must be logically consistent—this logical consistency is usually guaranteed by a mathematical and systematic expression of the propositions; and (*2*) the propositions must at some point be related to observations. This relationship is guaranteed by the appropriate "operational" definitions and measurements in terms of which the propositions are given physical meanings.

There are other criteria which satisfactory theories should fulfill. After we have studied some examples of theories, we shall consider these additional criteria. First, we shall consider the Ptolemaic theory of planetary motion, how it is related to observation, and how it was eventually replaced by the Copernican theory. By such a study, much can be learned about the nature and structure of theories in general, and the manner in which they are overthrown.

1. ARISTOTELIAN AND PTOLEMAIC THEORIES

We have seen how the Greeks decided that the shape of the earth was a sphere and how they had arrived at some respectably accurate measurements regarding the size of this sphere and also of the distance of the earth from the moon and from the sun.

If the earth is a sphere, the problem is at once posed as to what constitutes "down." (In later ages, this became a real problem—people could not conceive of anyone living on the underside of the earth at the antipodes—they would naturally fall off since the definition of "down" was in terms of gravitational attraction at the point where one was standing.) This problem was met by Aristotle's simple theory of gravitation; every object is composed of a mixture of four elements: earth, water, air, and fire. The heavier objects were composed of a larger proportion of "earth" than the lighter ones; and the natural place for earthy objects was the center of the sphere of the earth. Thus, heavy objects "sought" their center, namely, the center of the earth. The gravitational forces were explained in terms of an essential property of matter.

The natural terrestrial motion of heavy bodies was thus always directed toward the center of the earth. On the other hand, there was obviously another type of motion, that of the heavenly bodies in their courses. The natural motion for these bodies was conceived by Aristotle to be circular motion, since a circle was the most perfect figure. The stars, the sun, the moon, and the planets were all thought to move in circular orbits around the center of the earth. Only two types of motion were to be considered natural: terrestrial motion, which was in a straight line toward the center of the earth, and heavenly motion, which was motion in a circle about the earth as a center. All other motions were considered "violent motions," which would require some peculiar explanation, such as an impressed force or impulse, to explain why they existed at all.

Such was the basic scheme for explaining motion in the universe. It obviously, however, required considerable additional refinements to make it fit with the observations of astronomy. In particular it was well known that the sun, moon, planets, and stars all revolved about the earth with different speeds. To explain this, it was thought that the motions took place because the various heavenly bodies were embedded in a series of concentric spheres, each of which rotated at the particular rate peculiar to that sphere. The most distant sphere of the stars rotated the fastest, being kept in motion by God, the "Unmoved Mover," who dwelt in a sphere surrounding all the others. The spheres of the planets (including the sun and moon), nesting one within the other, were kept in motion by the frictional resistance between one sphere and the next.

This was, more or less, the groundwork scheme, which was made more precise and definite by Ptolemy, the Greek astronomer who, in his great compendium entitled "the Almagest," set forth the cosmology of the universe that held sway for well over ten centuries.

It is often the fashion, in discussing the ancient view that the sun, moon, and stars rotate about the earth, to dismiss such ideas as rather naive and not "in accordance with observed facts." The long and bitter struggle to establish the Copernican theory belies the idea that the geocentric theory was ever a foolish prejudice, as well as the fact that a great many well informed men such as Francis Bacon (sometimes cited as the leading spokesman of modern empirical science) rejected the idea that anything but the earth could be the stationary center of the universe.

Most certainly Ptolemy was well aware of the possible explanation of the motion of the heavenly bodies by the alternative hypothesis that the earth is both rotating on its axis and revolving about the sun instead of being stationary.

Certain thinkers . . . have concocted a scheme which they consider more acceptable and they think that no evidence can be brought against them if they suggest for the sake of argument that the heaven is motionless, but that the earth rotates about one and the same axis from West to East completing one revolution approximately every day . . .

These persons forget however that, while, so far as appearances in the stellar world are concerned, there might, perhaps, be no objection to this theory . . . yet, to judge by the [terrestrial] conditions affecting ourselves and those in the air about us, such a hypothesis must be seen to be quite ridiculous.[1]

He goes on with the argument (which was repeated many times in later days in the sixteenth century discussions) that if an object were thrown upward, since it would fall vertically downward toward the center of the earth, it would not fall into its original place, but would fall behind, since the earth would meanwhile have revolved through a certain distance.

Such arguments cannot be dismissed lightly. In fact to dismiss them at all requires the development of a very different type of physics from that of Aristotle, namely, Newtonian physics. (And the Newtonian physics also requires certain assumptions regarding absolute space which have been brought into question once again by the relativity theory. The relativity theory again poses the question: What do we mean when we say that the earth is rotating on its axis—rotating with respect to what?)

The Ptolemaic theory was indeed in agreement with both common sense and observation. Common sense (a rather vague idea that embraces both experience and unconscious assumptions) seems to indicate that the earth is at rest. Note that if we are on a merry-go-round we conclude that we are in motion and that the earth about us is at relative rest for a definite reason, namely, because we experience some effect of the rotary motion. Certainly common sense would tell us that we should experience *some* effect if the earth were rotating.

Second, we may consider the notion that "natural" motion must be motion in a circle somewhat naive and anthropomorphic. But this "naive" idea was a

[1] Sir Thomas Heath, "Greek Astronomy," *Library of Greek Thought,* Dent, London, 1932, pp. 147–148, quoted by Thomas S. Kuhn in *The Copernican Revolution,* Random House, New York, 1959, p. 86.

stubborn one, which certainly did not die until long after the geocentric cosmology had been displaced by the heliocentric one. Copernicus most certainly accepted circular motion as the natural one; Kepler was loathe to give up this idea until he came across evidence of motion almost as simple, namely, motion in an ellipse; Galileo would not even accept Kepler's evidence and remained convinced that the natural motion of bodies in the heavens was circular. If we stop to think about it, why is our notion that *straight line* motion is the natural motion, deviations from which must be explained by special outside forces, any more "natural" than is the Aristotelian circular motion? Have we not been as much dominated by our tradition as were the thinkers of past ages? The answer can only be given in terms of the structure and evidence which has supported our newer points of view.

Almost any theory runs into difficulties if it is to agree with *all* observations. This was certainly true of the Aristotelian theory of circular orbits as accepted by Ptolemy. In particular, observation showed that the planets do not maintain a *constant* eastward drift against the background of more rapidly moving stars, as would be the case if each planet were moving in a circle at a uniform rate slower than the rate at which the stars move westward. In fact it was observed that during certain periods of the year the eastward drift of planets is interrupted and, for a time, they move westward against the background of stars. How is this to be explained in terms of circular motion?

This so-called "retrograde" motion of the planets was explained within the Ptolemaic system by assuming a system of "epicycles." Each planet did move on a circle, but the circle was one that had as its center a larger circle whose center was the earth. See Fig. 4.1.

Even now with the modification introduced by the system of epicycles, the postulate that all heavenly bodies move with circular motion could not explain all the variations in the motions of all the planets, the sun, and the moon. The Ptolemaic system required other additional modifications. Before discussing these, however, it is wise to make a descriptive study of the data of astronomy as revealed by observation.

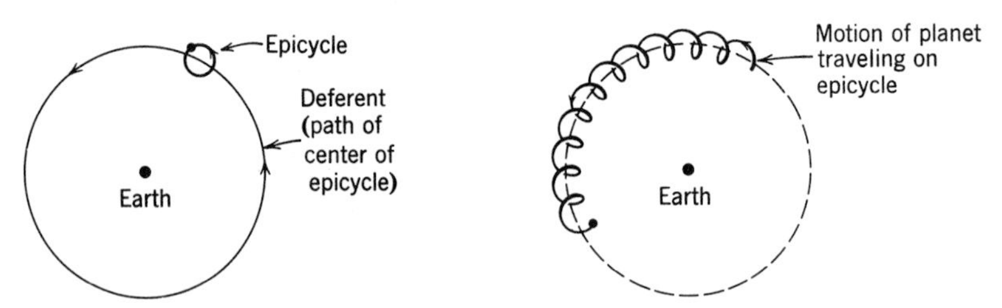

Fig. 4.1. Left, the epicycle is a circle on a circle. Right, retrograde motion occurs when the planet is at the bottom of each loop.

2. A SURVEY OF THE SKY; OR DESCRIPTIVE ASTRONOMY

For the most intimate understanding of a scientific theory it is best to see personally as much of the data supporting that theory as possible. For most theories studied outside of the modern laboratories this cannot be accomplished. However, in the case of the theory accounting for the motion of the heavenly bodies, we are in a position to see much of the supporting data directly by our own observation of the sky. (It is true that the *clinching* evidence in favor of the Copernican theory of a sun-centered rather than an earth-centered solar system depends on telescopic evidence.) At the very least we can achieve a first-hand appreciation of such aspects of astronomical phenomena as retrograde motion, nonuniform motion, changes in the brightness of the planets, etc., in terms of which we shall carry on the discussion of the relative merits of the Copernican and Newtonian theories over the Ptolemaic theories.

This section, therefore, will provide a brief guide to the study of the motions of the stars and planets. The necessity for brevity requires that anyone wishing to observe extensively for himself will have to depend on other sources for detailed information. Star charts and almanacs giving the positions of the planets during the year will be helpful. The charts found in Fig. 4.2 and in Appendix II give a few of the most prominent stars and constellations. An understanding of the use of these should enable the reader to find his way about the sky.

First of all we require a kind of "geography" of the sky so that we can locate the position of the stars, planets, moon, and sun, and relate them to positions of objects on the earth. For such a "celestial geography" we require some fixed points, directions, and a system of coordinates.

Consider first how we establish directions and coordinates on the earth. Since ancient times, even before the shape or size of the earth was known, one directional axis could easily and unambiguously be determined by a vertical shadow stick, or gnomon. Day after day, year after year, the direction of the *shortest* shadow cast by a vertical stick was invariably the same. The direction of north and south can thus be unambiguously and simply defined. At right angles to north and south, east and west are then specified. It happens that at sunrise and at sunset on two days of the year, the directions east and west are also marked by the direction of a gnomon's shadow.

Once the direction of north and south is specified, and once the spherical shape of the earth is established, the required coordinates, latitude and longitude, are simply specified in terms of which any position on the earth is uniquely located. (One arbitrarily selected point, zero longitude, is of course required. By international understanding this is now taken as the longitude through Greenwich, England.)

For "celestial geography" we need similar points and directions, preferably

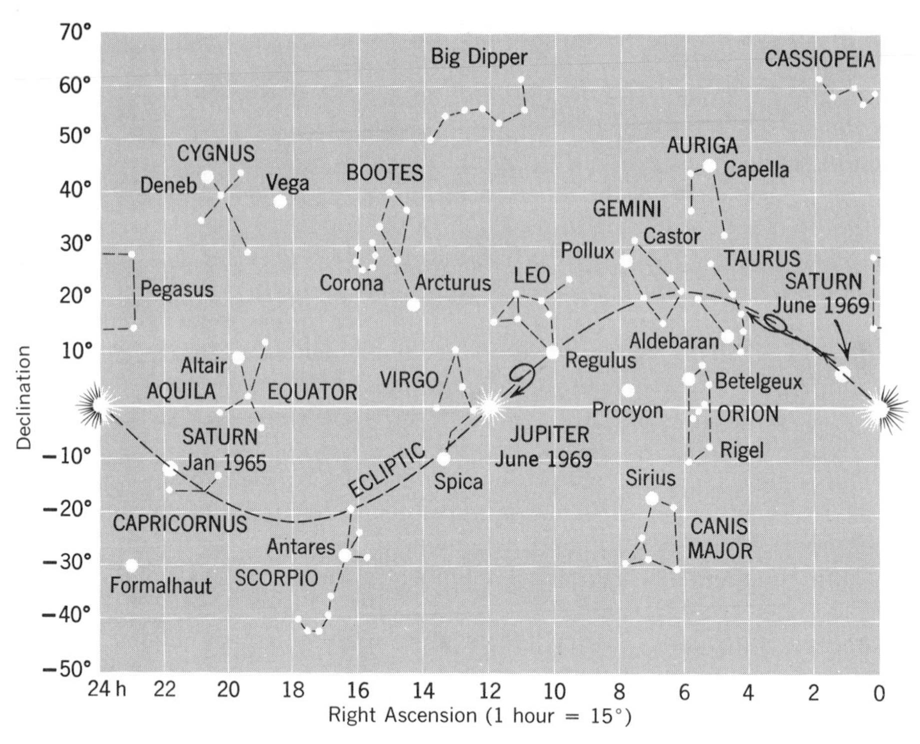

Fig. 4.2. A chart of the brightest constellations and stars. Note particularly the dotted line representing the ecliptic. The planets and the moon as well as the sun will travel against the background of stars along or close to this line. An observer should memorize the position (right ascension and declination) of a few bright stars. Note the position of Jupiter and Saturn as of June 1969. Note that Saturn has moved very little since Jan. 1965.

those which can be readily located by simple observation. If we watch the stars hour by hour, we see them move across the sky like the sun. Can we specify any unique direction? It turns out that we in the northern hemisphere can. If we look up at the constellations in the northern sky, we can see the star patterns rotating counterclockwise hour by hour about a central point. This central point is close to a bright star called Polaris or the North Star. See Fig. 4.3. This point viewed from any location in the northern hemisphere is in an invariable position season after season. Furthermore, its direction coincides with the direction of north as determined from the gnomon marking the shortest shadow of the sun. North and south as determined by observation of the sun coincide with north and south by observation of the stars. (Actually Polaris does not quite coincide with the north celestial pole, but is displaced from it by such a small angle that we can disregard it in naked eye observation.)

Fig. 4.3. A photograph of the northern sky taken by a time exposure of several hours duration. The circumpolar stars are shown leaving circular tracks centered at the north celestial pole. Polaris is shown to be very close to, but not quite identical with the north celestial pole.

This suggests a model to locate the stars. We place the stars on a celestial sphere, conceived of as concentric with the earth, with identical north and south axes, but infinitely far above the earth. There is, then, a celestial equator above the earth's equator, a celestial pole above the earth's north pole. See Fig 4.4. Corresponding to the angles of latitude which mark the angular distance of a point on earth from the earth's equator, there are the angles of declination of the stars with corresponding angular distance from the celestial equator (zero declination). This model is effective, regardless of whether it is believed, as the ancients did, that the celestial sphere rotates in its diurnal motion about the earth, or whether it is believed, as moderns do, that the earth rotates on its axis relative to the fixed celestial sphere.

It is easy enough to find the declination of any star. (Note: for practical observation, measure the angular distance of a star from Polaris and subtract this from 90°.) The problem of finding a "longitudinal" bearing of a star is more complicated. It is obvious that Greenwich cannot be taken as zero "longitude" for a star, since the celestial sphere and the earth rotate relative to one another.

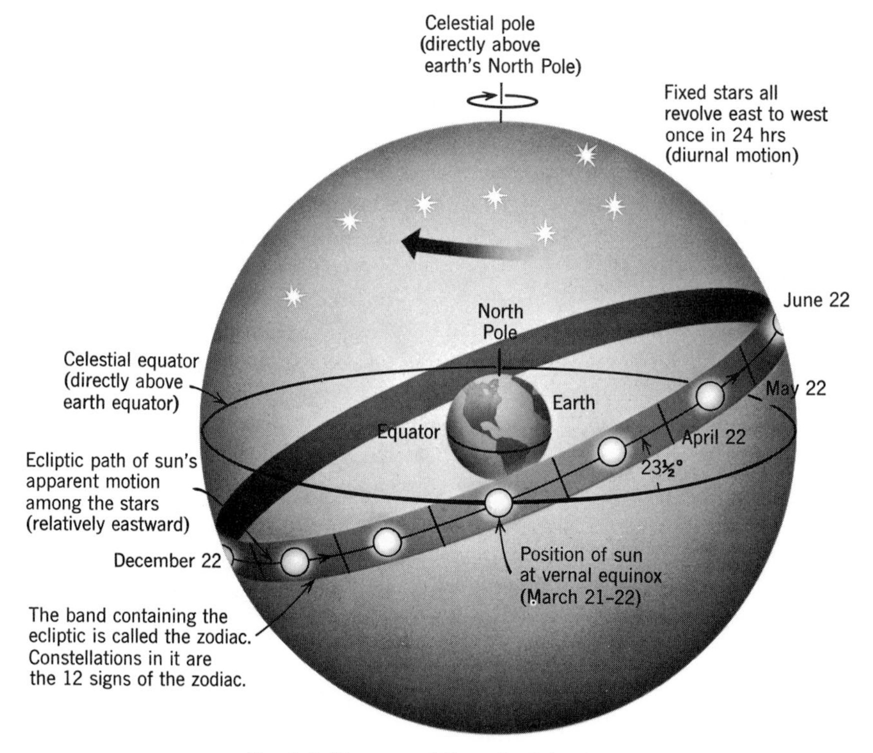

Fig. 4.4. Diagram of the celestial sphere.

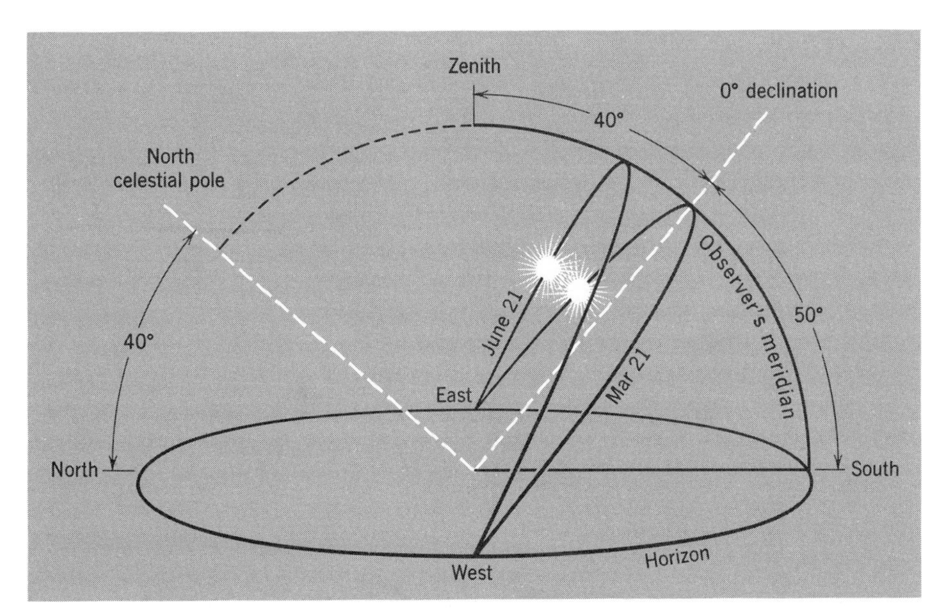

Fig. 4.5. Coordinates based on observer's horizon, drawn from north latitude 40°. The sun's diurnal motion is shown for March 21 and June 21. Note the altitude of sun as it crosses meridian. When the sun's declination is 0°, at vernal equinox, this altitude is 50°; at summer solstice it is $73\frac{1}{2}°$. Note also that altitude of north celestial pole is equal to latitude.

Some other fixed point must be chosen as the zero celestial meridian. This chosen zero point, called the first point of Aries and symbolized by ♈, is where the sun in its annual journey over the celestial sphere crosses the celestial equator at the vernal equinox. Let us see how this is done.

It is not possible to actually observe this motion of the sun relative to the stars, but a little thought leads both to the conviction that such motion does take place and to an understanding of how it takes place on the basis of a few simple observations. First of all, it is readily seen that regardless of the season, a star's declination does not change, thus implying that any particular star always crosses the local meridian (the north-south direction) at the same altitude above the horizon. (Trace the argument: Note Polaris from a given point on the earth has a fixed altitude above the horizon.) But the altitude of the sun above the horizon does vary according to season. Hence its declination changes. See Fig 4.5. Secondly, if one watches the motion of the circumpolar stars, the constellations surrounding Polaris, including the Big Dipper and Cassiopeia two kinds of counterclockwise motions are observed. Hour by hour these stars wheel about Polaris in their diurnal motion. But also, as the seasons change, there is an additional rotation. In October at 9 p.m. the Big Dipper is low on the horizon. Three months

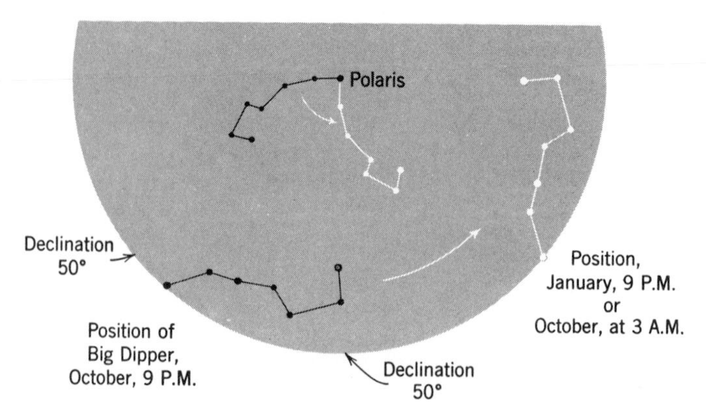

Fig. 4.6. Counterclockwise rotation of the Big Dipper and Little Dipper; diurnal and annual rotations take place at different rates.

later at the same time, it will have rotated through 90°. See Fig. 4.6. In a year's time it will have rotated through 360° in addition to the 365 diurnal rotations. It is only possible for this situation to arise if the observed rotation of the stars from east to west is more rapid than that of the sun. (By approximately how many degrees of arc, how many minutes of time each day?) The sun, therefore, relative to the stars is moving seasonally eastward. If we could observe the stars in daylight, we would see the sun against a different constellation each month, just as indeed we can observe the moon traveling eastward across the constellations from one night to another. The ancients named the series of twelve constellations through which the sun travels in its yearly course, the signs of the zodiac. The path of the sun they named the *ecliptic.*

Looking at Fig. 4.2 and at Fig. 4.4, you will see how the ecliptic path of the sun is related to the positions of the stars. The ecliptic is, of course, at an angle to the celestial equator. If this were not so, the sun's declination would not change during the year and there would be no seasons. At the point of highest and lowest declination the sun's declination angles are $+23\frac{1}{2}°$ and $-23\frac{1}{2}°$. At these points the noonday sun will be vertically above the Tropic of Cancer and the Tropic of Capricorn respectively. In the northern hemisphere a gnomon's shadow will be the shortest at local noon or longest at local noon, and these days will mark the summer and the winter solstice. When the sun's declination is zero, at the vernal and the autumnal equinox, the noonday sun will be vertically above the equator; hence, at these points, the ecliptic crosses the celestial equator. The point of crossing at vernal equinox is chosen as the zero point of celestial "longitude," named *right ascension.* In hours (or degrees) the longitudinal arcs are now marked off *eastward* (direction of sun's march across the sky) and each star now has its appropriate position on the celestial sphere designated by declination and right ascension. See Fig. 4.7.

For convenience of observation, the position of a few bright stars, easily located, will help fix the position of the lines of right ascension. Again refer to Fig. 4.2. The bright winter star, Sirius, is at about 7 hours right ascension. (Note: it will not be seen in June because the sun will be at 6 hours right ascension at summer solstice.) The bright summer star, Vega, is at about 18 hours right ascension. (At midnight, June 21, the sun will be at right ascension 6 hours and almost on the directly opposite side of the earth from Vega.)

Thus the "geography" of the celestial sphere is established on the basis of observation only. This geography was established by the Greek astronomers 2000 years ago on the basis of a model of an earth-centered universe; the stars, the sun, the moon, and the planets revolved about the earth at different rates and at different angles. But the same "geography" works as well in terms of our contemporary model of the earth's rotation on its axis and revolution about the sun, and hence the manner of assigning exact *location* to the stars, planets, sun, and moon in terms of declination and right ascension is retained in modern astronomy even though the "explanation" of the motion is very different. It is important to realize that stars remain very nearly "fixed" in their relative positions season after season, year after year. Their declination and right ascension do not

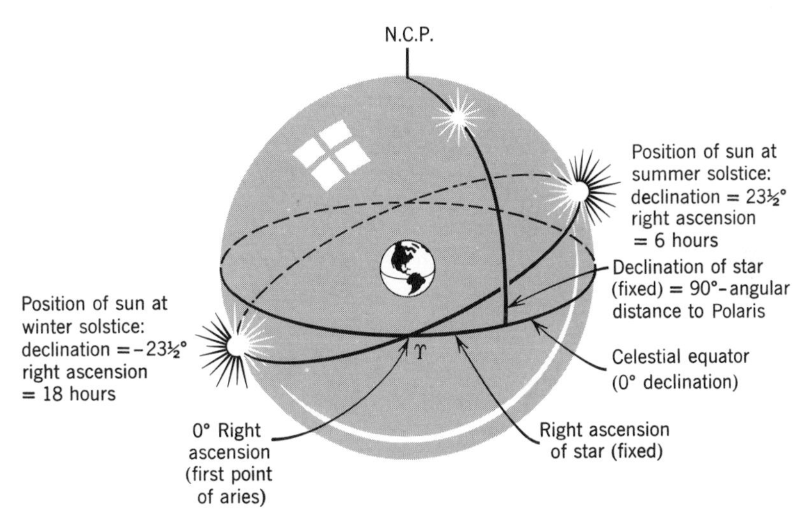

Fig. 4.7. Coordinates of the celestial sphere. The declination of a star is the angular distance from celestial equator. Right ascension is measured eastward from ♈, the intersection of ecliptic and celestial equator. Stars remain fixed, retaining same declination and right ascension. Sun, moon, and planets constantly change declination and right ascension, but all move along paths close to the ecliptic.

change; those of the sun, moon, and planets do change at varying rates, season by season.

For emphasis, the following major points about celestial geography are now summarized: (*1*) The North Pole of the celestial sphere is directly above the North Pole of the earth. (In the Northern Hemisphere, it is extremely convenient that the point on the celestial sphere, which is the North Pole, is marked within one degree by a star, called Polaris, or the North Star, which can easily be located.) (*2*) The equator of the celestial sphere is likewise directly above the equator of the earth. (*3*) *Declination,* the angle of a star above the celestial equator corresponds to degrees of latitude on the earth; a star at a declination of 90° means it is directly above the North Pole. (*4*) The earth is considered at rest, and stars travel in their diurnal motion (24-hr motion) about the earth in the direction shown by arrows, in Fig. 4.4, from east to west. (*5*) The "path of the sun"—the motion of the sun against the background of the stars in its yearly course through certain of the constellations—is called the *ecliptic.* This is shown as a line that makes a 23½° angle with the equator. This line runs through a band of stars called the stars of the zodiac. During each month, as the sun travels slowly eastward against the background of stars, it is located in a different constellation called a "sign of the zodiac." (Of course, this motion of the sun is very slow compared to its diurnal motion. It shifts its relative position by a very small angle every day.) The moon and the planets themselves all have paths that are very close to the ecliptic. They will, therefore, be always located also in one of the zodiacal constellations. (*6*) Where the ecliptic crosses the equator twice a year, the points of equinox are located. One of these (vernal equinox) is arbitrarily selected as the point of 0° longitude (just as Greenwich is arbitrarily selected as 0° on earth). This point is designated as 0° right ascension. Every 15° right ascension is designated as an hour. See Fig. 4.8.

Unfortunately, with respect to ease of understanding, we do not live at the North Pole of the earth, and the North Star, which marks the North Pole of the celestial sphere does not appear vertically above us, but appears somewhere below the zenith (point directly above us) and above the horizon, depending on our latitude. The dependence of the angle of elevation (angle above the horizon) of the North Star upon the latitude can be studied with the aid of the diagram in Fig. 4.9.

Figure 4.9 should be studied until the following relationships are clearly understood: since the declination of the North Star is 90°, and since the light rays from the North Star are all parallel (the distance of the North Star being considered to be practically infinite), it follows that if a person is standing on the earth at latitude of 40°, then the elevation of the North Star above the horizon will be 40° and the zenith angle of the North Star will be 50°.

(*Question:* What will the angle of declination of a star which is at the zenith

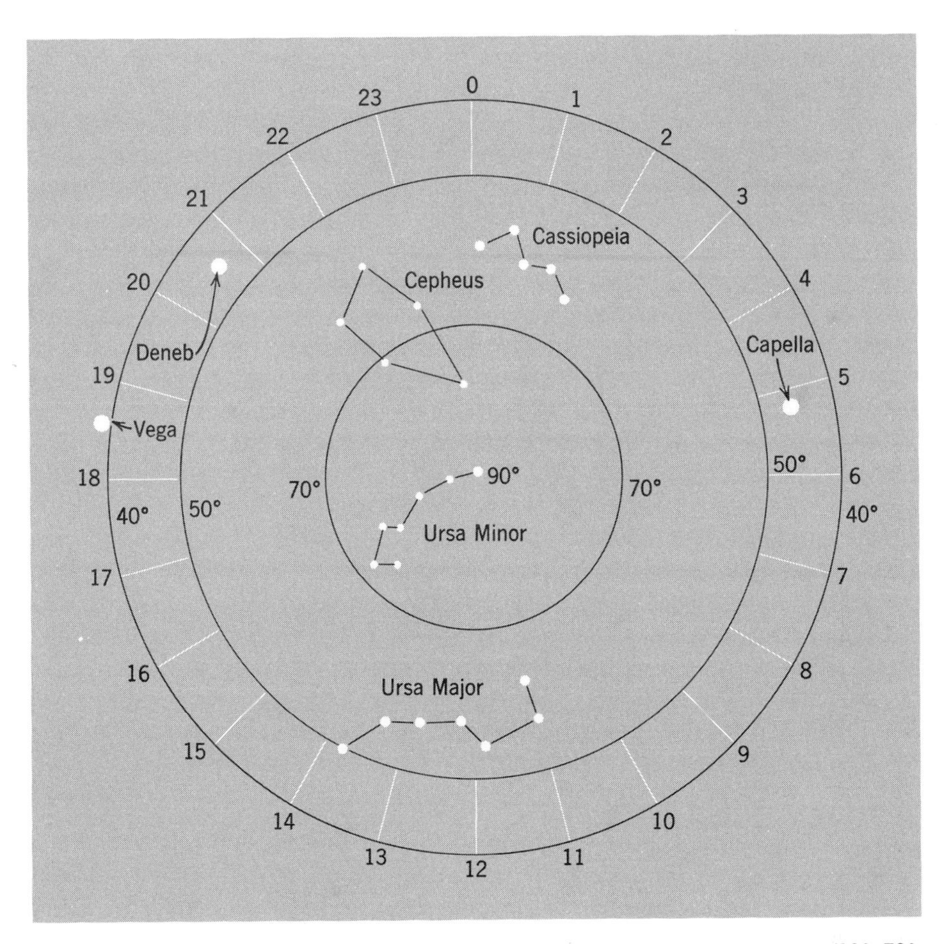

Fig. 4.8. Chart of the northern stars. In the chart, some angles of declination are shown (90°, 70°, 50°, 40°) and hour angles of right ascension (0 to 24). The constellations shown never go below the horizon when viewed from 40° latitude or above. A star such as Capella, Vega, or Deneb with declinations close to 40° will be almost directly overhead at some season, and will never sink far below the horizon. These are all first-magnitude stars. As Vega sets in the northwest, Capella is rising in the northeast. For location of stars, it is useful to note the hour angles 0 to 12 relative to the constellations Cassiopeia and the Big Dipper (Ursa Major), and the hour angles 5 to 18 relative to Capella and Vega.

be, if an observer is standing at 40° latitude? What will be the smallest angle of declination a star can have which will never sink below the horizon at the observer's latitude and will therefore be visible all year round?)

Next, let us consider the group of stars called the circumpolar stars. The first step in finding one's way about the stars is to be thoroughly familiar with this

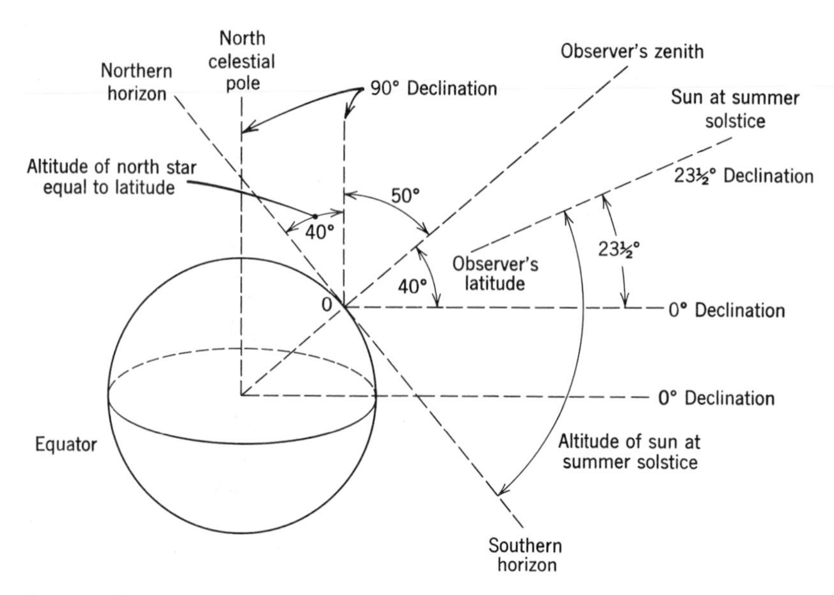

Fig. 4.9. Lines to zenith, horizon, 90° and 0° declination, drawn to a point on the earth at 40° north latitude.

particular group, how they move nightly and according to the season of the year.

First, let us study the "Big Dipper" or "Ursa Major." Probably most of us are very familiar both with this constellation and how it can be used to locate the North Star. However, it is well to review the way in which the constellation of the Big Dipper moves. In general, we can say that the Big Dipper revolves about the North Star, as do all the other circumpolar stars, once in 24 hours (very nearly) in a counterclockwise direction as we look northward. We do not keep track of time by the stars but by the sun. Therefore, the stars rotate in a counterclockwise direction a little more than 1 revolution of 360° every 24 hours. (Why? Can you determine how much more?) Hence, while the Big Dipper appears in the position shown at around 9 p.m. in an evening in October, with every ensuing 24 hours it will have rotated counterclockwise by a small amount and by the end of 3 months it will appear in the second position shown in the diagram in Fig. 4.6.

With the above observations in mind, let us study the star chart in Fig. 4.8, which lists some of the more important northern stars. A few first-magnitude stars, which have declinations somewhat less than 50°, are included as guide posts. They will be visible during some portion of the night at almost any season of the year.

One other point should be noted in the diagram in Fig. 4.8. There appear not only lines marking angles of declination, but lines marking what corresponds to longitudinal angles (instead of angles that divide the circle into 360°, the divisions are in hours, 1 hour corresponding to 15°, and 24 hours corresponding to 360°).

Another pertinent point is that although the circumpolar stars remain above the horizon, others do not. They rise and set just as the sun rises and sets, but some of them just skirt above the horizon in very short arcs, while others pass overhead, rising northeast and setting northwest. Those that describe the short segments of arcs, close to the southern horizon, have a negative declination, that is, they are below the celestial equator. Those that pass high overhead in latitudes near New York City, reach positions higher in the sky than the sun ever reaches. Since the sun at its *highest* point has a declination of 23½° at the beginning of summer (summer solstice, around June 22), these stars have declinations much greater than 23½°. In contrast, stars with a declination of less than −50° can never be seen in northern latitudes greater than 40°. The Southern Cross, for example, cannot be seen in the United States unless one is in the southernmost portion of the country. Those stars, on the other hand, with a declination of 50° or greater, are visible during all seasons of the year from latitudes of 40° or more. Those with a declination of 50° will, during certain seasons of the year, just skirt the northern horizon and, during other seasons of the year, they will appear at their highest point 10° from the zenith. (*Question.* In your locality, how close to the horizon will the end of the handle of the Big Dipper approach?)

Having learned the relative positions of the northern stars, these positions can be used to identify the other stars and constellations. All stars and constellations, of course, retain their relative positions, even though from hour to hour, and from season to season, they change their position relative to the directions of the compass and to the angle of elevation as observed by a person standing on the earth.

As a guide toward acquaintance with a few of the brighter stars and constellations, the seasonal maps shown in Appendix II should be studied. These maps record the approximate positions of the major stars at latitude 40°, as seen at 9:00 p.m. in the evening about the middle of the season stated for each map. (For earlier and later dates or hours, the maps should be rotated somewhat; that is, the meridian line, due south of the observer, will be passed by stars moving westward at the rate of 15° per hour and 30° per month.)

The first task in becoming familiar with the constellations is to note the very brightest stars, those of the first magnitude. In the latitude of about 40°, there are 15 first-magnitude stars (these are not all visible at once), some of which can be easily seen and located during each season.

For the location of the *planets,* Fig. 4.2 (a complete map of the stars of the

Northern Hemisphere showing the position of the *ecliptic*) will be useful. The ecliptic is the path of the sun against the background of the stars. It should be remembered, however, that this map is necessarily distorted, just as are maps showing all the continents of the earth. The distortion is inevitable because we are showing the positions of objects which are on a spherical surface, by points on a plane. In particular, just as the line of longitude on the maps of the earth are often represented as parallel lines on a plane surface, likewise our map of the sky shows the hour angles as parallel, when actually, in both cases, they of course meet at the North Pole of either the earth or the celestial sphere. One reason for studying the position of the ecliptic path is that the planets traverse paths very close to the ecliptic. See Fig. 4.10. Thus, any bright object in the sky, not identified as a well-known star, which is close to the line of the ecliptic is almost certain to be a planet.

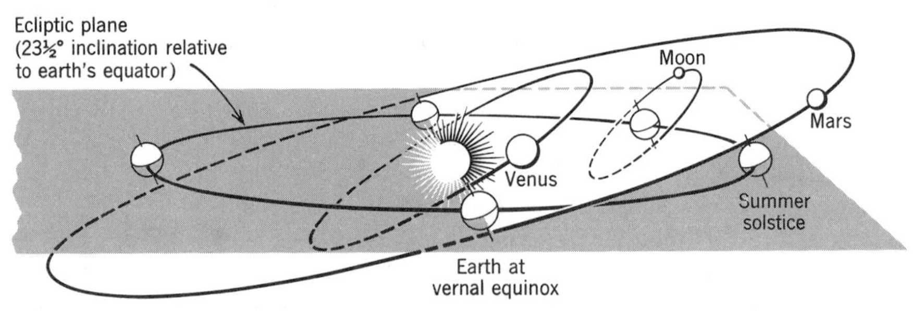

Fig. 4.10. Schematic diagram showing the ecliptic plane and how the orbits of Mars, Venus, and the moon are inclined to the ecliptic plane. (Angles, as shown, are greatly exaggerated for emphasis. The planets and the moon have orbits lying very close to the ecliptic plane. With the exception of Pluto the orbits of the planets are all inclined to the ecliptic plane at angles less than 5°.)

The positions of the planets cannot be permanently specified, since they continually change. But almanacs do show these positions for any date over a considerable number of years, and exact locations can be found this way. Or, more conveniently, the magazines such as *Sky and Telescope*,[2] which show positions of stars and planets month by month, can be consulted.

[2]Sky Publishing Co., Harvard Observatory, Cambridge, Mass.

However, since the positions of Jupiter and Saturn change relatively slowly, Jupiter making its rounds of the sky once every 12 years, and Saturn taking 30 years to complete its circuit, the positions of these two planets as of June, 1969, are shown in Fig. 4.2. In one year's time, Jupiter will have moved eastward by about 30 degrees and Saturn will have moved eastward but at a slower pace. The motion eastward will not, however, be a steady one, being interrupted by periods in which the motion is actually westward. Thus, the planets will appear to make loops in the sky when judged against the background of the fixed stars. This so-called *retrograde motion* was one of the big problems that any astronomical system had to explain. Note that neither the sun nor the moon has any retrograde motion during its course around the sky, although the motion of these bodies is not uniform. The nonuniformity of motion of the sun is a little difficult to observe but, in the case of the moon, a careful series of observations will disclose this nonuniformity.

Not only is this type of retrograde motion observed in the case of all the planets, but other phenomena are also readily detected, which must be accounted for in any systematic astronomical explanation. Among the most important are: the great variation in brightness of the planets; the fact that even during the eastward motion the planets do not move uniformly; and the fact that in the case of two of the planets, Venus and Mercury, they are never observed very far from the position of the sun. They are thus always seen only as either "evening stars," low on the horizon in the west shortly after sunset, or as "morning stars" in the east before sunrise.

3. THE COPERNICAN THEORY

Now that we have completed a brief survey of what might be termed descriptive astronomy, we return to the problem of a theoretical explanation of the large assortment of observations contained in that survey. Is there some simple set of principles by which we can bring together all the observations into a single system? Can we so develop such a system that it not only accords with the data of observation but will enable us to predict additional phenomena?

The Ptolemaic system was such an attempt, and it was accurate to so great an extent that it went unchallenged for over 1000 years. To repeat the three most basic principles that guided the Ptolemaic system: (*1*) the earth is at rest, and all other celestial bodies are in motion; (*2*) all celestial motions are in the form of circles; and (*3*) the motions in circles are uniform (that is, the speed of a planet, the sun, etc., as it describes its circle, remains constant). The major problem was to explain departures from uniform motion.

We have already seen how epicycles are needed to account for the so-called retrograde motion of the planets. Another modification, however, was also found necessary by Ptolemy to account for the observed fact that the moving bodies in the sky do not appear to move even in the epicycles with the constant motion required by principle. To explain this nonuniform motion, two additional devices were introduced: the eccentric, and the equant. These are diagrammed in Figs. 4.11 and 4.12 as they might be applied to the motion of the sun (which takes 6 days longer to pass from vernal equinox to autumnal equinox than from autumnal equinox to vernal equinox—which implies, of course, that the sun is moving somewhat more rapidly against the background of stars in winter than in summer).

To account for all the vagaries of planetary motion within the Ptolemaic system, it became necessary to make use of many combinations of these three devices: the epicycles, the eccentrics, and the equants. In fact, by the sixteenth century, there were in some systems as many as 64 epicycles as well as an assorted use of equants and eccentrics. Furthermore, since there was within this system no restriction on the radius of any epicycle or circle being used, many alternative combinations were suggested, and there was indeed no single Ptolemaic system universally accepted. Instead, there were many different variations of the system developed by different astronomers, all vying with one another, but

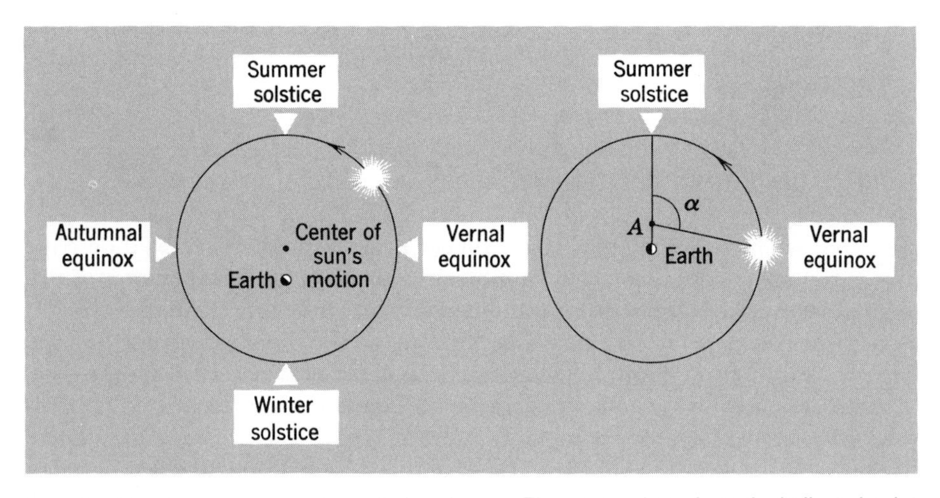

Fig. 4.11. Diagram of an eccentric applied to the sun. The sun resolves about the indicated point instead of about the earth as a center. Hence, at one point (just after winter solstice, actually in January), the sun is closest to the earth.

Fig. 4.12. Diagram of an equant as applied to the sun. This is used to explain the non-uniformity of motion. The sun moves in a circle about the earth. The angle, α, formed by lines drawn from the sun at different positions and properly selected equant point A, varies uniformly to account for the fact that with respect to the earth as center the angle would not vary uniformly.

all remaining within the framework of the three principles that we have listed.

In addition to these complications, during the span of thirteen centuries that had elapsed between the *Almagest* and the sixteenth century, deviations and inaccuracies that were too small to be apparent over short lengths of time, began to show up. (One of the tasks that the pope wished Copernicus to perform was that of reforming the calendar. The seasons and festivals were no longer in step with the appropriate dates in the year. Copernicus refused to undertake this task, considering that the then current astronomical theories could not provide an adequate base for an accurate calendar. The revised calendar, which is still in use today, was finally introduced among Catholic countries in 1583, but not until 1752 among the British dominions, and not until 1917 in Russia.)

Such was the background faced by Copernicus as he pursued his astronomical studies. He was offended both by the monstrous complexity of the Ptolemaic system as well as by the apparent inaccuracies. He therefore tried to develop a more elegant and orderly system for the explanation of the motions of celestial objects. It is interesting to read how he expressed his thoughts in the famous preface to his great work *De Revolutionibus Orbium Coelestium* (*On the Revolution of the Celestial Spheres*), published in 1543, the year of his death. In this preface, addressed to Pope Paul III, Copernicus first indicates that he is aware that his system may not be accepted at once, and then gives some of the main reasons which led him to propose the new system:

I may well presume, most Holy Father, that certain people, as soon as they hear that in this book about the revolutions of the spheres of the universe, I ascribe movement to the earthly globe, will cry out, that, holding such views, I should at once be hissed off the stage . . .

So I should like your holiness to know that I was induced to think of a method of computing the motions of the spheres by nothing else than the knowledge that the mathematicians [that is, the astronomers] are inconsistent in these investigations.

For, first, the mathematicians are so unsure of the movements of the Sun and Moon that they cannot even explain or observe the constant length of the seasonal year. Secondly, in determining the motions of these and the other five planets, they use neither the same principles and hypotheses nor the same demonstrations of the apparent motions and revolutions . . . Those again who have devised eccentric systems, though they appear to well-nigh have established the seeming motions by calculations agreeable to their assumptions, have yet made many admissions which seem to violate the first principle of uniformity of motion. [At this point Copernicus was objecting strongly to the use of equants; he himself was not averse to using eccentrics, but the equant appeared to him to be inconsistent with uniform motion.] Nor have they been able thereby to discern or deduce the principal thing—namely, the shape of the Universe and the unchangeable symmetry of its parts . . .

I pondered long upon this uncertainty of mathematical tradition. . . At last I began to chafe that philosophers could by no means agree on any one certain theory of the mechanism of the Universe, wrought for us by a supremely good and orderly Creator . . .

After quoting some Greek thinkers, who had considered the possibility that the earth moved around the sun, Copernicus continues:

> Taking advantage of this I too began to think of the mobility of the earth; and though the opinion seemed absurd, yet knowing now that others before me had been granted freedom to imagine such circles as they chose to explain the phenomena of the stars, I considered that I also might easily be allowed to try whether, by assuming some motion of the earth, sounder explanations than theirs for the revolution of the celestial spheres might be so discovered.[3]

Above, we have a summary of the motivating reasons for Copernicus's break with tradition. Notice that he maintains two of the three basic principles of the Ptolemaic system: (*1*) motion in a circle, and (*2*) the uniformity of the motion. May we emphasize again: one of the more impelling reasons for his break with Ptolemy was the fact that he felt that the several systems developed on the Ptolemaic assumptions did not actually retain the principle of uniform motion in its purest form. To preserve this principle, Copernicus abandoned the principle that the earth is stationary.

As a system for the *qualitative* explanation of the motions of the spheres, the Copernican system is simpler and also, for most scientists, more "elegant." (One criterion by which a theory takes precedence over another in the thinking of scientists is quite often a matter of aesthetic feeling or "taste"—for which it is as difficult in this field as in others to prescribe standards, but which must be taken into account. We shall encounter other examples later, in which such criteria have played a definite role in leading to the development of new theories.)

Let us look first at the explanation of retrograde motion on the basis of the Copernican theory. We shall use the motion of Jupiter as our example. Jupiter takes about 12 years to complete its circuit around the sun, although the earth takes 1 year. Therefore, the earth is moving with an angular velocity 12 times that of Jupiter. As the earth moves from A to B in its orbit in Fig. 4.13, Jupiter moves from a to b, and the image of Jupiter is seen against the background of fixed stars at a', and then at b', in which a' is west of b'. That is, as the earth is overtaking Jupiter, its motion against the fixed stars appears to be backward from east to west. However, after the earth has reached the point C and Jupiter has reached c, the image of Jupiter, c', now appears east of its former position, b', that is, the motion is once more in an eastward direction.

Next, let us examine the two alternative explanations of the orbits of Mercury

[3]Excerpts from Nicolaus Copernicus, *De Revolutionibus,* translated by John F. Dobson and Selig Brodetsky, printed originally as "Occasional Notes," Royal Astronomical Society, No. 10, 1947, quoted by Thomas Kuhn in *The Copernican Revolution,* Random House, New York, 1949, pp. 137–138. A full translation of *De Revolutionibus Orbium Coelestium* by Charles G. Wallis can be found in *Great Books of the Western World,* Volume XVI, Encyclopedia Britannica, Chicago, 1952.

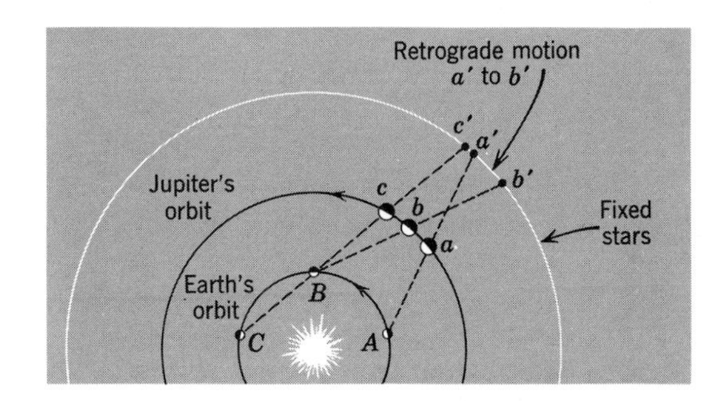

Fig. 4.13. Diagram explaining the retrograde motion of Jupiter.

and Venus. Here, the problem is to explain why, of all the planets, these two alone are never far from the sun in angular distance. In fact, the angle between the earth-to-sun line and the earth-to-Mercury line is never greater than about 28° (when this angle is maximum the planet is at its greatest elongation), which means that Mercury in this position sets about 2 hours after sunset, or rises about 2 hours before sunrise. It is therefore difficult to see except during these few days near greatest elongation. (Probably most modern-day persons have never seen Mercury in their lives!) Venus is much farther in angular distance from the sun and, consequently, it rises and sets considerably before and after the sun. When it is at an angle of 39° east or west of the sun, considerably before dawn or after sunset, it becomes the brightest object in the sky.

The two diagrams in Fig. 4.14 show the alternative explanations. While both are perfectly possible, it should be noted that the Ptolemaic explanation does have two rather artificial elements about it: (*1*) the center of the epicycles of

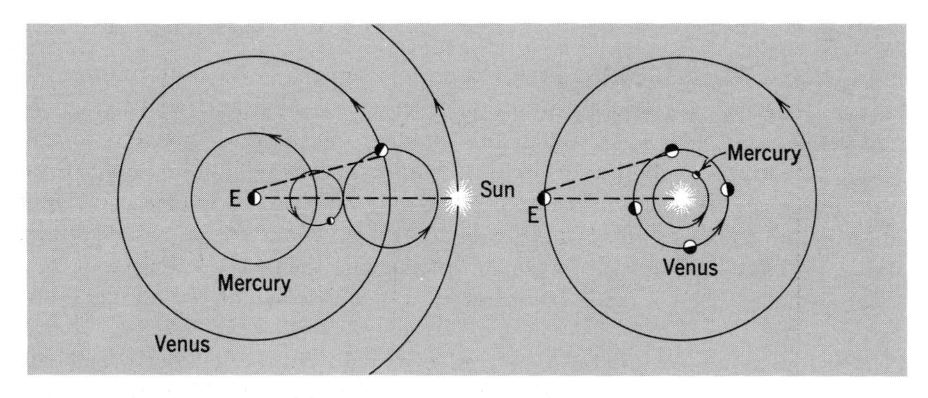

Fig. 4.14. Alternative explanations of the orbits of Venus and Mercury. *Left,* Ptolemaic system. The centers of the epicycles on which Venus and Mercury move must be in line with the sun. *Right,* Copernican system. The angular distance of Venus from the sun cannot exceed a maximum (at greatest elongation).

both Mercury and Venus must remain in line with the sun and therefore their period of rotation about the earth is identical with that of the sun; and (*2*) there is no convincing reason in this system for ascribing any order in the two orbits; either one could be closer to the earth. Indeed, some astronomers did consider Venus closer—others Mercury. Neither of these artificial-looking elements is present in the Copernican explanation. Mercury *must* be closer to the sun. Furthermore, as with the other planets, the periods of revolution of Mercury and Venus were in the same direction of magnitude as their distance from the sun. (From the Copernican system, it followed that the longer the period the farther away was the planet from the sun.)

Unfortunately, as Copernicus developed his system in later portions of his book, he was forced to introduce many of the same types of complexity that he had objected to in the Ptolemaic system. He was able to avoid the use of the equant, which he disliked so much, but to account for variations in the angular speed of the sun and the planets, he did have to introduce eccentrics and epicycles. In fact, the number of interlocking circles became more than thirty before he was through. Furthermore, his system did not turn out to be any more accurate than that of Ptolemy. It is not at all surprising that for many years his system was rejected not only by the theologians but by most astronomers. Why give up the traditional system for a new one, when the new theory was so contrary to the common-sense notion that the earth does not move, and when it contradicted the only system of physics that existed at that time? The only real advantage of the Copernican system was the qualitative simplicity of the explanations. This did appeal to a few astronomers, who began to make use of the system, often using it as a convenience without believing it to be "true." Galileo, however, with his great insight, accepted the Copernican system as valid even before any direct evidence appeared.

In addition to the failure of the Copernican system to be either more accurate or quantitatively simpler than the Ptolemaic system, there was one serious additional fault that was immediately pointed out. If the earth did revolve about the sun in an orbit known to have a radius of about 93 million miles, then, from two positions 6 months apart, a star should be observed to have a different angle. Although this shift of angle, called parallax, could not be observed from the ends of any base line on the earth, because the stars were known to be at a great distance, surely it could be observed from a base line of 180 million miles. See Fig. 4.15.

When no parallax could be discovered, even with this tremendously long base line, many considered this to be sufficient proof against the Copernican thesis that the earth revolved about the sun. Copernicus himself, however, pointed out that if the nearest star was at a great enough distance—millions of times farther than had been previously thought possible—then such parallax would be so small as not to be detectable by the instruments then in use.

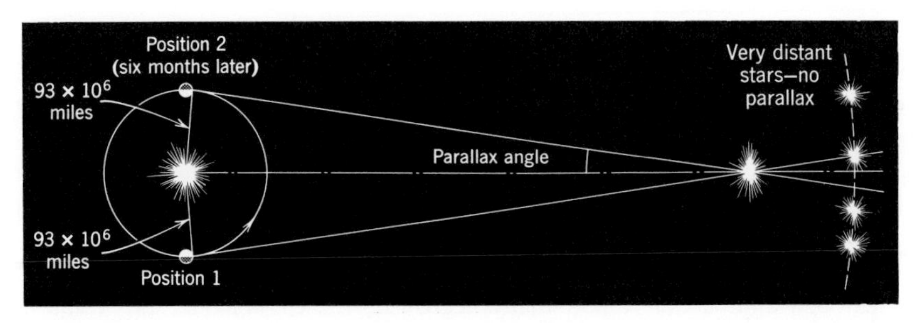

Fig. 4.15. Diagram explaining parallax. The farther away any star is, the smaller its parallax will be. The nearest star has a parallax angle of less than one second of arc (0.75″).

We now know that the answer of Copernicus was correct, but the parallax is indeed so small that it was not detected until very nearly 300 years after the time of Copernicus. In other words, although within this period indirect confirmation of the Copernican theory had grown to convincing proportions, the most direct and indisputable evidence in its favor was not found until three centuries after the theory was proposed.

It is not surprising that the contemporaries of Copernicus were reluctant to accept the tremendous enlargement of the scale of the universe that the theory entailed. The accepted cosmology placed the outermost sphere of the heavens at not too great a distance beyond Saturn. This was the abode of God and the angels. Nothing existed beyond. The new theory opened up all sorts of new possibilities, which ran counter to traditional thoughts and beliefs. We shall discuss further the controversy over the Copernican theory in the next chapter. Here, let us emphasize that failure to detect any parallax constituted an extremely serious, if not insurmountable argument against the theory that the earth revolved about the sun.

One more point should be mentioned. If we examine the diagrams in Fig. 4.14, showing the alternative explanations of the orbits of Venus and Mercury, one thing stands out. On the basis of the Copernican theory, the phases of Mercury and Venus should resemble closely the phases of the moon, proceeding from a thin crescent to practically a full phase. (A study of the diagram will show that the phase will never be quite full.) On the other hand, in the explanation of the Ptolemaic system, the phases of these planets, if they were planets reflecting only the light of the sun, would never reach beyond the crescent stage. Of course, before the advent of the telescope it was impossible to find out anything about the shape of the discs which constituted the source of light from Venus and Mercury. Nevertheless, we see here an important aspect of scientific theories. A theory should not only explain and bring into order what is known; it should make *predictions* that can later be tested. The Copernican theory, in this case, was the basis of a very definite prediction, a prediction that could not be true if the alternative theory were correct. Also, and importantly, the theory was vulnerable: it could be disproved if the conclusions predicted by it were not in fact found to be in agreement with observed data.

4. QUESTIONS

1. Through approximately how many degrees do the circumpolar stars rotate in 30 minutes?

2. Why cannot a lunar eclipse ever occur when the moon is not full?

3. Why are lunar eclipses observed more frequently than solar eclipses?

4. Can the existence of the seasons be accounted for as well by the Ptolemaic system as by the Copernican system?

5. The "arctic circle" is placed at 66½° latitude. What is the reason for the choice of this particular latitude? Explain the six months of night and the six months of daylight at latitudes north of 66½°.

6. Where will the North Star appear if you are standing on the equator? On the North Pole?

7. Does a star have the same declination and right ascension regardless of the season? Why is it possible to see certain stars in one season but not in others?

8. Sometimes a theory that is known to be invalid remains in use for certain purposes because of its simplicity. Navigators continued to use the scheme of a celestial sphere long after the Ptolemaic system had been discarded. Can you think of other theories or schemes known to be invalid which, nevertheless, continue to be used as a guide to action?

9. The earth does not travel about the sun in a perfect circle, but in an ellipse of very slight eccentricity. What evidence could you gather by direct and careful observation to indicate that this is true?

10. The earth is at its closest approach to the sun early in January. Why does not the warmest season of the year always arrive close to this time everywhere on the earth?

11. What are some of the ways by which you can recognize a planet?

12. Draw a diagram of the crescent moon as seen in the eastern sky and also as seen in the western sky.

13. How could you prove that the North Star is the closest bright star to the celestial pole? Polaris is actually about 1° in angular distance from the north celestial pole. What is its declination?

Problems

1. Prove that the altitude (angle of elevation above the horizon) of the North Star is equal to the latitude of the observer. Draw the necessary diagram, remembering that a star on the celestial sphere is so far distant that all the rays from such a star are parallel.

2. What is the declination and right ascension of the sun on December 21, March 21, June 21, September 21? If there is a full moon on June 21, what will the moon's right ascension be on that date? What will be its approximate declination? (*Hint:* recall that the moon's path lies within a few degrees of the ecliptic, as indeed do the paths of all the planets.)

3. What is the altitude of the sun as viewed from latitude 28° north, as it crosses the meridian on December 21?

4. From your latitude what is the smallest declination which a circumpolar star may have?

5. The nearest star is about 4 light years distant. What is the parallax angle? Is it surprising that it took so long to discover parallax?

6. A "parsec" is often used as a measure of a star's distance. It is defined as the distance at which a star will have a parallax angle of one second of arc. What is the equation relating the distance of a star in parsecs to the parallax angle?

7. How many light years is one parsec?

8. Can the moon ever have a retrograde motion?

9. Latitude may readily be determined by "shooting" the sun as it crosses the meridian, if its declination on that day is known. Can longitude be similarly determined if the right ascension of the sun is known on that day? What else is required?

Recommended Reading

Thomas S. Kuhn, *The Copernican Revolution,* Random House (Modern Library paperback), New York, 1959. This book provides an excellent and detailed discussion of all aspects of the Copernican revolution and should be consulted by anyone interested in this development.

E. Rosen, *Three Copernican Treatises,* Columbia University Press, New York, 1939.

Robert H. Baker, *Astronomy,* D. Van Nostrand, Princeton, N.J., 1955. This book may be valuable in aiding the reader to grasp the details of both descriptive and theoretical astronomy.

Henry M. Neely, *A Primer for Star-Gazers,* Harper, New York, 1946. This book will be helpful to anyone interested in learning the positions of the constellations.

Motz and Duveen, *Essentials of Astronomy,* Wadsworth Publishing Co., Inc., Belmont, Calif., 1967. This thoroughly up-to-date review of astronomy may be used as a reference source for this and other chapters. It covers the whole background of physics required for astronomy.

Edwin C. Kemble, *Physical Science, Its Structure and Development,* M.I.T. Press, Cambridge, Mass., 1966. Excellent and rigorous treatment of the development of the concepts upon which modern astronomy is founded. Highly recommended both because it includes topics not contained in this text and because it approaches some of the same topics from another point of view. Elementary treatment with historical survey.

chapter five

The significance of the Copernican revolution: Galileo and Kepler

Man's genius in inventing new systems of thought is matched by his genius in inventing methods of crushing opposing thoughts. In the ancient world, Socrates was forced to drink hemlock because he was "corrupting youth" —encouraging them to think a little deeper. A few centuries later, in 390 A.D., the extraordinary library in Alexandria, containing 440,000 manuscripts representing the pinnacle of ancient wisdom, was burned by order of a Christian bishop and, in this same period, the last Alexandrian mathematician, Hypatia, a woman renowned for her remarkable mathematical insights, was murdered in a most revolting manner by a crowd encouraged by the Christian authorities. Christian thought at that time identified learning with a heathen challenge to its authority. Learning had to be crushed; the so-called "Dark Ages" were ushered in.

About twelve centuries later, the story was repeated in a new form. By this time, Christian theology had been systematized by fusing the Judaic-Christian revelations with Aristotelianism. Upon this highly developed philosophy rested the church's authority. Galileo challenged one of the basic tenets, that is, that the earth stood still, and he was imprisoned and forced to recant. And Bruno, drawing implications from the same new Copernican theory, was burned at the stake.

If the thought should occur that such persecution of ideas could not take place in modern times, mention need only be made of the burning of the books and the banning of the teaching of Einstein's theories under the Nazis, in a country proud of its scientific achievements. Even in our own country, freedom in expressing ideas is not universally guaranteed, and certain parallelisms can be drawn between the Oppenheimer case and the treatment of Galileo.[1] Similarly, in Russia, under Stalin, the authorities persecuted biologists who would not follow the theories of Lysenko regarding genetics. While the modern methods by which authorities exert control over ideas are more subtle than the ancient ones, complacency is hardly warranted.

In this chapter we shall examine the whole controversy that raged for over a century around what is termed the Copernican Revolution. This should illuminate the relationship between philosophy and science as well as the process by which one scientific theory becomes replaced by another.

1. MEDIEVAL THOUGHT; SCHOLASTICISM AND OPPOSITION TO COPERNICUS

After about 400 A.D. interest in science among Europeans faded. For the next eight or nine hundred years the leaders of Western civilization were more interested in problems of social and political organization or of personal salvation than in those of understanding nature. St. Augustine, author of *The City of God* and *Confessions,* reflected the climate of thought. The Platonic theory of two worlds—the world of ideas and of the spirit, which constituted the basic reality, and the world of senses which was an unreal, flitting passage—became the central doctrine of the Catholic church. But it was a Platonism that was purged of any interest in mathematics. The mystical element was dominant.

In this situation, all the advances of Greece, in mathematics, art, science, and literature might easily have been lost forever except for one of those strange turns frequent in history. The Arabs were coming into contact with the remnants of Greek civilization as the vigorous movement of Mohammedanism spread westward. They became excited over all aspects of Greek learning and, quickly, they

[1]See the passage quoted subsequently on page 94 from *The Crime of Galileo,* by Giorgio de Santillana. See also *Lawrence and Oppenheimer* by N. P. Davis, Simon & Schuster, 1968.

translated into Arabic the works of Hipparchus, Galen, Archimedes, Aristotle, Euclid, Ptolemy, and others. Thus, some of these works became preserved for posterity. With the spread of the Arabs to Cordova and Sevilla, they brought their translations into Europe. In turn, many of them began to be retranslated into Latin around the tenth century. Not only did the Arabs thus redistribute Greek learning, but they also made their own significant contributions, particularly in mathematics, astronomy, medicine, and chemistry (then under the guise of alchemy).

As the works of Aristotle became available to European thinkers, a gradual shift took place away from the mystical Platonism toward the more naturalistic and empirical point of view of Aristotle. Finally, a systematic synthesis of Christian theology and Aristotelian science and philosophy was achieved in Thomas Aquinas's voluminous work *Summa Theologica,* which became and remained official church doctrine. In this scheme, the Ptolemaic theory fitted perfectly.

The resulting cosmology was a neat and imaginative one. The earth was at the center of the universe, set there purposely by God for the abode of man and as the stage for the whole drama of struggle and redemption. About the earth revolved the moon, the sun, and five other planets, as well as the stars, each set in "crystalline spheres" concentric about the earth. God and the redeemed dwelt in an outermost sphere, which constituted the heavens. God as the "unmoved mover" kept the crystalline sphere of stars in rotation, and the planets beneath moved in their courses by the frictional resistance between the nest of spheres. Hell was down toward the center of the earth. Purgatory was located at the antipodes opposite Jerusalem.

The universe was definitely finite, and the distance even to the sphere of stars was not unimaginatively far away. Dante's epic *The Divine Comedy* gives a poetic and dramatic picture of this cosmology. In the epic it was possible for the poet to travel through all the spheres in the course of seven days. The abode of the blessed was not too distant. (Since any possible base line on the earth would be too short for a determination of even a distance several times as far away as the sun, no serious effort was made by the astronomers to estimate the distance of the sphere of fixed stars. All that could be said was that it was outside the sphere of the most distant planet then known, Saturn, but the distance was still thought of as extraordinarily close when compared to modern conceptions.)

One other aspect of this cosmology should be mentioned, that is, that the planets and stars were all considered to be made of some pure substance of a nonearthly characteristic. They were all unchanging in composition. In particular, the stars were of an immutable nature, since they did not ever vary in brightness.[2]

[2]It had apparently been forgotten that the Arabs had already, in the medieval period, noted variations in Algol, the "devil star."

It was in this atmosphere of belief that Copernicus published his work *On the Revolution of the Spheres* in 1543. It caused no immediate stir. It was a difficult book and written primarily for technical astronomers. Even in the book's preface the suggestion was made (probably interpolated by Osiander, who saw the book through its publication) that the Copernican system could be regarded as an alternative method of calculation, and was not to be regarded as "philosophical truth." Certainly Copernicus did not concern himself with any of the implications of his work, and certainly did not consider it heretical.

It was not long, however, before the implications did begin to become clear. Most important was the idea that the earth no longer remained the *center* of the universe, but had become merely one of a number of planets revolving around the sun. Second, the sun was no longer regarded as being in motion—a theory which was in direct contradiction to a number of Biblical passages. Third, since Copernicus could explain the lack of parallax only by placing the stars at tremendous distances—millions of times farther than any previous estimates—the universe was now expanded to vast distances beyond what could be readily visualized, and God and the heavenly host were thus also now far removed. In addition, it now became perfectly possible to consider that these stars were not actually embedded in a sphere. Since they are so far away, under the new conception, they could be thought of as extending in a scattered fashion all the way to infinity.

This indeed was the conclusion drawn by Bruno, who traveled widely over Europe lecturing on the Copernican theory and propounding the heresy that the universe was infinite and that there were many other worlds outside our own upon which men might exist. Bruno was actually executed in 1600, primarily for his heresy regarding the doctrine of the trinity but, because of his espousal of the Copernican theory and its implications, his execution called immediate attention to the Copernican theory. Soon afterward the Church became really alarmed at the spread of Copernicanism, and took more and more violent action to prevent its spread. Copernicus's book was placed on the Index Expurgatorius in 1616, where it remained until 1822. The church refused to countenance any books that even cited the possibility of the earth's motion.

Thomas Kuhn has suggested that part of the Catholic church's reaction against Copernicus was the result of the pressure to take this position because of the strong anti-Copernican position of the protestants. The protestants regarded any theory of the earth's motion and the sun's stability as a challenge to the literal interpretation of the Bible. An insistence of the literal interpretation and the rejection of *any* interpretation by an organized church implied the rejection of any statement contrary to a Biblical passage. Thus, Martin Luther, in 1539, having heard about the new theory even before it was actually published, had this to say:

People gave ear to an upstart astrologer who strove to show that the earth revolves, not the heavens or the firmament, the sun and the moon . . .

This fool wishes to reverse the entire science of astronomy; but the sacred scripture tells us (Joshua 10:13) that Joshua commanded the sun to stand still, and not the earth.[3]

And other followers of Luther joined in the condemnation of Copernicus:

Now it is a want of honesty and decency to assert such notions publicly, and the example is pernicious. It is the part of a good mind to accept the truth as revealed by God and to acquiesce in it.[4]

In addition to the opposition from theologians, both catholic and protestant, it should be remembered that many others, often with excellent scientific training and insight, rejected the Copernican theory for two major reasons: (*1*) the failure to find any parallax displacement of stars; and (*2*) a lack of any physical explanation of how such a massive body as the earth could be kept in motion, or if it *were* in motion why it would not fly apart, or why a bird while in flight, for example, would not be left behind by the earth's rotation.

Tycho Brahe, one of the greatest of the observing astronomers, who laid the groundwork for later discoveries by his precise and careful observations, rejected the Copernican theory entirely. On his deathbed he pleaded with his assistant Kepler to maintain a modified Ptolemaic system and to give up interest in the obviously absurd Copernican theory.[5]

Similarly, Francis Bacon, who is occasionally referred to as the father of scientific empiricism, dismissed Copernicanism, even after hearing of telescopic evidence, with these words in 1622:

In the system of Copernicus there are found many and great inconveniences; for both the loading of the earth with a triple motion [rotation, revolution and changes in the tilt of the axis] is very incommodius and the separation of the sun from the company of the planets with which it has so many passions in common is likewise a difficulty and the introduction of so much immobility into nature . . . all these are speculations of one, who cares not what fictions he introduces into nature, provided his calculations answer.[6]

In the above quotation, we note how Copernicus is being criticized by Bacon for inventing "fictions," which are not based upon philosophical foundations, but are introduced to make his calculations come out right. We shall meet the same kind of objection again and again. Wherever hypotheses that are not directly

[3]Quoted by Thomas Kuhn, *The Copernican Revolution,* Modern Library, New York, 1959, p. 191.

[4]Ibid., p. 191.

[5]A fascinating account of Brahe's instruments and observations and of his part in establishing the basis for Kepler's later work is given in *The Life and Times of Tycho Brahe,* by John Gade, Princeton University Press, Princeton, 1947.

[6]Quoted by Morris Kline in *Mathematics in Western Culture,* Oxford University Press, New York, 1953, p. 117.

observed are introduced into the structure of scientific thinking, for a time these hypotheses are apt to be regarded as strange and artificial until they finally become absorbed into the intellectual tradition. For a long time, Newton's concept of gravity acting at a distance was not accepted as plausible. By the twentieth century, most people ceased to question it seriously but accepted it as an intuitively correct notion. Today the layman has difficulty in grasping the possibility that space is curved, or that time intervals may perhaps depend on the relative position and motion of the observer. In each case the method was based not upon establishing generalizations as a summary of raw observation, but upon the hypothetical-deductive method: the most valid explanation is the one from which consequences may be deduced that are in accord with the data of observation. "Fictions," as Bacon labeled them, introduced to "make the calculations come out right," are commonplace in modern science.

In summary, the strong objections to the Copernican theory can be seen to be founded on grounds which, although they may seem strange and foreign to us today, had enough solidity and persuasiveness that only the boldest thinkers were captured by the new point of view when it was first published.

2. EVIDENCE FROM TELESCOPIC OBSERVATION; GALILEO'S CONTRIBUTION

Until Galileo turned his telescope toward the night skies, the arguments in favor of the Copernican theory did not convince a wide audience. From 1610 when Galileo published his *Message from the Stars,* in which he announced some of his early observations with the telescope, discovery after discovery concerning the planets, moon, and sun, followed one another rapidly. These new observations, Galileo believed, should have convinced all honest persons of the truth of the Copernican system, and he proceeded to teach and to write accordingly. We list a few of these observations.

(*1*) On studying the planet Jupiter, Galileo saw in his telescope four moons, or satellites, circling the planet. This provided a model of a miniature solar system. Here could be seen celestial bodies, which were definitely *not* revolving about the earth.

(*2*) The surface of the moon was carefully examined. The mountains—very similar in appearance to mountains on the earth—could be discerned. Galileo actually measured the heights of these mountains with a surprising precision.

(*3*) The rotation of the sun upon its own axis was established through observations of the sun spots, which moved across the face of the sun.

(*4*) Most important of all, the phases of Venus became clearly visible. One will recall that according to the Copernican theory, Venus should show practically the same phases as our own moon, although the Ptolemaic system implied that only crescent phases would appear. See Fig. 4.14, p. 81. This discovery alone at once forced those astronomers, who maintained that the earth was central and immobile, to revise the Ptolemaic system drastically along lines that had been proposed by Tycho Brahe in which the earth remained central, but the planets other than the earth revolved about the sun. That such a revision was necessary opened another door to doubt regarding the Ptolemaic system.

However, even after all these telescopic observations, strong opposition to the Copernican theory continued. Often the evidence was rejected completely as an optical illusion. The Italian astronomer, Francesco Sizzi, for example, argued that there had to be exactly seven planets (including the sun and the moon) corresponding to the seven days of the week, to the seven metals of alchemy, to the seven openings in the head, and to the seven cardinal virtues. The satellites of Jupiter were invisible to the naked eye, and thus could have no influence on humans and, therefore, had no real existence.[7]

The controversy raged on. Galileo was warned in 1616, the year in which Copernicus' book was placed on the Index Expurgatorius, to cease advocating and teaching that the earth moves and not the sun. However, by 1632, after the election of Pope Urban VIII who previously as a cardinal had been the scientist's friend, Galileo decided to publish *Dialogues on the Great World Systems* in which he persuasively advanced the Copernican thesis.

The *Dialogues* became so widespread in influence that finally the Inquisition moved against Galileo. As an old man of seventy, he was finally forced to abjure his beliefs, after having been held under house arrest and threatened with jail. Upon his knees he was made to repeat his recantation, which went in part:

> . . . I wrote and printed a book in which I discuss this new doctrine [that the earth is not the center of the world and moves] already condemned and adduce arguments of great cogency in its favor without presenting any solution of these, I have been pronounced by the Holy Office to be vehemently suspected of heresy, that is to say, of having held and believed that the Sun is the center of the world and immovable and that the Earth is not the center and moves.
>
> Therefore, desiring to remove from the minds of your Eminences, and of all faithful Christians, this vehement suspicion justly conceived against me, with sincere heart and unfeigned faith I abjure, curse, and detest the afore-said errors and heresies and generally every other error, heresy, and sect whatsoever contrary to the Holy Church, and I swear that in future I will never again say or assert, verbally or in writing, anything that might furnish occasion for a similar suspicion regarding me . . .[8]

[7] Giorgio de Santillana, *The Crime of Galileo,* University of Chicago Press, Chicago, 1955, p. 13.
[8] Ibid., p. 312.

And so the then desperate authorities were able, although only temporarily, to crush freedom of thought and inquiry. Doubtless the story will be repeated again and again in the history of the future. Giorgio de Santillana comments:

> And lest our straining for beams make us overlook our own domestic gnats, we may perceive in the Oppenheimer case a parallel which is a shade too close for comfort. In such vastly different climates of time and thought, whenever the conflict comes up, we find a similarity of symptoms and behavior which points to a fundamental relation.
>
> True, the Oppenheimer case is very different from Galileo's as to context. Today there is a tendency not to suppress physics but rather to exploit it; a tendency to act not on deep philosophic differences but on mere issues of expediency. Yet, as the story unfolds before the public, the exact analogy in structure, in symptoms and behaviors, shows us that we are dealing with the same disease. Through the little that we are allowed to know, we can discern the scientific mind as it has ever been—with its free-roaming curiosity, its unconventional interests, its detachment, its ancient and somewhat esoteric set of values (it is the scientist, we may remember, who is reproached with having brought the concept of "sin" into modern contexts)—surprised by policy decisions dictated by 'Reasons of State' or what are judged to be such.[9]

On the other hand, and in a more optimistic vein, it is proper to note the modern attitude of the Catholic church. A Belgian priest, Abbé Lemaitre, the modern astronomer who first proposed the theory that the universe was born in a cataclysmic explosion several billions of years ago and has been expanding ever since, is quoted as saying:

> Religion has no bearing either on my theories or that of the steady state. I do not believe the Bible was intended to explain such things as cosmology. I do not believe that God ever intended to disclose to man what man could find out for himself.[10]

In the 17th century it is doubtful whether these sentiments could have been so freely expressed without arousing the wrath of organized religion.

3. KEPLER'S CONTRIBUTION TO THE COPERNICAN REVOLUTION

However powerful and persuasive Galileo's arguments in support of the Copernican System appeared to be, more conclusive evidence was still required to establish convincingly the superiority of this system over other possible ways of looking at the universe. Two additional developments were required: (*1*) the development of a new physics to replace the Aristotelian physics, and (*2*) a better-fitting description of the orbits of the planets than the Copernican theory could supply. We shall consider the first development in the next two chapters. Now we turn to the problem of the orbits.

[9] Ibid., p. viii.
[10] *Newsweek,* September 4, 1961.

Kepler, contemporary and friend of Galileo, reflected much of the same kind of faith in the harmony and simple mathematical orderliness of the universe as we have already encountered in Copernicus and Galileo. This innate sense of mathematical order was manifested by Pythagoras and Plato and is, in many ways, opposed to the more direct empiricism of Aristotle. A. N. Whitehead has said:

The Platonic world of ideas is the refined, revised form of the Pythagorean doctrine that number lies at the base of the real world. Owing to the Greek mode of representing numbers by patterns of dots, the notions of number and of geometrical configurations are less separated than with us. Also Pythagoras, without doubt, included the shape-iness of shape, which is an impure mathematical entity. So today, when Einstein and his followers proclaim that physical facts, such as gravitation, are to be construed as exhibitions of local peculiarities of spacio-temporal properties, they are following the pure Pythagorean tradition. In a sense, Plato and Pythagoras stand nearer to modern physical science than does Aristotle. The former two were mathematicians, whereas Aristotle was the son of a doctor, though of course he was not thereby ignorant of mathematics. The practical counsel to be derived from Pythagoras, is to measure, and thus to express quality in terms of numerically determined quantity. But the biological sciences, then and till our own time, have been overwhelmingly classificatory. Accordingly, Aristotle by his Logic throws the emphasis on classification. The popularity of Aristotelian Logic retarded the advance of physical science throughout the Middle Ages. If only the schoolmen had measured instead of classifying, how much they might have learnt![11]

Kepler, careful astronomer and mathematician though he was, remained continuously in the posture of a mystic—listening, at least with his mind, for the "music of the spheres." He sought constantly to find this harmony in some simple formula that would bring order into the multitude of observations on planetary motion. He was convinced of the truth of the Copernican theory, but felt it his duty to straighten out some of the difficulties that marred its beauty:

I certainly know that I owe it [the Copernican theory] this duty that as I have attested it as true in my deepest soul, and as I contemplate its beauty with incredible and ravishing delight, I should also publicly defend it to my readers with all the force at my command.[12]

He felt that he needed to bring about a more complete exposition of the harmonies of the system: the number, distances, and periods of the orbits of the planets became his chief concern. One of his early attempts was to show that the circular orbits of the planets had radii that were related to one another as were the five regular solids:

I undertook to prove that God, in creating the universe and regulating the order of the cosmos, had in view the five regular bodies of geometry as known since the days of Pythag-

[11] Alfred North Whitehead, *Science and the Modern World,* Macmillan, New York, 1928, p. 2.
[12] E. A. Burtt, *Metaphysical Foundations of Modern Science,* Doubleday, New York, 1954, p. 58.

oras and Plato, and that he has fixed according to those dimensions, the number of heavens, their proportions, and the relations of their movements.[13]

Kepler found great delight in this regularity, although it did not prove in the end to be very fruitful. In spite of all his deep desire to establish simple relationships, he was too good a scientist to maintain a theory that did not fit the observations satisfactorily, and his studies soon led him to conclude that the motions of planets were not well described by uniform motion in a circle. After studying the orbit of Mars for many years and groping for a better scheme, he finally came upon the discovery that the orbits of the planets could be described as ellipses, with the sun at one focus of the ellipse.

This discovery, called Kepler's first law, was in itself superb. Although the circle was now abandoned, at least the "next most perfect curve," one of great simplicity and symmetry, and one closely related to the circle, was substituted. Swept away were all the epicycles, deferents, eccentrics, and equants, which had so encumbered previous descriptions of planetary motion.

But Kepler could not remain satisfied, for just as important as the proper description of the orbits was the problem of the mathematical description of the motions of the planets. What kind of uniformity could be found? Again, to his delight, he discovered the secret: in equal intervals of time a line drawn from the sun to the planet will sweep out equal areas. This is known today as Kepler's second law. We shall examine its meaning below.

Even now Kepler was not satisfied: what could be the relation between the various orbits and the times of revolution of the planets in those orbits? After more years of study, he came upon the solution and announced in *The Harmony of the World,* which was published in 1619: ". . . after I had by unceasing toil through a long period of time, discovered the true distances of the orbits, at last, at last the true relation," namely, that for all the planets the squares of their periods of revolution were proportional to the cubes of their mean distances from the sun.[14] This is known as Kepler's third law.

Summarizing, the three laws are:

1. The planets travel in elliptical orbits with the sun at one focus.
2. For every planet, in equal intervals of time, equal areas are swept out by a line drawn from the sun to the planet.
3. Comparing the orbits of the several planets, it is found that the times of revolution are related to the distances from the sun by a proportionality: the *squares* of the periods of revolution are proportional to the *cubes* of their mean distances from the sun.

[13] Morris Kline, *Mathematics in Western Culture,* Oxford University Press, New York, 1953, p. 113.
[14] G. Holton and Duane Roller, *Foundations of Modern Physical Science,* Addison Wesley, Reading, Mass., 1958, p. 149.

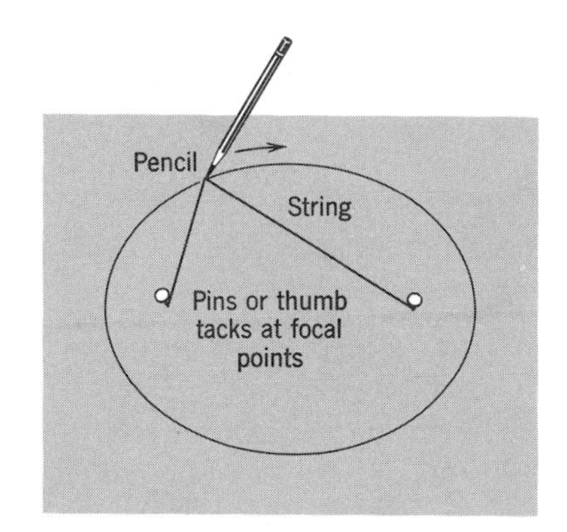

Fig. 5.1. How to draw an ellipse. The closer the thumb tacks are placed, the more nearly circular the ellipse will become. The orbits of the planets are almost circular.

Let us examine briefly the meaning of these laws.

Law 1. An ellipse can be studied by either geometric or algebraic methods. From its definition it is an easy figure to draw. It is the locus of all points the sum of whose distances from two given points is a constant. If one sticks two pins in a piece of paper, tying to them a single piece of string considerably longer than the distance between the pins, one can draw an ellipse by moving a pencil within the loop of the string in such a way that the string remains taut. This is illustrated in Fig. 5.1.

The construction in the figure satisfies the definition of the ellipse because every point on the curve is located in such a way that the sum of its distances from the two pins (called the focal points) is constant. See Fig. 5.2.

It will be noticed that as the two focal points are brought closer and closer together the shape of the ellipse becomes more and more nearly a circle until, when the two points coincide, a true circle results. In other words, a circle can

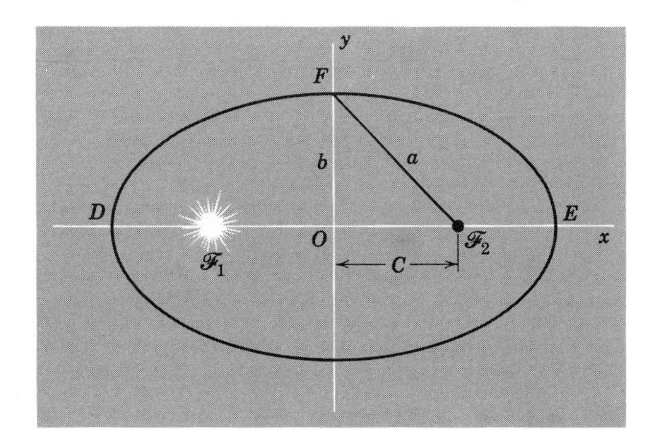

Fig. 5.2. Properties of an ellipse. The equation is $x^2/a^2 + y^2/b^2 = 1$. Note the following: If $a = b$, ellipse becomes a circle. The greater the inequality of a and b, the greater the eccentricity which is defined as $\epsilon = c/a$, where $a^2 = b^2 + c^2$. The semi-major axis is OE, the semiminor axis is OF. The *mean* distance of a planet from a focal point is equal to semi-major axis. If sun is at focus $\mathscr{F}_1$, D is perihelion point, E is aphelion point.

be regarded as a special case of an ellipse, that is, an ellipse in which the distance between the two focal points is zero. It happens that the planets, for the most part, have orbits that are very nearly circular. If the orbits are drawn to scale, their departure from circularity can hardly be detected by the eye. The comets and the planet Pluto have highly eccentric orbits. Halley's comet, for example, has an orbit that brings it closer to the sun than the earth at perihelion, and farther than the distant planet Neptune, at aphelion.

Law 2. Kepler establishes a relationship between the speed of a planet and its position in an orbit. This can best be understood by the diagram in Fig. 5.3.

In this figure, assume that the planet moves from position 1 to 2 in a given interval of time, say one month, and then later moves from position 3 to 4 in one month. Kepler's law states that the white areas of the diagram are equal. Without analyzing the meaning of this law (it will be discussed again later), it can be shown that an alternative statement of the law is that the product of the distance from the sun to the planet times its velocity remains constant. (The more enterprising of us may wish to give arguments in support of this by analyzing the diagram and noting that the white areas are approximately triangles.)

Law 3. This law of Kepler's can be neatly expressed as an equation: $T^2 = kR^3$ where T is the period of revolution and R is the mean distance of planet from the sun. The square of the time of revolution is thus found to be proportional to the cube of its mean distance from the sun.

These three laws finally brought simplicity into the Copernican system. No longer was there need of epicycles or eccentrics. Here was a group of easily expressed regularities that fitted all the motions of all the planets. While it is true that circular motion was abandoned, it is also true that an ellipse is almost as simple a figure as a circle. The only real source of difficulty remaining was to account for elliptical motion and, for this, a new physics was needed. The Aristotelian concept of the naturalness of circular motion had to be abandoned, unless some method could be found to explain forces which distorted circles into

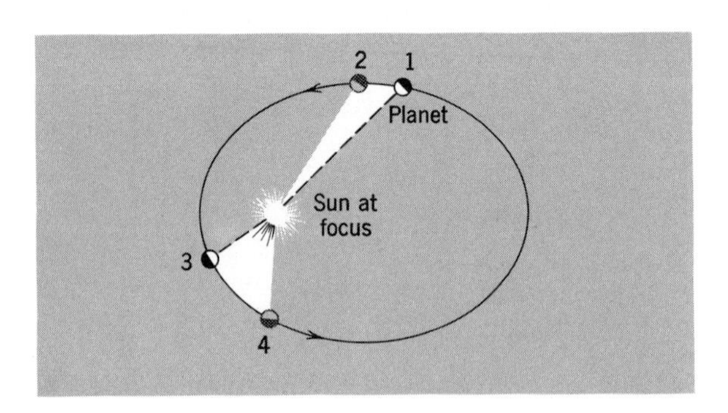

Fig. 5.3. Kepler's second law. The planets do not move with a uniform speed (as they would if they traveled in a circle). They move more rapidly at perihelion than at aphelion. The white areas are equal.

ellipses. The actual solution finally was developed by Newton in 1687, a solution we shall be studying in Chapter 7.

Kepler tried to find some "explanation" of why the orbits were as described, but he failed. His three laws, therefore, remained what may be termed "empirical" laws; that is, relationships (usually equations) that are established on the basis of observation but whose explanation, or connection with other basic principles, is not known. Often the first step in scientific analysis is to establish such empirical relationships, later seeking to show how several such relationships are tied together into a single theory, which theory is then often spoken of as an "explanation" of the empirical laws. When Newton developed his system, as we shall see, the three laws of Kepler were shown to be consequences derivable from the basic postulates of Newtonian theory. At that point an explanation for the laws was said to have been found.

4. QUESTIONS

1. Venus was once regarded as two separate stars, the "morning star" and the "evening star." Discuss the use of a "model" in identifying these two "stars" as one.

2. The synodic month is the period of time between two identical successive phases of the moon (e.g., between two full moons). It differs from the sidereal month (the period of time for a complete revolution about the earth relative to the fixed stars). Which of these months is longer? Draw a diagram to explain.

3. Why did Galileo's discovery of the moons of Jupiter and the sunspots render the Copernican theory more plausible? How does this "evidence" compare with the evidence entailed in the discovery of parallax, which came more than two centuries later?

4. Sometimes it is useful to refer to the north pole of the ecliptic. What is its relationship to the north celestial pole? If the planets are assumed to travel in circular orbits around the sun, they will all travel in a counterclockwise, nearly circular motion as reviewed from the ecliptic pole. As discussed in Chapter 4 the sun appears to move eastward along the ecliptic. Draw a diagram of the earth's orbit as viewed from the ecliptic pole and show the direction of rotation. Is it clockwise or counterclockwise?

5. Tycho Brahe, the astronomer under whose guidance Kepler worked and studied, was unwilling to accept the heliocentric model of the solar system. He introduced a revised system of the geocentric model to account for the phases of Venus which had been observed by Galileo. Can you think of how such a compromise model might be constructed?

6. If Kepler's third law is valid, what role, if any, does the *mass* of an earth satellite play in the determination of its orbit? In what way does this make problems of a "space platform" for assembly and refueling easier?

7. On the basis of Kepler's second law, and because the earth is at perihelion (closest distance to the sun) in winter, explain why the earth's orbital speed is greater in winter

than in summer. Explain why there are not an equal number of days between September 21 and March 21 and between March 21 and September 21.

8. Draw several ellipses of various eccentricities (the farther apart the two focal points, the more eccentric the ellipse). Explain how a circle and a straight line may be considered as limiting cases of an ellipse. Compare the equation of an ellipse, that is: $x^2/a^2 + y^2/b^2 = 1$ (where a and b are the semimajor and semiminor axes) with that of a circle $x^2 + y^2 = R^2$ and note why the equation of a circle is a special case of the more general equation of an ellipse. (Elliptical and circular orbits of satellites are similarly related. If a satellite does not have precisely the right velocity and direction of flight when it goes into orbit, its orbit will be elliptical and not circular.)

9. One of the objections to the Copernican theory that the earth was in rotation, was that an object dropped from a height would be "left behind" as the earth rotated beneath it and that it would therefore fall backward. Can you, at this point, develop a counter-argument?

10. In Kepler's third law, namely, that $T^2 = kR^3$, is the proportionality constant, k, the same for the moons of Jupiter as for the planets revolving about the sun and the satellites revolving about the earth? Cite evidence in support of your answer. Can you make any shrewd guesses at this point as to what data k might depend upon?

Problems

1. It has been proposed that three satellites might be placed in orbit about the earth in such a way that each would remain in a stationary position above the earth. In this manner, communication to all parts of the earth could be maintained by the bouncing of radio waves off the reflecting satellite. Calculate the radius of the necessary orbit and the orbital velocities. (Hint: Use Kepler's third law and your knowledge of the moon's period and distance from the earth.)

2. Calculate the value of k for Kepler's third law as applied to the solar system. (Use your knowledge of earth's period of revolution and distance to the sun.) Compare with k for satellite orbits.

3. The distance between the earth and the sun varies between about 9.2×10^7 miles and 9.4×10^7 miles. What per cent variation is there in the angular diameter of the sun as observed from the earth between these two positions?

4. Taking the distance between the sun and Venus as 7×10^6 miles, show that the angle of greatest elongation (greatest angular distance from the sun as observed from earth) is about 48°.

5. On the average by approximately how many degrees will the declination of the sun change during one day? When will the rate of change per day be the fastest? the slowest?

6. If an earth satellite is traveling in a circle about 1000 miles above the surface of the earth, what will its period of revolution be?

7. Using the fact that the period of Jupiter's revolution is about 12 years, find its mean

distance from the sun in astronomical units. (An astronomical unit is the mean distance of earth to sun.)

8. What method might have been pursued by Galileo in determining the height of a mountain on the moon? What measurements must be made? Can you surmise the approximate limits of precision?

Recommended Reading

Thomas S. Kuhn, *The Copernican Revolution,* Random House, New York, 1959. An excellent scholarly historical survey to which the author of this book is greatly indebted.

Angus Armitage, *Sun, Stand Thou Still,* Henry Schuman, New York, 1947.

Giorgio de Santillana, *The Crime of Galileo,* University of Chicago Press, Chicago, 1955.

Sir Oliver Lodge, *Pioneers of Science,* Dover Publications, New York, 1960 (chapters on Copernicus, Galileo, and Kepler).

I. Bernard Cohen, *The Birth of a New Physics,* Doubleday, New York, 1960.

Thomas S. Kuhn, *The Structure of Scientific Revolutions,* University of Chicago Press, Chicago, 1962. Another excellent book by Kuhn of a broader and more general nature than his *The Copernican Revolution.* Anyone interested in how revolutions in thought develop should study this work.

Arthur Koestler, *The Sleepwalkers,* Macmillan, New York, 1959. An authoritative and readable account of the lives of early astronomers through Newton, along with an analysis of the background in which they worked.

chapter six

The birth of experimental science

We have been discussing one aspect of a great revolution in thought that took place in the sixteenth and seventeenth centuries. Up to this point we have concentrated primarily upon systems for describing the motions of stars and planets, and we have seen how it became gradually more apparent that the Copernican view was more effective than the Ptolemaic one.

This shift in viewpoint brought out one problem clearly. The Ptolemaic system fitted well with Aristotelian physics, which gave a systematic explanation of motions. However, the Copernican description, particularly after Kepler's discovery of elliptical orbits, was in basic contradiction to the Aristotelian principle that there are two types of natural motion: the uniform motion of celestial bodies in circular paths, and the motion of earthly bodies "seeking" the center of the earth. The problem was compounded by the fact that no explanation could now be given as to any cause for the motion of the planets.

In this situation, a new physics was required, which would provide a more coherent explanation of the motion of all bodies whether terrestrial or celestial. While this new physics was not to be completely established until Newton published his great work in 1687, the groundwork was carefully and brilliantly laid by Galileo.

The work of Galileo is also important because it established a new method, a new direction for scientific investigations. A careful study of the methods he used will provide an illustration of the basic approach that has subsequently dominated the physical sciences. Since the concepts he used are closely connected to our everyday experience and the measurements he made are easily understood, Galileo's work forms an excellent springboard for our understanding of the more complicated structures that we shall be considering later.

Three aspects of Galileo's approach should be most carefully kept in mind as we examine his analysis of the motion of bodies: (*1*) the use of precisely stated hypotheses; (*2*) deductions, from these hypotheses, of consequences that can be checked by experiments involving measurements; and (*3*) analysis in terms of abstract, *idealized* situations, departures from which can be accounted for. (For example, by a "freely falling object" is meant one falling in a "perfect" vacuum. No such vacuum exists, even today, but the assumption of such an ideal condition is required for a proper statement of laws of falling objects.)

1. UNIFORM MOTION IN A STRAIGHT LINE

If we rely only on experience and observation, we must conclude with Aristotle that to keep an object moving on earth requires force. Every object in motion sooner or later comes to rest, unless an external force acts upon it. Galileo, however, was not content with such a vague generalization. He wished to know the numerical relationships that existed between measurable forces and amounts of motion. During his investigations in which he used smooth metal balls rolling down inclined planes, he finally came to the conclusion that, contrary to raw observation, a better hypothesis regarding motion is that "any particle projected along a horizontal plane without friction . . . will move along this same plane with a motion which is uniform and perpetual, provided the plane has not limits."[1]

This is obviously a hypothesis that cannot be reached by direct observation: there is no plane that extends without limit; there is no possibility of eliminating friction entirely. From where, then, does such a hypothesis come? How can it be tested?

First, we require a definition of what is meant by *uniform* motion. In the *Dialogues,* Galileo defines it this way:

[1]Galileo Galilei, *Dialogues Concerning Two New Sciences,* Dover Publications, New York, 1950, p. 244, copyright Macmillan Co., New York, 1914.

By steady or uniform motion, I mean one in which the distances traversed by the moving particle during any equal intervals of time, are themselves equal.[2]

Today we would say that by uniform motion, we mean one in which the speed is constant, speed being defined as the distance traveled divided by the time of traveling. (As we shall later see, velocity, strictly defined, includes direction as well as magnitude. A car turning a corner at constant speed is changing in its velocity. For the moment, we are disregarding changes in direction, hence also disregarding the distinction between speed and velocity.)

Galileo gave two arguments to support the validity of his hypothesis regarding the uniformity of motion. Here is his first argument:

Furthermore we may remark that any velocity once imparted to a moving body will be rigidly maintained as long as the external causes of an acceleration or retardation are removed, a condition which is found only on horizontal planes, for in the case of planes which slope downward there is already present a cause of acceleration, while on planes sloping upward there is retardation; from this it follows that motion along a horizontal plane is perpetual.[3]

The second argument is even more effective. First, it is shown that in the case of a pendulum, if a pendulum bob on the end of a string is released from rest, it will descend and again rise to the same horizontal height (very nearly) as it was originally. Furthermore, Galileo showed that if an obstruction such as a nail is firmly placed under the support of the pendulum, as shown in Fig. 6.1, at a point so as to be struck by the string, the bob will again rise to the original height, although this time, over a shorter arc.

Now Galileo states that the same thing would happen when a ball rolls down

[2]Ibid., p. 154.
[3]Ibid., p. 215.

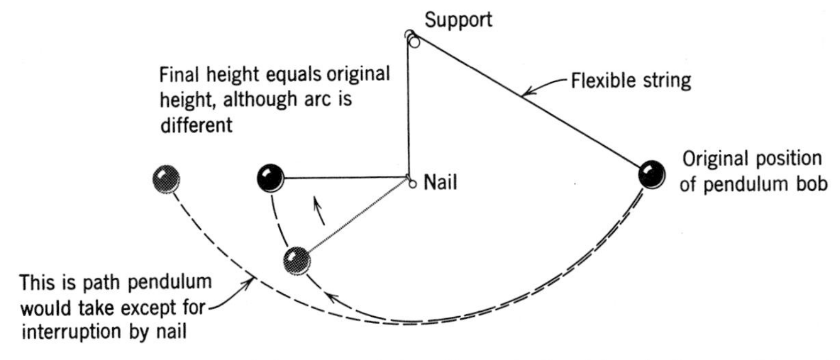

Fig. 6.1. If a pendulum bob, suspended on the end of a string, is drawn aside and released, it will rise to its original height. If a nail is interposed, as shown, the bob will still rise to its original height but along a different arc.

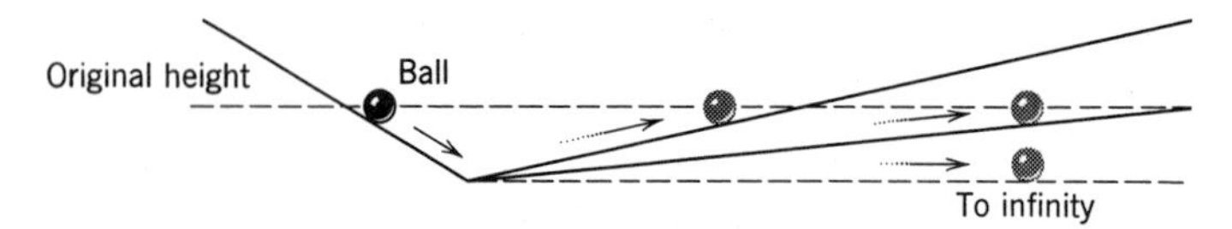

Fig. 6.2. An analogy to the pendulum motion of Fig. 6.1. A ball rolling down one inclined plane would rise to the same height on another inclined plane if it were not for friction. If the second plane is horizontal it would, according to Galileo, travel on forever. This suggests the idea of inertia.

one inclined plane and up another one, were it not for the loss of momentum which ensues as it reaches the joint at the bottom. Regardless of the slope of the second plane, it must eventually reach the same height *if friction* is disregarded, as shown in Fig. 6.2.

If the slope of the second plane is reduced more and more, the ball must roll a greater and greater distance to reach the original height. If the slope is reduced to zero, the motion would be perpetual and the ball would roll on forever.

We should notice how *idealized* are the conditions established for these "thought-experiments." There can be no friction between the ball and the plane, and there can be no air resistance. Neither of these conditions can be met in practice. On the other hand, if the hypothesis is correct, by reducing friction as much as possible, the consequences of the hypothesis are more and more closely approached. It is concluded that the force of gravity is not that which causes velocity, but that which causes a *change of velocity* either accelerating the ball, if the ball is moving downward, or retarding it if it is moving upward. Galileo could not eliminate the friction, he could only reduce it. After lining his inclined planes "with parchment, also as smooth and polished as possible, we rolled along it a hard, smooth, and very round bronze ball." The results then approximated those called for by his hypothesis.

In this way Galileo established, at least for terrestrial objects, a generalization which later became Newton's first law of motion: every object remains in a state of rest or in uniform motion unless acted on by an external force applied to it. This law carries with it the concept of *inertia,* or *mass,* which we may describe as an invariant, inherent property of matter which opposes any change in velocity.

2. LAWS OF ACCELERATION; THE VELOCITY OF A FALLING BODY IS INDEPENDENT OF ITS WEIGHT

Establishing the law of inertia is only one of the steps taken by Galileo in setting mechanics off in the right direction. Another big contribution was his mathematical formulation describing accelerated motion. Here again he departs from Aristotle both in conclusions and in his method of arriving at conclusions. It is important to appreciate the abstract, as well as the experimental nature, of his methods and to see how unerringly he attends to the significant points.

Much has been made of the fact that Galileo dropped stones of different sizes from a tower and refuted Aristotle by demonstrating that the stones reach the ground at the same time. This story is misleading on two counts. First, the stones do *not* reach the ground at *exactly* the same time. And, second, Aristotle had long since been proved wrong in his statements. (Aristotle had asserted that the heavier bodies fall from rest more rapidly than lighter bodies, and that the time taken to reach the ground is inversely proportional to weight. Thus, a three-pound ball should reach the ground in one third of the time it would take a one-pound ball. This proposition had already been proved incorrect by experiment as far back as the sixth century.)

Galileo proposed the hypothesis that *all* objects *in the absence of air* would fall to earth with the same speed and he convinced himself of the correctness of this hypothesis. How did he proceed? Certainly not *merely* on the basis of experiment, for although he found it true that objects such as a lead ball and a wooden ball fall with very nearly the same speed, the rate of fall, on the other hand, of an inflated bladder (which he also tried) is considerably slower. And other objects such as soap bubbles fall even more slowly. In Galileo's time vacuums were not available, and any experiment on falling bodies had to be conducted in air.

It is interesting how Galileo presents the problem and develops the argument for his hypothesis. The argument is presented in *The Dialogues Concerning Two New Sciences* in which three interlocutors discuss the basic problems of mechanics. One of the three characters of the dialogue, Simplicio, represents, as the name implies, a man of little perspicacity who has accepted the Aristotelian ideas and is impressed by their common sense. Salviati speaks for Galileo, and Sagredo represents a friend. (These are the same three interlocutors whom Galileo also used in his *Dialogue on the Great World Systems.* Simplicio is often treated with irony, bordering at times on contempt. It has been suggested that Pope Urban was told that Simplicio was supposed to represent His Holiness, and that part of his bitterness toward Galileo stemmed from hurt pride.)

Simplicio: . . . I shall never believe that even in a vacuum, if motion in such a place were possible, a lock of wool and a bit of lead can fall with the same velocity. . .

Salviati: A little more slowly, Simplicio. Your difficulty is not so recondite nor am I so imprudent as to warrant you in believing that I have not already considered this matter and found the proper solution. Hence, for my justification and for your enlightenment hear what I have to say. Our problem is to find out what happens to bodies of different weight moving in a medium devoid of resistance, so that the only difference in speed is that which arises from inequality of weight. Since no medium except one entirely free from air and other bodies, be it ever so tenuous and yielding, can furnish our senses with the evidence we are looking for, and since such a medium is not available, we shall observe what happens in the rarest and least resistant media as compared with what happens in denser and more resistant media... Let me once more explain that the *variation of speed* observed in bodies of different specific gravity *is not caused by the difference of specific gravity* [density], but depends upon external circumstances and, in particular, upon the resistance of the medium, so that if this is removed all bodies would fall with the same velocity...

... [Constant acceleration] you must understand, holds whenever all external and accidental hindrances have been removed; but of these there is one which we can never remove, namely, the medium which must be penetrated and thrust aside by the falling body. This quiet, yielding, fluid medium opposes motion through it with a resistance which is proportional to the rapidity with which the medium must give way to the passage of the body; which body, as I have said, is by nature continuously accelerated so that it meets with more and more resistance in the medium and hence a diminution in its rate of gain of speed until finally the speed reaches such a point and the resistance of the medium becomes so great that, balancing each other, they prevent any further acceleration and reduce the motion of the body to one which is uniform and which will thereafter maintain a constant value. *There is, therefore, an increase in the resistance of the medium, not on account of any change in its essential properties, but on account of the change in rapidity with which it must yield and give way laterally to the passage of the falling body which is being constantly accelerated.*

... Thus, for example, imagine lead to be ten thousand times as heavy as air while ebony is only one thousand times as heavy. Here, we have two substances whose speeds of fall in a medium devoid of resistance are equal: but, when air is the medium, it will subtract from the speed of the lead one part in ten thousand and from the speed of the ebony one part in one thousand, i.e., ten parts in ten thousand. While, therefore, lead and ebony would fall from any given height in the same interval of time, provided the retarding effect of the air were removed, the lead will, in air, lose in speed one part in ten thousand, and the ebony ten parts in ten thousand. In other words, if the elevation from which the bodies start be divided into ten thousand parts, the lead will reach the ground leaving the ebony behind by as much as ten, or at least nine, of these parts. Is it not clear then that a leaden ball allowed to fall from a tower two hundred cubits high will outstrip an ebony ball by less than four inches?[4]

We shall not further pursue Galileo's arguments in support of the hypothesis that if it were not for air, all objects would fall with equal velocities. The gist of his argument is that *very* light objects, such as "an inflated bladder," fall more slowly because in this case the *excess* of density over the density of air is much less than in the case of heavy objects, and hence the very light objects will be more affected by the resistance of the air than the heavy ones. (He also describes another series of experiments in which objects were allowed to fall through columns of water, and he showed that his general conclusion was valid for mediums other than air.) It is important to note that Galileo attended to the

[4]Ibid., pp. 72–76.

significant point, that is, that heavy objects even of different weights, fall *almost equally* fast. He then proceeded to explain the deviation from the *idealized* law of motion (which would be exactly true only in a vacuum), taking into account the resisting medium.

Having established the plausibility of the hypothesis, Galileo turns to another experimental test. In order to do so he used a second hypothesis, which is italicized in the quoted passage, that is, that the resistance of the medium is proportional to the velocity. If this second hypothesis is true, then if the fall of objects could be tested when they are traveling at low velocities the effect of resistance of air should be very much reduced. Galileo decided that this could be accomplished if instead of studying the vertical fall of objects, they were made to fall along an inclined plane, or along the arc of a pendulum (which can be considered to be a close approximation to an inclined plane if the arc is small compared to the length of the pendulum). If the hypothesis that all objects regardless of their weight fall at the same speed vertically is true, then it should also be true that objects of different weight, when suspended as pendulum bobs at the end of a string, should descend along identical arcs at the same speed. The advantage of the pendulum motion was that, since this speed was relatively slow, the resistance of the air would be reduced to a negligible amount. Galileo proceeded to the test, and found that a cork ball and a lead ball swinging through arcs at the end of a string of the same length traveled at the same speed.

In summary, Galileo's hypothesis that all objects would fall toward the earth with the same velocity and acceleration were it not for the presence of air is a statement about highly idealized conditions which certainly could not even be approached in Galileo's time when there was no possibility of achieving a good vacuum in which to drop objects. Instead, Galileo combines this hypothesis with the hypothesis that the resistance of the medium is roughly proportional to the velocity of an object falling through it. He then tries to see whether the *consequences* that can be drawn from these hypotheses are in accord with the facts. He finds that they are, both by the fact that objects which have nearly the same weight and are considerably heavier than air fall almost equally fast, and that objects describe equal arcs in equal times when used as pendulum bobs, even if their weights differ considerably.

3. LAWS OF ACCELERATION; THE VELOCITY OF A FALLING OBJECT IS PROPORTIONAL TO THE TIME IT HAS FALLEN

But the problem of how the velocity of falling bodies does vary remains. Thus far, all that has been shown is that the weight of an object is not a determining factor.

Now it would be perfectly possible to have bodies gain velocity in any number of ways. For example, the velocity might increase proportionally with the distance, or the square or cube of the distance. Or the velocity might increase proportionately to the time, the square of the time, or the cube of the time. Let us look at these possibilities in the form of equations:

$$v = ks \quad (1)$$
$$v = kt \quad (2)$$
$$v = ks^2 \quad (3)$$
$$v = kt^2 \quad (4)$$
$$v = ks^3 \quad (5)$$
$$v = kt^3 \quad (6)$$

(v = velocity; k = some constant (to be found by experiment); s = distance; and t = time.)

These are, of course, only a few of the infinite number of possibilities that might exist. Galileo, with his profound faith in the simplicity of nature began by considering the simplest possible alternative:

Finally, in the investigation of naturally accelerated motion we were led, by the hand as it were, in following the habit and custom of nature herself, in all her various other processes, to employ only those means which are most common, simple, and easy.[5]

On this basis, Galileo considered equations 1 and 2 to be equally simple in form. At first he was inclined to believe that velocity would increase in proportion to the distance traveled. However, upon examination, he concluded that this hypothesis led to an inconsistency.[6]

This left Galileo with the hypothesis, $v = kt$, or that the velocity of a falling object increases in proportion to the time it has fallen.

A motion is said to be uniformly accelerated if when starting from rest it acquires during equal time intervals, equal increments of speed.[7]

How could such a hypothesis be tested? Both because there were no instruments for measuring velocity directly, and because of the rapidity with which objects fall, a direct test of such a hypothesis was clearly impossible. Galileo resorted again to the indirect method of *deducing certain consequences* from the hypothesis and *testing these consequences* instead of the hypothesis itself.

One consequence that is mathematically deducible from the hypothesis of

[5]Ibid., p. 160.

[6]It has been pointed out by Mach and others that at this point Galileo erred in his analysis. It is true that his conclusions are valid. It can be shown (by using calculus and solving the differential equation $ds/dt = ks$) that equation 1 cannot hold for an object starting from rest.

[7]Reference 1, p. 162.

constant acceleration is that the *distance traveled* by a falling object is *proportional* to the *square* of the time during which the object has fallen. In other words, *if* the equation $v = kt$ (velocity is proportional to time) is valid, *then* the equation $s = Kt^2$ (distance is proportional to the square of time) is valid (where k and K are different constants, s is distance, v is velocity, and t is time). That is, the hypothesis $v = kt$ is to be tested indirectly by its consequence $s = Kt^2$.

Rather than interrupt the further statement of Galileo's experimental method, we shall postpone until the next section this derivation of $s = Kt^2$ from $v = kt$.

The problem has now been made a little simpler. It is no longer necessary to measure a relationship between velocity and time; instead, the more easily measurable relationship between space and time is now to be determined. Even so, there still remained a serious experimental difficulty in measuring time with sufficient accuracy because of the rapidity with which objects fall. Galileo surmounted this difficulty in an ingenious manner. In place of measuring distances and time with respect to objects falling vertically, the measurements were carried out on objects traveling down inclined planes. Thus, the effect of gravity on the speed of descending objects was diminished. (This, of course, introduced another subsidiary assumption, that is, that acceleration along an inclined plane is similar in nature and form to the acceleration of a vertically falling body. This assumption was carefully substantiated by Galileo who showed that the angle of inclination of the plane does not change the mathematical form describing acceleration.[8] He argued that as the angle of inclination approached 90°, the object on an inclined plane would have its velocity increased in an amount approaching that of a vertically descending object.)

We have already described the construction of the inclined plane down which Galileo rolled the balls. To show the further ingenuity of Galileo as an experimentalist, let us examine the manner in which he measured time.

For the measurement of time, we employed a large vessel of water placed in an elevated position; to the bottom of this vessel was soldered a pipe of small diameter giving a thin jet of water, which we collected in a small glass during the time of each descent, whether for the whole length of the channel, or for a part of its length; the water thus collected was weighed, after each descent on a very accurate balance; the differences and ratios of these weights gave us the differences and ratios of the times and this with such accuracy that although the operation was repeated many, many, times, there was no appreciable discrepancy in the results.[9]

[8] Since Galileo's time this statement is known to require slight modification. When a ball rolls down an inclined plane a small amount of energy is required to make it rotate. Since it does not rotate when it is falling vertically the laws governing free fall descent and descent along a plane are somewhat different. The modification introduced, however, is relatively negligible and Galileo was safe in disregarding it.

[9] Reference 1, p. 179.

With experiments such as these, Galileo indirectly established the validity and applicability of his hypothesis that all objects regardless of their weight will fall toward the earth with equal and constant acceleration under the ideal condition that there is no air resistance.

It is noteworthy that throughout these *Dialogues* Galileo avoids any discussion of the causes of motion. He carefully restricts himself to an accurate, mathematical description of how motion takes place. This approach in itself is a departure from Aristotelian physics in which a discussion of essential causes is given first emphasis. Many Aristotelians, for example, explained the increasing velocity of a falling object on the anthropomorphic basis that as an earthly object approaches its home, that is, the earth, it increases its speed of descent with "jubilation." The Aristotelians neglected to examine the way in which events occurred; instead, they concerned themselves with interpretation based on essences and final causes.

4. EXAMPLES OF ACCELERATED MOTION; DERIVATION OF $s = Kt^2$ FROM $v = kt$

It is essential that the concepts of velocity and acceleration be thoroughly grasped for any basic understanding of the physical sciences. First, keep clearly in mind what is meant by a *rate* of change. It is defined as the amount a quantity varies per unit time. If, *and only if*, the rate of change is *constant* is the varying quantity proportional to the passage of time. Thus, if velocity is constant, it is implied that the distance traveled during any interval of time, large or small, is proportional to the time taken to travel that distance, and we express this proportionality as $s = kt$; or, since k, the constant of proportionality, is the velocity, we may express the relation as $s = vt$. Obviously, the same relation can be expressed as $v = s/t$, the definition of velocity—distance traveled divided by the time it takes to travel that distance (assuming no change in direction). The units in which the velocity must be expressed then becomes units of distance divided by units of time, such as miles/hr, m/sec, and ft/sec.

Velocities are, however, rarely constant in the world we live in. Since *velocity is itself often a changing quantity*, to study how it changes, we must inquire as to the *rate of change of velocity*, that is, we need the concept of acceleration. In many cases, velocity changes uniformly with respect to time. *Under these conditions*, we then can say that velocity is proportional to the passage of time or $v = kt$ where we identify the constant k, as the acceleration, so that we can write $v = at$, or if we wish, $a = v/t$.[10] Acceleration is a measure of the rate of change of

[10]Note that it is just as possible for acceleration to be nonuniform as for velocity to be nonuniform. If acceleration varies, $v = kt$ will no longer be a valid equation. However, we shall only be concerned in this chapter with constant or uniform acceleration.

velocity with respect to time. But since velocity is a rate of change, *acceleration* becomes a *rate of change of a rate of change.* The units of velocity are those of distance divided by time, such as miles/hr. Hence the units of acceleration must be stated in the unfamiliar-looking form of units of distance divided by units of time, all divided by units of time, such as miles/hr/sec, or ft/sec/sec, or m/sec/sec.

Before considering the laws of acceleration, we need to have clearly in mind three different concepts of velocity: *average* velocity, *constant* or uniform velocity, and *instantaneous* velocity. Let us begin with constant or uniform velocity. For emphasis, we repeat that if any body moves in such a manner that equal distances are described in the same direction in equal intervals of time, *no matter how large or how small the intervals of time,* then the velocity is said to be uniform or constant. Only under these conditions is the distance traveled proportional to the time. Graphically the same relationship can be represented by a straight line, the slope of the line representing the velocity. See Fig. 6.3.

Speeds or velocities that remain constant over long periods of time are exceptional in the universe. When one travels by automobile between two cities 100 miles apart, one cannot hope to travel every mile in the same interval of time; the speed is not a constant. Under certain road conditions and under certain circumstances one may travel the distance in one length of time; under other conditions he may travel the same distance in another time length. The *total* distance traveled divided by the total time it takes to travel that distance is defined as the *average* velocity. (As we shall see later, speed and velocity are not strictly the same concept but, for the present, we shall not distinguish between them.) If a person travels 100 miles in 4 hours, the average velocity with which he has traveled is 25 miles an hour, regardless of whether he slowed down to 5 miles an hour or even stopped several times en route. He has not covered equal distances in equal intervals of time throughout his journey and hence has not maintained a constant velocity. If we represent average velocity by the usual convention, $\bar{v}$, it

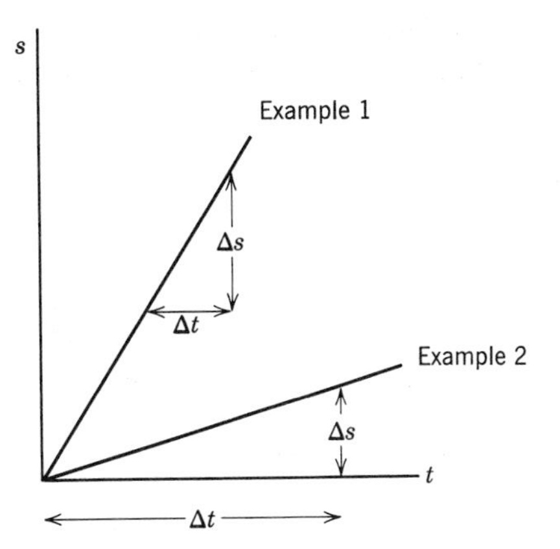

Fig. 6.3. Two examples of constant velocity. $s = vt$, where v is a constant and is given by the slope $\Delta s/\Delta t$ (read delta s divided by delta t; the Greek letter delta stands for a difference or increment). This is a change in distance divided by the change in time to cover that distance. Note how $\Delta s/\Delta t$ is determined in the diagram. For constant velocities, any two arbitrary points may be taken on the line, since the result will always be the same.

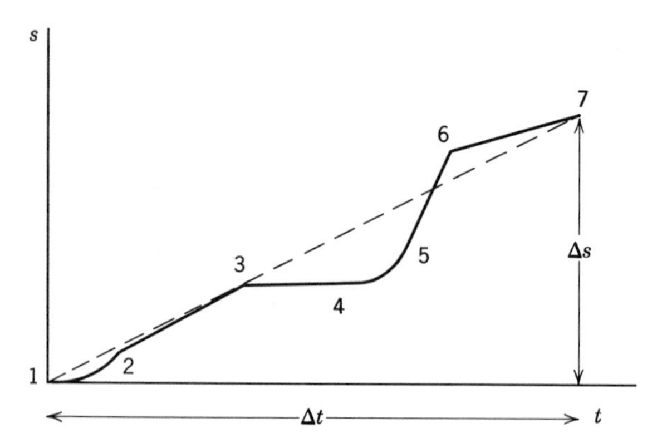

Fig. 6.4. Example of changing velocity. Note these points. (1) The average velocity $\bar{v}$ between starting point 1 and finishing point 7 is given by $\Delta s/\Delta t$ where, in this case, Δs represents the total distance traveled and Δt the time taken to travel that distance. (2) The velocity is constant between points 2 and 3. It is also constant between 3 and 4 (where it is zero) since, in this interval, distance does not change, and $\Delta s = 0$. (3) The instantaneous velocity is varying between 1 and 2, and between 4 and 5.

is valid to state the equation as $\bar{v} = s/t$, where s stands for the *total* distance traveled and t represents the total time required. But this equation expressed in the form $s = \bar{v}t$, although formally similar to $s = vt$, is very different in meaning. The latter equation is valid only for *constant* velocity. The former is valid even if the instantaneous velocity changes.

The distinctions between average, constant, and instantaneous velocities are illustrated in Figs. 6.3 and 6.4. In Fig. 6.4 the distances traveled by an object that starts from rest and moves with varying velocities is plotted against time. Both figures should be carefully studied until the distinctions between the three types of velocity are clearly in mind.

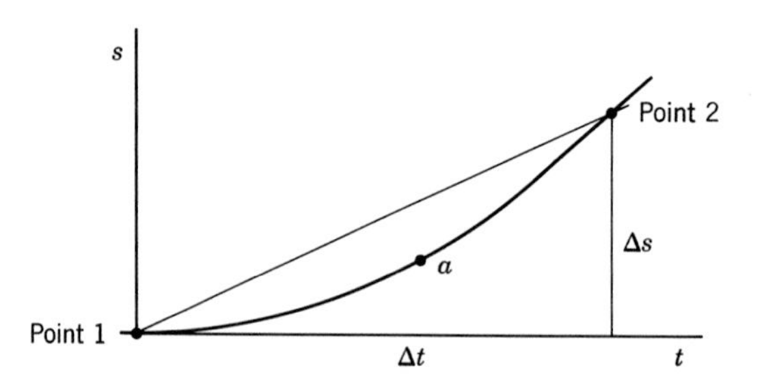

Fig. 6.5. Determination of average velocity during time t when instantaneous velocity is changing.

To study the meaning of instantaneous velocity, let us consider more carefully Fig. 6.4 with respect to points 1 and 2. To do this, it is convenient to enlarge the scale of the diagram, which has been done in Fig. 6.5.

Since the curve is no longer a straight line, the velocity is certainly not constant. However, the *average* velocity between points 1 and 2 may be calculated as before to give us $\bar{v} = \Delta s/\Delta t$ where Δs represents a change in distance and Δt represents a corresponding change in time.

Now suppose we take some intermediate point on the curve, such as a, lying between the points 1 and 2. In this case we can determine the *average* velocity from point 1 to a, and also from a to 2. It is at once apparent that the average velocity between points 1 and 2 is greater than the average velocity between 1 and a, and smaller than the average velocity between a and 2.

How can we give any meaning to the instantaneous velocity at point a? Fig. 6.6 shows how this may be done.

If we gradually reduce the distance between two points, for example a_1 and a_2 on either side of a, the average velocity can be computed over smaller and smaller time intervals. In particular, as the distance between the points a_1 and a_2 approaches zero, the line joining these two points approaches the tangent line. This tangent line is a unique one; it has a definite slope. We can use the slope of this tangent line to define what we mean by the instantaneous velocity at the point a. (If one wishes to be more careful in his statement of the above analysis, he can use the mathematical symbols and say the instantaneous velocity at a is defined as $v =$ limit (as Δt approaches zero) of $\Delta s/\Delta t$. This is the basic idea of the differential calculus, which consists of finding such limits for all kinds of changing quantities.)

At this stage, it would be well to draw and analyze a variety of curves representing various ways in which distance may change with time, in each case distinguishing carefully between constant, average, and instantaneous velocity.

We are now in a position to analyze what we mean by acceleration, which is, as we have indicated, a rate of change of velocity, or a rate of change of a rate of change.

First, if velocity is a constant then, of course, acceleration will be zero. If the velocity is constantly increasing at a uniform rate, the acceleration is then con-

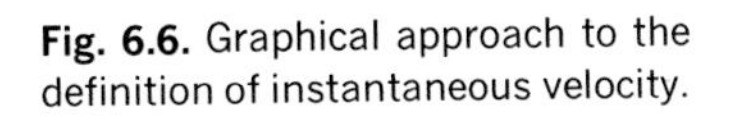

Fig. 6.6. Graphical approach to the definition of instantaneous velocity.

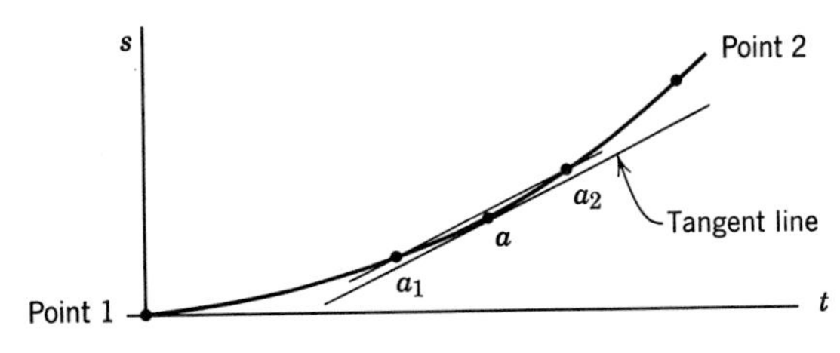

stant or uniform. Although there are many examples where acceleration is not constant (for example, in the motion of a pendulum), we shall be concerned primarily with cases of constant acceleration.

Let us use the same principles to analyze graphically the condition of *constant* acceleration (Fig. 6.7). When velocity (the rate of change of distance with respect to time) was constant, we found that we could graph the proportionality between s and t as a straight line with v as the slope of the line. Similarly, if *acceleration is constant,* since it represents the rate of change of velocity with respect to time, we can express this graphically as the slope of a straight line relating *velocity* and time.

Figure 6.7 and the equation in the caption represent Galileo's hypothesis that velocity changes uniformly (or acceleration is constant) and is, therefore, proportional to the time.

We now must consider the *consequences* of this hypothesis. In particular, we wish to find, under the condition that the velocity is proportional to time, what the relationship is between distance traveled and the time required to travel that distance.

Note carefully that even though acceleration is constant, since the velocity is changing from instant to instant, we *cannot identify v as being equal to $\bar{v}$.* But there is a relationship between them.

First, it is to be noted that a proportionality is closely related to an arithmetic progression. If we take the series, for example, 0, 3, 6, 9, 12, etc., this series of numbers may be plotted so as to fall on a straight line (Fig. 6.8).

It is apparent that the series can be expressed in an equation form, $y = kx$, where x represents the number of the term in the series minus one and where y represents the corresponding value. (Of course, in this case, x is restricted to whole numbers.)

If several such arithmetic series are examined, it will soon be observed that

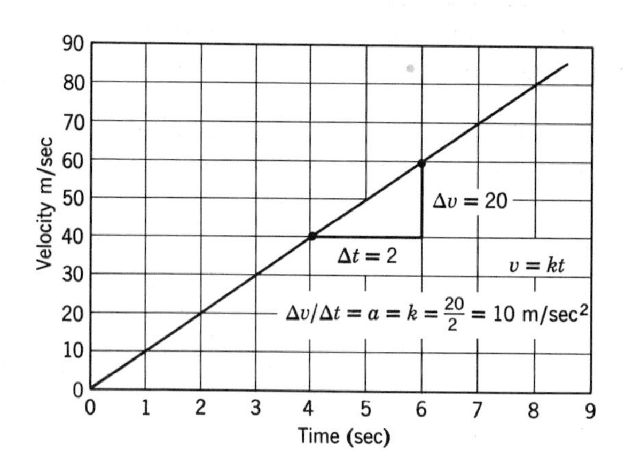

Fig. 6.7. Velocity changing uniformly with respect to time; thus acceleration is constant. $v = kt$; or $v = at$, where a is the constant k, or the uniform acceleration.

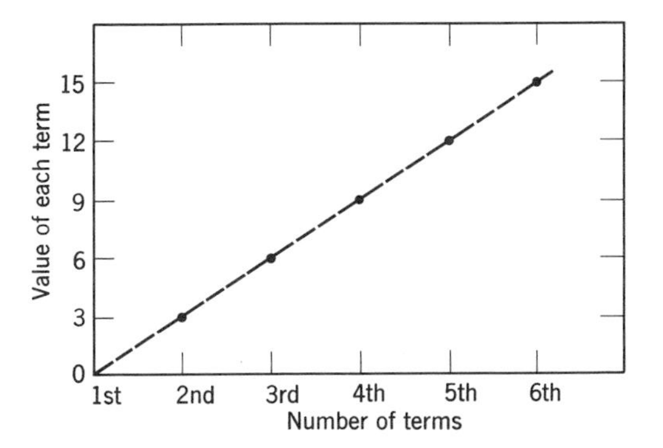

Fig. 6.8. The graph of an arithmetic progression. The term-by-term increase is constant. Therefore, it can be represented by a straight line graph. In this case, the average value of the numbers between two intervals is the sum of the first and last term divided by two.

the *average* of each series can always be expressed as the sum of the first and last term divided by two. For example, in the arithmetic series 4, 8, 12, and 16, the average value may be found by $(4 + 16)/2 = 10$. Can such a rule be proved to hold in general? Yes. The average value of such a series will always be given by $(a + L)/2$, where a is the first term, L the last term.

But since the average, by definition, is the sum S divided by the number of terms, or S/n, we see that the average value of the series is $S/n = (a + L)/2$, that is, the average is equal to the sum of the first term (which, of course, may be zero) plus the last term, divided by two.

It can be shown that this relationship will hold regardless of the number of terms, even though the number of terms is infinite. In other words, even if the series is composed of terms that are as close together as the points on a line, in which case the equation, $y = kx$, will cover *all* the points on the line, not just the integers, the average value of the whole series is also found by taking the sum of the first and last term and dividing by two.[11]

We can now state with simplicity the relationship between average velocity $\bar{v}$ and instantaneous velocity v under the condition of constant acceleration (that is, when $v = kt$). It is simply that the average velocity $\bar{v}$, over the time t, is the sum of the original velocity plus the final velocity (that is, the instantaneous velocity at time t) divided by two, or, if the original velocity is zero the average velocity becomes

$$\bar{v} = \frac{0 + v}{2} = \frac{v}{2}$$

Making use of this relationship, let us derive the equation tested by Galileo. We begin with three assumptions:

[11] Intuitively, the reader will probably appreciate the idea that for every point on a sloping line above the average value there lies a corresponding point below the average value.

$$\bar{v} = \frac{v}{2} \quad \text{(true under conditions of constant acceleration and zero initial velocity)} \tag{7}$$

$$\bar{v} = \frac{s}{t} \quad \text{(by definition of average velocity)} \tag{8}$$

$$v = at \quad \text{(the hypothesis to be tested, the acceleration } a \text{ being assumed constant)} \tag{9}$$

Combining equations 7 and 9, we have

$$\bar{v} = \frac{at}{2} \tag{10}$$

Combining equations 8 and 10, we have

$$\frac{s}{t} = \frac{at}{2} \tag{11}$$

Whence, on rearranging the terms, the required conclusion is reached

$$s = \frac{at^2}{2} \tag{12}$$

If the initial velocity is not zero but is v_0, then by a similar proof the equation is

$$s = v_0 t + \frac{1}{2} at^2 \tag{13}$$

It is essential to keep clearly in mind what has been proved. It has been shown that *if* the velocity is increasing uniformly with time, and *if* we define acceleration as the rate at which the velocity is increasing, and *if* the object starts from rest (zero velocity), *then* the total distance traveled by such an object will vary in direct proportion to the square of the time during which it has traveled, and the equation relating the distance traveled to the time of traveling will be given by $s = \frac{1}{2}at^2$.

We have already shown, in previous sections, why this is a relationship much more easily established by experiment than the direct verification of the law, $v = at$, where a is a constant. It is the *derived* relationship that Galileo verified experimentally.

Let us see how it applies to falling bodies.

The diagram in Fig. 6.9 shows approximately how an object will fall under the influence of gravity. (For the sake of convenience of calculation, we use the values for g, the acceleration of gravity, 32 ft/sec/sec, and 10 m/sec/sec, both accurate within 2 per cent, instead of using the somewhat more accurate figures of 32.2 ft/sec/sec and 9.81 m/sec/sec. The important point is that any specific

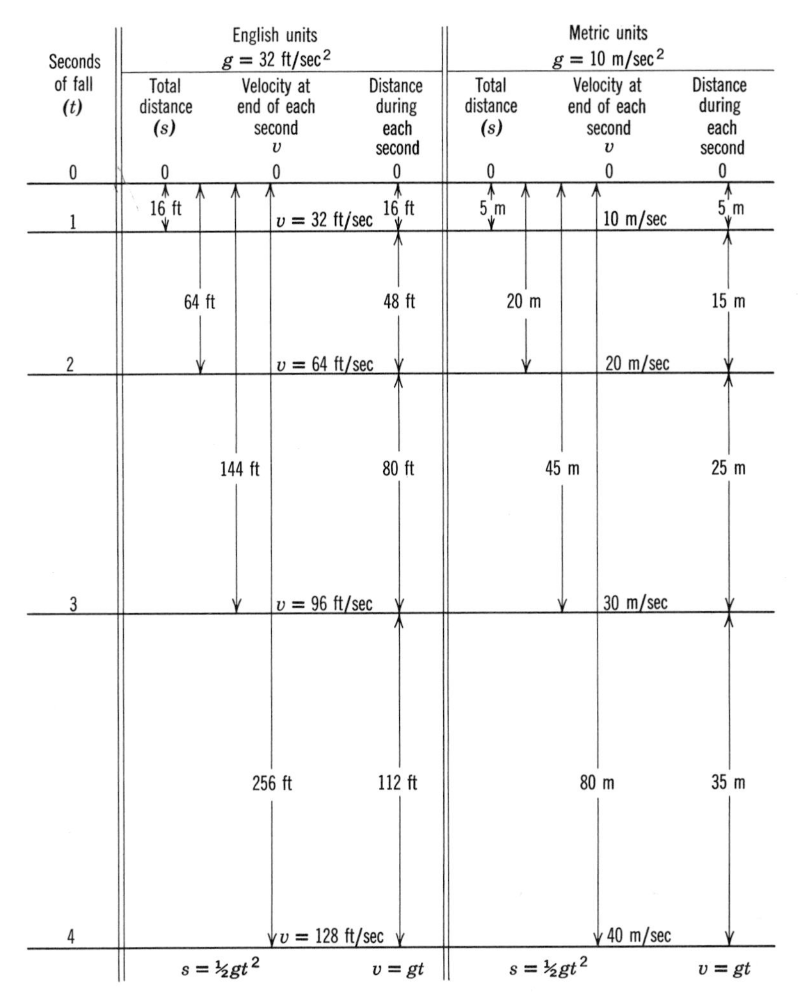

Fig. 6.9. Distances fallen under condition of free fall.

constant value of g is an approximation since, as we shall learn, the value of g varies with the location of the observer.)

One further step in the analysis needs to be carried out if we are to understand the full significance of Galileo's method. As we have mentioned, even to verify the uniformity of acceleration of falling bodies indirectly by measuring the distance fallen at various time intervals depends upon fairly accurate timing. Even a small error in timing causes too great an error in the results to enable satisfactory conclusions to be drawn. (One should determine for himself, on the basis

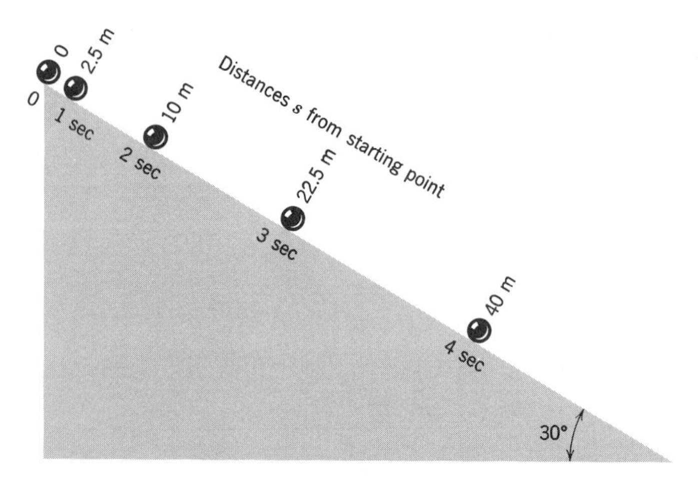

Fig. 6.10. Motion along an inclined plane of 30°. Acceleration = $g \sin \theta$. In this case $a = 5$ m/sec² since sin 30° = 0.5. Galileo showed that distances covered were proportional to the square of the time: $s = \frac{1}{2}at^2$, or in this case, $s = 2.5t^2$. The smaller the angle, the lower the acceleration, but the total distance traveled, as Galileo showed, is always proportional to the square of the time, regardless of the angle.

of the figures given in Fig. 6.9, about how much difference a one quarter-second error would make in the value of the distance at the end of three seconds.) Therefore, Galileo did not even attempt to verify the equation, $s = \frac{1}{2}at^2$ directly. Instead, he used a method by which the acceleration, and therefore the velocities and distances traveled, would be greatly reduced. To accomplish this, he used an inclined plane down which he rolled polished bronze balls.

To justify this procedure, *still another assumption* had to be introduced: the supposition that the acceleration would be diminished along an inclined plane in a way dependent only upon the angle of inclination of the plane. The relationship Galileo worked out was that the acceleration along the inclined plane would equal g, the acceleration of gravity acting on a freely falling body, *times* the sine of the angle of the inclined plane. *If* this hypothesis is true, *then for any given angle* of the plane, the distance traveled along the plane will also be directly proportional to the square of the time during which the object travels along the inclined plane. The only difference between this case and that of the freely falling body is that the constant of acceleration and, therefore, the distance traveled in a given interval of time will be much reduced. Consequently, the experimental verification of the law can be much more readily examined. Again, we show a diagram (Fig. 6.10) that will indicate the essential elements in the experiment.

5. DIRECTION OF MOTION

Up to now, we have not concerned ourselves with changes in the direction of motion. Galileo, as we have seen, showed that a force such as gravity causes a change in velocity *in the direction of the acting force.* If no force is acting on a body that is in motion, it will continue to move in a straight line at a constant velocity. *To change the direction of the motion* requires a force just as well as *to change the magnitude of the velocity.*

Here, the technical distinction between speed and velocity must be emphasized. *Velocity* implies direction: velocity is a measure of distance traveled in a *specified direction* divided by the time traveled. (A quantity that is directed is called a *vector,* and thus velocity is an example of a vector.) *Speed* is merely the *magnitude* of the velocity. The speedometer on a car does not record velocity but records speed. If one goes around a curve with the speedometer reading a steady 25 miles per hour, his speed is constant but the velocity is changing.

One of the triumphant results of Galileo's analysis was his explanation of projectile motion. We need not enter into the details of this analysis, but shall merely indicate the elements upon which it is based.

The question asked by Galileo was: If a body is in motion with a constant horizontal velocity, what is the effect of gravity acting upon that body at right angles to its motion?

He showed that, (*1*) since the action of the force of gravity is at right angles to horizontal motion, the magnitude of the horizontal velocity will remain unchanged; and (*2*) since there is a force acting vertically, there will be superimposed upon the horizontal motion a downward, vertical acceleration. The net result of these two independent components of the motion acting on a single body will determine the motion. Figure 6.11 should clarify these ideas. Three projec-

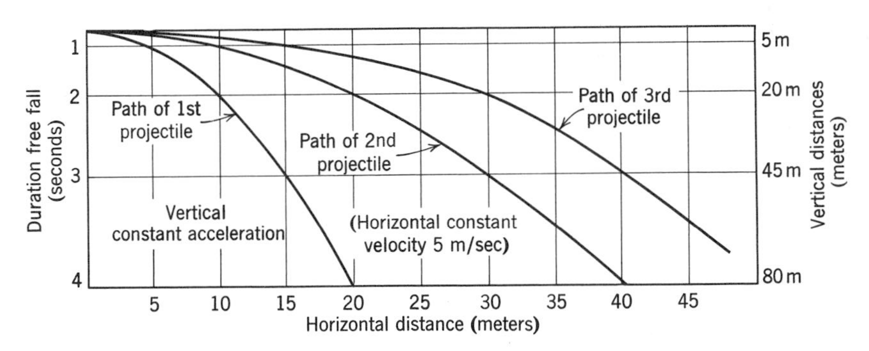

Fig. 6.11. Paths of projectiles (combination of constant horizontal velocity and vertical acceleration due to gravity). All projectiles will reach the ground at the same time, regardless of horizontal velocity.

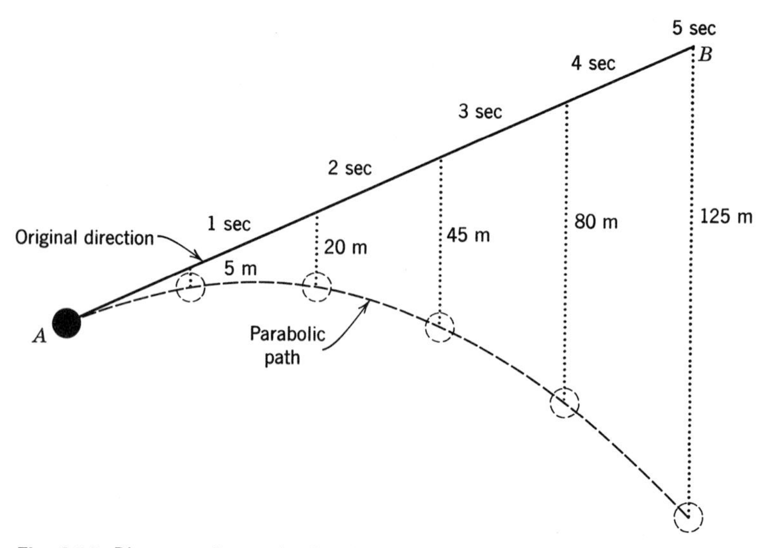

Fig. 6.12. Diagram of a projectile shot at an angle with the horizon. Without gravity the projectile would reach position *B* in 5 sec. Meanwhile, however, it is also falling by the vertical distances shown, and the actual path then becomes the dotted line.

tiles have been shot horizontally to the right with velocities of 5, 10, and 15 m/sec. They are each also being accelerated downward with accelerations of 10 m/sec/sec. The resultant motion is as shown.

Now consider the case in which the initial direction of the projectile is not horizontal, but is at an angle. The analysis is similar. The results are shown in Fig. 6.12.

The paths of projectiles as shown are in the forms of parabolas. (Actually, however, the vertical lines are not quite parallel, since the gravitational force is directed toward the center of the earth. A parabola is another special case of an ellipse—an ellipse with the focus at infinity. Since the focal point is actually the center of the earth, the parabolas are sections of ellipses with a focus very far away compared to the distances usually involved. If, however, a projectile is to travel distances comparable to the distance to the center of the earth, as in the case of intercontinental ballistic missiles, or artificial satellites, the paths must be considered elliptical.)

6. SUMMARY

Since the equations of motion are used so often it is best to master them thoroughly. Their derivation and interpretation should be understood. It is only by understanding the assumptions from which the equations derive that one acquires confidence in their application. The key equations are:

$$\bar{v} = \frac{s}{t} = \frac{1}{2}(v_0 + v)$$

where $\bar{v}$ is the average velocity, v_0 and v are initial and final velocity respectively. The latter equation *only* applies under conditions of constant acceleration. The same necessary condition applies to the following:

$$v = v_0 + at$$
$$s = v_0 t + \tfrac{1}{2}at^2$$
$$v^2 = 2as + v_0^2$$

You should be able to derive each of them from clearly stated assumptions.

Galileo's great contribution to science, in addition to the many particular discoveries he made, was to establish a method that combined hypothetical-deductive procedures involving idealized and abstract conditions with empirical verification through experiment. By establishing a mathematical and precise formulation of the laws of motion instead of beginning with generalizations and explanations in terms of anthropomorphic concepts such as "bodies seek their natural place," "nature abhors a vacuum," "water seeks its own level," and the like, the sciences were started in a new and much more fruitful direction. While it is true that among the Greeks, Archimedes had used the same method, he had little influence upon the further development of science until Galileo once more applied it. (Archimedes was first translated and published in Europe by the mathematician Tartaglia in 1543, and Galileo studied his works with great care and enthusiasm.)

In the next chapter, we shall see how the work that was begun by Galileo was extended to include motions of celestial bodies and, then, was incorporated into a single system by Newton. Subsequently, the whole development of physics and chemistry became more and more rapid, both in its theoretical development and in the practical applications to technology and industry. The myriad-minded Galileo had indeed built the foundations well.

7. QUESTIONS

1. Explain succinctly why average velocity and constant velocity differ. Under what conditions of acceleration are they the same?

2. Give an example of where it is important to distinguish between speed and velocity?

3. Can instantaneous velocity be measured by direct observation? Explain.

4. Consider a ball thrown vertically upwards which then reaches a certain height and afterwards falls back to the original level. Without using a mathematical equation, state the

logical arguments in support of the statement that the speed will be the same at the same point above the original level on its upward as on its downward flight. (Is this true of its velocity?) Now prove the validity of the same statement using mathematics. Also show how to find the distance at which $v = 0$.

5. Consider a pendulum. Would the equations of motion for falling bodies apply? Explain. State *two* factors which would make you conclude that the velocity of a pendulum bob is constantly changing. Is the acceleration constant? If not, where is it maximum in magnitude and where is it minimum?

6. In what way can you extend the arguments used regarding a ball thrown vertically upward to the case of a projectile? What general reasons would you give for predicting that a projectile fired at 45° will achieve its maximum range if air resistance is neglected? If air resistance is allowed for, can you surmise whether the reduction in range will be greater in the first portion of the flight or in the latter half?

7. Plot a graph of velocity against time and show that if acceleration is constant, the area under the plotted line will be equal to the total distance traveled. Can you extend the argument to the case where the acceleration is *not* constant?

8. Describe the types of motion represented by the following graphs:

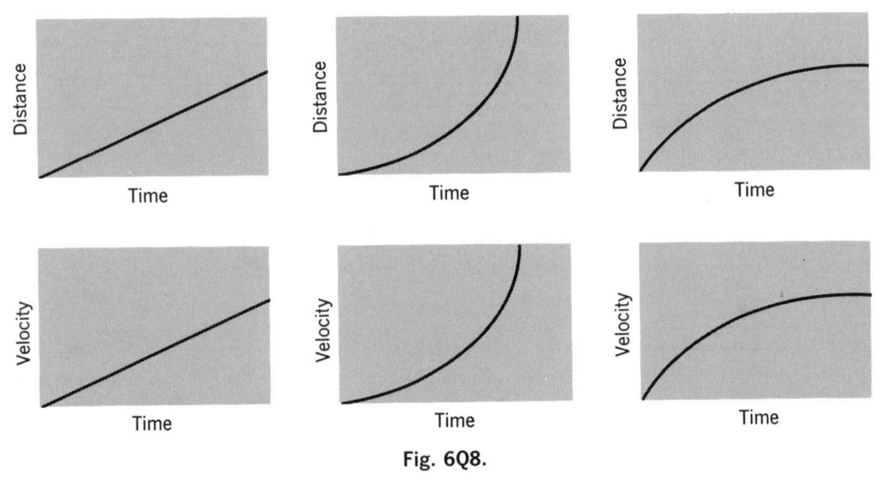

Fig. 6Q8.

9. Suppose a sprinter runs a straight 200 meters, slowing up slightly about midway and sprinting at the end. Construct a graph of distance against time and velocity against time. On the basis of your graphs, what is the approximate velocity?

10. What is the basic assumption used in analyzing the motion of a projectile? (Consider horizontal and vertical components.)

11. Explain clearly why acceleration has "clumsy" units, such as feet per second per second or meters per second per second.

12. Is compound interest mathematically analogous to constant acceleration? Explain, using a specific example.

Problems

1. The rate of 60 miles per hour is how many feet per second? How many meters per second?

2. What is the average acceleration of a car which starts from rest and is going 40 miles per second at the end of 3½ seconds? Does this average acceleration depend upon the way the instantaneous velocity changes during any one second?

3. What is the average velocity of a car if it travels for ten minutes at 30 miles per hour? What is the average velocity if it travels ten miles at 60 miles an hour and then ten miles at 30 miles per hour?

4. An object drops from rest at a height of 40 meters above ground. Neglecting air resistance, find: (a) the velocity with which it strikes the ground; (b) the time it takes to reach the ground; (c) the average velocity during the 2nd second of fall; (d) the instantaneous velocity at the end of the 1st and the 2nd seconds.

5. Neglecting air resistance, what is the velocity of an object which starts from rest after sliding down a "frictionless" 30° plane through a distance of 5 meters. Through what vertical distance would it have to fall to reach an equivalent speed?

6. A stone is thrown vertically upward to a distance of 100 ft. Neglecting air resistance, what is its initial velocity? What is its final velocity as it hits the ground? A big league pitcher can throw a ball about 90 mph. How high could he throw a baseball, assuming he could throw it upward at the same velocity?

7. Determine "g" at the surface of the moon from the following rough data obtained at an observation station set up by astronauts relating the distances fallen by an object dropped from rest and the time of falling:

s (meters)	1.0	1.9	3.1	5.0	7	10.1	13	17	21
t (seconds)	1	1.5	2	2.5	3	3.5	4	4.5	5

8. A stone is dropped down a well and, 10 sec later, a splash is seen. How deep is the well? Why can we neglect velocity of light in determining this depth? If, instead, the splash is *heard* after 10 sec, can the velocity of sound be neglected in finding the answer? (Take the speed of sound as 300 m/sec.) (*Hint.* To find the depth of the well in this case, set up simultaneous equations and use the formula for the solution of a quadratic equation, that is, if $ax^2 + bx + c = 0$ then $x = (-b \pm \sqrt{b^2 - 4ac})/2a$.)

9. If a .22 caliber bullet has a muzzle velocity of about the speed of sound (330 meters per second), what is its average acceleration if the rifle is one meter long?

Recommended Reading

Galileo Galilei, *Dialogues Concerning Two New Sciences,* translated by Henry Crew and Alfonso De Salvio, Macmillan, New York, 1914, reprinted by Dover Publications, New York. Of particular interest are pages 66–77, 153–175, 215–218, and 244–250. Galileo's style is elegant, even in translation, and should be sampled by all students.

I. Bernard Cohen, *The Birth of a New Physics,* Doubleday (Anchor Book), Garden City, N.Y., 1960.

E. A. Burtt, *The Metaphysical Foundations of Modern Science,* Doubleday (Anchor Book), Garden City, N.Y., 1954, pp. 91–104.

Sir Oliver Lodge, *Pioneers of Science,* Dover Publications, New York, 1960. This is an entertaining and instructive book containing both basic ideas and biographical data.

For further study of the laws of motion, any text in elementary physics will prove helpful. One that abounds with examples and clear illustrations is *Physics,* prepared by the Physical Science Study Committee, D. C. Heath and Company, Boston, 1960.

Another extremely useful but more advanced text written for the nonprofessional is Eric Roger's *Physics for the Inquiring Mind,* Princeton University Press, Princeton, 1960.

E. Pollard and D. C. Hutton, *Physics, an Introduction (Poet's Physics),* Oxford University Press, 1969, Chapters 3, 4, and 14.

L. Fermi and G. Bernadini, *Galileo and the Scientific Revolution,* Basic Books, Fawcett-World, Greenwich, Conn. 1961.

chapter seven

Newton's world system

Sir Isaac Newton, who was responsible for fully developing the conceptions and ideas of Copernicus, Kepler, and Galileo, was born in the year that Galileo died, 1642, almost exactly one hundred years after the publication of the great Copernican treatise *De Revolutionibus*.

Because of his experiments and studies in optics and his development of the calculus, Newton is entitled to the highest rank among scientists and mathematicians. But the crowning achievement of his career was the development of the laws of motion and the theory of gravitation, systematically presented in his *Mathematical Principles of Natural Philosophy* (otherwise referred to as the *Principia*), published in 1687.[1]

This work is generally regarded by scientists as the greatest single contribution to science that has ever been made. Certainly, few other publications have had such a profound influence on human thought and history. In the *Principia,* we find the extraordinary reach of an imaginative yet

[1] *Philosophia Naturalis Principia Mathematica.* There is a fine edition of this work translated by Andrew Motte, revised by Florian Cajori, published by the University of California Press, Berkeley, California in 1947. Every student of science should glance through this book and read some of its passages to "get the feel" of Newton's accomplishments and his style.

thoroughly disciplined mind, bold without measure in inventing hypotheses, yet cautious to an extreme in stating these hypotheses only after the most careful evaluation of the evidence.

The book sets forth both the framework for later analysis and specific suggestions, which kept the finest minds of mathematicians and scientists busy for two hundred years. We summarize a few of the topics that are treated.

In the first part (Book I) the three laws of motion and their consequences are established. The law of gravitation is set forth as a mathematical hypothesis from which the three laws of Kepler can be deduced as consequences. The orbits of two attracting bodies revolving about their common center of mass are successfully calculated. The vastly more complicated problem of the interaction of three bodies, each attracting the other two (the famous "three body" problem), is analyzed and, although the analysis is not complete, a good start is made in developing the theory of perturbations that is still used in modified form. The motion of the simple pendulum and of the cycloidal pendulum is thoroughly discussed.

In the second part (Book II) the effect of resisting mediums upon motions is considered in detail, and the science of fluid dynamics, including various studies of streamline effects, is established. The calculus of variations and the mathematical treatment of wave motion are inaugurated.

In Book III, the system of the world is set forth. The motions of all the planets, the satellites of Jupiter, and the orbits of the comets, are studied in detail and shown to fit the law of gravitation. The complicated motions of the moon, the precession of the equinoxes, and the tides are all examined in detail and systematically explained. Newton works out a prediction regarding the equatorial bulge of the earth, which was later found to be extremely accurate, and he estimates the density of the earth as between five and six times that of water (it was later found to be about five and one half times that of water.)

Contemplating the successive triumphs that marched forth one after another in the *Principia,* we are not surprised at the astonished wonder with which Newton's work was greeted, most effectively expressed perhaps by Alexander Pope's often-quoted couplet:

> Nature and Nature's laws lay hid in night,
> God said "Let Newton be!" and all was light.[2]

Having indulged in these superlatives with respect to Newton—all well deserved —it is necessary also to emphasize how much every scientist in his creative

[2]There is an amusing modern reply, almost as frequently quoted, by J. Squire:

> It could not last, the devil crying "ho!
> Let Einstein be," restored the *status quo.*

efforts depends on his predecessors. Newton was most aware of this when he said, "If I have seen a little farther than others, it is because I have stood on the shoulders of giants." Science fundamentally remains a co-operative affair. Newton would not have achieved his goal without the preceding work of Galileo, Kepler, Descartes, and the myriads of scientific workers, many obscure, to build upon. Rutherford has well expressed this aspect of science: "It is not in the nature of things for any one man to make a sudden violent discovery; science goes step by step, and every man depends on the work of his predecessors . . . Scientists are not dependent on the ideas of a single man, but on the combined wisdom of thousands of men."[3] Indeed, examples of how the development of new conceptions depend on the intellectual climate of the day can be found in Newton's own productions. While he was "inventing" that branch of mathematics called calculus, Leibnitz was independently developing the equivalent mathematical method in Germany. Even Newton's two basic ideas—that there must be a deflecting force that causes planets to pursue elliptical orbits, and that this deflecting force is inversely proportional to the square of the distance between the planet and the sun—were "in the air."

Here is an account of the events leading up to the publication of the *Principia,* as related by E. N. Andrade in his biography of Newton.

This question of the planetary motions was occupying the minds of many prominent Fellows of the Royal Society about this time. Early in 1684, the great astronomer Edmund Halley, whose name is known to most people from "Halley's comet," met with Sir Christopher Wren and Robert Hooke—probably in a coffeehouse—to talk about the matter. Hooke affirmed that the laws of the celestial motions followed from the inverse square law and that he had proved this. Sir Christopher, who was evidently a bit doubtful, said that he would give Hooke two months' time to produce the proof and would present him with a book costing 40 shillings if he did it. Hooke put him off: I believe that Hooke was convinced that the inverse square law was at the bottom of the matter, and had some kind of general indication and arguments, but no mathematical proof. He was not an oustanding mathematician: in any case he did not produce what was required. In the following August, Halley went to Cambridge to get Newton's views. He asked him what would be the orbit described by a planet supposing that it was attracted to the sun by a gravitational force obeying an inverse square law. Newton immediately answered *an ellipse.* "Struck with joy and amazement" to quote the record of the meeting, "Halley asked him how he knew it. Why, replied he, I have calculated it; and being asked for the calculation, he could not find it, but promised to send it to him." It is typical of Newton not to be able to lay hands on a little thing like that: he was not concerned about it at the time. Soon after he sent Halley two different proofs of the elliptic orbit, and, his interest having been revived by this, he wrote down the principles of mechanics in a little book *De Motu Corporum* (On the Movements of Bodies) which was founded on his lectures. This he sent to Halley at

[3]Quoted in Gerald Holton and Duane Roller's, *Foundations of Modern Physical Science,* Addison-Wesley, Reading, Mass. 1958, p. 166.

the end of the year. Halley at once saw the overwhelming importance of Newton's work and made up his mind to induce Newton to write a treatise setting out in detail his discoveries.[4]

1. COMMENTS ON NEWTON'S METHODS

In this chapter we shall examine the meaning and consequences of Newton's three laws of motion and the law of universal gravitation. Before going into detail, some general comments are in order.

Actually, Galileo had already implicitly established at least two of the three laws of motion that are named after Newton. But these were set forth in a somewhat discursive way in his *Dialogues*. It was Newton who set forth all three laws clearly and explicitly as basic propositions from which systematic deductions could be made to account for all terrestrial motions. Furthermore, by combining these three basic laws with a fourth hypothesis, the law of universal gravitation, Newton was able to establish a single systematic explanation that covered not only motions of objects on earth but the motions of all objects in the universe. Here, indeed, is a "system of the world," so broad in its range, so powerful in its methods of analysis, so elegant and simple in conception, that its dominance over the minds of men for many succeeding generations was inevitable.

Much can be learned from reviewing the way in which Newton became convinced of the validity and universality of the law of gravitation. Early in his scientific studies, he became familiar with various explanations that had been offered as to why the planets moved as they did about the sun. Kepler had suggested that the rays of the sun pushed the planets around in their course; by 1640, the famous mathematician and philosopher, Descartes, had developed a very elaborate theory of swirling vortices in an all-pervasive ether, which swept the planets around the sun. This vortex theory of Descartes was the one most generally accepted in Newton's day. In view of Galileo's proof of the law of inertia, however, Newton could not find it acceptable. He correctly surmised the proper approach was not to ask what kept the planets in motion, but what force could be acting that would *deflect* them from the straight-line motion that they would describe if there were no force acting upon them.

The story, which appears to be historically authentic,[5] indicates that Newton, as a young man of twenty three, was meditating in his garden over the problem of finding this deflecting force, when he heard the thud of an apple falling upon

[4]E. N. da C. Andrade, *Sir Isaac Newton,* Doubleday (Anchor Book), New York, 1958, pp. 67–69. Reprinted through the courtesy of Macmillan, New York, and W. Collins Sons & Co., Ltd., London.
[5]Unlike the story of Galileo dropping stones from the leaning tower of Pisa, which has never been authenticated, the story of Newton and the falling apple, told both by his early biographer and friend, Stuckley, as well as by his niece who lived with Newton, has every appearance of being valid.

the ground. Could it be that this same force that was acting on the apple extended out infinitely into space? Why should the force of gravity stop at any particular distance from the earth? Possibly, thought Newton, this gravitational force, although diminished by distance, acted on the moon and constantly deflected its linear motion, causing it to take a circular path. In a memorandum written years later, he refers to his speculations:

And the same year I began to think of gravity extending to y^e (the) orb of the Moon, and having found out how to estimate the force with w^ch (which) a globe revolving within a sphere presses the surface of the sphere, from Kepler's Rule of the periodic times of the Planets . . . I deduced that the forces wch keep the Planets in their Orbs must be reciprocally as the squares of their distances from the centers about wch they revolve: and thereby compared the force requisite to keep the Moon in her Orb with the force of gravity at the surface of the Earth, *and found them answer pretty nearly*. All this was in the two plague years of 1665 and 1666, for in those days I was in the prime of my age for invention, and minded Mathematics and Philosophy more than at any time since.[6]

Let us look at the procedure implied in the above passage. First, Newton surmised that the force of gravity is inversely proportional to the square of the distance between two attracting bodies. He tells us that he reached this conclusion by relating Kepler's third law to the equation that determines the amount of force required to keep a terrestrial object revolving in a circular orbit. The appropriate equation can be found either theoretically on the basis of the laws of motion, or experimentally. Undoubtedly, Newton had already derived the equation from Galileo's laws of motion, but it is to be noted that he refers to an *experimental* determination of the special case of circular motion, which he accomplished by measuring the force exerted on the inner surface of a hollow sphere in which a ball was made to roll. In either case the result is the same, the force required to keep a body moving in a circle will be given by

$$F = \frac{m4\pi^2 R}{T^2}$$

where R is the radius of the circle in which the object is traveling, m is the mass, and T is the time taken to complete one revolution.[7]

The above equation is a proportionality: for a given mass the force is proportional to the radius of the circle and is inversely proportional to the square of the period of revolution (time to complete the revolution). Since we are only concerned at the moment with the radius and the period, we can express the equation as

$$F = K\frac{R}{T^2} \qquad (1)$$

[6] Sir William Cecil Dampier, *A History of Science,* Macmillan, New York, 1949, pp. 150–51.
[7] The theoretical derivation of this equation from the laws of motion is given on p. 138.

But Kepler's third law is

$$T^2 = k\,R^3 \tag{2}$$

(where k is another constant).

If, therefore, a force is required in space to keep an object moving in circular motion in a manner similar to its action on terrestrial objects in circular motion, such a force would be a combination of equations 1 and 2, of the form

$$F = \frac{c}{R^2} \tag{3}$$

(where c is a third constant, namely, $c = K/k$, or, the force is inversely proportional to square of distance).

Thus, Newton arrived at the conclusion that the centripetal force acting toward the center of motion, which causes a planet to travel in its orbit, must obey the inverse square law. *If* his surmise was correct that the force that kept the planets in their orbits was the same as the force of gravity that made objects fall to the earth, *then* it should follow that the value of g (acceleration of gravity on the earth) could be calculated from a determination of the moon's acceleration, the moon's orbit being almost a perfect circle.[8] Newton carried out these calculations and found that the answers agreed "pretty nearly." He had found a beautiful verification of his theory that it was the same force, gravity, which both kept the moon in its orbit and caused the apple to fall from the tree. At last the laws of motion on earth and those of the celestial objects could be brought together under a single law applicable to both! One of the great "secrets of the universe" lay exposed! Discovered by a youth of twenty-three, who was "minding mathematics and philosophy!"

But Newton did not publish his results; he kept them to himself for over twenty years. Two possible reasons for this have been given (in addition to the fact that Newton was by nature shy, sensitive to criticism and, hence, unwilling to make any premature statement for which he could be criticized). Either explanation demonstrates an aspect of sound scientific method. One explanation is that, although the agreement with measurements of the moon's distance and period and the acceleration of gravity on earth agreed "pretty nearly," they still differed by enough to make Newton unwilling to give unquestioned acceptance to his hypothesis. The small discrepancy occurred because of inaccurate measurement of the earth's radius at that time. The discrepancy was later removed, in 1672, when a more accurate estimate was made by Piccard of the earth's radius (and hence of the ratio of the moon's distance to the radius of the earth, which

[8]The complete analysis is postponed until after we have considered in greater detail the laws of motion on page 140.

enters into the calculation). Another explanation is that Newton had not yet developed a way of showing that with respect to the gravitational force exerted, any spherical object such as the earth could be regarded as a point mass, that is, as if it were composed of a single particle with all the mass concentrated at one point. Probably what Newton needed to do was to develop and apply the calculus in order to convince himself that this was true. Until this was accomplished, no doubt Newton felt that his demonstration of the law of gravitation was not complete.

We should mention that in our discussion above, *circular* orbits are assumed. Does the same general law of gravitation explain *elliptical* orbits? Newton, depending on insight arising out of his knowledge of calculus, (which he does not, however, use in his *Principia*) was able to show many years later that not only for circular motion but for elliptical motion, an inverse square law of gravitation would explain the orbits of the planets. This was beyond the mathematical ability of Hooke and others who had guessed that an inverse square law held, but could not prove it.

2. NEWTON'S THREE LAWS OF MOTION AND THE LAW OF UNIVERSAL GRAVITATION

We repeat Newton's laws of motion as he gave them in the *Principia* along with his comments.

Law I. Every body continues in its state of rest or of uniform motion in a right line, unless it is compelled to change that state by forces impressed upon it.

Projectiles continue in their motions, so far as they are not retarded by the resistance of the air, or impelled downward by the force of gravity. A top, whose parts by their cohesion are continually drawn aside from rectilinear motions, does not cease its rotation, otherwise than as it is retarded by the air. The greater bodies of the planets and comets meeting with less resistance in freer spaces, preserve their motions both progressive and circular for a much longer time.

In equation form, we can express Newton's first law (or law of inertia):

$$\textit{if } F = 0 \quad \text{then} \quad a = 0$$

(where F is the *net* impressed force and a is the acceleration).

Law II. The change of motion is proportional to the motive force impressed; and is made in the direction of the right line in which that force is impressed.

If any force generates a motion, a double force will generate double the motion, a triple force triple the motion, whether that force be impressed at once, or gradually and successively. And this motion, being always directed the same way with the generating force, if the body was in motion before, is added to or subtracted from the former motion according as they directly conspire with or are directly contrary to each other; obliquely joined when they are oblique, so as to produce a new motion compounded from the determination of both.

Newton had already defined "quantity of motion" as the product of mass times velocity. Today this product is defined as *momentum*. Therefore, Newton's "change of motion" is the same as change of momentum. We can, therefore, restate Newton's second law as an assertion that change in momentum is proportional to the impressed force. Alternatively, if the mass involved is constant, we can express the law as an assertion that acceleration (change in velocity during time that the force acts) is proportional to the net impressed force and is in the same direction as that force. If we choose our unit of force properly, the proportionality constant is unity and we can sum up the law in the equation

$$F = ma$$

(where F is the net force acting on a body of mass, m, and where a is the acceleration in the direction of the force, F).[9]

Newton is careful to state that forces in the same or opposite direction are additive or subtractive respectively. In other words, they are quantities that are measurable by cardinal numbers in accordance with our discussion in Chapter 2. On the other hand, it is implicit in Newton's statement that if the forces do not act along the same straight line they must be added by the special rules of vector addition. However, we will not in this study do more than introduce vector addition.

Law III. To every action there is always opposed an equal reaction; or the mutual actions of two bodies upon each other are always equal, and directed to contrary parts.

Whatever draws or presses another is as much drawn or pressed by that other. If you press a stone with your finger, the finger is also pressed by the stone. If a horse draws a stone tied to a rope, the horse (if I may so say) will be equally drawn back towards the stone; . . .

Although the first two laws followed clearly from Galileo's analysis, this third law was explicitly stated for the first time by Newton. Today we would express the law by saying that to every *force* there is always an equal and opposite force instead of using the terms action and reaction. If the earth is pulled toward the sun by the force of gravity there is an equal and opposite force pulling the sun toward the earth. The centripetal force (force acting toward the center) of a piece of string acting on a body which is being whirled around in a circle is

[9] Mass in modern physics is no longer regarded as constant and independent of velocity when the velocity is close to that of light. Therefore, Newton's original statement of the law is superior to that given by the equation $F = ma$. However, we shall use the latter.

equal and opposite to the force acting on the string exerted by the whirling object. This is termed the centrifugal force, the force exerted by the mass pulling outward on the string, which is equal and opposite to the centripetal force, which is the force of the string acting on the mass.

In equation form the third law may be expressed

as

$$F_1 = -F_2 \qquad \text{(the two equal and opposite forces)}$$

or

$$m_1 a_1 = -m_2 a_2 \qquad \text{(the corresponding interacting masses, } m_1 \text{ and } m_2\text{, and their respective accelerations).}$$

The *Law of Universal Gravitation* was expressed by Newton in two propositions.

1. *That there is a power of gravity pertaining to all bodies, proportional to the several quantities of matter which they contain.*
2. *The force of gravity towards the several equal particles of any body is inversely as the square of the distance of places from the particles.*

This law can best be expressed in the equation

$$F = \frac{Gm_1m_2}{d^2}$$

(where m_1 and m_2 are any two masses, d is the distance between the two masses, F is the force of gravity between them, and G is a universal constant).

As expressed above, this is a law relating to attraction between particles, or point masses. However, Newton was able to show that spheres made up of particles can be considered as if they were point masses with all their mass concentrated at the center. For purposes of analysis of gravitational forces, therefore, we may regard any planet, including the earth, as being a point mass.[10] (For example, in calculating the force of gravity on a mass at the surface of the earth, its distance from the earth is taken as the earth's radius.)

[10]Newton comments on this point: "After I had found that the force of gravity towards a whole planet did arise from and was compounded of the force of gravity towards all its parts . . . I was yet in doubt whether the proportion inversely as the square of the distance did accurately hold, or but nearly so, in the total force compounded of so many partial ones; for it might be that the proportion which accurately enough took place in greater distances should be wide of the truth near the surface of the planet, where the distances of the particles are unequal and their situation dissimilar . . . But I was at last satisfied of the Proposition, as it now lies before us." (Comment after Proposition VIII, Book III.) Here is a good example of the caution and thoroughness with which Newton proceeded.

3. DEFINITION OF INERTIAL MASS FROM LAWS OF MOTION; HOW TO USE EQUATION F = ma

The first two laws of motion can actually be expressed in the form of a single law, namely,

$$F = kma$$

where, k, is a constant which can be made equal to one, if proper units for F and m are used. (Since $F = 0$ when $a = 0$ and vice versa, the first law is included under this law.)

We repeat the expression for the third law so that it may be kept in mind:

$$m_1 a_1 = -m_2 a_2$$

These equations, so fruitful in all they imply, appear beguilingly simple, as they indeed are when only considered mathematically. But, if they are to be applied to physical situations, we must have an operational definition for at least two of the three concepts—force, mass, and acceleration—which enter into the equations. We must, in other words, be able to prescribe unambiguous methods of measuring at least two of these quantities. We have examined already the definition of acceleration. This is defined in terms of measurements of length and time, both fully and unambiguously defined by specific operations of measurement.

Let us consider mass. How can we specify the proper definition of mass without using the concept of force, or of weight? The rough definition that mass is "quantity of matter" is of no help, because we cannot specify clearly how this "quantity" is to be measured. (We do not mean volume because we recognize how the volume of a substance may change with changes of temperature, pressure, etc. We do not mean weight because weight is a force, and if we define mass in terms of weight and weight in terms of mass, we should be going in a circle.)

The essential property of mass to which we must attend is inertia, or "resistance to change in velocity." But we need to specify the exact procedures for measuring such resistance without assuming that we have a definition of force (since we intend to define force in terms of mass and acceleration). Several suggestions as to how to escape the dilemma have been made.[11]

Probably the best approach in developing an operational definition of mass is to use the third law of Newton. This law states that to every force there is an equal and opposite force, but no mention is made of the magnitude of the forces involved.

Let us imagine we have two carts as shown in Fig. 7.1, with frictionless bearings.

[11] For a detailed and rigorous analysis of the problem, see Lindsay and Margenau's *The Foundations of Physics,* Wiley, New York, 1936, pp. 85–93.

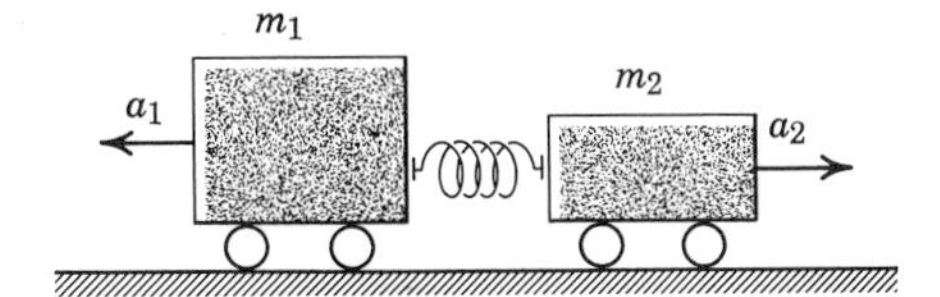

Fig. 7.1. An experimental way of determining relative masses. Since $m_1a_1 = -m_2a_2$, therefore $m_1/m_2 = -a_2/a_1$; if a_2/a_1 is measured, the ratio of the inertial masses m_1/m_2 can be determined.

They are separated by a spring that can be compressed. The spring is compressed and the carts are suddenly released. They fly apart with equal but opposite forces acting upon them. The third law then tells us that the magnitude of their relative accelerations will be in proportion to the masses.

Therefore, without knowing anything about the magnitudes of the force we may measure the ratio of the masses of any two objects by measuring the relative accelerations when they interact. Thus, if we take any particular body as a standard of inertial mass, we can determine the relative mass of any other body. In this manner we have achieved a way of measuring what we can term "inertial mass" by specific physical operations that can be performed.

Several aspects of our analysis need special emphasis. First, the quantity being measured is indeed "resistance to change in motion," and *has nothing to do with weight.* The same experiment conducted anywhere in the universe would yield the same relative measures of mass if the laws of motion are valid. Second, we have not required any specification of what we mean by a measurement of force. We may, therefore, define force in terms of the operationally defined quantities mass and acceleration.

If we take a standard kilogram as our unit of mass, and 1 m/sec/sec as our standard unit of acceleration, it is most convenient to define the unit of force in such a way that the equation $F = kma$ will be reduced to $F = ma$, that is, to take our unit of force so as to make the proportionality constant, k, equal to one.

Applying the above, the unit of force, *one newton,* is defined as that force that will give to *one kilogram* an acceleration of *one meter per second per second* in the direction in which the force is applied. (The term newton will be abbreviated to nt in what follows.)

Now it happens that all masses accelerate toward the earth at the same constant rate at the same location. This acceleration is so important that it is given the symbol g and, for the sake of simplicity, it is being taken here as being equal to about 10 meters/sec/sec. We should, however, keep clearly in mind that g is not a constant, and varies from location to location even on the earth. For more accurate work, g is taken as 9.81 m/sec/sec (which may be written as 9.81 m/sec^2) by engineers, unless still greater accuracy is required. For a more precise value, a calculation must be made that will allow for both height above sea level and for latitude. At sea level and at the equator the value of g is taken as 9.80616 m/sec^2; at latitude 45° at sea level it is 9.80665 m/sec^2.

The value of 10 m/sec^2 is, however, accurate within 2 per cent. Similarly, the value if 32 ft/sec/sec is accurate within 1 per cent.

What is the force of gravity acting on a mass of one kilogram? Since 1 kg is accelerated at the rate of 10 m/sec/sec, from the equation of motion the force of gravity must be 10 nt. A 2-kg mass will be accelerated at the same rate; therefore, the force of gravity on this mass will be 20 nt, etc.

It appears then that to the extent that g is constant, the force of gravity or weight of an object will be proportional to its inertial mass. We can always express weight by the equation $F = mg$ (F, being expressed in newtons, m, in kilograms, and g, in meters per second per second).

These relationships, while clear and unambiguous in themselves, are apt to become confused when different units are used. The only way they can be mastered is to keep the basic concepts very clearly in mind. Particularly important is the distinction between weight and mass, a distinction which in nonscientific terminology is often lost, since the same units such as kilograms are often used as a measure of weight as well as of mass. This can be done without disastrous effect on calculations only in those cases where motion is not involved. For example, in weighing an object on a balance, since weight is proportional to mass at the same location, we can express the relative weights (force of gravity) as equivalent to the relative masses in kilograms. To avoid any confusion, it is better in such cases to specify the unit kilogram weight, meaning the force that gravity exerts at a particular location on a kilogram.

A similar confusion exists in the English system. The unit pound is used sometimes as a unit of force or weight, sometimes as a unit of mass. Here, we shall almost always use the metric system. The data for any problem in accelerated motion should be translated from the English to the metric system before it is worked out. All that is needed is to recall that 1-kg mass *weighs* 2.2 lb, which is a satisfactory approximation for most purposes. Thus the *mass* of a 3300-lb automobile is $3300/2.2 = 1500$ kg.

4. THREE EXAMPLES: APPLICATION OF LAWS OF MOTION

A clear understanding of the implications of the laws of motion can be arrived at only by seeing how they are applied to specific situations. The three basic examples given in this section should be studied with attention to every step in the analysis.

Example 1 Given a cart, as illustrated in Fig. 7.2a, with a total mass of 50 kg, what horizontal acceleration will it undergo if it is pulled by a horizontal rope with a force of 4.2 nt? (The tension of the rope is then said to be 4.2 nt.)

Case 1 The frictional resistance is zero. To apply the equation $F = ma$, we

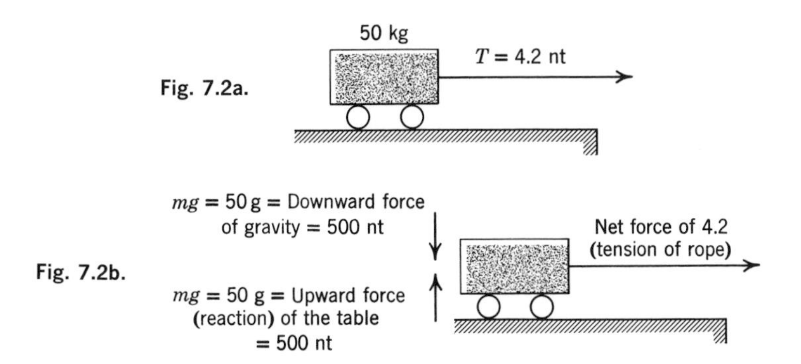

Fig. 7.2a.

Fig. 7.2b.

need to consider all forces acting on the cart and then determine the net force and its direction (Fig. 7.2b).

The upward and downward forces (50 g or 500 nt) exactly cancel and can be disregarded. The only force acting is the tension of the rope acting horizontally. It will, therefore, produce a horizontal acceleration. The problem then becomes

$$F = ma \text{ where } F = \text{net force} = 4.2 \text{ nt}; \; m = \text{total mass} = 50 \text{ kg}.$$

$$\therefore a = \frac{F}{m} = \frac{4.2}{50} = 0.08 \text{ m/sec}^2 \text{ in a horizontal direction.}$$

Case 2 Frictional resistance opposing the motion is assumed to be 0.3 nt (Fig. 7.3a).

The upward and downward forces again cancel. Now, however, the net horizontal force is $4.2 - 0.3 = 3.9$ nt.

$$\text{Therefore, } a = \frac{3.9}{50} = 0.078 \text{ m/sec}^2$$

Example 2 This problem is illustrated in Fig. 7.3b.

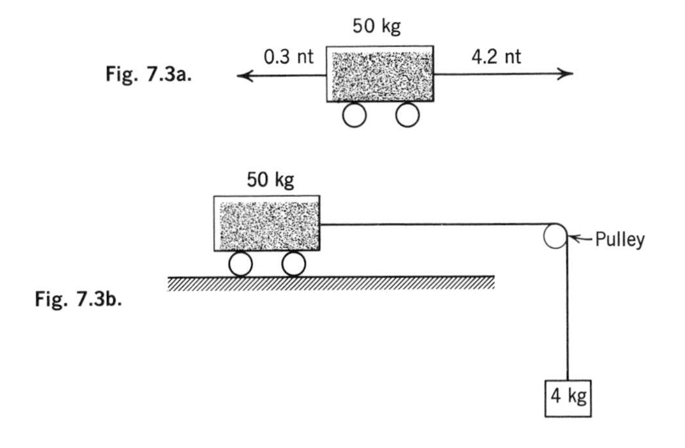

Fig. 7.3a.

Fig. 7.3b.

We again wish to determine the acceleration of the cart. In this case the tension or force is being produced by the force of gravity acting on the mass of 4.0 kg.

We shall assume absence of friction. What is the net force acting on the cart? Again the upward and downward forces *on the cart* are in balance. The pulley acts merely as a machine, which changes the direction of action of the tension of the rope. Therefore, the downward force of gravity acting on the 4 kg is producing a tension that is acting on the cart in a horizontal direction.

What masses are being accelerated? If we assume that the rope is not extensible, then both the 50 kg is being accelerated, and the 4 kg is being accelerated at the same rate (although in different directions). Our problem now can be analyzed:

$$F = \text{net force acting on the system} = \text{weight or force of gravity on 4 kg} = 4\text{ kg} \times g = 40\text{ nt}$$

$$m = \text{the total mass being accelerated is } 50\text{ kg} + 4.0\text{ kg} = 54\text{ kg}$$

Therefore, $a = \dfrac{F}{m} = \dfrac{40}{54} = 0.74\text{ m/sec}^2$

(Note that the answer is given to two digits or significant figures because the data in each case are given to an accuracy of two digits or significant figures.)

In this example the force of gravity, 40 nt, is accelerating *two* masses. If we analyze the forces acting on the 4 kg mass alone, we see that there are two forces acting in opposite directions: the downward force of gravity, 40 nt, and the tension of the rope that is acting upward (by the law of action and reaction the rope pulls upward on the 4 kg mass with the same force it pulls on the 50 kg mass). Thus, there is a net downward force equal to 40 nt $-$ T (tension of rope upward). If we wish, we can now determine this tension T as follows, using the acceleration we have determined.

$$40 - T = (4)(.74)$$

or

$$T = 40 - 2.96$$
$$T = 37\text{ nt}$$

To check whether this is the force necessary to accelerate the 50 kg mass at the rate of 0.74 m/sec^2, we note that if we attend *only* to the 50 kg mass, the only net force acting is the tension of the rope, namely 37 nt. Therefore in

$$F = ma$$
$$37 = (50)a$$

and again we find

$$a = 0.74\text{ m/sec}^2$$

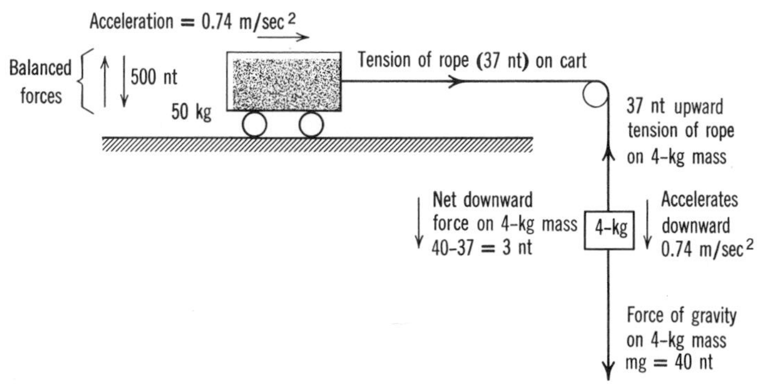

Fig. 7.4.

The complete solution has been diagrammed in Fig. 7.4, showing all the forces acting.

If such experiments are repeated with various combinations of masses and with various distances through which the masses move, and always yield the same result, it may be regarded both as a verification of the laws of motion and as a determination of the constant g *within the limits of accuracy* of the measurements. The experiments when performed will never yield exact results because the abstract conditions described cannot be fulfilled.

In addition to the rather specific experimental conditions that must be examined, there is a more general problem that for a long time was overlooked. In all the laws of motion *it is assumed that we can specify directions.* But directions with respect to what? Newton assumed the existence of absolute space. He also assumed that the so-called fixed stars were effectively at rest in absolute space. But if we assume that the earth is rotating relative to the fixed stars, will motion toward the center of the earth actually be straight-line motion? For the present we shall not go further into these problems except to point them out, and to state that, for many purposes, we have assumed as our frame of reference the earth itself (which implies that we consider it stationary). But we should be aware that we leave an important problem "up in the air." We shall return to this when we consider relativity.

5. MOTION IN A CIRCLE; CONNECTION BETWEEN MOTION IN A CIRCLE AND THE FORCE OF GRAVITATION; ILLUSTRATIONS OF PLANETARY AND SATELLITE ORBITS

Up to now we have been primarily interested in motions that take place in straight lines, except for the brief section in which we considered Galileo's analysis of projectiles. But the extraordinary contribution made by Newton to the science of mechanics was the development of an analysis that applied to *all* objects anywhere in space. By combining his three laws of motion with his law of gravitation, the basic manner in which all objects moved could be brought within a single system of analysis. Such a simple and elegant conception!

First, let us examine with some care the problem of motion in a circle. In Fig. 7.5 is shown a point mass P, which is being forced to move about the circle of radius R. We assume that the speed of P is constant or uniform. That is, the distances covered along the arc of the circle are always the same over equal time intervals, no matter how long or short the time intervals may be. Now, although the *speed* is *constant* according to this definition, since the *direction* of the motion is constantly changing, the *velocity* is *changing*. In other words, there must be some force that is constantly being exerted on the point mass P that will constrain it to move in the circle. (If no force were being exerted, the mass would fly off in the direction of the tangent line shown in the figure.)

To find the magnitude of this constant force acting upon P, we must determine the magnitude of its acceleration. This is equivalent to determining how far the mass P must "fall" (or depart from the straight line PA) in a given time if it is to remain on the circle. (Here is a problem whose elements are very similar to Galileo's analysis of projectile motion, which we reviewed on page 110. Another look at Galileo's analysis would be helpful before continuing.)

We use the geometric relationship, as shown in Fig. 7.5:

The angle OPA is a right angle (Why?)
Therefore, the distance $(h + R)^2 = R^2 + s^2$ (Why?) (3)
From algebra, we obtain

$$h = \frac{s^2}{2R + h} \tag{4}$$

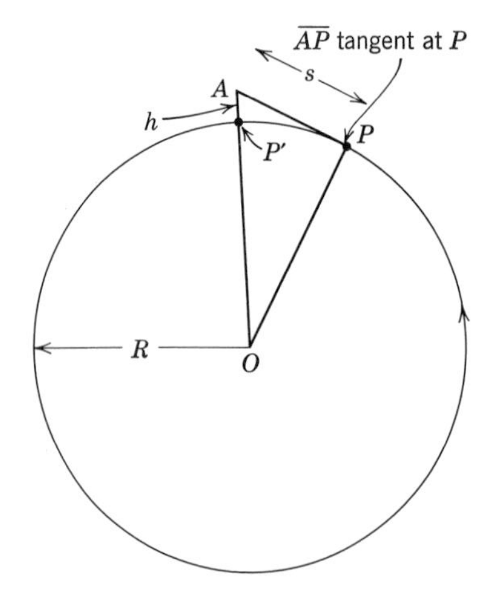

Fig. 7.5. Centripetal acceleration of an object moving in a circle. An object at P "falls" through distance h when it moves to P' because if it had continued in a straight line it would have reached point A.

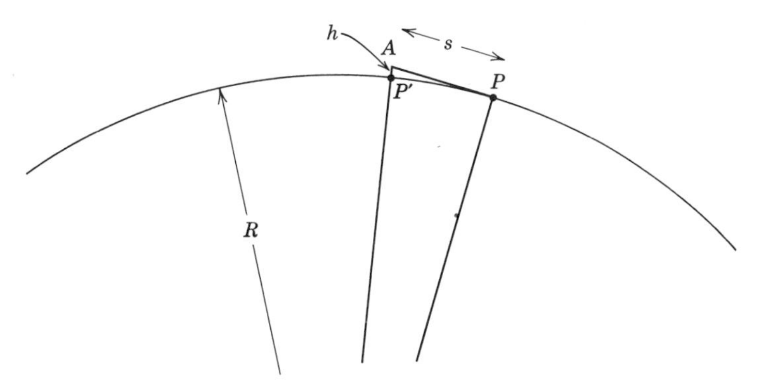

Fig. 7.6. Blown-up diagram of Fig. 7.5 to show how h diminishes relative to R when points P and P' are taken close to one another (that is, when the time interval of the motion is made small).

Now we need to use some more fancy mathematics, only rigorously justified by using the calculus. The essential ideas, however, should be intuitively clear if one considers the problem with care.

Suppose that we examine the motion of P over an extremely short time interval, as in Fig. 7.6.

Then two results transpire: the straight-line length PA becomes closer and closer to the arc length between P and P'; and the distance h becomes smaller and smaller compared to the length of the radius R.

As a specific example, suppose that we consider the motion of the point P through an arc that is 1/180 of the total circumference—in other words, through an angle of two degrees. Under these conditions, it is found that the ratio of the distance h to the distance R is less than 0.01%.

If the time interval over which we consider the motion is reduced sufficiently, we finally reach a point where the difference between $2R + h$ and $2R$ is so small that the effect upon the fraction of neglecting the value of h in the denominator is negligible. Under these circumstances we can appropriately restate equation 4 as

$$h = \frac{s^2}{2R} \tag{5}$$

Finally, since we assume uniform speed of the particle P, we can substitute for s, the value given by

$s = vt$ where v is the constant *speed* of the mass P

and arrive at

$$h = \frac{v^2t^2}{2R} \tag{6}$$

(It should be emphasized that v is used here as *speed* and not as *velocity*. The *speed* is constant, although the *velocity* is changing since direction is changing.)

Again, if we are assuming that the accelerating force directed toward the center of the circle is constant, we can use the relationship between the distance fallen and the time during which the mass falls, which is

$$h = \tfrac{1}{2}at^2$$

to get finally

$$\tfrac{1}{2}at^2 = \frac{v^2t^2}{2R}$$

which immediately reduces to

$$a = \frac{v^2}{R} \tag{7}$$

It will be seen that the centrally directed force, which is required to keep a mass m moving in a circle with uniform speed v, will thus become

$$F = ma = \frac{mv^2}{R} \tag{8}$$

Since $v = \dfrac{2\pi R}{T}$, we can also express the same equation as

$$F = \frac{m4\pi^2R}{T^2} \tag{9}$$

(This is the equation that Newton used to derive the inverse square law; see page 120.)

These equations for motion in a circle can be generalized (see Fig. 7.7) and

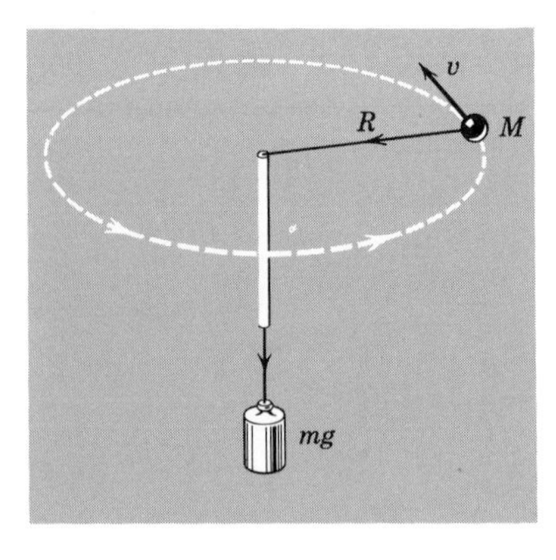

Fig. 7.7. A mass M, whirled in a horizontal circle of radius R, at the end of a string passing through a smooth (i.e., frictionless) tube, will be acted on by the tension of the string equal to mg. This force will keep the mass M moving in a circle of radius R according to the equation $mg = Mv^2/R$. (This result can be conveniently tested in a simple experiment by passing a string through a glass tube, with edges smoothed, and whirling the mass at various speeds and radii). The equivalent expression used by Newton is that the centripetal force mg is equal to $F = M4\pi^2R/T^2$.

it is found that very similar equations hold for motion in elliptical orbits. Probably Newton first derived these results through his use of the calculus, although the presentation of the equations in the *Principia* does not make use of calculus methods.

6. THE ACCELERATION OF THE MOON: SATELLITE ORBITS

By combining the equation derived above, for the force required to maintain an object in circular motion, with his equation of gravitation, Newton was able to show how the moon's acceleration was related to the acceleration of gravity on earth. He was also enabled to "explain" Kepler's three empirical laws pertaining to the motion of the planets.

Let us apply equation 9 for centripetal force to determine how fast the moon accelerates toward the earth.

$$F = \frac{M(4\pi^2 R)}{T^2} = Ma$$

where a is the moon's acceleration and M is the mass of the moon, R the radius of its orbit, and T its period of revolution.

The masses cancel, and we get

$$a = \frac{4\pi^2 R}{T^2}$$

In English units, as used by Newton

$$R = 2.4 \times 10^5 \text{ miles} = 1.27 \times 10^9 \text{ ft}$$
$$T = 27 \text{ days, } 7 \text{ hr, } 43 \text{ min} = 2.38 \times 10^6 \text{ sec}$$

This yields an acceleration of 0.0088 ft/sec/sec for the moon as it "falls" toward the earth, out of a straight-line path.

How does this compare with the acceleration of gravity on earth? How are they related? If Newton's inverse square law of gravitation is correct, since the radius of the moon's orbit is 60 times the radius of the earth, we can show that the acceleration of the moon due to gravity should be $1/60^2$ or $1/3600$ the gravitational acceleration of an object on the earth's surface. This may be shown as follows.

Force of gravity between the earth and moon $= G\,M_e M_m/R_m{}^2$ where M_e is mass of earth, M_m is mass of moon, and R_m is distance between their centers.

This force is equal to the mass of moon times its acceleration. Thus

$$G\frac{M_e M}{R_m{}^2} = Ma$$

which simplifies to

$$G\frac{M_e}{R_m{}^2} = a$$

On the earth's surface, gravity exerts a force on a mass m, which is

$$G\frac{M_e m}{R_e{}^2} = mg \qquad (10)$$

which simplifies to

$$G\frac{M_e}{R_e{}^2} = g \qquad (11)$$

where R_e is the radius of the earth.

We now can compare g and a by dividing equation 10 by equation 11

$$\frac{a}{g} = \left(\frac{R_e}{R_m}\right)^2 \quad \text{or} \quad \frac{a}{g} = \left(\frac{1}{60}\right)^2$$

thus

$$a = \frac{g}{3600}$$

Now g at the earth's surface is 32 ft/sec^2.

Therefore, $a = 32/3600 = 0.0088$ ft/sec^2, which agrees with our previous calculation. The hypothesis of the inverse square law of gravitation is thus verified. It is interesting that when Newton first made this calculation, the ratio of the radius of the moon's orbit to that of the earth was not very precisely known, and his calculations showed only an approximate agreement.

Finally, we can readily see that Kepler's third law is appropriately explained, by applying the law of gravitation to the attraction between the sun and a planet:

$$F = G\frac{M_s M_p}{R^2} = M_p a = M_p \frac{4\pi^2 R}{T^2}$$

yielding

$$\frac{GM_s}{R^2} = \frac{4\pi^2 R}{T^2} \quad \text{or} \quad T^2 = \left(\frac{4\pi^2}{GM_s}\right) R^3$$

where the bracketed symbols are a constant; thus

$$T^2 = kR^3$$

(where R is the orbital radius and T the period.)

It is of interest also to apply these equations to the orbits of artificial satellites. As in the case of the moon, any satellite circling the earth will have an orbit determined by

$$\frac{GM_eM}{R^2} = \frac{M(4\pi^2R)}{T^2}$$

where M is now the mass of any satellite, or

$$\frac{GM_e}{R^2} = \frac{4\pi^2R}{T^2}$$

At once we observe that the mass of the satellite does not enter into the equation relating the period of revolution to the radius of revolution. The speed of all satellites in the same orbit must be the same regardless of their mass. This enables space agencies to plan space stations with rendezvous among several different-sized satellites more easily. Regardless of mass, any satellite in the same orbit travels with the same speed.

Although we have not yet considered how G or M_e can be determined individually, we do have available a method of determining their product, GM_e, since

$$G\frac{M_em}{R_e^2} = mg$$

for any object of mass m at the surface of the earth (radius of the earth being R_e and mass of the earth being M_e), or

$$GM_e = gR_e^2$$

In the metric system, we are taking

$$g = 10\,\text{m/sec}^2 \text{ and } R_e = 6.4 \times 10^6 \text{ m}$$
$$GM_e = 4.1 \times 10^{14} \text{ m}^3/\text{sec}^2$$

On this basis, then, we can calculate the time of revolution of any satellite that orbits at a given distance above the surface of the earth.

For example, a satellite 6×10^2 km or 6×10^5 m above the earth (somewhat over 360 miles) will have a period of revolution given by

$$\frac{GM_e}{R^2} = \frac{4\pi^2R}{T^2}$$

where $R = 7 \times 10^6$ m (6×10^5 plus radius of earth), or

$$\frac{4.1 \times 10^{14}}{(7 \times 10^6)^2} = \frac{(4\pi^2)(7 \times 10^6)}{T^2}$$

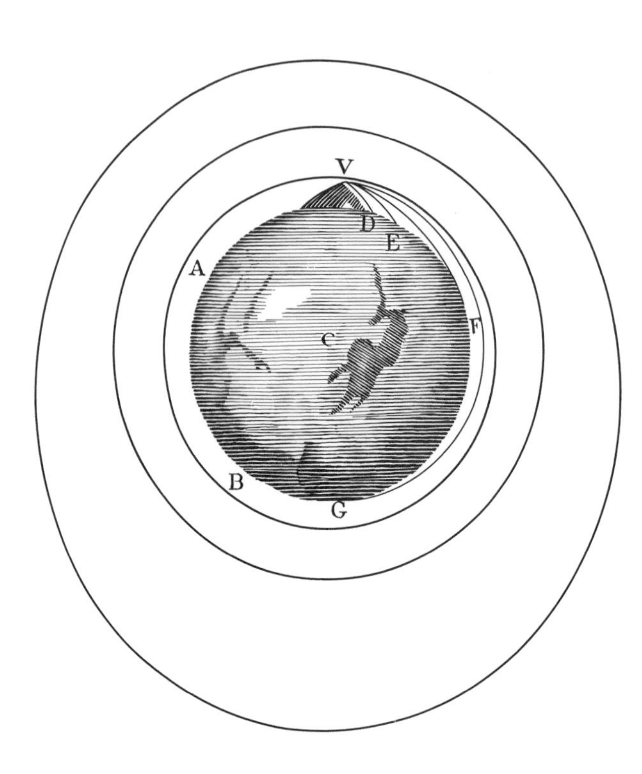

Fig. 7.8. A replica of a diagram appearing in Newton's *Principia,* which forecasts the possibility of launching satellites into orbit about the earth. Newton explained that if projectiles were launched from a high mountain at *V*, those projectiles with relatively low velocity would land at *D*, *E*, and *F*; a projectile with a higher velocity would go halfway around the world to point G; at still higher velocities, the projectile would go into orbit. This principle is used today in launching second-stage rockets with high velocities from points far above the earth's surface. (The drawing is reproduced from *Mathematical Principles of Natural Philosophy,* edited by Florian Cajory, University of California Press, Berkeley, 1947, p. 551.)

or

$$T^2 = \frac{(4\pi^2)(7 \times 10^6)^3}{4.1 \times 10^{14}} = \frac{1.4 \times 10^{22}}{4.1 \times 10^{14}} = 34 \times 10^6$$

or T is approximately 6×10^3 sec or about 1 hr, 40 min.

As an exercise, calculate the speed at which the satellite will travel in m/sec and in miles/hr. (Remember that the speed will be given by $2\pi R/T$.)

Figure 7.8 is a duplicate illustration of Newton's forecast of artificial satellites.

7. CAVENDISH EXPERIMENT TO DETERMINE VALUE OF G

In Newton's equation for gravitational force, $F = Gm_1m_2/R^2$, G is an extremely small quantity and, unless at least one of the masses is large, the force is so minute as to be difficult to measure. It was not until 1798, a century after Newton's publication of the law of gravitation, that G was measured with accuracy (and hence the value of the mass of the earth determined) by Henry

Cavendish, an Englishman of great wealth who devoted his life to scientific investigations, and after whom the famous Cavendish laboratory was named.[12]

Essentially the method consists of suspending from a thin wire a long slender rod with small spheres at each end, and measuring the deflection caused by the gravitational attraction to a pair of large massive balls brought close to the suspended pair (Fig. 7.9).

The approximate value determined by Cavendish and, since then, confirmed by other experimenters is $G = 6.7 \times 10^{-11}$ newton-meters squared per kilogram squared. (Can you explain these units for G?). From the value of G, the mass and density of the earth can readily be calculated as explained in the previous section. Consequently, Cavendish's experiment is sometimes referred to as "weighing the earth."

To appreciate the delicacy of the experiment, consider the apparatus used by Cavendish and the values from which his calculations were made. The larger lead balls each had a mass of 167 kg, the smaller 0.8 kg. The distances between the centers of the large and the small lead balls were 0.2 m in each case. Cavendish was able to measure the minute deflection (less than 1°) caused by the acting gravitational forces. From this, and knowing by independent measurements the amount of force required to give the observed deflection, Cavendish found the force between each pair of spheres to be 2.3×10^{-7} newtons.

It is not surprising that Cavendish had to take all kinds of precautions to prevent extraneous influences from affecting his results. In particular, he had to

[12] An account of Cavendish's experiment is given in the *Philosophical Transactions,* Vol. 17 (1798), p. 469, and is discussed in detail in *Great Experiments in Physics,* edited by Morris H. Shamos, Henry Holt, New York, 1959.

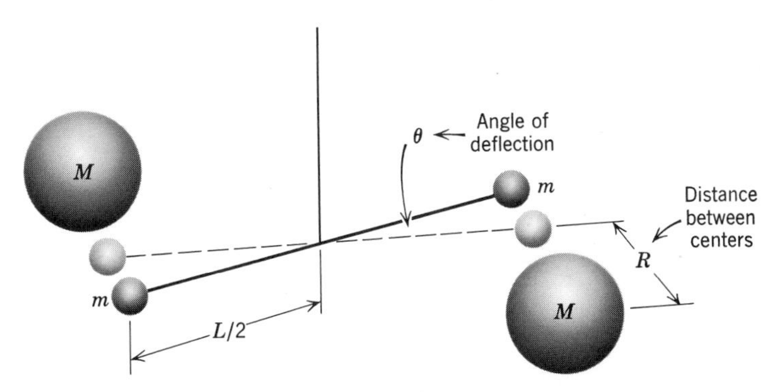

Fig. 7.9. Diagram of a torsional pendulum used in the Cavendish experiment to determine G. The slight deflection caused by the gravitational attraction between small masses *m* and large masses *M* is measured with great precision.

guard against slight temperature changes. Even though his apparatus was suspended in an air-tight box and he took his readings through a telescope outside the box, he discovered that very slight changes in temperature would set up convection currents in the air sufficient to cause small deflections in the suspension system. Nevertheless, he obtained a value for G which was within about one per cent of the values obtained by a later series of very carefully conducted investigations.

8. ELEMENTARY PRINCIPLES OF ADDITION OF DIRECTED QUANTITIES (VECTORS)

We have seen several examples of how directions as well as magnitudes must be considered in applying equations of physics. Let us review them. It was seen in Chapter 6 that in the case of a projectile moving with a given horizontal velocity, the force of gravity, being at right angles to the horizontal direction of motion, has no effect on any horizontal velocity. It merely imposes an additional vertical acceleration in the direction of the force of gravity. The combined effect of the vertical acceleration and the constant horizontal velocity was illustrated by diagrams of the resulting motion (see Fig. 6.11).

Similarly, in the last section, we saw that in the case of the moon, if we assume its orbit to be circular, the effect of gravitational attraction of the earth is not to increase or decrease the speed or tangential velocity of the moon in its orbit but only to cause a constant change of direction of the moon's motion. In other words, since the force of gravitation is at right angles to the tangential velocity at any point, the only acceleration present is at right angles to the orbit at every point.

Again, when discussing Galileo's inclined plane experiments we stated that an object on an inclined plane is subject to a force downward along the plane, which is considerably less than the force of gravity but is equal to the force of gravity times the sine of the angle of elevation of the plane. We shall see that this is a result of combining two forces at oblique angles: the force of gravity downward and the reaction of the plane.

Now, we shall consider a few of the basic principles used in combining quantities that have directions as well as magnitudes. These principles were developed systematically by Newton and are called the axioms of vector combinations.

To start with, we use, as an illustration (Fig. 7.10), the notion of a *directed* displacement (that is, a change of position by a certain amount in a given direction). Suppose that a person moves first 10 ft due north, then 25 ft due east. This change in his position is a displacement that can be expressed in one of

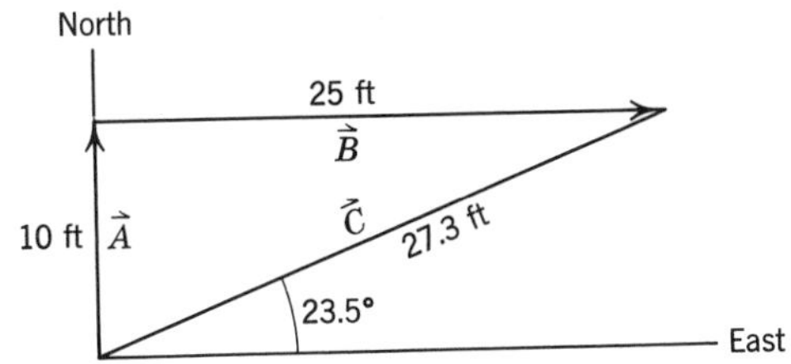

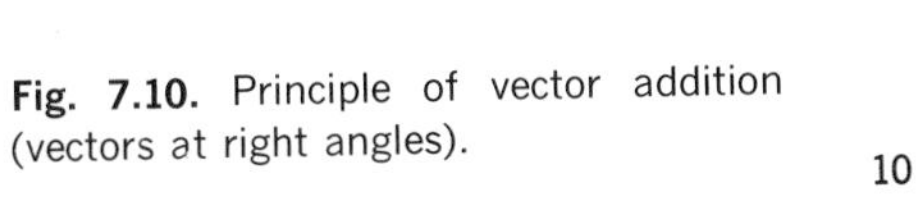
Fig. 7.10. Principle of vector addition (vectors at right angles).

two ways that are equivalent. We assume that he is moving on a plane surface. We can either say that he moves successively north 10 ft *then* east 25 ft, or we can say that he reaches a point that is 27.3 ft (from the Pythagorean theorem) from his starting point at the angle, 23.5° (from $\tan\theta = 0.4$) north of the eastward direction. We can, in other words, regard the displacement through 27.3 ft in a specified direction as equivalent to the successive displacements of 10 ft north plus 25 ft east. A single displacement is thus regarded as the *sum* of two displacements. (The two vector displacements $\vec{A}$ and $\vec{B}$, which are equivalent to the single displacement $\vec{C}$, are called components of $\vec{C}$.)

Now let us assume that after reaching the point P in Fig. 7.10, the person moves again 20 ft north (vector $\vec{C}$) followed by a displacement 15 ft east (vector $\vec{D}$). Again, it will be found (Fig. 7.11) that his new displacement from P to P' will be equivalent to a single displacement—in this case equal to 25 ft—at an angle of 53° north of the east direction.

If we combine all four displacements, we obtain a single displacement, OP', as shown in Fig. 7.12. In this figure we have shown the displacements A and B combined into OP, and the displacements C and D combined into PP'. We can think of OP' as being made up of the combination (or sum) of $OP + PP'$.

Figure 7.12, having to do with displacements, now suggests the following geometric representation and definition of addition of all vectors.

1. A vector may be represented by a line whose length indicates the magnitude of the vector and whose direction represents the direction of the vector.

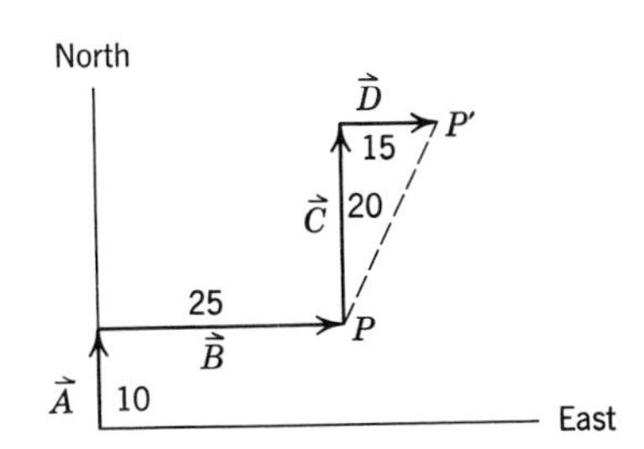

Fig. 7.11. Addition of vectors.

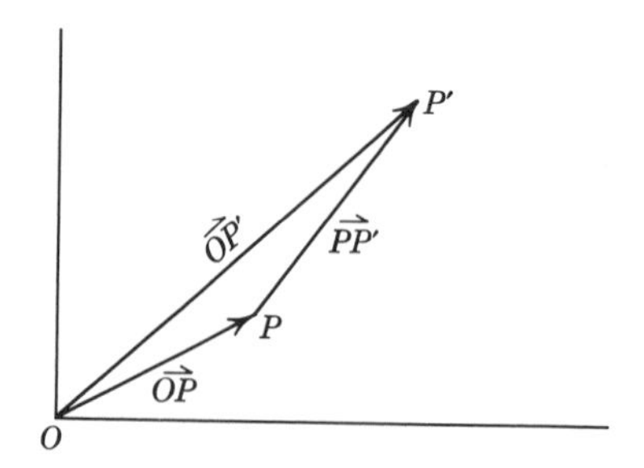

Fig. 7.12. Vector addition where vectors have an angle between them other than a right angle.

2. Two vectors $\vec{A}$ and $\vec{B}$ may be added by drawing vector $\vec{A}$ and then from the end of vector $\vec{A}$ drawing vector $\vec{B}$, the sum $\vec{A} + \vec{B}$ then becoming a vector drawn from the origin of $\vec{A}$ to the end of vector $\vec{B}$. Various examples of vector addition are illustrated in Fig. 7.13.

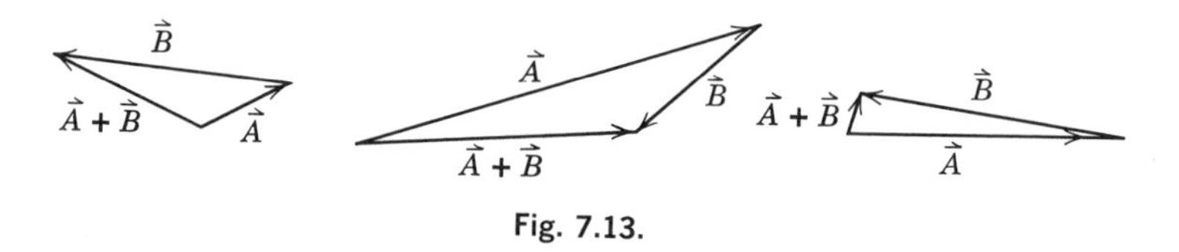

Fig. 7.13.

This figure involves displacements that take place successively. Other directed quantities in physics, however, often act simultaneously and it is convenient to show the combination of two vectors that act together starting from a point. We use velocity as an illustration, drawing on common experience. In Fig. 7.14 a man in a boat is rowing in a direction perpendicular to the bank of a river. If his velocity is 3 miles an hour and the river flows at 4 miles an hour, what will be the resultant velocity with which he travels? As he travels across, he is swept downstream and will land at point C.

Suppose now, instead of rowing directly across the stream, the course is made oblique to the bank. It will be found that the situation now becomes that shown in Fig. 7.15.

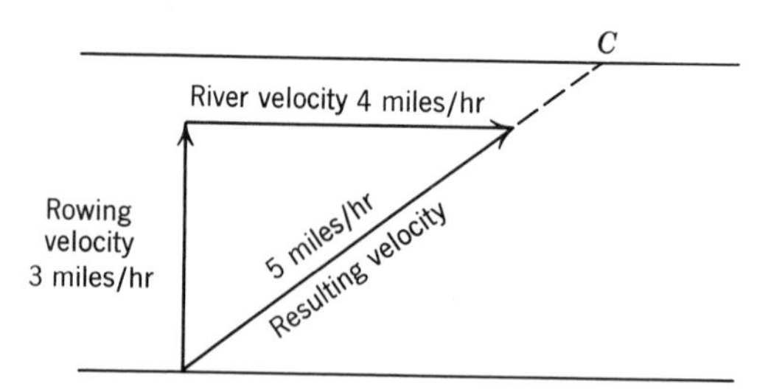

Fig. 7.14. Application of vector addition to velocities.

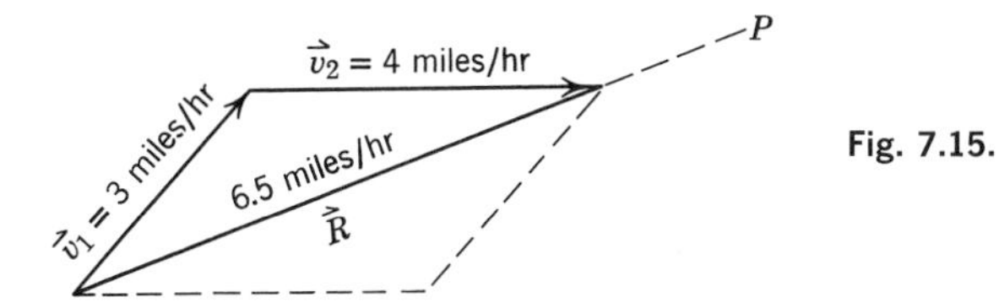

Fig. 7.15.

Where the resultant R can be regarded as the diagonal of the parallelogram, two of whose sides are given by vectors $\vec{v_1}$ and $\vec{v_2}$.

This manner of adding vectors, called the parallelogram law first formulated by Stevinus in 1586, was systematically developed by Newton and his successors. The law was stated by Newton, as applied to forces as a direct corollary of the second law of motion: "a body acted on by two forces simultaneously will describe the diagonal of a parallelogram in the same time as it would describe the sides by those two forces separately."

In the rowboat illustration the boat will reach the point P at the same time as it would if it first moved across the river at velocity $\vec{v_1}$ and then moved down the river at velocity $\vec{v_2}$.

In general the same rules of combination of vectors apply to forces, velocities, momentums, accelerations, and displacements. These are postulated to be vector quantities where we define a vector as a *line segment which has definite length and definite direction relative to a co-ordinate system and which combines with other vectors according to the parallelogram law.*

Up to this point, we have merely given some *illustrations* of how vectors are combined, drawing upon experience to suggest the rules. To be more formal, we may proceed to set up an algebra of vector addition. The law of combination has the same postulate as in the algebra that we have already considered. For vectors $\vec{a}$ and $\vec{b}$ and the parallelogram law of addition, the postulates are:

1. If $\vec{a}$ and $\vec{b}$ are vectors, then $\vec{a} + \vec{b}$ is a vector that is unique
2. $\vec{a} + \vec{b} = \vec{b} + \vec{a}$
3. $(\vec{a} + \vec{b}) + \vec{c} = \vec{a} + (\vec{b} + \vec{c})$
4. If $\vec{a} + \vec{b} = \vec{a} + \vec{c}$, then $\vec{b} = \vec{c}$
5. If $\vec{a} = \vec{b}$ and $\vec{b} = \vec{c}$, then $\vec{a} = \vec{c}$
6. There is a unique vector $\vec{O}$ such that $\vec{a} + \vec{O} = \vec{a}$
7. For every vector $\vec{a}$ there corresponds a vector $-\vec{a}$, such that $\vec{a} + (-\vec{a}) = 0$

There are also postulates for several types of multiplication of vectors. These we shall not consider, except to note the possibility of multiplying a scalar (ordinary number) times a vector. In such multiplication we postulate that the mag-

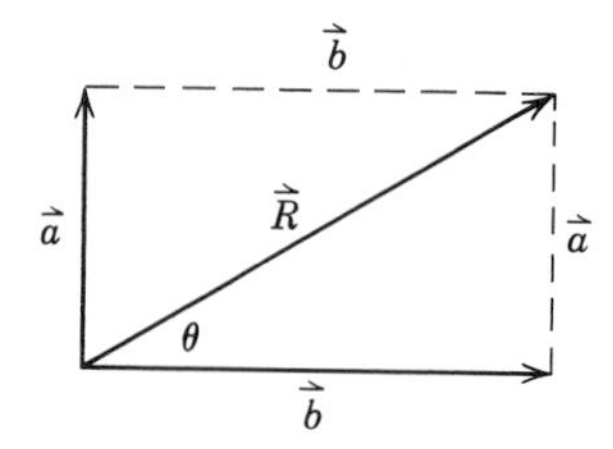

Fig. 7.16. Addition of vectors at right angle. $R^2 = a^2 + b^2$ and $\tan \theta = a/b$.

nitude of the vector is changed but not its direction, and that the algebraic axioms of multiplication, including that of distribution, remain valid. (*Note.* Multiplying a vector by a scalar greater than 1 implies stretching out the vector in that direction, etc.)

It is often necessary to find the actual value of the vector resulting from the addition of two vectors. If the two vectors are at right angles, the magnitude of the resultant is easily found by the Pythagorean theorem, its angle by the definition of the tangent (Fig. 7.16).

Often vectors are not at right angles. If one vector $\vec{a}$ is at an oblique angle with a second vector $\vec{b}$, then their resultant can be found by either of two methods as illustrated in Fig. 7.17.

Let us now consider the problem of finding the net force acting on a mass that is on a frictionless inclined plane. Since the plane is frictionless, there is no force on the mass *along* the plane. There are, however, two forces acting on the mass: the gravitational force downward, and the force of reaction by the plane

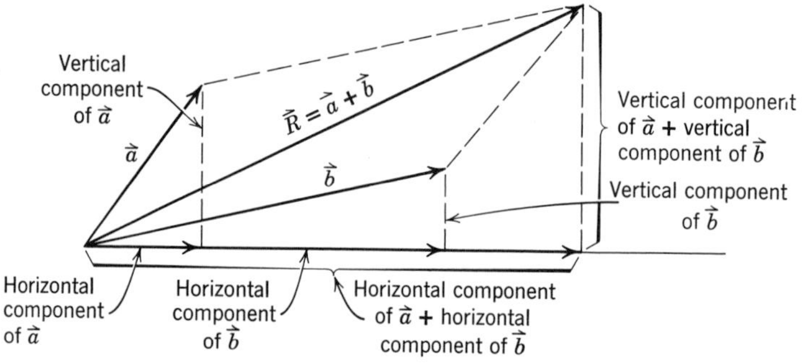

Fig. 7.17. Addition of vectors at an oblique angle. Two vectors, $\vec{a}$ and $\vec{b}$, not at a right angle, may be added by either of two methods to find the resultant $\vec{R} = \vec{a} + \vec{b}$. Either the diagonal of the parallelogram (in magnitude and direction) may be determined, or the vertical and horizontal components of $\vec{R}$ may be found by adding the vertical and horizontal components of $\vec{a}$ and $\vec{b}$ as illustrated.

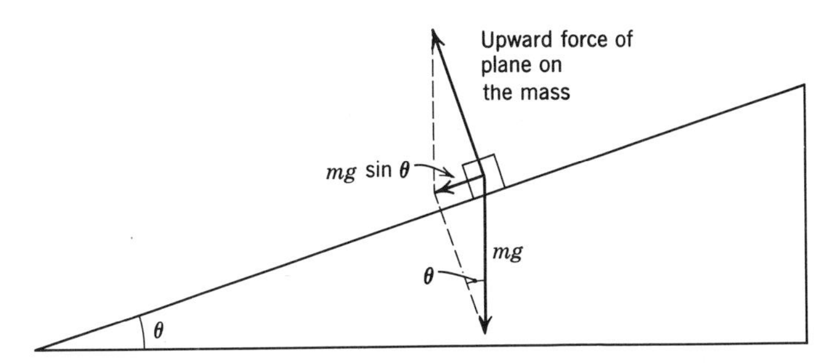

Fig. 7.18. Forces acting on a mass on an inclined plane. The resultant of the weight mg and of the upward reaction of the plane is a net force equal to $mg \sin \theta$ where θ is the angle of inclination of the plane.

pushing against the mass in a direction perpendicular to the plane. (This latter force is implied by Newton's third law. The mass must be pressing against the plane, therefore the plane is pressing against the mass.)

From a study of Fig. 7.18 it will be seen that the force along the plane is $mg \sin \theta$. If this force is set equal to ma the acceleration becomes $a = g \sin \theta$. (After any mathematical analysis of this kind it is good practice to consider some *physical* conditions, particularly of extreme cases. If $\theta = 90°$, then the object will drop with an acceleration g. Is this what you expect from the equation? At $\theta = 0°$ acceleration is 0. Does this agree with the equation? It is good practice to check your mathematical analysis, wherever possible, by considering extremes. Often, in fact, considering extremes can suggest the proper formulation of the mathematical analysis.)

9. SUMMARY OF LAWS

Motion. The first two laws of motion may be summarized by the equation:

$$F = ma$$

In using this equation three cautions should be kept in mind: (1) The force and the acceleration are vectors acting in the same direction; (2) the force F is the *net* or unbalanced force in the direction of the acceleration; (3) proper units must be used. Mass units and force units must not be confused. In the metric system the distinction is easily handled. Mass is expressed in kilograms and force in newtons. (In the English system mass is best expressed in slugs and force in pounds. We shall not use these units in equations of motion.)

Newton's Third Law.

$$F = -F$$

Action and reaction are equal and opposite. Note that the two forces act on *different* bodies. The sun pulls on the earth with a force equal and opposite to the gravitational force of the earth acting on the sun. The sun and the earth are each accelerated towards the center of revolution according to the equation:

$$F = \frac{m{v_\perp}^2}{R}$$

where R is the distance of the earth (or the sun) from this center of revolution (which lies within the surface of the sun).

Law of Gravitation.

$$F = G\frac{m_1 m_2}{d^2}$$

Note the units of G and its extremely small value, i.e. 6×10^{-11} in the KGM system. For spherical masses d is the distance between the centers of the two masses.

Central Acceleration.

$$a = \frac{{v_\perp}^2}{R}$$

The acceleration is directed towards the center of the circle, provided the tangential velocity or speed, v, is constant.

$$F = ma = \frac{m{v_\perp}^2}{R}$$

This equation expresses the centripetal force required to keep the object moving in a circle. See Fig. 7.19. Notice that a mass being swung in a circle at the end of a string is being accelerated by the tension force of the string. An equal and opposite force (centrifugal reaction) acts on the central body but not on the object.

Orbits of planets and value of g.

$$F = G\frac{m_s m_p}{R^2} = M_p\frac{v_\perp}{R} = M_p\frac{4R}{T}$$

where m_s is the mass of the sun, m_p is the mass of the planet and $v_\perp$ is the

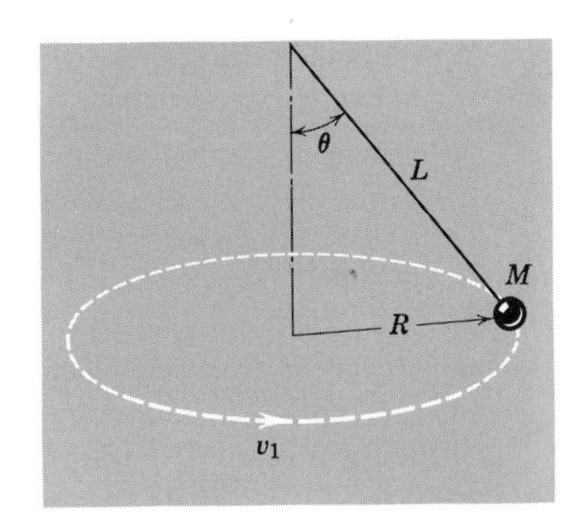

Fig. 7.19. The conical pendulum. A pendulum bob of mass M is suspended at end of string of length L. At any instant the tension of the string has two components: the vertical component Mg and the centripetal force component Mv^2/R. Thus for any given angle θ, $\tan\theta = Mv^2/R \div Mg$ or $\tan\theta = v^2/Rg$.

tangential speed of the planet. Note that the mass of the planet cancels out and that Kepler's third law can be derived from these equations.

$$F = mg = G\frac{mM_e}{R^2}$$

where M_e is mass of the earth and R is the radius of the earth. The value of g is thus determined and it is seen that it is independent of the mass of the accelerated object.

Tidal Force.

$$F = 2G\frac{M_m R}{d^2}$$

where M_m is the mass of the moon, d is the distance of the moon to earth, and R is the radius of the earth.

10. QUESTIONS

1. Distinguish clearly between mass and weight. Show that it follows from the definition of a newton that one kilogram weighs g newtons (approximately 10 newtons). Assuming that one kilogram weighs 2.2 lbs, what is the relationship between a newton and a pound?

2. What data are necessary to find the mass of the earth?

3. Why is it considerably more than "twice as dangerous" to go around a curve in a car at 30 miles per hour than around the same curve at 15 miles per hour?

4. What are two reasons for surmising that g is less at the equator than at the poles?

5. Why does the value of g vary by such a small per cent when measured first at the bottom and then at the top of a mountain?

6. Why do planets revolve around the sun in a plane? (Actually there is some slight deviation from plane motion. Can you account for it?)

7. Explain the principles of a centrifuge. Do you expect the lighter or the heavier fluids to collect near the center of rotation?

8. If a mass is suspended at the end of a spring and set into oscillation, is the acceleration constant throughout its upward motion? Explain. Can Galileo's equation $s = \frac{1}{2}at^2$ be applied?

9. If a Foucault pendulum is set in motion at the north pole, explain why its plane of oscillation will revolve by 360° in 24 hours? What will happen if it is mounted on the equator?

10. A heavy rope of uniform density is hung from a hook. Choose points at the bottom, top, and center of the rope and describe the tensions acting and the net forces.

11. Explain why a candle will not normally continue to burn in an orbiting satellite even if there is an adequate supply of oxygen.

12. In what way did the discovery of Uranus and of Neptune differ?

13. Will a person weigh more or less if he is placed on a planet of 5 times the mass and 5 times the radius of the earth?

14. A heavy mass is suspended by a string. Another string of the same breaking strength is fastened to the bottom of the mass. If the string at the bottom is given a strong enough downward jerk, which of the two strings will break? If instead, a steady force is exerted on the bottom string, which will break?

15. Does a planet farther from the sun have a higher or lower tangential speed than one at a lesser distance from the sun? Angular velocity?

16. Find the fault in the following analysis: two spheres of identical radius, one of lead, the other of wood, fall from rest in air from a certain height. If the resistance of air is proportional to the surface area, the frictional resistance will be equal on both spheres and since the constant of acceleration g is the same in both cases, they will reach the ground in identical times even in the presence of air.

Problems

1. What *force* is required to uniformly accelerate a car from rest to a speed of 30 mph in five seconds? The car *weighs* 2200 lbs. (It is sufficiently precise to assume 30 mph equals about 14 m/sec, and one kilogram *weighs* 2.2 lbs.) Express the answer in newtons and then convert to pounds.

2. What frictional force exerted by the roadbed on the tires is required to keep a car weighing 2200 pounds traveling at 30 mph in a circle of radius 60 yards (assume 1 meter = 3 ft)? Express your answer in newtons and then convert it to pounds.

3. Look up values for G, mass of the earth, radius of the earth and calculate the approximate value of g.

4. Assume that the earth is a perfect sphere. By what percentage will the value of g be less at the equator than at the poles because of the earth's rotation?

5. The mass of the moon is roughly ⅛ times the mass of the earth. The distance of the moon is about 60 earth radii. Where does the center of mass of the earth-moon system lie? What is the value of g on the moon if its radius is approximately 1000 miles? How high could you jump on the moon?

6. There are two methods of determining the mass of the earth, once the value of G is known. Look up the necessary data and calculate the mass of the earth by both methods.

7. A frictionless plane is inclined at the angle of 30° with respect to the horizontal. What is the acceleration down the plane?

8. An elevator of mass 1000 kg starts from rest and is accelerated upward uniformly so that it travels 8 meters in two seconds. What is the tension of the cable? Repeat the problem with the elevator accelerating downward.

9. If a plane travelling at 1000 miles per hour pulls out of a power dive in a circular path of one-half mile radius, to how many "*g*s" will the pilot be subjected? What force will he exert on the cushion of his seat if he weighs 220 lbs.

10. Show that the tidal generating force of the sun is somewhat less than half of the tidal generating force of the moon.

11. The length of a simple pendulum is 2 meters. The bob has a mass of 2 kg. If the pendulum is displaced at 30° from the vertical by a horizontal force applied to the bob, what is the tension on the string?

12. Apollo 8 circled the moon in about 120 minutes at a height not far, relatively, from the surface of the moon, whereas an earth satellite circles the earth at a height not far from the earth's surface in about 90 minutes. Using Newton's law of gravitation and assuming that in both cases the orbits are approximately equal to the radii, can you explain why it took longer to orbit about the moon whose radius is approximately ¼ that of the earth than it does to orbit about the earth? (*Hint:* consider density)

Recommended Reading

Sir Isaac Newton's Mathematical Principles of Natural Philosophy and His System of the World, translated by Andrew Motte in 1729, edited by Florian Cajori, University of California Press, Berkeley, Cal., 1947.

E. N. Da C. Andrade, *Sir Isaac Newton* (Anchor Book), Doubleday, Garden City, N.Y., 1958.

Sir Oliver Lodge, *Pioneers of Science,* Dover Publications, New York, 1960 (Lectures VII–IX).

E. A. Burtt, *The Metaphysical Foundations of Modern Physical Science* (Anchor Book), Doubleday, Garden City, N.Y., 1954.

Sir William C. Dampier, *A History of Science,* Prentice-Hall, Englewood Cliffs, N.J., 1957.

F. E. Manuel, *A Portrait of Isaac Newton,* Harvard University Press, Cambridge, Mass., 1969. The first full biography of Newton since L. T. More's biography of 1934, by an eminent scholar in the history of ideas. A sensitive and careful analysis of Newton's intellectual and personal background and motivation.

For a better understanding of the basic laws of Newtonian physics, simply presented with many illustrations and problems, see either *Physics* by the Physical Science Study Committee, D. C. Heath, Boston, 1960, or Eric Rogers's *Physics for the Inquiring Mind,* Princeton University Press, Princeton, N.J., 1960.

chapter eight

"The laws of nature and of nature's God"[1]

By the middle of the eighteenth century the tremendous success of the new approach to natural philosophy in unlocking the secrets of nature caught the imagination of men. This new vision of nature in turn inevitably colored men's thoughts about human institutions. A high optimism swept over Europe, England, and America that mankind could, by a proper combination of reason and observation, so order both society and nature that a happy and harmonious life on earth could be guaranteed.

Newton's *Principia* is not easy reading, but its main ideas, unlike some of the concepts of modern physics, were easily graspable. Fairly simple models could be provided. Consequently it did not take long for these ideas to spread throughout the world. Five editions of Benjamin Martin's *A Plain and Familiar Introduction to Newtonian Philosophy* and seven editions of James Ferguson's *Astronomy Explained Upon Sir Isaac Newton's Principles* were published by the middle of

[1] This phrase, which comes from the Declaration of Independence, is used as the title for this chapter, following the example of Carl Becker, in *The Heavenly City of the Eighteenth Century Philosophers,* Yale University Press, New Haven, 1932.

the century. Voltaire's *Elements de la Philosophie de Newton* published in French was translated into English, and Count Algorotti's *Newtonianism for the Ladies,* written in Italian, was available in French or English. Even a poetic statement of the Newtonian system was published in 1728, *The Newtonian System of the World the Best Model of Government,* by J. B. Desogulier.[2]

It should be remembered, of course, that familiarity with Newton's work was supplemented by a spreading awareness of new discoveries being made in rapid succession, both in science and in mathematics. The experiments of Boyle, Hooke, Huyghens, and Gilbert and the telescopic observations of Halley and other astronomers were fascinating to men of that age. (Their curiosity had not been dulled, as ours is apt to be today, by the overwhelming number of new discoveries that are made every month. We would probably be startled if more than one important new discovery were *not* made every year.)

Even the mathematical-deductive method of Descartes was widely used and understood.

> The mathematical interpretation of nature became so popular and fashionable by 1650 that it spread throughout Europe, and dainty, expensively bound accounts by its chief expositor, Descartes, adorned ladies' dressing tables.[3]

We shall, in this chapter, be concerned first with the impact of the Newtonian world view upon other aspects of thought and, second, with a more general consideration of the nature of scientific theory and its relationship to other social and philosophical theories.

1. NEWTONIAN INFLUENCE ON RELIGION

It was not many decades after Galileo's trial before the church's authority over man's outlook toward nature was seriously weakened. As new discoveries were made, the interest of men in nature slowly supplanted the sole and steady concern with the relationship of God to man which had hitherto marked the thought of the Middle Ages. Michelangelo's or Dante's conception—colorful, dramatic, powerful—of a cosmic drama with man at the center of the stage was slowly fading away. The possibility that man might control his own destiny within the limitation of natural law became the new vision.

The old authority of church and state was challenged on two grounds: on the basis of reason and on the basis of experience. Surely there were many inconsistencies in both church doctrine and in the Bible itself. If nothing else, Descartes' method emphasized consistency as a necessary foundation for truth.

[2]The authors and titles given above are all taken from Reference 1, p. 61.
[3]Morris Kline, *Mathematics in Western Culture,* Oxford University Press, New York, 1953, p. 107.

And, on the grounds of experience, the real existence of angels and devils could be seriously challenged. Furthermore, if the doctrine of natural law was valid, capriciousness in nature had to be ruled out. Miracles and the interference of God in human affairs for punishment or for reward could have no place in a system ruled by immutable and universal law.

The proper function of natural philosophy became the discovery of the basic regularities that existed in nature, not the enquiry regarding purposes, essences, occult qualities, or the organic relationship between individual entities. Instead of seeking an organic and final explanation of natural phenomena as sought by Aristotelians, the primary task became that of finding the general regularities, postponing inquiry as to "ultimate causes" until after the regularities were thoroughly understood.

It is true that, as Philipp Frank has pointed out, there remained a remnant of organismic philosophy in Newton's own thinking. Newton has been reported to have expressed his point of view as follows.

> The plain truth is that he believes God to be omnipresent in the literal sense; and that as we are sensible of objects when their images are brought home within the brain, so God must be sensible of everything, being intimately present with everything: for *he supposes that as God is present in space where there is no body, He is present in space where a body is also present.*[4]

The quotation is given not only to illustrate Newton's religious attitude but to emphasize the need felt by Newton to ground his idea of absolute space on a metaphysics. He was aware that he could not define absolute space in terms of any specific observation or measurement. (How can one specify any particular object as being at *absolute* rest, since the sun and the stars are all in relative motion with respect to each other?) This insight of Newton's into the problem of an adequate definition of absolute space soon became forgotten and was not raised again in a definitive form until the time of Mach and Einstein about two hundred years later.

Newton himself believed that God created the machinery and the laws by which the machinery operated. He was even inclined to doubt whether every few thousand years God would not have to interfere with the positions and orbits of the planets (in other words, fix up the machinery) because they might get into unstable positions and collide.

These questions and aspects of Newtonian thought gradually were dropped as the success of the system became more apparent. In time, scientists became more and more convinced that they could carry out the very program envisaged by Newton in the preface to his first edition of the *Principia:*

[4]From a diary by Newton's friend, David Gregory, quoted by Philipp Frank, *Philosophy of Science,* Prentice-Hall, Englewood Cliffs, N.J., 1957, p. 118.

I wish we could derive the rest of the phenomena of nature by the same kind of reasoning from mechanical principles, for I am induced by many reasons to suspect they may all depend upon certain forces by which the particles of bodies by some cause hitherto unknown, are either mutually impelled towards one another, and cohere in regular figures, or are repelled and recede from one another . . . I hope the principles here laid down will afford some light either to this or some truer method of philosophy.[5]

In time, scientists became more and more convinced of the validity of Newton's approach as applied to every possible phenomenon of nature. The harmony of nature became the harmonious workings of a vast machine. Symptomatic of the new view was the often-quoted remark of Laplace, celebrated mathematician, who extended Newton's scheme. Asked by Napoleon where God fitted into celestial mechanics, Laplace answered, "Sire, I have no need for such an hypothesis." A new dogma was replacing the old.

Almost aggressively hostile to the pretensions of the older absolutism, the would-be skeptic Laplace substituted one dogmatic creed for another. It was largely due to the success of his own celestial mechanics and his widely appreciated popular exposition of the mathematical consequences of Newtonian gravitation that a crude mechanistic philosophy afflicted nearly all physical scientists and many philosophers of the 19th century.[6]

These were the weapons: reason and experimental observation—together with an innate conviction that the universe had a basic harmony that could be discovered by man's own efforts—with which Voltaire, Diderot, Condorcet, Helvetius, Montesquieu, and the whole group of French philosophers of the eighteenth century attacked religious institutions and often corrupt authority that had indeed become reactionary impediments to social reforms so obviously required. A need for a new foundation for morality was widely felt. The philosophers exclaimed that "morals should be treated like all other sciences and that one should arrive at a moral principle as one proceeds with an experiment in physics."[7]

This brave although somewhat naive philosophy led to a rapid extension of both deism and atheism. Among our own founding fathers and the first presidents, deism probably was the creed of the majority. While many may deplore the adverse effect upon religion, deism's liberating influence was exhilarating, productive, and has lasted to the present day. Torture and punishment of heretics became absurd. If God did not intervene in the government or actions of man, one could hardly be aiding His cause by throwing heretics to the flame. Religious freedom (presumably, at least) became a fundamental right of man.

[5]Sir Isaac Newton, *Mathematical Principles,* translated into English by Andrew Motte in 1729, University of California Press, Berkeley, 1947, p. xviii.

[6]E. T. Bell, *The Development of Mathematics,* McGraw-Hill, New York, 1945, p. 363.

[7]Claude Helvetius (1776), quoted by Frank Manuel, *The Age of Reason,* Cornell University Press, Ithaca, N.Y. 1951, p. 39.

2. RATIONALISM AND SOCIAL THEORY

One of the dangers inherent in the discovery of any great idea is the temptation to extend it beyond its appropriate range. Later, we shall consider how Newtonian physics has failed to account for recent discoveries. At the moment we are concerned only with understanding Newtonianism and its connection with social theory. We cannot improve upon Whitehead's summary:

> Newtonian physics is based upon the independent individuality of each bit of matter. Each stone is conceived as fully describable, apart from any reference to any other portion of matter. It might be alone in the Universe, the sole occupant of uniform space. But it would still be that stone which it is. Also the stone could be adequately described without any reference to past or future. It is to be conceived fully and adequately as wholly constituted within the present moment.
>
> This is the full Newtonian concept, which bit by bit was given away, or dissolved, by the advance of modern physics. It is the thorough-going doctrine of "simple location" and of "external relations."[8]

John Locke, a contemporary and acquantance of Newton's, was developing the same ideas with respect to men. Men were no longer regarded as mere creatures bound into a social order or organism by God's decree, but were by nature free and distinct individuals. Government was not ordained but was created by the coming together of free individuals who entered into a social contract. There were natural laws in the field of social affairs just as in the field of nature, and a valid government would be based upon such laws which reflected the "natural rights" of citizens as individuals.

To determine the "laws of nature and nature's God," applicable to the affairs of men, could be just as effectively accomplished as it had been in the field of natural philosophy. This was the task to which Montesquieu and Diderot in France; Locke, Hume, and Adam Smith in England; and Jefferson and Franklin in America addressed themselves with fervid optimism. Even the methods were similar in many ways: a statement of principles or axioms followed by a deduction of their consequences. "We hold these truths to be self-evident . . ."

Here, also, can be found the origin of laissez-faire economics—man was an economic individual, acting according to economic laws just as the particles of physics moved according to physical laws.

Eventually, of course, there came a new orientation in thought. In America and in England the impact of Darwinism called attention of men to organic relationships once again (even if the "struggle for existence" was interpreted mechanistically for a long period), and the development of a critical metaphysics in the hands of Hegel encouraged the growth of new social theories such as

[8] Alfred North Whitehead, *Adventures of Ideas,* The New American Library (Mentor Book) 1958, pp. 160–161. Originally published by Macmillan, New York.

Marxism. Woodrow Wilson summed up the development of this political philosophy:[9]

The government of the United States was constructed upon the Whig theory of political dynamics, which was a sort of unconscious copy of the Newtonian theory of the universe. In our day, however, whenever we discuss the structure or development of anything, whether in nature or in society, we consciously or unconsciously follow Mr. Darwin; but before Darwin they followed Newton . . . Politics is turned into mechanics under his (Newton's touch) . . . the trouble with the theory is that government is not a machine but a living thing . . . Government is not a body of blind forces, it is a body of men, with highly differentiated functions, no doubt, in our modern day of specialization but with a common task and purpose . . . Living political constitutions must be Darwinian in structure and in practice.

Up to now we have been discussing the impact of a particular scientific body of theory, Newtonian Physics, upon social and religious philosophy. But a more general and more critical analysis of the interrelationship of scientific and social theories is required. Before entering into this analysis, however, a broader review of the nature of scientific theory itself, based upon the examples we have studied, will be helpful. We shall deal with this in the next two sections, after which, in the final section of this chapter, we shall consider again the connection between scientific theories and other social beliefs and attitudes.

3. THE DATA OF SCIENCE; PASSAGE FROM SENSED DATA TO CONCEPTS; PASSAGE FROM CONCEPTS TO THEORY[10]

The recurring phrase "laws of nature" demands an examination of its meaning. In this section we begin a further analysis into the nature of scientific laws and the problem of establishing their validity.

However, if we are not to be content with a superficial analysis, our inquiry must be broadened to include a consideration of that aspect of philosophy termed epistemology, that is, the study of the nature of knowledge in general, and its validity. How do we know? What is the basis and validity of our knowledge?

The general philosophical outlook of any age might be summed up in the answer given to three questions. "What is true?" "What is good?" "What is beautiful?" The answers to each, of course, are related to the answers to the other two. But, in the field of science, we restrict ourselves to the first of these problems.

[9]Woodrow Wilson, *Constitutional Government in the United States,* Columbia University Press, New York, 1921, pp. 54–55, quoted by F. S. C. Northrop, *The Meeting of East and West,* Macmillan, New York, 1952, p. 139.
[10]See the footnote to the heading of Section 5.

In the intellectual climate of today, we do not believe that we have now, or that we ever can have, any final and absolute answer to the problem of truth. We are historically minded and aware, perhaps a little too sensitively, of how past visions have failed because some significant aspect of reality has been overlooked. Hence we are skeptical about *absolute truth,* not only because of the failure of past science to come up with final answers (as was once considered possible) but also because we have seen how so many of the bright certainties of the past have had to be abandoned or modified. On the other hand, this very historically oriented point of view has also given rise to a faith in the possibility of approximating truth more closely through intellectual progress—progress being a theory that, in Bury's words "involves a synthesis of the past and a prophesy of the future."

We shall not try to handle all the complex considerations, both scientific and philosophic, which surround the problem of knowledge. But one interested in the foundations of science must come to some grips with this problem. Both the range of scientific method as well as its limitation will be more clearly appreciated if we proceed with an examination of what is involved. It is significant that so many of the great scientists of our own period, including Einstein, Bohr, Planck, Heisenberg, Bridgman, and Schroedinger, have been very much concerned with this problem, as were their immediate predecessors, such as Maxwell, Poincaré, and Mach.

The first consideration of any theory of knowledge is the passage from what may be called sense perceptions to the ideas or concepts under which the sense perceptions may be organized.

Knowledge of any kind is impossible unless we recognize repeated characteristics or patterns in our perceptions. Recognition implies the relative invariance of certain combinations of the sensed experiences. If every experience were completely different from every past experience, recognition would be impossible.

After obtaining knowledge of a sort through personal or subjective recognition, if we wish to communicate this knowledge or information, the first step is to *name* or give a symbol to the group of characteristics or the pattern that is being repeated. The name is said to denote or point to the pattern that is recognized. It is important to realize that in giving a symbol or name to some pattern we have already shifted from the purely sensory to an intellectual mode of identification. The thing symbolized and the symbol are certainly to be distinguished (which is the point that Juliet makes in her familiar line, "A rose by any other name would smell as sweet").

But for scientific knowledge, names are insufficient. The names must be related to one another. Furthermore, they must be related to each other in an orderly manner, so that the relationship will be clearly understood by all persons working within the subject matter of the science. Then the names begin to take on the role of carefully defined concepts. They now have a twofold possibility of reference: (*1*) *a denotation,* or direct reference to sensed data and (*2*) a reference to other concepts through logical specification such as definition, classification, description, or equational forms.

In the early stages of any science, it is difficult to select the most fruitful concepts, which completely meet both the requirement of exact denotative reference and logical relationship to one another. At this stage, a science may be characterized primarily by description and classification. Most of biology remained in this stage until quite recently. (The concept of heredity, for example—originally a qualitative concept of the existence of a group of unchanging characteristics passed from parents to offspring—is today more precisely defined in terms of the concept of the gene. Heredity is thus being specified more and more exactly by a precise denotative reference, which can be agreed upon by all workers in the field where the quantitative laws regulating genetics are being established.)

The primary reason that physics may be regarded as the fundamental science is because of the denotative precision with which all the basic concepts have been formulated. These *denotative references* are given by what we have termed *operational definitions,* in which every concept is related to sensed data through a complete specification of certain rules. These rules relate such concepts as mass, length, or time to the perceived pointer readings of the corresponding measuring instruments: balances, meter sticks, and clocks. Chemistry and biology make use of these basic definitions. The distinction between elements, for example, becomes one of mass; crystal structures or even organic structures are determined by a variety of X-ray diffraction measurements; neural impulses are measured in terms of electrical recording instruments, etc. While some biologists may properly claim that there is no proof that teleology (the doctrine that ultimate purposes act upon and influence organic processes) does not have its role in the interpretation of biological phenomena, practicing biologists continue to use only those concepts which have a well-defined operational reference. It may be philosophically wrong to deny the possibility of an *elan vital* (a spiritual force operating on organisms) but, until some operational definition identifying such a concept can be produced, it will not become an effective tool for scientific research.

The appreciation that ultimately all concepts must be defined in terms of precisely expressed operations has received emphasis in recent developments of physics, e.g., in quantum and relativity theories. For example, Newton's reference to absolute space: "Absolute Space, in its own nature, without relation to anything external, remains always similar and immovable," is rejected because no operational method has been found of determining the existence of such a space or frame of reference. Similarly the concept of a mechanically characterized ether (an all-pervasive fluid that was thought in the nineteenth century to be necessary to carry the waves of light) is no longer specified as an adequate concept because there is no prescribed method for its detection. Other examples could be given. It is interesting that this new emphasis in the science of physics has had the indirect result of giving rise to a similar emphasis in semantics, as found in the writings of Stuart Chase, Hayakawa, and others. The basic injunction or rule is the same: the definition of the terms must specify a definite and unambiguous reference, which all persons using the term agree upon. Part of the difficulty faced by such sciences as psychology, is failure to reach agreement upon an appropriate operational definition. Thus, for example, the concept of "abnormal" is defined in many different ways with no satisfactory agreement among psychologists as to an operational definition.

Objection has sometimes been raised against carrying this requirement too far. It is pointed out that the operations themselves are partly specified by our prior knowledge of the laws of physics. This is, of course, quite true, but it merely means that as our science advances the operational definitions themselves become modified and made more precise. In Bridgman's words, "at any moment our concepts are coextensive with the system of existing knowledge."[11]

Having considered the one aspect of concepts—involving their empirical content or their relationship to sense data—we should give equal attention to the other aspect of concepts, involving their relationship to each other. These latter relationships are established by propositions or statements, and the rules governing the relationships are those of logic and syntax. This twofold aspect of the conceptual elements of the physical sciences is well summarized by Einstein's statement:

> I see on the one side the totality of sense-experiences, and, on the other, the totality of the concepts and propositions which are laid down in books. The relations between the concepts and propositions among themselves and each other are of a logical nature, and the business of logical thinking is strictly limited to the achievement of the connection between concepts and propositions among each other according to firmly laid down rules, which are the concern of logic. The concepts and propositions get "meaning," viz., "content," only through their connection with sense-experiences. The connection of the latter with the former is purely intuitive, not itself of a logical nature. The degree of certainty with which this connection, viz., intuitive combination, can be undertaken, and nothing else, differentiates empty phantasy from scientific "truth." The system of concepts is a creation of man together with the rules of syntax, which constitute the structure of the conceptual systems. Although the conceptual systems are logically entirely arbitrary, they are bound by the aim to permit the most nearly possible certain (intuitive) and complete co-ordination with the totality of sense-experiences; secondly they aim at greatest possible sparsity of their logically independent elements (basic concepts and axioms), i.e., undefined concepts and underived [postulated] propositions.

[11]P. W. Bridgman, *The Nature of Physical Theory,* Dover Publications, New York, 1936, p. 9. For a more complete discussion of this point the reader may consult Philipp Frank, *Philosophy of Science,* Prentice-Hall, 1957, pp. 314–316.

> A proposition is correct if, within a logical system, it is deduced according to the accepted logical rules. A system has truth-content according to the certainty and completeness of its co-ordination-possibility to the totality of experience. A correct proposition borrows its "truth" from the truth-content of the system to which it belongs.[12]

A few remarks about the foregoing passage may be helpful. What Einstein terms the truth-content or meanings of the concepts, that is, their connection with sense-experiences, we have called the denotative reference of the concept. Einstein describes this connection between concept and sense-experience as "intuitive," but we have preferred to describe it as given in an operational definition.[13]

From what we have said, it is evident that any science consists of a large set of propositions, which may be thought of as the elements of the science and, further, that this set of propositions has a structure or order.

First, some, but by no means all, propositions are almost purely empirical; that is, they relate through the concepts involved very directly to the data of observation. "Iron melts at 1535°C" is a statement in which each term is operationally defined with considerable precision. (The warning, however, should be constantly kept in mind that these definitions may be modified with the further progress of the science. Iron, for example, may be found to be made up of more than one basic element—isotope—each of which has its own melting point. Or, greater precision in determining temperature may shift the temperature of 1535°C upward or downward.)

The proposition that iron melts at 1535°C can be regarded as a "fact," although "fact" is an exceedingly slippery term meaning that the proposition asserts that a well-defined relationship exists between the terms of the proposition which is identical to a similar relationship existing within the observed data to which the concepts of the proposition refer.

From such factual propositions, a mature science proceeds rapidly to generalizations of a higher and higher order. For example, we proceed from "iron melts at 1535°C" to "all metals melt at some temperature." This is a generality that is already postulational, since it was asserted even before all metals had been

[12]From Einstein's preface to *Albert Einstein: Philosopher-Scientist,* edited by Paul A. Schilpp, Library of Living Philosophers, Open Court Publishing Co., La Salle, Ill., 1949, pp. 11–13. This work is a collection of excellent essays by outstanding scientists, including Sommerfeld, de Broglie, Born, Heitler, Bohr, Margenau, and Bridgman, discussing the general philosophical and epistemological problems of modern physics.

[13]F. S. C. Northrop calls the relationship "epistemic correlation." Margenau uses the term "epistemic definition." A further elaboration of the nature of the relationship of concept to experience is beyond the scope of our study, but anyone with philosophic interest should refer to Margenau, *Nature of Physical Reality,* McGraw-Hill, New York, 1950, Chapter 12, to see how the idea of "operational definition" requires supplementation. See also the excellent critical analysis of several points of view in Ernest Nagel's, *The Structure of Science,* Harcourt, Brace & World, Inc., New York, 1961.

discovered. Next, we can state "all solids melt at some temperature" as another higher order of generality. This postulated generality, however, may and actually does involve many interpretations when applied to particular solids.

An even better example of an empirical generalization, that is, a generalization based upon direct observation, can be cited in Kepler's laws, which we have studied. Two aspects of these laws should be emphasized: (*1*) they were originally stated as generalizations unconnected with any other known phenomena and therefore led naturally to the inquiry as to how they could be connected or "explained" by a more general theory; and (*2*) although Kepler carried out his observations with respect to only a few planets, he had no hesitation in ascribing the same laws to all planets. He was thus generalizing upon the basis of few instances, by *no means* following a procedure of induction by enumeration, that is, of reaching generalizations only after an examination of a large number of specific instances. Kepler's instinct told him that some pattern, some regularity, must apply to planetary motion and that if this pattern could be found with respect to one planet, it would inevitably also apply to others. His instinct turned out to be correct. This procedure is typical of the physical sciences: a few instances lead to a postulated generalization, which is then tested in other circumstances.

Empirical generalizations such as these mark only the beginning of a mature science. They must be tied together into a structural pattern, in which each generalization is logically related to other similar generalizations. In other words, a consistent theory must be developed, with certain propositions given the same position as the postulates of a mathematical system. The various empirical generalizations are then shown to be consequences that can be deduced from a few basic or primitive propositions or postulates. We have seen an example of this in Newton's postulates, or laws of motion, combined with his law of gravitation. Kepler's laws can all be shown to be consequences of the Newtonian laws. In the next section we shall consider additional aspects of such basic theoretical laws of the physical sciences.

Before we turn to a further discussion of theoretical laws, we should note how much can often be gained by a proper formulation of empirical generalizations even before a mature theory has been developed. Just because the physical sciences almost invariably couch the expression of such laws in quantitative form, we should not conclude that any generalization otherwise expressed may not be fruitful as an important tool for extension of knowledge. Many of the generalizations of biology and physiology are extremely important, both in organizing knowledge and in enabling predictions to be made, even though they are neither included within a complete theoretical framework nor expressible quantitatively.

As a typical example, consider the argument found in Leonardo da Vinci's notebooks that certain mountains of Italy must at, one time have been covered by the sea.[14] His argument proceeds from a descriptive or classificatory generalization regarding certain types of marine animals: oysters and clams are found only in the sea. But their fossil remains have been found in layers on mountains far from the sea. Therefore, Leonardo drew the conclusion that, at one time, the mountains were covered by the waters of the sea. An unobserved phenomenon is thus deduced from a simple generalization. (Leonardo was careful in his argument to cover other possible explanations—showing each possibility to be contrary to other empirical generalizations.) A simple, nonquantitative generalization led to consequences that cannot be directly observed.

4. THE NATURE OF THEORY IN THE PHYSICAL SCIENCES

One of the difficulties that the nonprofessional encounters in trying to understand the nature of theoretical laws or postulates is the common misunderstanding that such postulates are necessarily implied by the evidence. This is far from the case. The postulates are most usually a free creation of the human intellect—an invention—which will enable the vast complexity of experienced data to be organized under a few basic ideas or assumptions. The theory is *not* deduced from the facts; it is either "verified" or contradicted by them. No theory can ever be regarded as representing the *final* truth.

Because the illuminating ideas or postulates that constitute the basis of theory are not directly deduced from any amount of factual information, "no *rule* can be given for bringing to birth in the brain a correct and fertile idea that may be a sort of intuitive anticipation of successful research."[15] We remember the story of how Newton, in a moment of contemplation, heard the apple fall and jumped to the theory of universal gravitation as a possible theoretical explanation. There are many similar examples. The resemblance between Coleridge's dream, which led to his poem *Kubla Khan,* and the inspiration of Kekulé by which he arrived at the postulate of the benzene ring which did so much to organize the science of organic chemistry is striking:

> I was sitting, writing at my textbook; but the work did not progress; my thoughts were elsewhere. I turned my chair to the fire and dozed. Again the atoms were gamboling before my eyes. This time the smaller groups kept modestly in the background. My mental eye, rendered more acute by repeated visions of the kind, could now distinguish larger structures, of manifold conformation: long rows, sometimes more closely fitted together; all twining and twist-

[14]The example is given, in part at least, to encourage acquaintance with Leonardo's extraordinary breadth of interest. It is taken from *Leonardo da Vinci's Notebooks,* arranged and translated by Edward McCurdy, Empire State Book Co., New York, 1923.

[15]Claude Bernard, *An Introduction to the Study of Experimental Medicine,* Dover Publications, New York, 1957, p. 133.

ing in snakelike motion. But look! What was that? One of the snakes had seized hold of its own tail, and the form whirled mockingly before my eyes. As if by a flash of lightning I awoke; and this time also I spent the rest of the night in working out the consequences of the hypothesis.[16]

Such was the origin of the theory of the molecular structure of the benzene ring. Even in mathematics, an intuitive idea that jumps well beyond the established bounds of the science very frequently occurs. Euler's intuitive hunch that the infinite series: $1 + 1/4 + 1/9 + 1/16 + 1/25 \ldots$ was equal to $\pi^2/6$ was a daring conjecture, without logical validity at the time it was made.[17]

Although the specific genesis of any theory of science is difficult to trace because it is of the form of a free creation of the intellect, we can find one point of view that is almost universal among scientists: the belief that nature can ultimately be understood in fairly simple terms and that there is an element of beauty in the simplicity and orderliness so discovered. This is an ancient doctrine asserted by Pythagoras and Plato, which is constantly reasserted by scientists throughout the ages. Here is the way a few scientists have expressed this sentiment:

Newton: For Nature is pleased with simplicity and affects not the pomp of superfluous causes.[18]

Poincaré: The scientist does not study nature because it is useful, he studies it because he delights in it and he delights in it because it is beautiful—I mean that profounder beauty which comes from the harmonious order of the parts and which pure intelligence can grasp.[19]

Einstein: Our experience hitherto, justifies us in believing that nature is the realization of the simplest conceivable mathematical ideas.[20]

Although in origin, any fundamental theoretical law of science may be the result of the free creative act of the intellect, its validity and acceptance must pass certain specific requirements. The criteria by which the postulates of science are judged are listed and discussed below.

(*1*) Every law of science must be *in accord with the facts.* This simple sounding statement may, in practice, be very difficult to apply. First, it requires that every concept used has an "operational definition," either direct or indirect which, at times, may be difficult to state with precision. Actually, many definitions are defined indirectly, instead of directly in terms of operations. For example, average velocity is defined as distance traveled divided by the time traveled. In this case,

[16] *What is Science?*, edited by James Newman, Simon & Schuster, New York, 1955, p. 183.

[17] How analogies have led to intuitive conjecture in mathematics is one of the main topics of G. Polya's book, *Induction and Analogy in Mathematics,* Princeton University Press, Princeton, 1954. The example given is taken from this book.

[18] *Principia,* the statement is made as an explanation of one of Newton's four rules of reason.

[19] H. Poincaré, *Foundations of Science,* Science Press, New York, 1946, p. 366.

[20] A. Einstein, *The World As I See It,* Covici Friede, New York, 1934, p. 36.

the "operational definition" applies only to measurement of distance and to the measurement of time, but not directly to the average velocity. More important, perhaps, is the fact that there is a sense in which every law, when tested, is found to agree with the facts only within a certain degree of precision. Thus, Kepler's laws or Newton's laws are valid only within a certain approximation. If any particular orbit of a planet is studied, it will not quite follow the specified orbit because of the influence of other planets. These interefere with the simple interaction between the sun and the particular planet being studied. Perturbations arise that mask the simple relationship between the two bodies. The application of given laws, such as the laws of motion and the law of gravitation, must specify what is relevant in a particular analysis and make allowance for deviations from the laws in question. This is much more difficult in the biological sciences than in the physical sciences.

(*2*) The theoretical or logical relationships must be *free from any inconsistency*. Again, one might believe that this criterion is a simple one, but the history of both science and mathematics is filled with examples of logical slips, due either to oversight or to making assumptions that have not been expressly stated and that later prove to be invalid.[21] The modern critical spirit of mathematics tends to reduce such errors, but the danger is always present.

(*3*) Between two competing theories, both of which are validated by the facts, that theory that is *more general* is to be preferred.

(*4*) We have already discussed the criterion of *simplicity*. But this criterion is, to some extent, an aesthetic one. It is very difficult to propose an objective measure of what is meant by simplicity. It certainly does not consist merely in citing the number of propositions that must be assumed in establishing a theory.

(*5*) Another criterion that is difficult to state with precision is that *no theory should introduce* ad hoc *assumptions* (that is, assumptions that are brought in arbitrarily to make the theory fit the facts). We shall find, however, that this criterion also is sometimes difficult to apply. For example, when we come to the study of atomic structure we shall study the Bohr theory of the atom in which it is assumed that no electron can pass from one orbit to another unless the difference in energy is a multiple of a certain quantity called a "quantum." This theory introduced an apparently artificial assumption whose sole justification was that it could explain all the facts of spectral analysis. While it was felt that the assumption was unsatisfactory as a final solution to the problem of orbits, it was accepted until a better explanation was finally forthcoming.

(*6*) Another very important criterion is that a theory not only should explain

[21]Two examples: Leibnitz, the co-inventor of the calculus, argued that the series $1 - 1 + 1 - 1 + 1 - 1 \ldots = \frac{1}{2}$. Even a student of elementary calculus would recognize this as false today. Einstein, in one of his papers, made a slip that amounted to a disguised division by zero. This was not discovered for some time.

facts as presently known, but that it should enable us to *make predictions* about phenomena that are not presently observed. Thus the orbit of Uranus, as calculated by Newton's theory, deviated considerably from the observed orbit. This led to the prediction of the existence of another planet, hitherto unseen. Two mathematicians, one in France and one in England, simultaneously carried out calculations required to specify the precise location of the undiscovered planet. When telescopes were trained upon the designated portion of the sky, the new planet, Neptune, was found almost exactly at the predicted point. Closely connected with this ability to predict unknown phenomena, another aspect of a fruitful theory is that it almost automatically suggests further questions to be investigated. In fact, a new direction of investigation arises out of the act of determining the range of application of a theory. Questions are asked which had not occurred to scientists theretofore.

(*7*) In the physical sciences it is most often regarded as essential that a theory be expressed in *quantitative* terms. This criterion amounts basically to two guarantees: first, it insures logical coherence within the theory (since mathematical mistakes are rarely made); and, second, if properly used, it insures a well specified operational definition of the terms entering into the theory.

But quantitative analysis may be misused and misleading. The success of quantitative measurements in the physical sciences has, indeed, led investigators in other fields to apply such measurements without careful enough prior analysis. Claude Bernard gives an effective example of such misuse in the application of mathematics:

> Another very frequent application to biology is the use of averages which, in medicine and physiology, leads, so to speak, necessarily to error . . . A startling instance of this kind was invented by a physiologist who took urine from a railroad station urinal where people of all nations passed, and who believed he could thus present an analysis of *average* European urine.[22]

Today, the social scientists are much more sophisticated in their use of statistics. Even so, the establishing of averages or quantitative correlations may be meaningless in themselves. Here is another effective example (from Morris R. Cohen) of a high correlation that has no meaning:

> In a book of mine published some years ago, I referred to Dr. George Marshall who found a correlation of 87 per cent between the membership of the Machinist's Union and the death rate of the state of Hyderabad. Many of my readers have since protested vigorously that such a high correlation extending over twelve years cannot be accidental or devoid of real significance. But the fact is that the correlation does not hold beyond the period taken.[23]

(*8*) Another aspect of successful scientific theories, which is closely allied to

[22]Claude Bernard, *Experimental Medicine,* Dover Publications, New York, 1957, p. 134.
[23]Morris R. Cohen, *A Preface to Logic,* Henry Holt, New York, 1946, p. 133.

predictive power already discussed, is that a theory should be *vulnerable,* that is, it should have specifically stated consequences that may be tested and proved wrong. We have seen that the Copernican theory predicted that Mercury and Venus would have phases similar to the phases of the moon; that parallax shifts should be detectable if instruments of sufficient precision were developed. If these predictions had not eventually been born out by observation the theory would have had to be abandoned or, at least, modified. We shall consider, later, many examples of what are called *crucial experiments* in which two theories, both in accord with the facts as known when they were developed, predict differing consequences that can be tested. (The classical example occurs in the theory of light. Two theories, the wave theory of light and Newton's theory that light was composed of particles, both could account for the phenomena of reflection, refraction, and dispersion. However, according to the wave theory, light would travel more slowly in a transparent material than in a vacuum but, according to the particle theory, the opposite would be true. The crucial experiment of measuring the velocity of light in water was finally performed, and thus established a basis of choice.)

It is possible to invent theories that cannot conceivably be proved wrong. Such theories cannot be regarded as scientific, precisely because they are not vulnerable. For example, the theory that the world was created in 4004 B.C. and that at the same time God created upon earth and in the sky all the evidence (fossils, radioactive material, etc.), which has led scientists to believe in a much older evolving universe, cannot possibly be disproved by any appeal to evidence. Such a theory cannot therefore be taken as scientific.

(*9*) It is usually an advantage, although not a necessary requirement, that a theory be able to be *visualized* in terms of some *model* that is familiar to common sense. Man's mind is so constructed that understanding is aided tremendously by such models. Unfortunately, the appropriate construction of models has become more and more difficult for modern scientific theories. Actually, some branches of modern physics were slow in developing only because the models of nineteenth century physics were so very effective and often misleading. It is still difficult to visualize an electron as being fundamentally different from a small, hard particle similar to the large-scale objects of common sense, but the essential properties of the electron in physics today are so different from ordinary particles that the model that can be visualized must be abandoned if the modern concept of an electron is not to be distorted.

This causes a part of the difficulty in presenting much of modern physics to the nonprofessional. To make a meaningful connection with his own experience, it is almost inevitable for him to use analogies. Even though the Bohr atom—viewed as a miniature solar system with electrons in orbit around a central nucleus—can only be regarded as farfetched and actually misleading, it is almost impossible to introduce the subject of atomic physics without, at least, beginning with the theory of the Bohr atom.

It remains, of course, true that as science expands and develops, the terms and concepts used become more and more familiar, and their strangeness disappears. For example, gravitational "action at a distance," which common sense repudiated for many years after Newton's time, became accepted as "philosophically" reasonable by the end of the nineteenth century, only to be brought into question again in the twentieth century. Or, again, when Einstein first showed the impossibility of defining absolute simultaneity, there was for a while the strong reaction of "common sense" against such a theory. Today it is no longer difficult to explain this aspect of his theory because it has become familiar to most persons in our generation.

(*10*) One final criterion by which scientific theories are judged by professional scientists is the most difficult of all to explain and illustrate, that is, the *aesthetic nature* of the theory. Possibly we have gained some appreciation of this quality from our comparison of the Ptolemaic with the Copernican theory.

5. NORMATIVE THEORIES AS THEY ARE RELATED TO SCIENTIFIC THEORIES[24]

We have now reviewed in some detail the nature of scientific theories and the manner of their verification. We wish tentatively to reconsider the problem (with which we began the chapter) of the relationship of scientific theory to political, ethical, and religious philosophy. Assuredly, one of the major purposes of studying the natural sciences is to understand how they are related to other attitudes and beliefs. Education must aim at making as clear as possible the grounds upon which ethical principles are founded and how they are related to the scientific generalizations of our age.

We have already remarked on the tremendous optimism that dominated the eighteenth-century age of enlightenment and how it was founded upon the vision of applying a new kind of Newtonian system to social affairs. This vision of the "heavenly city," which dominated the thinking of the philosophers of the

[24]Much of the material of this section and, to a lesser extent, the previous sections of this chapter is of a different nature, at least in degree, from the other portions of the text. It is interpretive in the broadest sense, and reflects the author's point of view about controversial matters. The excuse for incorporating it is that it might prompt the nonexpert to relate his understanding of science and its methods to his knowledge and point of view regarding other fields of interest. It is hoped that, whether he agrees or not, he will be forced to consider how the "climate of opinion" of any age is necessarily deeply affected by the science of that age. The point of view, rather sketchily stated here, derives primarily from Margenau, Northrop, Frank, and Whitehead, whose works are listed at the end of the chapter.

eighteenth century, has faded. We no longer believe that it is such a simple matter to organize society to the tune of nature's law; the "innate goodness" of man is questioned.

Let us look at the problem somewhat further. The Declaration of Independence was fundamentally an assertion of certain principles, declared to be "self-evident." What is meant by self-evident? What is the origin and justification of such principles?

First, we note, and this point needs emphasis, that the principles asserted are *normative,* that is, statements of what *should* be. That "all men are created equal" is not an expression of existing social fact. (To consider this a fact is an invitation to irony such as Orwell's: "all men are created equal, but some are more equal than others.") Jefferson was not naive; he was asserting a desirable goal, that is, that all men should, as a matter of right, be treated alike under the law, regardless of birth, position, property ownership, and the like.

This, then, is one mark of a social or political philosophy or ideology as contrasted with a scientific theory. Its principles do not attempt merely to organize *existing factual* material. They are not to be judged by the criterion of whether they accord with facts. A normative theory, or an ideology, asserts principles that reflect not what *is,* but *what ought* to be the case. Indeed a normative theory attempts specifically to change existing conditions.

In this sense, we certainly cannot judge an ideology, a political philosophy, an ethical system, or a religion on the same basis that we do a scientific theory. A scientific theory is judged wholly upon its ability to organize factual material effectively, but this can never be the sole criterion for an ideology.

We are left then with the twofold problem: where do normative theories come from, and how can we establish their validity?

To the first problem, there have been many different answers. Jefferson believed that certain principles were "self-evident." In light of the new attitude toward mathematical and physical generalizations and of modern philosophical criticism of both Plato and Kant, we can hardly accept the criterion today of "self-evidence" as an adequate basis for the assertion of *any* principle. Another suggested answer has been that the principles become known through religious revelation. However, because of the variety of revelations and interpretations, it is hardly possible to found a *universal* political ideology, let alone an ethical system solely upon such a basis.

What can we learn from a study of the nature of scientific theories that will throw light on social theories? If ideologies are normative, can anything in science affect them? If ideologies are neither to be regarded as self-evidently true nor to be verified in the manner of scientific theories, how can they be verified?

Consider, at least, several ideas that arise from a study of scientific theory and of the historical impact of past scientific theories.

First, as knowledge increases, it is of the nature of scientific theories and principles that they undergo modification and revision. New concepts are introduced and new meanings are given to old concepts. Our view of nature does not remain fixed.

Likewise, we should expect that the principles of social organizations should undergo continual modification as our knowledge of social relationships and of the nature of man changes. We should be cautious about accepting any fixed interpretation of any symbol or concept, however effective it may be in any given age as an integrating principle.

Second, we should recognize with Jefferson that ultimately the establishment of a good government must depend upon the discovery and effective use of a set of basic concepts and postulates that adequately organize all known facts about nature itself and about the nature of man. Although ideologies cannot find final verification in the social facts of a given age—since the ideologies are normative and seek to change the social facts—they must nevertheless be founded upon a concept both of nature and of the nature of man as most clearly revealed by the science of that age. (This is, of course, well recognized in many instances. For example, our government today will not torture the insane "to drive out evil spirits" because the older theory of insanity is contradictory to our scientific concept of the nature of man and his diseases. Similarly, slavery would find few defenders as an institution because the idea of slavery as a natural state is no longer tenable as a concept in the sciences of anthropology or economics.)

To recapitulate: any value of the system of government and organization under which we live is due to the recognition given to the fact that ideologies must be in accord with our knowledge both of nature and the nature of man, and that as knowledge is extended by all branches of science, the ideologies themselves must be modified. The basic attitude of Jefferson was thus appropriate. However, two errors were made: (1) the narrowly mechanistic interpretation of Newtonian physics was extended far beyond its proper range; and (2) the foundations of the political philosophy were taken to be "self-evident," after the manner of the interpretation of classical mathematics.

The result of the first error was to try to develop a kind of "particle" economics in which economic laws (law of supply and demand, etc.) were regarded as eternal verities. The philosophers of the eighteenth century and the social scientists of the nineteenth century took Newton's theory of nature too literally, and extended it far beyond its natural limitations. In the nineteenth century the doctrine that nature was a mechanical system of particles was extended to embrace individuals in an economic society. That such an extension was unwarranted became more and more apparent as the extent of human suffering under spreading industrialism became clear. The romantic poets raised their insistent voices against the working out of the laissez-faire, mechanical philosophy, and the conflict between machine and nature became clearer and clearer. Robert Frost in America echoed Wordsworth's sentiments when, finding a nest of eggs beside a railroad track, he expressed the conflict:

> The next machine that has the power to pass
> Will get this plasm on its polished brass.

Thus the narrow conception of all of nature and of society as a mechanistic interrelationship of independent entities, individuals, or particles, interacting only through external "forces" (in the one case the physical force of gravitation; in the other case, the force of psychological and economic wants) encouraged a type of social organization in the nineteenth century, which was ultimately repudiated. "Laissez-faire" economics did not bring about the harmonious and ideal social organization that had been anticipated. Too narrow an interpretation of the nature of man and of his social relationships had been the outgrowth of a philosophy uncritically based upon Newtonian physics. One of the natural reactions was Marxism, which did indeed provide a kind of dynamic interpretation to economic events lacking in classical laissez-faire economics.

As for the second error, that of regarding any proposition of political philosophy as "self-evident" and needing no further modification or testing, we should be just as willing here to modify such propositions as we are in the fields of natural science. Just as a scientist must be careful in not allowing the traditional success of a theory to close his eyes to new evidence, so the citizens and their leaders must be willing to consider new evidence and to reinterpret older ideologies.

Finally, there is another point in which a study of the sciences can be helpful in a study of political philosophy, that is, in the analysis of symbols and concepts. In this field the scientists, in erecting theories that organize experience and form the groundwork for predictions and the control of much of nature, have been forced into a very careful analysis of the meaning of all concepts that are used.

Symbols, slogans, and banners may often be necessary in a society to bring about decisive and coherent action when required, but they should nevertheless be recognized for what they too often are—shortcuts to careful thought and analysis. As in the sciences, a continual reexamination of the basic concepts is needed both with respect to their operational reference and with respect to the theories in which the concepts are embedded. Also, the rituals of our civilization need constant re-examination. Otherwise, we become an organism with automatic responses to symbols that no longer fit our present culture. We become subject to manipulation by those clever at handling the symbols. "Brainwashing" is more readily accomplished on those whose lives have already been governed by thoughtless response to symbols. All that is required is a substitution of a new set for the old (just as in the case of Pavlov's dog the sound of a bell could be substituted for the scent of food as the effective stimulus to action).

Furthermore, great care should be used in the spreading of "propaganda" merely to accomplish particular ends, lest we fall victims to our own propaganda. For the use of symbols for purposes of propaganda is an action completely contrary to the scientific use of symbols or concepts. When we criticize the Russians for their distortions and use of propaganda, internal and external, we should also be more critical of the growing carelessness with which slogans are used in our own society.

In this section, we have passed far beyond what is properly part of science. This has been done to create more consciousness of the powerful impact that scientific theories inevitably have upon social, political, and ethical theory. From what has been said, however, it should not be concluded that political or ethical theories should be patterned closely after the form of scientific theory. Again, we must emphasize that although a scientific theory is judged solely upon existing evidence, political and ethical theories are normative, and assert what *should* be the case, not merely what is the case. Therefore our analysis should not be construed to mean that a scientist as a scientist is in any better position to determine the ultimate ends and goals of society than the nonscientist. All that is claimed is that a *better* understanding of the nature of scientific theory can be helpful in analyzing and criticizing existing social and ethical theory. This is one of the basic reasons for including a course in science in a liberal arts curriculum.

6. QUESTIONS

1. In what way does a "scientific law" differ from a mathematical postulate?

2. What are some similarities and differences between scientific laws and laws of government?

3. Newton's laws "explained" Kepler's empirical laws. What does the explanation consist of? Does the law of universal gravitation need further or more basic explanation?

4. Scientists sometimes speak of "crucial experiments" or "crucial observations" on the basis of which a decision is made between alternative theories. Can a crucial experiment ever "prove" a theory, or does it not merely disprove one or more alternative theories? Are the number of possible alternative theories ever exhausted?

5. When the orbit of Uranus differed from its predicted path, why were astronomers so sure that another unobserved planet was affecting the orbit of Uranus instead of becoming convinced that Newton's law of gravitation required modification?

6. What evidence would you cite that convinces you that the moon is more nearly

spherical than hemispherical? (Disregard the evidence of recent photographs of the back of the moon.)

7. Most of us today are familiar with Foucault's pendulum and its purported "proof" that the earth rotates on its axis. The theory may be stated as follows. A pendulum, free to swing in any direction, is set into motion in a plane. By Newton's and Galileo's law it will continue to move in the same plane unless acted on by some external force at an angle to the plane. Since no force is exerted by the suspension system, its motion will continue in the same plane, even if the earth rotates with respect to it. Thus, at the North Pole, the pendulum will swing in a fixed plane in space with the earth rotating beneath it. It will "appear," therefore, to change its direction of swing relative to the earth by 15° every hour. Does this actually "prove" that the earth rotates on its axis, or does it merely show a rotation of the earth relative to the so-called "fixed" stars? Is it possible to conceive of the "fixed" stars as revolving, and of the earth as stationary?

8. In the history of thought, mathematics has sometimes been regarded as an empirical science; other times as transcendent philosophical truth. Discuss the manner in which Einstein's approach differs from both of these views.

9. Paradoxes have played a distinguished role in the history of ideas. If you are not already acquainted with them, look up Zeno's paradoxes and Bertrand Russell's discussion of one solution in his *Introduction to Mathematical Philosophy*.

Recommended Reading

F. S. C. Northrop, *The Logic of the Sciences and the Humanities,* Macmillan, New York, 1948.

Frank E. Manuel, *The Age of Reason,* Cornell University Press, Ithaca, 1951.

Gerald Holton and Duane Roller, *Foundations of Modern Physical Science,* Addison-Wesley, Reading, Mass., 1958, particularly Chapters 13–15.

Carl Becker, *The Heavenly City,* Yale University Press, New Haven, Conn., 1932.

Susanne Langer, *Philosophy in a New Key,* Mentor Press, New American Library, New York, 1948. A broad critical survey of the place of symbols in the arts and in science.

Norman Campbell, *What is Science?,* Dover Publications, New York, 1952.

Albert Einstein, Philosopher Scientist, edited by Paul A. Schilpp, Library of Living Philosophers, Open Court Publishing Co., La Salle, Ill., 1949.

Henry Margenau, *The Nature of Physical Reality,* McGraw-Hill, New York, 1950.

H. Poincaré, *Science and Method* and/or *Science and Hypotheses,* Dover Publications, New York, 1952.

P. W. Bridgman, *The Logic of Modern Physics,* Macmillan, New York, 1927.

Ernest Nagel, *The Structure of Science,* Harcourt, Brace & World Inc., New York, 1961. This volume is highly recommended to all serious students of the philosophy of science. It analyzes the significant problems with both scholarly care and imagination.

Arthur Danto and Sidney Morganbesser, editors, *Philosophy of Science,* Meridian Books, World Publishing Co., New York, 1960. Another book for the more advanced reader who wishes to pursue the many questions raised by modern physics.

Science and Culture, edited by G. Holton, Beacon Press, Boston, 1967. A series of essays examining relation of science to society from diverse points of view. Holton, Kepes, Dubos, Marcuse, Mead are all typically represented.

N. P. Davis, *Lawrence and Oppenheimer,* Simon & Schuster, New York, 1968. For those interested in investigating the interaction between scientists and government, this sensitively and carefully written book will provide much suggestive material. It illuminates the abiding dilemmas of modern political decisions regarding the use of ultimate weapons of destruction.

Jacob Bronowski, *Science and Human Values.* rev. ed. Harper & Row, Torchbooks, New York, 1965.

René Jules Dubos, *The Dreams of Reason: Science and Utopias.* Columbia University Press, New York, 1961.

chapter nine

Extension of Newton's system: conservation laws

The problem of precisely defining what is permanent amid change is a perennial one for all philosophy and science. It has been emphasized that the possibility of science, let alone recognition of any sort, implies an awareness of some more or less permanent or repeated pattern within the never ceasing flux of perceptions.

For this problem the ancient Greeks proposed several possible solutions, which are still being considered by Western scientists and philosophers. Some of the very early natural philosophers, among them Parmenides, accentuated permanence; others, such as Heraclitus, emphasized flux and process. One of the most influential philosophers of all ages, Plato, tackled the problem by asserting that ultimate reality consisted of permanent mathematical forms and that the senses merely give rise to an *illusion* of change.[1] This platonic doctrine has maintained its influence through the whole history of Western thought, reappearing, for example, in the interpretation of several twentieth century philosophers of science, among them

[1] This doctrine is given a dramatic and poetic expression in the famous allegory of the caves in Plato's *Republic,* Book IV.

Eddington and Jeans. This philosophy is often unconsciously or consciously echoed by leaders of religious thought as well as by some mathematicians and scientists.

The ancient solution to the problem, which is perhaps closest to the most widely accepted modern scientific answer is the one summarized in an extraordinary epic poem by Lucretius, *De Rerum Natura* (On the Nature of Things). We use one of his illustrations.[2] A flock of sheep grazing upon the green hillside is recognized as a white patch. The pattern moves about, the individual sheep moving in and out of the flock and changing places, causing many temporal variations, yet the pattern remains recognizable as the same white patch.

So, today, we consider an individual person to be made up of multitudes of cells, no one of which lives on for more than about a decade before being replaced. But the over-all pattern remains recognized as the same person, regardless of the fact that the composition of the whole has changed in all its constituent elements, and regardless of the fact that age changes even the pattern itself, psychologically as well as physically, leaving so little in the mature man that was once present in the child.

Science may be thought to consist of the study of those patterns that either repeat themselves, or are conserved amidst change. The importance of such aspects of permanence is given recognition by what are termed laws of conservation. In this chapter we shall study several of these laws: the laws of conservation of mass, of momentum, and of energy. In later chapters we shall examine other conservation laws, such as the laws of conservation of charge, of heavy particles, and of parity. Of the several conservation laws, we shall find that some of them require modification in light of recent analysis and experimental observation, but even with these modifications they represent important organizing principles required by all the physical sciences.

In addition to these general conservation laws, we shall find that certain quantitative relationships between measurable phenomena appear to remain invariant in our universe. These invariants are the universal constants, such as the constant of gravitation G and the constant velocity of light c. Just as π, e, 0, 1 (one), and i represent basic elements in the structure of mathematics, so a number of invariant constants remains basic to the structure of the physical sciences.

1. LAW OF CONSERVATION OF MASS

One of the earliest generalizations concerning invariance was proposed by some of the Greek natural philosophers: as a substratum underlying reality there is substance, stuff, or matter, which does not change. Change itself is accounted

[2] Lucretius, *On the Nature of Things,* translated by W. E. Leonard, Dutton, New York, Everyman's edition, Book II.

for by the fact that matter is composed of indivisible particles or atoms, which may be combined in a variety of ways, the combination giving rise to what appears to us as change. Referring again to Lucretius's flock of sheep, the flock seems to undergo changes because of the constant motion of the sheep that compose it, but the constituent particles, in this case sheep, remain individually unchanged.

By Newton's time this postulate of indivisible particles comprising the ultimate stuff of the universe was widely accepted. But there was the problem of what fundamental properties such atoms would have. Newton himself proposed as his hypothesis that

> All these things being considered, it seems probable to me, that God in the beginning formed matter in solid, massy, hard, impenetrable, movable particles, of such sizes and figures, and with such other properties, in such proportion to space, as most conduced to the end for which he formed them; and that these primitive particles, being solids, are incomparably harder than any porous bodies compounded of them; even so very hard, as never to wear or break in pieces; no ordinary power being able to divide what God himself made one in the first creation.[3]

The properties ascribed to these fundamental particles are extension in space, mobility, "massiveness" (by which is meant the inertia we have previously discussed), indivisibility, and impenetrability (or hardness). With respect to the last property (hardness or impenetrability) of the ultimate particles out of which all substances are made, it is noteworthy that the early explanation of the way in which one substance penetrates another was the same as today: the particles are pushed aside (when, for example, a stick is plunged into water), and one body can indeed penetrate into another by entering into the space between the particles.

From the postulate that all substances are made up of unchanging particles or atoms and that physical changes are changes in the position of the particles, it is a quick step to the postulate that in *all* changes organic and chemical, as well as physical, the total mass of the particles remains constant. Lavoisier was one of the first investigators to state this explicitly in 1789:

> We must lay it down as an incontestible axiom, that in all the operations of art and nature, nothing is created; an equal quantity of matter exists both before and after the experiment . . . and nothing takes place beyond changes and modifications of the combination of these elements.[4]

This postulate was one that was subject to more and more precise experimental testing, as instruments for weighing microscopic quantities became developed. A series of tests carried out by Hans Landolt from 1893 to 1909 showed that

[3]Newton, *Opticks,* 4th edition, London, 1730, p. 400. Reprinted Dover Publications, New York, 1952.
[4]Quoted by Gerald Holton and Duane Roller, *Foundations of Modern Physical Science,* Addison-Wesley Reading, Mass., 1958, p. 274.

there were no variations in weight greater than one part in over one million in any of 15 different chemical reactions studied. Similar tests in biology likewise showed the invariance of matter involved in organic changes.

The law of the conservation of matter seemed completely verified. However, as we now know, even this postulate or law has undergone modification. With the further development of nuclear and atomic physics, Einstein's prediction in 1905 that matter and energy may be transformed into one another has been borne out. Instead of the law of the conservation of mass, we have the law of the conservation of mass-energy ($E = mc^2$). We find that the energy given out or absorbed by chemical reactions is a result of a very minute change in mass of the constituents of the reaction. The change of mass is, however, so small that it cannot be detected in any reactions except in those in which the nucleus of the atom itself is involved. We shall return to this topic in a later chapter. For all phenomena, except for motions close to the speed of light and for nuclear reactions, the principle of conservation of matter is still used as an effective operating principle in analysis and experiment in both chemistry and physics.[5]

2. LAW OF CONSERVATION OF MOMENTUM

As our second example of the existence of permanence amid change, we consider another conservation law, that of the conservation of momentum which—unlike the laws of conservation of mass or conservation of energy both of which have been modified from their classical statement—still remains unmodified among the laws or postulate of modern physics.

Not long after Galileo successfully established his law of inertia (that is, that a body will continue in uniform rectilinear motion in a straight line) as an organizing principle, the law was extended and generalized. In particular, it was found useful to define the "quantity of motion" as the product of mass times velocity. Huygens,[6] a contemporary of Newton, used this concept to study the laws of collision; that is, the laws that govern the motion of two bodies before and after they collide with one another. He found that the total quantity of

[5]Discussion of the law of conservation of matter, which implies that matter is neither created nor destroyed (except, in the modern theory, through a transformation of matter into energy or vice versa), would be incomplete without some reference to the theory of H. Bondi and F. Hoyle, modern astronomers, who state that as space becomes emptied of matter through expansion of the universe, matter is automatically created in the empty space left. This cosmological theory will be discussed in a later chapter.

[6]Christian Huygens (1629–1695) was a Dutch physicist of a stature almost equal to that of Galileo and Newton. He submitted his definitive work on collisions before the Royal Society of which Newton was a member. His analysis of pendulum motion and of wave phenomena makes him one of the great physicists of all time.

motion (momentum), before and after a collision, remained constant, provided the correct signs denoting directions were given to the velocities.

Here, then, was a new invariance found in nature. Quantity of motion, which is the product of mass times velocity, is today defined as *momentum.* The new principle states that in any system composed of particles interacting in any fashion, that is, through collision or through attractive or repulsive forces, the sum of the momenta of all the particles remains constant.

This law is implied by Newton's three laws of motion. It will be recalled that according to the second law, when the proper units are used, the force exerted on any body causes a change in quantity of motion or momentum. If the masses are considered constant, the equation may be written as

$$F = ma$$

or

$$F = \frac{mv_2 - mv_1}{t}$$

or

$$F \cdot t = mv_2 - mv_1$$

where F is the average force[7] and $v_2 - v_1$ is change in velocity and $mv_2 - mv_1$ represents the change in momentum of the object acted on by the average force F. (The product $F \cdot t$ is given the name "impulse.")

Now if two objects collide, during the time of their impact (while they are interacting with one another), the mutual force that each exerts on the other must be equal but opposite in direction, on the basis of Newton's third law. It follows, therefore, that the change of momentum undergone by one body must be equal in magnitude, athough opposite in direction, from the change of momentum undergone by the second body. The total change of momentum is thus zero, or the sum of the momenta before and after collision is constant.

Momentum, since it is the product of mass times velocity, and since velocity is a vector or a directed quantity, must itself be a vector or directed quantity.

The principle will be more clearly understood if we apply it to specific examples.

Example 1 Suppose a bullet of 5-g mass is fired into a block of wood of mass ½ kg, the velocity of the bullet being 300 m/sec. What will be the velocity of the wood (plus the bullet) after impact? Assume that the wooden block is completely free to move (that is, assume friction is negligible). The analysis is made as follows.

[7] For this course we shall use the concept of *average* force. The force usually will vary with time. A complete analysis would require the calculus. But it can be shown that if the average is properly calculated, even if the force is varying, the equation given is correct. For those interested, the equation given may be written in terms of calculus as $\int f dt = mv_2 - mv_1$ where f is a variable force and where the sign $\int$ indicates that a sum is being taken.

Given:
M_b = mass of bullet = 5×10^{-3} kg
V_b = velocity of bullet before impact[8] = 300 m/sec
M_w = mass of wood block
V_w = velocity of block before impact = 0
$V_b' = V_w'$ = velocity of block *after* impact (the bullet is embedded in the block and they move with same velocity)

The velocities are all in the same direction before and after impact. Therefore, by the law of conservation of momentum

$$M_bV_b + M_wV_w = M_bV_b' + M_wV_w'$$

Since $M_wV_w = 0$

$$M_bV_b = M_bV_b' + M_wV_w'$$

Since $V_w' = V_b'$

$$M_bV_b = V_w'(M_b + M_w)$$

Using the given data we have

$$(5 \times 10^{-3})(300) = V_w'(5 \times 10^{-3} + 5 \times 10^{-1})$$

or

$$V' = 300/101 \text{ m/sec or approximately 3 m/sec}$$

Example 2 Assume the same mass of bullet and the same velocity as above. Calculate the recoil velocity of the gun which fires the bullet (assuming the gun is free to recoil and is not held firmly) if the gun has a mass of 2 kg.

In this case the momentum before firing is 0, therefore the total momentum after firing must be zero, or

$$M_gV_g' + M_bV_b' = 0$$

therefore

$$V_g' = -\frac{M_bV_b'}{M_g} = -\frac{(5 \times 10^{-3})(300)}{2} = -0.75 \text{ m/sec}$$

Note two points about this example. (*1*) The recoil velocity is negative, meaning that it is opposite in direction from the bullet's velocity. (*2*) It is assumed that the gun is free to move. If the gun is held, for example, against one's shoulder, the total mass set into backward recoil would have to include the mass of everything which moved backward, including the shoulder.

[8]This is approximately the velocity of a 22 caliber bullet, close to the speed of sound, or approximately 1000 ft/sec or 700 miles/hr.

Example 3 We may apply the principle of the conservation of momentum to the study of rocket propulsion. Fundamentally rockets are impelled in one direction by the recoil of a mass of gas ejected at high velocity in the opposite direction. The analysis of rocket propulsion is thus very similar to that of a gun subject to a recoil resulting from the ejection of a bullet. Just as the recoil of a gun will take place in a vacuum, that is, without any air to "push" against, so a rocket will be given an acceleration in a direction opposite from the direction of the expulsion of the gas, even if there is no surrounding atmosphere.

The only consideration that makes rocket propulsion more difficult to analyze is that instead of a single object of a given mass being shot out backwards once, the mass of gas being exhausted is forced out in a continuous stream at very high velocity. The rocket is thus subject to a continuous forward impulse and, at the same time, its mass is continuously being reduced by the loss of fuel that is being converted into gas.

Nevertheless, we can make some *approximate* calculations to illustrate the application of the principle of conservation of momentum. For simplicity, we shall neglect air resistance and the effect of gravity. For a satellite to circle the earth requires a velocity of somewhat over 7 km/sec, or a velocity of approximately 16,000 miles/hr. (See page 147 for the calculation of this figure.) We inquire as to the approximate amount of fuel required to give a satellite of 1 kg this velocity, after we have made some assumptions as to the velocity of gas being ejected from the rocket.

It should be clear that the higher the velocity at which the gas is expelled backward from the rocket the greater will be the forward impulse given to the rocket. At the present time, the limit on the velocity is set by the melting temperatures of the nozzle through which the gas escapes (the velocity of the gas being dependent on the temperature). A velocity of about 3 km/sec *relative* to the rocket may be assumed. But since we are interested in the velocity of the rocket *relative to the earth* and since the gas is being ejected from the rocket which is itself in motion, the velocity of the gas relative to the earth will be very much less than it is relative to the rocket. Calculation shows that a reasonable figure for the *average velocity* of the gases *relative to the earth*[9] will be no greater than approximately 0.3 km/sec. (If it seems puzzling that the *average velocity relative to the earth* is so drastically reduced, consider the fact that when the velocity of the rocket relative to the earth is greater than 3 km/sec, at this point the escaping gases have a velocity relative to the earth in the same direction as the rocket. The velocity of the gas after this point is then negative if we assume the original direction of the escaping gases to be positive when the rocket begins its flight.)

[9]For a more complete analysis (without the use of calculus) of the variation of velocity, see a good discussion of this point in *Physics* by the Physical Science Study Committee, D. C. Heath, Englewood, N.J., 1960, pp. 372–375.

The problem can now be easily solved by our momentum equation

$$M_gV_g - M_sV_s = 0 \text{ or } M_g = \frac{M_sV_s}{V_g} = \frac{(1)(7)}{0.3} = 27 \text{ kg}$$

where M_g is the total mass of the gas, V_g is the average velocity of the gas relative to the earth, M_s is the mass of the satellite, and $-V_s$ is the final satellite velocity (the negative sign is used because its velocity is opposite that of the gas).

In other words, about 27 kg of fuel is required to give 1 kg the necessary velocity. The actual amount of fuel required is, of course, very much larger than this since the rocket must also overcome air resistance and the force of gravity. It can be appreciated why such tremendous rockets are required to get satellites into orbit. The mass of the fuel is many times greater than the mass of the orbiting rocket.

We have in this section discussed only the principle that linear momentum is conserved. In addition to this principle, mention should be made of another principle derived from it: the principle of conservation of angular momentum. This law is derived from the law of conservation of linear momentum, but we shall omit the derivation. However, the principle is of sufficient importance that it should be stated. Angular momentum of a particle is defined as the following product: the mass of the particle times its distance from an axis of rotation times its tangential velocity. (Tangential velocity is the component of the velocity perpendicular to the line joining the particle with the axis of rotation. See Fig. 9.1.)

The principle of conservation of angular momentum states that in an isolated system the sum of the angular momenta of all the masses composing the system remains constant. (Dancers and figure skaters make instinctive use of this principle when they bring their arms down sharply as they are turning, with the result that their rate of rotation is increased. In this case, since the distance R of the mass of their arms is reduced, angular momentum can only be conserved if $V_\perp$, namely, the tangential velocity and, hence, the rate of rotation is increased.)

It may be shown that Kepler's second law, that is, that a line drawn from the sun to a planet sweeps out equal areas in equal intervals of time, may be de-

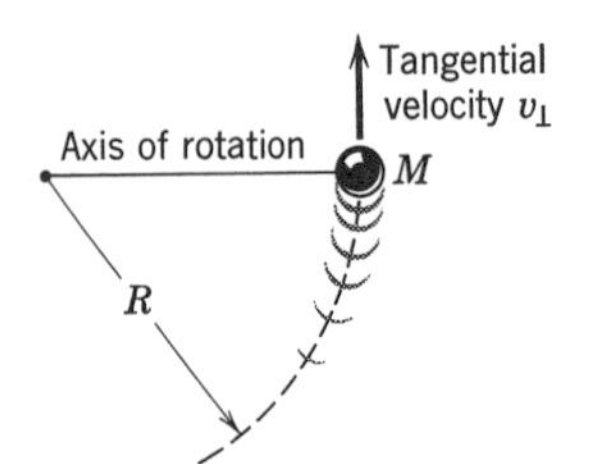

Fig. 9.1. Illustration of the angular velocity of a particle $= MRv_\perp = 2\pi MR^2/T$.

rived from the principle of the conservation of angular momentum. Thus, again, an empirical law is "explained" in that it is shown to follow from the three laws of motion.

The significance of this principle may be appreciated if it is realized that our standard for the measurement of time, which ultimately rests upon the assumption that the period of rotation of the earth or of the revolution of the earth about the sun is constant, in turn depends upon the conservation of angular momentum.

In modern physics, the law of the conservation of angular momentum has stood the test of time and remains one of the cornerstones upon which present theories of the structure of the atom are built.

3. LAW OF CONSERVATION OF ENERGY

In the search for what is constant or invariant amid the flux of natural events, one of the greatest triumphs of science has been the development of the generalization that a certain measurable quantity called energy is conserved in all changes involving material things. Today, the principle of the conservation of energy (when properly interpreted in light of the modern principle of the equivalence of mass and energy) remains the organizing principle that brings together all the various natural sciences in their interrelationships.

The emergence of the idea of energy has been more recent than that of either mass or momentum. It was not until the middle of the nineteenth century that the principle of the conservation of energy became dominant in the sciences. In fact, the wide use of the term "energy," so commonplace today, is comparatively modern. To understand the concept and its present significance, it will be helpful to trace its development.

Galileo had the germ of the idea of energy when he concluded that a pendulum bob when released from any given original height would swing through an arc, reach a maximum velocity at the bottom of its swing, and again reach very nearly the same height it had originally. That it did not quite attain the original height was properly ascribed by Galileo to the resistance of the air. He argued that if no air were present, the pendulum bob would have the "power" (using Galileo's term) because of its velocity to rise again to the original height.

A pendulum, swinging back and forth in an endless repetition of a pattern of movement, leads naturally to the inquiry as to what property abides in the system which remains constant throughout the motion. There is a constant exchange repeated again and again between the spatial configuration of the system and the velocity of the swinging mass.

When we consider the problem of relating spatial configuration to velocities, it is natural to be reminded of how velocities were found to be related to time through the concept of force. In the preceding section we saw how the product of a force times the *time* during which the force acts may be equated to a change in momentum.

$$F \cdot t = mv_2 - mv_1$$

In 1695, Leibniz proposed an analysis very similar to the above, but he asked how a force acting over a *distance* (instead of over an interval of time) would affect velocity. He established the relationship

$$F \cdot s = \frac{mv_2^2}{2} - \frac{mv_1^2}{2}$$

The term $mv^2/2$ was called by Leibniz *vis viva* or vital force. Today it is given the designation kinetic energy, while the term on the left-hand side of the equation is called work.[10]

Let us see how this expression may be derived.

Suppose we have a constant force F, which is giving an acceleration a to the mass m. For simplicity, we assume that the mass is at rest when the force is first applied.

$$F = ma$$

Multiplying both sides of the equation to obtain the desired product of force times distance we have

$$F \cdot s = ma \cdot s \tag{1}$$

But, since the force is constant, the acceleration is constant

$$v = at \qquad \text{or} \qquad t = \frac{v}{a} \tag{2}$$

We have from the equations for constant acceleration

$$s = \tfrac{1}{2}at^2 \tag{3}$$

Hence, by equation 2

$$s = \tfrac{1}{2}\frac{v^2}{a} \tag{4}$$

Substituting in equation 1, we obtain

$$F \cdot s = \tfrac{1}{2}mv^2$$

[10] It was not until 1807 that the term energy was used. Thomas Young was the first to give *vis viva* that designation. The term work was first used by J. Pomelet in 1826.

(*Note.* We have taken the original velocity as 0. If the original velocity is taken as v_1 and the final velocity as v_2, then the equation becomes $F \cdot s = \frac{1}{2}mv_2^2 - \frac{1}{2}mv_1^2$.)

We saw in the previous section how Huygens analyzed the problem of motion of an isolated system of masses in collision, and found that in all cases momentum was conserved. Is this also true of kinetic energy? Here Huygens found a somewhat different kind of result. Instead of a *general* law, which would require that the total quantity $\frac{1}{2}mv^2$ remain constant before and after collisions, he found that this constancy is true only for certain types of collisions involving very hard objects, such as steel balls or billiard balls (objects that are said to be elastic[11]). And even in these cases the kinetic energy is not quite conserved, but only nearly so.

Let us examine rather closely an example based upon fairly common experience. Suppose we have two billiard balls, one of which is at rest and the other in motion, and the moving ball makes a "head on" collision with the one at rest. It will be found in this case that after collision the moving ball comes to complete rest (very nearly) and the second ball takes up the velocity of the first (very nearly). Can we account for this result from the two laws—the law of conservation of momentum and the law of conservation of kinetic energy? The analysis is given as follows.

It is assumed that the balls have an equal mass.

The mass of the moving ball is designated M_1; the mass of the ball originally at rest is designated M_2.

The velocity of the first ball before impact is designated V_1 and after collision as V_1'. The velocity of the second ball before collision V_2 is 0 and after collision is V_2'.

From the law of conservation of momentum

$$M_1V_1 + 0 = M_1V_1' + M_2V_2' \tag{5}$$

If kinetic energy is conserved, then

$$\tfrac{1}{2}M_1V_1^2 + 0 = \tfrac{1}{2}M_1[V_1']^2 + \tfrac{1}{2}M_2[V_2']^2 \tag{6}$$

From equation 5, since the masses are equal,

$$V_1 = V_1' + V_2' \tag{7}$$

[11]To give a clearer idea of elasticity, it may be pointed out that a measure of elasticity is given by noting, for example, how high a steel ball will rebound when dropped from a height onto a heavy steel plate. If the ball bounces as high as the point from which it was dropped, then the steel would be said to be perfectly elastic. *Perfect* elasticity is never found among objects of ordinary size. We shall find later that the collisions between atoms are assumed to be perfectly elastic.

From equation 6

$$[V_1]^2 = [V_1]^2 + [V_2']^2 \tag{8}$$

If we square equation 7 and substitute in equation 8, we arrive at

$$[V_1']^2 + 2V_1'V_2' + [V_2']^2 = [V_1']^2 + [V_2']^2$$

or

$$2V_1'V_2' = 0$$

From this it follows that *either* V_1' or V_2' is zero.

Since the second ball is certainly not at rest after the collision, we conclude that the first ball must be at rest after the collision, implying from equation 5 that the second ball has a velocity after collision exactly equal to that of the first ball before collision. Now in experimental tests of colliding objects these analytical results are *almost* but not quite realized. We conclude that in certain types of collisions (in elastic collisions such as between billiard balls) kinetic energy is approximately conserved, whereas in others (in inelastic collisions, such as a ball of putty colliding with a billiard ball) kinetic energy is not even approximately conserved. See Fig. 9.2.

Until it was discovered how kinetic energy was dissipated into heat, the value or fruitfulness of the concept of energy was necessarily greatly restricted. It was not until fairly recently (the middle of the nineteenth century) that it was found

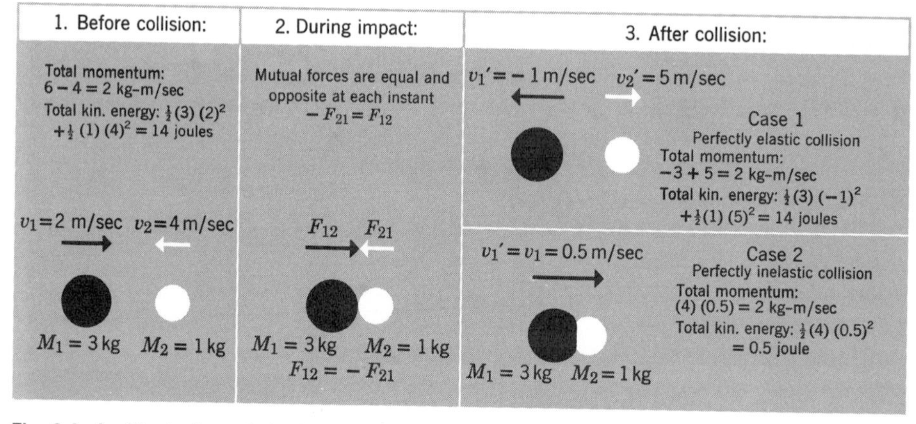

Fig. 9.2. An illustration of the limiting cases of collision (perfectly elastic and totally inelastic). In a perfectly elastic collision the kinetic energy is conserved. In a totally inelastic collision the masses remain coupled together and move with a common velocity after the impact. Momentum is conserved in both cases. The total kinetic energy is conserved for elastic but not for inelastic collisions. In this illustration, note how much kinetic energy is lost (usually in the form of heat) in the inelastic collision. Also note that momentum is a directed quantity (a vector) and we must attend to the signs, whereas kinetic energy is not a vector and is independent of the velocity. Although *collisions* have been used for illustration, *all* interactions obey the same laws.

the energy did not disappear but was merely changed into nonmechanical forms (heat and sound for most collisions). Newton himself made no use of the concept of energy because of its restricted applications.

Up to now we have been considering the term on the right-hand side of the equation $F \cdot s = \frac{1}{2}mv^2$, that is, the kinetic energy. In the collision problem that we have examined, it has been assumed that any force such as the force of gravity acting on the two bodies before and after collision is at right angles to the motion of the bodies and, therefore, has no effect on that motion. Let us see how the concept of mechanical energy may be applied to an object falling from rest, in which the force acting on a mass is in the same line as that of the motion.

If an object of mass m is lifted to a height h above ground level and then dropped from rest, since the downward force acting on the body is mg (the weight of the mass) and acts through the distance h, the equation tells us that the quantity mgh, as determined at the beginning of the motion, is equal to $\frac{1}{2}mv^2$ at the instant the mass strikes the ground.

Otherwise stated, we see that when $s = 0$ the kinetic energy is $\frac{1}{2}mv^2$, and when $s = h$ the original height, then $v = 0$ and the kinetic energy is 0. There is thus a quantity mgh which at the beginning of the motion is equal in value to the kinetic energy which appears at the end of the motion. We term the quantity mgh, the potential energy (or energy of configuration) of the system. Thus, the original potential energy mgh is transformed into kinetic energy $\frac{1}{2}mv^2$ as the system changes from its first state to its second state.

Analysis shows that if we take any intermediate point between the level of the ground $s = 0$ and the original height $s = h$ we always find that the sum of the terms $mgs + \frac{1}{2}mv^2$ remains a constant. Throughout the motion, therefore, we can say that the total mechanical energy is conserved. (*Note.* Strictly speaking, we have to make two assumptions that in practice are valid for relatively small velocities and small distances s, that is, that the resistance of the air is negligible and that g is constant.)

Another way of stating the analysis is to say that if we do work by lifting a mass vertically through a certain height h, we give to the mass a certain amount of potential energy mgh which, in turn, can be transformed into kinetic energy by allowing the mass to drop again under the force of gravity. It will be seen that the units of energy and the units of work are the same, and are the product of force times distance. The fundamental unit that we shall most often use is the newton-meter (newton times meter) which, for convenience, is called a joule.[12]

[12]Other common units are the erg, which is a very small unit (10^7 ergs equal 1 joule), and the foot-pound (1 joule equals 0.738 foot-pound, or approximately $\frac{3}{4}$ of one foot-pound).

The principle of the conservation of mechanical energy for frictionless systems can easily be extended to cases where the motion is not in the same direction as the force. For example, let us consider what happens when a mass is pushed up along a frictionless inclined plane through a distance s. The work done in this case is given by $F \cdot s = (mg \sin \theta) \cdot s$ since the force along the plane is now $mg \sin \theta$ (see page 155). But if the height of the plane is h, we see at once that the work done or the potential energy at the top of the plane is equal to mgh, regardless of the angle of the plane. Now this is a very interesting and significant result. It tells us that regardless of how a mass is lifted to a certain elevation, it will have the same potential energy. (This can be shown to be true for curved paths as well as for paths along a straight line.) Further analysis shows that the potential energy is completely transformed into the same amount of kinetic energy by allowing the mass to be accelerated downward along any path until zero elevation is reached. Hence, we conclude (as Galileo, indeed, first showed) that the velocity will be the same regardless of the path followed, if the change in elevation is the same. This is often a very useful principle, and is illustrated in the following examples.

Three cases in which potential energy is transformed into kinetic energy are shown in Fig. 9.3.

It should be noted that although the magnitude of the velocities is the same in the three cases illustrated, the direction of the velocity is different. We conclude that energy is a nondirected magnitude, that is, a scalar rather than a vector quantity.

We also see that in the case of the pendulum there is a continuous repetition of transformation between potential and kinetic energy and back to potential energy. In this case the mechanical energy is conserved over a long period of time (that the energy is finally dissipated is, of course, due to the frictional resistance of the air and the frictional resistance at the support). In the other two cases, motion ceases when the mass strikes the ground. The energy is, at this point, converted into nonmechanical forms such as heat.

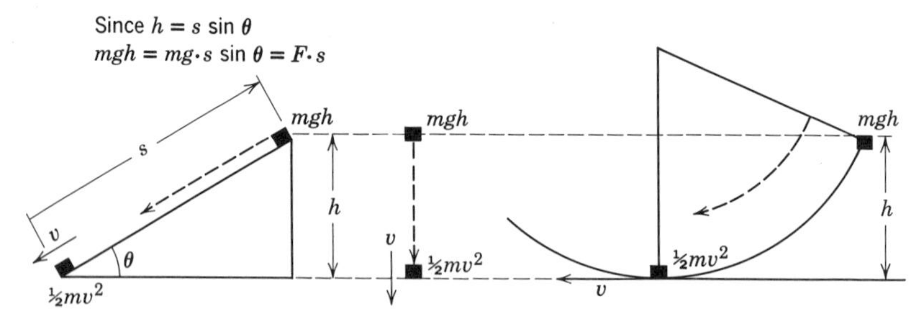

Fig. 9.3. The potential energy mgh in each of the above cases is the same, since the elevation h is the same and the mass is the same. The kinetic energy at ground level is the same; therefore, the speed or magnitude of the velocity is the same in each case.

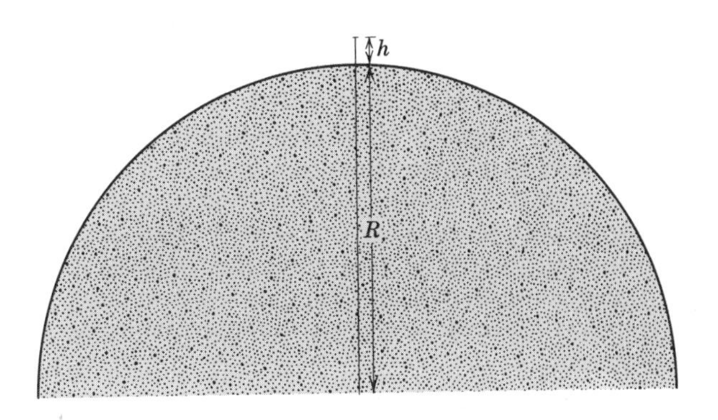

Fig. 9.4. Potential energy in a nearly constant gravitational field. If the height h above the surface of the earth is very small relative to R, the radius of earth, then force of gravity is nearly constant because $F = GM_em/R^2$ is very nearly equal to $GM_em/(R + h)^2$. Therefore the potential energy difference between a mass m at a point on the surface and the mass at a point at distance h above the surface can be taken as P.E. = $GM_em/R^2 \times h$ which is equal to gmh.

It is necessary, particularly in discussing potential energy, to be very clear on one point. In almost all physical phenomena we are interested in the *difference* between the energy of two distinct states of a system and not in the absolute amount of internal energy contained in the system. As we shall find out later, there is an equivalence between mass and energy. Therefore any mass, no matter in what state, has more than zero energy. This particular consideration need not concern us now. But it is important to realize that when we speak of the potential energy of an object, we are really referring to its potential energy *difference* with respect to *some* arbitrarily selected zero point. Thus a mass m lifted to a height h above the surface of the ground is said to have potential energy *relative to the ground* of mgh. Obviously if a well were dug below ground and the mass could fall towards the center of the earth it would acquire velocity and hence kinetic energy. Thus the mass at ground level has potential energy relative to the earth's center. For most purposes, where the ground level or zero point is easily defined the problem causes little difficulty. This is particularly true if the potential field of force is more or less uniform as it is near the earth's surface. See Fig. 9.4.

In other cases, it is more difficult to establish an adequate zero point potential energy. In the case of the gravitational field about the earth at large distances from the surface of the earth, it is most convenient to take a point at infinity as the zero point. At this point the gravitational attraction between any mass and the earth is zero, but since the potential energy of a mass increases as the distance from the earth increases, the absolute potential energy of an object at a distance d (less than infinity) from the earth must be negative. It can be shown that it follows from the inverse square law of gravitation that the difference of potential energy of a mass m between two points at distances d_1 and

d_2 from the earth's center is given by

$$G\frac{mM_e}{d_1} - G\frac{mM_e}{d_2}$$

where d_2 is greater than d_1. Therefore if d_2 is taken at infinity and d_1 equals R, the radius of the earth, the expression

$$-G\frac{mM_e}{R}$$

represents the potential energy of the mass m at the surface of the earth. See Fig. 9.5.

The implication of this analysis is that it takes a certain potential energy equal to:

$$G\frac{mM_e}{R}$$

to "lift" the mass m to infinity, i.e., to a distance so great as to be outside the earth's field. Hence any mass m that is given a kinetic energy

$$\frac{1}{2}mv^2 = G\frac{mM_e}{R}$$

will escape the earth and the corresponding velocity will be what is termed the escape velocity of the earth (or, with appropriate substitutions for mass and radius, for any other planet).

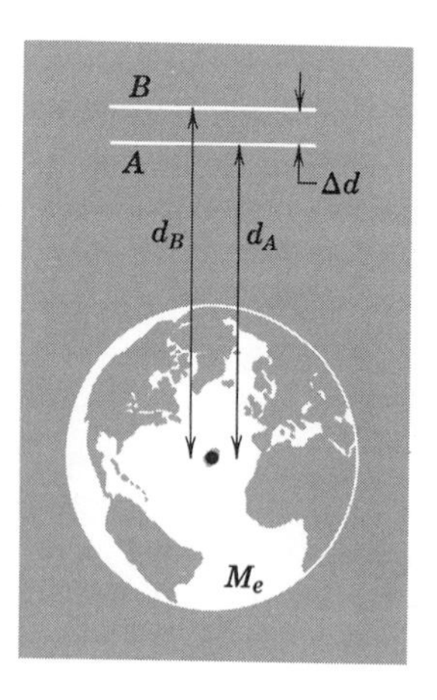

Fig. 9.5. Potential energy in a varying gravitational field. At considerable distances above earth's surface, gravitational force changes significantly with distance. Change in potential energy from point A to point B can be thought of as the *average* gravitational force times the distance $d_B - d_A = \Delta d$. The average force exerted on mass m can be taken as GM_em/d_Ad_B. Therefore the work (or increase in potential energy) is

$$[GM_em/d_Ad_B](d_A - d_B) = GMm(1/d_B - 1/d_A).$$

If zero level is taken arbitrarily as "at "infinity," the potential energy at point A becomes $-GMm/d_A$.

4. TIDAL THEORY

Before leaving the subject of dynamics we refer briefly to the theory of tides which weaves together so many of the basic principles of physics. Gravitational forces, centripetal acceleration, angular momentum, energy, and power, all play their role. The complex periodicities of the tides have fascinated poets over many generations—have often affected the course of history.

It was Kepler who made the first well-formulated guess that it is a force exerted by the moon that causes tides. Interestingly, Kepler's brilliant contemporary, Galileo, rejected this idea. The explanation smacked too much of "occult" forces and astrology for the skeptical mind of Galileo. So the analytical solution of tide generation had to wait for Newton. In the Principia he was able to show that tides are a consequence of the law of universal gravitation when that law is combined with the laws of centripetal acceleration. Newton found the answer in the *differential* gravitational effect of the moon (or sun) on different points of *the extended* body of the earth.

Consider as an example the earth-moon system. As we have presented it, the moon is assumed to revolve about the earth as a center. Actually both revolve about an axis situated at about 3,000 miles from the earth's center. If every particle composing the earth and the ocean were at the the same distance from the moon, there would be no relative force on the solid earth different from the force on the fluid oceans. The gravitational force of the moon would affect all particles alike and the earth would move as a point mass around the mutual center of mass. There is, however, a departure from this "average" gravitational force of the moon on each particle that depends on that particle's distance from the moon. On the side of the earth facing the moon there is a small *additional* gravitational force compared with the force acting on the center of the earth. On the side of the earth away from the moon there is a similar differential force which is *smaller* than the force acting on the earth's center. These two departures from the average force on the earth's center are nearly equal and oppositely directed. In the one case the differential force urges the oceans, free to move, to pile up on the side of the earth facing the moon; on the opposite side, the oceans, being less forcefully attracted than is the earth's center, are caused to flow and to pile up in the opposite direction. (The earth is, so to speak, "pulled away" from the oceans on the side which is opposite the moon.) Finally, since the earth rotates with respect to the moon with a period of about 25 hours, we should expect high tide to arrive at periods of a little over 12 hours.

The sun also and in a similar manner causes tides. Although the sun's gravitational force is greater than that of the moon upon any single particle of the earth, it is found that the *differential* of the force on particles on opposite sides of the earth is much smaller. The tide-generating force of the sun is about half that of the moon. The reason is not hard to find. Whereas gravitational forces vary inversely with the square of the distance of the mass causing the force on a single particle, it can be shown that the *differential* or tide-generating force varies inversely with the *cube* of the distance to the mass causing the tide.

Consider mass M exerting a gravitational force at a distance d from the earth's center upon a small particle of mass m. If the radius of the earth is R, we have for the *difference* of the force on two masses situated on opposite sides of the earth, that is, at d and $d + R$, given by

$$\Delta F = G\frac{mM}{d^2} - G\frac{mM}{(d+R)^2} = GmM\left(\frac{1}{d^2} - \frac{1}{(d+R)^2}\right) \tag{9}$$

where M is the mass of the moon. Now if R is *much* smaller than d, this expression approximates closely to

$$GmM\left(\frac{2Rd}{d^4}\right) = \frac{2GmMR}{d^3} \tag{10}$$

(As an exercise, show that in the case of the moon, this approximation leads to a negligible error.) See Fig. 9.6. This is an inverse cube law expressing the difference of the gravitational attraction of the same mass M on small particles of mass m on different sides of the earth. If it is found that when approximate values for G, M, and R, and d are substituted in equation 10, the sun's tide-generating capacity is about one-half that of the moon.

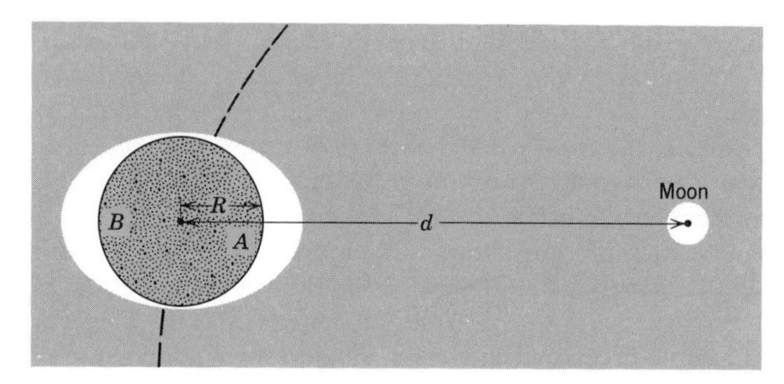

Fig. 9.6. Schematic representation of tidal action, simplified and exaggerated. The moon's gravitational force on a unit mass m situated at point A nearest the moon is $GMm/(d - R)^{2'}$ where M, is the mass of moon. Since R is very small compared to d, the difference between this force and the force at the center of the earth is very nearly $2GMR/d^3$. Likewise at B the differential force per unit mass is $2GMR/d^3$ less than at earth's center. Ocean waters at point A are attracted more strongly towards the moon and at point B less strongly towards the moon, than is the earth's center. Waters thus tend to pile up at A and B.

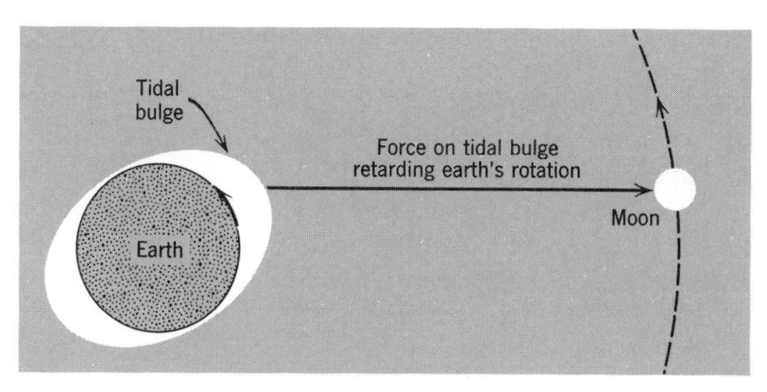

Fig. 9.7. Earth-moon tidal system. Because of the rotation of the earth, "tidal bulges" are carried forward as shown. Friction then causes the earth's rotation to slow down somewhat. To preserve total angular momentum of the system, the angular momentum lost by earth in its rotation requires gain in angular momentum of moon.

Upon this analysis, we would expect and indeed we do find that when the sun, earth, and moon are in line, we have the "spring tides." Under those conditions the new moon (or the full moon) acts in line with the sun. When the moon is at its first quarter or third quarter, the effects of the sun and moon partly cancel out and we have "neap tides."

Such is the idealized, abstract analysis. Applied to actual conditions along the shores of the continents, the situation becomes vastly more complicated. The earth is not a solid overlaid with a single homogeneous level of frictionless fluid, free to move under forces. The shores along the continents are marked by bays, inlets, coves. Wave propagation and resonances give rise to great local anomalies depending on local irregularities.

Other important and interesting consequences flow from tidal theory. The energy of the oceans in motion must be in a process of constant dissipation, kinetic energy eventually turning into heat. This implies that the kinetic energy of rotation of the earth is decreasing. The earth's rotation must therefore be slowed. See Fig. 9.7. In turn it follows that, as the earth's rotation slows, its angular momentum is lost. But angular momentum of an isolated system is never lost. Hence the angular momentum of the moon must take up the angular momentum lost by the earth, and to do this the moon must recede slowly but inexorably year by year from the earth until at last the period of revolution of the earth will correspond to the period of revolution of the moon. The moon long since has gone through this process. The moon's tides, i.e., the slight fluctuations of the moon's solid surface (caused by the earth) slowed it down until it now always presents the same face to the earth.

When the time is finally reached that the earth's day is a month long, the tidal generating force of the sun will then become predominant. Now the earth's rotation relative to the *sun* will slow down, this time through the friction of *solar* tides. Again because angular momentum must be conserved, the earth's orbit about the sun will become larger, and the year longer. The day eventually will become longer than the month. The whole earth-moon system will lose angular momentum. The moon will again approach the earth. If no other solar-system catastrophe has intervened, if the sun has not already blown up into a red giant and scorched the earth, the moon will eventually approach (in perhaps 8 billion years) a distance of about 10,000 miles. It will then be burst asunder by the tremendous gravitational force exerted by the earth.

From the considerations of this section we see that, again, we have arrived at another pattern of invariance. It is, however, much restricted by the condition that frictional resistance must be absent. The principle of conservation of *mechanical* energy would appear to be too restricted to be of great practical significance. However, the very statement of this ideal principle does give rise to the problem of what happens to kinetic energy when it disappears due to friction. The answer was forthcoming in the middle of the nineteenth century. It was then found that a more general principle of conservation could be established: that in any isolated system energy is always conserved if we include not only *mechanical* energy but other types of energy including heat, electromagnetic, sound, and chemical energy.

5. POWER: RATE OF CONVERSION OF ENERGY

Although we shall not consider in detail here the matter of conversion of the chemical energy of fuels to the mechanical energy that generally drives the machinery of our civilization, it is well to bring out the concept of the time rate at which such conversion takes place. For example, a small gasoline motor in one automobile may have just as great an over-all efficiency as one in a second automobile with respect to the total energy conversion of the fuel consumed, but the first motor may not be able to make the conversion as rapidly as the second. This relative rate of conversion is measured by the respective power of the two motors, where *power is defined as the time rate of doing work.*

In the metric system, power is expressed in *joules of energy per second.* For convenience, we call one joule per second one *watt.* In the English system of units, the unit of power is called a *horsepower,* 1 hp being defined as a rate of doing work that is equivalent to raising 550 lb vertically through 1 ft in 1 sec. The unit of 1 hp is equivalent to 746 watts.

For example, if an automobile weighing 2200 lb goes up a hill at a constant rate of speed and reaches a point 400 ft above the starting point in 20 sec, we can calculate the power required (neglecting all frictional wastes and inefficiences) as

$$\frac{2200 \times 400}{20} \times \frac{1}{550} = 80 \text{ hp}$$

As a practical matter, of course, more horsepower than this is required because of the frictional resistance that needs to be overcome.

6. SUMMARY OF EQUATIONS AND CONCEPTS

$$\textit{Momentum}, \text{ by definition, } = mv \tag{11}$$

It has the units kg · meter/sec in the MKS system.

The *law of conservation of momentum:*

$$m_1v_1 + m_2v_2 = m_1v_1' + m_2v_2' \tag{12}$$

This applies for all types of interactions whether elastic or inelastic.

If the collision is elastic, the condition also holds that:

$$\tfrac{1}{2}m_1v_1^2 + \tfrac{1}{2}m_2v_2^2 = \tfrac{1}{2}m_1v_1'^2 + \tfrac{1}{2}m_2v_2'^2 \tag{13}$$

Work by definition equals force times distance (the unit for work is the joule). This applies only when the force is acting in the direction of the displacement (otherwise work equals force times distance times the cosine of the angle between the direction of the force and the direction of the displacement). Work is *not* a vector; it is a special kind of scalar product between two vectors, force and displacement, namely $W = F \cos \theta \cdot d$ where θ is the angle between vectors F and d

Mechanical energy is divided into two types: *Potential* and *kinetic.*

Potential energy of a mass lifted to a height h above zero reference level

$$\text{P.E.} = mgh \tag{14}$$

where the units are in joules for the MKS system.

Kinetic energy arises from the motion of a mass

$$\text{K.E.} = \tfrac{1}{2}mv^2 \tag{15}$$

where the units are in joules for the MKS system.

Conservation of mechanical energy:

$$\text{P.E.} + \text{K.E.} = \text{constant} \tag{16}$$

Power by definition is the rate of doing work or expending energy. Therefore

$$\text{Power} = \frac{\text{Energy or Work}}{\text{Time}} \tag{17}$$

where the MKS units are joules per second or, more commonly, watts.

7. QUESTIONS

1. Explain why, by definition, momentum is a vector. Then illustrate by an example drawn from your own experience the directional nature of momentum.

2. If a golf ball is bounced against a solid wall and then a piece of putty of the same mass is thrown at the wall with the same velocity, which will exert the greater average force, if it is assumed that the "duration" of the impact is the same?

3. Two men are standing, one at each end of a cart equipped with frictionless wheels. They toss a heavy medicine ball back and forth. Describe the motion of the cart.

4. A mason jar with a fly perched in it is balanced on a very delicate analytical balance. What happens when the fly starts to fly around? Does the mason jar weigh more, or less?

5. From a consideration of the principle of conservation of energy alone, at what point in its elliptical orbit do you conclude the earth to be traveling at its maximum orbital speed? How does this explanation compare with the explanation based upon Kepler's second law? Which is more general?

6. An isolated system preserves angular as well as linear momentum. If the moon and the earth are considered an isolated system and the earth's rotation is slowed by tidal friction, what must happen to the moon's orbit?

7. A stone is dropped from the west side of a tall tower and then from the east side. Will the motion be identical?

8. How fast must you climb a flight of stairs 30 feet in height to develop 1 horsepower?

9. It is asserted that if a perfectly elastic collision were possible, a steel ball dropped from a height of 10 ft onto a large steel plate would rebound to exactly the same height. Explain why this is not *precisely* accurate.

10. What is the connection between the law of conservation of energy and the impossibility of constructing perpetual motion machines?

11. Look up definition of "coriolis forces" and relate to wind direction in a cyclone.

Problems

1. Contrast the results of the following: (a) a ball of clay is thrown at an object resting on ice (i.e., we assume frictionless conditions) and sticks to it; (b) a tennis ball, of the same mass is thrown at the same object with the same velocity and bounces backward. Show which will impart the greater velocity to the object.

2. Assuming the principle of conservation of energy, prove the equation $v^2 = 2gh$ for a body falling from rest from a height h. Neglect frictional resistance of air.

3. A mass is raised to a height h above ground level: first, by sliding it up a frictionless plane at 30°; second, by raising it to the same height vertically. Is the potential energy the same in each case? Explain why. Show that mgh is the potential energy in both cases by using the definition of work = force × distance. Also show that this must be true under the

assumption that all the potential energy is changed to kinetic energy if the mass is dropped vertically.

4. A simple pendulum is composed of a bob of mass 5 kg suspended from a string 2 m in length. It is released from a horizontal position (i.e., 2 m above the lowest point of swing). What is the velocity it will have at the lowest point? What is the tension (force) exerted by the string. (Don't forget gravity.)

5. A model T Ford is loaded with stones so that it has the same mass as a Mustang. If they both climb the same hill and travel over the same distance, the model T Ford takes ten minutes and the Mustang takes two. Disregarding frictional losses, is the energy expended the same? What is the difference in the two cases? Can the difference be expressed quantitatively on the basis of the data supplied?

6. Where does the momentum go when a billiard ball strikes a cushion of a billiard table and bounces straight back? Is not the change in momentum in the billiard ball offset, according to conservation laws, by a change of momentum elsewhere? Give an explanation with an approximate quantitative analysis. (*Hint:* Consider the *whole* system involved, including the earth.)

7. A certain rocket ship at launching weighs 50 long tons (that is, a mass of 50,000 kg). How many kilograms of gas must be exhausted per second at a velocity of 6000 mph relative to the rocket to give it an initial upward acceleration of 6 g? (The unit 6 g is close to the maximum acceleration that a human can withstand.) Remember that the weight of the rocket must be taken into account.

8. Show that no matter how hard you hit a golf ball, the velocity of the ball just after impact cannot be more than twice as great as the velocity of the clubhead just before impact. This is an exercise in using extreme conditions to obtain significant limiting results. Assume perfect elasticity. For ease of analysis, assume that the ratio of the masses is k. In your final equation for the ratio of velocities, let k become infinite (meaning the mass of the clubhead is extremely large compared with the mass of the ball), and see what happens.

9. What power is required to drive a 3300 lb automobile at a constant velocity of 13 m/sec (about 30 mph) up a road that rises 2 ft for every 100 ft along the road. Disregard friction.

10. An isolated system preserves angular as well as linear momentum. If the moon and the earth are considered an isolated system and the earth's rotation is slowed by tidal friction, show analytically that the moon's orbit must acquire an increasingly larger radius.

11. Outside the surface of the earth the gravitational force on a unit mass m is proportional to $\frac{M_e}{d^2}$ where M_e is the mass of the earth and d the distance of the unit mass from the earth's center. Do you expect the same law to hold true if the unit mass is beneath the surface of the earth? Explain. Note again the limiting conditions. If the law was universally true, what would happen to the force as d became close to zero (i.e., at the center of the earth)?

12. If a tunnel were bored through the earth from opposite points on the earth and evacuated, can you give a general description of how an object would move if it were dropped from either end of the tunnel?

13. The moon's mass is $\frac{1}{81}$ times that of the earth, its radius is $\frac{1}{4}$ that of the earth. Show that the escape velocity from the moon is about $\frac{1}{4.5}$ that of the earth, or about 2.38 km/sec (1.5 miles per second).

Recommended Reading

There are numerous elementary physics texts in which problems of Newtonian mechanics are worked out in detail. The selections below are listed in increasing order of difficulty, but all contain many illustrative problems without use of any advanced mathematics.

Physics, prepared by the Physical Science Study Committee, D. S. Heath, Boston, 1960. Chapters 20, 23, 24, and 25.

Eric M. Rogers, *Physics for the Inquiring Mind,* Princeton University Press, Princeton, N.J., 1960. This is an elegant volume, which abounds in illustrations. It was prepared specifically for nonscience majors.

Jay Orear, *Fundamental Physics,* Wiley, New York, 1961. Both this book and its companion by the same author, *Programmed Manual for Students of Fundamental Physics,* Wiley, New York, 1962, provide good examples and exercises on conservation laws. See, particularly, Chapters 3 and 5 of the latter.

chapter ten

Fluids and heat

We have watched the development of the basic concepts of science from Copernicus through Newton. In general, the phenomena studied have been on a scale that we can directly experience. It is true that the positions and motions of astronomical bodies have already forced us to use indirect methods of measurement. But where this has been done it has not been difficult to extend and enlarge ideas founded on everyday experience. We can discuss the solar system in terms of easily understood models.

We now begin a study that will take us into the realm of the submicroscopic, in which we discuss particles and their properties. These cannot be observed directly, but must be studied entirely through indirect evidence. And although by now we are familiar with the manner in which science departs from "common sense," through the study of the Copernican system, we shall find, as we go along, that the departure from common sense becomes more and more drastic, even to the point, in modern physics, where it is difficult to construct any models or analogies that both convey meaning and do justice to all the evidence.

We have been concerned only with that branch of physics called mechanics or the study of the motion of massive bodies acted on by forces. We have already seen that this study needs extension. In the first place, the law of conservation of mechanical energy breaks down whenever friction is present. Can the concept of energy be broadened so that its conservation is universally valid? Second, we have carefully restricted ourselves to "mechanical" forces, such as the force of gravitation or the forces arising when bodies are in physical contact. What about other forces such as those arising out of magnetism or electricity? How do we handle such apparently massless phenomena as light radiation?

In this chapter we make the first extension of the concept of energy, and shall end it by finding that *mechanical energy* can be converted into a different form called *heat energy* and vice versa by a specific conversion factor. This discovery was not made until about the middle of the nineteenth century. From now on the details of the historical development will not be pursued with the same thoroughness that has marked previous chapters, but we shall refer to and make use of the now discarded caloric theory of heat, both for the additional light it throws upon the nature of theories in science, and because the theory can be used to formulate some of the basic concepts in terms of which heat phenomena are explained.

1. THE STATES OF MATTER

Before undertaking a study of heat itself it is best to review some of the observed facts about the distinction and properties of solids, liquids, and gases. The idea that all matter is made up of atoms or small indivisible particles—a theory that can be traced back to ancient Greece—was undoubtedly developed, at least in part, by observing the penetrability and flow of liquids. The Greek philosopher Democritus asserted that liquids flow easily because the atoms composing the liquids are smooth, but the atoms composing solids have interlocking projections that cause them to hold together. Today, our explanation is on the basis of the intermolecular forces between the particles composing any material substance. In solids, the forces hold the particles together in definite patterns that cannot be readily disrupted although the individual particles (atoms or molecules) may vibrate about some central position. In gases, our present theory asserts that the particles are almost completely free of any mutually attracting forces. They move about at rapid speeds colliding with each other, bouncing away again, and moving apart as a result of their haphazard collisions until they once again collide with one another or with the containing wall. In the intermediate liquid state, which is not as well understood as either the solid or gaseous state, the particles are bound together by forces that are certainly greater than the forces between the particles composing a gas and less than those composing a solid.

The explanation of the three states of matter—solid, liquid, and gaseous—and of their distinctive properties depends on an adequate theory of heat. For this explanation, two basic definitions are required, namely, *density* and *pressure.* The density of a substance is defined as the mass per unit volume. The pressure exerted on a substance is defined as the force exerted per unit area taken perpendicular to the direction of the force. In fluids at rest (the term fluid embraces both liquids and gases in scientific terminology) pressures are of two kinds: (*1*) the pressure due to the weight of the fluid, each layer of the fluid having to "carry" the weight of all the layers above it, and (*2*) external pressures that are transmitted equally to all parts of the fluid. Thus, the pressure at the bottom of a column of liquid is the sum of the liquid pressure that is proportional to the depth plus the pressure on the surface of the liquid. The latter pressure is ordinarily due to the weight of the atmosphere above it. See Fig. 10.1.

The unit of density most commonly used in the metric system is a gram per cubic centimeter. Since water at 4°C has a density of 1 g/cc, and since specific gravity is the relative density of any substance compared to water, it follows that in the metric system density and specific gravity have the same numerical value. In the English system a common unit of density is pounds per cubic foot. Water has a density of approximately 62.5 lb/cubic ft.

The unit for pressure that we shall use most often is a newton per square meter in the metric system, and a pound per square inch in the English system. However, it is often found convenient to refer to atmospheric pressure by reference to barometric readings.

Although it had been known for a long time that air was a material substance, it was not until 1643 that Torricelli, with the invention of the barometer, proved that air has mass or weight. The simplest type of barometer is illustrated in Fig. 10.2.

Torricelli showed his barometer to Galileo and they were led to the hypothesis that the reason the column stood at the height it did was that there was a "sea of air" the weight of which pressed down on the surface of the mercury in the bowl, and that the pressure resulting from this weight of air was equal to the

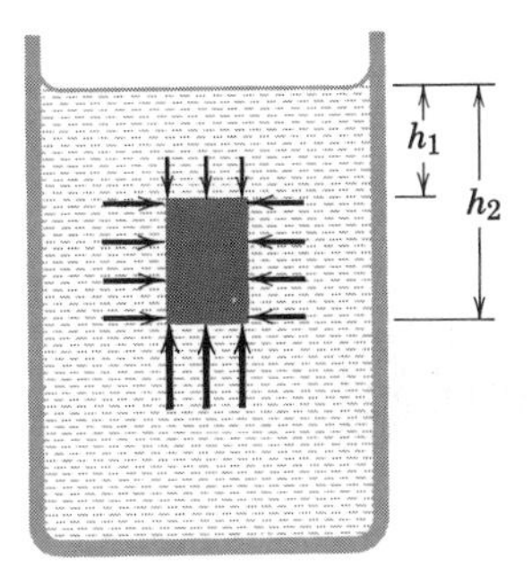

Fig. 10.1. Archimedes' principle. Liquids, such as water, being incompressible and unable to sustain shear stress, exert a pressure at all points and equally in all directions in proportion to depth. Difference in pressure at top and bottom of submerged object shown above is $(h_2 - h_1)dg$ where d is the mass density. (Horizontal pressures are equal and opposite at every level.) Difference in force on top and bottom is difference in pressure × cross-sectional area or area $(h_2 - h_1)dg$. This is equal to weight of fluid displaced. Simple analysis shows that the principle is independent of the shape of the submerged object.

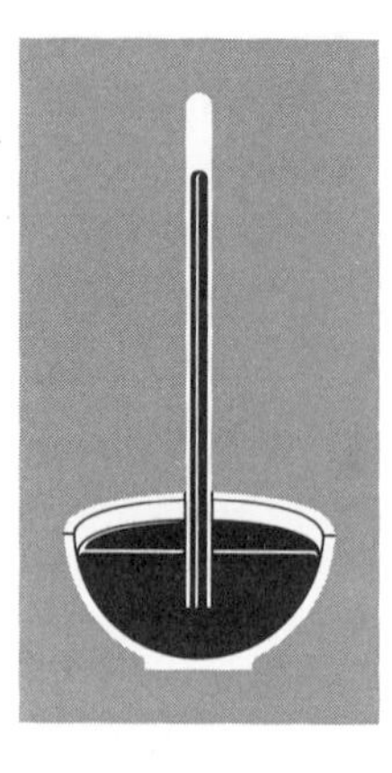

Fig. 10.2. A simple barometer. A glass tube, over 76 cm in length, is filled with mercury. It is then inverted into an open bowl of mercury. The mercury column then will fall to around 76 cm or 30 in.

pressure of the mercury column. To test this hypothesis, Torricelli took his barometer to the top of a mountain and, as the hypothesis suggested, the column of mercury dropped to a lower point, since the "depth" of the air above it and hence its weight was reduced.

It is readily seen that atmospheric pressures can be recorded in barometric units as well as in units of newtons per square meter or lb/sq in. When the mercury barometer stands at 76 cm, the pressure recorded is 15 lb/sq in. (Can you show that this is true? Try to find the pressure in newtons per square meter, using for the density of mercury 13.6 g/cm^3.)

2. HEAT AND TEMPERATURE

Although the use of heat from fires goes back to the earliest civilization, an understanding of what heat consists of and a statement of adequate quantitative laws describing heat phenomena is quite recent in the development of science. It turns out that the measurement of heat is surrounded by difficulties. Much easier is the measurement of what is called temperature. It is necessary to keep these two concepts, heat and temperature, clearly distinguished. Temperature, we shall find, merely gives an indication of the direction in which what we call heat will "flow"; it does not itself measure the amount of heat in a material body.

The idea of temperature arises directly from our sensations. An object is usually said to be "hot" or "cold" depending upon whether it feels hotter or colder than our own skin. From ancient times, doctors were helped in making diagnoses by judging whether a person felt warmer than normal. The implicit assumption made here is that there is a specific property of bodies that may be said to be equal. But how is such equality of this property, which we call temperature, to be judged accurately? We all know how difficult it is to judge equalities of temperature on the basis of our own personal sensation. Just as the primitive notion of degree of magnitude of forces, which arose out of muscular sensations, had to be refined by inventing instruments that could make accurate and unambiguous measurements of force, so also a method had to be devised for measuring degrees of hotness and coldness that would be independent of individual sensations.

The first such instrument was designed by Galileo in 1603. (Note again the broad range of interest of this man of genius.) He used the property of the expansion of gases when heated. The instrument[1] as sketched and described by one of Galileo's pupils, is illustrated in Fig. 10.3. As we can see, it looks somewhat like a barometer but, in this case, there is a large glass bulb filled with air instead of a thin column containing a vacuum.

Apparently the thermometer was used by doctors to determine whether a patient had a fever. If the air expanded beyond a certain point when the bulb was placed under the patient's arm, fever was indicated.

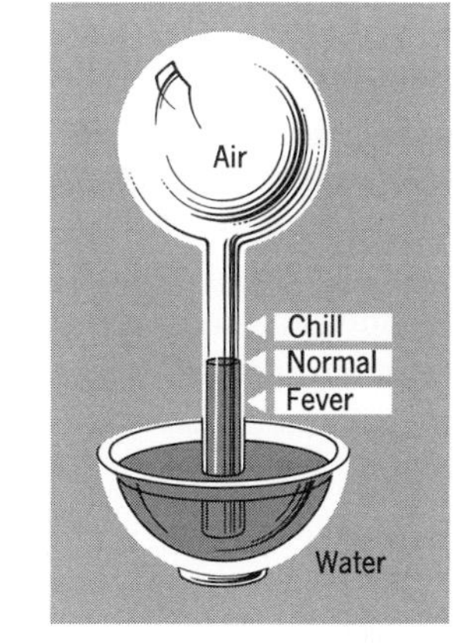

Fig. 10.3. An air thermometer used in Galileo's time.

Such a thermometer could yield only a rough indication of degree of temperature. For more accurate purposes a scale was required with two fixed points and a system of intermediate divisions. The two fixed points chosen for the centigrade scale are the temperature of melting ice and of boiling water. These were designated as 0° and 100° respectively, with the intermediate range divided into 100 parts.[2]

[1] Henry Crew, *Rise of Modern Physics,* Williams and Wilkins, Baltimore, 1935, p. 196.

[2] In the Fahrenheit scale, 0° was chosen as the point determined by melting ice in brine, 100° as the normal human temperature. Later, two points were standardized; 32° being the freezing point of water, and 212° the boiling point. There are thus 180 equal divisions or degrees between the two fixed points corresponding to 100° of the centigrade scale. Can you develop the conversion formula between degrees centigrade and degrees Fahrenheit?

The most common thermometers of today covering the ranges of temperature met on the surface of the earth are constructed by enclosing a liquid, usually mercury, in a glass tube. With an increase in temperature the liquid expands and rises. The question then is asked, "Does it matter what liquid is used?" The answer is yes. It is found that even if two fixed points, 0 and 100, are made to agree, when the distance between these points is divided into 100 equal parts, the thermometers using different liquids do not agree at these intermediate points (although the departure is often small). The problem thus becomes that of selecting one definite scale out of the enormous number of possibilities, corresponding to all possible liquids. Such a selection appears to be arbitrary because there is no way of determining for what liquid, if any, the expansion is proportional to the temperature, if temperature itself is to be defined in terms of the expansion of liquids.

Fortunately, it was found by the investigations started by Robert Boyle in the seventeenth century and carried on by his followers in the eighteenth century, particularly by Gay-Lussac and Jacques Charles, that for gases maintained at constant pressure, the rate of expansion is so nearly the same for all gases that an ideal temperature scale can be constructed.[3] This may be seen in Fig. 10.4, showing the variation of volume of any gas with temperature, the pressure being kept constant. It is found that for a given mass of gas a line drawn between the points representing the volume at two fixed temperature points (determined by boiling water and melting ice) will, when projected backward, cross the zero point of volume at −273°C.

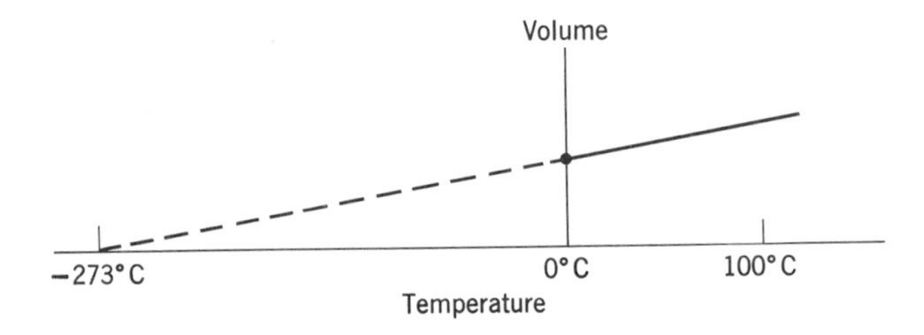

Fig. 10.4. Relationship between the volume of a given mass of gas and its temperature (pressure being kept constant). Absolute zero is taken as −273°C.

To make the absolute temperature scale as simple as possible, the point −273°, where the volume theoretically reaches zero, is chosen as absolute zero temperature.[4] The temperature of melting ice becomes 273° Absolute (referred to as

[3]Although most gases expand in almost identical fashion as temperature rises, particularly if the pressure is low, an even better scale of absolute temperature was devised by Lord Kelvin on the basis of an analysis of the efficiency of heat engines. This scale is completely independent of the properties of any substance, but is in essential agreement with the temperature scale based on gas laws.

[4]Note that the volume of any real gas never actually falls to zero. A change of state (condensation) takes place long before such a point is reached.

Fig. 10.5. Boyle's law, showing the inverse proportionality between pressure and volume of an ideal gas. The two graphs shown are equivalent. The advantage of the second straight-line graph is that the constant k can be read directly from the graph.

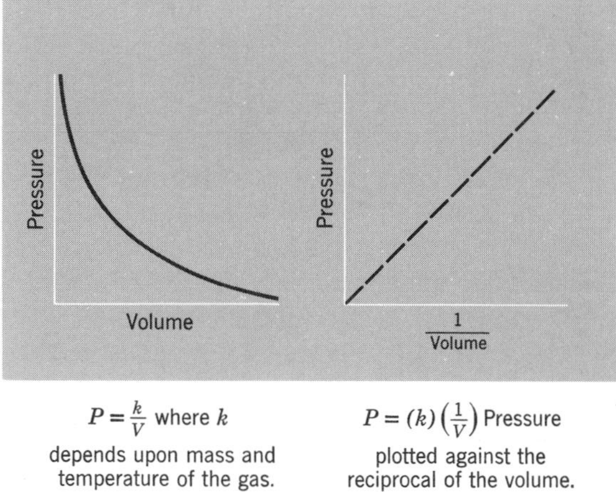

$P = \frac{k}{V}$ where k depends upon mass and temperature of the gas.

$P = (k)\left(\frac{1}{V}\right)$ Pressure plotted against the reciprocal of the volume.

273°K in honor of Kelvin) and the temperature of boiling water becomes 373°K. Gases on this scale expand in direct proportion to the absolute temperature, provided the pressure is kept constant. Expressed as an equation we have

$$V = cT$$

For pressure changes, Robert Boyle had already found another empirical law in 1662. This law, called Boyle's law, relates volumes and pressure changes in gases rather simply. At *constant* temperature the volume of a given mass of gas varies inversely with the pressure. See Fig. 10.5

$$P = \frac{c'}{V}$$

This law combined with the law for temperature changes, results in a simple combined law of gases

$$PV = kT$$

where k depends on the mass and composition of the gas. (This equation will later be replaced by the more general form, namely,

$$PV = NRT$$

where R is a *universal* constant, independent of the type of gas, and where N is a measure of the mass of gas under consideration, namely, the number of "moles"—an expression which will be explained later.)

The gas law, when first discovered and used, was an empirical one based on measurement and observation. It was not "explained" until a full analysis was carried out in the nineteenth century on the basis of the kinetic theory of gases. At that point an elegant synthesis was made between several separate branches of physics: mechanics, thermodynamics (which may be defined as the study of the mechanical aspect of heat phenomena), and atomic theory.

Having established an operational definition for the meaning of temperature, we turn next to the problem of defining *heat.* The distinction becomes clear if we consider a few examples that were studied and interpreted by Joseph Black in 1803.

First, it is found that a small flame, as from a match, may be enough to bring a very small quantity of water (a few drops) to a boil, but it will not raise the temperature of 1 kg of water by as much as 1 degree. Thus, the amount of heat in a substance must in some way depend upon its mass. A further experiment may be carried out by mixing, for example, 1 kg of water at 60°C with 1 kg of water at 30°C. It will be found that the resulting temperature of the final mixture will be 45°C. However, if 2 kg of water at 60°C is mixed with 1 kg of water at 30°C, the resulting temperature is 50°C. It is then apparent that the product of the mass times the temperature loss of the 2 kg of water is equal to the mass times the temperature rise of the 1 kg of water. This suggests the general law that the quantity of heat gained or lost is proportional to both mass and change of temperature. We may express the results of our experiment above in equation form

$$\text{Heat gained by 1 kg of water} = (m_1)(\Delta t_1) = (1)(50 - 30) = 20$$
$$\text{Heat lost by 2 kg of water} = (m_2)(\Delta t_2) = (2)(60 - 50) = 20$$

or

$$(m_1)(\Delta t_1) = (m_2)(\Delta t_2)$$

However, as Black found when two *different* substances, for instance mercury and water, are used in such an experiment, the results are not the same. In this case, it is found that if equal masses of mercury and of water, each at a different temperature, are brought together, the resulting temperature is not halfway between the two original temperatures but very much closer to that of the water. He concluded that the mercury has a much smaller capacity to absorb or to give up heat. For example, it will be found that if 2 kg of mercury at 61°C is mixed with 2 kg of water at 30°, the resulting temperature is not 45.5° (as it would be if both samples were water) but is only 31°C. In other words, when two different substances are used, we must modify the application of the general equation (that is, heat gained by one substance equals the heat lost by the other) by inserting proportionality constants that depend on the substances involved:

$$\text{Heat gained} = \text{heat lost}$$
$$C_1 m_1(\Delta t_1) = C_2 m_2(\Delta t_2)$$

(C_1 and C_2 now being proportionality constants). Applying this to the experiment of mixing mercury and water we get

$$C_1(2)(31 - 30) = C_2(2)(61 - 31)$$

which leads to

$$C_1 = 30C_2$$

Thus, we find that the proportionality factor, which measures the capacity of a substance to absorb or give up heat, is 30 times as great for water as for mercury. Now it is convenient to choose the proportionality factor for water as 1. For mercury, then, the factor becomes 0.033. This proportionality factor, which is a measure of the capacity (relative to water) of a substance to absorb or give up heat, is called the *specific heat* of a substance.

Furthermore, it is convenient to use water as a substance for the definition of a unit of heat. One Calorie (one kilocalorie), the unit of heat, is defined as the amount of heat required to raise 1 kg of water 1°C.[5]

We have now established both units of temperature, which define the *direction* in which heat flows, or the intensity of heat, and units of heat as a quantity. But we have not considered what heat actually consists of. The earliest carefully stated theory of heat, which was in agreement with the experimental findings of the day, was the caloric theory of heat.

3. CALORIC THEORY OF HEAT

Even though today we regard heat and electricity as forms of energy, we speak of them as if they were fluids that flowed from one point to another. It is not surprising to find that early investigators were led to believe that heat must be some kind of a fluid substance, differing from other material substances only in being imponderable, that is, without weight. Lavoisier in his *Elements of Chemistry*, published in 1789, which gives the first careful definition and listing of chemical elements, included "caloric" or heat fluid as an element.

Caloric was thought of as a form of substance that flowed from one body to another whenever they were at different temperatures. It was believed to consist of particles that mutually repelled one another but that were attracted to other material particles. Because of the mutual repulsion between the particles of caloric, there would be a natural tendency for them to disperse and separate

[5]The Calorie defined here is the large Calorie used also in dietetics as distinguished from the small calorie defined as the amount of heat required to raise 1 g of water 1°C. (Note the capital C for large Calorie, contrasting with small c used for the small calorie.)

from one another, that is, heat would naturally flow away from a point of greatest concentration. Furthermore, the fact that a gas tends to be warmer when compressed could be explained by saying that the caloric particles were now being forced into greater concentration causing an increase in the heat of the gas and also an increase in pressure because of the tendency of the particles to separate from one another under forces of mutual repulsion.

There is another phenomenon, closely associated with heat, which was studied by Black and was adequately explained on the basis of the caloric theory, involving changes of the state that substances undergo when heated or cooled. If water is slowly cooled, its temperature drops until it reaches 0°C, at which point the water begins to freeze, but the temperature of the water-ice mixture remains constant at 0° until all the water has been frozen. Black found that during this process of freezing, heat had to be taken out of the water at the rate of 80 Cal/kg. Similarly, this same amount of heat was required to melt 1 kg of ice at 0°C. Black explained that the heat absorbed during melting is not lost but remains latent in the liquid. He termed this quantity of heat the latent heat of fusion, or melting. A similar situation was found to hold for the change of state from liquid to vapor, and the latent heat of vaporization for water was determined to be 540 Cal/kg. (The latent heats of fusion and vaporization differ among various substances.)

To explain these transfers of heat during changes of state, it was postulated by the theory that the element caloric combined with other substances to form a chemically new compound. The element caloric remained constant in quantity but was latent when combined with other substances with no temperature change to be expected as the result of the combination.

It is interesting how well the caloric theory was able to account for the observed heat phenomena. One significant difficulty presented itself: if caloric was a substance, it should have mass. But the most careful weighing of the same mass of a hot or cold substance showed no difference in weight. Thus, supporters of the caloric theory were driven to the postulate that the fluid caloric was imponderable. To us this may appear strange, because it appears to be contrary to common sense and our general experience with material substances. However, proposing such a postulate was sound scientific theory construction. We should not be misled into thinking that the postulates of a theory must themselves be either in accord with experience or easily grasped by analogy. If from the postulates we can derive results that are in accord with observation, the theory remains verified, until such time as a deduction is found that is in contradiction with the facts. Many of the postulates of modern physics appear "unreasonable," but that in itself is no reason for abandoning those theories as long as they provide the most adequate basis for a systematic explanation or description of observed phenomena.

That heat was an imponderable fluid was made more plausible by the contemporary theories of electricity and magnetism in which other imponderable fluids were being postulated: electricity and magnetism were both regarded as fluids without weight for many years.

Another postulate of the caloric theory, that is, that the fluid was indestructible and could neither be created nor destroyed, was necessary to account for all the phenomena of heat. This constituted another conservation law: the law of conservation of caloric. This was the postulate that ultimately proved to be the vulnerable point in the whole theory.

4. HEAT AS A FORM OF ENERGY

It is a mistake to think of a theory of science as one that either is suddenly erected or, if it is later found inadequate, is suddenly abandoned. The whole process of theory construction, modification, and replacement is a slow one, often occupying many generations of investigators who are continuously making tentative surmises, expanding the scope of the theory, and modifying or replacing it as it is tested against observations and experiments suggested by the theory. So it was with the caloric theory of heat. Many scientists, among them Boyle and Newton, were inclined to consider that heat would ultimately be found explainable in terms of the motion of particles composing all matter. But the success of Black's work in expanding the caloric theory was so great that in time the caloric theory found general acceptance. And even after the work of Count Rumford, which we next describe and which brought strong evidence against the theory, it was not abandoned overnight. Indeed, for many purposes, we still find the concept of heat as a fluid a suitable model with which to describe heat phenomena, even though we no longer regard the theory as valid (just as navigators often use a modified form of the Ptolemaic theory as a more suitable model for navigation purposes than the more valid Copernican theory).

Count Rumford, a colorful figure in the annals of science, was a man who pursued a wide range of activities, but who had a continuing interest in the field of physics. At one point in his varied career he was supervising the boring of cannons at Munich and was surprised at the large amount of heat generated by friction. "If heat cannot be created," he asked himself, "whence can such a large amount of heat arise?" He was familiar with the explanation of the caloric theory that heat is generated by friction by "rubbing or pressing out" the caloric fluid that exists in all substances. But, in this case, the amount of heat appeared too large to be accounted for in this way. He therefore arranged an experiment to test the amount of heat that could be produced.

A hollow cylinder of iron with a very closely fitted iron plunger was surrounded by a water jacket in which a thermometer was placed. The cylinder was then turned by horsepower while the plunger was held stationary, causing a great deal of friction between the two, with the change in temperature noted. We quote from Rumford's essay[6] that was read before the Royal Society of London in 1798.

The result of this beautiful experiment was very striking, and the pleasure it afforded me amply repaid me for all the trouble I had had in contriving and arranging the complicated machinery used in making it. The cylinder had been in motion but a short time, when I perceived, by putting my hand into the water, and touching the outside of the cylinder, that heat was generated.

At the end of one hour the fluid, which weighed 18.77 lbs., or 2½ gallons, had its temperature raised 47 degrees, being now 107 degrees.

In thirty minutes more, or one hour and thirty minutes after the machinery had been set in motion, the heat of the water was 142 degrees.

At the end of two hours from the beginning, the temperature was 178 degrees.

At two hours and twenty minutes it was 200 degrees and at two hours and thirty minutes it ACTUALLY BOILED!

By meditating on the results of all these experiments, we are naturally brought to that great question which has so often been the subject of speculation among philosophers, namely, What is heat—is there any such thing as an *igneous fluid?* Is there anything that, with propriety, can be called caloric?

We have seen that a very considerable quantity of heat may be excited by the friction of two metallic surfaces, and given off in a constant stream of flux *in all directions,* without interruption or intermission, and without any signs of *diminution* or *exhaustion.* In reasoning on this subject we must not forget *that most remarkable circumstance,* that the source of the heat generated by friction in these experiments appeared evidently to be *inexhaustible.* (The italics are Rumford's.)

From our modern vantage point, it might appear that Rumford's work would have given the death blow to the caloric theory of heat, and that it would also provide a point of departure by relating heat to the amount of work required to produce it. However, it actually took over 50 more years for the conception that heat is another form of energy, convertible in definite amounts to mechanical energy and vice versa, to become generally accepted. A significant essay by Julius Mayer in 1842 asserted the general equivalence of all forms of energy, but it went without notice; and, at first, the series of experiments carried out by James Prescott Joule—which indicated that there was a constant ratio between the amount of work required to produce heat and the amount of heat produced (a ratio today called the mechanical equivalent of heat), in the compression of gases and in the stirring of liquids—were not given much attention.

However, by 1850, not only because of the many experiments by Joule and others, but also because of the increasing use of the steam engine, which obviously made use of chemical change (burning of fuel) and heat to obtain mechanical

[6]Quoted by John Tyndall, *Heat a Mode of Motion,* London, 1870, pp. 56–57.

work,[7] the scientific community was ready to accept the extension of the concept of energy to embrace other than mechanical phenomena. In 1850, Joule presented a summary of his conclusions before the Royal Society of London and these were now accepted as significant.[8]

In accordance with the pledge I gave the Royal Society some years ago, I have now the honour to present it with the results of the experiments I have made in order to determine the mechanical equivalent of heat with exactness. . . .

For a long time it had been a favorite hypothesis that heat consists of "a force or power belonging to bodies," but it was reserved for Count Rumford to make the first experiments decidely in favor of that view. . . .

"It appears to me," he remarks, "extremely difficult, if not quite impossible, to form any distinct idea of anything capable of being excited and communicated in the manner the heat was excited and communicated in these experiments, except it be motion."

One of the most important parts of Count Rumford's paper, though one to which little attention has hitherto been paid, is that in which he makes an estimate of the quantity of mechanical force required to produce a certain amount of heat. Referring to his third experiment, he remarks that the "total quantity of ice-cold water which with the heat actually generated by friction, and accumulated in $2^h\ 30^m$, might have been heated 180°, or made to boil, = 26.58 lb." In the next page he states that "the machinery used in the experiment could easily be carried round by the force of one horse (though, to render the work lighter, two horses were actually employed in doing it)." Now the power of a horse is estimated by *Watt* at 33,000 foot-pounds per minute, and therefore if continued for two hours and a half will amount to 4,950,000 foot-pounds, which, according to Count Rumford's experiment, will be equivalent to 26.58 lb of water raised 180°. Hence the heat required to raise a lb of water 1° will be equivalent to the force represented by 1034 foot-pounds. This result is not very widely different from that which I have deduced from my own experiments related in this paper, viz., 772 foot-pounds; and it must be observed that the excess of Count Rumford's equivalent is just such as might have been anticipated from the circumstance, which he himself mentions, that "no estimate was made of the heat accumulated in the wooden box, nor of that dispersed during the experiment. . . ."

In 1834 Dr. Faraday demonstrated the "Identity of the Chemical and Electrical Forces." This law, along with others subsequently discovered by that great man, showing the relations which subsist between magnetism, electricity, and light, have enabled him to advance the idea that the so-called imponderable bodies are merely the exponents of different forms of Force.[9]

[7] A patent for a steam engine was granted to James Watt, engineer and scientist, in 1769. Not, however, until the early part of the 19th century was the engine developed sufficiently for widespread use. Watt not only perfected the engine but conducted experiments to determine an appropriate unit of power or the rate of doing work. The term horsepower and its definition as being equivalent to 33,000 foot-pounds of work per minute results from his investigations. The unit watt, equivalent to one joule per second, is named in his honor.

[8] James P. Joule, "On the Mechanical Value of Heat," *Philosophical Transactions,* Vol. 140 London, 1850. Joule's paper can be found in *Great Experiments in Physics,* edited by Morris Shamos, Henry Holt, New York, 1959, with annotations and comments by the editor.

[9] Today we refer to forms of *energy* instead of forms of *forces.*

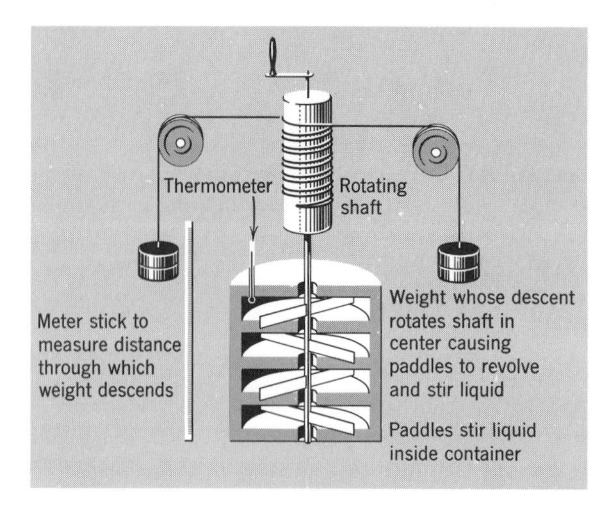

Fig. 10.6. Diagram of one piece of an apparatus used by Joule to determine the mechanical equivalent of heat by the stirring of liquids.

At the end of his paper, Joule summarized his quantitative findings, that is, that 772 foot-pounds of work are equivalent to the heat required to cause a rise of temperature of 1°F in 1 lb of water. The accepted modern value for the mechanical equivalent of heat is 4185 joules/Cal.[10] This constant conversion factor is usually symbolized by J in honor of Joule.

One piece of an apparatus used by Joule is shown in Fig. 10.6.

5. CONVERSION OF HEAT INTO MECHANICAL ENERGY

Joule's experiments were primarily directed at the investigation of the conversion of mechanical energy into heat. The possibility of the reverse transformation had already been clearly demonstrated by Watt's steam engine. However, it was soon found that although it is possible to convert mechanical energy completely into heat, no heat engine can ever recover more than a small portion of heat energy when transforming it into mechanical energy.

Every heat engine, including reciprocating steam engines, steam turbines, and internal combustion engines, utilizes the principle that gases expand when heated.[11] In the expansion that is allowed to take place at high pressure, the gas pushes against a moving piston or against the moving blades of a turbine, and work is performed while the gases give up heat during this expansion. Unfortunately, in this process, in addition to this partial conversion of heat into

[10] Recall that we have defined a joule as a newton-meter. It is equal to approximately $\frac{3}{4}$ foot-pound. We shall use the approximate conversion factor 4200 joules/Cal.

[11] More strictly the gases suffer a pressure-volume change according to the gas laws. In a heat engine, the gas is allowed to expand during the power stroke, and the pressure as well as temperature then falls.

mechanical energy, a large quantity of heat is inevitably lost to the surroundings, and is wasted. No heat engine can be designed that is more than about 40% efficient, the upper limit to the efficiency being determined by the difference of temperature of the hot gas and the temperature of the exhausted gas. Only if gases could be exhausted at absolute zero could a heat engine be 100% efficient.

A machine is a mechanism that transforms energy from one form into another. For the reasons stated above, *even in the absence of friction* a heat engine is inefficient. But friction is always present, and additional losses that convert mechanical energy into heat energy are also inevitable, however much they may be reduced to a low level through proper lubrication and well-designed bearings. The overall efficiency of any mechanism is measured by the ratio of the useful work output to the total energy input. For example, a modern automobile engine has an approximate efficiency of 20%. This means that of the total energy of gasoline that is in the neighborhood of 30,000 kcal/gal, only about ⅕ or 6000 kcal/gal is used in the useful mechanical work of transportation (the work in this case consisting of overcoming air resistance, road resistance, and other frictional resistances within the car—a total amount of work equal to roughly 25×10^6 joules/gal of gasoline).

That no device, no machine, can achieve 100% efficiency has long been known and expressed as the impossibility of perpetual motion. When, therefore, we speak of the law of conservation of energy (sometimes referred to as the first law of thermodynamics when heat energy is involved), we should always remember that although total energy is conserved, much of it is in a form that can never be recovered for useful purposes.

6. THE SIGNIFICANCE OF ENERGY AS A UNIFYING CONCEPT FOR ALL SCIENCES

Any good idea in science almost always appears to take on a life of its own. It grows vigorously in every possible direction. So it has been with the concept of energy. As soon as it was shown that heat could be regarded as a form of energy, the idea of energy was extended and generalized rapidly to include electrical, chemical, and light and sound energy, each with its own conversion factor. All events in nature, biological as well as physical and chemical, were at last found to involve one process in common, that is, energy transformations. Even though large gaps remain in the knowledge of the exact mechanism by which such transformations take place, at least a unifying principle and method of approach has been developed.

Of course, it is now common knowledge that both the concept of mass and that of energy as defined in the nineteenth century were too narrow. Since the turn of the century the concept of energy and the concept of mass have each been expanded and related to one another. We know today that mass and energy can be transformed into one another and that the laws of conservation of mass and conservation of energy have become one single and more general law, that of conservation of mass-energy. And, just as the transformation of one form of energy into others was a major preoccupation of scientists in the latter half of the nineteenth century, so in the second half of this century scientists are concentrating heavily on the transformation of mass into energy.

Our understanding of the universe has also been widely expanded. For example, the ancient puzzle as to the source of energy of the sun is now solved. The sun's heat which has remained very nearly constant over billions of years, is explained in terms of the conversion of its mass into energy according to the well-known equation of Einstein that energy may be produced from mass by a nuclear reaction, the quantity of energy being equal to the loss in mass times the square of the velocity of light, expressed in appropriate units. It is estimated, on this basis, that the tremendous outpouring of energy from the sun results in a daily loss of mass of about 2×10^{11} tons. (Because of the enormous mass of the sun, however, this is little cause for concern!)

7. ILLUSTRATIVE EXAMPLES OF THE APPLICATION OF PRINCIPLES

To make the discussion more specific, it is well to study three examples, which illustrate the application of one or more basic principles.

Example 1 A 0.5 kg mass of ice at 0°C is placed in a vessel containing 9 kg of mercury, which has a temperature of 60°C. The ice and mercury come to what equilibrium temperature when mixed together? (*Note.* If the problem is to be solved at all, as in so many similar cases in science, certain simplifying assumptions must be made. As the analysis becomes more sophisticated, these simplifying assumptions may be modified. In this case, we assume that the ice-mercury mixture is isolated, that is, no heat is gained or lost from or to the surroundings.)

The basic principle used is that total heat energy remains constant, or that heat gained by water (originally in the form of ice) is equal to heat lost by mercury. We use two constants

The latent heat of fusion of water = 80 kcal/kg
The specific heat of mercury = 0.033 kcal/kg °C

Heat necessary to melt ice = (.5)(80) kcal.
Heat necessary to raise water to final temperature t is $(0.5)(t - 0°)$ kcal.
Heat lost by mercury = $(9)(0.033)(60 - t)$ kcal.

Total heat remains constant. Therefore, heat gained by ice and water equals the heat lost by the mercury, or

$$(0.5)(80) + (0.5)(t - 0) = (9)(0.033)(60 - t)$$

from which the final temperature t is readily determined.

Example 2 An automobile weighing 3000 lb traveling at 60 mph (approximately 27 m/sec) is brought to complete rest by braking it. What is the total heat in Calories generated by the over-all friction and the application of the brakes?

The principle used is that since all the kinetic energy is converted into heat, the total kinetic energy is first calculated and then its equivalent in Calorie units of heat is determined.

$$\text{Kinetic energy} = \tfrac{1}{2}mv^2 = \tfrac{1}{2}\left(\frac{3000}{2.2}\right)(27)^2 \text{ joules}$$

using $J = 4200$ joules/Cal, the total heat becomes

$$\left(\frac{1}{2}\right)\left(\frac{3000}{2.2}\right)(27)^2\left(\frac{1}{4200}\right) \text{ or about 120 Cal}$$

Example 3 Assuming Niagara Falls to be about 250 m in height, how many degrees warmer will the water be at the bottom of the falls than at the top? (Joule proposed a similar problem and, *on his honeymoon!* visited a waterfall and tested its temperature to confirm his expectations.)

In this problem we do not appear to be given enough data. How can we calculate energy, which is obviously involved, without knowing the mass? Often in such cases, it is enough to assume a given mass M and, in the analysis, the factor M is then found to drop out. Thus

$$\text{Potential energy of a mass, } M\text{, at top of the fall} = (M)(g)(250) = 2500\, M \text{ joules}$$

Assuming this energy is converted to kinetic energy as the mass descends and then converted to heat at the bottom, we have

$$2500\, M \text{ joules} = \frac{2500\, M}{4200} = 0.6\, M \text{ Cal}$$

0.6 M Cal will raise the temperature of a mass M by 0.6°C.

8. QUESTIONS

1. Does Archimedes' principle apply to: (a) bodies floating in water, (b) bodies floating in calm air (e.g., balloons), (c) bodies submerged in water?

2. The product of pressure of a gas times change in volume may be expressed in units of: (a) force, (b) energy, (c) momentum, (d) energy per unit volume, (e) none of these.

3. In a fluid pressure varies directly with depth. Discuss why this same relationship does not hold for air.

4. To how great a depth must one descend in water to double the pressure found at the surface?

5. Is a temperature measurement cardinal or ordinal?

6. An iron ship floats in a lake. The sea cocks are opened, and the boat sinks. Will the water level of the lake go up, down, or remain the same? Verify your answer by a simple experiment.

7. Another classical problem in physics is the following. Canals sometimes have been carried across roads by bridges. A barge weighing 50,000 lb floats across the bridge. As it passes across, approximately how much additional weight must be carried by the bridge?

8. Estimate roughly your own volume in terms of: (a) cubic feet, (b) cubic meters, (c) cubic centimeters. (*Hint.* Use Archimedes' principle and your experience in floating.)

9. One form of the general law is given as

$$\frac{P_1V_1}{T_1} = \frac{P_2V_2}{T_2}$$

Explain why pressure and volume units can be arbitrary (as long as they are used consistently on both sides of the equation) whereas the temperature scale used must be absolute or Kelvin scale.

10. Trace the energy of each of the following through its transformations from its source in the sun: (a) electricity generated by a power station, (b) a waterfall, (c) water lifted to the top of a tall tower through a windmill pump.

11. Two metal springs are dissolved in two identical acid baths, one of the springs first having been compressed, the other not compressed. What happens to the potential energy of the first spring? How could you verify your prediction? Make some reasonable assumptions and some rough calculations as to the apparatus required.

12. Which contains more heat, 1 kg of steam at 100°C, or 1 kg of water at 100°C?

Problems

1. From the fact that mercury has a mass density of 13.6 grams per cubic centimeter, and assuming that a barometric reading stands at 0.76 meter, show that atmospheric pressure is approximately 10^5 newtons per square meter. Convert this to pounds per square inch.

2. Calculate the maximum height to which a suction pump can lift a column of water?

3. Estimate the pressure in pounds per square inch to which a submarine is subject at a depth of 8000 feet. (Take the density of sea water as 64 lbs per cubic foot.)

4. There is a simple linear equation expressing the relationship between centigrade and Fahrenheit temperature. Develop the equation from the following assumptions: (*1*) boiling point of water is 100°C and 212°F; (*2*) freezing point of water is 0°C and 32°F; the equation is linear, of the form $y = mx + b$ where y is Fahrenheit, x centigrade temperature, and m and b are constants to be determined.

5. An iron kettle containing 1 quart of water at 20°C (normal room temperature) is placed on a stove and brought to a simmer. The kettle weighs 1 lb. How many Calories of heat are required? The specific heat of iron is 0.11. Use the approximation that 1 pint of water weighs 1 lb.

6. A pendulum with a bob of mass 1 kg and length 1 m is set into oscillation with an initial amplitude of 30°. After 600 oscillations it comes to rest because of friction. How many Calories of heat have been given off to the surroundings?

7. If a car weighing 2200 lb is brought to rest from 60 miles an hour in 2 seconds by the application of brakes, how much heat must be dissipated?

8. If a cube 10 cm on a side with specific gravity of 8.0 is floated on mercury, how much of it will appear above the surface?

9. A girl's spiked heel has a diameter of ¼ in. As she walks across a rug, all her weight (120 lb), is, for a moment, concentrated on that heel. Calculate the pressure in tons per square inch. (If you can estimate the weight of an elephant, and the size of his foot, make the obvious comparison.)

10. Estimate the mass of the earth's atmosphere, using your general knowledge.

11. A lead bullet (specific heat 0.03) of mass 20 g strikes a stone wall at 1000 m/sec. If the collision is completely inelastic (that is, the bullet does not ricochet from the wall), and if 80% of the energy is dissipated in heating the bullet, will the bullet melt? (The melting point of lead is 327°C.)

12. Using your general knowledge (or looking up the data), make a very rough estimate of the number of Calories required to keep a person's heart beating for 24 hours.

13. Show that a kilowatt-hour is a unit of energy, and determine its equivalence in Calories and joules.

Recommended Reading

Gerald Holton and Duane Roller's *Foundations of Modern Physical Science,* Addison-Wesley, Reading, Mass., 1958, contains much historical material as well as an excellent treatment of the concepts of heat and the kinetic theory.

Eric Rogers, *Physics for the Inquiring Mind,* Princeton University Press, Princeton, 1960. This book, to which reference has been made before, contains the principles of heat and kinetic theory and many illustrations of the application of the principles.

John Tyndall, *Heat as a Mode of Motion,* Longman's Green & Co., London, 1863. See the comment in the suggested reading list for Chapter 11.

Hans Thirring, *Energy for Man,* Harper & Row, New York, 1958. For those interested in a survey of the energy resources—from windmills to nuclear reactors. It summarizes the historical development of power production and relates it to broad social problems.

Morton Mott-Smith, *The Concept of Energy Simply Explained,* Dover Press, New York, 1964. A nontechnical, largely historical, review of thermodynamics. For an elementary, nonmathematical exposition of the main topics of heat and energy, it is a clear and readable introduction.

chapter eleven

Heat as a mode of motion

A great step forward was made when it was discovered that mechanical energy could be transformed into heat energy and vice versa, but there still remained much to be explained. What is the process of the transformation? What is the nature of heat if it is not a fluid? Can the empirical equations of the gas laws be explained in terms of more fundamental conceptions?

One of the great triumphs of nineteenth century science was the finding of answers to these questions by an extension of the fundamental laws of Newtonian physics. This was accomplished largely by the development of a new and powerful means of analysis, the statistical method. In addition, as the mathematical analysis was extended, more physical facts were discovered regarding the fundamental particles, molecules, and atoms, which the chemists were investigating during the same period. Success was so overwhelming that by the end of the nineteenth century, scientists developed great confidence with respect to the "reality" of these particles and, perhaps of greater consequence, felt a new fearlessness in "inventing" hypothetical entities of all kinds, which were not directly observed but were useful in explaining what could be observed in or out of the laboratory.

Francis Bacon, as early as the sixteenth century, had considered that heat might be thought of as a motion of the small particles composing material substances.[1] Certainly by the time of Joule's work, the possibility that heat consisted of the kinetic energy of molecules was "in the air," and occupied the thought of many scientists. Indeed, a qualitative explanation for the law of gases had already been reached: the molecules of a gas were conceived of as miniature spheres, all in rapid motion, colliding with each other and with the containing walls, and rebounding in each collision with no diminution of energy (such collisions are defined as "elastic," that is, no mechanical energy is lost). Pressure was interpreted as the sum of the forces caused by the change of momentum undergone by each molecule as it struck the wall and rebounded. That the pressure was not irregular was explained by the existence of myriads of minute molecules striking the walls successively with great rapidity. Furthermore, the temperature of a gas was thought to result from the energies of the molecules composing it. As the temperature increased, the energies, therefore the velocities, of the molecules increased, causing an increased change of momentum at each collision with the containing wall and, thus, a change of pressure. To sum up, a good qualitative description of all the phenomena summarized in the gas laws appeared at hand. Finally, this descriptive picture was beautifully confirmed by a more direct type of evidence discovered in 1827 by the botanist Robert Brown. As he was examining under the microscope some small particles suspended in a solution, he noted that the particles were in constant irregular motion. The same phenomenon (called Brownian movement) can be observed if particles of smoke suspended in air are illuminated and observed through a microscope. The particles seemingly are engaged in an endless dance. At first, and certainly to Brown, explanation was lacking, but with the development of the kinetic theory of gases the cause of this motion could be surmised. The small particles observed were being bombarded by very much smaller and invisible molecules traveling at high velocities. The momentum of the small molecules was sufficient to transmit a visible velocity to the suspended particles. The Brownian movement thus provided an early and very plausible confirmation of the kinetic theory.

But pleasing as such a descriptive explanation might be, there remained the problem of developing a quantitative kinetic theory of heat that would relate pressure and temperature to the concepts of Newtonian physics. To accomplish

[1]Bacon's language is interesting: "Heat is a motion, expansive, restrained, and acting in its strife upon the smaller particles of bodies." Also: "Whether the particles of a body work inward or outward, the mode of action causing heat is the same." *Novum Organum,* second book, Spedding's translation, D. C. Heath, London, 1857.

this, a new viewpoint was required, one incorporating probability and statistical considerations. This analysis was carried forward and developed by both mathematicians and physicists in the latter half of the nineteenth century.[2]

1. STATISTICAL ANALYSIS IN PHYSICS

The new face that modern physics presents to the world has many profiles. We have hinted at a few. (*1*) "Laws" are no longer regarded as "absolute truths" as they once were, but rather as postulates whose function it is to organize our present knowledge of nature and to suggest further problems for investigation. (*2*) Scientific concepts are no longer based upon *a priori* or intuitive ideas but rather upon a community of agreement determined by operational definitions, (*3*) Value judgments of what "ought to be" are distinguished from judgments of "what is." (*4*) Scientific theory may depart considerably from "common sense." *Visualized* models are no longer always necessary, although they remain helpful and suggestive.

Another aspect of modern science that has become more and more important is the statistical approach to predictions, which replaces the more deterministic approach of the past. Here, we encounter some problems in which both science and philosophy meet. At many points questions of a fundamental nature arise, still unsettled, and about which there is controversy among the theoretical scientists and among the critics of science.

Modern physics and chemistry, especially in that branch called quantum mechanics, makes an extended and effective use of the concept of probability. In many cases it is asserted that although one cannot predict the occurrence of a single event (for example, the time during which one atom of radium will disintegrate into its daughter elements), one can safely make certain statistical predictions (for instance, that out of a billion atoms of radium a given number will disintegrate within a year). Although there are novel elements in the modern application of probability to physical phenomena (such as the "uncertainty principle," which will be discussed in a later chapter), the most basic ideas of the statistical approach can be traced back to the application of the mathematical and statistical analysis of heat phenomena, particularly the law of gases.

It should be clear that Newtonian physics can hardly be applied in its

[2]Clausius' work in 1858 is generally considered the first satisfactory statement of the kinetic theory, although an elementary statement of this type of theory appeared in a work on hydrodynamics by the mathematician, J. Bernoulli, in 1738. The work of Clausius was extended rapidly by C. Maxwell, L. Boltzmann, and J. W. Gibbs. Gibbs is the outstanding American scientist of the nineteenth century.

original form to any very large assemblage of particles such as are assumed to exist in gases. The laws of motion may indeed be valid for each individual molecule, but if there are billions of such molecules all in various conditions of motion, it is impossible even for the modern computer to cope with and find solutions for the vast number of equations of motion involved, even if it were possible to collect and record all the relevant data applying to each particle (including all positions and velocities). The solutions can be found in principle, but not in practice.

Although we cannot trace the individual motions of the molecules, the very fact that the numbers involved are so large enables us to find statistical regularities upon which we can depend, just as life insurance companies can profitably predict that out of a very large number of persons a certain number will die at every age, although no prediction can be made regarding the life span of any single individual. Thus, statistical determinism replaces the older determinism applied to individual entities. Laws and regularities are discovered that apply to aggregates, even if it remains impossible to determine the precise future of any one of the individuals making up the aggregate. Therefore, necessity and chance, each within limits, appear to rule over the affairs of both man and nature.

Statistical theory had its beginning in the seventeenth century. To offset the temptation to believe that all scientific knowledge originated and developed from purest motive, it is only necessary to cite the somewhat disreputable origin of probability theory. In 1654, Chevalier de Mere, an aristocrat addicted to gambling, proposed to the mathematician Pascal a problem involving probabilities occurring in certain games of chance. Pascal became intrigued with the mathematical analysis. He and his friend Fermat worked on the problem and founded the theory of probability upon which modern statistics is largely based. At the same time that Pascal and Fermat were concerning themselves with the mathematics of probability, the earliest serious analyses of social statistics were being undertaken.

The dual origin of probability theory and statistics—analysis of games of chance and the study of data such as frequencies of death—brings out the twofold nature of probability and statistics, which has come down to us today. There is, on the one hand, a mathematical calculus of probability and, on the other, a method of measurement that assigns probability on the basis of frequency. The relationship between the two aspects of probability theory is not always clear, and controversy has continued to rage between those who have emphasized one or the other as being more fundamental. We shall not go into the details of the controversy, but shall merely point out some of the useful and well-established results of both approaches, especially as they can be applied to the kinetic theory of heat.

The earlier mathematical investigators, among them Laplace, whose work on celestial mechanics has already been mentioned, took the definition of probability from games of chance: the probability of a specified event is the ratio of the number of ways that a specified or "favorable" event can take place to the total number of all *equally possible* events that can occur. For example, to draw an ace of hearts out of a deck of 52 cards yields a probability of $1/52$, it being assumed that there is an *equal possibility* of drawing any one of the 52 cards. It does not take much acumen to realize that this definition is circular, for it implies that we have already established the "equal possibility" of drawing any one of the 52 cards. To determine this we require a test based upon frequency (or a considerable experience of analogous situations, which ultimately depends upon some frequency test).

The mathematical analysis of probability, however incapable it was of establishing an effective definition of probability for one event, did yield a calculus of *combining* probabilities, which is used in all applications, whether probability is defined on the basis of frequency or by some other method. (This is not surprising, and is analogous to all scientific application of mathematics: we do not get from mathematics our definition of the measurement of time or distance, but once the definition has been established these terms can be combined in various ways by the application of mathematical analysis.)

The important axioms of the calculus of probability are:

1. Probability can range from 0 to 1 (0 indicating impossibility, 1 indicating certainty, that is, the sum of all the probabilities of a set of independent events = 1).
2. The probablity that two independent events will occur is the product of their individual probabilities.
3. The probability that one or the other or both of two independent events will occur is the sum of their probabilities minus the probability that both occur.

Whatever method is used to assign individual probabilities, the above axioms of the calculus are always used in statistical analysis.[3] (Examples can most easily be cited from games of chance. For example, if the probability of any single number of one die appearing on any one throw is assumed to be $1/6$, the probability of two sixes being thrown in two throws is $1/6 \times 1/6 = 1/36$. The probability that either a one or a six is thrown in a single throw is $1/6 + 1/6 = 1/3$. Note here that the probability of both a six and a one being thrown on a single throw is zero.)

The most common empirical method of assigning a probability is to use the frequency of the event. If, after examining enough cases—for example, of the incidence of measles among American school children—we find that in any given year one child out of five gets the measles, we assert that the probability

[3]This is not strictly the case. Von Mises' definition of probability as a limit of relative frequency in a collective modifies the axiom that certainty = 1.

of getting the measles is 0.2. (The objections and difficulties of defining probability on this basis are undoubtedly realized by one who has been exposed to the many discussions of sampling procedures, but we shall not consider these objections.) Suppose, similarly, that the probability of the same child getting a cold during the year is 0.5. The axioms of the calculus of probability would now tell us that the probability of a child getting both the measles and a cold is 0.1, and that the probability of getting either a cold or the measles or both is $0.7 - 0.1 = 0.6$. We see, in this example, one way in which the mathematical method is applied to combinations of probabilities, each of which is measured by frequency.

Of greater interest, for our purposes, is the development of the concept of a normal distribution curve. The normal curve arose out of the observation that usually a series of measurements of any property in an aggregate of individual entities tends to cluster about a central value and fall off rapidly in both directions. Gauss and Bernoulli, making certain mathematical assumptions, were able to define a specific form of this curve, which is found to fit a very wide variety of phenomena.[4]

To understand the meaning and significance of this curve, let us take a specific example. Look at the way in which the height of 10,000 adult males of the United States might be distributed. On the basis of experience, we would expect the number of men having various heights to be represented by a bell-shaped curve such as is shown in Fig. 11.1.

[4]The assumptions may be stated as follows. (*1*) The curve has a maximum point that is the arithmetic mean of all values. (*2*) There is equal probability that a measurement will be greater or less, by a given amount, than the arithmetic mean. (*3*) The probability of any measurement becomes less as the departure from the mean increases. (*4*) The three axioms of the calculus of probability, already stated. From these assumptions the normal, or Gaussian curve of error, that is, $y = (N/\sigma\sqrt{2\pi})e^{-(x^2/2\sigma^2)}$ is derived, σ represents the standard deviation.

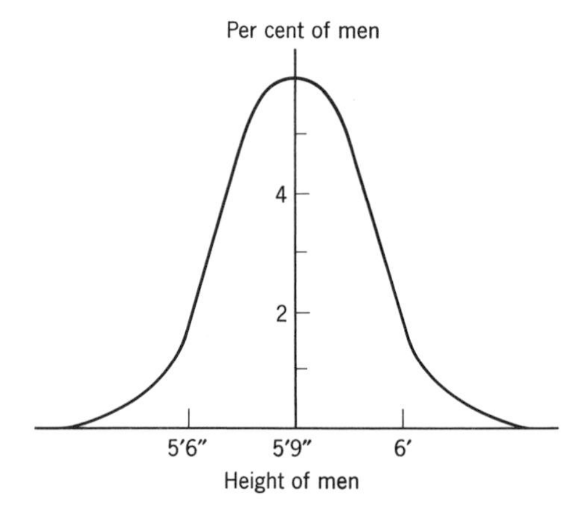

Fig. 11.1. Hypothetical distribution of the height of 10,000 men (normal or Gaussian curve).

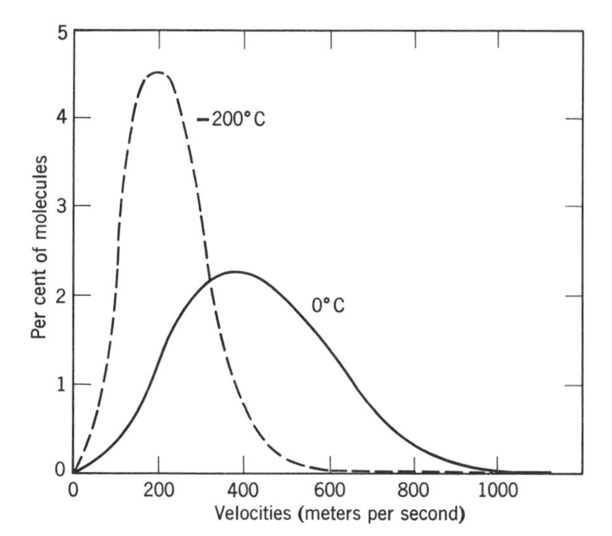

Fig. 11.2. The Maxwell-Boltzmann distribution of velocities of molecules.

An extraordinarily wide variety of phenomena fall into a distribution pattern such as the one shown in the figure. The curve may, of course, vary considerably in its sharpness. The more sharply it rises to its maximum the more closely we expect any given measurement to approximate the average value. The relative width of the curve thus can be used to obtain an estimate of the error involved in finding an average on the basis of a large series of readings. But we shall not pursue these details further.[5]

The application of a modified normal curve to the velocities of the molecules of a gas was carried out by Maxwell and Boltzmann in the latter half of the nineteenth century. This resulted in the reduction of the gas laws to a statistical explanation in terms of Newtonian concepts, as we shall see in the next section. Maxwell's analysis showed that the distribution of the velocities can be represented by a curve that differs slightly from the normal curve, primarily in that the curve does not extend out to the left in a manner symmetric to the right-hand side. This is obviously because molecules cannot have a negative velocity. The curve in Fig. 11.2 shows the Maxwell-Boltzmann distribution of velocities at a given temperature.

In this distribution curve, molecules subject to random collisions are divided into groups, each group moving within a small velocity range. The percentage of each group moving at various velocities is shown by the vertical height. The maximum percentage marks the most probable velocity, and there are very few molecules that move at more than three times this velocity. (If this were not the

[5]Those interested can refer to any text on statistics for details, such as John Kenney's *Mathematics of Statistics,* Van Nostrand, New York, 1950.

case the atmosphere of the earth would long ago have disappeared, since the velocities would have been greater than the so-called escape velocity—the escape velocity on earth is about 7 miles/sec; any object traveling at this velocity would escape permanently from the earth's gravitational field.)

It is interesting that if any mass of gas is suddenly disturbed so that its distribution is momentarily other than that shown, after a time the Maxwell distribution will again be reached, usually however with a different maximum point. This tendency for the molecules to end up in the most probable velocity distribution is often referred to as *entropy*. We shall consider its nature in Section 4, after we have completed the mathematical analysis of the kinetic theory of heat.

2. MATHEMATICAL STATEMENT OF THE KINETIC THEORY OF HEAT

It is readily apparent that the kinetic theory of gases provides a good qualitative basis for explaining many phenomena such as changes in volume, pressure, and temperature: compression or rarefaction occurs when the gas molecules move closer together or farther apart; pressure rises or drops as the number of collisions increases or decreases; and temperature changes occur when the average velocity of the molecules change. But to be really satisfactory the theory must be established on such a mathematical basis that all the quantitative gas laws can be derived from the theory.

The assumptions required for the simplest form of the theory are as follows.

1. The small particles (molecules) composing a gas each have the same mass and move in a completely random fashion, that is, as many are moving, on the average, in any one direction as in any other with the same random velocity distribution in each direction.

2. The molecules themselves are very small compared to the distance between them, and as a first approximation the molecules are therefore regarded as having negligible dimensions.

3. The forces of attraction or repulsion between the molecules are also so small as to be neglected. The only interaction between the molecules that is considered arises out of collisions between them.

4. The collisions between the molecules either with one another or with the containing walls are elastic, that is, no kinetic energy is lost in any collision. (If this were not true, the motion of the molecules would eventually cease as the energy was dissipated.)

5. The gravitational force of the earth acting on the molecules is assumed to have a negligible effect on the randomness of the motion of the molecules.

From these five postulates many conclusions can be drawn that apply to what is termed an "ideal gas." The ideal-gas laws thus derived are closely approximated by actual gases, particularly where the range of pressure is not high. It is possible to modify some of the assumptions by removing certain restrictions (such as the assumption that the space occupied by the molecules is negligible) and then obtain, by derivation, modified gas laws, which are even better approximations to the actual behavior of all gases over all ranges of temperature and pressure. However, here, we shall present the analysis and derivations based only upon the first approximations contained in the assumptions made above.

First, we calculate the pressure exerted on one face of a cubic box caused by the successive collisions of a *single* molecule traveling back and forth between the faces of the box, as in Fig. 11.3.

In this figure, the velocity $\vec{v}$ is a vector having three perpendicular components, which we designate as $\vec{v_x}$, $\vec{v_y}$, and $\vec{v_z}$. From vector analysis[6] we can write $\overrightarrow{v^2} = \overrightarrow{v_x^2} + \overrightarrow{v_y^2} + \overrightarrow{v_z^2}$. (Also, by hypothesis, in the case of many molecules the *average* velocities are the same in all directions. It can be shown, therefore, that we can write an equation for the *averages* of the velocities as follows:

$$\overline{v^2} = \overline{v_x^2} + \overline{v_y^2} + \overline{v_z^2} = 3\overline{v_x^2} \tag{1}$$

where the lines above the velocity symbols indicate that averages are being used.)

Hence, for randomly moving molecules, the average of the squares of the velocities in any one direction is one third the average of the squares of all the velocities or

$$\overline{v_x^2} = \tfrac{1}{3}\overline{v^2} \tag{2}$$

Now analyze the momentum change that the molecule undergoes upon striking the right-hand face of the box at each collision:

$$mv_x - m(-v_x) = 2mv_x \tag{3}$$

[6]Remember that the Pythagorean theorem applies to three as well as two dimensions. The diagonal of a rectangular box of dimensions a, b, and c will be given by $d^2 = a^2 + b^2 + c^2$, and the vector equation shown above similarly follows.

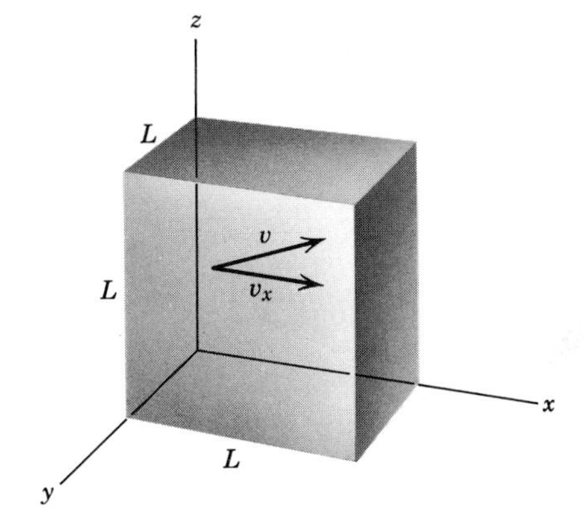

Fig. 11.3. A molecule traveling in a cubical box with velocity v. The components of the velocity are v_x, v_y, and v_z.

(remembering that momentum is a vector, and only the component of the momentum in the x direction will give rise to a force on the right-hand face of the box).

Next, apply Newton's law of momentum, which states that the *average force* exerted during a time interval Δt is equal to the change of momentum divided by the time interval during which the momentum is changed:

$$F = \frac{2mv_x}{\Delta t} \tag{4}$$

Take as the time interval the time between successive collisions on the right-hand face, which will be the time it takes for the molecule to travel across the box and back or the distance $2L$. Therefore

$$\Delta t = \frac{2L}{v_x} \tag{5}$$

from which

$$F = \frac{2mv_x}{\Delta t} = \frac{2mv_x}{2L/v_x} = \frac{mv_x^2}{L} \tag{6}$$

But the pressure exerted is the force divided by the area:

$$P = \frac{F}{A} = \frac{F}{L^2} \tag{7}$$

Thus, the average pressure resulting from the collision of one molecule on the right-hand face of the box will be

$$P = \frac{mv_x^2}{L^3} \tag{8}$$

But L^3 is the volume V of the box. Therefore

$$P = \frac{mv_x^2}{V} \tag{9}$$

If we now consider what happens when we have n molecules in the box, the average pressure becomes the sum of the following n terms,

$$P = \frac{m}{V}\left(v_{x_1}{}^2 + v_{x_2}{}^2 + v_{x_3}{}^2 + \cdots + v_{x_n}{}^2\right) \tag{10}$$

Using the concept of the *average of the squares* of all the n velocities, we may rewrite the equation as

$$P = \frac{mn\overline{v_x{}^2}}{V} \quad \text{(the line above the squared velocity indicating an average)}$$

But since all the velocities are randomly distributed as to directions, we can write for $\overline{v_x^2} = \frac{1}{3}\overline{v^2}$ and the equation finally becomes

$$PV = \frac{nm\overline{v^2}}{3} \tag{11}$$

Now this is quite a striking result, both because of the form of the equation that recalls the empirical gas law equation $PV = KT$, and because of the expression $m\overline{v^2}$ that obviously appears related to a kind of average kinetic energy, which would have the form $\frac{1}{2}m\overline{v^2}$. Let us multiply and divide the right-hand side of the equation by 2, in order to make the relationship appear explicitly:

$$PV = (\tfrac{2}{3})(n)(\tfrac{1}{2}m\overline{v^2}) \quad \text{or} \quad PV = (\tfrac{2}{3})(n) \times (\text{the average kinetic energy}) \tag{12}$$

As a final step, we identify the average kinetic energy of the gas as proportional to the absolute temperature:

$$PV = nkT \tag{13}$$

where k is a proportionality constant known as Boltzmann's constant, which is expressed in units of energy per degree of temperature.

When this step is taken we have achieved a formal definition of temperature in terms of the average kinetic energy of particles of a gas under standard conditions of pressure. If an independent determination can be made of the average kinetic energy, then a definite value of k can be found. Evaluation of k and n becomes the next topic for our attention.

3. CONSEQUENCES OF THE KINETIC THEORY; AVOGADRO'S NUMBER; VELOCITY OF MOLECULES

The kinetic theory as outlined is confirmed by the usual phenomena relating to gases, and embraced under the empirical gas laws discussed in the previous chapter. This is not surprising because the theory was constructed to explain those phenomena. In addition, however, the theory leads to many other consequences and predictions that can be tested by experiment. Again we see an example of how a good theory can guide further experimentation and search. One of the more significant results derived from the kinetic theory is called Avogadro's law, which asserts that the number of molecules in a given volume of any gas, under the same conditions of temperature and pressure, will be the same.[7]

This conclusion follows from the last equation of the previous section. If the volumes, pressures, and temperatures of two gases are the same, then the equa-

[7]Avogadro actually surmised that this was true on the basis of chemical evidence, so that the prediction on the basis of kinetic theory may be regarded as a strong confirmation of the hypothesis.

tion implies that there must be an equal number of molecules present in each volume of gas.[8] We shall shortly examine the method whereby the actual number of molecules for a given volume of gas can be determined. What began as a qualitative surmise that temperature is related to the average energy of motion of the particles of a gas now takes on a definite quantitative formulation.

Since equal volumes of any two gases, under the same conditions, contain the same number of molecules, it also follows that the ratio of the weight of the two volumes will be equal to the ratio of the weights of the single molecules of the two gases. This fact enables chemists to construct a table of the relative molecular weights (or masses) of any two substances that can be put into the form of a gas. Hydrogen is the least massive of the gases, and it is convenient to express all molecular weights in terms of ratios to the weight of the molecule of hydrogen. However, instead of taking the weight of hydrogen as one, it is actually taken as two, since it has been determined that a molecule of gaseous hydrogen is composed of two smaller particles called atoms, closely bound together by interatomic forces. The molecular weight of hydrogen being taken as two (the atomic weight, as we shall later see, being taken as one), the molecular weight of oxygen is, for example, found to have the value 32, which merely indicates that a single molecule of oxygen has a mass 16 times as great as a molecule of hydrogen gas.

A convenient unit of mass of any substance is the gram-molecular mass sometimes called a mole which is *that mass of any substance, expressed in grams, equal to its molecular weight.* Thus, for example, 1 gram-molecular mass of hydrogen is equal to 2 g of hydrogen, and 1 gram-molecular mass of oxygen is equal to 32 g of oxygen. Experimentally, 1 gram-molecular mass of any gas under standard pressure and temperature conditions occupies a volume of 22.4 liters. (Standard pressure is normal atmospheric pressure. Standard temperature is 0°C.)

The number of particles in 22.4 liters of any gas is called a mole, and is designated as Avogadro's number. It is an extremely large number and has been determined as about 6×10^{23}. (*Note.* 10^{18} would represent a billion times a billion.) The same value has been reached by many independent investigations, and may be regarded as a firmly established constant of physics and chemistry.

One elegant method of determining Avogadro's number was carried out by Jean Perrin in 1909. He applied the hypothesis that all particles in a mixture at the same temperature will have the same average kinetic energy as the energy of the moving particles suspended in a liquid (we recalled that the Brownian

[8]Allowance must, of course, be made for statistical fluctuations. It should not be forgotten that the results obtained require the use of statistical assumptions and, hence, one should not expect perfect or exact agreement at all times. The statistics of large numbers, however, indicate that the percentage fluctuations are extremely small—far under even $1/100$ of 1%, if the volume of gas being considered is as much as 1 cc at standard pressure.

movement is that endless dance of suspended particles caused by their being bombarded continuously by the molecules of the fluid in which they are suspended). Perrin carefully prepared a suspension of very small particles of gambose (a kind of gum, appropriate for the experiment) in a known quantity of water. Through a microscope the movement of the particles was carefully observed and measured. A good estimate of the average mass of the particles was easily obtained by counting the number that occupied a known volume, and by using the total weight of the gambose in suspension. The determination of the average velocity of the particles, needed for finding their average kinetic energy, required a more indirect approach and the application of statistical analysis.

The statistical generalization, which Perrin used, comes from an analysis classically referred to as the "random walk" problem. It can be shown that any object that moves in a series of short movements or steps of average length L, such that at the end of each step the direction abruptly changes in a random fashion, will move a total distance D from its starting point after N steps, according to a statistical equation: the most probable value[9] of D will be equal to $L\sqrt{N}$. Using this equation and measuring the time it took for particles to travel over different distances, Perrin was able to determine the average velocity of the particles. Knowing the average mass and the average velocities, the average kinetic energy could then be calculated. At 0°C, this is found to be approximately 5.65×10^{-21} joules. See Fig. 11.4.

If the kinetic theory is correct, this average kinetic energy should also be the average kinetic energy of any particle in any fluid (gas or liquid) which is at this same temperature. Therefore, the molecules of all gases should have an average kinetic energy of 5.65×10^{-21} joules, at 273°K. We can substitute this value for the average kinetic energy per particle in equation (12):

$$PV = (2/3)(n)(1/2 m\overline{v^2})$$

If we use 22.4 liters for the volume of the gas, and the standard atmospheric pressure at 273°K, the number n can be readily calculated:

$$n = \frac{3}{2}\frac{P_0 V}{5.65 \times 10^{-21}}$$

where P_0 is the standard atmospheric pressure and V is the volume of one mole, both expressed in appropriate units.

The number n, or Avogadro's number, is determined as 6×10^{23}. (See if you can carry out this calculation in detail. You will find it helpful in the handling of units as well as a review of the use of powers of 10. P_0 is atmospheric pressure, which must be expressed in the correct units. One liter is 10^3 cc.)

[9] An amusing derivation and discussion of this famous problem is given in George Gamow's *One, Two, Three—Infinity,* Viking Press, New York, 1948, p. 199 ff.

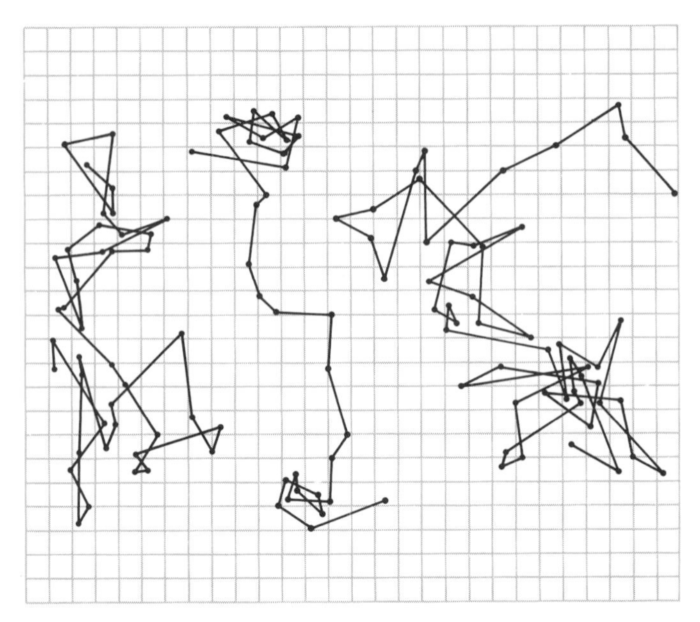

Fig. 11.4. A reproduction of Jean Perrin's diagram of Brownian movement. Microscopic study of this and similar timed plots of the movements of particles suspended in solution enabled Perrin to obtain a value for Avogadro's number. In Perrin's words: "Three diagrams are shown, the scale being such that sixteen divisions represent 50 microns." (A micron is defined as 10^{-6} meter.) "The consecutive positions occupied by the same mastic grain (radius equal to .53N) were marked every 30 seconds." Perrin from the velocities thus obtained computed N_0 as 6.4×10^{23} (compared with modern value of 6.02×10^{23}). Figure taken from *Atoms* by Jean Perrin, Constable & Co., London, 1923, p. 115.

Once Avogadro's number has been determined, many other values follow. For example, it is possible to determine the mass of any molecule merely by dividing the mass of 1 mole by Avogadro's number. In this way, 1 molecule of hydrogen is found to have a mass of about 3.3×10^{-24} g, and an atom of hydrogen to have a mass of one-half this amount.

Many other consequences have been drawn from the kinetic theory of heat, including a quantitative explanation of specific heats, but we give only one other example, selected for its interest and simplicity. This is the determination of the average velocities at which the molecules of gases travel at different temperatures.

The equation, derived in Section 2, which leads most simply to the calculation of velocity is

$$PV = \frac{1}{3}nm\overline{v^2}$$

Since the velocity term in this equation is the average of the squares of the velocities, we may, by taking its square root, find the value of an average velocity, which is called the root-mean-square average. (If a frequency distribution is symmetric, the arithmetic mean, the most probable value, and the root mean square are identical. Since the velocity distribution curve is not quite symmetric, the root-mean-square velocity is slightly higher than both the arithmetic mean and the most probable value.) Solving the equation for the root-mean-square velocity, we arrive at

$$\bar{v} = \sqrt{\frac{3PV}{nm}}$$

Since nm is the product of the number of molecules times the mass of each molecule, it represents the total mass of the gas. We can measure the values of P and V for any given mass at a particular temperature and, hence, the average velocity can be calculated for that temperature.

Some typical values for these average velocities at room temperature (20°C) are

Hydrogen: 18×10^2 meters/sec (over 1 mile/sec)
Helium: 13×10^2 meters/sec
Oxygen: 4.6×10^2 meters/sec
Water vapor: 6.2×10^2 meters/sec

(Can you calculate these approximate values? Try it by using appropriate values for a mole of gas at standard pressure and temperature.)

The significance of figures for velocities lies, at least in part, in the fact that they are subject to an experimental check. Experimental measurements of velocities of many substances have been carried out, and the results predicted by the kinetic theory have been confirmed. Confirmations such as these, along many different lines, have given to the kinetic theory of heat the status of an established scientific generalization. But remember that the verification is indirect, since it is quite impossible to confirm the assumptions or postulates of the theory by direct observation.

4. ENTROPY AND THE SECOND LAW OF THERMODYNAMICS

The concept of entropy, already referred to in the first section of this chapter, is one that cannot be precisely defined except in technical and rather complex terms. However, it is a concept of such importance, and one so often referred to, that it is well to gain a sense of its meaning and significance. Entropy is closely related to the ideas of order and disorder when these are defined in terms of the probability of particular statistical distributions of assemblages. Some examples will make this clear. If one layer of fine white sand is placed in a large jar and on top of it another layer of fine black sand, and if the jar is then tumbled about, one correctly expects the black and white sand to become thoroughly mixed in time, and the color of the mixture to turn grayish. And, no matter how much shaking we do, it is most unlikely that we can recapture the original arrangement of the two separated layers of black and white sand. Why is this? The statistical answer is that there are many millions of ways more in which the grains of sand may be intermingled than the number of ways in which they can be found in two distinct layers. The original orderly arrangement has become disordered in the process of shaking (which gives the grains of sand random motion); and the less probable distribution has been supplanted by a more probable one. We do not say that further shaking *might* not in a long enough period of time reproduce once again the original ordered arrangement in two distinct layers; we only say that this is *highly* improbable. The greater the number of grains of sand in the assemblage, the more improbable it becomes that the original order can be recaptured. (Now consider a smaller assemblage: a deck of cards which, when bought, is arranged in suits, and then becomes disordered in the process of shuffling. The chance of reproducing the original order through further shuffling, although small, is still large enough for it to occur occasionally. Ever so often there are reports of a bridge hand being dealt in which each person holds 13 cards of the same suit.)

In considering assemblages of molecules, we refer to extremely large numbers. The number of molecules in 1 cc of gas is over 10^{19}, which is a number larger than all the grains of sand on all the beaches of the world. If we connect two jars—one a vacuum and one filled with a gas—the gas will quickly fill the empty jar through the random motion of the molecules of the gas. Theoretically, because the random motion continues, it is possible that all the molecules might at some future time return to the original jar, leaving the other a vacuum once again. There is nothing in Newtonian dynamics that would prevent such a possibility. But the statistical probability against it happening is so overwhelmingly large that it is regarded as impossible. (A calculation of the time it would take for such an improbable occurrence is many, many billions of times longer than the billions of years of the universe's existence.)

The measure of this tendency for large assemblages of molecules in random motion to go from a less probable configuration (an orderly arrangement) to a more probable configuration (more disorderly) is called entropy. We may, therefore, *define entropy* as a measure of disorder. The tendency of all physical changes to go in the direction of a more probable configuration until a state of equilibrium is reached that corresponds to the maximum probability is called the *second law of thermodynamics*. Stated another way, it says that any spontaneous change in a system of particles (that is, any change that takes place in the system without a gain or loss of energy to the outside) will be in the direction of increasing entropy and that entropy will be maximized when equilibrium is reached.

It is the operation of this famous law that determines the direction in which heat flows. Two isolated bodies in contact, which are originally at different temperatures, will ultimately reach the same equilibrium temperature, at which point the energies of the molecules composing the two bodies will have a distribution of maximum probability. (Only very minute and normally undetectable fluctuations from this maximum will occur.) Because of this law, the vast amount of heat energy in the ocean, for example, is unavailable. Only if we could arrange for this heat energy to flow from the ocean could we obtain useful energy, and this would be possible only if we could allow it to flow to some other body at a lower temperature.

In this connection, the idea of the "heat death" of the universe, about which philosophers and scientists sometimes speculate, may be mentioned. It has been argued that because of the apparently inviolable second law of thermodynamics the universe must inexorably run down to a position of maximum entropy in which all physical bodies are at an even dead level of energy equilibrium. When this time comes, many billions of years hence, nothing new or significant can occur because there will be no energy differentiation anywhere in the universe. Impressively scientific and depressing as such a prognosis may seem, it is actually a rather naive speculation. We know such a little portion of the universe and have observed it for such a minute period of time that extrapolating our knowledge to this extent hardly seems warranted except as a flight of speculative fancy. Nothing can either be "proved" scientifically or asserted (except hypothetically) about the universe as a whole on the basis of our limited and finite knowledge. It is true that, as far as we know, the second law of thermodynamics has never been violated. It remains an empirical generalization, a useful directive to further scientific investigation, yet it is no more of an absolute certainty than is the law of conservation of matter which was so widely accepted a brief generation or two ago. It hardly seems wise to base a philosophic outlook on the meaning of life upon the generalities of contemporary science, no matter how well founded these may appear to be at the moment.

An additional comment may be made regarding the second law of thermodynamics as it is related to biological phenomena. Sometimes it is alleged that the second law of thermodynamics—the tendency toward statistical disorder as measured by increasing entropy—is suspended in the case of living organisms. Such a conclusion is, however, scarcely warranted. All we can say is that *locally* because of the extraordinary stability of highly complex molecules found in living cells, particularly the genetic molecules, orderliness is maintained and then extended by extracting, so to speak, "order" from the surrounding environment. ("Order" as used here refers not only to structural complexity but also to energy-level differences. If two bodies are at different energy levels or at different temperatures they constitute a more orderly system, in this case, than as if they are at the same temperature.) On an over-all basis, the second law of thermodynamics is not suspended in living processes, any more than it is in the case of a refrigerator that produces ice (a more ordered structure than water). In both cases, the total entropy, if we include the entropy of the environment, is increased. This principle is sometimes stated by saying that a living organism feeds upon the "negative entropy" of its surroundings, even while the over-all entropy of organism plus environment increases. An organism, from this point of view, is an island of locally decreasing entropy. It should be added, however, that although the laws of thermodynamics do not appear to be suspended in biological phenomena, this does not imply that the peculiar ability of living organisms to maintain and even increase order and complexity on a local level can be "explained" on the basis of the laws of physics.

5. THE SIGNIFICANCE OF THE STATISTICAL APPROACH

From one point of view the results of the analysis pursued in this chapter appear to be merely an extension of Newtonian dynamics. Thermodynamics, which covers all processes involving heat, can be reduced to the same principles of energy that appear in the earlier physics, provided use is made of statistical laws that were absent in Newtonian dynamics. However, from another point of view, it is apparent that a new approach is introduced. Although Newtonian dynamics still provides a basis for making exact and determinate predictions with respect to aggregates containing large numbers of particles, it is impossible to make exact predictions regarding the movement of the individual particles composing those aggregates. Our predictions, in other words, are based upon the statistical laws pertaining to large numbers. Thus, a wedge has been opened in what might be called the determinism of the Newtonian system. This opening is not wide, since *theoretically,* at least, it would be possible, if we could determine the position, mass, and velocity of every particle in a system of particles, to predict the future configurations of those particles, using myriads of equations. In theory, then, Newtonian determinism is still valid—even if it has no operational meaning because of the impossibility of making all the measurements necessary.

This new approach, the statistical one, has received a profound extension in modern physics with the development of quantum theory. In quantum mechanics, as we shall see, it is no longer *even theoretically* possible to measure both the position and the velocity of the smallest particles at the same time. Therefore, even the theoretically possible application of Newtonian dynamics to particles is swept away, and statistical laws replace the dynamical laws of classical physics in the explanation of events on the subatomic level. The opening wedge provided by the statistical methods developed in the nineteenth century has been widened significantly in our own century.

6. QUESTIONS

1. Explain from the kinetic theory of heat, why evaporation cools a liquid.

2. What are some of the reasons why the ideal gas law is not strictly obeyed, particularly at high pressures?

3. If a perfumer is interested in speed of response at a distance, should he select a scent with a high or a low molecular weight?

4. Does doubling the absolute temperature of a gas lead to a doubling of the average velocity of the molecules if the same pressure is maintained?

5. Ordinary thermometers will melt at very high temperatures. What meaning can be given to the term temperature when we talk of the temperature of the sun's gases as being thousands of degrees?

6. In a mixture of gases, would you expect the average kinetic energy of molecules of the different gases to be the same? Would their velocities be the same?

7. In separating the two isotopes of uranium (elements identical in chemical properties, but of different atomic weights) a gaseous diffusion process is one method used. A gaseous compound of the mixed uranium istotopes is made to flow through porous material. Explain why the lighter gas would flow through with a selectively higher velocity.

8. Air, unlike water, is compressible. Show why the pressure of the atmosphere is therefore not proportional to the depth of atmosphere. Draw a curve representing the general shape of a graph of pressure against height above sea-level.

9. Do you expect the root-mean-square velocity of a gas to be higher or lower than the arithmetic mean velocity?

10. Viscosity is a measure of the frictional resistance of a fluid to motion. The viscosity of gases in general increases with temperature. Can you think of an explanation in terms of the kinetic theory?

11. What is the explanation for the total lack of atmosphere on the moon?

12. The average kinetic energy of translation of all gases, monatomic, diatomic, and triatomic is proportional to the absolute temperature. But the specific heats of monatomic, diatomic, triatomic gases differ. What reason can be given for this difference? (*Hint:* Consider types of possible kinetic energy other than those due to the translational velocity of the molecules as a whole.)

Problems

1. If the general gas law is written $PV = NRT$ where N is the number of moles, R is a universal gas constant, and T is the absolute temperature, in what units is R expressed?

2. The gas law can also be expressed as $PV = nkT$ where n is the number of molecules, k is then known as Boltzmann's constant. What are the units of k? What is the relationship between k and R? How is k related to the average kinetic energy?

3. Are there more or fewer molecules in a gram of steam than in a gram of oxygen? What is the number of atoms of hydrogen in a gram of water?

4. By approximately how much does water expand when it turns into steam at 373°K and one atmosphere? (Recall that the density of water is 1 gram/cm^3 and one atmosphere is 10^5 newtons/m^2.

5. When gases, such as air are a mixture of elements, then the total pressure is the sum of the partial pressures. Show why this follows from the kinetic theory of gases.

6. Derive the convenient equation $P_0V_0/T_0 = P_1V_1/T_1$ (where the subscript 0 refers to the original state of the gas and 1 to the final state) from the general gas law. Then explain why this equation holds regardless of the units in which pressure and volume are expressed (provided they are the same on both sides of the equation), but that the temperature *must* be expressed in Kelvin (absolute) degrees.

7. A spherical bubble of air rises from the bottom of a pond 5 meters below the surface. If the radius of the bubble is originally 1 cm, what is the radius as it breaks out of the water?

8. Consider the equation $u = \sqrt{3PV/nm}$ where u is the average velocity of the molecule of gas, P the pressure, V the volume, n the number of molecules, and m the mass of the molecule. If we take a liter (1000 cubic cm) of one gas and compare it to a liter of another gas at the same pressure and temperature, the only quantity that differs in the equation is nm, which is equivalent to the mass of a liter of the gas. A liter of oxygen weighs 16 times as much as a liter of hydrogen. How does the average velocity of an oxygen molecule compare with that of a hydrogen molecule?

9. At normal pressure and temperature, one liter of hydrogen has a mass of approximately 9×10^{-2} g. Calculate the approximate average molecular velocity of hydrogen. *Remember to use the correct units in the equation.*

10. Perrin, from his studies of the Brownian movement, determined that the average kinetic energy of a molecule of hydrogen was 5.65×10^{-21} joule. Using the data and results of the preceding problem, determine the approximate mass of one molecule of hydrogen. (*Note.* As we shall see, one *molecule* of gaseous hydrogen is composed of two *atoms,* hence one *atom* will have one-half this mass.)

11. Using the results of Problem 6, calculate the approximate number of molecules in a liter of hydrogen at standard conditions. How many molecules will there be in 2 g of hydrogen gas (which will occupy 22.4 liters under standard conditions). This is "Avogadro's number," which we shall refer to again and again. Can you show, on the basis of kinetic theory, that it is the same number as the number of molecules in 32 g of oxygen (which also occupies 22.4 liters)?

12. Many of the closer artificial satellites circle the earth in about 90 minutes. Com-

pare the speed with which they travel with the average speed of a molecule of air at room temperature (use 300°K). Since nitrogen is the major constituent of air you may take the molecular mass of air to be 28.

13. By modern vacuum pumps pressures as low as 10^{-11} newtons/m^2 are obtainable. What fraction of normal atmospheric pressure does this represent? In such a vacuum how many molecules are there in a cubic meter? In a cubic centimeter? Compare these figures with the number of molecules in interstellar space, where the vacuum is still "more complete." When the density of the "dust" reaches 10 atoms per cubic centimeter, it appears as a dense cloud through the telescope, because of the scattering of light.

Recommended Reading

John Tyndall, *Heat as a Mode of Motion,* Longmans Green & Co., London, 1863. Although written a century ago, this still remains a very readable and entertaining book, based on lectures given by the author for popular audiences.

Gerald Holton and Duane Roller, *Foundations of Modern Physical Science,* Addison-Wesley, Reading, Mass., 1958.

George Gamow, *One, Two, Three—Infinity,* Viking Press, New York, 1948. Chapter VIII has a delightful summary of the ideas behind the kinetic theory. The chapter on the "Law of Disorder" treats of the concept of entropy.

Albert Einstein and Leopold Infeld's, *The Evolution of Physics,* Simon & Schuster, New York, 1942, contains a good nonmathematical treatment of much of the material of Chapters X and XI of this text.

Henry Margenau, *The Nature of Physical Reality,* McGraw-Hill, New York, 1950. Chapter 13, entitled "Probability," is an excellent review of the relationship between the mathematical definition of probability and the frequency definition.

Norbert Wiener, *The Human Use of Human Beings,* Houghton Mifflin, Boston, 1950. The connection between entropy and information theory is handled in an elementary but illuminating manner.

Erwin Shrödinger, *What Is Life?* Macmillan, New York, 1946. A physicist examines biological phenomena, particularly as related to the second law of thermodynamics and quantum theory.

Niels Bohr, *Atomic Physics and Human Knowledge,* Science Editions, Inc., New York, 1961. This book, particularly the chapter on "Biology and Atomic Physics," is quite meaningful to those acquainted with modern atomic physics.

Stanley W. Angrist and Loren G. Helper, *Order and Chaos,* Basic Books, Inc., New York, 1967. For those interested in a nontechnical nonmathematical discussion of entropy and its relationship to other laws of physics and chemistry, this amusing and instructive book will be found rewarding.

D. K. C. MacDonald, *Near Zero,* Doubleday, Garden City, N.Y., 1961. A survey in an elementary fashion of the major recent developments of low temperature physics which has been omitted from our own discussion of heat.

chapter twelve

Molecules and atoms chemistry before the Bohr atom

Although the development of physics was dependent upon the discovery of the appropriate organizing principles and theories, it would be a mistake to assume that all branches of science originate in theoretical speculation. In particular, the growth of chemistry as a science was preceded by many centuries in which knowledge was slowly accumulated by skilled artisans and technologists. They had perfected many complex processes with little guidance by anything resembling a scientific theory.

As far back as 2000 B.C. the Egyptians and Mesopotamians had mastered the applied chemistry required for the working of gold, silver, lead, copper, iron, and bronze. Glass factories had been developed. Dyes of great variety had been prepared, necessitating the bringing of dyes into solution by the chemical process called reduction. The fixing of dyes by use of metallic compounds was known. Skilled artisanship and a high degree of competence in chemical processing had been developed, the knowledge being passed down from generation to generation.[1]

By the time Alexandria was founded in 322 B.C. by Alexander the Great as a port at the mouth of the Nile, this artisanship of the Egyptians had been brought to a high degree of perfection. Within the next two centuries Alexandria became the greatest seat of trade, learning, and culture of the ancient world. It was a meeting place of the ancient civilizations: Greek, Semitic, Egyptian, and Arabic.

It was inevitable that the Greek speculative philosophers with their theory of the atoms and of the four elements should try to apply these theories to the chemical processes practiced by the Egyptians. Here, once again, the problem of what is invariant in a world of change and flux presents itself. If *everything* changes in time, how can there be any orderliness in the universe, since the rules of order are subject to change? However, if there is a permanent order in the universe, how can change be accounted for? We have already examined physical changes that are accounted for in terms of the locations and movements of mass particles. But the changes into which we now inquire are those that apparently involve much more than mere location in time and space: in a chemical process the properties of color, taste, smell, and texture are all involved. What regularities, if any, are involved in such changes?

One of the earliest theories, traceable to the Egyptians (1500 B.C.) but refined by Aristotle and the Greeks, was that all matter is composed basically of four elements: earth, water, air, and fire. The element fire, combining with wood, can cause visible changes in the properties of wood, giving off smoke (a kind of gas or air), causing drops of moisture (water), and leaving an ash (earth). The forms of matter were thus subject to change or transmutation, most often accomplished by the action of the element fire. These four elements, themselves, remain unchanged, the apparent and visible changes arising out of the many possible mixtures and separation of the elements. According to Greek thinking:

> Birth (that is, change) properly speaking, does not occur in anything mortal nor does any mortal end in a corrupting death. Only commingling takes place and the separation of the commingled.[2]

From this point of view, matter is permanent and unchanging, and only the body or outward forms change and undergo transmutation. An ancient Egyptian-Greek symbol of the cosmos as a serpent swallowing its own tail

[1] J. R. Partington's *A Short History of Chemistry,* Macmillan, London, 1937, reprinted as a Harper Torchbook, Harper, New York, 1960, gives a brief but interesting account of the origins of chemistry. Partington accepts the idea proposed by Berthelot that the term "chemistry" originally stemmed from a reference to the Egyptian art, or the art of the black earth "Khem," found in Egypt.

[2] H. Diels, *Fragmente der Vorsocratiker,* Berlin, 1934, quoted by Andrew G. Melsen, *From Atomos to Atom,* Harper, New York, 1960, p. 23.

with the enclosed words "all is one" symbolizes the ultimate unity of matter, and expresses the old doctrine of a substratum of permanent reality that could be operated on in a way that resulted in only apparent change.

Out of this background of Greek speculation and Egyptian artisanship arose the promise of finding quick methods of causing transmutations of matter, that is, the hope of finding the "philosopher's stone," which could cause desirable changes in metals, and the "elixir of life," which could produce desirable changes in human bodies. In this way, there arose what appears to us to be the false theory of alchemy—naive, compounded of elements of mysticism and magic, which motivated the widespread experiments carried on by the Moslems, who conquered Alexandria in 640 A.D. For several centuries these Arabic alchemists preserved and extended the experimental knowledge of chemistry, as well as mathematics and medicine, and finally, once again transmitted this vast lore to Europe, especially through Spain. Without these techniques, the classification of the substances, and the exhaustive study of reactions by the alchemists, the modern science of chemistry would hardly have come into being in the seventeenth century.

1. THE BIRTH OF MODERN CHEMISTRY; DEFINITION OF AN ELEMENT

In 1661, Robert Boyle, after whom one of the gas laws is named, published a book with the significant title *The Sceptical Chymist.* In it, Boyle expresses his dissatisfaction with the ancient doctrine of the elements because of the traditional inability to specify clearly any empirical method of determining what the elements are. He says,

> I have not yet, either in Aristotle, or any other writer, met any genuine and sufficient diagnostic for discriminating and limiting the species of things.[3]

Boyle defined as an element "those primitive and simple or perfectly unmingled bodies; which not being made of any other bodies . . . are the ingredients of which (all other bodies) are immediately compounded." But Boyle took only a tentative first step toward an empirically precise definition of a chemical element. He gave no prescription for distinguishing elements from compounds and provided no list of such elements. These steps awaited the work of the great French scientist, Antoine Lavoisier.

Lavoisier, political liberal, social reformer but orthodox financier, competent lawyer and brilliant scientist, was executed in the aftermath of the French Revolution in an act of extreme irresponsibility on the part of the revolutionary tribunal whose president declared, "The Republic has no need of savants."

[3] Robert Boyle, *The Sceptical Chymist,* London, 1674, p. 356.

But Lavoisier had already made a significant scientific contribution with the publication of *Elements of Chemistry* in 1789, in which he not only gave a fully empirical definition of the elements but a tentative table that listed them. Quoting from his book:

If we apply the term elements . . . to express our ideas of the last point which analysis is capable of reaching, we must admit, as elements, *all substances into which we are capable, by any means, to reduce bodies by decomposition.* Not that we are entitled to affirm, that these substances we consider as simple may not be compounded of two, or even a greater number of (elements); but since these (elements) cannot be separated, or rather since we have not hitherto discovered the means of separating them, they act with respect to us as simple substances, and *we ought never to suppose them compounded until experiment and observation has proved them to be so.*[4] (Italics supplied.)

At last, a definite, although tentative, list of elements was available for investigators to work with. Furthermore, Lavoisier and his contemporaries were using quantitative methods in their experiments, and were making careful records of the weights of the substances being combined and of the products of the chemical changes. Chemistry was at last on a firm scientific foundation; the organizing principles now began to be discovered with great rapidity.

2. THE OXYGEN THEORY OF COMBUSTION

Before considering the sequential advances in the science of chemistry it is worthwhile to pause in this historical era to look into the development of the modern theory of combustion, a theory largely attributable to Lavoisier and intimately associated with his concept of chemical elements.

Until late in the seventeenth century, the fundamental ideas about matter remained essentially Aristotelian in nature. The *elements* air and water and certain *principles* such as inflammability were considered to be the basis of chemical form and change. The German chemists Becker and Stahl advanced the hypothesis that in the *combustion* or burning process, the principle of inflammability or the fact that *phlogiston* (the Greek word for flame) is liberated by the burning material and absorbed by air. In the somewhat similar process of calcination, e.g., rusting of iron or the change of mercury into a red powder, it was also believed that phlogiston was released by the metal as it underwent chemical change. The presence of air was essential to the theory, for it was observed that in a vacuum neither combustion nor calcination could occur.

The theory, seemingly in agreement with many observations, gained wide

[4]Antoine Lavoisier, *Elements of Chemistry*, translated by Robert Kerr, Edinburgh, 1790, reprinted for the University of Chicago by Edward Brothers, Ann Arbor, Mich., 1940. page xxiv.

support. However, several facts stubbornly resisted satisfactory explanation by the Becker-Stahl hypothesis. One fact related to the mass of phlogiston: In the calcination process the product was heavier than the original metal, suggesting that phlogiston had a negative mass, whereas in the combustion process the resulting ash was lighter than the wood, suggesting a positive mass. Positive or negative? The theory gave no adequate answer. A second fact concerned volume: If the combustion and calcination processes were carried out in closed vessels, the volume of air *after* completion of the process was less than the original volume. This was even more difficult to reconcile with known physical principles. Finally, we mention the lack of reversibility of the combustion process. Calcination was indeed reversible, that is, the *calx* or powder could be transformed back into the original metal (a process usually called *reduction*); according to the Becker-Stahl theory this was supposed to occur as a result of absorption by the calx of phlogiston from the air. However, it is not possible, as we all know, to recover wood from ash in an analogous manner.

As chemical science progressed, it was necessary to modify the theory of combustion in order to incorporate new facts, just as it had been necessary to modify Ptolemaic astronomy earlier. It became obvious to a small group of chemists of especially keen insight that the whole edifice would have to be abandoned and replaced. Among them was Lavoisier. The focus of his scientific thinking was the concept of elements defined in an operational sense. The Aristotelian systems were of little use to him because they were not directly verifiable by experimental methods. What Lavoisier was seeking was a chemical element of ordinary matter which would be absorbed, for example, in the combustion and calcination processes. Furthermore, it must be a constituent of air because the latter was necessary to both processes.

The element *oxygen* had already been discovered by the English chemist Joseph Priestly and independently by Carl Scheele, a Swede. *Discovery* is used here in the sense that these investigators produced an elemental substance which they recognized as having unique properties not previously known. In particular, oxygen (called dephlogisticated air by Priestly) supported combustion exceptionally well. According to the Becker-Stahl theory, it must be air from which phlogiston had been removed. Scheele also subscribed to the phlogiston theory and reached essentially the same conclusions as Priestly.

After Priestly's discovery came to the attention of Lavoisier, he recognized oxygen as a chemical element which participated in the combustion process. Exhaustive experimentation soon verified his theory. Without going into great detail, we can summarize his work as follows. Mercury was heated in a closed atmosphere of air as in Fig. 12.1 until the red calx appeared on the surface; eventually its formation ceased. Weighing of the air left in the chamber revealed that it was about ⅘ as heavy as the original amount. Subsequently, the red calx

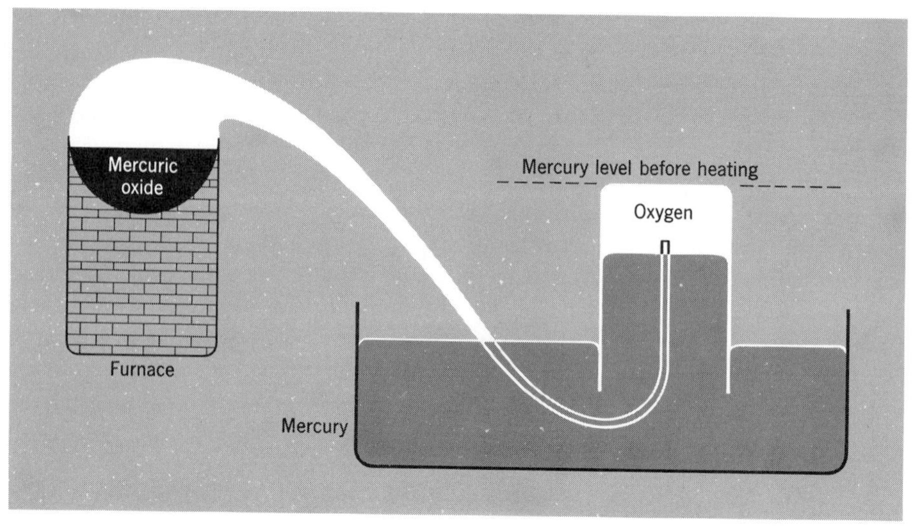

Fig. 12.1. Lavoisier's preparation of oxygen by heating mercuric oxide.

was heated and the mercury was recovered; in this process an amount of gas equal to that lost by the air in the calcination reaction was liberated. When this gas was mixed with the residual gas left from the original calcination, the product was indistinguishable from ordinary air. Thus, reasoned Lavoisier, calcination and reduction consist simply of the absorption of oxygen by mercury and its release by the red calx, respectively. The extension of the theory to include combustion is obvious. In the burning process, the wood reacts with the oxygen in the air to produce the ash. How then can we account for the apparent loss of weight in going from wood to ash? Additional gases are liberated, in particular (this was discovered later) carbon dioxide and water vapor.

Although this theory provided a very simple and all-encompassing explanation of combustion, calcination, and respiration processes, it was not immediately accepted. In fact it was never fully accepted by the older generation and its ultimate adoption depended on the younger, more receptive chemists—a recurring feature of the human drama that is not confined to science, though perhaps most conspicuously present there. For example, Scheele and Priestly, although contributing enormously to the development of the oxygen theory, never supported it, just as, in an earlier age, Tycho Brahe who contributed so much to establishing the Copernican theory, as modified by Kepler, never supported that theory.

The next step in the historical development of the basic chemical concepts, was the atomic theory of John Dalton.

3. DALTON'S ATOMIC THEORY; LAW OF MULTIPLE PROPORTIONS; AVOGADRO'S HYPOTHESIS

The chemical elements listed by Lavoisier were thought to be composed of different kinds of atoms, but Dalton saw the necessity of ascribing to the atoms of each element a definite relative weight. In his *A New System of Chemical Philosophy,* he states his purpose:

> Now it is one great object of this work, to show the importance and advantage of ascertaining *the relative weights of the ultimate particles, both of simple and compound bodies, the number of simple elementary particles which constitute one compound particle, and the number of less compound particles which enter into the formation of one more compound particle.*[5]

Dalton built his atomic theory largely upon the basis of the law of definite proportions, that is, when elements combine to form compounds they do so only in fixed proportions. Consequently, water is composed of eight parts by weight of oxygen to one part by weight of hydrogen. (Dalton, more of a theorist than a meticulous experimenter, found the ratio to be 1 to 7 "nearly.") If we assume with Dalton that compounds are formed by the junction of integral numbers of atoms of the elements, we conclude that an atom of oxygen is eight times heavier than an atom of hydrogen, *provided we also assume that the simplest combination—one atom of each*—gives rise to the compound water.

We need not consider all the details of Dalton's theory. Its importance is that he realized if the theory was to be valid with respect to chemical composition, then an integral number of atoms, each with a specific atomic weight, must form all known compounds. Reminiscent of the quotation from Empedocles regarding birth, death, or change, Dalton said:

> All the changes we can produce consist in separating particles that are in a state of cohesion or combination, and joining those that were previously at a distance.

The atoms themselves were permanent and indivisible; changes of properties, recognizable to our senses, occurred only with the combinations and separation of the particles.

One noteworthy feature of Dalton's system of atomic combinations, which soon had to be modified, was the assumption that atoms will combine in the simplest manner possible. Thus Dalton assumed that a compound particle of water was composed of one atom of hydrogen and one atom of oxygen. Unless evidence to the contrary could be found, this appeared to be a reasonable assumption, reflecting the traditional point of view of all scientists that nature is basically simple. However, his contemporary Gay-Lussac had discovered another

[5] John Dalton, *A New System of Chemical Philosophy,* Manchester, 1808, reprinted by the University of Chicago Press, 1940, p. 213.

quantitative and empirical generalization with respect to gases, that is, that gases combine with one another in simple ratios *according to volumes*. Thus two volumes of hydrogen combine with one volume of oxygen to form two volumes of water vapor. Similarly, three volumes of hydrogen combine with two volumes of nitrogen to form two volumes of ammonia gas, and one volume of hydrogen plus one volume of chlorine form two volumes of hydrochloric acid gas.

To explain these combinations by volume, Avogadro hypothesized that there are an equal number of particles in every volume of gas under the same conditions of temperature and pressure. But, if such a hypothesis were true, it would contradict the simplest form of the atomic theory, which assumed that the fundamental particles of gases were single atoms. (For example, 100 atoms of oxygen would then combine with 200 atoms of hydrogen to form 200 atoms of water vapor, which would imply that each of the 100 atoms of oxygen would have to be split into two parts, each half combining with an atom of hydrogen. This is contradictory to the atomic hypothesis that the atom is indivisible.)

To explain the puzzling contradiction, Avogadro proposed the hypothesis that the particles, or molecules, of many gases, were themselves each composed of two atoms. On this basis, everything known experimentally would fall into place. For instance, in the case of two volumes of hydrogen and one volume of oxygen combining to form two volumes of water, the hydrogen and oxygen molecules are each conceived to be composed of two atoms. The molecules of water are each composed of two atoms of hydrogen and one atom of oxygen. This is illustrated in Fig. 12.2.

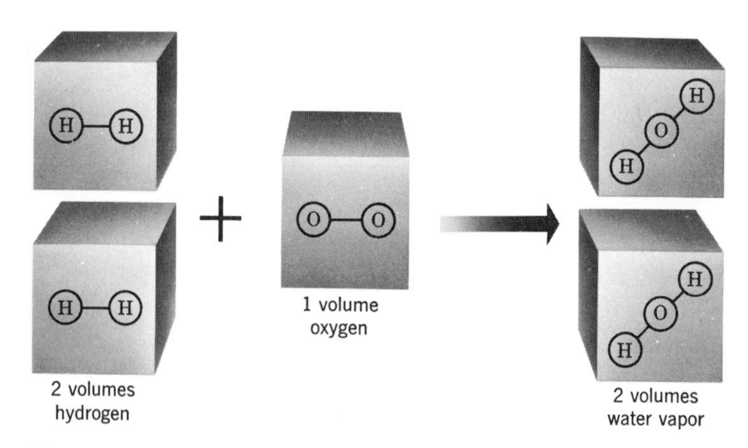

Fig. 12.2. Illustration of chemical combination of gases. Two volumes H plus 1 volume O give 2 volumes water vapor. (Can this be explained if molecules are made up of *single* atoms?)

Without pursuing all of the detailed developments and analysis that grew out of Avogadro's hypothesis, we point out that the basic structure of matter was now susceptible to at least a partial explanation on an atomic basis in a manner that seemed to fit most of the important facts known by experiment and observation and that provided firm guiding principles for further investigations and development. It was true that many serious problems remained. There was, for instance, the puzzle of why two atoms of hydrogen or oxygen or many other gases would combine with themselves to form molecules, as assumed by Avogadro. This could not be explained until the twentieth century, and did indeed prevent the full acceptance of Avogadro's theory for several decades. The answer could finally be found only when the atoms were at last discovered to have an internal structure, which produced an explanation of why exactly two, no more no less, of some atoms such as hydrogen combine with one another to form a molecule of gas.

4. CHEMICAL SYMBOLISM; QUALITATIVE AND QUANTITATIVE EQUATIONS

Lavoisier's instinct, Dalton's speculations, and Avogadro's logic, working upon the growing body of chemical knowledge, combined to form a synthesis that appeared to fulfill many of the early and apparently visionary dreams of the Greek atomists. Once more, invariance and regularity were being found amid the obvious changes to which matter is subject. To express this knowledge and compress it into a form easily grasped and manipulated, a symbolism had to be invented. The role of a good symbolism in any branch of science can never be exaggerated: Mathematics could hardly have progressed beyond the geometry of ancient Greece if it had not been for the invention of zero and the symbolic expression of algebraic equations; the advance of physics to its present state would not have been possible without its compressed symbolism; also a symbolism expressing the essential aspects of chemical changes was needed if chemistry was to advance as a science instead of remaining a collection of special recipes for accomplishing specific results.

The basic modern symbolic scheme that has evolved for the expression of chemical change is simple to grasp. It is as follows.

1. Each element, of which slightly over 100 have now been identified, is given a letter (sometimes two letters) usually the first letter of its English or Latin name. Thus, H stands for hydrogen, O for oxygen, Fe for iron (from the Latin name ferrum), and the like. These symbols can represent either the element or one atom of the element.

2. A compound molecule, consisting of a combination of atoms is indicated by the atoms of the elements of which it is composed. A subscript number shows the number of atoms of each element making up the molecule. For example, H_2O (water) has a molecule made up of two atoms of hydrogen and one of oxyygen, and H_2SO_4 (sulfuric acid) has a molecule made up of two atoms of hydrogen, one of sulfur, four of oxygen. These two compounds are noteworthy not only because of their general importance but also because of their *names.* The name *water* tells us nothing about the composition of the substance, but *sulfuric acid* does. Although the naming of compounds is a rather complicated matter, it is desirable to point out a few general guide posts, realizing that there are many exceptions to what follows. If two elements form only one compound, the name of the latter consists of the name of one element followed by the second which carries the suffix, "ide," e.g., *sodium chloride.* If a metal is one of the elements, its name always appears first. If the two elements form more than one compound, two conventions may be employed: The first element may carry the suffix "ous" (or "ic") to indicate the molecule with the smaller (or greater) relative number of atoms of this element. Examples are ferrous oxide, Fe_2O_3 and ferric oxide, FeO. In an alternative convention, a prefix (mono, di, tri, tetra, etc.) is added to the name of the second element to indicate the number of atoms of that type. Examples are sulphur dioxide, SO_2, and carbon tetrachloride, CCl_4. In more complex molecules, it frequently happens that a group of atoms behaves like a unit within the molecule. These units are called *radicals* and bear names which appear like names of elements in two-element molecules. Examples are the nitrate radical, NO_3, the sulphate radical, SO_4, and the hydroxyl radical, OH. Accordingly $NaNO_3$ is called sodium nitrate and $K(SO_4)_2$, potassium sulphate. An important class of compounds which departs from this convention is the *acids,* all of which contain the element hydrogen (Note, however, that most of the hydrogen-bearing compounds are not acids). For example, when the compound HCl (hydrogen chloride) is dissolved in water it forms *hydrochloric acid.* On the other hand, we do not call HNO_3 *hydronitric acid,* but simply nitric acid. In the case of acids containing radicals, the prefix *hydro* is customarily omitted.

3. In a chemical reaction the compounds or elements entering are shown on the left of the arrow, the products on the right. Because atoms are indestructible, the equation must balance, that is, the same number of each element must appear on the left-hand as on the right-hand side. The number preceding each molecule shows the number of molecules entering the reaction. Thus

$$2H_2 + O_2 \longrightarrow 2H_2O$$

(*Note.* It would be incorrect to write $H_2 + O \longrightarrow H_2O$ because hydrogen and oxygen in the gaseous stage have molecules composed of two atoms each.) Another example is

$$2HCl + Zn \longrightarrow ZnCl_2 + H_2$$

(This represents the reaction between hydrochloric acid and zinc. The zinc is dissolved, combining as a salt [zinc chloride], and free hydrogen is liberated. The reason that zinc chloride molecules contain two atoms of chlorine will be discussed under valence in 5 below.)

4. Since the relative atomic weights of all the elements in the particular chemical reaction being considered are known, it is easy to calculate quantities involved in the equation. (As an example, how much zinc and hydrochloric acid is required to produce 22.4 liters [2 g of hydrogen] given the atomic weights $H = 1$, $Cl = 35.5$, and $Zn = 65.4$? Since one atom of chlorine is combined with one atom of hydrogen in hydrochloric acid, a total of 72 g of hydrochloric acid is required [2 g of hydrogen are combined with 2×35.5 g of chlorine], and 65.4 g of zinc are required according to the reaction under 3.)

5. The number of hydrogen atoms with which one atom of an element can combine is called its *valence* number; for example, the valence number of oxygen is 2 (recall H_2O). It frequently happens that an element will not combine with hydrogen, in which case its valence is determined indirectly. A good example is the element sodium which, like hydrogen, combines with oxygen in the ratio 2 to 1, with NO_3 in the ratio 1 to 1, etc. Hence, the valence number of sodium is 1. Calcium on the other hand forms CaO, $Ca(NO_3)_2$, etc., and has the valence number 2. In addition to its magnitude, the valence number bears a sign. Hydrogen and elements which behave like hydrogen in compounds have *positive* valence number and elements and radicals which combine with them have *negative* valence number. Hence the valence number of sodium is $+1$, whereas that of chlorine is -1. Many elements have multiple valence numbers, rendering the whole subject quite complex. As it is best understood from the standpoint of the modern theory of atomic and molecular structure, we defer further discussion to Chapter 19.

Although the chemical equations such as those shown are of great value in chemical reactions, too much significance should not be read into them. Setting up a *possible* equation does not guarantee that a reaction will actually take place. Furthermore, a great amount of previous knowledge is generally required to determine the molecular formula for a given compound, even if the elements of the compound are known. For example, there are two gases that are composed of carbon and oxygen: carbon monoxide, a lethal gas, symbolized CO; and carbon dioxide, one of the less harmful products of combustion of carbon, symbolized CO_2. In addition, the equations specify only the mass relationships entering into chemical reactions. Nothing is indicated with respect to energy exchanges. In some cases energy is absorbed; in others, energy is given out.

To handle these and other problems, a great deal of knowledge of the properties of the elements and how they combine is necessary. For a complete explanation in terms of a few fundamental laws, an understanding of the internal structure of the atom is required. This has only become available in our own century. We shall review a few more of the basic ideas in terms of which chemistry has been organized and, in later chapters, we shall show how the theory of the internal structure of atoms and molecules has developed.

5. WEIGHT RELATIONSHIPS IN CHEMICAL REACTIONS

One of the oldest known laws governing chemical reactions is that of conservation of mass. It was a recognized principle as far back as the days of the Becker-Stahl theory of combustion in which it was central to the paradox of the mass of phlogiston. However, it could not be employed in quantitative analysis until the concept of *elements* in the sense of Lavoisier's definition was established on a firm basis. Further, it is necessary to know the atomic weights of the elements. Once these had been determined by Dalton and his successors, all the knowledge needed to calculate the quantities of elements and compounds participating in a chemical reaction was available.

For example, let us look at the reaction

$$S + O_2 \longrightarrow SO_2$$

from the standpoint of the masses of each substance involved. Suppose we start with one gram atomic weight (called a *mole*) of sulphur—32 grams. If this combines with oxygen molecules in the ratio of 1 atom of sulphur to 2 atoms of oxygen as prescribed by the equation, the amount of oxygen necessary to carry the reaction to completion is one gram *molecular weight* or 32 grams. According to the law of conservation of mass the yield must be 64 grams. Since one atom of S and one molecule of O_2 yield one molecule of SO_2, the molecular weight of the latter should be 64 grams; the same result is obtained by adding the atomic weight for each atom in the sulphur dioxide molecule:

$$\underset{\text{sulphur}}{32\text{ g}} + \underset{\text{oxygen}}{2 \times 16\text{ g}} = \underset{\text{sulphur dioxide}}{64\text{ grams}}$$

With the foregoing as a preliminary, we can now proceed to a more complicated problem in which the initial quantities of the reactants have arbitrary values. Consider the reaction

$$Fe_2O_3 + C \longrightarrow Fe + CO_2$$

The first step is to *balance* the equation so that we have the same number of atoms appearing on each side. There are no rigorous rules for doing this but depends rather on good judgment and practice. The balanced form of this particular reaction equation is

$$2Fe_2O_3 + 3C \longrightarrow 4Fe + 3CO_2$$

The next step is to specify the quantities of the reactants involved and then to compute the quantities of the products plus the residual, if any, of the reactants. Suppose we start with 32 grams of Fe_2O_3 and 5 grams of carbon. Now, one easily arrives at the gram molecular weight of the iron oxide; it is $(2 \times 56 + 3 \times 16)$ or 160 grams. Two of these will combine with three gram atomic weights of carbon or $3 \times 12 = 36$ grams. In other words 320 grams of iron oxide combine with 36 grams of carbon. However, only 32 grams of the former and 5 grams of the latter are available. Let us assume that all the iron oxide reacts and compute the amount (X) of carbon necessary for the reaction. The relevant mathematical equation is

$$\frac{\text{X g of carbon}}{\begin{array}{c}\text{36 g mole}^{-1}\\ \text{of carbon}\end{array}} = \frac{\text{32 g of iron oxide}}{\begin{array}{c}\text{320 g mole}^{-1}\\ \text{of iron oxide}\end{array}} \quad \text{or X} = 3.6 \text{ g}$$

Hence all the Fe_2O_3 is consumed and $5 - 3.6 = 1.4$ grams of carbon are left. The next and final step is to determine the quantities of the products. The total weight must of course be the sum of the weights of iron oxide and carbon which react or $32 + 3.6 = 35.6$ g. The relative weights of iron and carbon dioxide evolved are in the ratio of the number of molecules multiplied by the molecular weight or 4×56 g of iron to $3 \times (12 + 32)$ or 3×44 g of carbon dioxide. Thus the 35.6 g is divided between Fe and CO_2 in the ratio $X/Y = 4 \times 56/3 \div 44 = 56/33 = 1.70$, where X and Y are the actual weights of Fe and CO_2 produced. But we have already seen that $X + Y = 35.6$. Substitution of $X = 1.70Y$ yields $Y + 1.70Y = 35.6$ or $Y = 13.2$ g and $X = 22.4$ g which is the result desired.

All of the important weight relations involved in a chemical reaction are illustrated by this example.

6. CLASSIFICATION OF THE ELEMENTS

After the promulgation and general acceptance of Dalton's atomic theory, the number of known and identified elements increased rapidly. (Today, 103 elements have been identified, including several that do not occur in nature but have been created by man through various types of accelerators, such as the cyclotron, bevatron, and the like.) It soon became apparent that it would be useful to group certain elements having common properties into families and, then, try to work out relationships between them. Many suggestions for such classifications were made. Finally, in 1869, the Russian chemist Dmitri Mendeléeff developed a powerful method of classification, known as the periodic table of the elements.

If the elements are arranged in the order of their atomic weights it is found that certain properties repeat themselves regularly and periodically. To quote from Mendeléeff:

> This is the fundamental idea which leads to arranging all the elements according to their atomic weights . . . if all the elements be arranged in the order of their atomic weights, a periodic repetition of properties is obtained. This is expressed by the law of periodicity; the properties of the elements, as well as the forms and properties of their compounds are in periodic dependence or (expressing ourselves algebraically) form a periodic function of the atomic weights of the elements.[6]

Mendeléeff, following this idea, developed a system which is still used today. Because the periodic classification scheme is best understood from the standpoint of the atomic theory developed many years later, we defer presentation of the complete periodic table and detailed discussion of the modern theory until Chapter 18. However, from Table 1 which gives a portion of the periodic table, the general principle involved is immediately evident. Under the vertical columns (groups I–VIII), elements are found, all having similar properties. The last group (called the zero group) was added after Mendeléeff's time, since these elements, the stable or inert gases, were discovered after Mendeléeff's work was finished.

The periodic table provided both a successful summary classification of existing chemical data and also a challenge to investigators to find more basic principles in terms of which the regularities could be explained. Because emphasis has been given to the deductive method in the development of physics, we should not be misled into thinking that this is the only scientific method. There are indeed a number of scientific methods, each appropriate to a particular discipline or to a stage of inquiry that has been reached by that discipline at any time. Without an adequate system of classification based upon clearly stated criteria, chemistry could not quickly or easily have advanced to its modern form. (Nor could the science of biology have passed beyond a natural history stage before a broad base of empirical observations had been brought into an organized form by the classification system inaugurated by Linnaeus.[7])

[6] Dmitri Mendeléeff, *Principles of Chemistry,* 1871, Chapter XV, reprinted from the English translation of 1897 in John Knedler, *Masterworks of Science,* Doubleday, New York, 1947, p. 552.

[7] Linnaeus, in *System of Nature,* published in 1737, established the system of classification of genus and species. This classification was later extended by Buffon and Cuvier, and formed the background upon which Charles Darwin could develop his theory of evolution, published in 1859.

TABLE 1. A PORTION OF THE PERIODIC TABLE (LIGHTER ELEMENTS)

I	II	III	IV	V	VI	VII	VIII	Zero
Hydrogen $_1H^1$ [a]								Helium $_2He^4$
Lithium $_3Li^7$	Beryllium $_4Be^{9.4}$	Boron $_5B^{11}$	Carbon $_6C^{12}$	Nitrogen $_7N^{14}$	Oxygen $_8O^{16}$	Fluorine $_9F^{19}$		Neon $_{10}Ne^{20}$
Sodium $_{11}Na^{23}$	Magnesium $_{12}Mg^{24}$	Aluminum $_{13}Al^{27.3}$	Silicon $_{14}Si^{28}$	Phosphorous $_{15}P^{31}$	Sulfur $_{16}S^{32}$	Chlorine $_{17}Cl^{35.5}$		Argon $_{18}A^{40}$
Potassium $_{19}K^{39}$	Calcium $_{20}Ca^{40}$	$_{21}(\ )^{44}$ [b]	Titanium $_{22}Ti^{48}$	Vanadium $_{23}V^{51}$	Chromium $_{24}Cr^{52}$	Manganese $_{25}Mn^{55}$	Iron (Fe) Cobalt (Co) Nickel (Ni)	
Copper $_{29}Cu^{63.5}$	Zinc $_{30}Zn^{65}$	$_{31}(\ )^{68}$	$_{32}(\ )^{72}$	Arsenic $_{33}As^{75}$	Selenium $_{34}Se^{79}$	Bromine $_{35}Br^{80}$		Kryton $_{36}Kr^{84}$
R_2O [c]	RO	R_2O_3	RO_2 H_4R	R_2O_5 H_3R	RO_3 H_2R	R_2O_7 HR	RO_4	

[a] The subscript to the left of each element indicates its serial or atomic number. The superscript indicates its approximate atomic weight.

[b] Blank parentheses represent elements that had not been discovered when Mendeléeff first published his table. He predicted the properties and atomic weights of these unknown elements, which have since been discovered.

[c] The last two rows indicate the manner in which any element R combines to form an oxide or hydride. Thus lithium in group I combines with oxygen to form Li_2O, oxygen combines with hydrogen to form H_2O.

Mendeléeff's perspicacity and intuitive sense of significant relationships, coupled with the most careful and patient review of the known facts which led him to the ordered system of elements, remind us of the same qualities exhibited by Kepler as he sought for and found an order in the orbits of the planets. (However, there was nothing in Mendeléeff's make-up of the mystic "sleepwalker," which marked Kepler's personality.) In both cases, an extraordinarily fruitful generalization was reached which, in time, led to basic principles or laws of science as we know them today.

The full significance of the periodic system, depending as it does upon the serial ordering of the elements according to their atomic weights, requires a full understanding of the principles and methods used in determining both atomic weights of the elements and the structural composition of the compounds.

7. QUESTIONS

1. Previously, the two ideas—continuity and discontinuity—have been used but not closely examined. Try to define what we mean by continuous, both in abstract (or mathematical) terms and as a physical operation implying "infinite divisibility." Is a line infinitely divisible? Is space or time infinitely divisible? Aristotle, objecting to the atomic theory of Democritus, declared that matter and substance were infinitely divisible, much as space and time. Discuss what is meant by the infinite divisibility of matter.

2. An effective analogy that has been used to explain the type of quantitative analysis carried out by Dalton goes like this: you are given ten glasses of water in which sugar has been dissolved and are asked whether the sugar came in lumps (cubes) or whether it was powdered or granulated. You evaporate the water and find that in each glass the following weights of sugar were dissolved:

Glass	*Weight (in grams)*	*Glass*	*Weight (in grams)*
1	1.8	6	3.6
2	5.4	7	5.4
3	1.8	8	1.8
4	7.2	9	3.6
5	1.8	10	5.4

What do you conclude as to the probability that the sugar came in lumps or in granulated form? What hypothesis can you make as to the weight of one lump? Can you be as sure that this is the weight of one lump as you can be that the sugar came in lumps?

3. Several different gases are all found to be composed of various combinations of nitrogen and oxygen. Equal volumes of each gas (specifically 22.4 liters at standard pressure and temperature) are quantitatively analyzed with the results shown in Table 2. What is a simple hypothesis to explain the ratios between the quantities of nitrogen and oxygen found

TABLE 2. SOME COMPOUNDS OF NITROGEN

Compound	Mass of gas (g)	Mass of nitrogen (g)	Mass of oxygen (g)
Nitrous oxide	44	28	16
Nitric oxide	30	14	16
Nitrogen dioxide	46	14	32
Nitrogen tetroxide	92	28	64
Nitrous anhydride	76	28	48
Nitric anhydride	108	28	80

in the combinations? Can you think of any alternatives? Write out possible chemical formulas for each of the listed gases.

4. Because Dalton always found that there was a mass of eight parts of oxygen to one part of hydrogen in water, he concluded that oxygen had an atomic weight eight times as great as that of hydrogen. State the argument, on the basis of the laws of combination of gases, explaining why oxygen is now given an atomic mass *sixteen times* that of hydrogen.

5. Explain why molecules instead of simple atoms are thought of as the ultimate particles of gases.

6. Why is it that chemists are able to use weight instead of mass as their basic unit of measurement?

7. In the text we mentioned that both carbon dioxide and water vapor are the products of the combustion of wood. However only carbon dioxide is produced when charcoal is burned. Explain this difference in terms of the Lavoisier theory of combustion.

8. Besides the concepts of continuity and discontinuity which become more important as we approach the study of modern scientific developments, periodicity also assumes a larger and more significant role. What is meant by periodicity? Consider both the examples from previous chapters and the partial periodic table of elements (Table 1) presented in this chapter. What is there in common? In the periodic table, it is generally found that properties of elements (boiling points, specific heats, densities, and chemical activities) have approximately average values between those of the adjacent elements in the same row and same column. Mendeléef used a general formula for a quantitatively stated property, such as density, as follows: density of unknown element = ¼ (density of element in column in the periodic table to right and that of element to left plus density of element above and that of element below). On this basis, calculate what the density of the element magnesium in Table 1 will be, looking up the necessary data for the neighboring elements, and compare it with actual value.

Problems

1. The concise symbolism of chemistry (as in the case of all scientific symbolism) is extraordinarily effective in summarizing (or mapping) many occurrences in nature, not only qualitatively but quantitatively. How many quantitative as well as qualitative conclusions

can be drawn from the following equation (combined with data about atomic weights, valences, etc., of the elements)?

$$C + O_2 \longrightarrow CO_2 + \text{heat (7000 calories per kg of carbon)}$$
$$2H_2 + O_2 \longrightarrow 2H_2O + \text{heat (34,000 calories per kg of hydrogen)}$$

2. In one of the above exothermic reactions (that is, reactions that give off energy in the form of heat) translate the energy given off by the chemical reaction into the loss of mass that has taken place according to Einstein's equation: energy = change of mass × (velocity of light)^2. Actually all exothermic or endothermic chemical reactions involve a change in mass. Can you explain why such a change in mass is not detectable?

3. What *volume* of oxygen under standard conditions is required to completely oxidize 10 grams of carbon? How much volume will the resulting carbon dioxide occupy if the temperature has increased to 300°C and the pressure remains constant? How many molecules of carbon dioxide will be present?

4. Balance the following chemical equations

a. $NH_3 + H_2SO_4 + H_2O \longrightarrow (NH_4)_2SO_4 + H_2O$
b. $Fe_2O_3 + CO \longrightarrow Fe + CO_2$
c. $SiCl_4 + H_2O \longrightarrow H_2SiO_4 + HCl$
d. $HNO_3 + HCl \longrightarrow NOCl + H_2O + Cl_2$
e. $Cu_2O + H_2 \longrightarrow 2Cu + H_2O$

5. Write the names of the following compounds

Na_3PO_4
Na_2HPO_4
CS_2
$Ca(OH)_2$

6. In process 4b, one starts with 100 kg of Fe_2O_3 and the reaction goes to completion. How much CO is required and what are the quantities of the reactants produced?

7. In process 4d, one starts with 150 g of HCl and produces 36 g of H_2O. How many grams of HNO_3 were necessary to produce this quantity of water, how many grams of NOCl and Cl_2 were produced and how many grams of HCl are left?

8. If reaction 4e starts with just enough reactants to produce 50 g of copper with no reactants leftover, how many g of Cu_2O and H_2 were initially present? How much water was produced?

9. Two oxides of an element can be formed. In the first oxide the ratio by weight of oxygen to the other elements is $\frac{2}{27}$. In the second compound the ratio is $\frac{1}{9}$. What is the ratio by weight of the oxygen in the first compound to the oxygen in the second?

Recommended Reading

J. R. Partington, *A Short History of Chemistry,* Harper, New York, published as a Harper Torchbook, 1960.

Andrew Melsen, *From Atomos to Atom,* New York, 1960. This book contains both a historical review and a philosophical interpretation of the development of atomic theory.

James R. Newman, Ed., *What is Science?* Simon and Schuster, New York, 1955. The chapter on chemistry by John Read is an excellent presentation of the basic ideas of chemistry, descriptive rather than quantitative.

Dmitri Mendeléeff *Principles of Chemistry,* English edition 1897, Chapter XV, reprinted in *Masterworks of Science,* edited by J. W. Knedler, Doubleday, Garden City, N.Y., 1947.

G. Holton and D. Roller, *Foundations of Modern Physical Science,* Addison-Wesley Publishing Co., Reading, Mass., 1958. Chapters 22, 23, and 24 contain an expanded version of much of the material of this chapter.

Marie Boas Hall, "Robert Boyle," *Scientific American,* August 1967; page 96 is an excellent account of Boyle's contributions to chemistry and to pneumatics.

Henry M. Leicester, *The Historical Background of Chemistry,* Wiley, 1956. Chapters 3–15 are concerned with the early history of alchemy and chemistry and with the later important work of Boyle, Lavoisier, and other investigators of the late seventeenth, and eighteenth centuries.

chapter thirteen

Elementary electric phenomena

A fascinating, dramatic aspect of the history of science and thought is the surprising cross-currents that mark its progress. A period of quiescence, in which older ideas are developed, consolidated, and exploited to their fullest, is suddenly interrupted by new discoveries, often apparently completely unrelated to the old ideas, but which upon examination throw a bright new light upon the approach to questions that have remained troublesome within the framework of existing theories and beliefs. One new discovery often sets in motion a series of investigations that, eventually, bring about a complete reorganization of the old ways of thinking.

In this chapter we shall consider many examples of how this process works. All kinds of observations had been collected since the time of the ancient Greeks relative to what is included under the term electrical phenomena. Magic and mystical ideas surrounded many of the reports. There were few philosophers or thinkers who paid careful attention to these mysterious observations. Certainly none connected them closely to the other branches of scientific knowledge, which we have been studying. Astronomy (under Newton and those following him), mechanics, and chemistry already seemed on their way toward becoming "completed sciences." Newtonian physics, in particular, seemed to lack nothing other than a filling in of the gaps. Chemistry, however, involved a good many puzzles. There was no easy explanation of why elements had particular properties such as affinities and valences, or why reactions took place as they did. It was thought that the answer to these puzzles would be found in the further extension of Newtonian concepts. Upon this quiet, even development of old ideas, new discoveries, however, began to intrude. A physician, William Gilbert, in 1600 had begun a study of magnetic and electric phenomena, and published an organized summary of his carefully made observations. This stimulated many partially haphazard experiments and observations, which continued with greater and greater frequency. One hundred and fifty years later, Benjamin Franklin, a man of extraordinary versatility, became excited over the many experiments that he could perform with simple apparatus. His work stimulated an even more widespread interest. Then the Italian physician, Galvani, in 1780, found that a frog's leg would go into convulsion when its nerves were touched with a scalpel under the right conditions. The electric cell was the outcome of Galvani's discovery and, from this point on, the vast science of electricity opened up and developed with extraordinary rapidity, affecting every other branch of knowledge. The puzzling problems of chemistry were explained in the laws of electricity. Atomic theory and the structure of matter were intimately tied up with electrical and magnetic phenomena. By the end of the nineteenth century, even the laws of optics were believed reducible to those of electricity and magnetism.

Also, the study of electricity began to have a profound effect upon the general point of view regarding the fundamental laws of nature. As the investigations progressed and understanding grew regarding the laws governing electrical and magnetic phenomena, it became more and more clear that the Newtonian conception of a universe composed of independent particles affecting one another by "action at a distance" had to undergo serious modification. The new discoveries required that much greater attention be paid to the *medium* or the *intervening space* between particles. Within the century after the discoveries of Franklin and Galvani, Michael Faraday, Clerk Maxwell, and their followers had formulated a new point of view labeled "field physics" in which emphasis was shifted from the configuration of particles to a study of the properties of space itself. The story, even today, is by no means ended. Although we are now acquainted with the nature of the propagation of electromagnetic waves (light, radiant heat, and radio waves) at high but finite velocity, and can express such properties effectively by mathematical equations, the possibility of gravitational waves and their mathematical formulation remains a problem for modern physics. The shift in viewpoint from classical Newtonian physics has by no means run its full course.

1. EARLY DISCOVERIES AND THEORIES IN ELECTRICITY AND MAGNETISM

Of the three types of attractive forces that may be made observable directly by our senses—gravitational, electrical, and magnetic—the gravitational force between any two small objects is so slight that it cannot be detected except by extremely delicate instruments, although both magnetic and elecrical forces have effects that are easily observed. Under these circumstances, it seems curious that the quantitative law of gravitation was established and used by Newton and his followers as an effective working hypothesis more than a century before similar laws were established for electrical and magnetic forces, which can so easily be produced in varying quantities in a laboratory. The very variety of the effects, magnetic and electric, may have hindered the formation of ideas needed before measurement could be undertaken.

Phenomena that we now include under electricity and magnetism have been known and observed for several millenia. The Greeks noticed that amber, when rubbed, would attract to itself bits of chaff, thread, and the like (the very term electricity derives from the Greek word "elektron" for amber). They also noticed the attraction of several types of minerals (lodestones) for iron. In both cases, they attempted an explanation, although it was often ambiguous. Some speculators insisted on an anthropomorphic view: a "sympathy" existed between the bodies, causing them to be attracted to one another. Others suggested that a kind of emanation spread outward from the attracting bodies. For example, Lucretius explained magnetic attraction in this way:

> First, stream there must from off the lodestone, seeds
> Innumerable, a very tide, which smites
> By blows that air asunder lying betwixt
> The stone and iron. And when it is emptied out
> This space, and a large place between the two
> Is made void, forthwith the primal germs
> Of iron, headlong slipping, fall conjoined
> Into the vacuum.[1]

In other words, the lodestone's emanations affect the space around it, and it is this distortion of spatial surroundings that causes the motion of the iron. This is the primitive germ of the idea of a "field" surrounding a magnet (even though as Gilbert—the first true experimenter in the investigation of magnetism—points out in his quotation of the above passage in *De Magnete,* no "vacuum" is cre-

[1] Lucretius, *On the Nature of Things,* translated by W. E. Leonard, Everyman's Library, Dutton, New York, 1957, Book VI, p. 289. The cited passage is quoted by William Gilbert in *De Magnete* and criticized in Book II, Chapter 2.

ated by the "magnetic rays," [as Gilbert calls them] which are subtle enough to penetrate the most dense materials).

In spite of many references to electrical and magnetic phenomena, no really systematic work was done in this field until the time of Elizabeth I. In 1600, William Gilbert, personal physician of the English queen, published his famous treatise on the magnet after 18 years of research and experiment. In this work, which is the first important publication in English on the physical sciences, Gilbert not only investigates with great thoroughness the properties of magnets and electrified bodies but proposes a new doctrine for the pursuit of "natural philosophy:" it must be founded upon direct experiment and observation, and not upon ancient authority. In the preface to his treatise, Gilbert explains his effort:

> To you alone, true philosophers, ingenuous minds, who *not only in books, but in things themselves* look for knowledge, have I dedicated these foundations of magnetic science—a new style for philosophizing . . . (Italics supplied.)
>
> This natural philosophy is almost a new thing, unheard of before . . .
>
> Our doctrine of the lodestone is contradictory of most of the principles and axioms of the Greeks. Nor have we brought into this work any graces of rhetoric, any verbal ornateness, but have aimed simply at treating knotty questions about which little is known in such a style and in such terms as are needed to make what is said clearly intelligible. Therefore, we sometimes employ words new and unheard of, not (as alchemists are wont to do) in order to veil things with a pedantic terminology and to make them dark and obscure, but in order that hidden things with no name and up to this time unnoticed may be plainly and fully published.[2]

Gilbert was attempting to accomplish by experiments in magnetism and electricity what his contemporary, Galileo, in Italy, was accomplishing so successfully in his analysis of gravitation. Francis Bacon, influential English philosopher, belittled Gilbert's work and, yet, in his writings on the experimental method, Bacon copied many passages almost directly from it. Although Gilbert's studies were purely qualitative, he was careful and precise in his observations and experiments, and set the right course for future investigators. He had little use for authority that was not backed by observation:

> Our generation has produced many volumes about recondite, abstruse, and occult causes and wonders, and in all of them amber and jet are represented as attracting chaff; but never a proof from experiments, never a demonstration do you find in them. The writers deal only in words that involve in thicker darkness subject-matter; they treat the subject esoterically, miracle-mongeringly, abstrusely, reconditely, mystically. Hence such philosophy bears no fruit; for it rests simply on a few Greek or unusual terms—just as our barbers toss off a few Latin words in the hearing of the ignorant rabble in token of their learning, and thus win reputation—bears no fruit, because few of the philosophers themselves are inves-

[2] *Preface to De Magnete,* by William Gilbert, translated by P. F. Mottelay in *Great Books of the Western World,* Volume 28, pp. 1–2, published by Encyclopaedia Brittanica, 1952.

tigators, or have any first-hand acquaintance with things; most of them are indolent and untrained, add nothing to knowledge by their writings, and are blind to the things that might throw a light upon their reasonings. For not only do amber and (agates or) jet, as they suppose, attract light corpuscles (substances): the same is done by diamond, sapphire, carbuncle, iris stone, opal, amethyst, vincentina, English gem (Bristol stone, *bristola*), beryl, rock crystal. Like powers of attracting are possessed by glass, especially clear, brilliant glass; by artificial gems made of (paste) glass or rock crystal, antimony glass, many fluor-spars, and belemnites.[3]

To carry out his investigations, Gilbert used an instrument that he called a "versorium," instead of the chaff and threads commonly used as objects of attraction to amber, which was charged by being rubbed with silk. The versorium (Fig. 13.1) was made of a simple, light metal pointer mounted on a pin, probably suggested by the compass, which had already been invented. This pointer or rod rotated freely and, thus, moved under very small attractive electric forces.

Gilbert explains his versorium:

Now in order clearly to understand by experience how such attraction takes place, and what those substances may be that so attract other bodies (and in the case of many of these electrical substances, though the bodies influenced by them lean toward them, yet because of the feebleness of the attraction they are not drawn clean up to them, but are easily made to rise), make yourself a rotating-needle (electroscope—*versorium*) of any sort of metal, three or four fingers long, pretty light, and poised on a sharp point after the manner of a magnetic pointer. Bring near to one end of it a piece of amber or a gem, lightly rubbed, polished and shining: at once the instrument revolves. Several objects are seen to attract not only natural objects, but things artificially prepared, or manufactured, or formed by mixture. Nor is this a rare property possessed by one object or two (as is commonly supposed), but evidently belongs to a multitude of objects, both simple and compound, e.g., sealing-wax and other unctuous mixtures.[4]

(It is simple to make your own versorium. You can do it by folding a long strip of tinfoil into a rod, then mounting it on the point of a pin stuck through

[3] Ibid., p. 27.
[4] Ibid., pp. 27–28.

Fig. 13.1. Gilbert's versorium.

a piece of cork. It will operate very successfully to demonstrate attraction of rubbed pieces of plastic or glass brought near one end.)

Perhaps, because the phenomena of both electricity and magnetism in spite of their variety were specialized and not as universally present and uniform as that of gravitational attraction, quantitative studies were postponed, while all sorts of speculations arose as to why the phenomena existed at all.

Gilbert speculated as to the cause of both electrical and magnetic attractions. He surmised that from an electrified body (an electric, as he termed it)[5] there extended an emanation or an "effluvium" but, unlike the previous accounts of Lucretius and others, this effluvium was not a sudden emanation, causing a vacuum into which objects would be sucked. Instead:

> The effluvia spread in all directions . . . they are as it were material rods which hold and take up straws, chaff, twigs till their force is spent or vanishes (i.e., the force becomes weakened with distance). . . . electric bodies attract the electric only and the body attracted undergoes no modification through its own native force, but is drawn freely under impulsion in the ratio of its matter (composition.)[6]

Gilbert's experiments and discoveries stimulated many investigations and observations over the next 200 years. During this period no quantitative work was done but the basic qualitative observations were being accumulated upon which, later, the quantitative theories could be constructed. Many ingenious devices were built for generating and storing up electric charges. Popular interest grew rapidly, as dramatic exhibitions were given. A machine was developed by a seventeenth-century associate of Kepler, Otto von Guericke, for producing charges. The machine was improved upon until, by 1750, a static electric mechanism was developed for producing charges by rotating frictional devices. Benjamin Franklin and others also devised various means of storing large charges by such arrangements as Leyden jars (Fig. 13.2). These could be suddenly discharged to produce strong and dramatic sparks (or shocks to persons).

I. B. Cohen makes this comment about the Leyden jar:

> In France the new device provided a means of satisfying simultaneously the court's love of spectacles and great interest in science. . . . Seven hundred monks from the Convent de Paris, joined hand to hand, had a Leyden jar discharged through them all. They flew up into the air with finer timing than could be achieved by the most glorified corps of ballet dancers.[7]

[5] Although Gilbert distinguished between "electrics" and "nonelectrics," that is, those bodies that could by friction be given attractive properties similar to those given to amber, the term "electricity" was not coined until 1646 by Thomas Browne in his fascinating book, *Vulgar Errors,* which recounts and attacks many of the pseudoscientific stories surrounding attractive properties.

[6] Reference 2, p. 34.

[7] "Benjamin Franklin," by I. B. Cohen, in *Lives in Science,* Simon & Schuster, New York, 1957, p. 121.

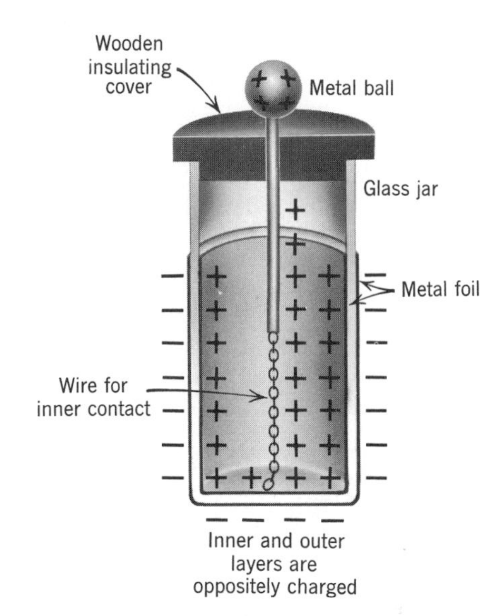

Fig. 13.2. The Leyden jar (a type of condenser). A large accumulation of charges may be stored in metal foils, since the opposing charges on inner and outer layers attract one another. Powerful shocks or sparks may be produced by this device.

As experimental investigations continued, the nature of electricity became more clearly defined. Instead of the rather vague conception of effluvia or emanations mysteriously surrounding an electric, there gradually arose an idea that there was a property, an "electric virtue," which was inherent in certain substances such as glass. However, when it was found that this "electric virtue" could be transmitted over metal wires from place to place, the idea that electric virtue was a kind of fluid began to develop. It had by now also been discovered that there were two kinds of electric charges, and that when they were alike they repelled one another; when they were unalike, they attracted one another.

In light of our present knowledge, we are apt to ask: why did not these early investigators apply measurements, and start building up a mathematical theory of the "laws" governing electrical phenomena? The simple answer is that measurement and quantitative analysis cannot be applied until an adequate conceptual scheme has been developed at least to that point where the properties to be measured are clearly and unambiguously identified.

But, finally, by the middle of the eighteenth century—150 years after Gilbert's publication on magnetism and electricity—a theory of electricity was emerging that could be put into a form where measurements could be taken. In the years 1747 to 1752, Benjamin Franklin published his *New Experiments and Observations on Electricity,* in which he outlined a fluid theory of electricity, and suggested the law of the *conservation of electrical charges.* He insisted that electricity is not created by rubbing silk on glass, but that the electric fluid is made to pass from the silk to the glass. The glass then acquires a superfluity $+e$ of electricity and the silk a deficiency $-e$ of electricity. The terms positive charge and negative charge, which are used to this day, stem directly from Franklin's original conception. Electricity itself became regarded as an imponderable fluid, similar in many ways to "caloric," the heat fluid, which was discussed in an earlier chapter. Franklin described this fluid as composed of "particles extremely subtile, since it can permeate common matter, even the densest metals, with such ease and freedom as not to receive any perceptible resistance."[8]

The fluid was sometimes spoken of as "electric fire," in obvious reference to the theory of heat or caloric as a kind of fluid. Caloric was considered to be composed of "elastic" particles, which repelled one another. On this basis, heat would flow from a greater concentration of particles to a lesser concentration, that is, from a hot body to a cool one. Similarly, a greater concentration of electric particles would flow toward a point where there was a deficiency. But, in one respect at least, the electric charges differed from the heat particles, and this was due to their ability to attract or repel each other over a distance.

The time was now ripe for an attempt at measurement of both the amounts of "charge" (that is, quantity of fluid particles according to the theory) and the forces exerted between the charges. Franklin's experiments were being repeated everywhere, and the basic instruments for producing and storing charges had become common knowledge.[9]

2. QUANTITATIVE MEASUREMENTS OF CHARGE: COULOMB'S LAW

By the end of the eighteenth century, the problem of a quantitative statement of the laws of attraction and repulsion between charges had been "in the air" for some time. It was inevitable that the suggestion should be made that an inverse square law would be found to apply to electrical forces as well as to gravitational forces. This was already suspected by the mathematician Daniel Bernoulli in 1760. Clear argument was presented by Joseph Priestley in 1767. Priestley had been following suggestions for an experiment designed by his friend Franklin. He found that when a hollow metallic vessel was electrified there was no

[8] Sir Edmund Whittaker, *A History of the Theories of Aether and Electricity,* Thomas Nelson & Sons, 1951. Harper Torchbook edition, 1960, Vol. I, p. 48.

[9] Benjamin Franklin's writings on electricity established him as one of the foremost scientists of the day. Translations into French, Italian, and German were widely read. It is interesting how this extraordinarily versatile American was able in such a short period—between his duties and interests as a public servant—to establish himself as an internationally recognized scientist (natural philosopher). In his voyage to France in 1776, as a representative of the newly founded nation, Franklin carried with him a letter from his government asking that, in case of capture by the English, he be given his freedom, since he was a man of science and "his contribution to humanity should be respected even in war."

electric force in any direction detectable on its interior. Priestley was well aware of Newton's analysis of gravitation, in which it had been shown that a hollow spherical mass would have no gravitational effect on its interior, *provided the inverse square law of gravitation was correct.* By analogy, therefore, Priestley argued in 1767:

> May we not infer from this experiment that the attraction of electricity is subject to the same laws with that of gravitation, and is therefore according to the squares of the distance, since it is easily demonstrated that were the earth in the form of a shell, a body inside of it would not be attracted to one side more than another?"[10]

This was indeed a very powerful argument in support of an inverse square law for electricity, but still it was an indirect one. It can be stated this way: (*a*) *if* the charges are *uniformly* distributed over the surface of a hollow sphere and (*b*) *if* they repel one another with a force inversely proportional to the square of the distance, *then* (*c*) there should be no net force in any direction upon any charge inside the sphere. Experimentally, condition *c* was found to be true. This does not *prove* that the inverse square law holds, but does give strong substantiation to the hypothesis.

In the above argument we have another example of the power of the formalism of mathematics when properly applied to concrete problems. Although it was suspected that gravitational and electrical forces acted in a somewhat similar fashion, there did not seem to be any reason to be sure that the forces would follow an exactly similar mathematical formulation. But when Newton's law of gravitation was applied to a spherical hollow shell of uniform mass, formal deductions showed that no net gravitational force in any direction would result within the shell. *This could not be tested experimentally* in the case of gravitational forces because of the minute forces involved for any shell devised by man. But experiments on electricity showed that a hollow charged sphere produced no net force in any direction within it. At once, by formal analogy, Priestley (and Cavendish and Coulomb, independently) concluded that the force must be similar in mathematical form to that of gravity. Again and again, examples of formal analogies are found leading investigators to fruitful conjectures. In this manner, Maxwell was led to his supposition that light is an electromagnetic phenomenon; de Broglie was led to his guess that electrons have wave properties; Einstein and contemporary physicists have been led to speculate on the existence of gravitational waves traveling with the velocity of light (a conjecture, however, which is still not experimentally established, and may not be for some time, because of experimental difficulties in testing this hypothesis).

[10] *The History and Present State of Electricity, with Original Experiments,* J. Priestley, London, 1767, p. 732, quoted by Whittaker in *The History of the Theories of Aether and Electricity,* Thomas Nelson & Sons, London, 1951. Harper Torchbook edition, New York, 1960.

The direct proof that the inverse square law holds true for electric charges was finally provided by the French engineer and physicist, Charles Coulomb, in 1784.[11] The experimental problem of measuring with precision the minute forces between charges was solved by using a torsion balance. Essentially, Coulomb's procedure was to use the principle of Gilbert's versorium (already discussed), but instead of supporting the rod on a point that gives it complete freedom to move in every direction, the rod was suspended by a fiber. Under such an arrangement a rod then will find its equilibrium position in one direction. If twisted out of that direction, it will be subject to a restoring force that is *proportional* to the angle through which the rod is turned. Thus, a precise measurement of the angle through which the rod is displaced also yields a precise measurement of the force causing that displacement.

Coulomb used the "torsion balance" principle (Fig. 13.3) to measure forces and, by measuring a variety of distances through which these forces acted, was able to show with considerable precision these two things:

> (1) The repulsive force between two small spheres charged with the same type of electricity is inversely proportional to the square of the distance between the centers of the two spheres, and
>
> (2) . . , the attractive force between the charged balls, one charged positively and the other negatively, is inversely proportional to the square of the distance between the centers of the balls.[12]

These laws of attraction and repulsion can be expressed in the equation form

$$F = \pm \frac{k}{d^2}$$

the sign depending on whether the charges are alike or different.

[11] Coulomb was a contemporary of Lavoisier but, unlike the latter who was executed in the French revolution, he left Paris in 1789 after the storming of the Bastille. He returned to Paris after Napoleon had established his consulate in 1799. A fine translation of Coulomb's memoir of 1781, with annotations by M. H. Shamos, may be found in *Great Experiments in Physics* edited by M. H. Shamos, Holt, Rinehart & Winston, New York, 1959.

[12] Ibid., pp. 63–66.

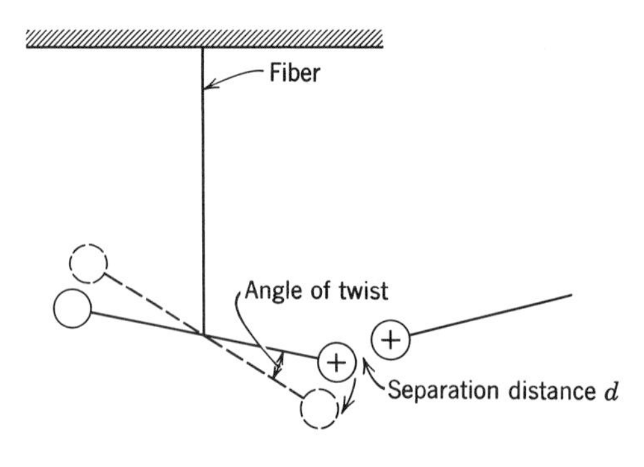

Fig. 13.3. Schematic diagram of Coulomb's torsion balance. The force of replusion is proportional to the angle of twist. Compare with Fig. 7.9 on Cavendish experiment.

Once the inverse square law of electrical forces was firmly established by experiment, the analogy between electrical and gravitational forces almost at once suggested a form similar to that of Newton's equation for gravity.[13]

Coulomb showed that if the distances are kept the same, the electric force varies as the product of the "electrical masses of the two balls." What he termed "electrical mass," emphasizing the analogy to gravitational mass, we now term *charge.* The equation for attraction and repulsion becomes

$$F = c\frac{q_1 q_2}{d^2} \tag{1}$$

where c is a constant similar in form to the constant G in Newton's law of gravity, that is,

$$F = G\frac{m_1 m_2}{d^2} \tag{2}$$

In the case of gravitational attraction, the units for m_1, m_2, and d were arbitrarily selected as the kilogram and meter, and the unit of the force, the newton, was defined by the equation $F = ma$.

In the case of Coulomb's law (equation 1), it would have been possible to choose some arbitrary charge as a unit charge. Instead of multiplying the number of arbitrary standard units, it was found convenient to define a unit charge in terms of equation 1, using familiar units of force and distance. The electrostatic unit of charge has been defined as a charge that repels a like charge at a distance of 1 cm with a force of 1 dyne (a dyne, it may be recalled, is 10^{-5} nt). Such a definition has the advantage of making c in the equation equal to 1. However, this unit of charge is so small that it is impractical when considerations of current (that is, the flow of charges) are studied. Instead, the more practical unit of charge, which we shall use, is the coulomb (coul). This is now defined in terms of the equation

$$F = \frac{9 \times 10^9}{k}\frac{q_1 q_2}{d^2}$$

where F is in newtons, d is in meters, and q_1 and q_2 are in coulombs. The k in the denominator is equal to 1 in a vacuum, but has other values when the charges are separated by a medium. It is called the dielectric constant. It is apparent that a coulomb of static charge is extremely large and cannot actually be accumulated on a small sphere without discharging (or "sparking"). (Two charges, each of 1 coul, if they could be accumulated, would repel or attract

[13] It is interesting that Newton's law, already firmly established for 100 years, was not *experimentally* verified until 1798 by Cavendish, since the gravitational force between two masses of a size which can be manipulated in a laboratory is miniscule.

each other in a vacuum with the tremendous force of 9×10^9 nt if placed 1 m apart.) Furthermore, the most elementary charge, the charge on an electron, is only 1.6×10^{-19} coul. When, therefore, we are only dealing with static charges, a coulomb would seem much too large for a useful unit. However, the *flow* of charges and the flow of 1 coul/sec (one amp) is in the useful range, and is then used as a practical unit.

Not only did Coulomb determine the mathematical form of the law of force between electric charges but, by using his torsional balance, he also demonstrated that a similar law could be established for magnetic forces:

$$F = \mu \frac{p_1 p_2}{d^2} \tag{3}$$

where p_1 and p_2 are the "magnetic mass" or pole strengths of two poles, and μ is still another constant.

The striking similarity between equations 1–3, governing electric, gravitational, and magnetic forces, has of course stimulated many theoretical physicists to try to find some single underlying law of which each of these equations would merely represent one aspect of the universe. This search for a more general "explanation" is typical of the continuous hunt that takes place in all science to find fundamental laws or principles of which the many more particular laws are merely illustrations or applications under special circumstances.

As we shall see later, there is a close connection between electricity and magnetism. But the relationship of mass to these phenomena still escapes us. It might not be an exaggeration to say that the modern physicist regards electromagnetic phenomena as better understood and probably more fundamental than those involving masses.[14]

3. THE FLOW OF CHARGES

Although Coulomb took the first step in the development of quantitative laws in electricity, further development of the science had to await a method of producing a constant and reliable source for the flow of charges from one point to another.

By the end of the eighteenth century it was common knowledge that charges did flow, although it was uncertain whether there was one fluid or two, and the fact that certain substances, particularly metals and even damp strings, were

[14] In this connection, we mention a suggestion to take length, time, charge, and magnetic flux as the fundamental units of physics, allowing mass to be defined as a derived unit. This proposal by the Russian physicist, P. Kalantaroff, in 1929, has received more attention in European than in English and American technical journals.

conductors, while other substances such as glass and dry organic materials, were nonconductors, was also well known. But it was difficult to store or produce a large steady flow of charge. Franklin arranged an ingenious device for collecting and storing electric charges with which he could experiment. He extended an iron rod from the roof of his house and connected this to a bell by means of a wire. Another bell was grounded by a wire leading to the earth. Between them was suspended a metal ball hanging from a silk thread. Whenever the rod was charged with electricity the charge passed down to the bell, drawing aside the metal ball to strike the bell. Franklin explained,

"I had given orders in the family that if the bells rang when I was away from home, they should catch some of the lightning for me in electric phials" (the electric phials were condensers such as the Leyden jars mentioned earlier in which the charges could be stored).[15]

In this ingenious way, he accumulated the electric charges with which to carry on his experiments. Obviously more certain methods of producing a copious supply of electric charges were needed.

In 1780, Luigi Galvani, a professor of anatomy, was studying the nerves of frogs in a laboratory that happened to contain a frictional electrical machine for producing charges. One day he had, in his words,

. . . dissected and prepared a frog and laid it on a table, on which, at some distance from the frog, was an electric machine. It happened by chance that one of my assistants touched the inner crural nerve of the frog, with the point of a scalpel, whereupon at once the muscles of the limbs were violently convulsed.[16]

He at once became interested, and conducted a long series of experiments to determine the conditions under which the convulsion of the frog's legs occurred. He suspected that the contractions were due to atmospheric electricity stored up in the frog, and decided on a series of experiments to test his theory. "For it is easy in experimenting," he wrote, "to deceive ourselves, and to imagine the things we wish to see."

Finally, Galvani found that the convulsion could be brought about if the nerve and muscle were touched at two points by dissimilar metals, themselves connected. He then proposed a theory of a new kind of electricity called *animal electricity* or *galvanism.* Although his theory was shortly discarded, his discovery did lead to a widespread interest and many further investigations.[17] Later, the correct solution was found by Volta, that is, that the electricity is generated by the connection of two dissimilar metals separated by any moist body, not neces-

[15] Cohen, I. B., editor, *Benjamin Franklin's Experiments,* Harvard University Press, 1941.

[16] Reference 8, p. 67.

[17] Galvani himself became another victim of the French Revolution. He lost his chair at the University of Bologna and died in poverty and disgrace because he refused to take an oath of allegiance to the Cisalpine Republic. Ibid., p. 70.

sarily organic. In 1800, Volta found that if he made a pile of metal disks, alternately silver and zinc, separated from one another by layers of a moist substance such as a moistened cloth or blotting paper, and connected the two ends of the pile, a very strong galvanic effect could be achieved. A touching of the two ends with the moistened fingers would produce a strong continuous sensation. Furthermore, sparks could be drawn if the two ends were connected by a metal wire.

> Volta inferred that the electric current persists during the whole time that communication by conductors exists all round the circuit and that the current is suspended only when this communication is interrupted. "This endless circulation or perpetual motion of the electric fluid," he says, "may seem paradoxical and may prove inexplicable, but it is nonetheless real and we can, so to speak, touch and handle it."[18]

A steady source of current from voltaic piles, or what are now called batteries, made possible the tremendous development in the field of current electricity. The century from 1800 to 1900 was to be marked as much by the vast development of electrical theory and its applications as by the development of chemical theory and its applications. These, together with the further development of heat engines, an efficient form of which was constructed by Watt in 1769, led to the radical and breath-taking changes in civilization with which we are all familiar.

4. UNITS OF ELECTRIC CIRCUITS: VOLT, AMPERE, OHM, WATT

One of the earliest needs for the development of the science of current electricity was the establishment of a system of units, precisely defined in terms of laboratory operations, in which various quantitative laws could be expressed. Taking the *coulomb* as a unit of charge, current is defined as the number of charges flowing past a given point in one second. This unit is given the name *ampere* (amp).[19] One ampere is the flow of 1 coul/sec past a given point in a circuit.

The flow of current must be brought about, of course, by some condition in the circuit. Just as a mass, when free to move, will fall from a higher point to a lower point under the influence of gravity, so the charges will move between two points, the positive and negative terminals of a battery, if they can find a path over which to move. We can, if we wish, describe the two heights in the gravitational field as points of different *potential*, meaning by this that the same mass at these points will have different potential energies. The difference in height could be termed a potential difference. Similarly, we can define the dis-

[18] Ibid., p. 73.

[19] After Andre-Marie Ampere, whose investigations of the force of attraction and repulsion between two current-carrying parallel wires suggested an operational definition of a unit of current.

similar condition existing between the terminals of a battery as a *potential difference.* To measure this potential difference, the unit of *volt* (V) is used. The volt is then defined as being a potential difference between two points such that 1 joule of work is required to bring 1 coul of charge from one point to another. When a potential difference between two points exists, charges will flow, provided that there is a conducting material between them, and will continue to flow until the potential difference falls to zero.

(Sometimes the volt is referred to as a unit of electromotive force. Strictly speaking, this is inaccurate. Although a potential difference between two points results in a flow of charges when the points are properly connected, the forces causing this flow are the forces of attraction and repulsion between the elementary charges, and the difference in potential is no more a force than is the difference in elevation through which a mass will fall under the influence of gravity.)

One of the many useful features of the volt as a unit of measurement is that when current flows through a metallic conductor, the amount of current flowing in amperes is proportional to the voltage difference applied between two points on the conductor. This is known as Ohm's law and can be expressed as

$$V = RI$$

where V is the potential difference expressed in volts, I is the current in amperes, and R is a proportionality constant which depends on the conductor. (R is not strictly speaking a constant, but depends upon other conditions such as temperature.) This proportionality constant is given the name resistance, and is measured in ohms. One ohm is defined as the resistance of a conductor such that 1 amp will flow in that conductor between two points when there is a potential between the two points which is equal to 1 volt.

Finally, there is another very useful connection between the units of electricity and those of mechanics. Since 1 V is 1 joule/coul and, 1 amp is 1 coul/sec, 1 V $\times$ 1 amp becomes 1 joule/sec or 1 *watt.* And, it also follows that electrical energy is conveniently measured in watt-seconds or multiples thereof such as kilowatt-hours.

Without going further into the laws of direct current electricity, let us concentrate on a summary of the units.

Coulomb. A unit of charge, defined in terms of newtons and meters, by the equation

$$F = 9 \times 10^9 \frac{q_1 q_2}{d^2}$$

If $F = 1$ nt, $d = 1$ m, and $q_1 = q_2$, then the charge on q_1 and q_2 is 1 coul on each. This equation must be modified by a constant if the charges are separated by a medium other than a vacuum.

Ampere. A unit of current equal to 1 coul/sec. (Actually, an ampere can be more easily measured in the laboratory than a coulomb. Thus, an ampere may be defined in terms of the magnetic effect produced by a wire carrying a current. A coulomb, then, is defined as an ampere-second.)

$$I = \frac{Q}{t} \qquad \text{amperes} = \frac{\text{coulombs}}{\text{seconds}}$$

Volt. A unit of potential difference (sometimes loosely called emf or electromotive force) equal to 1 joule/coul.

$$V = \frac{E}{Q} \qquad \text{volts} = \frac{\text{joules}}{\text{coulombs}}$$

Ohm. A unit of resistance. If 1 amp flows through a conductor when there is a 1 V potential difference between the ends of the conductor, it has a resistance of 1 ohm.

$$R = \frac{V}{I} \qquad \text{ohms} = \frac{\text{volts}}{\text{amperes}}$$

Watt. A unit of power. A watt is 1 joule/sec and, since a volt is 1 joule/coul, and an ampere is 1 coul/sec, a watt is equivalent to an ampere-volt.

$$P = IV \qquad \text{power in watts} = \text{amperes} \times \text{volts}$$

Watt-Second. A unit of energy equal to 1 joule.

$$W = VIt \qquad \text{joules} = \text{volts} \times \text{amperes} \times \text{seconds}$$

5. INDUCTION; ELECTRIC AND MAGNETIC FIELDS

The similarity of the mathematical form of the attractive and repulsive forces exerted between electric charges and between magnets strongly suggests that there may be some relationship between electricity and magnetism. Even more strongly is this connection suggested when the variety of phenomena included under the term *"induction"* is studied.

Why is it that a charged body will attract an uncharged body? Why is it that a magnetized piece of iron attracts one that is not magnetized? The reason is given by the theory of induction: if a positively charged sphere is brought close to one that is not charged, the theory asserts that the two types of charges on the uncharged sphere, positive and negative, which are equal in number and intermingled, will separate, the negative ones being drawn in the direction of the positively charged sphere, the positive ones being repelled to the opposite side, as shown in Fig. 13.4. No charge has flowed from one sphere to the other. There is no physical connection between them. But the rearrangement of the charges on the second uncharged sphere has been brought about over the intervening space, and this process is called induction.

Similarly, if a permanent magnet is brought close to a piece of iron, without touching it, the piece of iron becomes magnetized by induction, as is also shown in Fig. 13.4. The induced magnetic pole in the iron is, as in the case of the induced charge, opposite to the magnetic pole closest to it, and hence the piece of iron is attracted by the magnet even though the iron was originally not a magnet itself, just as an uncharged body is attracted by one that is charged.

This similarity strengthened the suggestion that there might be a connection between electrical and magnetic effects, but the precise connection escaped investigators for a long time.

The exciting discovery was finally made after a steady source of current became available in the voltaic pile or battery. Oersted, a Danish physicist, noticed that sometimes compass needles moved erratically during thunderstorms. He, therefore, tried to find a deflection of a compass needle when a wire bearing a current was placed over it. For a long time he was unsuccessful, until one day in 1820 he placed the wire parallel to the compass needle instead of at right

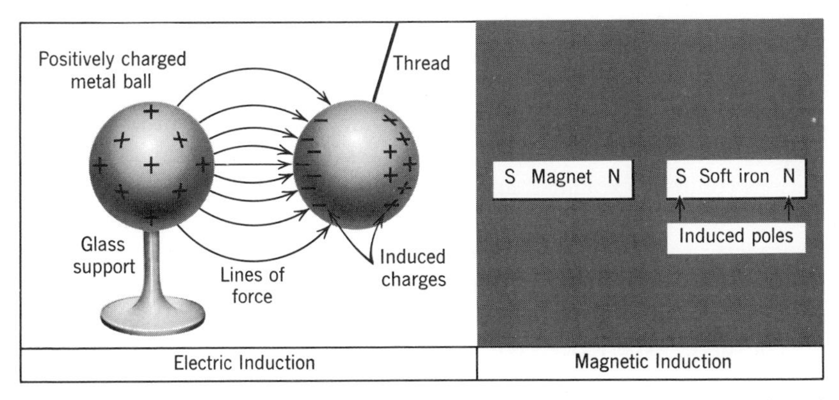

Fig. 13.4. Electric and magnetic induction. Neutral objects have an equal number of positive and negative charges, a haphazard orientation of small magnetic fields. When a charged body is brought close to a neutral body, the electric charges are attracted and repelled as shown. Similarly, poles are induced on a piece of soft iron when a permanent magnet is brought close to it. In both cases, attraction takes place.

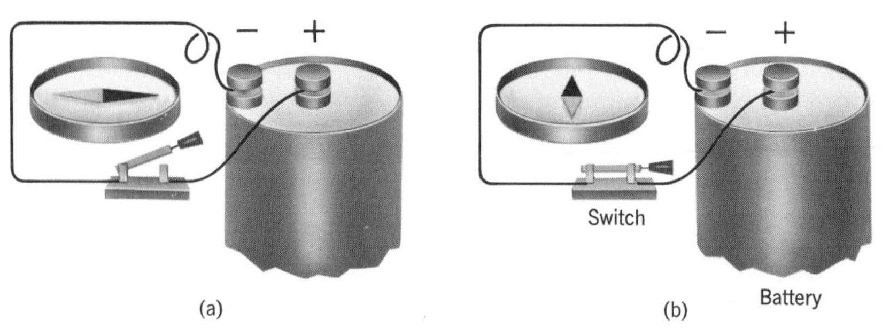

Fig. 13.5. An electric current generates a magnetic field, directed at right angles to the flow of current.

angles, as he had done previously. A distinct movement of the needle resulted (Fig. 13.5).

Once it was demonstrated that an electric current will "cause" magnetism (or, in more sophisticated scientific language, is associated with magnetic effects) the reverse problem suggested itself: can electricity be produced from magnets? For a decade after Oersted's discovery, investigators on both sides of the Atlantic Ocean tried to find evidence for the reverse phenomena. Again and again, the attempt was made to detect electric currents flowing in conductors placed near magnets, all with negative results. Finally, Michael Faraday discovered the answer: an electric current is produced if a magnetic field is *moved* when it is near a coil of wire.[20] A stationary magnet produces no effect, the *current* arises only when *the conductor and the magnet or magnetic field are in relative motion.*

The basic experimental discoveries had now been made: charges in motion produce a magnetic field; and a magnetic field in motion produces an electric field, which causes any charges that are present to move. The problem now was that of erecting a systematic theory on the basis of which the many experimental observations could be explained and summarized. Faraday tackled this task.

He was a man of extraordinary imaginative ability and curiosity, with the nimble fingers of an experimentalist, but with little formal training in mathematics. The models he used to "explain" or formulate the regularities that he observed were those that could be easily visualized and understood in terms of concepts of direct experience. The modern physicist tends to express the same laws by means of more abstract mathematical sets of equations, but it is to Faraday's lasting credit that he developed a clear pictorial representation of the

[20] Joseph Henry, an American, was undoubtedly the first to notice and identify that an electric current can be caused by a changing magnetic field, but he did not publish his observations until after Faraday had done so, and thus Faraday is given credit for the discovery.

complex relationships existing between electric and magnetic fields. Upon his preliminary visualization, the later mathematical formulation by Maxwell was erected.

It had already been observed that if iron filings are scattered on a sheet of paper covering a magnet, they will arrange themselves in a distinct pattern; a series of fairly well-defined curves will be described by the iron filings (Fig. 13.6).

From these curves, Faraday developed the idea of lines of magnetic force. They were conceived as filling all the space surrounding a magnet. The lines of force form closed curves passing through the interior of the magnet as shown in the figure. They never intersect, but are more concentrated in certain portions of space than in others. In those places where they are more concentrated, the magnetic field is stronger.

Faraday said in 1851:

> I cannot refrain from again expressing my conviction of the truthfulness of the representation which the idea of lines of force affords in regard to magnetic action.[21]

A current of electricity in a wire was conceived by Faraday to be produced whenever a wire was moved in such a way as to cut the lines of force, whether the lines were cut perpendicularly or at an angle. Furthermore, the voltage induced, and hence the current, was found to be directly proportional to the number of lines of force cut per second. Mathematically this statement may be expressed as

$$E = -\frac{\phi_1 - \phi_2}{t}$$

where $\phi_1 - \phi_2$ is the change in the number of lines of force or *flux* linked by the

[21] Reference 8, p. 172.

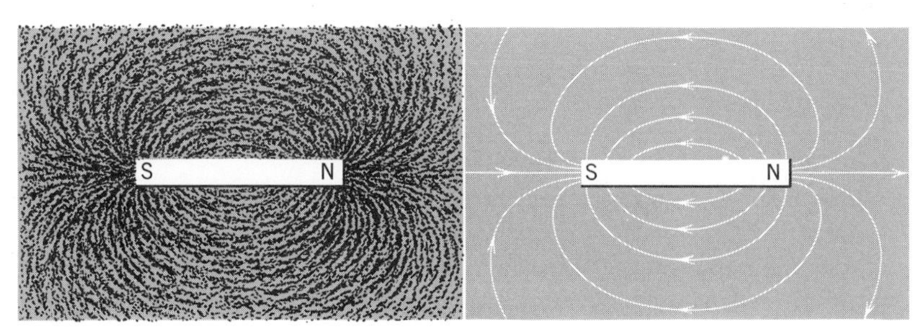

Fig. 13.6. *Left:* Iron filings scattered over a paper, covering a magnet, will distribute themselves in a pattern such as that shown. *Right:* The lines of force are suggested by the pattern of iron filings; the arrows show the directions of the lines of force.

circuit, t is the time during which the change occurs, and E is the average induced voltage. The negative sign indicates that the magnetic field of the induced current tends to oppose the flux change which causes it.

It made no difference whether the wire was moved across the lines of magnetic force, or whether the magnet was moved in such a way that its lines, moving with it, were cut by the wire. "In this view of the magnet," wrote Faraday, "the medium or space around it is as essential as the magnet itself."

Thus "action at a distance" between two distant bodies was abandoned as an explanation. To trace the manner of how one body could affect another, it was necessary to see how the space between them was effected. Also this transmission of the force through space would take time. If a magnet or an electric charge is moved, the pattern of the lines of force surrounding the magnet or the charge is disturbed, and this disturbance is carried outward into space at a finite velocity. Faraday's model laid the groundwork for the later theory, developed so elegantly by Maxwell, that light itself is nothing but the radiation of such an electromagnetic disturbance.

The concept of lines of force lends itself very readily to a qualitative explanation of the force between two current-carrying wires. Consider the wires with opposed currents shown in Figure 13.7 to be infinitely long and infinitely far

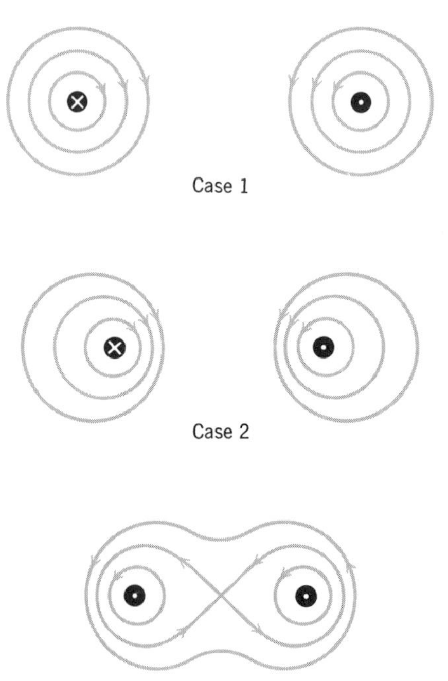

Fig. 13.7. Magnetic fields of long, parallel current-carrying wires: case 1, opposed currents, wires infinitely far apart; case 2, opposed currents, wires close to each other; case 3, currents in the same direction; ⊙ denotes current flowing out of page, ⊗ current flowing into page.

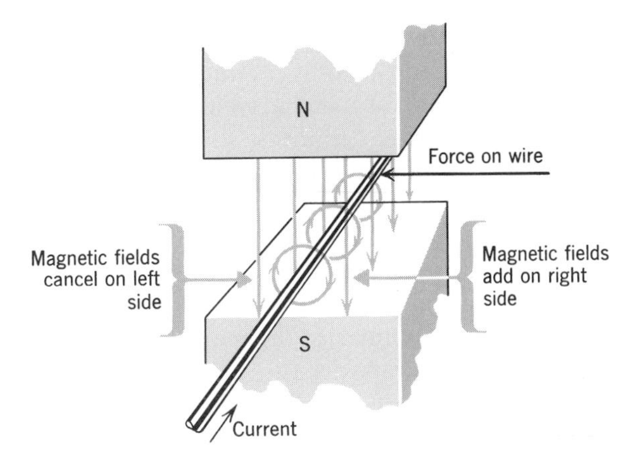

Fig. 13.8. A current-carrying wire in a uniform magnetic field.

apart. The magnetic lines of force in this case are circles which are centered at the wires. Now suppose the wires are brought close together and oriented parallel to each other. In the space between them, the lines of force have the same direction and are thus denser than before. Because the latter try to "avoid" each other they become distorted as shown in Figure 13.7 Case 2. This leads to a repulsive force between the two wires because the lines of force are, in effect, under compression. If the wires are now released, they will move apart until the circular lines of force again become concentric with the wires. Finally, consider the case in which the currents are parallel and in the same direction. In the space between the wires the magnetic fields are opposed and the lines of force tend to cancel each other, thus forming the pattern shown in the final case of Fig. 13.7. In this case they are under tension and can attain the equilibrium configuration of concentric circles only if the wires coincide. Hence, the force is an attractive one. The operation of all electric motors is based upon the interaction of current-carrying wires with magnetic fields. This is illustrated in Figure 13.8 which shows a current carrying wire in a uniform magnetic field. The two magnetic fields add on one side of the wire and cancel on the other. The wire tends to move in the direction of the weaker field.

One of the reasons why the "field concept" was developed in connection with electrical and magnetic phenomena before it became associated with those of gravity is probably that the *direction* of interaction is not as simple in the case of electromagnetic forces as it is in the case of gravity. Two isolated masses exert a mutual force directly along the line connecting them. In contrast, a wire carrying a current when placed near a magnet is acted on by a force that is not, as one might expect, in the line of direction between the north and south poles of the magnet, but is at right angles to the lines of magnetic force. Let us look at the matter a little more closely.

A *stationary* electric charge produces no magnetic effects. But if such a charge moves, a magnetic effect is produced. The *magnitude* of the effect depends both on the magnitude of the charge and the velocity with which it moves, but the *direction* of the effect cannot be simply stated, since it is not in the direction of the movement of the charge but is at right angles to the direction of motion. Suppose a conducting wire is formed in a loop, as shown in Fig. 13.9, and that charges flow around the loop. Then magnetic lines of force are produced whose directions are shown by the small closed loops typical of the many lines of resulting magnetic force.

We can now summarize the major developments that led Faraday, and after him Maxwell, to consider the "field" or the "properties of space" surrounding magnets and electric charges as equally important as the magnets and charges themselves.

(*1*) Magnets and electric charges do not interact except when they are in relative motion.

(*2*) The direction of their interaction, that is, the forces that they exert upon one another, is not simply in the same direction as their motion. The changing pattern of lines of force surrounding an *electric* charge *in motion* gives rise to a set of *magnetic lines* of force in a direction at right angles to the motion of the

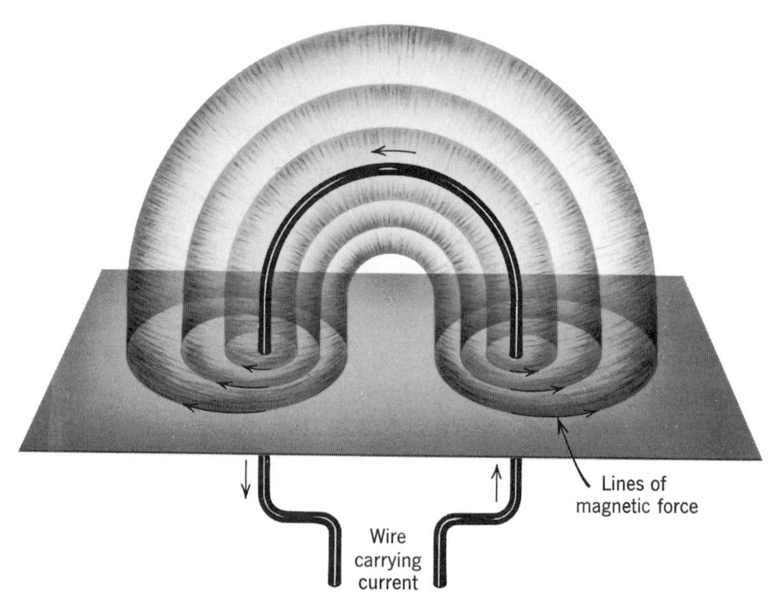

Fig. 13.9. Direction of lines of magnetic force produced by a current flowing through a loop of wire.

electric charge. Similarly, a magnetic field in motion brings about a force on a charge that is at right angles to the magnet's motion.

(3) It takes a finite time for the changing pattern of lines of force to spread from one point in space to another. An electric charge in motion will set up a magnetic field, but this magnetic field is not instantaneously built up through all of space. Instead, it spreads out into space with a definite velocity. (The *charge* may come to rest or be neutralized by an opposite charge and, yet, the effect of its original motion continues to travel outward through space, even if the source of this disturbance is no longer present.)

The significance of these conclusions is precisely this: even so-called empty space has certain physical properties to which attention must be paid when electromagnetic phenomena are being considered. The simple Newtonian picture of objects "acting at a distance," in a mathematical space devoid of any physical properties has to be abandoned.[22]

When dealing with fields of forces which vary from point to point depending on the distribution of the sources of the force field, be it gravitational, electric, or magnetic, it is often convenient to use a new unit: the field strength or intensity at a point.

In Section 2 we examined the field of force extending from spheres, which carried charges on their surface. We found that the force on a charge Q_1 placed at a distance d away from the center of a sphere carrying the charge Q_2 would be given by

$$F = c\frac{Q_1 Q_2}{d^2} \quad \text{where } c \text{ is } 9 \times 10^9$$

It is convenient, here, to introduce a new unit of field strength: electric field strength measured in newtons per coulomb. This unit is designed to define precisely what is meant by *field strength* at a point, in terms of the force exerted on a small test charge, say Q_1 placed at the point in question. From the equation above for Coulomb's law, we can derive the necessary concept, that is, the field strength E at a point d meters away from a second charge Q_2. Defining E as the force per unit charge Q_1, we arrive at

$$E = \frac{F}{Q_1} = c\frac{Q_2}{d^2}$$

[22] Suppose the sun, and all its mass, disappeared "in an instant," would the effect on the earth's motion be instantaneous and "at the same instant" or would the gravitational force continue to spread outward in space at some velocity, such as the speed of light, and be felt only (let us say) 8 minutes later? Under the Newtonian conception where gravity is associated only with masses acting in a mathematical space without properties, the effect would be instantaneous. Under "a field concept" the effect would take some finite time (presumably the 8 minutes required if gravitational effects are transmitted at the same velocity as electromagnetic effects).

Should this unit at first appear strange, consider how the strength of a gravitational field can be specified in a precisely analogous manner. The gravitational force exerted by the earth on a mass m, it will be recalled, is given by $F = G(mM_e/d^2)$. Since the weight of a mass (i.e., the gravitational force acting on it near the earth's surface is given by $F = mg$, it follows that $g = G(M_e/d^2)$ where g is expressed in newtons per kilogram and specifies the gravitational field intensity at a point d in distance from the mass M_e. The general equation for $g = G(M/d^2)$ for any mass expresses the intensity of the gravitational field at a point of distance d from the mass in question. Because in most cases of common application we do not move far from the earth, g may be taken as approximately constant and equal to 10 newtons per kilogram. In contrast when we deal with the field of an electric charge, very small distance variations are significant because the charges are concentrated in small regions and the electric field intensity in newtons per coulomb may vary rapidly from point to point. The analogy between $E = c(Q/d^2)$ and $g = G(M/d^2)$ is clear. Furthermore just as $F = mg$, so $F = Eq$, where m is the mass subject to a gravitational field strength g, and q is a charge subject to electric field strength E.

One effective geometric method by which the field strength E at a point may be conveniently represented was proposed by Faraday. In this representation, the field is thought of as being threaded by many lines of force. Field strength E can, in this way, be measured by the number of lines of force perpendicularly crossing an area of unit cross section.

In this representation, the lines of force will spread out radially and uniformly from an isolated charged sphere. The total number of lines of force will remain constant in any volume surrounding the charged sphere (they will only be interrupted if they terminate on an oppositely charged body). Hence the number of lines of force crossing any perpendicular unit area will become less and less as the distance from the sphere increases. Since the number of lines of force remains constant, whereas the total spherical surface at a distance from the center increases as the square of the distance, the intensity E, represented by the number of lines of force per unit area, must fall off inversely as the square of the distance from the center. However, if an oppositely charged body is introduced in the neighborhood of the charged sphere, the lines of force will now be distorted, bending toward the oppositely charged body, with many terminating on that body (Fig. 13.10A).

Up to now we have considered the field of force exerted by charges on the surfaces of spheres. Analysis shows how the field of force between two flat oppositely charged plates can be calculated. Without going into the details of proof, the plausibility of the results can be demonstrated by examining the three diagrams in Fig. 13.10. In Fig. 13.10A, the lines of force between two spheres

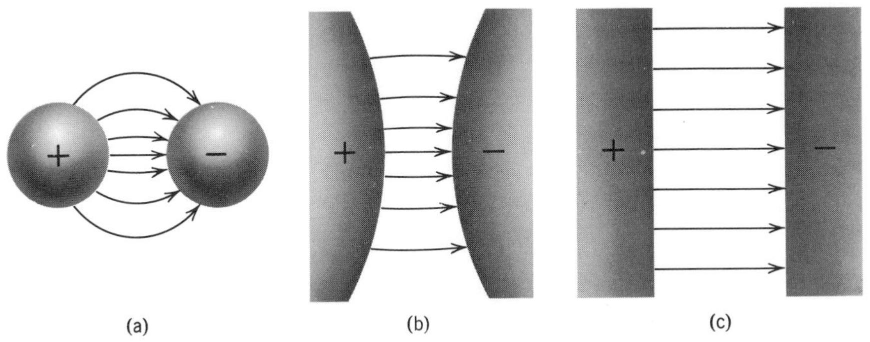

Fig. 13.10. (a) the lines of force between two spheres; (b) the lines are more nearly parallel as the radii of the spheres are increased; and (c) the lines are parallel, since the surfaces of the two spheres have become plane surfaces as the radii increases to infinity.

are shown. In Fig. 13.10B, the radii of the spheres are increased (the spheres are being flattened out.) In Fig. 13.10C, the radii of the spheres become infinite and the surfaces are now flat plates. Under the third condition (parallel plates), the lines of force are parallel and the field between the two plates is uniform.

The net result of the analysis shows that the field does not vary either in magnitude or direction anywhere between the charged plates (an exception being very close to the edge). A charged particle, such as an electron, therefore, if placed between the electric plates will be acted on by *a constant force.* In other words, it will be *uniformly accelerated,* just as a mass is accelerated uniformly under the force field of gravity near the earth's surface.

Furthermore, it is simple to measure experimentally the magnitude of the intensity of the field between the plates. It will depend directly upon the voltage difference between the plates. This is shown as follows.

The *work* done in moving a positive charge Q from the negative plate to the positive plate will be equal to the uniform *force times the distance* through which it is moved. Now the *force* is the electric *field intensity* E (in newtons per coulomb) *times the charge* Q (coulombs). The work is therefore

$$W = EQd \tag{4}$$

where d is the distance between the plates.

But, by using the definition of potential difference, we know that the *work* can also be expressed as *the potential difference* (measured in volts, which are defined as joules per coulomb) *times the charge* moved between the two points of potential difference. Thus

$$W = VQ \tag{5}$$

Therefore, equations 4 and 5 yield

$$EQd = VQ$$

or

$$E = \frac{V}{d} \tag{6}$$

Equation 6 expresses the field intensity very simply in volts per meter, where V and d are easily measured.

There is a similar unit of magnetic field strength, which is not quite so simply expressed because a magnetic field commonly arises out of a flow of charges, and because no such thing as an isolated unit pole, except as an abstraction, has been observed. The unit of magnetic field strength can be considered to arise out of the equation expressing the force of attraction between two parallel wires of small elements of length ΔL_1 and ΔL_2 separated by the distance d:

$$F = \frac{k(I_1\Delta L_1)(I_2\Delta L_2)}{d^2} \text{ where } k \text{ is } 2 \times 10^{-7}$$

and where I_1 and I_2 represent the current flowing in the two wires. The force on the wire carrying current I_1 due to the magnetic field produced by current I_2 can be thought of as arising out of a magnetic field defined by $B = kI_2/d^2$, and the equation then becomes $F = ILB$ where B,[23] the magnetic field intensity, is in units of newtons per ampere-meter. Sometimes in applying this equation, as we shall find in the next chapter, the force will be exerted on a free charge q, such as an electron, traveling through space with a velocity v. In such a case qv is equivalent to IL (explain why) and the equation becomes

$$F = qvB$$

We will return to this important equation in Chapter 14 where the discussion of determining the nature of the elementary charge on an electron is undertaken in connection with cathode ray tubes.

6. ELECTROLYSIS AND THE HYPOTHESIS OF AN ELEMENTARY CHARGE

During the very period that the laws of electricity and magnetism were being formulated, it was found that a new understanding of chemical theory and of the structure of matter also resulted from a knowledge of electrical laws. The

[23] In this book no distinction is made between H and B. Usually B and H are only identified in vacuo. H is considered to be the intensity of magnetization produced by an electric current.

unity of science is beautifully illustrated by the contributions made to chemistry growing out of discoveries in electricity.

Sir Humphrey Davy, the English chemist and world renowned scientist, gave some public lectures in 1812 which were attended by young Michael Faraday who was at that time a journeyman bookbinder. Faraday took careful notes on the lectures and applied successfully for a position as Davy's laboratory assistant. Although their work in the laboratory was primarily in chemistry, as the newest discoveries in electricity became known, Davy and Faraday became keenly interested in the new discoveries and their relationship to chemistry.

Shortly after the voltaic pile was put into use, it was noticed that if a current of electricity was made to travel through solutions, bubbles of gas arise at the electrodes (that is, the portion of the conductor which was immersed in the solution). Examination soon revealed that the liquid through which the current is being passed decomposed. For example, if current is passed through water, which is made sufficiently conductive by the addition of a slight amount of acid, bubbles of oxygen will collect at the positive pole, and hydrogen at the negative. Similarly with other solutions. In general, hydrogen and all metals that are dissolved in the form of a salt travel toward and are deposited on the negative electrode.

This type of action is of course very different from what occurs when electricity flows over a metallic conductor. In the latter case the metal itself is not affected, and retains the same chemical composition.

These observations soon led Davy and others to consider the possibility that the chemical composition could be explained in terms of electrical forces. "Chemical and electrical attractions," Davy said in 1826, "are produced by the same cause." Berzelius used and extended Davy's explanation of chemical affinity as arising from the play of electric forces originating in the electric charges within the atoms.

By the aid of this conception Berzelius drew a simple and vivid picture of chemical combination. Two atoms, which are about to unite, dispose themselves so that the positive pole of one touches the negative pole of the other.[24]

The beginning of an explanation for chemical affinities or valences among the elements which had remained a mystery to Mendeléeff was now taking shape.

Furthermore, as Faraday picked up the trail and made quantitative observations relating the amount of current to the mass of the elements being deposited at the electrodes, the groundwork was being laid for the theory that *there is an elementary atom, or "package" of electrical charge.* He found that invariably the same total amount of electricity measured in coulombs deposited the same mass of hydrogen at the negative electrode, regardless of any variation in volt-

[24] Reference 8, pp. 76, 78.

age, temperature, speed, or chemical composition of the solution. For example, 1 g of hydrogen would always come out of solution for each 96,500 coul of electric charge carried through the solution. Additionally, he found that 35.5 g of chlorine, 80 g of bromine, 127 g of iodine, and 108 g of silver were always deposited by this same quantity of charge.

These figures are exactly the same as the atomic weights of the elements involved. It is natural to conclude that each atom therefore carries the same charge as an atom of hydrogen. Faraday also found that for those atoms which combine with two atoms of hydrogen the charge required to liberate 1 g atomic weight is *one-half* the amount of 96,500 coul (that is, 8 g of oxygen, whose atomic weight is 16, is deposited by the 96,500 coul). Thus atoms of elements such as oxygen were considered to carry two electric charges when in solution, and so forth. Since oxygen was found to travel toward the positive electrode, these charges were considered negative.

The laws established by Faraday inevitably led to the conclusion that there must exist in nature an elementary indivisible electric charge. The name "electron" for this elementary charge was first coined and universally accepted in 1874, long before any additional direct proof of the existence of the electron was forthcoming. (Here, we have another illustration of how "hypothetical" elements are often called for or suggested by theory, long before their "reality" is more directly substantiated by experimental observation. The same story repeats itself throughout science, again and again: the atom, the electron, the photon, the positron, and the neutrino, in physics and chemistry, and the gene in biology were postulated as hypothetical entities, helpful in the development of theory long in advance of their actual and more direct "discovery." The "graviton" belongs in this theoretical limbo today, awaiting possible direct confirmation.)

Even before any direct measurement of the value of the elementary charge could be made, the theoretical structure for its evaluation was present, and some early attempts were made to arrive at a rough approximation. We have seen that Avogadro's number (the number of atoms in a gram molecular weight of a substance) can be approximately calculated from the kinetic theory of gases. If this number of atoms in a gram of hydrogen is taken as 6×10^{23}, and if it is further assumed—on the basis of Faraday's work—that in a conducting solution each atom of hydrogen carries one unit of the smallest charge, then, because 96,500 coul of charge liberate 1 g, the charge on an electron must be

$$96{,}500 \times \frac{1}{6 \times 10^{23}} = 1.6 \times 10^{-19} \text{ coul}$$

There is a beautiful relationship, which continuously displays itself between various branches of theory and a variety of independent observations, each correcting, modifying, and substantiating one another. Avogadro's number 6×10^{23}

has been determined by a variety of different methods. The kinetic theory of gases provides one method of evaluation. Another is to use the mass of a single atom as determined by modern electronic devices. Still another is to determine interatomic spacings by means of X-ray diffraction. Finally, if the charge on an electron is independently determined with precision by means of Millikan's oil-drop experiment (to be discussed in the next chapter) the resulting value can be used to calculate Avogadro's number.

7. SUMMARY COMMENTS

With the discoveries discussed in this chapter, it is quite apparent that a new dimension has been added to the physical sciences, and it is well to pause and take stock.

Modern science emerged and established itself as being of prime importance in the affairs and thinking of men during the sixteenth and seventeenth centuries. During this period, extending from Copernicus through Galileo and Newton, the combination of carefully formulated logical theories with well-defined experimental methods and observation replaced an earlier anthropomorphic point of view. Science before this had been often tinged with haphazard generalizations and willful thinking, fashioned primarily to fit into the dominant philosophy and theology of the age.

With Galileo's and Newton's work, a vision grew of a world of nature subject to regularities that could be formulated and understood and used for prediction. However, so great was the success of the Newtonian formulation in terms of a few ideas, such as mass, force, gravitational laws, velocity, and acceleration that the focus became narrow, and the laws soon were taken as absolute and unchanging truths.

In the eighteenth and nineteenth centuries, a wide assortment of new discoveries, observations, and experiments began to be made. Chemistry, thermodynamics, electricity, and magnetism in the physical sciences, and physiology, genetics, and bacteriology in the biological sciences were being formulated. At first the various threads of knowledge did not appear closely related, but as regularities were studied more carefully the interrelationships began to indicate themselves.

And, as the newer fields of study became incorporated into "natural philosophy," new methods and new points of view were required. Of particular importance was the need for creating new models and new mathematical methods to replace the older and simpler ones used to describe matter in motion. It was not difficult for a layman in the eighteenth century to visualize the workings of Newton's laws, and even to observe them in operation as they found illustration in the operation of the solar system or in the way machines could move masses that were directly visible.

Much of this simplicity, this ease of visualization, together with relatively simple algebraic laws, disappeared as the new sciences developed. An atom cannot be seen. There exists such a myriad of them in the smallest amount of visible matter that the number is inconceivably large.[25] Even the atom could be, at first, regarded as a small sphere of matter obeying the same general laws as those of larger bits of matter. (Only in the twentieth century was this simple model to evaporate.) But during the nineteenth century, the laws of electricity and magnetism dealing with invisible lines of force having complicated patterns of direction required new and more subtle models and, indeed, a much more complicated mathematical formulation.

Finally, by the middle of our own century the general structure of science has departed so significantly from the earlier Newtonian world picture that we are living through a new "revolution in science" as significant in its effect upon the general outlook of man as the earlier revolution that took place in the days of Copernicus, Kepler, and Galileo.

8. QUESTIONS

1. Define precisely what is meant by "potential difference." What unit is used and how is it defined?

2. The Cavendish apparatus to "weigh the earth" (that is, to determine the value of G) and the apparatus that Coulomb used to establish Coulomb's law were very similar in construction. In use, however, there was one important difference. Cavendish had an independent method of measuring masses (by weighing). Coulomb could not measure his charges independently. He could, however, arrange to make the pair of repelling charges equal. Explain how he could do this. Also, if he used two additional identical spheres, he could reduce the charge on each to one-half the original charge. Explain how this was done, stating the theoretical assumptions used.

3. Whereas an electric charge can be isolated, it is generally believed that magnetic poles always occur in pairs. What evidence can you cite to support this conclusion?

4. If a stream of electrons flows from point A to point B as shown below, what direction will the magnetic field take at the point P?

$$A \rightarrow \overset{P\,\circ}{\longrightarrow} B$$

[25] Possibly Lord Kelvin's example of the numbers involved is as good as any other. If all the molecules of a glass of water could be painted bright red and such a glass of water were dumped into the ocean and evenly mixed in all the oceans of the world, each glass of water dipped out of any sea would then contain several hundred red molecules. Another illustration, perhaps more intriguing to some, is that with each breath we take, we inhale many of the same molecules of nitrogen once breathed by Cleopatra.

5. Can an electric charge and a magnetic field interact? If not, explain; if so, explain under what conditions.

6. The electric field intensity everywhere inside a hollow charged shell is zero. Can you suggest a plausible argument why this would be so? Would you expect it to be true as well for the gravitational field inside a hollow shell? How does this lead you to think that the equation for the acceleration of a body falling inside the earth differs from that outside the earth?

7. In what characteristics does a magnetic field produced by a current differ from a gravitational field produced by a mass?

8. Suppose you have two long iron bars, one of them known to be magnetized, the other not magnetized. How can you determine which is which without using any apparatus other than the bars themselves?

9. If you reverse the direction of a current, what happens to the surrounding magnetic field?

10. Although the direction of a current is commonly spoken of as going from positive to negative, most often it is the negative charges (electrons) which are moving. In which direction do these negative charges move when the current flows from positive to negative?

11. Define a Faraday (unit of charge). Trace the arguments that led Faraday to guess that there was an "atom" of electric charge.

12. Set up analogous equations between field strength at a point, that is, the force exerted on a unit mass placed at that point, due to the gravitational attraction of a spherical mass (the gravitational field strength of the earth is g in newtons per kilogram) and field strength at a point due to an electric charge on a sphere. For many purposes, g can be regarded as a constant. Why cannot the field strength outside of a charged sphere be similarly taken as a constant?

13. Study Fig. 13.7 on page 290. If two parallel wires were carrying current in the same direction, would the magnetic force between them be one of attraction or repulsion? It is found convenient to measure the unit of current, the ampere, *directly* instead of in terms of coulombs flowing past a point in one second. The ampere can in this way be defined by the equation $F = k(2I_1I_2l/d)$ where F is the force of attraction (or repulsion) between two parallel wires of length 1 meter carrying currents I_1 and I_2 separated by a distance d. When F is in newtons, I_1 I_2 in amperes, d in meters the equation becomes $F = 10^{-7} \times 2I_1I_2l/d$ (where the constant $k = 10^{-7}$ is exact by definition).

14. An ammeter is a current measuring device which depends upon the magnetic effect of the current carried by a coil of wire. A very simple type of ammeter can be constructed by a circular coil of wire with a compass at its center, the compass needle pointing north when no current is flowing and in the plane of the coil. The tangent of the angle of deflection of the needle is then found to be proportional to the current. Can you think of a way of calibrating such an ammeter? (*Hint.* Consider Faraday's laws of electrolysis.)

15. As we shall consider in the next chapter, units of magnetism (or magnetic field intensity) can be developed by defining them in terms of the effect of a standard current passed around a coil. For field intensity, H, there is a variety of units in use. However, magnetic field intensity gives rise to different effects depending on the medium. The effect is

increased thousands of times when iron is present over what it is *in vacuo*. The effective magnetic field (taking into account the medium) is then designated as B and is measured either in units called gauss or in units called newtons per ampere-meter based on the effect of a standard current circuit. One newton per ampere-meter equals 10^4 gauss. To give you a conception of the magnitude of these units, the earth's magnetic field is somewhat less than one gauss but varies from place to place. Assuming B of the earth is ½ gauss at a certain point, what will be its value in newtons per ampere-meter? A strong electromagnet may easily produce a field strength 3×10^4 times as great as that of the earth. How many newtons per ampere-meter will result from such a magnet? If you multiply units of B by units of charge and by units of velocity (as we shall do in the next chapter) what kind of unit results?

Problems

1. When current flows through a resistance wire electrical power is converted to heat. Show that the heat given off per second is proportional to the square of the current in amperes and to the resistance in ohms. Calculate the heat energy given off by a wire carrying 4 amp for one-half hour if its resistance is 10 ohms. Give your answer in joules, kilowatt hours, and Calories.
2. How many electrons flow past a given point each second in a wire carrying a current of four amperes?
3. Prove that one volt times one ampere is equal to one watt.
4. A very small unit of energy often used in electronics and in atomic and nuclear physics is the electron-volt (ev). Calculate the equivalent energy of 1 billion ev in joules (recall that 1 volt is 1 joule per coulomb).
5. Compare the gravitational and the electrical forces between two electrons at a distance 10^{-9} m apart. (This distance is approximately 10 atomic diameters.)
6. Small electric motors are usually about 80 per cent efficient (that is, about 20 per cent of the input power is wasted in heat and not available to do useful mechanical work). How many amperes will be drawn by such a motor connected to a 110-volt line if it has an effective output of ½ horsepower?
7. If two resistances R_1 and R_2 are connected in parallel, and if the currents through the resistances are I_1 and I_2 caused by a potential difference of V, prove that the total resistance is given by

$$\frac{1}{R} = \frac{1}{R_1} + \frac{1}{R_2}$$

(*Hint:* use ohm's law plus the fact that the total current, $I = I_1 + I_2$. Why?)

8. How many coulombs of charge deposited on the earth and on the sun in equal amounts, but of different sign, would be required to hold the earth in its orbit if there were no gravitational force?
9. If a small object carrying a charge of one-millionth of a coulomb "falls" through a potential difference of one thousand volts, starting from rest, how much kinetic

energy would it acquire? If it had a mass of a millionth of a kilogram what would its final velocity be?

10. Silver has a valance of one and an atomic weight of approximately 108. How much silver will be deposited by electrolysis in five minutes if the current is three amperes?

Recommended Reading

William Gilbert, *De Magnete, Great Books of the Western World,* Encyclopaedia Brittanica, Chicago, 1952.

Michael Faraday, *Experimental Researches in Electricity,* originally published in 1839; reproduced in many places, among them by John Warren Knedler, Jr., in *Masterworks of Science,* Doubleday, Garden City, New York, 1947.

John Tyndall, *Faraday as a Discoverer,* Crowell, New York, 1961. A reprint of the original edition of 1868 (Longmans, Green) with an introduction and notes by Keith G. Irwin.

Morris H. Shamos, editor, *Great Experiments in Physics,* Henry Holt, New York, 1959. This carefully annotated and edited selection of significant experiments, as originally reported by the experimenters, should be consulted repeatedly by those who wish to see the great scientists at work in their laboratories. The selection especially recommended as pertinent to this section of the text is Chapter 5 on the "Laws of Electric and Magnetic Forces" by Charles Coulomb.

Gerald Holton, and Duane Roller, *Foundations of Modern Physical Science,* Addison-Wesley, Reading, Mass., 1958, Chapters 27 and 28.

Jay Orear, *Fundamental Physics,* Wiley, New York, 1961, Chapters 7–9. For a mature and fairly complete presentation of the fundamental ideas of electricity (noncalculus approach).

K. R. Atkins, *Physics,* Wiley, New York, 1966. Although designed for a straight physics course, this text contains excellent illustrations which help to make many of the basic concepts clearer. The section on Electricity and Magnetism, pp. 215–353, is particularly recommended although the student should be warned that the system of units used is different from that used in this text; hence, the corresponding equations may sometimes look different.

chapter fourteen

The electron: experimental determination of charge and mass

As our study moves into the realm of modern physics, one feature becomes more and more prominent. The basic concepts of classical physics (that is, physics up to the twentieth century) either can be directly defined in terms of a specific observation or laboratory measurement or can be analytically defined by simple equations. Thus mass, length, time, temperature, and even charges, can all be defined in terms of easily understood laboratory procedures. Other concepts such as energy, power, force, velocity, and acceleration can be defined in terms of straightforward equations.

In modern physics the problem of stating the relationships between "objects" becomes very difficult for three reasons. First, the size of the objects is so small that they can be investigated only by extremely indirect methods. Second, because of these extremely small sizes, most phenomena under investigation usually involve billions of objects. Instead of tracing a single object, as we can do when examining macroscopic phenomena, we must use methods involving statistical aggregates. (This procedure had already been used as an effective analytical tool in thermodynamics before the twentieth century and is now being significantly extended.) Third, when we pass from the macroscopic realm (involving masses, velocities, forces, etc., which are on a human scale of magnitude) to the microscopic (in which the magnitudes are so small that they escape all possibility of observation by our senses), it is often found that we have no right to extrapolate the "laws" so carefully established in the macroscopic to the realm of the microscopic. New and often startlingly different laws and regularities reveal themselves. Sometimes these new regularities are of such a different nature from those in terms of which we are accustomed to think, that "models" based on visual experience cannot be developed at all, and the only model that can be established is a mathematical one, often involving a highly sophisticated type of mathematics developed specifically for that purpose.

However, these considerations having been pointed out, we should also realize that basically most of modern physics and chemistry uses the same hypothetical-deductive approach that was used in classical physics. This approach is retained even if we renounce the old attitudes and direct ideas so firmly based on visualized experience. Postulated theories are still set up in such a way that a wide variety of deductions can be drawn from these theories. The deductions are then tested against observations or experiment. The most significant differences between the modern postulated theories and the older classical theories are that the new postulates have much less visualizable content, the deductions are extremely elaborate, and the experiments are often of tremendous complexity and refinement.

The modern approach, however, is also marked by a somewhat greater flexibility of imagination than occurred in classical physics. Frequently, today, postulates shock common sense in a more profound way than they did in the past. The test of a good theory in modern science is "does it work," does it lead to good ideas even if it "does not make sense," or "cannot be easily understood." One might say that philosophy has not yet caught up with modern science.

It is convenient to think of modern physics as beginning in 1900, for within five years each side of the turn of this century the face of experimental and theoretical physics was radically changed. Roentgen discovered X-rays (1895), Becquerel discovered radioactivity (1896), Planck proposed the quantum theory (1900), Thomson identified and made the first measurements on the electron (1897), and Einstein propounded his theory of relativity and the quantum theory of photoelectricity (1905).[1]

The identification and measurement of the properties of the electron is the

[1] In a series of three papers, all published in 1905, Einstein established his reputation. Any one of these would have brought fame. He actually received the Nobel Prize for his work on photoelectricity and not for his theory of relativity for which he is better known. His third paper was a definitive study of Brownian motion.

subject of this chapter. The transition to modern physics is begun. At first, there is no sharp break with the past. In fact, the first model for the electron is classical, and it obeys the usual laws of classical physics. Only in the indirect instrumentation is the subject typical of the work in modern physics: we extend our senses through instruments that involve an elaborate theory of operation.

1. IDENTIFICATION; EXPERIMENTAL DETERMINATION OF THE RATIO OF THE CHARGE ON AN ELECTRON TO ITS MASS

Previously we have presented the investigation of the conduction of electricity through metals and liquids. Conduction by gases was discovered in the late nineteenth century, and still another line of investigation was opened up. The basic instrument for these investigations was a glass tube containing two metal plates, called a cathode and an anode, the cathode being connected to the negative terminal of a voltage source, the anode to the positive. The glass tube could be filled with gases of various types, and the conduction of electricity between anode and cathode could be studied. Many interesting phenomena occur when electricity passes through gases at various pressures. The discovery of the greatest importance is that electricity passes between the anode and the cathode, *even if the glass tube is completely evacuated* (Fig. 14.1).

In 1897, J. J. Thomson undertook to find out the nature of this process. It was already known that if the evacuated cathode ray tube was arranged as shown in Fig. 14.2, some kind of mysterious rays seemed to emanate from the cathode passing directly through a slit in the anode and striking the glass at the

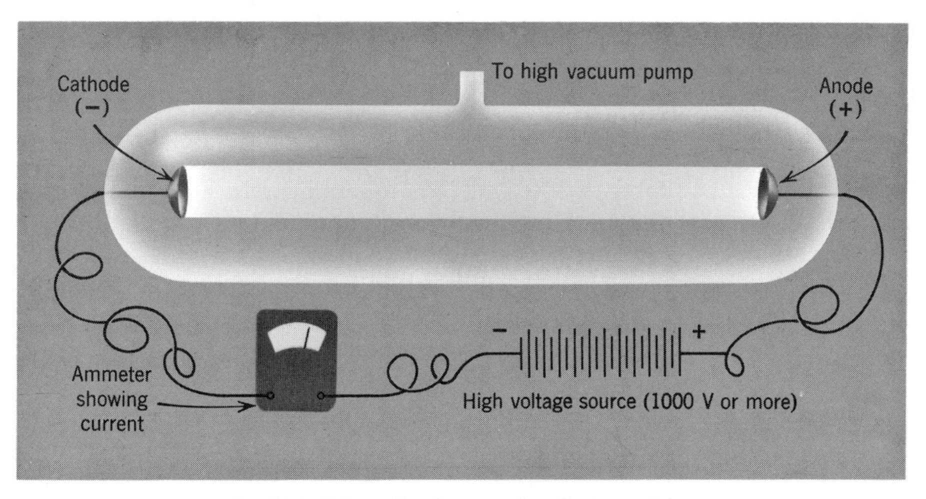

Fig. 14.1. Schematic diagram of cathode-ray tube.

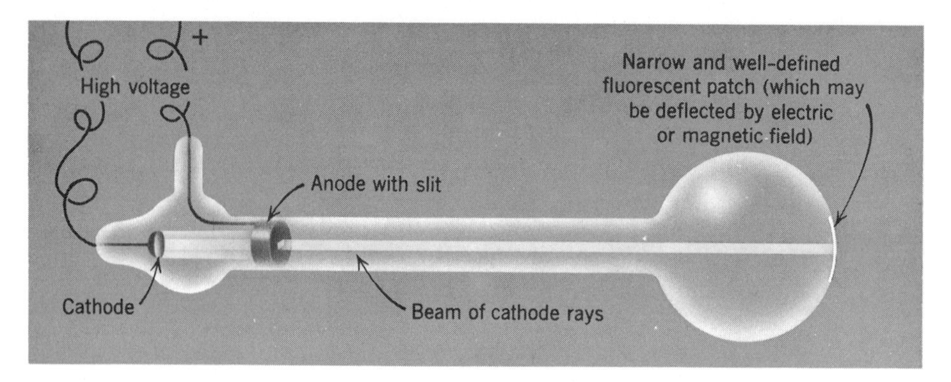

Fig. 14.2. Schematic diagram showing evidence of a cathode beam.

other end of the tube, causing fluorescence. What kind of beam was shooting across the length of the tube?

Two hypotheses had been proposed to explain the nature of these strange "cathode" rays as they are named. The German physicists looked upon them as a kind of *electromagnetic radiation,* similar to light. English physicists leaned toward the hypothesis that these were *particles* that carried negative electric charges from the cathode to the anode.

Thomson settled the problem by his series of experiments. He showed that the particles had mass, although it was extremely small compared to the mass of an atom. Furthermore, he measured the *ratio of the amount of charge to the amount of mass with considerable precision,* finding that his ratio was constant regardless of the material of the cathode acting as the source of the rays. Thus, the existence and some of the properties of what we now call electrons became experimentally determined. The elegance and simplicity of these determinations are worth careful study. With these experiments, begun by Thomson in 1897, was ushered in the period of modern electronic devices.

Two types of apparatus were used by Thomson in his quantitative studies, both depending on the possibility of deflecting the beam of particles from the straight-line path. In one case, a pair of electric plates is introduced into the cathode ray tube (Fig. 14.3a). In the other, a magnetic field at right angles to the position of the electric field is produced (Fig. 14.3b). In both cases a deflection is produced: downward by the electric field as shown, and upward by the magnetic field as shown. An analysis of these deflections lead to the value of e/m.[2]

[2]Originally Thomson used several independent methods to determine e/m, described in his paper in the *Philosophical Magazine,* Vol. 44, Series 5, 1897, page 239, which is conveniently reprinted in *Great Experiments in Physics,* edited by M. H. Shamos, Holt, Rinehart & Winston, New York, 1959. The procedure we explain is outlined in *The Corpuscular Theory of Matter* by J. J. Thomson, Archibald Constable, London, 1907, pp. 2–9.

In Fig. 14.3a, the amount of deflection from the straight-line path as the electron passes between the electric plates can be determined. The electron would travel with a constant velocity toward the right, if it were not for the presence of the electric plates. However the plates cause a downward force and, hence, superimpose a downward acceleration upon the electron. (The problem for analysis becomes very similar to the problem of finding how far a projectile traveling horizontally will fall under the influence of gravity. If we know the force of gravity and the velocity at which it travels, the distance it will fall vertically relative to the horizontal distance it travels at the same time can be easily calculated. See Chapter 6.)

Note that the *gravitational* force acting on the electron is so small as to be entirely negligible. Experimentally, this is verified by the fact that without any field the electron beam travels in a straight line from the slit in the anode to the target point.

If the ratio of charge to mass, i.e., e/m, is to be determined, it is necessary to find the horizontal velocity of the electrons being deflected by the known force.

The procedure to find this velocity is simple. The electric and magnetic fields are "crossed," that is the forces produced are arranged to offset one another. Since each field can be controlled, the electric field by the voltage difference across the deflecting electric plates and the magnetic field by the amount of

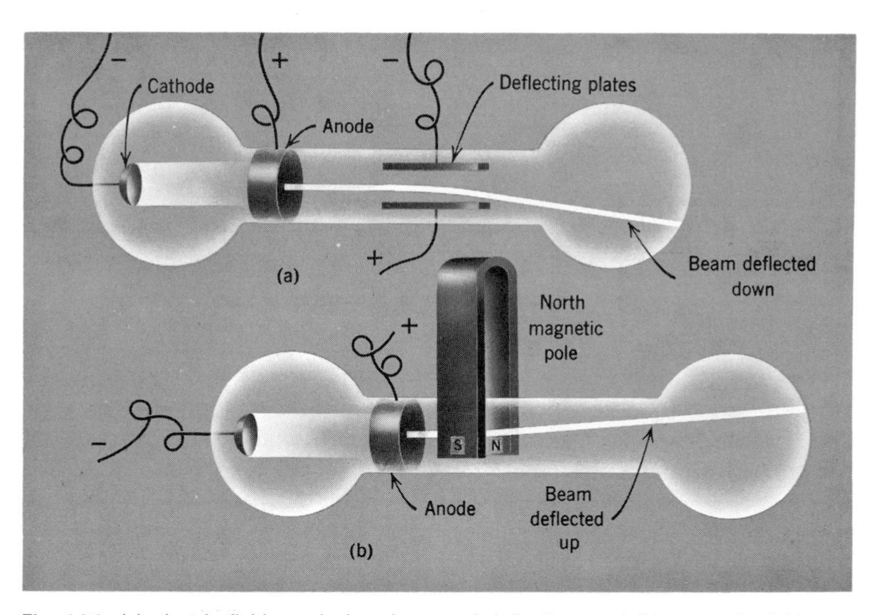

Fig. 14.3. (a) electric field, producing downward deflection; and (b) magnetic field, producing upward deflection. See Fig. 14.4 for explanation of magnetic deflection.

current flowing through deflecting magnetic coils, it is possible to balance the two force fields in such a way that there is a null deflection.

Under these conditions there is an equality of two forces:

$F = Ee$ (1) Where E is the measurable electric field intensity

$F = Bev$ (2) Where B is the magnetic field intensity and v is the horizontal velocity (see Chapter 13, Section 5)

Therefore $Ee = Bev$ (3)

and the velocity becomes

$$v = E/B \tag{4}$$

where both E and B are measurable. Thus the horizontal velocity of the electrons can be precisely measured. This velocity is determined entirely by the voltage difference between the anode and the cathode. Once the electron has passed through the anode, as illustrated in Fig. 14.3 there is no further accelerating force in the direction of its motion. The only forces now acting are the vertical force of the electric field when it is turned on and the force of the magnetic field, both of which operate at right angles to the instantaneous direction of motion, as we shall see.

Once the velocity is obtained by this method of crossing the electric and magnetic fields, the ratio of e/m can be found by using the electric field *alone*. In this case the distance through which the electron drops under the force of the electric field is measured while it travels a known distance. Or else the ratio e/m can be found by using the known magnetic field and finding the radius of the circular path the electron travels over. Thomson studied both. We will consider one method and show how the ratio is obtained when the electron is subject to the magnetic field alone. This is reminiscent of the method used in finding the mass of the earth from the radius and the speed of the moon.

It will be recalled that a planet is kept in its circular orbit by a gravitational force exerted by the sun, producing a central acceleration as follows:

$$F \text{ (the force of gravitation of the sun)} = \frac{m{v_\perp}^2}{R}$$

where $v_\perp$ is the speed or "tangential velocity" of the planet and R the radius of the orbit. In this way it is possible to "weigh" the earth.

Similarly a magnetic field acting at right angles to the motion of the electron will force the electron into a circular path:

$$F \text{ (the force of the magnetic field)} = \frac{m{v_\perp}^2}{R}.$$

We must use the measurable unit for the intensity of the magnetic field strength, similar to the electric field strength, E. This *magnetic field strength B,* is controlled and determined both by the amount of electric current flowing through a coil producing the field and by the number of turns of wire, and the radius of the coil.

It was shown in Chapter 13 that a force is exerted on a wire carrying a current if the wire is appropriately placed in a magnetic field. The reason for this is that a charge *in motion* with a velocity v perpendicularly crosses magnetic lines of force at right angles to its motion. The action of the force is shown in Fig. 14.4.

Therefore in the cathode ray tube, the deflecting force acting at right angles to the direction of motion of an electron is given by

$$F = Bev \tag{5}$$

Since the action is *always at right angles* to the direction of the velocity of the particle, the particle will tend to be driven into a circular motion, similar to a planet moving in a circular orbit. Just as a planet is kept in that path by the gravitational force of the sun always acting at right angles to its instantaneous velocity, so the electron may be kept in a circular orbit by a magnetic force.

This force is Bev. Therefore

$$Bev = \frac{mv_{\perp}^{2}}{R} \tag{6}$$

From the amount of deflection of the cathode beam, the radius R of the partial circle being followed by the electrons can be measured. Since v is already known from equation (4) and B can be measured, the ratio of e/m can now be determined, and is found to be 1.7×10^{11} coulombs per kilogram.

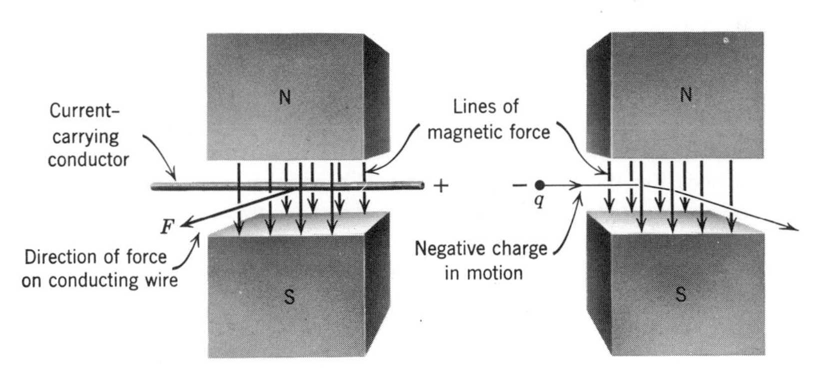

Fig. 14.4. Illustration of the direction of the force exerted by a magnetic field on a conductor carrying a current (*left*), and on a charge in motion (*right*). (cf. Fig. 13.8)

What is the significance of this ratio e/m? It is twofold. (*1*) Since Thomson used many different cathode materials and *always found precisely the same ratio,* the hypothesis arose that there was *some basic particle of matter that always carried the same negative charge* and that this particle of matter existed identically in all substances regardless of their chemical composition. (*2*) If the above value of e/m for the electron were used in conjunction with the estimate of the value of e/M (as determined on the basis of electrolysis), where M is the mass of a hydrogen atom, it follows that *there exists a particle more than 1000 times smaller in mass than the lightest atom previously known.* When Thomson read his paper at a Friday evening discourse at the Royal Institution in London on April 30, 1897, at least one person in the audience said later that he thought the lecturer was pulling his leg.[3]

Quoting Thomson's own observations on the significance of his results:

This constant value is, when we measure e/m in the CGS system of magnetic units, equal to about 1.7×10^7 (or 1.7×10^{11} in the MKS system). If we compare this with the value of the ratio of the charge to the mass of electricity carried by any system previously known, we find that it is of quite a different order of magnitude. Before the cathode rays were investigated the charged atom of hydrogen met with in the electrolysis of liquids was the system which had the greatest known value for e/m, and in this case the value is only 10^4; hence for the corpuscle of the cathode rays the value of e/m is 1700 times the value of the corresponding quantity for the charged hydrogen atom. This discrepancy must arise in one or other of two ways, either the mass of the corpuscle must be very small compared with that of the atom of hydrogen, which until quite recently was the smallest mass recognized in physics, or else the charge on the corpuscle must be very much greater than that on the hydrogen atom. Now it has been shown by a method which I shall shortly describe that the electric charge is practically the same in the two cases; hence we are driven to the conclusion that the mass of the corpuscle is only about $1/1700$ of that of the hydrogen atom. Thus the atom is not the ultimate limit to the subdivision of matter; we may go further and get to the corpuscle, and at this state the corpuscle is the same from whatever source it may be derived.

Furthermore, Thomson's conclusions showed that these "corpuscles" are widely distributed:

It is not only from what may be regarded as a somewhat artificial and sophisticated source, viz., cathode rays, that we can obtain corpuscles. When once they had been discovered it was found that they were of very general occurrence. They are given out by metals when raised to a red heat: you have already seen what a copious supply is given out by hot lime. Any substance when heated gives out corpuscles to some extent; indeed, we can detect the emission of them from some substances, such as rubidium and the alloy of sodium and potassium, even when they are cold; and it is perhaps allowable to suppose that there is

[3]A. E. McKenzie, *The Major Achievements of Science,* Volume I, Cambridge University Press, New York, 1960, p. 283.

some emission by all substances, though our instruments are not at present sufficiently delicate to detect it unless it is unusually large.

Corpuscles are also given out by metals and other bodies, but especially by the alkali metals, when these are exposed to light. They are being continually given out in large quantities, and with very great velocities by radioactive substances such as uranium and radium; they are produced in large quantities when salts are put into flames, and there is good reason to suppose that corpuscles reach us from the sun.

The corpuscle is thus very widely distributed, but wherever it is found it preserves its individuality, e/m being always equal to a certain constant value.

The corpuscle appears to form a part of all kinds of matter under the most diverse conditions; it seems natural, therefore, to regard it as one of the bricks of which atoms are built up.[4]

Finally the very magnitude of the ratio of the charge on an electron to its mass needs emphasis. This ratio is 1.7×10^{11} in coulombs per kilogram—a large ratio indeed! Recall that a coulomb is itself a large unit of electricity. If you could accumulate two like static charges of a coulomb upon each of two spheres one meter apart, the force of repulsion would be 10 billion newtons, almost a million tons! Therefore an electron is subject to a vast force relative to its inertial mass when placed in an electric or magnetic field. This means it can be accelerated or decelerated with extraordinary ease by small fields over short distances. An electron can acquire enormous velocities (thousands of times as great as those of a bullet) over a distance of a few inches when subject to any small electric or magnetic field. This is the secret of the speed and effectiveness with which electronic circuits operate. In a TV set electrons arriving at the fluorescent screen with extremely high velocities may be deflected one way or another by slight voltage changes in a matter of microseconds. Modern electronic computer and communication technology depends upon the large ratio of the electric charge on an electron to its inertial mass.

2. DETERMINATION OF THE CHARGE OF AN ELECTRON; MILLIKAN'S OIL-DROP EXPERIMENT

After Thomson's investigations indicated that an electron was an elementary particle having both a charge and a very small mass, the absolute determination of the amount of the charge became most important for further advances in the field of atomic structure. Once the charge could be determined precisely, many other important physical constants such as Avogadro's number, the mass of a hydrogen atom, and the mass of all atoms could also be pinned down with precision.

[4] J. J. Thomson, *The Corpuscular Theory of Matter,* Archibald Constable, London, 1907, pp. 10–11.

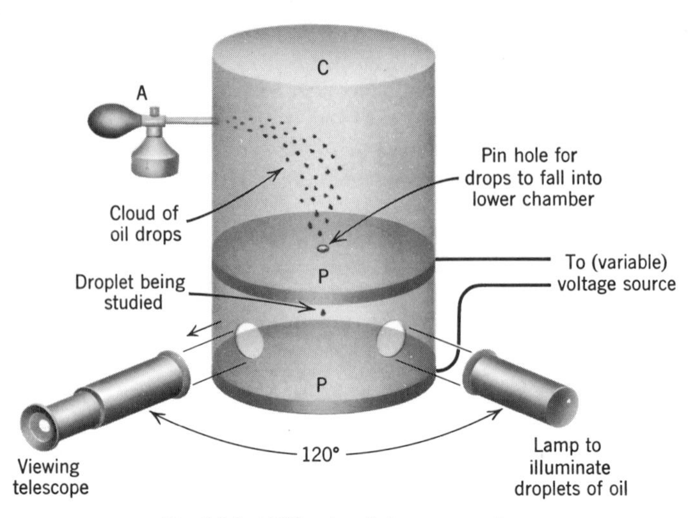

Fig. 14.5. Millikan's oil-drop apparatus.

In 1906, Robert Millikan, the American physicist, undertook the task of making this precise determination. Finally, he succeeded in measuring the charge as 1.602×10^{-19} coulomb in 1911, and in demonstrating that this value was precise within 0.1%. The experiment is beautiful in conception and simple to understand. Like Thomson's experiment, it is typical of modern procedures in its dependence on instruments that measure indirectly quantities in the submicroscopic realm.[5]

A schematic drawing of Millikan's apparatus is shown in Fig. 14.5. By means of an atomizer A, a cloud of fine droplets of oil is blown into the enclosed chamber C. These droplets settle, a few of them passing through a pinhole at the center of the upper plate P. The two plates P and P′ are made of brass, and may be charged to a potential difference by a battery or other source. The electric force field produced between the plates is uniform, and may be regulated by varying the potential difference (voltage) between the plates. Even the direction of the field may be changed by an appropriate switch that can reverse the potential. Glass windows are arranged as shown. Through one window a bright beam of light enters and continues out through the opposite side. The third window serves, as Millikan said, "for observing, with the aid of a short focus telescope placed about two feet distant, the illuminated oil droplet as it floats in the air between the plates. The appearance of this drop is that of a brilliant star on a black

[5]It is difficult to give any meaningful conception of the microscopic nature of either the charge or the mass of the electron. There are more than a billion times a billion electron charges flowing past a given point in a current of 1 amp flowing in a wire.

background."[6] The droplets appear somewhat as do motes of dust dancing in a beam of sunlight.

The observing telescope contains a system of cross hairs, accurately calibrated, so that, as a droplet moves either downward, under the influence of gravity, or upward because of the superimposed and greater effect of the electric field, the distances traveled in a given time may be measured with precision.

When the droplets emerge from the atomizer, a few of them will carry a charge. If a particular droplet carries a negative charge, an upward force will be exerted on this charge by the electric field. According to the same equation that we have studied, this upward force will be

$$F = Eq$$

(where E is the field strength and q is the charge carried by the droplet).

Now it is possible to adjust this field strength to such degree that the upward force on the droplet is just sufficient to offset the downward force of gravity. Under these conditions,

$$Eq = mg \tag{7}$$

(where mg is the effective weight of the droplet, taking account of the buoyancy of air); or

$$q = \frac{mg}{E} \tag{8}$$

The charge q on the droplet can now be determined *if the mass can be measured.* The mass of a spherical droplet of oil can be expressed as

$$m = \tfrac{4}{3}\pi R^3 d \tag{9}$$

(where d is the density of the oil).[7]

The radius of the droplets of oil is too small to be determined by direct measurement. But there is a well-known relationship, called Stokes' law, between the terminal velocity of a small spherical drop falling in a medium (in this case air) of known viscosity and the radius of the drop. So Millikan merely allowed the drop to fall under the influence of gravity when there is no charge on the horizontal plates, and timed the fall of the drop over a known distance. Since the density of the oil was known all the values on the righthand side of equation (9) could be measured. The charge on a single drop could thus be determined with great precision.

[6] R. A. Millikan, "The Isolation of an Ion, A Precision Measurement of Its Charge," *The Physical Review* (1911), Volume 32, p. 35.
[7] The more precise equation is $m = \tfrac{4}{3}\pi R^3(d - \rho)$, where ρ is the density of air.

Millikan, using the ideas outlined above, measured the charges on droplets of oil and noted that, although the charges varied considerably, they were always *an exact multiple of the same figure,* that is, 1.64×10^{-19} coul. Millikan concluded that a droplet of oil becomes ionized or charged by the addition of or subtraction of one or more fundamental packages or quanta of charge—the charge of an electron.

Millikan summarizes:

> With the aid of this [apparatus] it has been found possible to catch upon a minute droplet of oil and to hold under observation for an indefinite period of time one single atmospheric ion or any desired number of such ions between 1 and 150.
>
> In the interval between December 1909 and May 1910, Mr. Harvey Fletcher and myself took observations in this way upon hundreds of drops which had initial charges varying between the limits 1 and 150, and which were upon as diverse substances as oil, mercury, and glycerine and found in every case the original charge on the drop an exact multiple of the smallest charge which we found that the drop caught from the air. The total number of charges which we have observed would be between one and two thousand, and *in not one single instance has there been change* which did not present the advent upon the drop of one definite invariable quantity of electricity or a very small multiple of that quantity.[8]

(The *changes* in charge mentioned by Millikan arise spontaneously during the experiment, when the oil drops are occasionally ionized by radioactivity or cosmic rays always present.)

He concludes that:

> . . . all electrical charges, however produced are exact multiples of one definite, elementary electric charge, or in other words, that an electric charge instead of being spread uniformly over the charged surface has a definite granular structure, consisting, in fact of an exact number of specks, or atoms of electricity, all precisely alike, peppered over the surface of the charged body.[8]

3. DETERMINATION OF PHYSICAL CONSTANTS, USING MILLIKAN'S VALUE OF e: MASS OF ELECTRON, AVOGADRO'S NUMBER, MASS OF HYDROGEN ATOM, AND ELECTRON VELOCITIES

Once an accurate determination of the charge on an electron was made, many other important physical constants were also accurately evaluated. The mass of an electron followed immediately from the knowledge of the ratio of e/m determined by Thomson's experiment. Using Millikan's value of e, and Thomson's value of e/m, the value of the mass of the electron is 9.11×10^{-31} kg. Today,

[8] Reference 6, p. 39.

this value is known to a precision better than 0.01%. The mass referred to here is the so-called rest mass of the electron. As we shall see later, the mass of any object increases when it is given very high velocities (approaching the speed of light) and, since electrons can be given such velocities by modern accelerators, their mass may be made to increase by a significant amount.

Second, Avogadro's number (the number of atoms in a gram-atomic weight) can be determined with great accuracy using Millikan's results in combination with Faraday's laws of electrolysis. When electricity is passed through certain liquids called electrolytes, the passage of the current causes the conducting solution to be decomposed. In Chapter 13 we saw that Faraday, studying this decomposition quantitatively, found that 96,500 coul of electricity always liberated a gram-atomic weight of a monovalent element at either electrode. (It will be recalled that a gram-atomic weight of an element or compound is its atomic weight expressed in grams.) Since a gram-atomic weight of any element always contains the same number of atoms, the total charge that causes the deposit of a gram-atomic weight of a monovalent substance at an electrode must be equal to the number of such atoms times the charge on each atom deposited. (If an element such as sodium or chlorine has a valence of one, then in solution the ionized or charged atom carries only *one* charge, positive or negative. The total charge then is made up of the number of charges carried between the electrodes, one for each atom. If an element has a valence of *two,* each ionized atom carries two charges and, therefore, the total charge is the product of *one-half* the number of ionized atoms times the elementary charge.) It follows, therefore, for monovalent elements (as was shown in Chapter 13, Section 6) that

$$96{,}500 \text{ coul} = N_0 e$$

where N_0 is Avogadro's number, and e is the elementary charge. Since $e = 1.6 \times 10^{-19}$ coul, N_0 is 6.03×10^{23}. This is the most accurate method of determining Avogadro's number, and has been found to agree with other independent methods of calculation. Importantly, the significant constants of physics are usually measured by many independent methods, each of which strengthens our acceptance of the accuracy of the determined values. For example, Avogadro's number has been determined by studying the Brownian movement (see p. 242). Other methods can also be used, including a recent one that amounts practically to a direct count. Helium atoms are emitted by radium, and may be counted by a Geiger counter. Knowing the number of atoms per second emitted by a gram of radium and also knowing the volume or mass of helium emitted, Avogadro's number is readily calculated. Altogether about 15 different and independent methods of determining Avogadro's number have all yielded the same result within the expected experimental error. These confirm both the accuracy with which Avogadro's number is known and the accuracy of Millikan's method.

The accurate knowledge of Avogadro's number enables us to determine the mass of any atom. For example, since 16 g of oxygen (1 gram-atomic weight) contains 6.03×10^{23} atoms, the mass of 1 atom becomes 16 divided by $6.03 \times 10^{23} = 2.66 \times 10^{-23}$ or 2.66×10^{-26} kg.

On this basis, we may also calculate what is defined as an atomic mass unit (amu). The relative atomic weight of oxygen is arbitrarily taken as precisely 16.00; therefore, one-sixteenth of the mass of an oxygen atom is taken as an atomic mass unit, and is determined to be 1.66×10^{-27} kg (more accurately $1.66035 \pm 0.00031 \times 10^{-27}$ kg). The mass of any atom can then be found by multiplying the atomic mass unit by its atomic weight. For example, the hydrogen atom has a mass of approximately 1.67×10^{-27} kg. Other determinations of great precision yield results in agreement with these calculations. (In particular, by using especially designed tubes similar to that of Thomson, a ratio of e/m for charged atoms of many elements has been determined with extraordinary precision). Note that the mass of the hydrogen atom is almost 2000 times the mass of an electron.

It is also interesting to calculate some typical electron velocities. When electrons "fall" between two charged plates, they are subject to an accelerating force that depends on the voltage or potential difference between the plates. The work done on one electron as it falls between the cathode and anode is equal to the voltage difference V times the charge e (joules per coulomb $\times$ charge in coulombs yielding the answer in joules). But this work is equal to the kinetic energy attained by the electron as it reaches the anode. Thus

$$Ve = \frac{1}{2}mv^2$$

Knowing V, e, and m, the velocity of the electron is easily calculated. Such electron bullets travel at extremely high velocities. For example, if an electron starts at the cathode and moves through a potential difference of 2000 volts, its velocity is calculated as follows:

$$Ve = 2 \times 10^3 \times 1.6 \times 10^{-19} = 3.2 \times 10^{-16} \text{ joule}$$

Therefore

$$\frac{1}{2}mv^2 = 3.2 \times 10^{-16} \text{ joule}$$

$$v^2 = \frac{6.4 \times 10^{-16}}{m} = \frac{6.4 \times 10^{-16}}{9.1 \times 10^{-31}} = 7 \times 10^{14}$$

$$v = 2.7 \times 10^7 \text{ m/sec or about 16,000 miles/sec.}$$

Much higher velocities of both electrons and protons can be obtained today in modern accelerators such as cyclotrons. For the highest velocities—close to the speed of light—a change in the mass of the accelerated particles must be taken into account, as we shall show when Einstein's theory is discussed.

Because electrons are so commonly accelerated by letting them fall through a potential difference, another convenient measure of energy has been established, that is, the *electron-volt.* An electron-volt is the energy acquired by an electron, which has been accelerated by a one-volt potential difference. Mev and Bev, in this terminology, designate one million electron volts and one billion electron volts, respectively. Accelerators are now in operation, capable of producing more than 10 Bev. These are used to study the internal structure of the atom, the electrons (or positive ionized atoms) being used as "bullets" that will disrupt atoms themselves. (Recently because of the different meaning of "billion," as used in the United States and in Europe, it has been internationally agreed to use the term Gev, Giga electron volts, for 10^9 electron volts instead of Bev.)

4. ELECTRONS USED IN AMPLIFICATION

Electrons are of interest not only as part of the basic structure of the universe, but also as extraordinarily useful tools for further studies. Through light rays entering our eyes, we learn most about the macroscopic aspect of nature (that is, events taking place on our own scale or larger), but they do not suffice when we study the smallest microscopic dimensions (that is, dimensions that are so small that we cannot detect differences by our senses). In this case, we must use electrons and instruments designed to receive and translate beams of electrons into visual and auditory sensations.

The general principles underlying the use of electrons can be well illustrated by considering two devices, the diode vacuum tube and the triode vacuum tube, and the use of the latter in what is known as amplification.

When a metal plate is heated to a high temperature, the particles composing the metal are set into rapid kinetic motion. Some of the electrons in the metal then gain sufficient thermal or kinetic energy so that they are "boiled" out of the metal. Ordinarily, they do not stray far from the metal electrode out of which they come, since, upon their departure, the electrode is left positive, and they are quickly recaptured by the attractive force. If, however, the electrode is made negative and there is another metal plate nearby which is made positive, the electrons that are boiled out of the negative electrode (cathode) travel rapidly to the positive plate. This is the principle of the simple diode tube, which is sketched in Fig. 14.6.

The triode is a modification of the diode tube. A third electrode (hence the name triode) is introduced between the cathode and the plate. This third electrode is called the *grid* because it does not consist of a solid metal plate

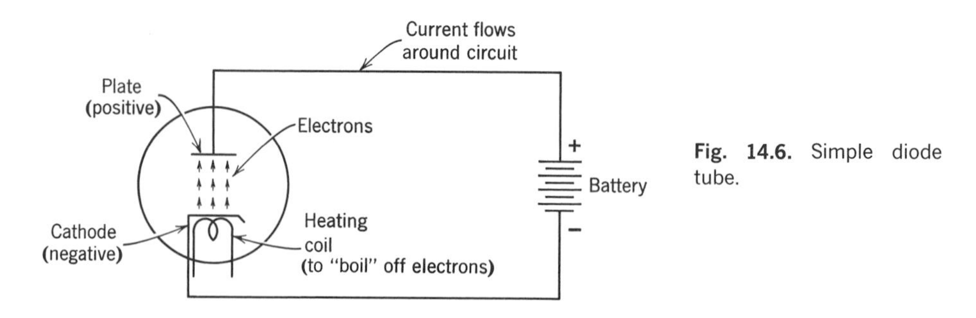

Fig. 14.6. Simple diode tube.

but of a screen of fine wire through whose meshes electrons can easily pass. The grid, if it is made negative, will impede the free flow of electrons from the cathode to the plate. The beauty and effectiveness of the grid lies in the fact that an *extremely small variation* in the charge on the grid will give rise to a *large variation* in the number of electrons that pass from the cathode to the plate. This fact enables us to reproduce a small variation as a large one copying the form of the small variation with great exactness (Fig. 14.7).

Figure 14.8 shows how a very small variation (input signal) in the potential difference applied to the grid is reproduced on a magnified scale as a large variation (output signal) in the potential difference in the plate circuit. (*Note:* when the current varies in the plate circuit, the voltage must correspondingly vary, according to Ohm's law $V = IR$.)

That which is defined as the voltage amplification of the triode tube is given as

$$\text{voltage amplification} = \frac{\text{voltage variation of output signal}}{\text{voltage variation of input signal}}$$

Several such tubes can be arranged in steps, the "output" of one tube being fed into the "input" or grid circuit of another. In this way, very large amplification

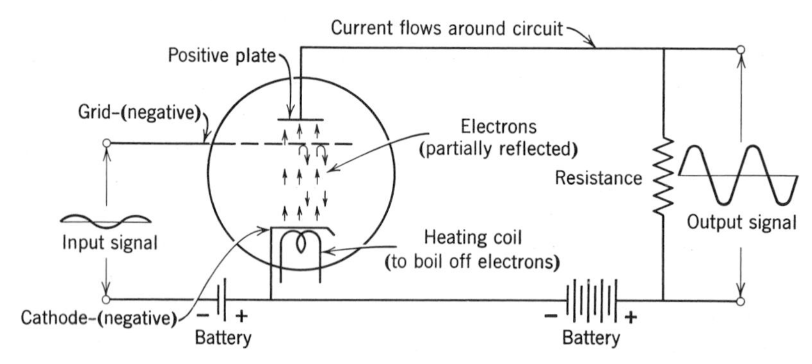

Fig. 14.7. Simple triode tube. The grid impedes the free flow of electrons from the cathode to the plate.

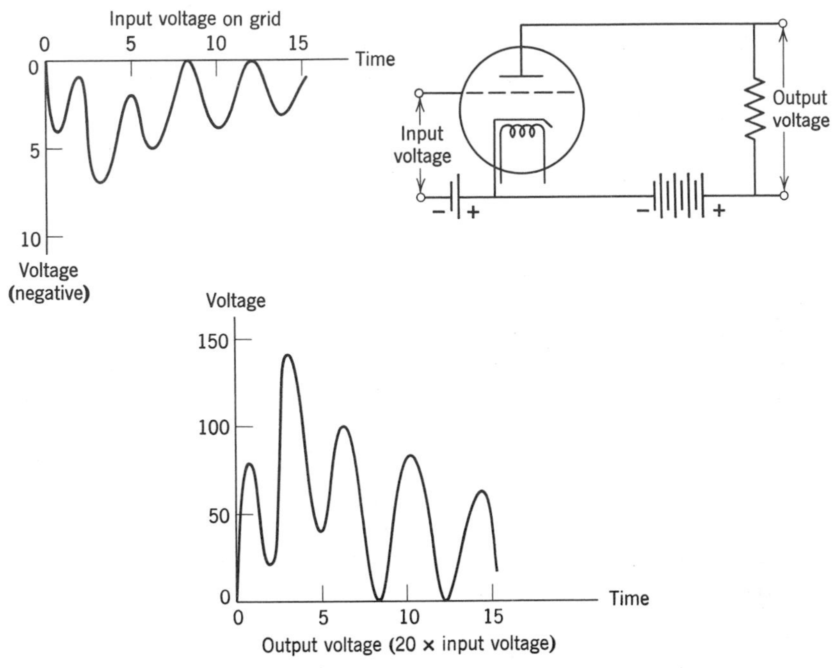

Fig. 14.8. The principle of amplification. Variation in the grid potential generates a magnified variation in the plate circuit.

can be obtained. A radio signal, for example, when picked up by an antenna, may be in the order of microvolts. This may easily be built up into a signal of more than one volt, which may be used to activate the speaker in a radio receiver.

There are so many triumphs as well as problems, and so many technical aspects involved in the use of electronic devices that space does not allow us to go into further detail. The general significance of these developments, however, is important. We have now developed a new and distinctly different method by which the universe may be explored. And new problems of interpretation have arisen. Modern electronic technology has given us what amounts to a new sense organ. This expansion of our senses is significantly different from previous devices. The microscope and the telescope were able to extend our vision by what was essentially a process of *enlarging* the *spatial* and geometric relationships in a simply understood and quite direct manner. With electronic *amplification,* patterns of *temporal* variation are enlarged, and this is done in several steps with intermediate steps that are often indirect and complicated. In both cases, reproduction is the aim: the microscopic photograph *maps* the minute cell being examined, enlarging it to the scale of our senses; the microphone and associated electronic instruments reproduce the energy variations produced by an orchestra hundreds of miles away. However, reproduction is only one of the significant aspects of the electronic devices. Often the output pattern or "signal" must be interpreted by an expert, who is guided by theory to determine the nature of the "input signal." In a way, this is similar to a doctor's process of diagnosis: from an examination of the pupil of the eye, a skilled physician attempts to determine physiological disturbances; from a Rorschach test a psychiatrist attempts to determine the nature of hidden psychoneuroses. So does the physicist, examining an arrangement of blips or spots on a photographic film caused by electrons, attempt to arrive at some conclusion regarding the inner structure of the atom which is so far beyond the reach of our senses (but where physicists are inclined to arrive at a common agreement, psychiatrists do so only rarely). Interpreting the output signal to determine the unseen input signal is one of the modern scientist's tasks, requiring great experience, theoretical knowledge, and skill.

The various electronic devices available to mankind today enable us to acquire and interpret with extraordinary rapidity vast quantities of information that we could not handle before. Thus, in a real sense, we have added fast-operating receptors to the ones that nature endowed us with. As Norbert Wiener, who has had so much to do with the design of modern computers, has expressed it:

> It is my thesis that the operation of the living individual and the operation of the newer communication machines are precisely parallel. Both of them have sensory receptors as one stage of their cycle of operation: that is, in both of them there exists a special apparatus for collecting information from the outer world at low energy levels, and making it available in the operation of the individual or of the machine. In both cases these external messages are not taken *neat,* but through the internal transformation powers of the apparatus.[9]

Out of the rapid growth of electronic devices not only has a need grown to fit the new knowledge gained into our picture of the external world, but also an awareness has developed that communication of meaning in general needs re-examination. Both "Semantics" and "Information Theory" have arisen as critical studies of how symbolic content can be handled and transferred by "machines that think" and this in turn has led to new appraisals of the more general problems connected with transfer of information from one person to another.

5. QUESTIONS

1. In Millikan's experiment, if there is no charge on an oil droplet, will it fall with an acceleration or with a constant velocity? Explain. If an electron is added, but there is no charge on the plates, will the rate of fall be any different? Explain.

[9]Norbert Wiener, *The Human Use of Human Beings or Cybernetics and Society,* Houghton Mifflin, Boston, 1950, p. 15. In this book, Wiener discusses for the nonexpert some of the significant ideas that he developed in his more famous and technical *Cybernetics.*

2. The acceleration of gravity is taken as nearly constant, namely, as 32 ft/sec^2. Would you expect a positively charged sphere to exert an electrostatic attraction to produce a *constant* acceleration on a negatively charged sphere? (Consider dimensions.)

3. For a time it was claimed by another physicist that Millikan did not discover the most elementary charge, but that there were "sub-electrons" of lesser charge than he claimed for the electron. Discuss this possibility, suggesting what kind of evidence might substantiate such a theory. Cite the experimental evidence and arguments which support the view that the charge on the electron represents the smallest discernible charge in nature.

4. Analyze the forces acting on a man descending in free fall after he has jumped from a plane at a high altitude, showing the reason why he will ultimately reach a terminal velocity. (Assume that the frictional force of the air on the man is proportional to the velocity of descent.) Will a child reach the same terminal velocity as the man? Will he reach it more rapidly or more slowly? Explain.

5. It is known that a helium atom has a mass four times as great as the mass of a hydrogen atom. However, the ratio of e/m for an alpha particle (an ionized helium atom) is one-half the ratio of e/m for a hydrogen ion. What can we therefore conclude regarding the charge on an alpha particle?

6. Explain how a charged particle moves if its original motion is perpendicular to the lines of force due to a uniform magnetic field.

7. Why did not Millikan have to take account of the effect of the earth's magnetic field in his experiments?

8. Once in a while, Millikan noted as he watched the drops of oil, that after falling with a constant velocity for some time, a drop would suddenly reverse its motion and ascend. What is the explanation? Would it ascend with a constant velocity?

9. There are two reasons why it is more difficult to give a high velocity to positively charged atoms than to electrons. One of these is connected with relativity theory. Can you think of another on the basis of classical physics?

10. The inside of a hollow spherical metal shell, which is evenly charged with an excess of electrons, has *no net* electric field in any direction. (For this reason, you are safe from lightning in a closed modern car during a thunderstorm.) Can you explain what the connection is between this fact and the inverse square law? (*Hint:* Draw a diagram of a sphere and, taking a point within it at random, consider the area which two cones with apex at the point would subtend at the intersection of the cone with the surface of the sphere.) By analogy, what could you conclude about the gravitational field at any point in the interior of a hollow spherical shell? Can you go on from here and show *qualitatively* that the gravitational force within the surface of the earth does not vary inversely with the square of the distance from the center but according to some other law? Refer to Fig. 21.3 for a geometric diagram of inverse square law.

11. Review the units for B, the magnetic field strength. Show why the units, by definition, must be newtons per ampere-meter.

Problems

1. Gravitational field intensity close to the earth's surface is constant and represented by g in newtons per kilogram. Electric field intensity, which is analogously expressed in newtons per coulomb, is constant in the region between parallel charged plates, but de-

pends on the potential difference between plates (voltage) and the distance between the plates. Show that in the equation $E = V/d$, the units (volts per meter) are equivalent to newtons per coulomb. Calculate the acceleration of an electron leaving the negative plate of a condenser (pair of parallel plates) if the plates are 2×10^{-2} m apart and are at a potential difference of 100 v. (*Hint:* Use $F = ma$ as well as $E = V/d$; also use values for the charge and mass of an electron.)

2. What is the velocity of an electron that has "fallen" through a potential difference of 500 v? (Even higher and more dangerous voltage differences are common within TV sets, where electrons must be shot out at high velocities within the main picture tube.) What energy does such an electron possess in joules? In electron volts?

3. Compare the acceleration due to the earth's gravity of a mass of 10^{-6} kg. with the acceleration of the same mass if it carried a charge of 10^{-6} coulombs due to a constant electric field of intensity $E = 10$ volts per meter.

4. From magnitudes for e/m and e as determined by Thomson and Millikan, determine the mass of an electron.

5. A proton has a positive charge equal in magnitude to the charge on an electron. Its mass is about 2×10^3 times as great. Compare the velocities attained by an electron and a proton when subject to a constant field of $E = 10^3$ volts/meter over a distance of 10 cm. How do these velocities compare with the velocity of light?

6. The *gauss* is commonly used as a unit of magnetic field intensity. It is equal to 10^{-4} newtons per ampere-meter. If an electron has been accelerated by a potential of 500 volts and is deflected by a magnetic field of 1000 gauss what is its radius of curvature?

Recommended Reading

Morris H. Shamos, editor, *Great Experiments in Physics,* Holt, Rinehart & Winston, New York, 1959, Chapter 18 on Millikan's experiment (see the comments on this book under the heading Recommended Reading in Chapter 13).

Robert A. Millikan, *Electrons* (+ *and* −), University of Chicago Press, Chicago, 1947 (revised edition). Those portions of this book dealing with the oil-drop experiment will reveal how beautifully theory and experiment are combined in modern research.

F. W. Constant, *Fundamental Laws of Physics,* Addison-Wesley, Reading, Mass. 1963. This book contains a thorough analysis of the quantities discussed in this chapter. Treatment is fairly advanced, but calculus is not used.

Eric Rogers, *Physics for the Inquiring Mind,* Princeton University Press, Princeton, 1961.

chapter fifteen

Periodicity; waves; sound and music

In previous chapters we have studied patterns and the quality of invariance that enable us to recognize and deal with those patterns. Conservation laws of mass, momentum, and energy have been grouped together in the general attempt to find what remains constant within the flux of nature. But we can recognize at once that patterns may be continuously changing in time, slowly or fast, and yet may be changing in such a way that they repeat themselves. During these transitional changes, nothing seems to remain constant; the original pattern reveals itself, however, if enough time is allowed to elapse.

Actually, this is the basis of time's operational definition. Even the lowest animals have an unconscious recognition of the rhythm that marks most changes in nature. The beating of the heart, the flowing and ebbing of the tides, the waxing and waning of the moon, the rhythm of day and night, of the seasons, of hunger, of sex, and of sleeping and awakening—these and the legions of other repetitive patterns are woven into the web of our existence, yielding both extraordinary variety and recognizable familiarity. Which of these rhythms is to be taken as a standard of elapsed time? Which is the most universal, basic, easily understood, and uniform? The choice is almost inevitable, even if arbitrary: the successive passages of the sun through its highest point in the sky; here is a universal, easily understood standard for the unit of time.

Once this choice is made, however, difficulties ensue. One problem is that seasons have a rhythm of their own. Unfortunately, the repetition is not easily expressible in units of the day, since the length of the year is not exactly divisible by the length of the day. And, from a scientific point of view, even worse consequences follow: if the passage of time is identified and defined by the day, then such apparently uniformly rhythmic repetitions as the swing of the pendulum become subject to small but significant variations. In terms of the sun's day, the pendulum sometimes beats faster and sometimes slower. Similarly, if the day is used as a standard of time, the passage of stars across the meridian is sometimes faster, sometimes slower.

Eventually, it was found that more rhythms were in agreement with sidereal (star) time than the solar time, and the former was chosen. More recently, it has been judged that the rhythm found in the vibrations of atoms is the best available standard, and this has now been accepted. On this basis, variations of sidereal, solar, lunar time, variations in the seasons, and all other variations are regarded as departures from the rhythm chosen as the standard.

1. MATHEMATICAL EXPRESSION OF PERIODICITY

Once the operational definition of time has been clearly determined, it is not difficult to give a mathematical expression to the meaning of rhythm. Suppose that the quantity y is a variable which is a function of the time t. What we mean by a function is that for each value of t, y has itself a uniquely determined value. We express this as $y = f(t)$, leaving undetermined the exact form of the function. For example, $y = 3t$, $y = 6t^2$, or $y = 2/t$ are each special cases of the function $y = f(t)$. Now, to express the property of rhythm, we need a special kind of function, that is, one whose value repeats itself again and again. We need a function, in other words, that has the property expressed mathematically as follows: $y = f(t) = f(t + nP)$, where n is any whole number and P is a constant called the period of the function. For such a function, the value of y of the function at a given time t repeats at a later time (t + the period, or t + twice the period, etc.). Fortunately, we have two well-known functions that have this property: the sine and the cosine of angles. To understand this use of the trigonometric functions, we must extend or generalize them. We have defined the sine and cosine of angles of less than 90° (see Chapter 2). These functions can be easily extended to angles greater than 90°, or even 360°. The manner in which this is done is shown in Fig. 15.1.

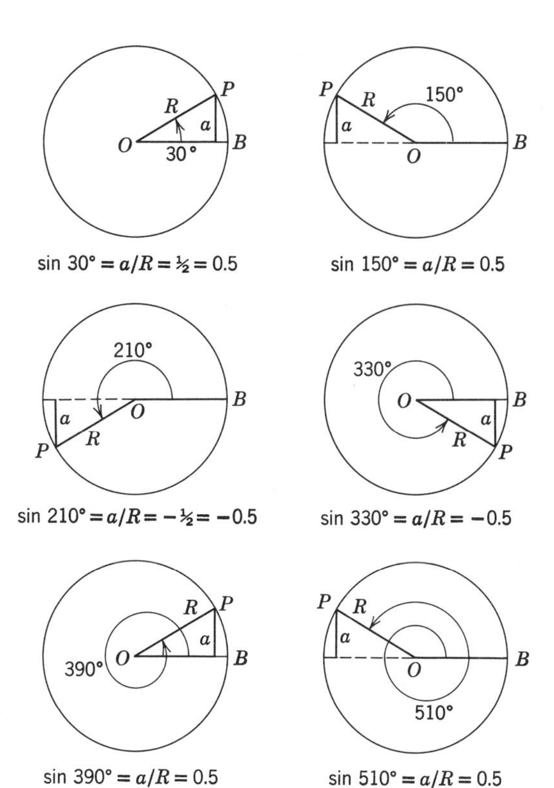

Fig. 15.1. The sine defined for angles greater than 90°.

As the point P moves around a circle of radius R, OP making an angle with the zero reference line OB, a perpendicular dropped to the diameter of the circle as shown will vary in length. The ratio of a/R is defined as the sine of the angle. When the point P is below the diameter, the value of a is taken as negative. When a cycle of 360° has been completed, the sine of the angle then repeats the values originally traced out.

We see that, with this definition, the graph of the function $y = \sin x$ will have the appearance shown in Fig. 15.2.

The value of y, it will be seen, repeats itself as x increases. The same value always reappears after 360° and, therefore, 360° is called the period.[1] The mathematical expression then becomes $y = \sin x = \sin [x + n(360°)]$ where n is any whole number.

Many modifications of this function are possible. For example, $y = 2 \sin x$

[1]Mathematicians and scientists prefer using another measure for angles, that is, radian measure. One radian is defined as that central angle that is subtended by an arc equal in length to the radius. In this measure, $360° = 2\pi$ rad. 2π may thus replace 360° if radian measure is used.

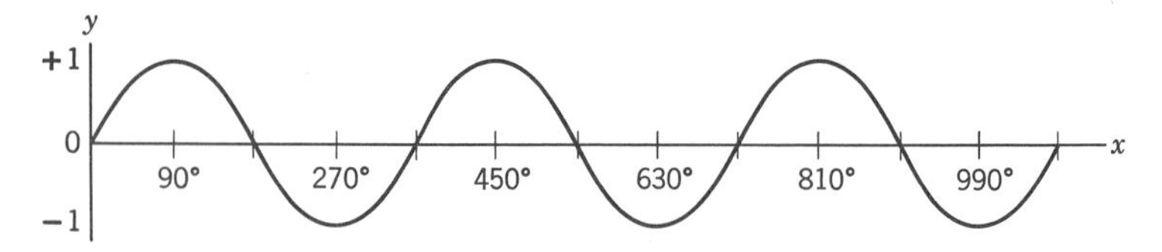

Fig. 15.2. Sine curve.

will be identical in form to $y = \sin x$, except that the y values are all doubled. The number 2 is called the *amplitude* of the periodic function, and represents the maximum departure of y from zero (Fig. 15.3).

Another modification will be apparent if we draw the graph of $y = \sin 2x$, as shown in Fig. 15.4.

In this case, the period of the function is reduced to one-half the original value. In general, for $y = \sin kx$, $360°/k$ is the period.

In addition to these modifications, there is the possibility of combining, by addition, several forms of the function. The results are often irregular-looking graphs, which are still periodic. An example is indicated in Fig. 15.5.

We have reviewed these mathematical functions to illustrate how a changing pattern, which is periodic, can be represented mathematically. In fact, it is not an exaggeration to say that almost any type of periodic variation can be represented by the sum of many properly chosen sine curves.[2]

[2]Joseph Fourier, a French mathematician, developed this theorem in 1807.

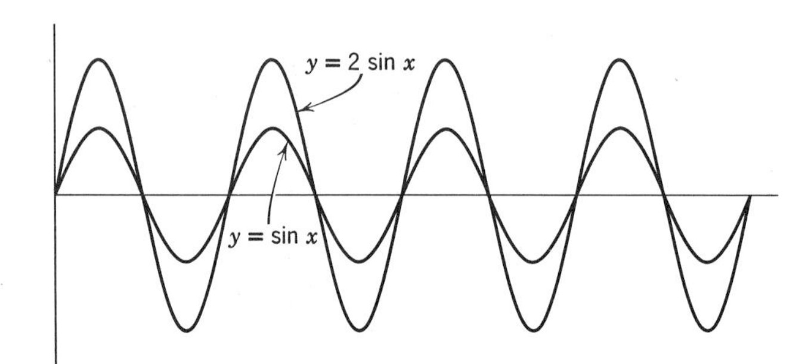

Fig. 15.3.

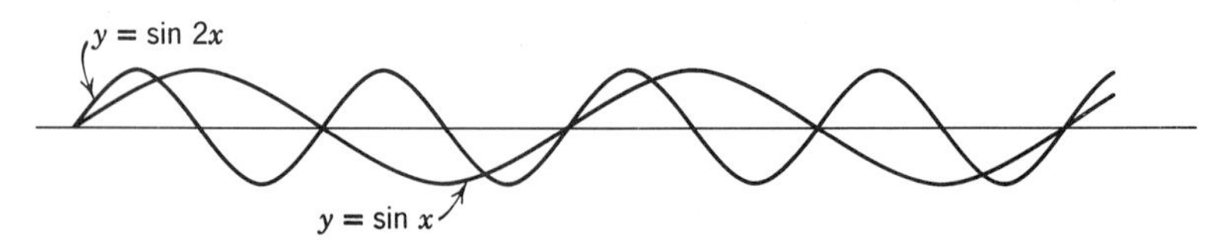

Fig. 15.4.

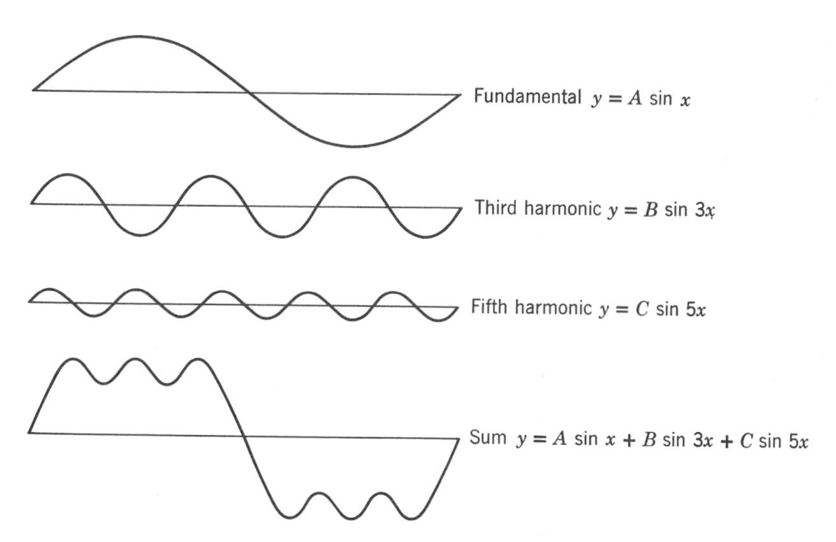

Fig. 15.5. Illustration of harmonics. The sum of several harmonics can produce complicated curves such as the one shown in the final illustration.

2. THE NATURE OF WAVES AND WAVE PROPAGATION

Rhythmic vibrations, whose mathematical expression we have been studying, are important because they give rise to "waves." It is through sound and light waves that most of our knowledge of the external world comes. A wave may be broadly defined as a state of disturbance that is propagated from one point to another at a definite velocity. Up to now, we have omitted any detailed examination of the question of how an action taking place "here and now" can lead to a reaction "elsewhere and later," except in so far as particles of mass move through space. But it is a matter of common knowledge and experience that a disturbance at one point in water, for example, gives rise to a disturbance that travels through the water to another point, although the particles of water themselves are not moving in the direction or with the velocity with which the disturbance is moving. Since this traveling disturbance bears energy, we must examine this energy-carrying process, which differs significantly from the mechanical energy transmitted from point to point by a moving mass.

The most important type of wave is the harmonic or sinusoidal one, and we shall restrict ourselves to this type.[3] These waves originate in the motion of par-

[3]It should not be forgotten that a sudden and single pulse, such as a "shock wave," is a kind of discontinuous disturbance that travels with a definite velocity and often carries tremendous energy. These wave pulses are also analyzed under the general term "wave propagation," but are omitted in our discussion.

ticles of a specific nature, called simple harmonic motion, defined mathematically by the sine function: $y = A \sin kt$, discussed in the previous section. A typical example of such motion, drawn from our experience, is that of a crankshaft attached to a wheel moving with a uniform angular velocity (strictly speaking, this is not quite a simple harmonic motion but closely approximates it). The piston moves up and down with a motion that is graphed as shown in Fig. 15.6. Note that its amplitude is A (the farthest departure from its original position, the position when $t = 0$, and hence $\sin kt = 0$), and that this point is reached when kt is equal to 90° (or $\pi/2$ rad). Furthermore, it will return to its original position *and* will be going in the same direction when $kt = 360°$. This, then, defines the period of one complete cycle, that is, $P = 360°/k$.

Now let us construct a model to aid our analysis. We visualize the end of the piston on our crankshaft attached, in some fashion, to the end of a string stretched horizontally. The end of the string will now be put into simple harmonic motion, and will drag the particles of the string next to it upward and downward. In Fig. 15.7, the positions of several connected particles are traced out, and the resulting positions of the particles are shown.

It will be seen that at the point where the left-hand particle (source) has completed a cycle, another particle down the length of the string is beginning to move upward. The wave disturbance has just reached this point which is at a distance λ, called a wavelength, away from the source of the disturbance. The velocity of the wave is therefore $v = \lambda/P$ (or, if we use the frequency f defined as the reciprocal of the period or the number of vibrations per second, we may express this as $v = \lambda f$). If no energy were lost internally within the string, each particle of the string would, in turn, be sent into vibration with the same amplitude and the same frequency—but at different times—as the source particle at the left end. Energy, therefore, would be propagated[4] down the string with a definite velocity v.

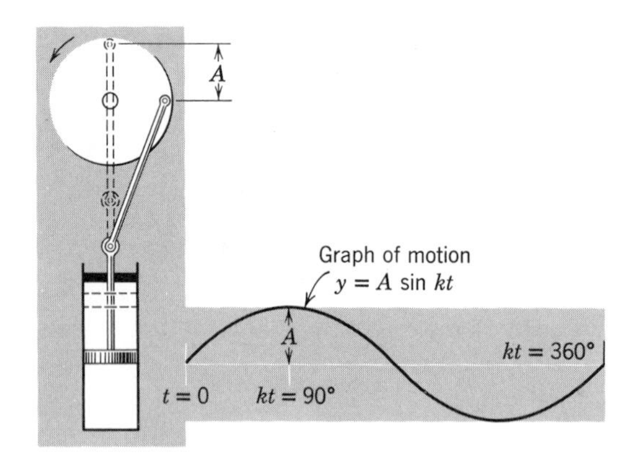

Fig. 15.6. A crank shaft. The piston's motion approximates that of a sine curve.

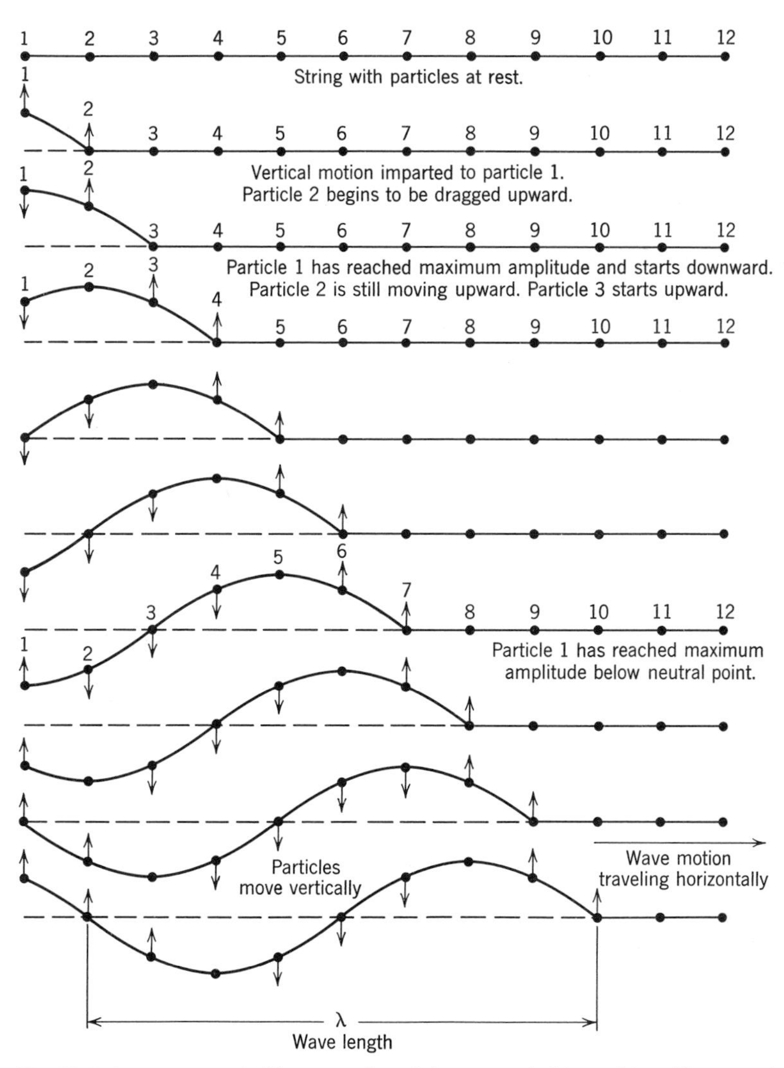

Fig. 15.7. A wave generated in a row of particles connected by a string. The up and down simple harmonic motion impressed upon the first particle generates a simple harmonic motion to the other particles at a later time (or phase).

[4] The full wave equation pictures the disturbance as periodic both in time and in space and, since each particle will vibrate with the same configuration of motion but at a differing time, the equation may be expressed as $y = A \sin [(360/p)(t - x/v)]$, which gives the displacement of any particle at distance x from the origin at the time t.

In our model, the particles are in motion at right angles to the direction of propagation. Such waves are called *transverse* waves. There are, however, other types of waves in which the particles move back and forth about a central position with a motion in line of the direction of propagation. Sound waves are an example of such *longitudinal* waves. All the basic properties involving amplitude, energy, wavelength and velocity, however, take the same form whether the waves are transverse or longitudinal.

In addition, although in our example the wave travels in a single linear direction, waves more often travel outward from the source in a circular (or more commonly spherical) pattern, unless there is some obstruction in the medium. A plunger, moving up and down in water, will act as a source of ripples which spread outward in a circular pattern. In such a case, since each wave front becomes larger as it departs from the source, the energy also must be spread out through the whole wave front. From similar considerations, it will easily be appreciated that for spherical waves, the amount of the energy that flows past or through a given surface area must fall off with the square of the distance from the source (since the same total energy at any distance is spread over a spherical surface whose area is $4\pi R^2$).

This mechanism of the propagation of sound waves is explained in the kinetic theory of gases. A rapidly vibrating source of sound, such as a tuning fork, alternately causes a compression and rarefaction of the contiguous layers of gas. This compression disturbance then spreads to the next layer and so on outward. The process can be diagrammed (Fig. 15.8) where the small dots represent gas molecules. The positions where the dots are closely spaced are to be thought of as points of compression, and the arrows show the *average* direction of the motion of the molecules at that point. In the bottom part of the figure, the compression configuration is shown to have moved to the right. As the wave moves past any point, the *average* velocity of the molecules is alternately to the right and to the

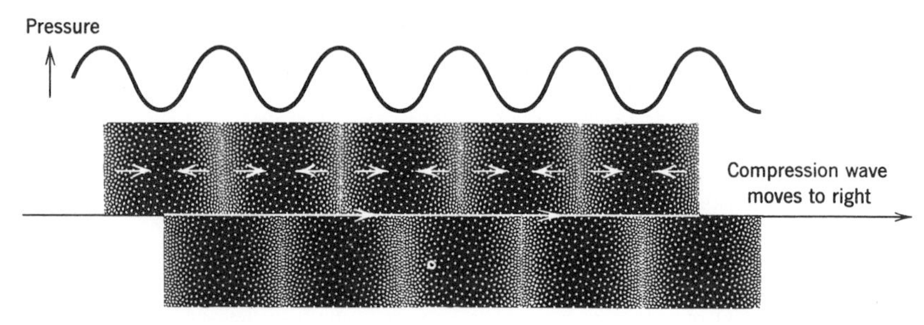

Fig. 15.8. A longitudinal compression wave in a gas moving to the right. Pressure maxima at the start are shown above the diagram. Dots represent gas molecules and their concentration.

left, but the *over-all velocity* of the sound continues on toward the right at a constant velocity. Thus, the sound can be thought to be a wave of changing pressure and density moving out from the source. When this wave strikes an object, such as, for example, the eardrum, the variations in pressure will cause the object itself to be set into vibration, and the sound energy is at this point absorbed or transformed into mechanical energy once more.

3. WAVE PROPERTIES; INTERFERENCE AND DIFFRACTION

With some wave patterns, such as those in water, which are directly perceptible, it is not difficult to understand and measure such characteristics as amplitude, velocity of propagation, wavelength, and frequency. But, with other types of waves, such as sound and light, although their effects are sensed directly, the wave patterns cannot be seen, and must be established indirectly. Two properties, common to all waves, that is, interference and diffraction, are invaluable in measuring some of the other characteristics such as frequency and wavelength.

Probably everyone, when watching ripples travel across water, has noticed how two series of ripples, starting from different sources, can travel across and through one another without losing their characteristic patterns after they have crossed. Over the space where they meet, the patterns mingle, resulting in complex shapes, but after they have crossed, the two wave fronts continue independently with the original pattern re-established.

In that portion of space where the two waves are superimposed on one another, the resulting disturbance is called an *interference pattern.* Here, the particles of the medium are subject to a combined motion, due to the influence of each constituent wave. Mathematically, the analysis is quite simple. The resulting pattern may be represented by the sum of constituent sine curves. Where two sine functions of the same period and same amplitude are superimposed, they reinforce each other, if they are "in step" (that is, have the same phase angle); or, if they are exactly "out of step," they destroy one another. See Fig. 15.9. (If their frequencies and amplitudes are different, the result is a pattern similar to that shown as the final illustration in Fig. 15.5.)

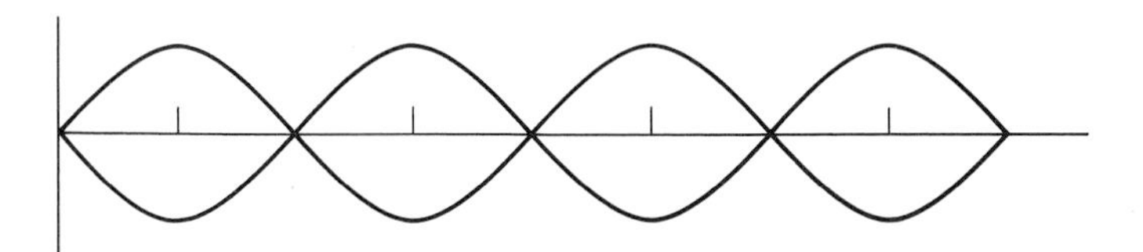

Fig. 15.9. Addition of waves "out of step" by 180° gives 0.

Now suppose that two sources are each producing a train of waves that are being propagated outward with the same amplitude, period, and velocity. At those points of the surrounding region from which the distance to the two sources differs by one-half a wavelength, there will be *destructive interference,* the crest of one wave train arriving just as the trough of the other wave train arrives at the same point, the two waves at all times canceling one another. At those points where the distances to the two sources are either the same or differ by a whole number of wavelengths, there will be *constructive interference.* In the case of sound, for example, if two speakers are giving off the same note, as shown in Fig. 15.10, any point, such as P_1 on the perpendicular bisector, will be a point of constructive interference; but at another point, such as P_2, there will be destructive interference, and a minimum of sound will be heard. (Theoretically, this would be a point of silence. Actually, because of reflection of sounds, because neither the speakers nor the eardrums are points, and because the amplitudes of the sounds may differ, complete silence is only approached.)

As an example, suppose that the distance from A to P_2 is 1½ m longer than the distance from B to P_2, then the wavelength is 3 m. If we take the velocity of sound as about 300 m/sec, the frequency is now determined to be 100 cycles/sec, obtained from $v = \lambda f$.

Of course, the frequency of sound waves is more easily determined from a study of the source of vibration, the vibrations in the medium corresponding to those of the source. But, if there were no such direct way of making this determination, it still would be possible to measure the wavelength by interference methods. In the case of light, as we shall see later, where the frequency is extremely high and the wavelengths very short, interference measurements must be used to give us accurate determinations of frequency and wavelength.

There is yet another property of waves that has played a significant role in the understanding of natural phenomena, that is, the ability of waves to spread

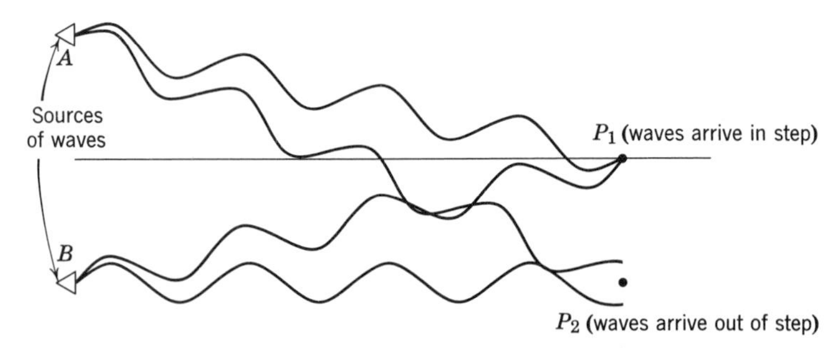

Fig. 15.10. The principle of interference. Waves generate in phase and with same amplitude at A and B arrive at P_1 in step, and reinforce each other. At P_2 they are out of step, and interfere.

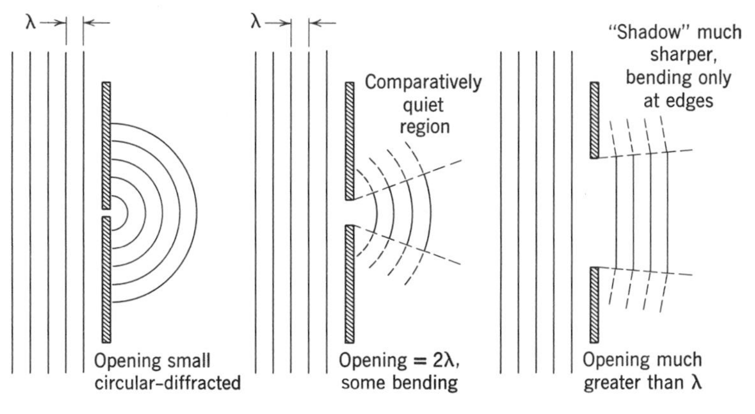

Fig. 15.11. Diagram of diffraction for waves of differing wavelengths.

out or bend around corners, which is referred to as *diffraction.* Possibly those of us who have lived by the water and watched the spreading of waves, both of long and short wavelength, will have noticed that when a wave front meets an obstacle, such as a breakwater, if the wavelength is long, the wave bends around the object much more effectively than if it is short (see Fig. 15.11). The obstacle will cast a fairly sharp "shadow" if the waves are short ripples, otherwise waves will be propagated "around the corner." Similarly, if there is an opening in a breakwater that is small relative to the wavelength, the waves will spread out circularly from the opening as a center. If the opening is large relative to the wavelength, then the opening will cast a fairly sharp shadow.

This explains why persons with keen ears may be able to detect the tonal difference in a hi-fi instrument when listening from another room. Since the longer wavelengths (or lower frequency) bend around corners more easily than the shorter, there is a tendency for the lower notes to be relatively accentuated, the higher ones to be subdued. Proper placement of speakers reproducing the high notes is more critical than for the lower notes.

We need to examine the mathematical analysis showing why short waves cast sharper shadows than the longer. But one reason for emphasizing this difference arises from the controversy that arose in Newton's time over the nature of light. Newton believed that light was composed of particles, since the shadows cast by it are sharp and clear, and light rays give little evidence of bending, as they should if wavelike in nature. The eventual answer was found to be that light waves have an extremely short wavelength and, consequently, the bending around corners is slight, and shadows are sharply defined.

4. MUSIC

Pythagoras discovered many numerical relationships between notes of music and the lengths of plucked strings that emit those notes. He was so impressed with these and other simple numerical relationships that he declared the fundamental stuff of the universe is number. (In more recent times, the Pythagorean doctrine that the ultimate reality of the world is mathematical is echoed in the writings of such scientists as Eddington and Jeans.)

Although few would go this far in speculation, it is certainly true that the sensations of music may be analyzed in terms of mathematical form. That which we distinguish with our ears as notes is determined by the number of vibrations per second (frequency) of the sound waves reaching our ears. These frequencies are determined by the vibration of the source of sound. A plucked string may be used as a typical illustration. If a stretched string fastened firmly at each end is put into vibration by plucking at its center, a wave disturbance (wave pulse) is sent out in each direction symmetrically. The wave is reflected back at each end and then continues to travel back and forth. The net result is to set up in the string what is called a standing wave. The whole string vibrates, as shown in Fig. 15.12a. Instead of the string vibrating in a single segment, however, it may be caused to vibrate in two or more sections (Fig. 15.12b, c) with one or more *nodes* or positions of no vibration at intermediate points.

If the wavelength of the standing wave is determined, the frequency is also

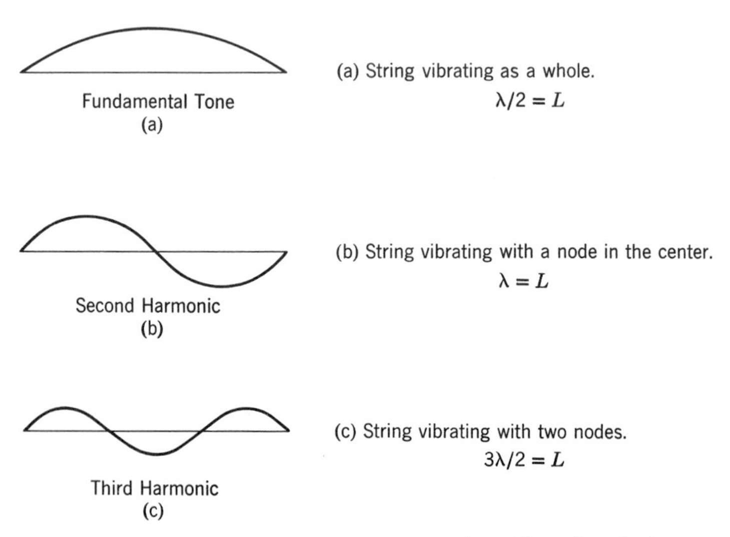

Fig. 15.12. Diagram of standing waves in a string. All modes of vibration may be present at the same time if properly plucked or struck.

determined for a particular string, since the velocity of the wave disturbance, which is moving back and forth between the ends, is fixed by the tension and the weight of the string. Since $f = v/\lambda$, a string may vibrate with any of the frequencies determined by $f = v/2L$, $2v/2L$, $3v/2L$, $4v/2L$, etc. Usually, it will vibrate with several of these frequencies mingled, the quality of the sound being determined by which frequencies or harmonics are present.

The notes on a musical scale are, of course, arbitrarily fixed by convention. It is usual to take A as a standard pitch of 440 vibrations per second. The major scale (Table 1) is then represented by frequencies, bearing a simple ratio to one another.[5]

The notes that, played together, yield harmonious sounds to our senses bear the simplest numerical ratios, as Pythagoras pointed out. The octave C and C' has a ratio of 2:1; the fifth G to C has the ratio of 3:2; the major third E to C has the ratio 5:4; etc. It is extraordinary how our minds and hearing organs take more sensuous pleasure in consonances, which always have simple mathematical ratios. (This is more difficult to explain than the pleasure enjoyed by perception of various rhythms, since the latter are directly sensed in the natural operations of the body: the beating of the heart, the act of breathing, and the natural swing of arms and legs in walking.)

The superimposition of many tones together does, of course, yield an infinite variety of possible patterns. Some instruments accentuate certain of the harmonics which others do not. With modern instruments, such as the oscilloscope, the analysis of the component waves which make up the resultant pattern can be studied in detail. As an example, we quote from Lemon and Ference's book:

Considerations of tone quality (i.e., presence of harmonics) should help in blending instruments of different families. The combination of piano and string quartet to form a quintet is, at best, unfortunate because the piano is essentially a percussion instrument. Replace the

[5]In order to accommodate the minor scales and the various keys, which would introduce too many notes in one octave for keyboard instruments such as piano and organ, an even-tempered scale has been developed, giving twelve notes per octave, each note bearing the same frequency ratio to the preceding note, that is, $2^{1/12}$. For a discussion of the even-tempered scale, see H. B. Lemon and M. Ference's *Analytical Experimental Physics,* University of Chicago Press, Chicago, 1943. Chapter 37 gives an excellent survey of the principles involved.

TABLE 1. THE MAJOR SCALE

Notes	C		D		E		F		G		A		B		C′
Frequency	264		297		330		352		396		440		495		528
Ratio of frequencies		$\frac{9}{8}$		$\frac{10}{9}$		$\frac{16}{15}$		$\frac{9}{8}$		$\frac{10}{9}$		$\frac{9}{8}$		$\frac{16}{15}$	

piano by a clarinet, and we have a combination rich in overtones, which has been sadly neglected by composers—the clarinet quintet [examples cited by the authors are Brahms Opus 115 and Mozart's K 481]. The partial tones of the clarinet are sufficiently like those of the strings so that homogeneity of tone is preserved when needed; on the other hand the clarinet tone is sufficiently different to enable its melodic line to be discerned with ease in an intricate developmental section or in a quiet melodic section. Similar analysis should be applied by composers to other possible combinations.[6]

5. ENERGY PROPAGATION; INTENSITY OF SOUND

The transmission of energy from one point in space to another, without the necessity of an accompanying transmission of mass, is perhaps the most significant aspect of wave motion. It is largely through such processes that the physical unity of the universe is maintained, in spite of the diversity of its massive parts. Through energy propagation, both information and interactions can take place between widely separated masses. "Action at a distance" begins to lose some of its mystery.

Upon what factors does the energy of a wave depend? Instinctively those who have traveled on the sea recognize that high waves (large amplitude) coming with high frequency are more dangerous than those that are small and long. Analysis shows that the energy per unit volume of a wave is proportional to both the square of the amplitude and the square of the frequency. We cannot present a rigorous proof of this in these pages, but the plausibility of this conclusion can be easily indicated.

Consider a mass moving in simple harmonic motion (the type of vibration that we are considering). If the mechanical energy is considered to be conserved (and, in this analysis, we omit consideration of loss of energy through heat dissipation), its energy at the point of greatest amplitude, where the velocity is zero, is all potential energy. The total amount of energy is constant throughout the cycle, changing to kinetic energy at the bottom of the swing as previously discussed. To determine the relation between amplitude and energy, we shall reconsider the pendulum as an illustration of oscillatory motion.

At its greatest amplitude the pendulum has potential energy of mgh (see Fig. 15.13). We inquire first into the relationship between h and the amplitude A, when the angle of displacement is small (only for small angles does the pendulum oscillate with simple harmonic motion). In the figure, it is seen that L is the hypotenuse of a right triangle with legs A and $(L - h)$.

[6] H. B. Lemon and M. Ference, *Analytical Experimental Physics*, University of Chicago Press, Chicago, 1943, p. 468.

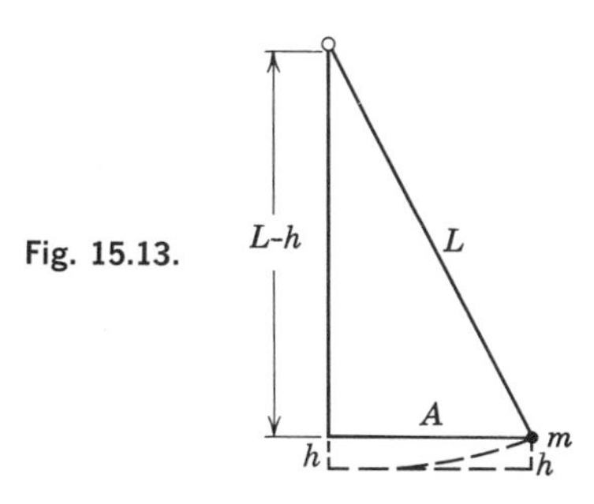

Fig. 15.13.

Therefore

$$L^2 = A^2 + (L - h)^2$$

or

$$L^2 = A^2 + L^2 - 2Lh + h^2$$

whence

$$h(2L - h) = A^2$$

Now, if the angle of swing is very small, as we assume, h will be very small compared to L, and the equation may be very closely approximated[7] by

$$2hL = A^2$$

or

$$h = \frac{A^2}{2L}$$

Since energy of the mass is mgh,

$$\text{energy} = \left(\frac{mg}{2L}\right)A^2$$

or

$$\text{energy} = KA^2$$

that is, the energy of the particle m is proportional to the square of the amplitude.

Carrying the analysis a step farther, since the period of a pendulum is given by

$$P = 2\pi\sqrt{\frac{L}{g}}$$

[7]Without rigorously justifying this approximation, it essentially amounts to this: to the extent that the pendulum approximates simple harmonic motion, to the same extent the above approximation is valid.

and, since the period and frequency are reciprocals,

$$\frac{1}{f} = 2\pi\sqrt{\frac{L}{g}}$$

whence

$$L = \frac{g}{4\pi^2 f^2}$$

Therefore, substituting in the energy equation above, the energy of the mass is

$$mgh = \frac{mgA^2}{2} \cdot \frac{4\pi^2 f^2}{g} = 2\pi^2 m f^2 A^2.$$

Thus, the energy per oscillating particle is proportional to the square of both the frequency and of the amplitude. If there are n particles per volume of the medium through which a sinusoidal wave is traveling, the energy per unit volume becomes

$$E = 2\pi^2 d f^2 A^2$$

where $d = nm$ is the density of the medium.

Now we are in a position to specify the intensity of a wave, which is defined as the rate of energy flow per unit time in the direction of propagation per area of surface perpendicular to the direction of propagation. We measure intensity in *watts per square meter*. (The intensity equation then becomes $I = 2d\pi^2 f^2 A^2 v$, where v is the velocity of the wave. Note that the energy per unit volume must be divided by time to give power per unit *volume*, and multiplied by the distance traveled by unit time to give power per unit area. But distance traveled divided by time yields velocity.)

It is interesting to see how extremely small is the power per square meter (or intensity) of sound at the lower limit of audibility. Using appropriate figures for air, this threshold intensity of sound is about 10^{-12} watts per square meter. This is an extremely small rate of energy flow (10^{-12} watts is millions of times less than the power used by a small insect in flight). Our ears are thus extremely sensitive organs for the detection of energy flow.

Can you show that the amplitude of a spherical wave must decrease inversely as the radial distance from the source? Can you then plot a sinusoidal wave showing a decrease in amplitude?

The physiological sensation of loudness does not increase in simple ratio with the intensity of sound as measured in watts per square meter. Instead, the level of loudness increases as an exponential power of the intensity. Thus a scale of loudness has been developed with the bel as its unit. The threshold of hearing (about 10^{-12} watts per square meter) is assigned a loudness value of 0 bels. Ten times this intensity has a value of 1 bel (or, in more convenient units, 10 decibels). If the intensity is 10^2 times as great as at the threshold, the loudness level is 2 bels (20 decibels), and so forth. At an intensity of about 10^{12} times as loud as the threshold of intensity, when the decibel level has reached 120, the sounds begin to cause pain. Ordinary conversation is carried on at a level of about 40 decibels.

This wide range of intensity, which the ear can accommodate, is matched by the wide range of frequencies to which the ear is sensitive. The normal ear can hear notes ranging from as low as 20 vibrations/sec up to about 20,000 vibrations/sec. (Many animals and insects emit and respond to sounds of still greater frequency. The bat uses the echoes of the sound it emits to guide it in flight, somewhat as radar is used. It is advantageous to have short wavelengths for this purpose, and the frequency of the bat's cry is high and well above the audible range for humans.) We shall see, in our study of light, that the ear has a much greater range of response to frequency than the eye, which responds to the frequencies of vibration of light in terms of color. The eye can perceive only approximately one octave of frequencies against the many octaves perceptible by the ear.

6. WAVE TRANSMISSION OF INFORMATION; EXAMPLE OF SEISMIC WAVES

Intimately associated with the wave propagation of energy is wave transmission of information. By analyzing waves, information can be gathered both about objects emitting radiation and about the medium through which the radiation passes. Modern scientific instruments have so extended the range and the power of discrimination that, for special purposes, they outperform even the eye and the ear—those amazingly flexible and sensitive receptors of waves. Scientific instruments can measure wider ranges than biological sensors, and can better discriminate minute variation of amplitude, frequency, and source direction. They often respond more rapidly. In addition, whereas biological receptors depend upon radiation spontaneously emitted or reflected from the objects perceived, scientific receptors may be used in conjunction with companion instruments which emit radiation. (The bat with its echo-sounding squeaks and the deep-ocean lantern fish that illuminates its surroundings, are exceptions in nature.) Radar, sonar, X-rays, electron-microscopes, laser beams, artificially produced seismic waves—these are but a few of the wave-emitting mechanisms that are in use today as probes into nature.

One illuminating and easily grasped use of wave analysis is well-illustrated by the geologist's use of seismic waves to learn more about the structure of the earth. Information so gathered is, of course, coordinated with other data being accumulated, including data about such diverse phenomena as the earth's magnetic field, variations in g, the orbits of artificial satellites, the abundance of radioactive materials. By a combination of these studies the geologist can "see" into the center of the earth and delineate its structure with a considerable degree of confidence.

Modern geological surveys have confirmed the fact that the earth's center is much denser than the surface rocks. More than this, they have shown that the central core, most probably composed of dense silicates such as those of iron and nickel, is a molten fluid. It is this ferromagnetic fluid, constantly kept in circulation by the earth's rotation, that is thought to explain the earth's magnetic field.

The seismograph is the geophysicist's instrument for recording the amplitude, the frequency, and the time of arrival of tremors of the earth due to earthquakes or to induced shocks. Today these instruments are widely scattered. The arrival of waves generated by an earthquake is recorded at many stations. The velocity and nature of the wave and the source of the earth's local rupture can then be determined. In addition, with enough stations established, the attenuation of the waves, i.e., the decrease in amplitude with distance from the earthquake's focus, can be measured.

From an analysis of these data it has been found that three kinds of mechanical waves are simultaneously generated by an earthquake:

(1) *P* or primary waves. These are compression or longitudinal waves and travel the fastest within the earth.

(2) *S* or shear waves. These are transverse waves and arrive at a seismograph station later than the *P* waves.

(3) Surface waves. These have complicated motions and travel slowly in upper parts of the earth. Ocean waves are a type of surface wave.

These three types of waves are easily distinguished on a seismogram.

Perhaps the most fundamental information concerns the state of the material within the earth. It was recognized early in this century that for a given earthquake, the kinds of seismic waves and their velocities recorded at various seismograph stations varied considerably with the location of the station with respect to the earthquake focus. For example, the velocity of *P* waves increased, the deeper they penetrated before reaching the station. This revealed a fairly consistent increase of velocity with depth in the earth. Since the velocity of a wave depends on the density of the medium, an increase of density with depth must follow. See Fig. 15.14.

However, more importantly, scientists noticed that for many stations only *P* waves were recorded; the *S* waves somehow were disappearing between the

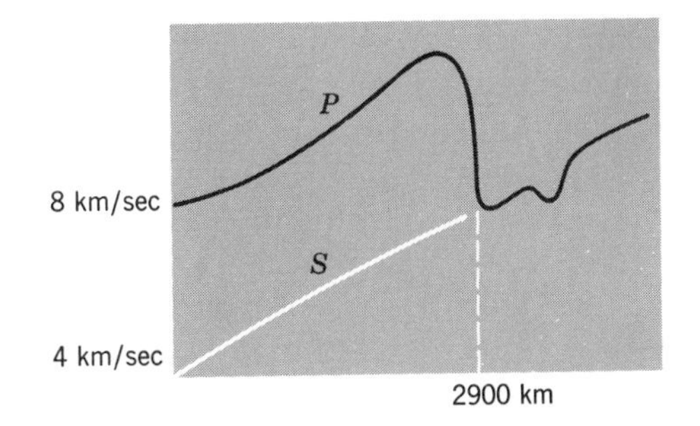

Fig. 15.14. Chart of velocity of *P* and *S* waves plotted against depth. At 2900 km *S* waves disappear and velocity of *P* waves suddenly drops. This point marks the boundary of the earth's core. (Another anomaly shown suggests another boundary between an inner and outer core.)

focus and some of the seismograph stations distributed around the earth. As seen in Fig. 15.15 this distribution indicates that there must be an inner region of the earth (the core) of radius approximately 2160 miles which does not transmit shear (transverse) waves. It is known that solids transmit shear waves, but that liquids do not; the conclusion follows that the core of the earth is liquid. Additional geophysical evidence concerning the thermal and the magnetic properties of the earth fortifies the conclusion.

In this geophysical theory of the earth's core, we have another illustration of how the grand theories "about the environment of man" are built. Many independent studies and different approaches to the same subject eventually interlock and support each other. The geophysicist studying changes in the earth

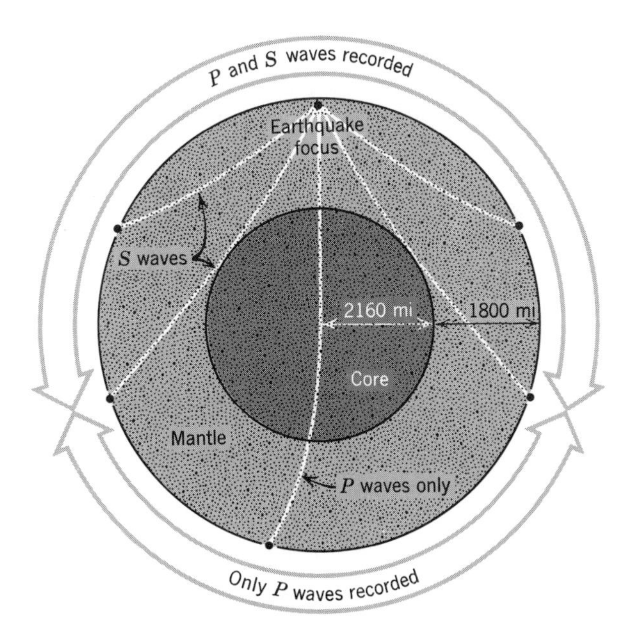

Fig. 15.15. *P* and *S* waves in the earth. The wave paths are bent (refracted) because seismic velocity increases with depth. Since *S* waves are not seen for paths that pass through the earth's core, the core is thought to be liquid.

confirms what Newton surmised three centuries earlier. It was Newton, studying, not the earth beneath, but turning his eyes heavenward and analyzing the motion of the moon, who first concluded that the central core of the earth must be much denser than the surface. He calculated the mass of the earth from the moon's orbit. Knowing the earth's radius, he found that the average density of the earth was much greater than that of the surface rocks. Modern geology by another route has now confirmed Newton's conclusion.

7. QUESTIONS

1. State, in your own words, as clearly as possible what is meant by periodicity. Give several examples based upon your own observations. Now study the mathematical definition of periodicity, and see how the definition applies to these examples. Is a cardiograph a good example of periodicity? What is the difference between "noise" and a musical note?

2. Hold a tuning fork up to your own ear and slowly rotate it as it sounds. What happens and what is the explanation?

3. Can you think of a qualitative reason for expecting the speed of sound to increase with temperature? (Consider the kinetic theory of gases.)

4. Do you expect the speed of sound to be greater or less in hydrogen than in air? (Again consider the kinetic theory of gases.)

5. Will a tuning fork vibrate for the same length of time when its base is placed on a sounding board as when it vibrates (with the same original amplitude) when not so placed. Explain in terms of energy dissipation.

6. State a common experience that shows that sound must travel at a constant velocity regardless of frequency. (Consider orchestral music, listened to at different distances from the orchestra.)

7. Why is it that in a well-designed piano, the hammer strikes each string at a slightly different position for each separate note? (*Hint:* A node cannot exist at the point where the hammer strikes the string.)

8. Why is it that solids, but not liquids or gases, transmit both transverse waves and compression waves? Gases and liquids cannot transmit transverse waves except at interfaces. (This principle is important for geophysicists who study the structure of the earth through seismographs.)

9. If a person fills his lungs with helium and then sings or talks as he expels the gas his voice changes in pitch. This is because the resonant frequencies are changed. Explain why this takes place. (*Hint:* Use the fact that the size of the vocal cavities remains unchanged.)

10. A "whispering gallery" has an ellipsoidal dome. Why are the sounds emitted at one focal point reflected to the other focal point?

11. If a particle moves in simple harmonic motion, where will its acceleration and velocity be at a maximum and at a minimum? Where will its kinetic energy be the maximum?

12. How would you define the intensity of a sound? In what units is intensity expressed?

13. If the velocity of sound in water is approximately 1500 meters per second, compared with the velocity of sound in air which is about 330 meters per second, contrast the relative amplitude of sound waves in air and water when the intensity and frequency are the same. Can you think of a general reason for this difference in amplitude?

Problems

1. On studying ocean waves an observer notices that when the crest of a wave is located at one buoy, the trough is located at another buoy which is 50 meters distant from the first in the direction in which the waves are travelling. If the crests pass the first buoy at the rate of 12 per minute, what is the velocity of the waves?

2. Show how the velocity, frequency and wavelength of waves must be related by considering the dimensions of each.

3. What is the amplitude, the period and the frequency of $y = 5 \sin 4t$?

4. Draw a curve of $y = 4 \sin x$ and $y = 2 \sin x$, and then draw a single curve showing the addition; that is, $y = 4 \sin x + 2 \sin x$. Do the same for $y = \sin 2x + \sin x$. Do the same for $y = \sin 2x + \sin 3x$. What is the amplitude and period of each?

5. The Greek philosopher and mathematician Pythagoras found that when the ratio of two notes of frequency f_1 and f_2 was equal to a ratio of two integral numbers $n_1 \div n_2$ such that n_1 and n_2 were between 1 and 6, the notes sounded together formed pleasing harmonic combinations. In the diatonic musical scale (differing slightly from the "even-tempered" scale) the frequencies of the eight notes in the octave have ratios to "do" of 1, $9/8$, $5/4$, $4/3$, $3/2$, $5/3$, $15/8$, 2. Which of these notes combined will form two-note combinations that are pleasing according to the Pythagorean principles?

6. The wavelength of a certain musical note is $3/4$ meters. What is the frequency? (Speed of sound taken as 330 meters/sec.)

7. The speed of sound can be shown to be equal to the square root of the ratio of pressure to density. Show that this statement is dimensionally correct. Then, using the ideal gas law, show that the speed of sound must therefore be proportional to the square root of the absolute temperature.

8. If the speed of sound at 0°C is 330 meters per second, what will the speed be at the high altitudes at which modern jet planes fly when the temperature is −100°C. (See problem 7.)

9. Two similar waves are generated in phase at points A and B and arrive at P_1 as in Fig. 15.10. Under what condition is maximum intensity recorded at P_1?

10. How many octaves can be heard by the normal ear? A bat emits sounds of wavelength somewhere between 0.1 and 0.2 inches. Why are these called ultrasonic notes?

11. If the A string of a cello (frequency of 220 vibrations per second) is bowed and, at the same time, touched lightly in the middle by a finger, the string will vibrate in two sections. What will be the frequency of this note? If by touching it at another point it vibrates in three parts, what will be the frequency? What will the emitted note be? How is the length of a vibrating string related to the frequency?

12. If you are familiar with logarithms and slide rules, state the way in which a piano resembles a slide rule. Also, taking C as 264 vps, show that E has a frequency about 1 per cent different on the even-tempered scale from what it has on the natural diatonic scale. Some persons can actually detect this difference rather easily if they first sing the notes do, re, mi, and then strike E on the piano, comparing the pitch of mi, as sung, with the note E as sounded on the piano.

13. If the threshold of sound has a power intensity of 10^{-12} watts per square meter, what is the approximate power intensity of the sound of an airplane motor, which has an intensity of 100 decibels?

14. What is the approximate amplitude of the particles of air molecules vibrating with a sound of 264 cycles per second at an intensity level of 20 decibels. Take the density of air to be 10^{-3} gm/cm^3. (Note how small this amplitude is and consider the delicacy of the response of the human eardrum!)

15. A contralto is singing at about 60 decibels in a closed room of dimensions $3 \times 3 \times 3$ meters. Assuming that all of the energy is absorbed and retained by an acoustic layer of material 0.01 m in thickness, and that this layer has a density of 0.5 g/cc and a specific heat of 0.01 cal/kg/°C, how much does the temperature of this layer of material rise after one minute of singing?

Recommended Reading

Alfred North Whitehead, *An Introduction to Mathematics,* Oxford University Press, New York, 1958, Chapters 12 and 13. This small volume written for the nonexpert by the distinguished mathematician and philosopher was first published in 1911, but is now available as a paperback. It is beautifully written, and should be read by all students interested in science. Chapter 12 is devoted to periodicity and is followed by a chapter on how trigonometry is applied to periodicity.

Willem A. von Bergeijk, John R. Pierce, and Edward E. David, Jr., *Waves and the Ear,* Doubleday, Garden City, N.Y., 1960 (an Anchor book). An interesting elementary presentation, covering both the physics of sound and the physiology of hearing.

James R. Newman, *The World of Mathematics,* Simon & Schuster, New York, 1956. Part XXIV, an essay on the mathematics of music by Sir James Jeans.

Hermann von Helmholtz, *Popular Scientific Lectures,* Dover Publications, New York, 1962, pp. 22–58, "On the Physiological Causes of Harmony in Music." This essay is a classic, and should be read by all students of music.

Harvey B. Lemon and Michael Ference, Jr., *Analytical Experimental Physics,* University of Chicago Press, Chicago, 1946, Chapters 34–37. A more advanced treatment of wave motion. The chapter on musical sounds and the scientific aspects of musical art should be studied by those interested in the structure of musical scales.

O. G. Sutton, *Mathematics in Action,* Harper, New York, 1960, Chapter 4 on "An Essay on Waves." For those who are familiar with calculus, this is an excellent treatment of the significance and application of wave analysis.

chapter sixteen

The nature of light

Mankind is haunted by an impulse to find unity amid the obvious multiplicity of observed events. The problem of "the one and the many" no less than that of "permanence amidst change" has, for centuries, been a traditional one for philosophers. The prophets' proclamation that there is but *one* God is an assertion both of the unity of nature and the unity of man. As the poet declares that "no man is an Island, entire of it selfe" and seeks to portray the involvement of every individual in humanity at large, so the scientist by announcing *laws* connecting separated events seeks to show the unity of nature by establishing the relationship between the event "here-now" to the event "there-not-now." In the modern terminology of science these connecting laws are formulated as laws of energy transformation.

The transmission of energy and, consequently, the transmission of information from place to place has been studied in two forms: kinetic energy in which particles move with a given velocity from one place to another, and wave energy in which mass is not transferred, but in which the energy (such as sound) is transmitted as a disturbance propagated through a medium. For man, the most important source of information about the external world is the flow of messages brought to him by light.

The story of the development of our modern ideas about the nature of light is a fascinating one; like light itself, it is full of color and change. New theories have been erected again and again and, one after another, a particular theory has had its day, only later to be found wanting in some fundamental aspect. Even old theories once discarded and spurned, often find themselves reinterpreted and reinstated.

Many of the ancient speculators assumed that when we "see" an object, a kind of tentacle reaches out from our eyes to the subject of our attention. "We cast our gaze" upon objects, reaching out in some manner to touch them. Today, we regard this as a metaphorical way of speaking, and it is generally accepted that the eye is the receiving instrument, accepting and interpreting a flow of energy from the object toward which it is turned. But such a statement tells us very little about how this process takes place. Does light consist of particles of some kind, either shot off or reflected by a luminous object? If so, how fast do such objects travel? Or, is light, like sound, a kind of wave propagation? If so, how can we explain the fact that although a bell in a vacuum cannot be heard, the same bell remains clearly visible in an evacuated glass container? Waves, as we ordinarily experience them, always appear to require a material medium, yet light can flow millions of miles through the empty space between the stars. These and many other puzzles have plagued the investigators seeking to pin down the nature of light. Theory has followed theory. Newton's theory that light is composed of particles took the place of Descartes' belief that light is a kind of pressure, and the particle theory was supplanted by a wave theory, the wave theory in our own century again being modified to come closer again to Newton's particle theory. In science, few developments illustrate more effectively the nature of theory construction and change.

1. THE VELOCITY OF LIGHT

It is hard for us to put ourselves into the position of the natural philosophers of a few centuries back. We are brought up with the "knowledge" that "light travels," and accept this quite naturally and without question. But, if we examine our experience, precisely what leads us to believe that light is, indeed something that takes time to travel? As Galileo put the problem, over 300 years ago, in his *Two New Sciences.*

> Everyday experience shows that the propagation of light is instantaneous; for when we see a piece of artillery fired at a great distance, the flash reaches our eyes without lapse of time; but the sound reaches the ear only after a noticeable interval.

This experience is familiar to us all. Furthermore, if we look about us in a quiet room and ask ourselves *what* is moving, it takes considerable sophistication to answer that *light* is moving, unless the source of the light is itself in motion.

It is not surprising, therefore, to find that many early scientists were wedded to the belief that light was some form of instantaneous connection between source and illuminated object. Descartes, for example, argued that light was a kind of pressure instantaneously transmitted through the "ether" of space from the sun and fixed stars. Vision is compared to the sense of pressure transmitted to a blind man using a cane to find the position of objects.[1]

We are aware now, of course, that light does travel, but at such an extraordinarily high velocity (more than seven times around the earth in one second) that no simple means can reveal that it is not instantaneous. Here, again, we notice how the great minds of science are willing to go beyond experience of the senses and entertain the nonobvious possibilities. Galileo, in his *Two New Sciences,* continues:

Sagredo: Well, Simplicio, the only thing I am able to infer from this familiar bit of experience (observing the flash of a cannon and later hearing the sound) is that sound, in reaching our ears, travels more slowly than light; *it does not inform me whether the coming of the light is instantaneous, or whether, although extremely rapid, it still occupies time.*

In addition to grasping the possibility that light traveled at a finite velocity, Galileo also proposed a method for deciding the issue. It was as follows. Two men are placed on mountaintops at some distance apart, holding lanterns, which may be covered by blinders. The first man opens his lantern and as soon as the second man sees it, he opens his lantern. When this light is seen by the first man, he knows that the light has traveled forth and back between the two mountaintops and, by knowing the distance and measuring the time, the velocity can be calculated. We smile today at this crude method, but we should remember that the first careful formulation of a problem is often a great leap forward in the making of discoveries. Obviously, such a method is hopelessly inadequate because of the high velocity of light, but the important point is that *the problem has been formulated.* When the Italian scientists of one of the oldest scientific societies, established in the time of the Medicis, did follow Galileo's suggested experiment with as much care and precision as they could, they concluded that there was no reason to believe that light was not instantaneous. However, two centuries later, this general procedure was used with very delicate mechanical refinements, and the velocity of light was successfully determined.

But, meanwhile, and from an entirely different direction, there came substantial evidence that light did travel with a definite and finite (although extremely high) velocity. A Danish astronomer, Roemer, had been studying the periods of rotation

[1] This instantaneous transmission of light pressure through "space" or "ether" became modified, according to Descartes, when the light pressure met substances such as air or glass. He believed that changes in this velocity in transparent substances of different densities give rise to refraction. According to his theory, as with Newton's, the velocity of light is greater in the denser medium.

of the inner satellite or moon of Jupiter. He noted that although the *average* period of rotation as measured by successive eclipses (and in this way very accurately timed) was 42¾ hours, the *period varied with the time of the year.* Such a variation, if taken as accurately established, would be directly contrary to Kepler's laws which required each period to be identical. Here, we see the natural and necessary conservatism that marks the methods of science, as well as other human endeavors. Roemer hardly considered the possibility that Kepler's laws were proved invalid, but he sought another explanation. He hit upon the explanation accepted today. In a paper read before the Paris Academy in 1675, he declared that the observed variation in the periods was caused by (*1*) the motion of the earth at different times of the year either toward or away from Jupiter, and (*2*) the *finite* velocity of light.

It was known that the earth travels in the neighborhood of 19 miles/sec. In the period of one revolution of Jupiter's moon, the earth travels a few million miles. However, the earth and Jupiter travel in different orbits; the earth, at a greater speed, overtakes Jupiter, approaching and passing it. When we are approaching Jupiter at maximum speed, the period of the satellite decreases by 15 sec, there being about 3 million miles shorter distance for the light to travel; when we are departing at maximum speed, it increases by a similar amount. Roemer, therefore, concluded that the velocity of light was in the neighborhood of 200,000 miles/sec, since the distance the light must travel is decreased by about 3 million miles.[2]

Thus, by the time that Newton published his second great work *Opticks* at the beginning of the eighteenth century, the scientific community was well aware that light travels with a very high but nevertheless finite velocity. Other astronomical methods, which we need not examine, confirmed Roemer's supposition, and the velocity of light in space was calculated with considerable precision by Bradley in 1728, two years before Newton revised and published his final edition of *Opticks.*

2. THE READILY OBSERVABLE FACTS ABOUT LIGHT; RECTILINEAR PROPAGATION; REFLECTION; REFRACTION

To know the *velocity of light* does not help very much in discovering the *nature of light.* What kind of messenger is this that travels so swiftly? Light had fascinated natural philosophers and scientists for more than 20 centuries, but little progress

[2] The figures used are illustrative. Roemer's estimates of the variations in time were somewhat high. Nevertheless, this early estimate of the velocity of light was extraordinarily accurate. The problem is not quite as simple as presented: (*1*) the surface of Jupiter is not regular and timing of eclipses with accuracy is difficult; and (*2*) there are other secondary effects that cause irregularities in the period. Here, again, note how the scientist must *idealize* and *abstract* to make any significant advance. He almost always "follows a hunch" in hitting on a solution, which is seldom really completely clear in the light of evidence.

had been made in this time beyond what was known by Euclid and Ptolemy who both wrote treatises on optics.

In the Renaissance, artists, pursuing a new natural realism, were pressing the mathematicians and natural philosophers for better answers to problems of perspective and color. In the seventeenth century the greatest minds were struggling to understand light phenomena: Descartes, Fermat, Hooke, Newton, and Huygens, among others, all studied and wrote about optics, examining and advancing various theories of explanation. Great progress was made at that time based upon observations readily available.

Let us start with a few of the simplest empirical observations that can be made without complicated apparatus, and then look for the simplest explanation.

1. Light appears *to travel in straight lines.* Shadows cast by an object placed before a point source of light (a very distant source, for example) are sharp and clear and follow geometric patterns. We have seen how the Greeks used shadows cast by vertical sticks placed in the sun to determine the circumference of the earth through analysis based on similar triangles. This assumes that we can represent rays or beams of light by straight lines. This assumption was followed from antiquity onward whenever optics was studied.

2. When a ray of light is *reflected* from a surface such as a mirror, *the angle of incidence* (the angle between the normal or perpendicular to the surface and the incoming ray) *is equal to the angle of reflection* (the angle between the perpendicular and the reflected ray). This law was also known to the Greeks and was clearly stated by Ptolemy.

3. Furthermore, it is a common observation that when light travels through a surface that lies between transparent substances of different densities (for example, the interface between water and air), it is *bent.* (This shows most clearly when we look at a stick that is half-submerged at an angle in water.) The way in which the light rays are bent (refracted) is more complicated than the way in which they are reflected. Qualitatively, it has been long known that a ray of light, upon entering a more dense medium, is *bent toward the normal* or, on coming from the more dense medium into a less dense medium, is bent away from the normal.

4. In the seventeenth century, Snell found the *quantitative* law describing *refraction,* that is, that for any two transparent substances of different densities, the ratio of the sine of the angle of incidence to the sine of the angle of refraction is constant, regardless of the angle of incidence of the light. This ratio is called the *index of refraction.* As an example, a ray of light entering from the air into water at any angle will be bent in such a way that its index of refraction (see Fig. 16.1a):

$$\mu = \frac{\sin i}{\sin r} = 1.33$$

where

$$i = \text{angle of incidence}$$
$$r = \text{angle of refraction}$$

If, on the other hand, the ray of light is coming from under the water out into the air (see Fig. 16.1b):

$$\frac{\sin i}{\sin r} = \frac{1}{1.33}$$

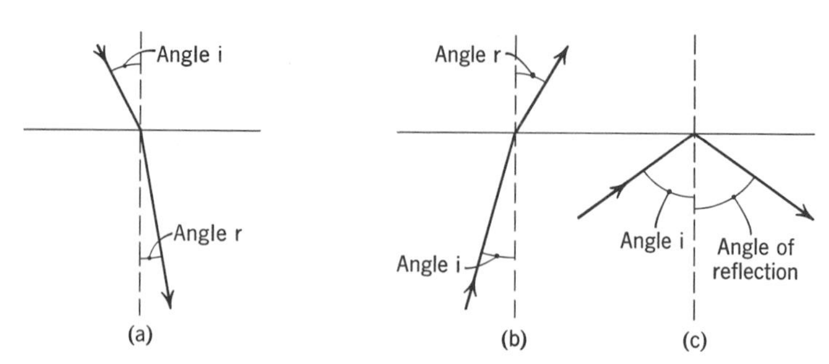

Fig. 16.1. (a) Ray entering from air into water is bent toward the normal. (b) Ray entering from water into air is bent away from the normal. (c) Total internal reflection. When the incident angle is too large, a ray cannot penetrate from water into air, and is reflected.

(However, since the sine of any angle can never be greater than one, if the angle of incidence of the ray coming out of the denser medium is so large that it would require the sine of the angle to be greater than one, the ray does not penetrate through the surface at all but is reflected into the denser medium. This is called total internal reflection.)

Lens making and the construction of telescopes and microscopes had been pursued as skilled crafts before the discovery of Snell's law of refraction, but once this law was discovered, technology rapidly entered into a precise and effective stage. The story of the development of what is often termed *"geometric optics,"* in which the paths of light rays through air and other transparent mediums are studied as geometric lines, is a fascinating one, both for itself and for its relationship to the technical problems of painters.[3]

Since our interest is primarily in following the development of an adequate theory of the nature of light, we are more concerned with the problem of explaining why refraction and reflection take place than in describing how these take place (this constitutes the subject of geometric optics, which we shall not pursue).

5. Finally the significance of one additional elementary observation should not be overlooked, that is, that two beams of light (as from two flashlights) can pass through one another without any mutual interference.

Such are a few of the most elementary facts of observation that any theory of light must account for. Later, we shall consider other and more complicated observations whose discovery reintroduced the whole baffling problem (even to this day) of deciding precisely *what* travels.

[3]The Renaissance painters, seeking a new descriptive realism in place of the conceptual emphasis of the earlier artists of the Middle Ages, studied the mathematical principles of perspective, and undertook to solve many problems and these, in turn, helped lead to the mathematical development of projective geometry. Leonardo and Durer were two among many of the artists who concerned themselves with such problems. Later, even problems connected with the distortions caused by lenses were of concern to Vermeer, the Dutch painter, who used lenses to form images on a screen. The so-called "barrel distortion" of a lens showed up in his earlier paintings but gradually disappeared as better lenses were developed. Those interested in art should refer to Chapters X and XI in *Mathematics in Western Culture* by Morris Kline, Oxford Press, New York, 1953.

3. ALTERNATIVE EXPLANATIONS OF LIGHT PHENOMENA: NEWTON'S THEORY OF PARTICLES; HUYGEN'S WAVE THEORY; COLORS AND POLARIZATION

Whenever the frontiers of science are being extended and new phenomena are being studied, there usually emerges a long period in which incomplete theories of explanation vie with one another for supremacy. Intuition, imagination, speculation, logic, and attention to details—all play their role. Only rarely does one theory suddenly stand out supreme and incontrovertible. In this respect Newton's formulation of his system of mechanics was exceptional. Even here, the earlier development of the right ideas in the work of Galileo and Kepler was necessary groundwork. Nevertheless, Newton's world system, embracing his theory of gravitation, was a brilliant and finished piece of work—a complete structure. Very different, and for many of us more fascinating, was Newton's work in optics. Here, we see a brilliant mind at work, trying to fit together phenomena, as then known, into a consistent whole. Newton recognized how unfinished his explanatory theory was and, therefore, stated many of his speculations as "queries." There are few better instances where the mind of a great scientist reveals its method of working, its tentative groping for the right directions, than in the third book of Newton's *Opticks,* which includes 31 queries covering not only optics proper but many related topics.

It was quite natural for Newton, puzzling over the nature of light, to find the best and simplest explanation for "what travels" in the idea that it is composed of particles. There was considerable evidence that led him to this conclusion. Newton, however, also realized that a simple "particle" explanation was not sufficient to account for *all* the phenomena that he studied and, therefore, he was quite willing to modify his theory to whatever extent necessary by associating "vibrations" with the particles of light themselves. When we examine Newton's famous study of colors and of the phenomenon termed "Newton's rings," we recognize his desire to combine particle theory with vibration theory, certainly not to his own complete satisfaction, but as a tentative speculation for further investigation.

The theory that light consisted of particles emitted by hot or radiating bodies appealed to Newton as the simplest and most effective theory, primarily because *light travels in straight lines* and casts *sharp shadows.* He said:

If it [light] consisted in Pression or Motion, propagated either in an instant or in time, it would bend into the Shadow. For Pression or Motion cannot be propagated in a Fluid in right Lines, beyond an Obstacle which stops part of the Motion, but will bend and spread every way into the quiescent Medium which lies beyond the Obstacle . . . The Waves on the Surface of stagnating Water, passing by the sides of a broad Obstacle which stops part of them, bend afterwards and dilate themselves gradually into the quiet Water behind the Obstacle. The Waves, Pulses or Vibrations of the Air, wherein Sounds consist, bend manifestly, though not so much as the Waves of Water. For a Bell or a Cannon may be heard beyond a Hill which intercepts the sight of the sounding Body, and Sounds are propagated as readily through crooked Pipes as through straight ones. But light is never known to follow crooked Passages nor to bend into the Shadow.[4]

So much for the basic reason for rejecting the wave theory. In addition, Newton felt that all the other experimental phenomena that he observed so carefully over a long period of years could be accounted for if light were conceived as consisting of particles, that upon striking material media could set up some kind of vibrations within that media. He continues,

Are not the Rays of Light very small Bodies emitted from shining Substances? For such Bodies will pass through uniform Mediums in right Lines without bending into the Shadow, which is the Nature of the Rays of Light. They will also be capable of several Properties, and be able to conserve their Properties unchanged in passing through several Mediums, which is another Condition of the Rays of Light. Pellucid Substances act upon the Rays of Light at a distance in refracting, reflecting, and inflecting them, and the Rays mutually agitate the Parts of those Substances at a distance for heating them; and this Action and Re-action at a distance very much resembles an attractive Force between Bodies. If Refraction be performed by Attraction of the Rays, the Sines of Incidence must be to the Sines of Refraction in a given Proportion, as we showed in our Principles of Philosophy; And this Rule is true by Experience. The Rays of Light in going out of Glass into a *Vacuum,* are bent towards the Glass; and if they fall too obliquely on the Vacuum, they are bent backwards into the Glass, and totally reflected; and this Reflexion cannot be ascribed to the Resistance of an absolute *Vacuum,* but must be caused by the Power of Glass attracting the Rays as their going out of it into the Vacuum, and bring them back.[5]

(Note Newton's explanation of total internal reflection, and his comment on the basis of attraction of the glass to particles.)

Newton had discovered that just as white light could be separated into all the colors of the spectrum by means of a prism, so the various colors of the spectrum could be brought together once again to form white light. He, therefore, concluded that each color consisted of a separate kind of particle and that the combination of such particles reaching our eyes gave rise to a sensation of whiteness:

Nothing more is requisite for producing all the variety of Colours, and degrees of Refrangibility, than the Rays of Light be Bodies of different Sizes, the least of which may take violet the weakest and darkest of the Colours, and be more easily diverted by refracting Surfaces from the right course; and the rest as they are bigger and bigger may make the stronger and more lucid Colours, blue, green, yellow, and red, and be more and more difficultly diverted. Nothing more is requisite for putting the Rays of Light into Fits of easy Reflection and easy Transmission, than that they be small Bodies which by their attractive Powers, or

[4] Sir Isaac Newton, *Opticks,* Dover Press, New York, 1952, p. 362.
[5] Ibid., pp. 370–377.

some other Force, stir up Vibrations in what they act upon, which Vibrations being swifter than the Rays, overtake them successively, and agitate them so as by turns to increase and decrease their Velocities, and thereby put them into those Fits.[6]

Before discussing two other phenomena that Newton was aware of and studied—polarization and diffraction—let us examine the rival theory of light: that it is composed of waves instead of particles. This theory was first set forth by Newton's contemporary, Huygens (a brilliant Dutch physicist, whose reputation would have been even greater if he had not lived in the same period and been overshadowed by Newton).

By the seventeenth century, it was well known that *sound* is a wave disturbance that travels at a finite and measurable velocity. It was inevitable that the possibility that light was also a wavelike phenomenon should be considered. Huygens, in his "Treatise on Light" presented before the Paris Academie in 1678, pursued this idea with a brilliant mathematical analysis, showing how reflection and refraction can be readily explained on the theory that light is composed of longitudinal vibrations similar to sound waves, instead of particles as suggested by Newton.

The principal reason for Huygen's belief in the wave theory was the fact that *beams of light may cross each other without any apparent interference.* If light were composed of a series of particles traveling through space, then surely some of these particles would strike one another and be deflected from their paths. (Newton did not answer this argument directly. Apparently he considered light particles to be so small that swarms of them could cross one another with little chance of collision, just as a stream of bullets from two machine guns fired at an angle will cross through each other's paths without interference.) Furthermore, Huygens asked: Why should all colors travel with the same velocity? If they are different-sized particles, why is it that light of different intensities travels at the same velocity? These aspects of light transmission are more easily explained if light is a wavelike disturbance than if it is composed of particles.

To explain how waves could travel in empty space, Huygens felt compelled to assume, as others had done to "explain" gravity, the existence of a very subtle medium that was called the *ether.* The ether was thought to extend everywhere, not only between the sun and the earth but even within the spaces between all particles or atoms of which material substances were composed. Just as air is needed to carry waves of sound, and water is needed to carry its waves, so it was felt that a medium was needed to carry light waves. Although the assumption of the existence of such a peculiar material as the ether, pervading all of space (frictionless because it produced no opposition to the planets traveling through it, and weightless—but still with sufficient firmness and elasticity to

[6] Ibid., pp. 372–73.

propagate waves) appeared highly artificial, scientists, even then, were unafraid to postulate the existence of unseen entities (such as atoms), which were needed to explain observed phenomena. Equally artificial, according to Huygens, was Newton's theory of a multitude of tiny particles each corresponding to a specific color.

To emphasize how slow, confused, and baffling is the attempt to build up a good scientific theory, we cite the explanation of *polarization,* known to both Newton and Huygens. It is fascinating to find that Newton felt that the phenomenon of polarization buttressed his own theory of particles, although 100 years later, it was cited as very strong evidence for the wave theory.

Although Newton and Huygens discussed polarization in terms of the phenomena of double refraction, which had been noticed in certain crystals brought to England from Iceland in the seventeenth century, it is simpler to discuss the phenomena in terms of polarizing crystals (or the "polaroid" glass, common in today's market) and in terms of the modern wave theory of light. We then can trace better the argument between Newton and Huygens.

The modern theory of polarization may be briefly summarized as follows.

(*a*) Light consists of *transverse* waves (similar to water waves) in which the vibrations are at right angles to the direction of the velocity or progress of the wave.

(*b*) Ordinary light, as from the sun, consists of unpolarized light in which there are many vibrations at all angles, but all in a plane perpendicular to the direction of the wave (Fig. 16.2).

(*c*) If these vibrations in a plane are restricted *to one direction only, the light* is said to be *polarized.* This can be accomplished by passing it through certain crystals or through a polarizing medium such as polaroid. Such a medium transmits only rays whose vibrations are in one direction—the direction being determined by the orientation of the crystal axis.

(*d*) If, as in Fig. 16.3, an unpolarized beam of light passes first through a polarizing crystal oriented in one direction and then through a second crystal similarly oriented, all the light is transmitted. If, on the other hand, the beam of light passes through the first crystal and is polarized vertically, let us say, and then strikes the second crystal when its axis is so oriented that it allows only horizontal vibrations to pass, then no rays of light can penetrate through the second crystal.

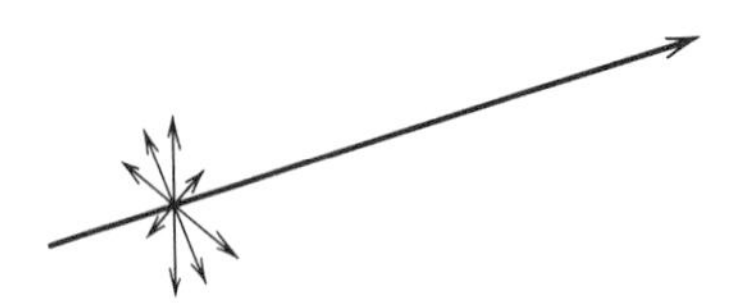

Fig. 16.2. Transverse vibration in a plane, perpendicular to the direction of propagation.

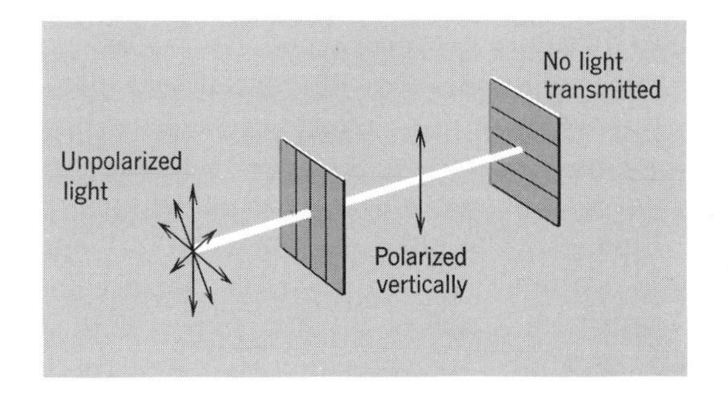

Fig. 16.3. Principle of polarization. Light rays that are unpolarized pass through a crystal that eliminates all rays except those polarized vertically. When these rays meet a second crystal, which eliminates all rays polarized horizontally, no light is transmitted.

Such, briefly, is the modern theory accounting for polarization. (The so-called "polaroid glasses," which aim at reducing glare, depend on the fact that much reflected light from a shining surface is polarized in one direction and may be mostly eliminated by the glass, which only allows light polarized in another direction to be transmitted. Reflected "glare" is thus reduced.)

Now, Newton and Huygens, when they considered the hypothesis that light consists of waves, *only thought of longitudinal waves* (those in which the vibrations are in the direction of the wave propagation, such as sound waves or other compression waves). If this were true, then it would be difficult indeed to explain polarization on the basis of a wave theory. The only escape from the dilemma would be to assume the *two* ethers with different properties. Newton comments:

> But how two *ethers* can be diffused through all Space, one of which acts upon the other, and by consequence is reacted upon, without retarding, shattering, dispersing, and confounding one anothers' Motions, is inconceivable. And against filling the Heavens with fluid Mediums, unless they be exceeding rare, a great Objection arises from the regular and very lasting Motions of the Planets and Comets in all manner of Courses through the Heavens.[7]

Newton had accomplished more than any other scientist of his time in the explanation of various phenomena connected with the transverse waves on water. While we may be surprised that such an obvious hypothesis—that light waves might be *transverse*—never occurred to Newton or Huygens, such oversights occur frequently in science. Ideas that later seem obvious and simple, even clamoring for attention, are often overlooked by the most penetrating minds.

Newton, therefore, rejected the explanation of polarization on the basis of the wave theory. Instead, he extended his particle theory. Just as particles might have different sizes and, therefore, set up different-sized vibrations accounting for colors, so the particles could have different shapes on their sides. "Whereas by those experiments it appears, that the Rays of Light have different Properties in their different sides," said Newton. Rays, therefore, having one shape would

[7] Ibid., pp. 364–365.

pass through a crystal oriented in one direction; those of another shape would pass through a crystal oriented in a different direction.

Finally Newton, in a long series of experiments on the colored rings seen in thin films (such as soap bubbles), developed a theory to account for a whole series of observed phenomena. He speculated that particles of light set up vibrations within the film that were of such a "bigness" that they varied in "fits of easy transmission or reflection." Some colors entered and were reflected out from the rear side of the film; others with the wrong "fit" were not reflected. Consequently, we see variegated patterns of colors. This explanation is not far from the explanation in terms of interference; yet it is based on a particle theory. Once again, we find that insights—"hunches," if you will—developed tentatively by a brilliant mind are taken by later generations to contain all of "truth" in a more literal fashion than the originator of these hunches had intended. Newton wrote:

> Tis true that from my theory I argue the corporeity of light; but I do it without any absolute positiveness . . . I knew, that the *properties* which I declared of light were in some measure capable of being explicated not only by that, but many other mechanical hypotheses.[8]

Finally, before we look at the more modern theories of light, another comment should be made. Just as today, Joseph Wood Krutch, the naturalist scientist and student of the poetry and beauty of nature, objects to the abstract formal approach to nature, reducing effectively so much of phenomena to the mathematical abstract relationships, so likewise, not long after Newton, the observers of nature found objection to the attempt to formulate the immediate experience in terms considered cold, forbidding, abstract and, therefore, distorted. In particular, Goethe, not only a poet but a nature scientist, objected to the dissection of nature, and felt impelled to erect an alternative theory of color closer to the immediately experienced (thus, anticipating the outraged cry of Wordsworth, early in the nineteenth century, who charged that in science "we murder to dissect.") Goethe said,

> The physicist also wins mastery over natural experience, fits and strings them together by artificial experiments . . . but we must meet the bold claim, that this is *Nature,* with at least a good-humoured smile and some measure of doubt. No architect has yet had the notion of passing off his palaces as mountains and woods.[9]

[8]One of Newton's replies to the criticism of Hooke, a contemporary of Newton who expounded the wave theory. Quoted by Holton and Roller, *Foundations of Modern Physical Science,* Addison-Wesley, Reading, Mass., 1958, p. 554.

[9]For this quotation and for an interesting discussion of the conflict between the approach of a keen observer of nature, a poet, and that of an abstract scientist, see Werner Heisenberg's lecture on "The Teachings of Goethe and Newton on Color," in *Philosophic Problems of Nuclear Science,* Chapter V, Pantheon Books, New York, 1952, p. 63.

Goethe, in 1790, developed an alternative theory of color that was most fruitful in aesthetics, physiology, and art. It was a conceptual theory which, as Heisenberg remarks,

> . . . brings together into a unified orderly whole the many effects of colours in our world of the senses rather by way of a guiding idea, based not on reason but on experience. The harmonious arrangement laid before us by Goethe's colour theory gives even the smallest details a living content and comprises the whole range of objective and subjective colour phenomena.[10]

Thus, while Newton was laying a brilliant foundation for the association between colors, which appeared to be pure *qualities* and the abstract *numbers,* the recognition that the full immediacy of experience was not exhausted by any advance in science was already widespread. It foretold the later reaction of poets and philosophers, even to this day, against science which was, by its very success, misleading people into the belief that *all* problems, natural, social, ethical, and religious could be solved by the abstract mathematical approach alone, that science exhausts the nature of reality.

Goethe's theory of color is also pertinent, not because it had any direct influence on the development of optical theory but because, as so often happens, a poet and nature scientist, by his influence, gave a different direction to the atmosphere in which the traditional development of abstract science had been proceeding. Goethe's thoughts, at least, encouraged those who questioned the traditional Newtonian theory of particles. Physiologists and artists as well as physicists were interested in color, and Newton's corpuscular theory had no appeal for them. The pressure to find a more complete explanation was continued.

4. ASCENDANCY OF WAVE THEORY OF LIGHT IN THE NINETEENTH CENTURY

In the eighteenth century the mechanics founded by Newton flourished and reached its culmination in the works of Laplace and Lagrange, whose analysis of celestial and terrestrial motions of bodies formed what appeared to be a complete system. Meanwhile, however, little progress was being made in extending the understanding of light and optical phenomena.

Thomas Young, an English doctor, with one of those extraordinary minds that at an early age seem to develop in all directions at once,[11] was interested as a doctor in the problems of the physiology of both optics and sound. This led

[10] Ibid., p. 62.

[11] Young had mastered seven languages by the age of fourteen. He had read the Bible through twice by the age of four.

to his making a connection between the two, using a wave theory. He explained satisfactorily why light appears to travel in straight lines. If light waves were extremely short, he argued, they would appear to travel in straight lines. To show clearly the interference pattern and to measure the corresponding wavelengths, Young designed the experiment as diagrammed in Fig. 16.4 based on principles already discussed in connection with sound interference.

In this figure, the positions of the light and dark bands correspond to positions where the waves reach the screen either in step or out of step. By measuring the distances between the slits, the distance between the first and second screen, and the position of the first light band, the wavelength of light can be calculated. (From the dimensions and from the fact that the two rays of light differ by *one* wavelength at the first light band or fringe, can you find the equation for wavelength?) Young calculated the wavelength of light to be about one-millionth of a meter. This is so small compared with the size of ordinary visible objects that Young showed that light thus composed would travel in straight lines and cast relatively sharp shadows. He also showed that the very slight bending close to the edge of opaque objects, that is, diffraction, could more simply be explained by the wave theory than by Newton's theory. (Newton's explanation was that particles of light suffered a very small attraction when very close to the material edge of the object.)

At first, Young was as puzzled about explaining polarization on the basis of wave theory, as his predecessors had been before him—all of them assuming that the waves were *longitudinal* vibrations. Suddenly, however, in 1817, it occurred to him that if the waves were *transverse* (that is, if the vibrations were perpendicular to the direction of wave velocity, as those on water or those

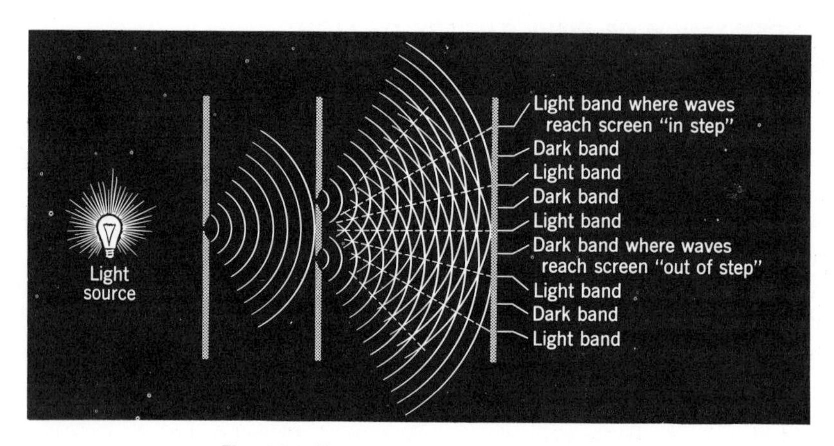

Fig. 16.4. Young's double-slit interference.

on a stretched string), an explanation would be possible. He suggested this hypothesis to his friend, Arago, who mentioned it to a young civil engineer, Fresnel. Fresnel made this assumption the basis of a prize essay, presented to the Paris Academy of Science, in which he showed *that all known optical phenomena* could be explained if light were considered to consist of *transverse waves.*

But there remained many who rejected even this form of the wave theory. In particular, the concept of an "ether," the medium for carrying waves, caused serious objection. "Common sense" experience demands that waves must have a medium for travel. Our intuition rejects the idea of waves without a medium as an artificial construction (much as the "grin of the cheshire cat," amusing and intriguing though it may be, cannot be seriously considered as a real possibility after the cat disappears). It was even more difficult to visualize the kind of ether that is required for *transverse* waves because such waves can normally only be propagated in a medium with properties of elasticity usually associated with very rigid bodies. How could there be a *frictionless* medium, all pervasive, that had a rigidity greater than steel, as the theory required?

In addition, a vituperative reaction developed in England against Young's new theories because of the preeminence of Newton's reputation. Newton believed in light particles, therefore all other theories were absurd. Somewhat reminiscent of those who refused to look through the telescope in Galileo's day to see for themselves the moons of Jupiter, were those in the early nineteenth century who refused to repeat Young's experiment on interference. One of Young's detractors, later to become Lord Chancellor of England, wrote:

> The fact is, we believe the experiment was inaccurately made; and we have not the least doubt, that if carefully repeated, it will be found that the rays when inflected, cross each other and thus form fringes (Newton's theory) . . . or that in stopping one portion, Dr. Young in fact stopped both portions; a thing extremely likely, where the hand had only one-thirtieth of an inch to move in, and quite sufficient to account for all the fringes disappearing at once from the shadow.[12]

Young did, in fact, admit his great indebtedness to Newton even as he saw that more emphasis should be placed upon Newton's recognition that there is some kind of periodicity associated with the particles that constitute light. Young, in reply to his critics, expressed himself in the spirit of most creative scientists:

> Much as I venerate the name of Newton, I am not therefore obliged to believe he was infallible. I see . . . with regret that he was liable to err, and that his authority has perhaps even retarded the progress of science.[13]

[12] A. E. E. McKenzie, *The Major Achievements of Science,* Cambridge University Press, London, 1960, p. 156.
[13] Quoted by Stephen F. Mason in *History of Science,* Abelard-Schuman, Ltd., London, Collier Books Edition, New York, 1962, p. 468.

But Young's ideas, as developed and elaborated mathematically by Fresnel, drew the attention of many of the European mathematicians. One of these, Poisson, who believed in the longitudinal vibration of the waves, gave an analysis to show that the transverse wave theory cannot be correct because this would lead to the "absurd" consequence that the shadow cast by a circular disk in monochromatic light would have a central bright spot. This was at first taken as a proof of the invalidity of the theory of transverse vibrations, until an investigator tried out the test experimentally and found that there was indeed a faint but distinct bright spot at the center of such shadows. Thus, again, an observation easily made had been overlooked until sought for carefully.

Finally, near the middle of the nineteenth century, the wave theory of light received a decisively important experimental confirmation. It will be recalled that on the basis of the particle theory refraction was explained by the supposition that the velocity of the particles of light would *increase* as they entered the denser medium, due to the greater attraction of the substance of that medium; although, on the wave theory, the velocity of light would *decrease* as it entered the denser medium. It was not until 1848 that instruments were developed that could measure the velocity of light with sufficient accuracy for such a decision. In that year, however, this was finally accomplished, and the velocity of light in air was found by H. L. Fizeau to be 1.3 times that of the velocity in water, in exact agreement with what was called for by the wave theory of light. The wave theory at last came into its own, accepted by all scientists. Ironically, 50 years after its unanimous acceptance, it was once again to be questioned.

5. MAXWELL'S THEORY: LIGHT AS AN ELECTROMAGNETIC VIBRATION

The crowning achievement of what is now regarded as classical physics (roughly the body of physics that was developed up to about 1900) was the unification achieved when light, radiant heat, X-rays, and radio waves were all found to consist of electromagnetic waves, differing only in wavelength and frequency. It was an achievement not only in the development of physical theory but in mathematical analysis. The full power and scope of the differential calculus invented by Newton and Leibnitz in the seventeenth century came into its own as a tool for analyzing extremely complicated problems of continuous variation.

Clerk Maxwell was one of those scientists in whom a profound insight into the physical mechanics of a problem was blended with an exceptional mathematical talent. His early work at Cambridge University consisted in trying to systematize the work of Faraday, Oersted, and Ampere by the construction of an adequate model susceptible to organized mathematical analysis. Faraday's ideas of lines of force extending from electric charges and from magnets was the starting point for Maxwell. More emphasis was placed on the intermediate space or medium between the magnets and charges, either at rest or in motion, than upon the magnets and charges themselves. Such an approach contrasts with the particle approach of Newton, in which only the masses, their position, and their distance apart had to be specified to define the phenomena of gravity. No analysis was required regarding what happens *in between* the particles or masses. Under Faraday's leadership, a concept of a *field* was developing in which the intervening, continuous medium, whether of "empty" space or of material substance, required analysis. Newton purposely avoided such an analysis, thinking it went too far beyond what is known.[14] Now with the introduction of a "field" arises the recurrence of the ancient idea that *natura non saltus facit* (nature makes no jumps) and that there is a continuous subtratum to all events, reminiscent of those Greek philosophers who felt that basically the universe is made up of a fluid. There is no "jump" in a fluid; pressures and velocities vary in a continuous manner and by slight gradations.

In addition to the aspect of continuity, it often occurs that forces applied in *one direction* in a substance may very well give rise to results in a *different direction.* Pushing a piston down in a cylinder of water gives rise to pressure not only in the direction of the push, but in directions at right angles to the push. In gravitational attraction, the force of attraction is always directed in a straight line between the attracting masses, but in electromagnetic interactions, as we have seen, a magnetic field exerts a force on a moving charge not in the direction of the magnetic lines of force but always at right angles to them. Charges and magnetic fields, as they change and interact with one another, must therefore be expressed by some method of mathematical analysis that not only connects magnitudes but connects the directions involved. This requires a new type of mathematical analysis in which the mathematical symbols refer not only to magnitudes but to directions as well. The *vector* differential calculus gave expression to this sort of interrelationship, and such a calculus was developed during the nineteenth century.

Let us review briefly some of the important facts of electromagnetism, which we have examined in Chapter 13. First, if a current flows in a conductor, which means that charges are moving, this will give rise to a magnetic force field. This means that at each point near the conductor there exists a force that would cause a tiny test magnet placed at that point to move in the direction of the force. The direction of this force field is circular about the conductor as shown in Fig. 16.5. The intensity of this force field at any point varies both with the num-

[14]Newton's often-quoted phrase *hypothesis non fingo* (I make no hypothesis) referred strictly to the question as to how gravitational forces extend across empty space.

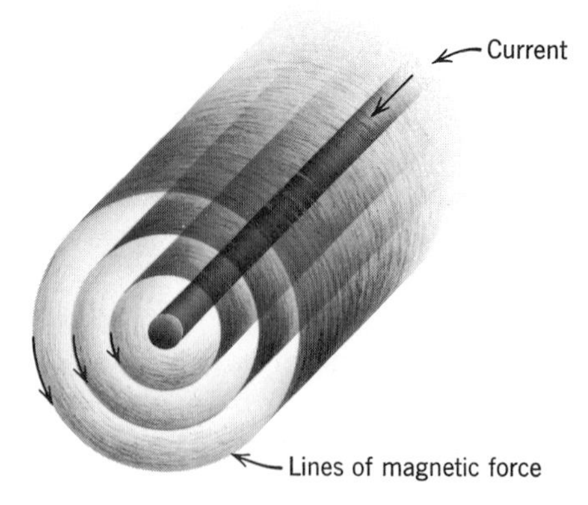

Fig. 16.5. Direction of magnetic lines of force generated by a current.

ber of charges flowing in the conductor per unit time and inversely with the distance from the moving charges. (What is meant by a *field* now becomes clearer, that is, that at any point in space there is both a magnitude and a direction of force assigned to that point and that there is a "continuous" variation in both magnitude and direction as one passes from one contiguous point to another.)

Likewise, a symmetrical relationship exists between moving magnets and charges. If we think of a group of *magnetic lines* of force enclosed by a circular conductor, Faraday's laws state that when the number of such magnetic lines is changed (either by increasing or decreasing the "magnetic strength" or by changing the direction of the lines), then an electrical force will be set up that will "induce" electric charges to flow in the conductor (see Fig. 16.6).

Such were the experimental facts, known to Maxwell and his contemporaries, indicating the relationship on the one hand between moving electric charges and the induced magnetism, and on the other between changing magnetic lines and induced currents. We shall find that there were concise mathematical equations that were able to represent these relationships effectively. To understand the significance of these equations, however, it is necessary first to consider an

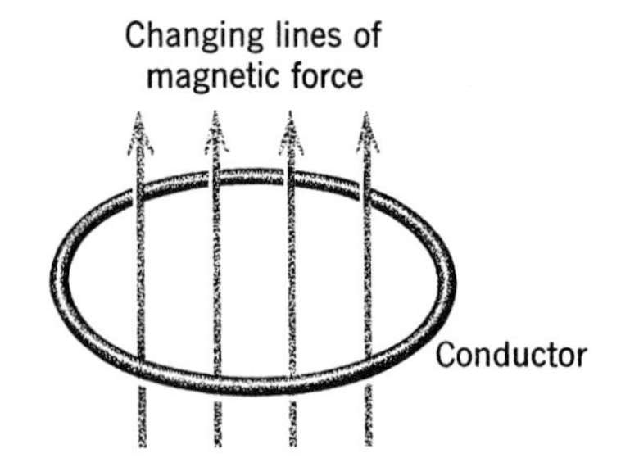

Fig. 16.6. Changing lines of magnetic force generate a current in a conductor.

important and original contribution to theory made by Maxwell, which led him to his final and powerful generalizations.

In Chapter 13, in which we were dealing with electricity, we attended primarily to circuits that were closed, that is, in which a continuous conductor provided a complete circuit for the charges to flow from the source of current around the circuit and back. The vector equations expressing the magnetic effect of such currents are *not* difficult to establish. However, consider the sequence of events in a circuit, which is first closed in which the battery is connected through a switch to a pair of parallel plates separated by an insulating medium. What happens when the switch is opened? See Fig. 16.7.

According to the fluid theory of electricity, positive electricity or charges would "pile up" on the positive plate and negative electricity upon the negative plate until an equilibrium was established, but no charges or fluid would flow across the insulating medium or the "empty" space between the plates. In other words, for a very short time charges would move to each plate until electrostatic equilibrium was re-established.

Now Maxwell found, in trying to extend the electromagnetic equations to cover the space between the plates, that the consistency of the equations was violated, unless it could be assumed that something occurred in this "empty" or insulating space that acted analogously to current. He termed this a "displacement current," indicating that although no actual physical charges were flowing, a kind of "strain" or "displacement" was occurring in the medium between the plates that has the same effect (so far as changing the magnetic field is concerned) *as if* a true current were flowing, or *as if* charges were moving.

Maxwell's extraordinary physical intuition, guided by his creative mathematical abilities, enabled him to construct a dynamical model that carried him far into an understanding of radiation phenomena and into predictions of consequences that later were to be proved valid. It is true, as in many such instances, that the original model, analogy, or mechanism upon which he built was later discarded—much as a scaffolding—leaving the abstract relations or equations that he developed as his contribution. (We shall see later how Niels Bohr similarly developed a model of the atom predicting many relationships between observed phenomena. The model he used to visualize what was taking place within the atom has been abandoned, but many equations, developed from that model, remain.)

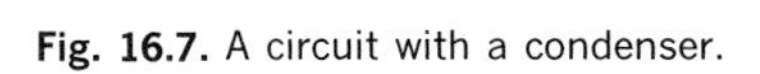
Fig. 16.7. A circuit with a condenser.

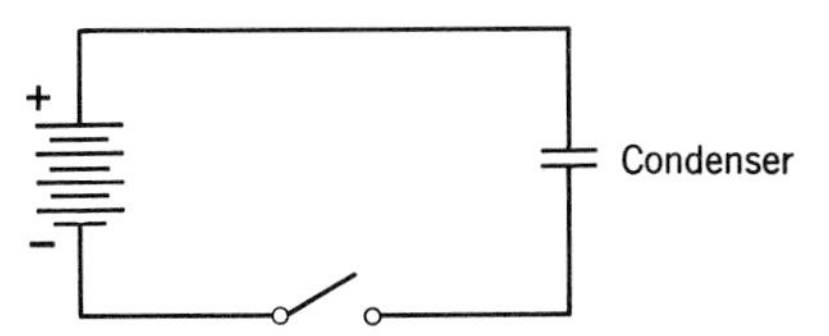

In 1864, Maxwell presented to the Royal Society his famous paper "On a Dynamical Theory of the Electromagnetic Field," in which he set forth his four, now famous, equations governing all electromagnetic radiations. Before examining the meaning of the equations themselves, let us look briefly at certain aspects of the model, that is, the analogy upon which Maxwell developed these equations.

We have seen how, in a pendulum system, there is a continued alternation between potential and kinetic energy. If friction did not enter, the alternation would continue forever. Similarly, in material waves, there is an alternation between kinetic energy and potential energy. In a water wave, for example, a mass particle of water is first displaced from its equilibrium position, reaching maximum displacement at the crest where it has maximum potential energy. This energy is then converted into *kinetic* energy as it returns to its equilibrium position. In the case of a wave pulse, the energy of one particle is transmitted to the contiguous particles of matter which are, in turn, set into motion, gaining kinetic energy. Maxwell concluded that a similar series of events occurred in the ether; if the ether were strained or displaced by an electric force, this would store up potential energy; as the displacement was restored, this would give rise to the kinetic energy of the magnetic field and, in turn, the kinetic energy of the magnetic field would again be transformed into potential energy by setting up a displacement or strain in the ether equivalent to an electric field. The alternations of electric fields and magnetic fields then become propagated as waves, carrying energy through the ether.

Although Maxwell's model (the mechanism in terms of which he first conceived of this transfer of energy) was soon abandoned, the equations that he developed on the basis of this model are still valid. Fifty years later, the assumption of a material ether was dropped, although the equations remained.

These equations state that at each point of space, as radiant energy flows through it, an oscillation of an electric force is coupled with an oscillation of a magnetic force at right angles. Furthermore, the directions of these oscillating force vectors are at right angles to the direction in which the energy is traveling as a wave. We can represent the two oscillations, one a vector representing the electric intensity E, and the other a vector representing the magnetic intensity B as shown in Fig. 16.8.

From here on, as we follow Maxwell's analysis, we see an example of how mathematical abstractions in the hands of men of great insight seem to take on a life of their own, and lead to quite extraordinary generalizations pointing the way to new and unexpected consequences. Up to this point, Maxwell had been struggling with a precise formulation of what goes on in an "open" electric circuit containing a condenser when the current varies. He wished to include all the experimental laws of electricity developed by Faraday, Oersted, Ampere, and others, and to include in his summary equations not only the space surrounding the conductor, but the empty space within the condenser. He was forced to con-

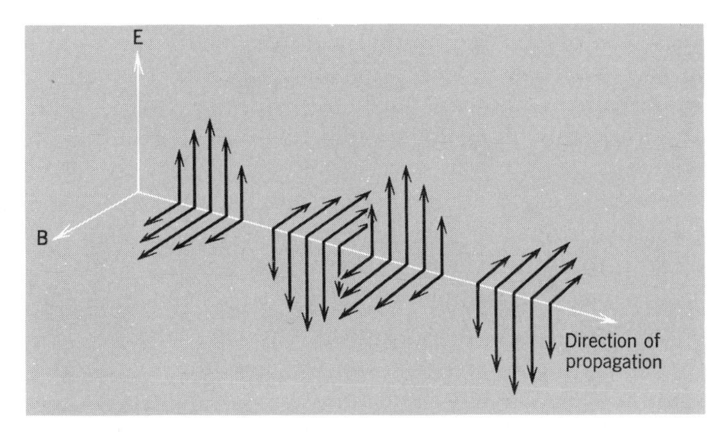

Fig. 16.8. Illustration of the relationship between the variation in electric intensity and magnetic intensity in an electromagnetic wave.

clude that some kind of wave phenomena, both electrical and magnetic, must exist in the *space or insulating material of the condenser itself,* and that the energy of this wave must be traveling through this space with a finite velocity. From the equations describing such wave propagation, which had *not even been detected at that time,* how could Maxwell jump to the conclusion that *light* was a form of electromagnetic wave propagation? Let us look at the equations to get a "feel" of how this insight developed.

Even though their full meaning, power, and elegance may not be grasped without training in more advanced mathematics, we list the four equations that Maxwell developed, both as an illustration of the extraordinary conciseness of the language of mathematics, which can pack into a simple equation (sentence) a tremendous number of ideas, and to show how the equations led Maxwell to suggest that light is an electromagnetic phenomena.

The four equations are these:[15]

$$\text{div } E = 0$$

$$\text{div } B = 0$$

$$\text{curl } E = -\frac{1}{c}\frac{\partial B}{\partial t}$$

$$\text{curl } B = \frac{1}{c}\frac{\partial E}{\partial t}$$

[15] The cited equations are for empty space, free of matter, charges or magnets. The more general equations developed by Maxwell are not here discussed because they are somewhat more complicated without adding anything relevant. For a fairly elementary discussion, see *Fundamental Physics* by Jay Orear, Wiley, New York, 1967.

These equations, which appear so simple, carry with them a vast amount of information, and imply consequences of great significance. Without entering into all the rigorous mathematical complications (which are, indeed, necessary if the consequences deduced by Maxwell are to be followed in detail), let us see what the equations say.

E and B are vector quantities, that is, they have both direction and magnitudes at every point in space. E is the intensity of the *electric* field at any point. In other words, it represents the force that would act on a very small charge if placed at that point. B is the intensity of the *magnetic* field, or the force that would act on a small electric charge moving through that point. E and B vary, of course, from point to point in space, and they also may vary in time.

The first two equations contain the expression "div," which stands for *divergence.* When a divergence is equal to zero it means that the number of lines of force (which represents the field intensity, either electric or magnetic) entering a very small volume of space are equal to the number leaving it. (This expression may be made a little more meaningful perhaps if we think of an analogous expression, $\text{div } v = 0$, in the case of an incompressible fluid. In this case, the total mass of water entering a very small volume is equal to the mass leaving it. Thus, in a pipe varying in internal diameter through which water is flowing, although the velocity v changes at various points in the pipe, and although the over-all velocity may change, if the water is incompressible at any one point, the amount of water entering one particular volume is the same as the amount leaving.)

The third equation

$$\text{curl } E = -\frac{1}{c}\frac{\partial B}{\partial t}$$

is a concise way of stating Faraday's law of induction. Again, a great deal of meaning is packed into this equation, and for those trained in mathematics, it also carries with it a visual picture of what is happening physically, which would take many paragraphs to state *completely* in words. Let us look at the terms of the equation. On the right-hand side, the expression $\partial B/\partial t$ refers to the rate of change in the magnetic field B. (It must be remembered that B is a directed or vector quantity and, therefore, such a change may be either an increase or decrease in the magnitude of B or a change in direction). The term c is a constant of proportionality, which becomes extremely significant as the analysis continues. We shall look at it shortly. On the left-hand side of the equation, we find the expression "curl E." It symbolizes the fact that the electric forces "curl" around the changing magnetic lines of force. If a wire conductor in the form of a circle were placed around the magnetic lines of force (Fig. 16.9), and these lines of force either changed in magnitude or direction, a current would be induced

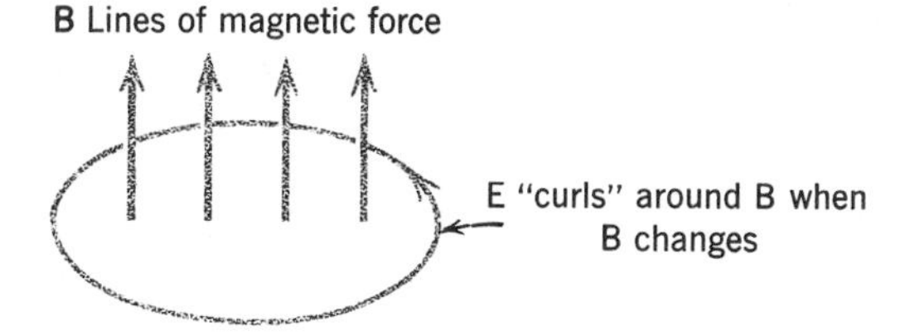

Fig. 16.9. When the magnetic intensity B changes in empty space, an electric intensity E is generated, even in the absence of a conductor.

in this wire conductor. Even without the presence of a wire able to conduct actual charges, an electric force is set up in space, which curls around the changing magnetic lines of force. The equation states that the *limiting value* of the electric force (per unit area) thus induced is negatively proportional to the rate of change in the magnetic field, and that the factor of proportionality is $1/c$. The fourth and final equation

$$\text{curl } B = \frac{1}{c}\frac{\partial E}{\partial t}$$

was Maxwell's contribution.[16] Note the symmetry between E and B (except for the negative sign whose significance we shall not discuss).

This fourth equation affirms that a changing electric field gives rise to a magnetic field in a manner similar to the generation of the electric field by a changing magnetic field (even though such a magnetic field had never been detected within a condenser).

Two interesting results follow from the above four equations.

(*1*) From the equations a simple derivation gives rise to another very significant equation:

$$\frac{\partial^2 E}{\partial x^2} = \frac{1}{c^2}\frac{\partial^2 E}{\partial t^2}$$

We shall not interpret the meaning of this equation other than to state that it is recognizable by mathematicians and scientists as the archetype of a wave equation, indicating that the electric intensity is propagated with the velocity c.

(*2*) Furthermore, Maxwell noted that this constant, c, which mathematically represented the velocity of a wave, happened to be identical with the ratio of two other constants connected with electricity and magnetism, which had already been measured in the laboratory. We have already noted that the electro-

[16]Originally, the differential equation that related the generation of a magnetic field by an electric current was given by the equation, curl $B = 1/c\ J$, where J was the density of current defined in terms of electrostatic units. Maxwell saw that conditions of continuity required that this equation be rewritten as curl $B = 1/c(J + \partial E/\partial t)$. Therefore, in empty space where $J = 0$, he concluded that curl $B = 1/c\ \partial E/\partial t$.

magnetic repulsion between two parallel wires of length ΔL_1 and ΔL_2 meters separated by a distance d and carrying currents I_1 and I_2 have an interacting force in newtons given by

$$F = K_\mu \frac{(I_1 \Delta L_1)(I_2 \Delta I_2)}{d^2}$$

We have also noted that two charges Q_1 and Q_2 in coulombs, separated by a distance d in meters interact with a force in newtons represented by

$$F = K_\epsilon \frac{Q_1 Q_2}{d^2}$$

The ratio of the two constants K_ϵ to K_μ is 9×10^{16} meters²/sec², which is exactly the square of the velocity of light.

Maxwell, noting this identity, surmised that light is an electromagnetic wave. He predicted, therefore, the existence of radio waves later found by Hertz, and he also was able to relate the optical properties of transparent substances to their magnetic and dielectric properties.

This unexpected fusion of two apparently unrelated branches of physics was a tremendous triumph for abstract mathematical analysis. It opened up many new avenues for exploration. Maxwell predicted the possibility of producing "radio waves," which would differ from light waves only in their wavelength and frequency. He even made suggestions as to the type of apparatus that might be used to test his theory. Finally, in 1888, ten years after Maxwell's death, Heinrich Hertz produced electromagnetic waves (now identified as microwaves) of much longer wavelength than light, and showed that these waves could be reflected, refracted, and polarized in the same way as waves of light. Those skeptics among scientists who had, up to 1888, refused to follow Maxwell's theory were now convinced.

As in all cases when a big leap forward has been made by bringing together two previously unrelated sciences into a single theoretical system, each of the separate sciences, in this case, optics and electricity, becomes much more thoroughly understood. A new technology rapidly came into being. Within a century after Maxwell's speculations, the communication system that we know had developed as a monument to his creative insight and the bold pursuit of logic in his mathematical equations.

6. THE ELECTROMAGNETIC SPECTRUM

We return briefly from the abstract world of the physicist to the colorful world of our direct experience. Just as each musical note distinguished by the ear can be assigned a different number, that is, the number of vibrations per second,

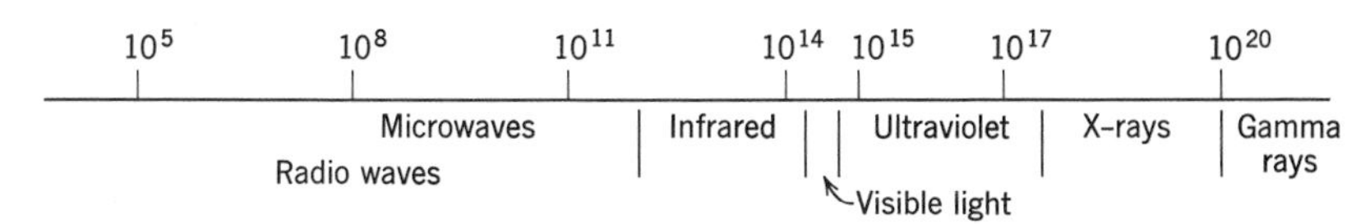

Fig. 16.10. Electromagnetic spectrum. Frequencies are shown in cycles per second.

similarly each color perceived by the eye is associated with a number, again the number of vibrations per second. Just as the human ear responds to sounds varying in pitch only within a certain range (normally from about 15 cycles/sec to about 18,000 cycles/sec, a range of about ten octaves), so the eye can respond to electromagnetic vibrations within only a certain range of frequency.

We have seen that for light to travel in straight lines the wavelengths must be extremely short. From the relationship that holds for all waves, that is, that the velocity equals wavelength times frequency, together with a knowledge of the velocity of light (determined by either of several direct experimental methods, or by Roemer's method from observations on Jupiter's satellite) and of the wavelength (by Young's double slit interference method), the frequency may readily be calculated. Upon this basis, colors range from a frequency of about 4×10^{14} cycles/sec for red (longest visible wavelength) to about 7×10^{14} cycles/sec for violet (shortest visible wavelength). In contrast with our hearing ability, our visual perception thus extends over less than one octave of the range of electromagnetic vibrations. Immediately above the range of 7×10^{14} cycles/sec is the ultraviolet range to which our skin and photographic plates are sensitive; immediately below 4×10^{14} cycles/sec is the infrared and radiant heat range to which also our skin is sensitive. Still lower frequencies (longer wavelengths) correspond to radio waves; higher frequencies correspond to X-rays.

The total range of frequencies is called the electromagnetic spectrum, and is charted in Fig. 16.10. Note what a very small proportion of the radiations constitutes visible light. Our new artificial sense organs, such as radio receivers, photographic plates, and other electronic detectors are now picking up many messages from the universe around us to which we have been blind in the past.

7. THE LASER

A modern device, the laser, has recently been developed which gets around many of the disadvantages of using other types of electromagnetic vibrations in the investigation of nature. Because in most cases the electromagnetic waves are emitted by a myriad of sources—each source vibrating with its own frequency (within limits) and phase, each with its own plane of vibration—the resulting wave disturbance is jumbled and lacks "coherence." This confusion arises from the addition of millions of wavelets of energy slightly different in frequency and not in phase ("in step"). The result is that no matter how well focused, the beam "spreads" rapidly. The laser so controls the emission of very short pulses of light that all the emitting sources are "in step" and emit waves of exactly the same wavelength and frequency. See Fig. 16.11.

Three types of lasers have been developed: (1) the ruby laser, (2) the gas laser, and (3) the semiconductor laser. Although the mechanical operation of the lasers are different, the principles are much the same.

The ruby laser is typical. It is composed of a ruby rod about ⅝ inch in diameter and ten inches long (see Fig. 16.12). The ruby is flooded with radiation from a flash lamp. The ends of the ruby rod are silvered so that light is trapped for a microsecond or so, traveling back and forth along the axis as it is reflected by the mirrors at the ends. One end, through which the laser beam eventually emanates, is only lightly silvered; the other end is almost opaque. Once the beam attains a certain energy level, it is suddenly released as a laser beam. In this way a sudden pulse of energy is released and is transmitted in a beam of small

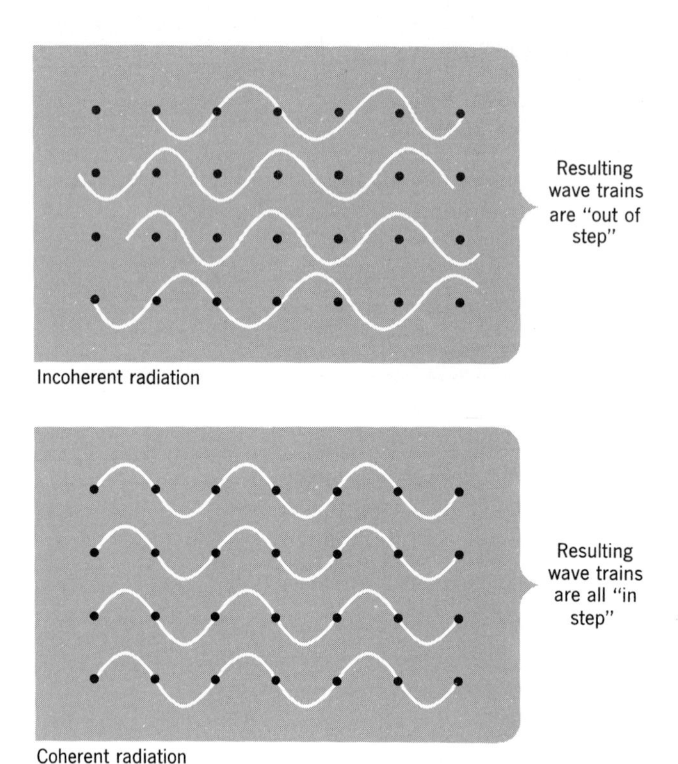

Fig. 16.11. Schematic diagram of coherent and incoherent radiation. With the more commonly occurring incoherent radiation, many sources emit electromagnetic waves which are "out of step," (as well as being, very often, of a wide range of frequencies.) In case of lasers, each of the myriads of small sources emits light waves of the same frequency and all are "in step." This is a type of "coherent radiation" in which the "rays" or photons do not interfere with one another and therefore do not "spread out" as do incoherent rays.

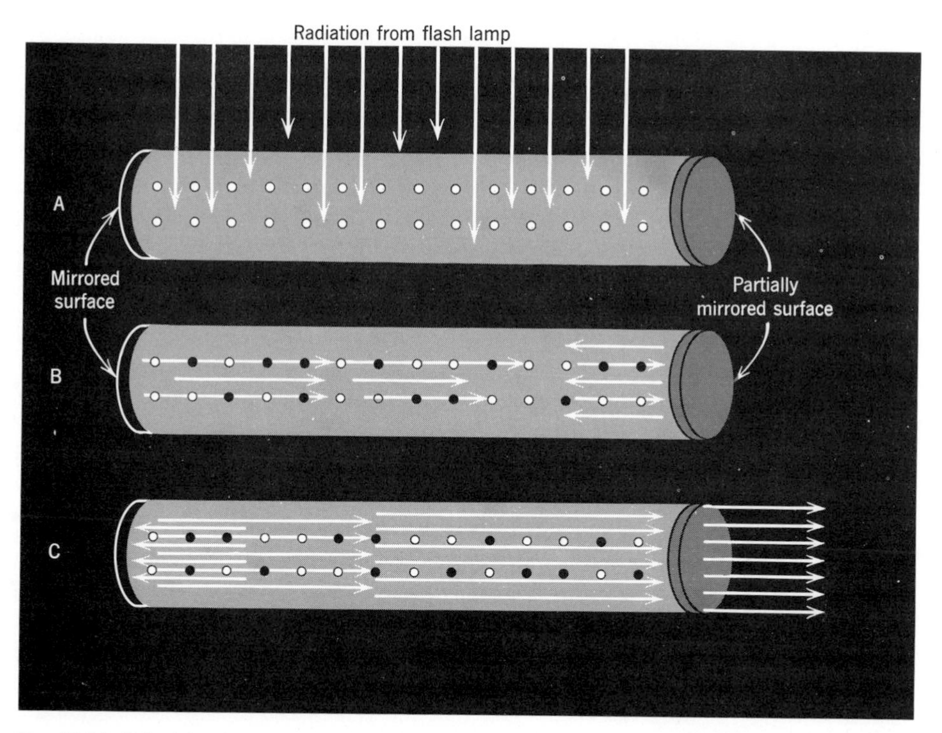

Fig. 16.12. Principle of ruby laser. Radiation from flash lamp raises energy level of atoms to an excited state. In the excited state an atom can be triggered into emitting a photon of a specific frequency. When sufficient atoms have been raised to excited state, a cascade of photons travels back and forth between the mirrored surface on left and the partially mirrored surface on the right until finally a burst of rays breaks out on the right.

diameter. The emission level is as high as 50 megawatts in 5 to 10 nanoseconds (10^{-9} sec). Such a beam will travel a thousand miles in a vacuum without "spreading out" to a beam larger than half an inch!

The theory and principles of laser operation depends upon quantum theory. Therefore further discussion will be postponed until Chapter 19.

8. QUESTIONS

1. Why is the index of refraction of a substance always greater than one?

2. Huygens assumed that light waves were longitudinal. Why was this a natural assumption? What phenomena indicate they are transverse?

3. Explain why there can be no total internal reflection when light passes from a less dense into a more dense medium.

4. In view of our earlier discussions about the nature of scientific theories, review the evidence supporting Huygen's wave theory and that supporting Newton's corpuscular theory. What type of evidence finally convinced physicists of the validity of Huygen's theory? Was the theory proved?

5. The atmosphere surrounding the earth is of greater density than that of "space" beyond the atmosphere. A ray of light from a star or the sun will, therefore, be bent. When will it be bent the most? Draw a diagram showing how the apparent position of the sun on the horizon is shifted from its true position.

6. How do we know that the velocity of light of every color in vacuum is precisely the same within at least one part in 10^{10}? Recall the similar problem with respect to the velocity of different notes of music. In the case of color, consider astronomical data, particularly eclipsing satellites of Jupiter and eclipsing stars. If colors traveled at different velocities, what would happen as a moon of Jupiter emerged from behind the planet after an eclipse? Is the velocity of light of different colors the same in materials other than a vacuum?

7. Compare the number of "octaves" of light visible to the eye with the number of "octaves" to which the ear responds.

8. Why does a star twinkle, although a planet does not?

9. Does a moving charge always radiate an electromagnetic wave? Explain.

10. When a light beam hits a mirror obliquely, the angle of incidence equals the angle of reflection. Compare this with a particle bouncing off a heavy wall (assuming no spin). If the collision is completely elastic (no energy is lost) the angle of incidence and angle of reflection will be equal and the speeds before and after collision will be equal. With reflected light, some energy is always absorbed by the reflector. Is the speed of light, nevertheless, the same before and after reflection? (*Hint:* If not, what effect would it have on the angle of reflection? Pursue the argument to see that light is not reflected in the same way as are particles, but is absorbed and then re-emitted.)

11. The emission of electromagnetic radiation depends on the oscillation of charged particles. The frequency depends on the speed of the oscillating particle and the distance covered between oscillations. As one might expect, the equation $f = v/d$ is a good approximation, where d is the distance traveled in each cycle and v is the average speed. Show that $\lambda = \frac{c}{v} d$ and relate this to the generalization that the emitted wavelength must be longer than that of the emitting system. Explain why a radio system cannot emit light waves or X-rays.

12. Why does the sky appear blue?

Problems

1. The velocity of light in vacuum is close to 3×10^8 m/sec and is only insignificantly less in air. Convert to miles per second. If you time the difference between the sighting of a lightning flash and the sound of the thunder, you can determine how far away the lightning struck. If the interval is 5 sec, *approximately* how far away will the lightning be? (Take the speed of sound as approximately 1000 ft/sec—about the same velocity as that of a 22 caliber bullet.)

2. Referring to the limits of frequency of visible light, how many Angstrom units are there in the corresponding wavelengths? (One Angstrom unit is defined as equal to 10^{-10} meters.)

3. The astronomer Roemer calculated the speed of light in 1675. Another astronomer Bradley in 1727 used a method depending on a phenomenon called aberration. If raindrops are falling vertically with velocity c and you are moving forward with velocity v, show that the angle θ (angle of aberration) to the vertical, from which the raindrops appear to come is given by the equation $\tan\theta = v/c$. Sketch the orbit of the earth moving about the sun with a speed v. Show how the light from a given star will be seen through a telescope which is set at a certain angle of aberration. The angle depends on the relative direction of the earth's motion and the light of the star. Where will the angle be maximum? Where will it be zero?

4. Using the law of reflection, determine the length of a plane mirror that a man 6 ft tall would require to see his complete image, assuming his eyes are 3 in. below the top of his head.

5. The index of refraction of water is about 4⁄3. What is the speed of light in water?

6. The visual threshold of the normal eye is around 10^{-22} watts per square meter. Compare the power sensitivity of the eye to that of the ear (disregarding any difference in cross sectional area between eardrum and cornea). (See problem 13, Chapter 15.)

7. Instead of using just two slits as in Young's experiment to get interference patterns, a diffraction grating is now commonly used in spectroscopic work. Instead of the two slits there are many thousands of slits through which the light passes. See Fig. 16.13 below.

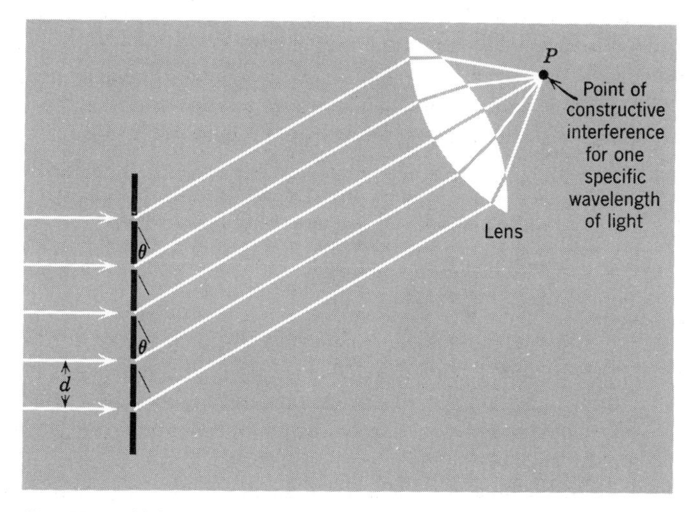

Fig. 16.13. Light of one wavelength passing through the slits of diffraction grating will be "in step" at a certain angle θ and will be focused by the lens at point P. For other wavelengths the angle will be different and will come to a focus at a different point.

Derive the equation relating λ, θ, and d.

8. Standard radio broadcasts have a range of frequency of about 1 megacycle. The very high frequency band of TV is in the range of 50 to 150 megacycles. How do the associated wavelengths compare with the wavelength of an X-ray with a frequency of 10^{18} cycles/sec?

9. An interferometer is an instrument that separates two beams of light so that they travel over two different paths at right angles to each other and then the two beams are brought together again. When viewed through a telescope, bands of bright and dark light can be detected showing the positions where the two beams of light interfere constructively and destructively. As one of the path lengths is changed by extremely small amounts, the interference bands move across the field of view. They can be counted as they move across and, in this way, the number of wavelengths by which the two paths differ can be determined. Such an instrument is used to determine lengths with extreme precision. For this reason the wavelength of certain frequencies of light is chosen as a new standard unit of length. How many wavelengths of light of frequency 4.5×10^{14} cycles/sec (in the red range) are there in a centimeter?

10. As light spreads out from a point source, its intensity (that is, the energy flow per unit time across a unit area perpendicular to the direction of motion of the beam) varies inversely as the square of the distance from the source, provided the light is being emitted equally in all directions. Draw a diagram of two concentric spheres about a source of light to illustrate why this must be true (assuming, of course, that no energy is absorbed as light travels through space). If a point source of light is emitting at a rate of 1 watt, how far away can it be detected if no energy is lost. Use the threshold sensitivity of the eye as 10^{-22} W/m^2.) *Note:* Of course, if light is not emitted in all directions equally, its intensity does not vary inversely with the square of the distance. A parabolic reflector sends out a beam of light which does not spread out as much as ordinary light. However, since no ordinary source of light can be emitted from a point, there is no way of keeping the beam from spreading. The advantage of the newly developed laser (light amplification by stimulated emission of radiation) is that all the light is coherent, intense, and travels in one direction.

11. The equation for a diffraction grating (see Problem 6) is given by $n\lambda = d \sin \theta$ where n is a whole number. (If n is one, the point of maximum intensity is called the point of the first order image, etc.) If the second order image is formed at 30° by a diffraction grating of 4000 lines per centimeter, what is the wavelength of the light?

Recommended Reading

Sir Isaac Newton, *Opticks,* with foreword by Albert Einstein, introduction by Sir Edmund Whittaker, preface by I. Bernard Cohen, Dover Publications, New York, 1952. This great classic of science, unlike the *Principia,* can be read with ease and enjoyment by the non-scientist. The preface by Cohen of Harvard, a historian of science, places the work in modern context. Newton's queries in Book III show the mind of a great scientist probing into and speculating about problems whose solutions were not yet at hand.

Physical Science Study Committee: *Physics,* D. C. Heath and Company, Boston, 1960, Part II. A careful presentation with many illustrations of optical and wave phenomena.

Sir Oliver Lodge, *Pioneers of Science,* Dover Publications, New York, 1960, Lecture X on the velocity of light.

James R. Newman, *Lives in Science,* (A Scientific American Book) Simon & Schuster, New York, 1957, pp. 155–183. An essay on James Clerk Maxwell.

Hermann von Helmholtz, *Popular Scientific Lectures,* Dover Publications, New York, 1962. Lectures on "The Recent Progress in the Theory of Vision" and "On the Relation of Optiks to Painting." These lectures, like the one on music, are classical essays on the broader relationship between physics and perception.

James R. Newman, editor, *The World of Mathematics,* Simon & Schuster, New York, 1956, Part XXI, A Mathematical Theory of Art. For those interested in an attempt by a modern mathematician to extend mathematics into the field of aesthetics.

F. Woodbridge Constant, *Fundamental Laws of Physics,* Addison-Wesley, Reading, Mass., 1963, Chapters 11 and 15. Two excellent chapters on wave propagation and Maxwell's equations, presented simply but thoroughly. A mature knowledge of mathematics (but not calculus) is required for mastery of the ideas.

chapter seventeen

Elements of quantum theory

The end of the nineteenth century was marked by an extraordinary confidence that most of the basic problems of nature were close to a final solution. Helmholtz's prediction that the vocation of science "will be ended as soon as the reduction of natural phenomena to simple forces is complete" appeared to be coming true. Many investigators felt that there was little left to be done except to fill in a few gaps in knowledge, and to tidy up a few ragged edges. There were only faint hints of the tremendous new era in science about to dawn with the new century.

In physics, the dynamical laws of motion and the electromagnetic laws of radiation, both operating within the framework of the laws of conservation of energy and of matter, provided, it seemed, a foundation upon which all phenomena of nature could eventually be explained. Even the biologists believed that their own subject matter would be reduced to physical determinism. "The great abstract law of mechanical causality now rules the entire universe as it does the mind of man," wrote Ernest Haeckel, the biologist, in his *Riddle of the Universe.* The fundamental particles of matter were located in a continuous space describable by the "true" axioms of Euclidean geometry and subject only to various forces of attraction and repulsion; the transmission of energy between these atoms took place according to well-known laws.

In mathematics, a firm and rigorous foundation to the calculus had been established, overcoming the weaknesses that had plagued its earlier formulations. The nature of continuity was at last understood, and that marvelous tool of analysis, the differential equation, could be applied with extraordinary success to all kinds of spacio-temporal phenomena: electromagnetic radiation, heat flow, fluid motion, gravitational effects, and forces acting between masses. Although non-Euclidean geometries had been explored by the middle of the nineteenth century, they were not considered very significant by either mathematicians or scientists. Furthermore, the modern development of mathematical logic with all its implications still awaited the work of Whitehead and Russell (in 1905) to accomplish the first significant departure from the Aristotelian logic, which had held sway for so many centuries. In other words, nineteenth century mathematics was marked by consolidation, rather than by striking innovation.

In chemistry, also, most scientists probably felt that the basic laws had now been discerned and formulated as a result of the triumphant development of atomic theory of that day. It was true that tremendous work still needed to be done, particularly in organic chemistry, in analyzing and synthesizing the myriads of compounds that could be formed out of the 90-odd known elements recognized by 1900, but no revolutionary discoveries were anticipated.

Even in the biological sciences, the revolutionary developments in genetics awaited the turn of the century. It is true that Mendel's paper on genetic variation foreshadowed the later work, but deVries' paper in 1900 was the first publicized recognition that evolution does not proceed by slow continuous steps but takes place in "quantum jumps" or mutations. Finally, in psychology the break came at the turn of the century when the older view of the mind of man as one of simple rationality suddenly came into question with Freud's "discovery" of the unconscious and its role in behavior.

The new era in the physical sciences, which was ushered in with the twentieth century, comprises three closely related branches: the quantum theory, atomic and nuclear physics, and the relativity theory. Not merely was a technological revolution introduced. Each of these three developments has had significant implications because of its impact on the biological and social sciences, and on man's fundamental attitudes toward himself and his destiny.

1. PLANCK'S QUANTUM HYPOTHESIS

In our previous discussion of light waves, little was said either about their origin or about their interaction with matter. Maxwell's theory, however, both as he formulated it and as it was later more fully developed and extended by Hertz and his followers, embraced not only what took place in the medium through which the radiant energy traveled, but its source and its ultimate absorption.

It was evident on the basis of Maxwell's theory that an electric charge when set into accelerated motion would originate an electromagnetic wave, since an acceleration of a charge implies a changing electric field, which gives rise to a changing magnetic field, etc. (If the electric charge oscillates about some central point, the wave given off is sinusoidal.)

Now, whenever any solid is raised to a high enough temperature, the atoms composing that solid are set into very rapid kinetic motion of an irregular pattern. The atoms, as we now know, contain charged particles, and these then undergo rapid but irregular acceleration, and a great mixture of electromagnetic waves of many frequencies in the range of visible light are given off. If gases, however, are heated, the atoms are much farther apart, and certain sharp frequencies of light dominate the spectral range. Furthermore, each element is found to give off a characteristic pattern of lines of specific frequencies. Thus, it was realized during the last half of the nineteenth century that much information about how atoms differed from element to element could be gathered by a study of these lines. Consequently, a vast amount of data on spectra was collected and classified which was to become one of the foundations of a new theory of the atom.

The problem of the *energy distribution* of light emitted by heated bodies was also emerging as significant. To study this energy distribution, an arrangement is used called (somewhat misleading) a "blackbody radiator." By a blackbody is meant one from which no light, or practically no light, is emitted or reflected. A hollow box with a small hole in it approximates this condition. Very little light from the outside can be reflected through this hole. If, therefore, the whole box is heated to a very high temperature, and the resulting light coming from the interior and given off through the hole is studied, the condition of blackbody radiation is approximately achieved (Fig. 17.1).

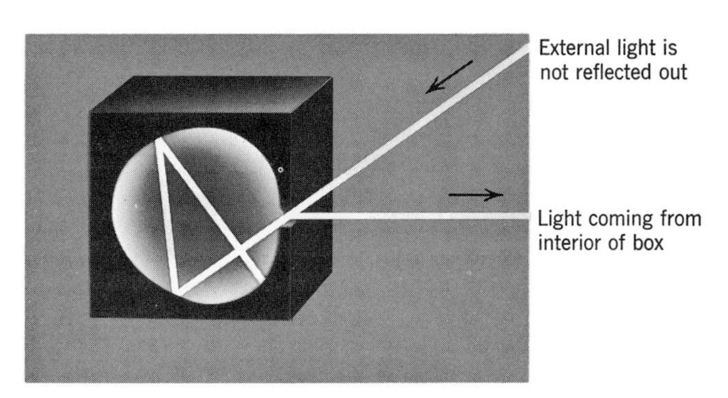

Fig. 17.1. Blackbody radiator.

When classical physics is used to analyze the energy of this light, as it is given off at different frequencies, a most amazing result occurs. The actual energy given off and the energy predicted on the basis of all known physical laws do not agree. At higher frequencies—in particular, toward the violet and ultraviolet range of the spectrum—classical physics predicts that energy should increase, although actually it decreases. This contradiction is sometimes referred to as the "ultraviolet catastrophe."

Since classical physics was so overwhelmingly successful in all its other branches, this particular and rather specialized puzzle was largely disregarded until Max Planck undertook to find a solution. Planck, a very competent mathematician, as well as a physicist, studied the data and the empirical curve (a bell-shaped curve showing that the maximum radiation energy is found not at the highest frequencies but at intermediate ranges) and, then, sought an explanation, or a set of assumptions from which the shape of the curve could be deduced. In other words, he was attempting to find a more basic theoretical explanation, just as Newton had proceeded to do when he had sought to find an explanation for Kepler's empirical laws of elliptical orbits. As Newton had proposed the postulate of universal gravitation (the inverse square law), together with the three laws of motion, as sufficient to explain the orbital motions of all planets, so in 1900 Max Planck proposed a basic postulate to "explain" the empirical data of radiation. See Fig. 17.2.

This postulate was a fantastic one, according to all known laws of science, and Planck himself, who was by nature a conservative thinker, was loathe to accept it. The assumption required that bodies can radiate energy only in bundles or quanta instead of continuously. Nature, according to this postulate, not only "made jumps" but, in the process of energy emission, it could do nothing else

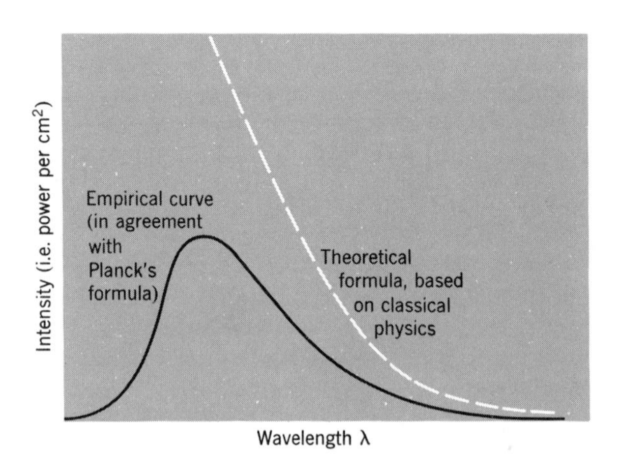

Fig. 17.2. General shape of the radiation intensity curve emitted by blackbody radiators. The observed empirical curve does not agree with classical prediction but fits Planck's equation. In the classical equation, the intensity should increase as wavelength decreases, i.e., as frequency increases. This contradiction between experiment and classical theory was known as "the ultraviolet catastrophe."

but make jumps. Planck postulated that energy was radiated in "packages" or quanta—the energy of which always is equal to a constant, h, (now called Planck's constant) times the frequency of radiation, f. Just as Millikan had proved that electric charges are always an integral number times a basic constant (the charge on an electron), so Planck was forced to the conclusion that solid matter can only radiate an integral number of quanta of energy, equal to $h \times f$.

It is told by Planck's son that during a walk in the Grunewald in Berlin in 1900 his father announced to him, "Today I have made a discovery as important as that of Newton."[1] Try as he would, Planck found no way of explaining the experimental data other than on this revolutionary concept of a basic discontinuity in nature. At first, other scientists remained skeptical, unwilling to accept such a departure from the well-established principle of continuity of energy. It was not until Einstein showed the application of Planck's hypothesis to photoelectric effects and Bohr successfully applied the quantum idea to atomic structure that the resistance to the conception of the existence of a quantum of energy evaporated. It was the youngest scientists in particular who picked up, applied, and extended Planck's hypothesis rapidly. Planck made a telling comment in his autobiography:

> A new scientific truth does not triumph by convincing its opponents and making them see the light, but rather because its opponents eventually die, and a new generation grows up that is familiar with it.[2]

This is but another illustration that the scientist, rigorously grounded in his subject, is often as conservative as investigators and thinkers in other fields. Even Lord Kelvin, an innovator and pre-eminent scientist, in an earlier period had affirmed in his often-quoted words of 1884 regarding Maxwell's explanation of light:

> I am never content until I have constructed a mechanical model of the object I am studying. If I succeed in making one I understand; otherwise I do not. Hence, I cannot grasp the electromagnetic theory of light. I wish to understand light as fully as possible, without introducing things I understand still less.

The whole idea first introduced by Planck, that energy might be discontinuous, was so much more radical and so hard to represent by any model (let alone a *mechanical* model such as Kelvin and an earlier generation of physicists had required) that it is little wonder that scientists were reluctant to accept the idea. However, even more radical ideas were soon to develop in the early twentieth

[1]Quoted by A. E. E. Mckenzie in *Major Achievements of Science,* Cambridge University Press, London, 1960, p. 316.
[2]Ibid., p. 317.

century, of such a nature that the whole structure of classical physics was to be brought into question. Relativity theory, the quantum theory, the uncertainty principle, and the problem of the nature of space, all combine to form the scientific revolution in which we now live. Planck's theory of quanta, of 1900, was but the beginning of a spectacular growth of many new and radical concepts.

2. THE PHOTOELECTRIC EFFECT; EINSTEIN'S APPLICATION OF QUANTUM PRINCIPLES

In 1905, the young Albert Einstein, working in the patent office at Berne, wrote three short scientific papers, any one of which would have assured him lasting fame in the annals of science. One of these announced his theory of relativity, which will be discussed in a later chapter; another was on Brownian motion; and the third was on the photoelectric effect, which established Planck's quantum theory as a principle to be construed even more broadly than Planck had thought necessary.[3]

The photoelectric effect, familiar in the operation of automatic electric doors and in the translation of visual images into electrical impulses in TV transmitters, consists in the fact that light upon striking certain metals causes electrons to be given off, transforming the radiant energy absorbed by the metal into kinetic energy of emitted electrons.

Now, if such an effect is to be understood on the basis of classical physics, it is to be expected that as the intensity (thus energy) of light increases, the kinetic energy of the electrons emitted will increase. This would imply that, as the intensity of the beam of light increases, the velocity of the electrons released would increase (since the classical view requires that the kinetic energy equals $\frac{1}{2}mv^2$, where m is the mass of the electron, and v its velocity).

Experimentally, however, this is not the case. Instead, the *maximum* velocity given to the electrons is constant, *regardless of the intensity of* light, but varies in direct proportion to the *frequency*. Note that we refer to the *maximum* velocity of any emitted electron for the following reason, according to the theory developed by Einstein.

Electrons are embedded in the atoms of the metals and are held there by attractive forces. When light is absorbed of sufficient energy, the energy is used in two ways: first, to overcome the attractive forces holding the electron to the metal atoms and, second, if there is any energy left over, to impart that energy

[3] Max Planck, while convinced that matter could only *give off* or *absorb* radiation in bundles or quanta, did not wish to go on to the next step, that the radiation *itself* must be in the form of quanta.

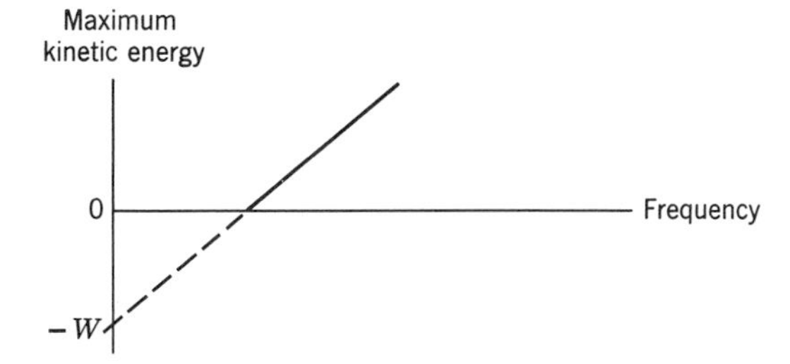

Fig. 17.3. Photoelectric relation between maximum electron energy and frequency.

to the electrons as kinetic energy. Therefore, for those electrons near the surface of the metal, less energy is required to overcome the "binding force," and there is more kinetic energy available. The "surface" electrons therefore acquire the maximum kinetic energy. The formula developed by Einstein was $hf = W + \frac{1}{2}mv^2$ (see Fig. 17.3).

Here, W is the work required to free the electron from the metal. It is seen that there is a linear relation between the frequency and the maximum kinetic energy of electrons emitted. Furthermore, there is nothing in this equation relating to the *intensity* of the light causing electron emission. However, although the *velocity* of the electrons does not depend in any way upon the intensity of light, the *number* of electrons emitted does so depend. From this, Einstein, following Planck's lead, concluded (and subsequent investigations confirmed) that the light is made up of quanta or, as Einstein called them, *photons*. The energy of a single photon is determined only by hf. The intensity of the light affects, not the energy of the emitted electrons, but only the number emitted. The process of photoelectric emission of a single electron depends upon the absorption of a single photon. If the number of photons is increased, the number of emitted electrons is increased, but no change occurs in the maximum kinetic energy and hence the velocity of electrons. The peculiar constant h, called Planck's constant, is the same for all metals. W, the threshold energy required to release electrons, varies from metal to metal, being so large in the case of some metals, for example, platinum, that no photoelectric effect occurs.

Now all of these phenomena are impossible to account for on the basis of the classical wave theory of light developed in previous chapters. It will be recalled that the energy of a wave of a given frequency is dependent on its amplitude. On a seashore, for example, during a storm in which the incoming waves are high, rocks immovable by small waves may be carried far up the beach. The kinetic energy given the rocks, in other words, is directly related to the intensity of the approaching waves. Similarly, if light consisted of waves, one would predict that the kinetic energy given to the electrons would depend on the amplitude of the light waves. The prediction failed to hold true in the photoelectric effect.

The Einstein-Planck theory of quanta (in the case of light, called photons) seemed indeed to reinstate Newton's corpuscular theory. Light now became regarded as composed of photons, each photon having a certain amount of energy depending on its frequency (color). Furthermore, it can also be shown that each photon has a certain momentum, equal to hf/c, and that as the photons strike a substance and are reflected, the change of momentum causes a small but measurable pressure. (However, it should be noted that under the older wave theory also, light had momentum and caused pressure.)

Has a full circle now been described in the development of a theory of light—from a particle theory to a wave theory and back again to a particle theory? This might seem so, if we consider only certain phenomena such as the photoelectric effect. But just as the wave theory cannot explain photoelectricity and similar quantum phenomena, so the quantum theory cannot explain many phenomena of interference and diffraction. For a while this dilemma appeared to be a contradiction inherent in modern physics. As expressed in the often-quoted words of Sir William Bragg: "On Mondays, Wednesdays, and Fridays, the physicist believes and acts on the basis of wave theory; on Tuesdays, Thursdays, and Saturdays he follows the particle theory of light." Only recently has a reconciliation of a sort been possible with the development of what is called quantum mechanics. But the cost has been great. An adequate, easily understood model in common-sense terms of experience at present seems to be impossible. If Kelvin found it difficult to comprehend Maxwell's theory because it was too abstract and nonvisualizable, the difficulty today of "understanding" quantum mechanics is much greater for scientists and nonscientists alike. Before considering even the elements of quantum mechanics, another surprising discovery of the twentieth century must be reviewed, that is, that *just as light waves have properties reminding one of particles, so material particles have properties reminding one of waves*. The idea at first seems even more startling than the discovery of light quanta. The disconcerting aspect of the idea is mitigated only by the symmetry of nature that is now revealed.

3. THE WAVELIKE NATURE OF ELECTRONS

The creative innovations in science, as in the arts, are often found in the metaphors or analogies constructed by man in his attempt to understand the world about him. However, although metaphors and analogies in the arts and literature are suggestive and often loosely drawn, those in science are meticulously developed and are subject to a precise and "objective" mode of criticism compounded both of logic and accepted experimental procedures. The very word "rationalism" stems from "ratio," which implies an analogy. Scientific reasoning has come to depend more and more upon the establishment of ratios, that is, analogies between aspects of physical reality and the number system of mathematics. A ratio might be called half of a metaphor: it is a fraction, that is, a relation between two numbers; the other half of the metaphor is the subject to which the ratio refers. Thus, a measurement is an analogy, as we indicated earlier, where it was asserted that a given length is to a standard unit as one number is to another.

As science has expanded, its metaphors and analogies have become more and more abstract and divorced from immediate experience, but also deeper in meaning, generality, and effectiveness. For example, the analogy Maxwell found between fluid flow and certain electromagnetic phenomena is not at all easy to visualize. It can only be expressed through mathematical equations applicable to both.

A profound and fortunate analogy, which was to lay the foundations of modern quantum mechanics, occurred to Louis de Broglie while working in 1924 on his Ph.D. thesis. The topic that he chose involved an attempt to synthesize the two aspects of light: the wavelike characteristics, and the quantum characteristics. This was two decades after Einstein and Planck had established the quantum theory and after Einstein had expounded his special theory of relativity. By this time also, Bohr had proposed a new theory of the atom. These three theories were pulled together by de Broglie to construct his wave theory of matter.

Let us examine de Broglie's comment reviewing the history of his discovery, as given in his address delivered in Stockholm in 1929 when he received the Nobel prize for his contribution:

Some thirty years ago, then, Physics was divided into two camps. On the one hand there was the Physics of matter, based on the concepts of corpuscles and atoms which were assumed to obey the classical laws of Newtonian Mechanics; on the other hand there was the Physics of radiation, based on the idea of wave propagation in a hypothetical continuous medium; the ether of Light and of electromagnetism. But these two systems of Physics could not remain alien to each other: an amalgamation had to be effected; and this was done by means of a theory of the exchange of energy between Matter and radiation.

. . . When I began to consider these difficulties I was chiefly struck by two facts. On the one hand the Quantum Theory of Light cannot be considered satisfactory, since it defines the energy of a light-corpuscle by the equation $W - hf$, containing the frequency f. Now a purely corpuscular theory contains nothing that enables us to define a frequency; for this reason alone, therefore, we are compelled, in the case of Light, to introduce the idea of a corpuscle and that of periodicity simultaneously. On the other hand, determination of the stable motion of electrons in the atom introduces integers; and up to this point the only phenomena involving integers in Physics were those of interference and of normal modes of vibration. This fact suggested to me the idea that electrons too could not be regarded simply as corpuscles, but that periodicity must be assigned to them also.[4]

[4]Louis de Broglie, *Matter and Light,* Dover Press, New York, reprint of W. W. Norton and Co. New York, 1939, pp. 166, 168–169.

In his attempt to bring into a synthesis optics and mechanics, de Broglie, was led to propose this fantastic analogy: just as light waves have a particlelike character, so mass particles such as electrons must have a wavelike character. This wavelike character of massive bodies escapes notice only because the wavelengths are so short. The shorter the wavelength of any wave, the more difficult it becomes to detect it. A wave does not easily bend or diffract if the wavelength is very short. This is basically the reason that light appeared to Newton to be composed of particles. It traveled in straight lines and did not seem to spread out as it hit obstructions, and appeared to be reflected from substances much as a bullet would be reflected. De Broglie assumed this to be the case with electrons. They seem to act like bullets or particles only because their wavelength is extremely small, much smaller in fact than the wavelength of visible light. Hence, the phenomena of diffraction and interference escape our attention.

The idea was profound and simple, and even portions of the mathematical analysis that lead to consequences testable by experiment can be followed in simple equations.

We have already mentioned that, according to classical electromagnetic theory, light has a momentum per unit volume, P, given by the equation:

$$P = \frac{E}{c}$$

where E is the energy per unit volume and c the velocity of light.

But by Planck and Einstein:

$$E = hf$$

Therefore

$$P = \frac{hf}{c}$$

We know that λ, the wavelength, for all waves is given by

$$\lambda = \frac{c}{f}$$

Therefore, for light, we have the relationship

$$P = \frac{h}{\lambda}$$

or

$$\lambda = \frac{h}{P}$$

that is, the wavelength of light is equal to Planck's constant divided by momentum.

De Broglie suggested that a similar relationship should be found for *electrons,* if they are wavelike in character:

$$\lambda = \frac{h}{P}$$

Furthermore, since an electron of mass m traveling at velocity v (small compared to the velocity of light) has a momentum

$$P = mv$$

then

$$\lambda = \frac{h}{mv}$$

This is an extraordinary equation, telling us that any mass m traveling with a velocity v must have associated with it a "de Broglie wavelength," as it is termed. But when we look at the equation closely we can see how small the wavelength must be because h is such a small number ($h = 6.6 \times 10^{-34}$ joules-sec). Even electrons with their extremely small masses have wavelengths that are very much shorter than the wavelengths of light.

Let us calculate, as an example, the wavelength of an electron traveling at a velocity that is given by an "electron gun," namely, by being accelerated in an electric field with a potential difference of say 100 volts. We remember that a volt is a joule per coulomb, that is, 1 joule of energy is given to a coulomb when it falls through a potential difference of 1 volt. If we multiply the potential difference by the charge of an electron, we obtain the kinetic energy that it will acquire in falling through this potential difference. Thus

$$\tfrac{1}{2}mv^2 = Ve$$

hence, the momentum

$$mv = \sqrt{2Vem}$$

Our wavelength then becomes

$$\lambda = \frac{h}{\sqrt{2Vem}} = \frac{6.6 \times 10^{-34}}{\sqrt{(2)(100)(1.6 \times 10^{-19})(9 \times 10^{-31})}}$$
$$= 1.2 \times 10^{-10} \text{ meter}$$

This is a wavelength shorter by 1000 times than the wavelength of visible light. This being the case, a beam composed of electrons would not show the properties of waves as easily as beams of light. Diffraction and interference, by which waves can be most easily identified, would not be revealed even under the finest apparatus designed to demonstrate these phenomena with light rays. Nevertheless, within two years after de Broglie had speculated that electrons were accompanied by waves, evidence for the predicted interference and difffraction of electrons was experimentally found.

4. EXPERIMENTAL EVIDENCE FOR WAVE NATURE OF ELECTRONS

The de Broglie wavelength of electrons, as calculated above, is in the same range as those of X-rays that are also thousands of times smaller than light. They are, therefore, also too short to be subject to the usual optical methods of demonstrating interference. The finest instruments cannot make lines or slits small enough. However, nature provides a natural apparatus, that is, crystals in which layers of atoms are so closely and regularly spaced in a lattice formation that they can be used to show the interference effects. When X-rays are reflected off crystals at certain angles, the typical diffraction pattern appears on a photographic plate, just as a diffraction pattern appears when light is reflected off a thin film. This was a phenomenon that had been analyzed and used at the time of de Broglie's proposal.

To understand this process, let us first examine the way in which light is reflected from thin films (such as soap bubbles). In Fig. 17.4, a ray of light striking a thin film of thickness d is partially reflected at the front surface (Ray 1) and partly at the back surface (Ray 2). Now if these two rays are brought together, they may be exactly "in step" (that is, the optical distance covered by Rays 1 and 2 differs by *an integral number* of wavelengths) and they will reinforce each other. But if they are out of step (differing by *one half a wavelength*), they will cancel each other. Whether they are in or out of step for a given transparent medium depends on the thickness of the film, the wavelength, and the angle of the incident beam of light. Thus, at different angles, the reflected rays show a typical interference pattern of bright and dark regions. These phenomena are very often observed. The colors seen, when light is reflected off a soap bubble or off a film of oil spread out over water, are due to the interference between the light rays reflected from the front and back surface of the film. Each color has a different wavelength; therefore, at different angles, the light of particular

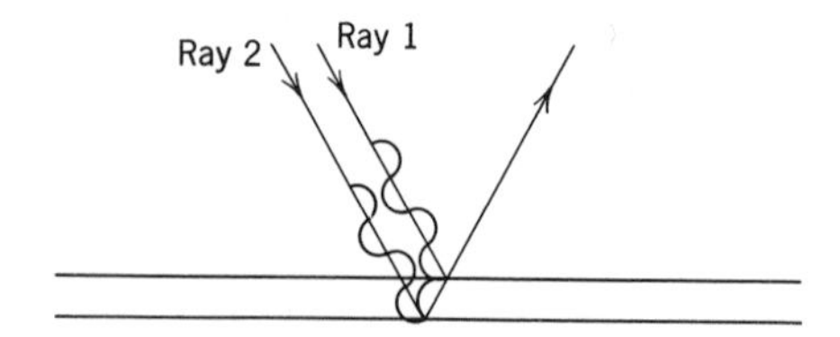

Fig. 17.4. Thin film interference. Ray 1 is reflected at the upper surface, Ray 2 at the bottom surface. When they meet after reflection, they may be either in or out of step, depending on the angle, thickness of the film, and wavelength.

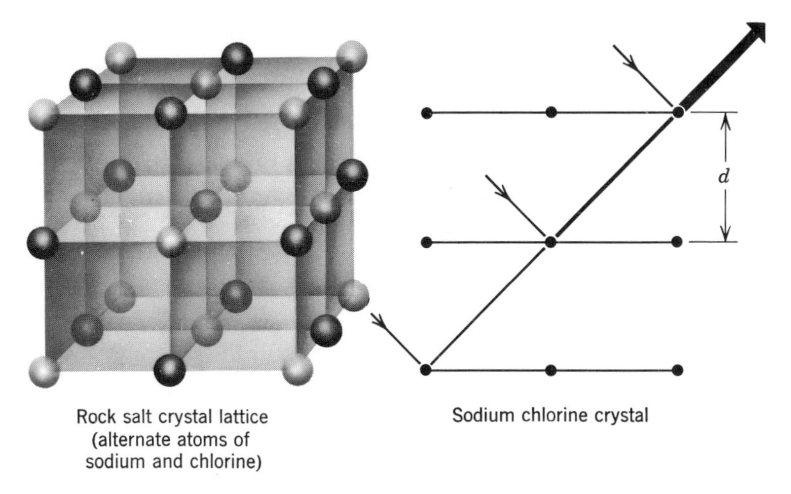

Fig. 17.5. Diagram of X-ray diffraction by crystals.

wavelengths is reinforced and, at other angles, is diminished. The colors, therefore, become separated into distinct regions or bands.

After X-rays were discovered, near the beginning of the twentieth century, it was suggested that they might also be reflected by thin films to show similar interference patterns. However, no ordinary film is thin enough. The thickness of any material film is many thousand of times the wavelength of an X-ray.

In 1912, Van Laue suggested that crystals in which the atoms are arranged in definite planes might be used instead. When this was done, the characteristic diffraction pattern showed up for X-rays. Measurements performed on these diffraction patterns then allowed investigators to calculate either the wavelength of an X-ray beam if the crystal structure is known, or the crystal structure if the wavelength of the X-ray being used is known (Fig. 17.5).

As we said, these and many other interference and diffraction phenomena for X-rays were well known when de Broglie suggested his wave theory of the electron. It happened that at about this time, two American physicists, Davisson and Germer, were experimenting with electron beams, bouncing them off various substances to study the angles at which they would be reflected. They had never heard of electron diffraction and were puzzled to find that, at certain specific angles, the reflected electron beams were intense, but dropped off almost to zero at other angles. In 1926 they presented some of their data to an international conference of physicists without an explanation. It was suggested to them that possibly their results were an example of electron interference. They at once began checking their data against de Broglie's equation for the electron wavelength, and found almost perfect agreement. (It happens that electrons accelerated by the usual apparatus attain velocities that make their de Broglie wavelengths close to those of X-rays. Therefore, the crystals reflecting beams of electrons give off electron diffraction patterns similar to those of X-rays.)

Such was the first more or less accidental confirmation of the existence of waves associated with electrons as foreseen by de Broglie. Other more carefully planned confirmations followed until, today, there is little doubt that not only electrons but neutrons and protons and all other elementary particles of physics have properties most effectively explained by wave characteristics. There is one method, used—double slit interference—that is worthy of analysis for its general significance.

Just as light from a distant source (implying that the rays of light are parallel), upon passing through two narrow slits in a screen, forms an interference pattern on a more distant screen, so does a beam of electrons. It is interesting that the electrons passing through one slit are *affected* in the path they take by whether the other slit is open or closed. Figure 17.6 illustrates electron interference.

The interference pattern found is precisely similar to that formed by light passing through a double slit. Again, may we emphasize that when one slit is closed and the number of electrons reaching a particular point is high, the opening of the other slit will often cause a *reduction* in the number of electrons reaching the same point. In other words, the path of the electrons through one slit is affected by whether the other slit is open or closed. The conclusion is inevitable that electrons act very much as waves, and not just as "bullets" or particles. The physical interpretation of this duality will be considered later, but

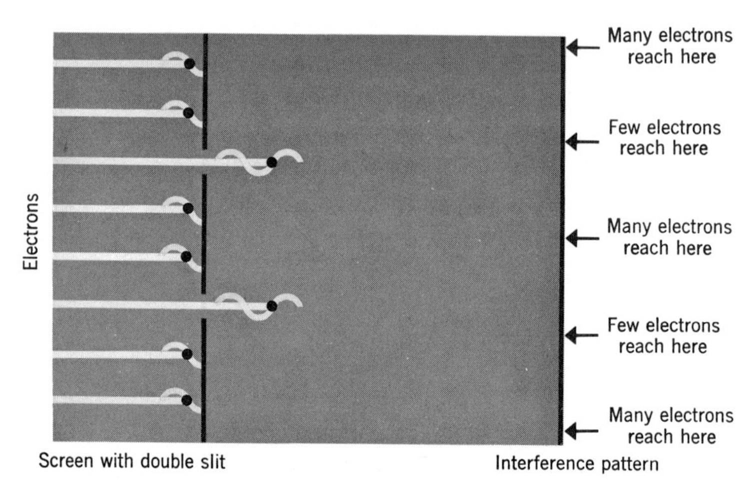

Fig. 17.6 Double-slit interference of electrons (compare with Figs. 16.4 and 15.10).

one aspect is mentioned briefly now; that is, that the intensity of the interaction of an electron with the detecting screen is a measurement of the *probability* of the electron arriving at that portion of the screen. This means that although we cannot predict with certainty the path of a single electron, we can assign a probability, and arrive at "statistical certainty" regarding the distribution of the paths that a large number of electrons will take under a given set of circumstances. The laws of quantum mechanics, therefore, become statistical laws. Wherever large numbers of particles are being handled, we can make statistical predictions with the same kind of certainty as we obtain, for example, in the case of gas laws where millions of atoms are involved. When we are dealing with ordinary masses of a scale visible to our senses, we are dealing with billions of particles. The laws of quantum mechanics then become the same as those of classical mechanics, and statistical uncertainties become so small as to be negligible.

5. THE UNCERTAINTY PRINCIPLE

Stemming out of quantum mechanics, there has grown one of the most controversial and difficult-to-interpret principles of modern physics, called the uncertainty principle (or, sometimes, the indeterminancy principle). A great deal has been said about this principle by scientists, philosophers, theologians, and social scientists. Undoubtedly, more time will be needed before a final interpretation will be reached that is generally satisfactory to philosopher and physicist alike. What is presented here should therefore be regarded with caution, and not be taken as such a final interpretation. We shall examine both the genesis of the principle, as it arose out of the experimental and theoretical structure of quantum mechanics, and some of the consequences that follow from it.

The uncertainty principle, first enunciated in 1926 by Heisenberg, one of several creators of what is now referred to as quantum mechanics, derives from the dual nature of the elementary particles of physics. These particles appear both as waves, which by their nature have a certain *"spread"* or extension, and also as a concentrated bundle or package identified in some sense as a particle with a defined position. It is difficult to "visualize" this duality on the basis of our ordinary experience, or even on the basis of classical physics that has seeped into the very mode of our thought processes, without a feeling of uncomfortable contradiction.

In Newtonian mechanics, any particle of mass m necessarily has a well-defined position in space, and also a well-defined momentum $P = mv$. Even if we modify the Newtonian mechanics to take into account the consequences of the special theory of relativity, the position and momentum relative to a given frame of reference can both be exactly and simultaneously defined. Furthermore, it follows from Newtonian mechanics that although a system may be extraordinarily complicated because it is made up of many particles in motion, and although our methods of measurement may be always subject to some inaccuracy *due to the specific measuring process* used, nevertheless, it is theoretically possible to ascribe a definite momentum to every particle in a system. And, from Newtonian laws, it follows that once these momenta and positions are all established (however difficult it may be as a matter of practice to do so), then all future motions of the particles will be completely and accurately determined once and for all. It was on this basis that Laplace was able to assert the complete mechanistic determinism of the universe as so often quoted:

> . . . An intellect which at any given moment knew all the forces that animate nature and the mutual positions of the beings that compose it, if this intellect were vast enough to submit its data to analysis, could condense into a single formula the movement of the greatest bodies of the universe and that of the lightest atom: for such an intellect nothing could be uncertain; and the future just like the past would be present before its eyes.

In contrast to the above, Heisenberg's uncertainty principle asserts the following. If you measure the position, x, of an electron (or other particle) and the momentum, p, of the same particle, there will be a certain amount of inaccuracy in the measurement of its position, which is designated as Δx, and a certain amount of inaccuracy in the measurement of its momentum, designated as Δp. By the uncertainty principle, when a simultaneous measurement of both position and momentum is undertaken, it is impossible to reduce the product of the inaccuracies, that is, $\Delta x \times \Delta p$ below the quantity h. *This is a theoretically irreducible minimum and does not depend, as in classical theory, upon the inaccuracies of either observation or of the experimental apparatus used.* The other way of putting the results is that the more precisely we specify the position of a particle, the less precise becomes the specification of its momentum and vice versa. We cannot specify both with an accuracy beyond a well-defined limit.

The immediate consequence that flows from this principle, if it is accepted as a final principle upon which nature operates, is to undercut Laplace's whole argument (which is basically the nineteenth century argument for "mechanistic" determinism), since classical physics demanded exact specification of both position and momentum of all particles in a system for the exact specification of the future of the system. The narrow determinism based on a Newtonian physics has indeed been dealt a death blow if Heisenberg's principle is accepted. This does not necessarily mean that *all* determinism is ruled out of science, but that the old mode of interpreting determinism must be re-evaluated. Before we consider this aspect of the problem more fully, however, let us examine the uncertainty principle more closely and weigh typical evidence adduced for its support.

First let us consider again what is meant by a wave that is traveling. Those waves that have the simplest mathematical expression are called simple harmonic waves, and can be represented by a sine function. They spread out evenly and continuously without end. Now the important waves of nature are never of this simple idealized form, but come as a pattern with a beginning and an end—a bundle of waves, so to speak. And, in particular, many wave bundles have a single maximum crest fading off to zero in either direction. A flash of light, or a sudden noise may be thought of as a wave pulse with a fairly well-defined maximum. Such are the waves usually associated with an electron, and such types are much more complicated to analyze mathematically than a simple sine wave. Let us assume that the wave pulse associated with an electron looks something like that shown in Fig. 17.7.

The problem now is to define both the position of the electron and its momentum. Where do we locate the electron? Certainly not just at the single point representing the maximum of the wave, for the electron is "guided" by and causes effects over the whole range L. One answer is to say the electron has an uncertainty of position Δx equal to L.

Second, what is the *momentum* of the electron? This involves specifying the wavelength. But the wavelength is no longer a simple matter to specify by the same mathematical method that we apply to a sine curve where the wavelength is measured from crest to crest and never varies. In the present case we must resort to a mathematical trick to which we have previously referred, that is, Fourier analysis. *Any* complicated function existing in an interval and zero outside, such as the one in Fig. 17.7, can be analyzed into a sum of many component sine waves such that the sum of these waves is zero outside the interval. (It may seem unrealistically abstract to "decompose" a unity in this manner and, in a sense, it is. However, analysis through "mathematical" abstractions is not perhaps so fundamentally different from a more "physical" analysis. A note struck by an instrument is a unit as heard by the ear; a color seen by the eye is a unit; but, in both cases, the note or color may be reproduced indistinguishably if the right combination of frequencies and amplitudes of component waves is produced. The determination of whether, by mathematical or physical analysis, the decomposition of a unit into its component parts aids or defeats understanding or appreciation depends upon the aim in view.)

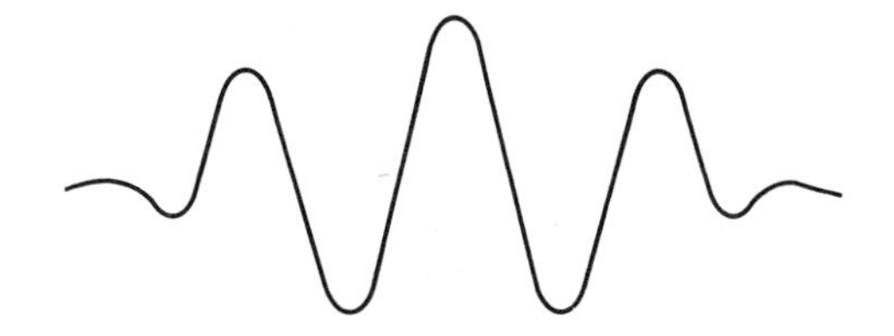

Fig. 17.7. A wave pulse that falls to zero outside a given interval.

For instance, let us see how we could decompose the wave pulse shown in Fig. 17.8 into at least two simpler sine waves. What is required is that at either end of the two components of the wave pulse, the waves must interfere with one another in such a way that their sum is zero at this point. This will be the case for two waves with wavelength λ_1 and λ_2 such that the number of waves in the interval of length L differs by a whole number, say 1. For example, if the number of waves in the interval is 3 for the first wave and 4 for the second wave, that is

$$\frac{L}{\lambda_2} = 4 \quad \text{and} \quad \frac{L}{\lambda_1} = 3$$

then

$$\frac{L}{\lambda_2} - \frac{L}{\lambda_1} = 1$$

Such a combination is shown in Fig. 17.8 within the interval L.

Now let us consider what wavelength we must ascribe to the wave pulse as a whole. This wavelength cannot properly be defined uniquely. Instead, we can only speak of a "group" wavelength, which lies somewhere between λ_1 and λ_2. If this is true, then we can ascribe momentum of the wave as given by the de Broglie equation, that is, $\lambda = h/mv$ or $\lambda = h/P$, as being between $P_1 = h/\lambda_1$ and $P_2 = h/\lambda_2$. Now we assign the uncertainty of the momentum P, the value $P\Delta$, as follows.

$$\Delta P = \frac{h}{\lambda_2} - \frac{h}{\lambda_1} = h\left(\frac{1}{\lambda_2} - \frac{1}{\lambda_1}\right)$$

but the length L of the interval is the "uncertainty" or "spread" Δx, of the wave.

Combining the two equations, and recalling that $L = \Delta x$, we find

$$\Delta P \cdot \Delta x = h\left(\frac{L}{\lambda_2} - \frac{L}{\lambda_1}\right) = h$$

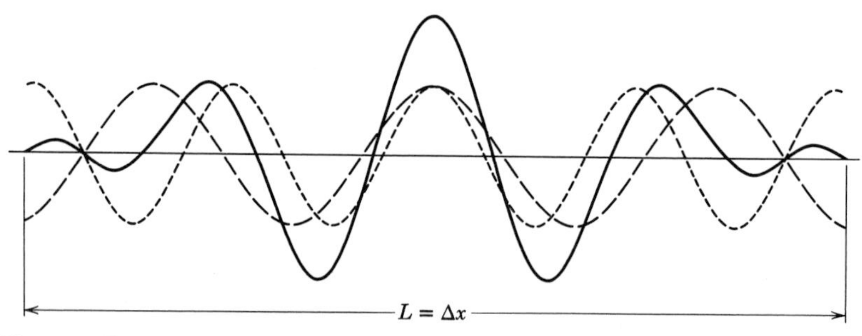

Fig. 17.8. The sum of two sine waves can yield a wave pulse that falls to zero at the boundary of a given interval.

which is Heisenberg's famous uncertainty relationship, expressing the fact that the product[5] of the uncertainty in position times the uncertainty in momentum is equal to h.

6. COMMENTS ON IMPLICATIONS OF THE UNCERTAINTY PRINCIPLE

Although the full scope of the wave-particle duality of subatomic particles will not become clear until after the next chapter, it is well at this point to make our first observations regarding the philosophical implications of Heisenberg's principle of uncertainty. A great deal has been said about this principle, which appears at present to be an integral part of the foundation of modern physics.[6]

It is commonly held that with the introduction of the uncertainty principle, the laws of causality have been suspended, and a place for "free will" has been again introduced into science itself. We shall later examine more carefully the nature of physical causation, but at present we should look more simply at the way in which the uncertainty principle has wrought a change in the basic approach of physicists to the realm of the subatomic.

Throughout this book we have stressed that science, however formal its structure and abstract its concepts, eventually identifies elements in that structure in an unambiguous fashion through carefully defined processes of measurement. These are the operational definitions. In classical, Newtonian physics, material particles were the basic elements, whether the constituents made up stars or atoms. They were defined as masses with positions in space and in time that interacted with one another through laws involving momenta. There was no doubt in the minds of the classical physicist that both the position and momenta

[5]It may have been noted that although the two waves cancel at both ends of the wave pulse, they will not do so outside the interval. In a more complete analysis, additional component waves must be added so that the sum of all the waves outside the interval will remain zero. Such an analysis is easily developed, again by Fourier's theorem. As a consequence, the spread of the wave length and hence the momentum is even larger, and the uncertainty relationship then becomes

$$\Delta P \cdot \Delta x > h$$

h still remains, in any case, the lower limit to the product of the uncertainty of the position and the momentum.

[6]However, among physicists themselves, questioning voices have been raised. Planck, Schrödinger, and Einstein, among the older generation of modern physicists, were all convinced that a newer and deeper penetration would reveal that the uncertainty principle and the probabilistic nature of physical laws had an explanation at a more fundamental level. "I do not believe that God plays dice with the universe," was Einstein's comment. de Broglie and David Bohm have, more recently, attempted critiques of the uncertainty principle and have suggested ways of circumventing it; and Heisenberg, himself, in his *Physics and Philosophy* subjects the principle to searching question.

of atoms had a clear meaning, at least theoretically measurable, in terms of the same general kind of operation performable in a laboratory. Classical determinism, then, consisted in the affirmation that if at any given moment the initial state of any system of material particles is completely fixed in terms of the positions and velocities of the particles, the future configurations are inexorably determined by the basic laws of physics.

In the application of the uncertainty principle, however, it is often said that you cannot simultaneously measure both position and velocity of a particle. Hence, the future action of the particles themselves remains open to uncertainty. A careful criticism of what this means, however, reveals not so much that a new kind of "freedom of action" has been granted to particles, as that the model of our scientific description has broken down. We are mistaken, in fact, in attempting to apply the concepts of definite position and velocity, drawn from our experience with materials and from our measuring processes met in the laboratory, to particles on the atomic or subatomic level. An electron does not have a position and does not have momentum, in the same well-defined sense that a billiard ball does. It is an error at the very beginning to attempt to ascribe the same type of "deterministic" behavior to these particles that classically has been ascribed to larger masses in motion.

The instruments and materials that we use in our laboratories, even when we experiment on subatomic particles, are, however, still subject to the classical laws of physics. The temptation, then, is always to build our models, even in this indirectly reached realm, in terms of our directly experienced conceptions. And when, as often happens, the logical structure (explanations) is similar in many ways to what we see visibly before us in the laboratory, we impose the visible model upon the conceptions themselves. For example, as we shall see when we consider experiments on the interaction between charged alpha particles and charged nuclei of atoms, the mathematical law best describing this interaction is similar in form to Coulomb's law of electrostatic repulsion. It is an easy step, therefore, to assume that since this mathematical expression leads to results confirmed in observation of larger charges, there must indeed be "particles" present carrying charges, and interacting exactly as do the charged pith balls that we play with in the laboratory. The *formal* similarity leads us to conclude that we are dealing with objects similar to those that we see, but on a vastly smaller scale.

This temptation becomes even more powerful when we find that in certain experiments with electrons they act as if they were particles traveling through space in trajectories and give up their kinetic energy upon collision according to classical laws of mechanics; although, at other times, they act as waves that interfere in patterns according to laws of optics. Both alternative pictures fit so well, *each under the appropriate conditions,* that they are regarded, in Bohr's terms, as "complementary" to one another. We then are tempted to say that the electrons are both waves *and* particles, yet they are in fact known to be neither.

Since they are essentially neither waves nor particles in the classical sense, we should not expect the same deterministic laws of classical physics to apply to subatomic particles. But saying this does not mean that on the subatomic level "Whirl is king" or that there is any arbitrary suspension of physical law.

The modern physicist tends to avoid the term "cause" except in informal discourse; instead, he refers to functional relationships, specifying these, when he can, by equations.[7] However, because the terms "cause" and "causation" are still widely used in the social and biological sciences as well as in philosophy, we shall analyze their meaning briefly in the next section, taking into account the modern developments in physics.

7. CAUSATION AND DETERMINISM

The nature of causation is one of the fascinating and perennial problems of philosophy. Among the ancient natural philosophers, Plato, Aristotle, and Lucretius grappled with it; among the ancient dramatists, it was portrayed in its widest scope through the working of destiny and fate. The poet Virgil wrote "Happy is the man who knows the causes of things." Down through the ages, interpretation has followed interpretation as to what we mean by "cause." In modern times, Hume and Kant narrowed the problem somewhat by discussing its relation to science.

Still later, as the success of Newtonian mechanics grew, the meaning of causation became even more narrowly prescribed: the limited doctrine of "mechanical causation" became dominant, and was applied dogmatically to all human and natural events. Effective criticism of the narrow nineteenth-century attitude is perhaps nowhere more delightfully expressed than in the chapter on Romantic Reaction, in Whitehead's *Science and the Modern World,* where he discusses the basic inconsistency in our inherited civilization:

> A scientific realism, based on mechanism, is conjoined with an unwavering belief in the world of men and of higher animals as being composed of self-determining organisms. This radical inconsistency at the basis of modern thought accounts for much that is half-hearted and wavering in our civilization.[8]

Whitehead goes on to discuss the insight expressed by poets regarding the nature of man—a deeper insight than that of those scientists or philosophers who were pressing the narrow interpretation to its limits. To those who consider that science always takes the lead in human thought, affecting all the arts and humanities,

[7] If and when time enters into a functional relationship, the notion of cause is carried by a differential equation in which time is the independent variable but does not occur explicitly in the equation. See H. Margenau, *The Nature of Physical Reality,* McGraw-Hill, New York, 1950.

[8] A. N. Whitehead, *Science and the Modern World,* Macmillan, New York, 1925, p. 110.

Whitehead's comments on the role of the nineteenth century poets in leading the revolt against scientific mechanism is a refreshing antidote.

In our own time, the problem of scientific causation has received a new twist with the emergence of the *uncertainty principle* which we have presented. In seeking to evaluate the significance of this principle we reiterate our earlier warning. It is even now too early to give it a definitive interpretation. Certainly the principle has been used too loosely in the pulpit and elsewhere, as the glib answer to the problem of "free will" in a world "governed" by "scientific laws." Even today, over 40 years after Heisenberg announced it, there are serious questions among both scientists and philosophers regarding its interpretation.[9]

Without going into a lengthy analysis of the nature of causation, let us pin down as closely as we can the meaning of this term as we shall use it. We shall assume that *strict physical causation* must fulfill the following conditions.

(*1*) It is a relationship between two completely specified states of a system: state A at time t_1, and state B at time t_2. Although, in ordinary language, a causal connection is often thought of as a relationship between *two* objects or *two* events, the scientist is aware that there are a great number of elements entering into any causal relationship, and thus he either specifies all of them or assumes that it is clearly understood that they are present. To specify the state of a system might mean, for example, stating all the values of momenta and the position of all of the particles in the system. These are called the state variables of the system.

(*2*) The relationship is invariable, that is, the occurrence of A at time t_1 is both a necessary and sufficient condition for the occurrence of state B at time t_2. Ideally, the relationship is expressible as a mathematical function relating the state variables.

(*3*) Either A and B are contiguous in space and time, or else there is a regular sequence of intermediate states between A and B, which are contiguous in space and time. In other words, if A and B are spatially or temporally separated, there is a well-understood chain of events that can be formally specified if necessary.

Such a definition of causation is admittedly abstract and ideal, and all the conditions cannot be met within practice, but some such limitation must be placed on the term "cause" if it is not to remain hopelessly vague. It follows from the definition that causation can only occur in *isolated systems,* since the state of

[9]Einstein, to the end, never could accept the uncertainty principle as more than a temporary expedient. De Broglie, Bohm, and others, have attempted ways of getting around the uncertainty principle. Those interested can refer to David Bohm, *Causation and Chance in Modern Physics,* Harper Torchbook, Harper, New York, 1959. For criticisms of the attack made by Bohm and others on the orthodox "Copenhagen interpretation" of the uncertainty principle, see Norwood Hanson's article in the *American Journal of Physics,* Volume 27, No. 1, pp. 1–15 (1959). An intermediate position is contained in a significant and penetrating analysis by Werner Heisenberg in his *Physics and Philosophy,* 1958, Harper Torchbook, Harper, New York, 1962. This book, written in nontechnical language, is invaluable for philosophers and nonprofessionals seeking to understand the significance of the uncertainty principle and modern physics in general.

the system must be completely specified. In actuality no system is ever completely isolated. The motion of a pendulum, for example, is slightly affected by the changing positions of the distant stars relative to the earth. In physical and chemical systems, relative isolation can be more readily achieved than in the biological sciences. In general, although the biologist seeks to find causal laws to explain phenomena, he usually still finds it necessary to complement his explanations by *teleological* considerations, that is, by considering the organic function of, for example, the kidney in preserving the over-all stability of the organism.

If we insist on the condition of asymmetry, that is, that the cause A precedes the effect B, the majority of physical laws cannot be considered strictly causal laws at all, since most laws of physical dynamics can work backward as well as forward. We can, from determining the position of the moon today, calculate its position yesterday as well as its position tomorrow. It is partly for this reason that the term "cause" is not often seen in a book on mathematical physics. Since, in discussing causation, the matter of predictability is the central issue, and the dynamical laws of physics are generally regarded as prototypes of causal laws, we have not included the condition of asymmetry in time of states A and B.

Departing from the abstract concept of causation and looking at the manner in which laws are used to make predictions, let us examine how modern and classical physics differ. In classical physics the state of a system was considered to be specified if the observed momenta and positions of all the particles of the isolated system were specified. With such a specification of the state variables, a prediction could be made of the values of the state variables at any future time. To the extent that the system was isolated from outside disturbing influences, the predictions would be fulfilled with a precision limited only by the inaccuracies of observation of the state variables.

Before considering the difference in approach in quantum physics, we point out that in spite of the successful application of classical physics to a large variety of changing systems, the conclusion implied by Laplace is hardly warranted that all events are strictly and mechanically determined, just because it is conceivable that some superintelligence might know all the state variables of the whole system of the universe and, therefore, be able to predict all future states. We repeat that there is no real meaning to postulating the existence of a function of infinite and arbitrary complexity covering the infinite number of state variables that would have to be specified. Unless such a function can be produced and handled in such a way that predictions can be made, it is meaningless to postulate its existence, except as a tautological statement to the effect that "what is" somehow arose from the past. The success of classical physics with its apparently deterministic predictions depended upon the ability to find "simple" functional relationships between state variables in comparatively isolated systems where it is easy to distinguish between relevant and irrelevant influences.

Turning to modern physics, particularly quantum mechanics with its uncertainty principle, in what sense is it true that causal laws have been suspended? At once we see that classical mechanical causation, as previously conceived, cannot be applied at the microscopic level, if for no other reason than that ascribing momentum and position to an atomic particle has no meaning. The classical variables defining the state of a system cannot be used to predict what will happen to an electron. We can make an even stronger assertion on the basis of our present knowledge. No other state variables have been suggested, which can be used to make predictions about individual particles. On the other hand, by means of the square of the Schrödinger psi function, we can obtain a probability norm that allows us to make definitive predictions not about individual particles but about very large assemblages of particles. In other words, if we allow probability to enter into our equations of state in a manner not allowed in classical physics, the relationship between states A and B could still be described as one of causal relationship. It has been suggested that this type of causation involving probability be called "weak causation" in contrast to the "strong causation" of classical physics.

The older problem of philosophy, that is, to reconcile teleological functions or "goal directed" activities with the "stronger" more deterministic type of causation has now, therefore, been replaced, to some extent at least, by the problem of a reconciliation with the "weaker" type of causation implied by the uncertainty principle. Since it is now apparently impossible to attribute a rigorous determinism to atomic events, some have argued that because many of the activities that take place in the interior of cells are on a microscopic, or even atomic level, possibly new light will be thrown upon living processes.

Whether one refers to the consequence of the uncertainty principle as a "breakdown" of causality is a matter of definition. We can, however, conclude that it is impossible to predict the result of an observation on *a particle* with certainty. In this sense, the type of determinism suggested by Laplace is not even conceivable. The argument about "free will" has been thrown into a somewhat different focus. However, we shall not pursue this topic which would carry us too far into speculative philosophy.[10] But even if the general principle of causality is given up as a philosophic tenet, it will still have its role to play as *a regulative principle* guiding the objectives of scientific research which is to discover which factors are relevant in determining the course of events in a given situation.

8. QUESTIONS

1. Explain what a "black body" radiator is. Why is it used in the study of the relationship between frequency of light emitted and temperature?

[10] See Ernest Nagel, *The Structure of Science,* Harcourt Brace, New York, 1961, for a thorough analysis of this problem.

2. What are the metric units in which Planck's constant h are expressed? (Sometimes these units are called units of "action.")

3. As one moves into the study of atomic and nuclear physics, the unit of energy of an electron-volt becomes more convenient than a joule. (Just as, on the other end of the scale of physics, a light year is a more appropriate unit of measurement of astronomical distances than a mile or a kilometer.) Review the meaning of an electron-volt. How many electron volts are there in a joule? What potential difference is required to give an electron a kinetic energy of 10^3 eV? With what velocity will it be traveling? If an electron has a kinetic energy of 10^6 electron volts, or one MeV, how many joules of energy does this represent? Cosmic rays (usually protons) may have energy equal to 10^9 electron volts. How fast is such a 1.0 BeV (or in more modern terminology a GeV) proton traveling?

4. When a proton strikes an electron, they both recoil somewhat as particles do. (The recoil of the photon is called "scattering.") But the speed of the photon is unchanged, even with the transfer of kinetic energy to the electron. What happens to the frequency of the photon?

5. The "stopping potential" in photoelectricity is the potential difference in volts between the metal and any collecting plate receiving the electrons of precisely that magnitude which will prevent electrons emitted by the metal from reaching the negatively charged collector. Explain why this "stopping potential" may be used to measure the maximum kinetic energy of the electron. Show that the stopping potential $V = hf/e - W/e$.

6. Explain why the threshold frequency for the photoelectric effect is more apt to lie in the ultraviolet than in the visible range of light.

7. There is an optical toy (often called a "radiometer"), resembling a miniature windmill which is composed of four vanes, each of them black on one side, mirror-faced on the other, and mounted on a vertical shaft in an evacuated glass tube. The vanes revolve when set in a strong light. Interestingly enough, instead of revolving in such a way that the mirrored surfaces (which receive a greater light pressure than the blackened sides and should therefore move away from the direction of the source of light) they move in the opposite direction from that expected. Can you explain why? (*Hint:* This is only true because the tube is not completely evacuated. Consider the kinetic theory of gases. If the tube is completely evacuated, the vanes rotate as expected. See how far you can carry the analysis quantitatively.)

8. An X-ray apparatus can be considered to be a reversal of the photoelectric effect. High-speed electrons are shot at a metal plate and X-ray protons of a very high frequency are given off. If the *velocity* of the electrons is increased, would you expect *the number* of photons given off to increase? Would you expect the wavelength of the X-rays to increase or to decrease as the velocity of the electrons is increased?

9. A photon's energy and momentum both depend on its frequency. How are they related? Place in proper order of magnitude the energy and momentum of infrared waves, X-rays, radio waves, and ultraviolet rays.

10. A baseball, like "elementary particles" such as an electron, should have its own frequency and a wavelength if the quantum theory is completely general. Why are not the wave properties of a baseball apparent?

11. The "resolving power" of a microscope is a measure of the smallest distance between two points which can be distinguished. The limit is reached when the distance between the

points is approximately the same as the wavelength. Why does an electron microscope have a much higher resolving power than an ordinary microscope? (Note: An X-ray microscope is not practical because X-rays cannot be focused by a lens. They show almost no refraction. The "focusing" of electrons can be accomplished by appropriate magnetic and electric fields.)

Problems

1. Find the ratio of the number of photons emitted in one second from a radio antenna broadcasting at one megacycle to the number of photons emitted in one second by a source of visible light of frequency 5×10^{14} cycles per sec., assuming that they both radiate at the same level of power.

2. If a human eye can just detect 10^{-18} watts entering the pupil, approximately how many photons per second must enter the eye?

3. A 10-eV photon is absorbed near the surface of a metal. If the "work function" is 2 eV, what will be the maximum velocity of the emitted electrons (assuming that the frequency of the photon is higher than the photoelectric threshold.)

4. If the momentum of an electron is reduced by one-half, how is its de Broglie wavelength affected?

5. What is the de Broglie wavelength of an electron with a velocity corresponding to a kinetic energy of 1 MeV.

6. At room temperature (300°K) what is the average de Broglie wavelength of a helium atom?

7. Planck's equation for blackbody radiation is: energy at wavelength λ, $E_\lambda = \dfrac{2\pi hc^2}{\lambda^5(e^{hc/\lambda kt} - 1)}$. Show that this can be reduced to $E_\lambda = \dfrac{c_1}{\lambda^5(e^{c_0/\lambda} - 1)}$. Now present an argument that E_λ will fall to zero when λ approaches either zero or infinity. (*Hint:* Consider what happens to the denominator as λ varies between 0 and infinity.) From this, present an argument that a blackbody radiator will emit maximum energy at some frequency between 0°K and infinite °K, regardless of the temperature. (Hence, the "ultraviolet" catastrophe predicted by classical principles is avoided.)

8. Plot a graph similar to that in Fig. 17.3 (using electron-volts for the ordinate) for both cesium and copper which have "work functions" (the threshold energy, W, required to release an electron from the metal) of 1.8 eV and 4.3 eV, respectively. Can visible light eject electrons from copper? What is the maximum velocity of the electrons emitted from cesium if light of a frequency of 5×10^{14} cycles/sec is used?

9. Using the de Broglie relationship $\lambda = h/P$, calculate the momentum of a photon with a frequency of 5×10^{14} cycles/sec. What units is your answer expressed in? Note carefully the units for h.

10. Photons have a zero rest mass, and the equation for the energy of a particle, $E = hf$ holds true only for particles with zero rest mass. However, the photons have an inertial (sometimes called a relativistic) mass, which can be calculated from the momen-

tum equation $P = mv = mc$, where c is the velocity of light. Find the inertial mass of the photon of frequency 5×10^{14} cycles/sec. How does the mass vary with frequency?

11. Light exerts a small but measurable pressure upon the surfaces that it strikes. Assume that the energy E is completely absorbed by a surface (a case approached when the surface is black) during a time t. Then the momentum for each photon would be $P = E/c = hf/c$ as already shown. If, on the other hand, the light is completely reflected (a case approached when the surface is mirrorlike) the momentum $P = 2E/c = 2hf/c$. Explain the difference, referring to previous discussions of momentum in elastic and inelastic collisions. Now assume that n photons strike 1 square meter of surface in 1 minute, and that the energy flux (that is, the power crossing a unit surface) is 10^3 watts per square meter (which is an intensity of the same order of magnitude as the intensity of sunlight in summer). The energy then delivered in 1 minute to a small mirror, 1 cm square, would be $E = (10^3)(10^{-4})(60)$ joules. What would be the momentum delivered to the mirror? What would be the force on the mirror? (*Hint:* Remember that Ft = momentum delivered, therefore F = momentum/t.) What would be the pressure? Compare these values to a rough estimate of the force exerted by the weight of a fly on a horizontal mirror.

12. Another effect that bears a formal similarity to the photoelectric equation is the *Compton effect.* When a photon strikes an electron, it recoils (is scattered) from the collision, and the electron is given additional kinetic energy. If the additional kinetic energy of the electron is $\frac{1}{2}mv^2$ the original frequency is f_0 and the recoil energy of the photon is hf, write the basic equation for the Compton effect. How does the new frequency f compare with the original frequency f_0?

13. Calculate the de Broglie wavelength of an electron traveling at one-half the velocity of light; at one tenth the velocity of light. Compare this with the wavelength of a 0.2 kg baseball traveling at 30 m/sec. Why are the wave properties of objects on "our scale" of observation never detected?

Recommended Reading

Albert Einstein, and Leopold Infield, *Evolution of Physics,* Simon & Schuster, New York, 1942. A nonmathematical but authoritative summary of developments that led up to the quantum theory.

Physical Science Study Committee: *Physics,* D. C. Heath and Company, Boston, 1960, Chapters 33 and 34. Many illustrations and problems are included in this text, designed primarily for advanced high school students. Simple algebra and graphs are used.

Jay Orear, *Fundamental Physics,* Wiley, New York, 1961, Chapter 12. A short but excellent presentation of quantum theory and the uncertainty principle. For readers who wish to pursue their own more detailed study of modern physics (as well as classical), the manual by Orear, *Programmed Manual for Students of Fundamental Physics,* Wiley, New York, 1967, is highly recommended.

Eric Rogers, *Physics for the Inquiring Mind,* Princeton University Press, Princeton, 1960, pp. 718–727. This excellent volume prepared for mature students with little technical mathematics has already been referred to. It abounds with illustrations, exercises, and problems to aid the reader to grasp the fundamental ideas of modern physics.

For those interested in the philosophical implications of the quantum theory, the following books are recommended, several written by the originators or developers of various aspects of the quantum theory.

Louis de Broglie, *Matter and Light,* Dover Publications, New York, 1939. An early and authoritative exposition of the development of the conception of wave-particle duality by its originator. The uncertainty principle and its implications are discussed. The book is highly rewarding both for its historical and philosophical speculations, but should be supplemented by the later works of the same author.

Louis de Broglie, *The Revolution in Physics,* Noonday Press, New York, 1958.

Louis de Broglie, *Physics and Microphysics,* Harper Torchbooks, Harper, New York, 1960.

Werner Heisenberg, *Physics and Philosophy,* Harper Torchbooks, Harper, New York, 1958. A thoughtful and excellent discussion of the problems arising out of quantum mechanics by the originator of the uncertainty principle. The critical foreword by F. S. C. Northrop should be read with attention.

Neils Bohr, *Atomic Physics and Human Knowledge,* Science Editions, New York, 1961. This volume contains an illuminating discussion with Einstein on the uncertainty principle. In other essays, Bohr suggests broadening the scope of the complementarity principle, which he developed.

Erwin Schrödinger, *What Is Life?,* Macmillan, New York, 1947. Every biologist should read this book containing the speculations of the renowned physicists who did so much to develop the mathematics of quantum mechanics.

Erwin Schrödinger, *Science, Theory, and Man,* Dover Publications, New York, 1957. Nine broad essays on the implications of modern science, including one on indeterminism and one on the fundamental ideas of wave mechanics.

David Bohm, *Causality and Chance in Modern Physics,* Harper Torchbooks, Harper, New York, 1961. An interpretation of the uncertainty principle is proposed, alternative to the Copenhagen interpretation. Bohm's work has been widely discussed by physicists and philosophers.

Ernest Nagel, *The Structure of Science,* Harcourt Brace & World Inc., New York, 1961. No student of philosophy who wishes to understand the critical issues raised by modern science should fail to study this scholarly work with care.

Philipp Frank, *Philosophy of Science,* Prentice-Hall, Englewood Cliffs, N.J., 1957. A critical, scholarly work on the philosophy of science by one of the members of the Vienna circle of logical positivists.

B. H. Hoffman, *The Strange Story of the Quantum,* Dover Publications, New York, 2d edition, 1959. Modern science as a creative activity is dramatically, if somewhat loosely, presented. The book will appeal to those who seek a lively, nonmathematical, but authoritative story of how the physicist has arrived at his ideas about the structure of matter.

chapter eighteen

Internal atomic structure

The second step in the development of modern physics is the new and revolutionary conception of the atom, which led to the destruction of Hiroshima and to many technological changes that distinguish our civilization from earlier ones. By the end of the nineteenth century, the atomic theory was well established in physics and chemistry. But there was no hint that the atom was radically different from substances experienced on the macroscopic or human level. The atom was conceived as a hard, indivisible material (a small billiard ball was a familiar analogy), whose size and mass could be closely approximated, even if its shape was not completely known. It had properties of elasticity and mass, and it obeyed all the same laws of classical physics, just as larger observable masses do. Thus, the same physical laws governing matter extended from the smallest known particle, the atom, to the largest known masses, the stars.

As late as 1892, Ensign Albert Michelson, a young man of twenty-eight, the very physicist whose experiments even then were laying a foundation for the radical theories of Einstein, was saying:

It seems probable that the grand underlying principles (of physical science) have been firmly established and that further advances are to be sought chiefly in the rigorous application of these principles to all phenomena . . . An eminent physicist has remarked that the future truths of Physical Science are to be looked for in the sixth place decimals![1]

We have already alluded to the extraordinary decade of discovery and expansion that began only three years after Michelson's confident pronouncement. For emphasis we list once more the amazing series of discoveries and new theories that originated between 1895 and 1905: the discovery of X-rays, radioactivity, and photoelectricity; the identification of the electron and, hence, the birth of electronics; the creation of new and radical theories of quanta, of relativity, and of internal atomic structure. In every case, these discoveries and theoretical developments, each in a specialized branch of physics or chemistry, cross-fertilized all other branches, and an extraordinary development took place with consequences still unknown and still to be explored.

The developments, so significant in their effects on the mode of living and on the philosophy of the twentieth century, came about with very little fanfare or publicity. For example, the great Cavendish laboratory in England, where so much was being done in establishing the foundations of modern atomic physics, had a budget so meager that it acquired the nickname "string and sealing-wax laboratory." Madame Curie and her husband worked in an abandoned shed, summer and winter, to separate by hand a minute amount of a peculiarly active substance from tons of pitchblende, while Einstein worked as a lesser official in the Berne patent office. The scientists of that day, watching the unfolding of a totally new conception of the universe, had little to reward them except the respect of their colleagues and the excitement of new discoveries.

1. RADIOACTIVITY AND ITS SIGNIFICANCE

In 1896, Roentgen found that under the proper conditions an evacuated tube would emit a new and very penetrating type of ray whose nature at first was unknown and, hence, was given the name X-ray. X-rays were soon identified as very short electromagnetic waves similar to light but with shorter wavelengths and much more penetrating power. X-rays, in themselves, were so little different from light that their discovery did not contribute directly to any great expansion of ideas. However, it stimulated J. J. Thomson in England and Henri Becquerel in France to inquire further into the nature and effect of these rays. These were the investigations that opened up new vistas. We have already covered the work of Thomson, who established the nature of an electron as a charged particle of a given mass in 1897. We now turn to the story of radioactivity.

Becquerel's discovery of radioactivity in 1896 is another of many examples of

[1]Quoted by Eric Rogers in *Physics for the Inquiring Mind,* Princeton University Press, Princeton, 1960.

how an intensive inquiry in one direction often leads to wholly unexpected results in another. Becquerel, on considering the nature of the newly discovered X-rays and hearing that these rays seemed to be emitted from that portion of the glass X-ray tube that fluoresces, decided to see whether other phosphorescent materials might also give off X-rays. Among several substances, he tried a uranium salt, which may be made to phosphoresce when exposed to light. He wrapped some uranium salt crystals, which had been exposed to sunlight, in opaque paper and found by photographic means that they did give off penetrating rays similar to X-rays. As he reports his findings:

> If I placed between the phosphorescent substance and the paper a coin or a metallic screen pierced with an open-work design, the image of these objects appeared on the negative.[2]

While conducting these experiments, a series of cloudy days intervened, and Becquerel put away in the dark some of his prepared samples and photographic plates. A few days later, to test the freshness of the plates, he developed one of them expecting to find at most a feeble image, since the phosphorescence of the uranium should by now have disappeared. Instead, to his amazement, he found that the images were very intense. "I thought at once," he said, "that the action might be able to go on in the dark" (that is, without previous exposure to sunlight). Tests quickly showed his surmise to be correct. Radioactivity had been discovered only two months after the discovery of X-rays! The nature of these so-called uranium rays now became a matter of exciting study to the scientists on the continent and in England.

Other radioactive elements were soon found. Madam Curie undertook as her doctoral thesis in 1898 a systematic investigation of a large number of minerals to see which might be radioactive. The announcement of the discovery of another radioactive element, polonium, was made in 1898, soon followed by the discovery of radium in 1902. The story of the laborious task of separating by physical and chemical means one-tenth gram of radium from six tons of pitchblende, a brown ore that was delivered in coal sacks at the abandoned and unheated shed in which Madame Curie worked, is dramatically told by her daughter, Eve, in the biography of her mother. Madame Curie received a Nobel prize in 1903 and another in 1911.

Under the hands of many investigators, particularly Rutherford in England, many startling facts about radioactivity quickly accumulated, which inexorably led to the conclusion that the atoms of the elements were disintegrating. First, it was found that this radioactive process is not an ordinary chemical one, but continues spontaneously with an intensity proportional to the mass of the element regardless of how it is combined with other elements, and regardless of any change in the surroundings such as high or low temperatures. Second, it was found that three types of rays are given off, which were originally given the

[2] W. F. Magie, *A Source Book in Physics*, McGraw-Hill, New York, 1935, p. 610.

neutral names alpha, beta, and gamma rays because their nature was unknown. The alpha ray is the least penetrating, and can be stopped by a few centimeters of air or by a thin sheet of metal. It was soon identified as an atom of helium (actually, a doubly ionized helium atom carrying two positive charges; in modern terminology this would be regarded as the helium nucleus). The beta ray became identified as an electron, and the gamma ray as a very short X-ray. Additionally, the study of radioactivity revealed that the amount of energy being yielded by the changes taking place within the atom was many thousands of times greater per weight of substance than any chemical change previously known. (Radium, for example, yields 0.135 Cal/g/hr for a period of centuries.)

Rutherford and Soddy, in 1902, explained that radioactive elements were disintegrating into other elements. A given atom of radium, for example, will after sufficient lapse of time suddenly explode to yield an atom of another element, radon, an atom of helium, several electrons, and high-frequency X-rays. The old, indivisible atom, thus, was recognized as having component parts. Furthermore, the ancient dream of the alchemists, the transmutation of one element into another, was now revived in a new form. Here was the transmutation process taking place by itself, at its own pace, and without external cause. Could such a transmutation be induced artificially? After many attempts, Rutherford in 1919 tried bombarding nitrogen gas with the alpha particles given off by polonium. He found that the alpha particles acted as high velocity bullets which, striking the atoms of nitrogen, shattered them, changing a few nitrogen atoms into atoms of other elements. The alchemists's dream was accomplished, even though on a miniscule scale and at high cost!

All of these facts and the developing theory of radioactivity brought about an entirely new conception of the chemical atom, which is considered in the next section. However, before leaving the subject of radioactivity, there are several interesting and significant points that should be covered regarding the *rate of radioactive decay* and its measurement.

Each radioactive substance decays at its own specific rate, but the general law of this decay is the same—a fundamental statistical law. It will be recalled from Avogadro's hypothesis that the number of atoms in a gram of radium is

$$\frac{6 \times 10^{23}}{222 \text{ (atomic weight in grams)}}$$

or approximately 2.8×10^{21}. Out of the large number of atoms present at any time in a mass of radium a certain specific proportion will decay in every ensuing interval of time; that is, the number of explosions per second is proportional to the number of atoms present. As a result of this decay law, it is most convenient to indicate the rate of decay of an element by its *half-life.* The half-life of an element is the period of time that it takes for the amount of the element to fall to one-half its original amount. Thus, for radium, the half-life is the time

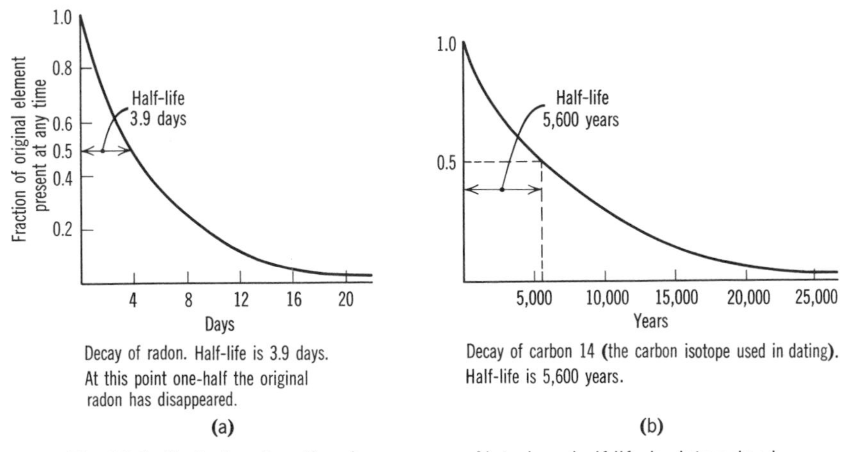

Fig. 18.1. Typical radioactive decay curves. Note how half-life is determined.

it would take for the number of atoms of a gram of radium to be reduced from 2.8×10^{21} to 1.4×10^{21}. (It does *not* mean that in two half-lifes the element will have disappeared. Rather, what is meant is that in two half-lifes there will remain one-quarter of the original amount, in three half-lifes there will remain one-eighth of the original amount, etc.) The half-life is used as a convenient measure because the time it would take for *all* the atoms to disintegrate cannot be determined. This is true because when the number of atoms left becomes very small, the statistical laws, which depend on very large numbers, can no longer be applied. Out of a billion atoms of radium we can predict with great precision how many will decay in an hour; out of ten atoms we can make predictions of probability only. Figure 18.1 indicates the curves of decay of two radioactive elements, each with its own half-life.

This is called an exponential decay law and may be represented[3] by a very

[3]The companion equation $y = Ce^{kt}$ is often called the natural curve of growth, and is of widespread significance in all the sciences. In biology, for example, in a culture of bacteria, the rate of reproduction is proportional at any time to the number of bacteria present (until autointoxication, lack of nourishment, or some external effect sets a limit). The curve of such growth is sketched below:

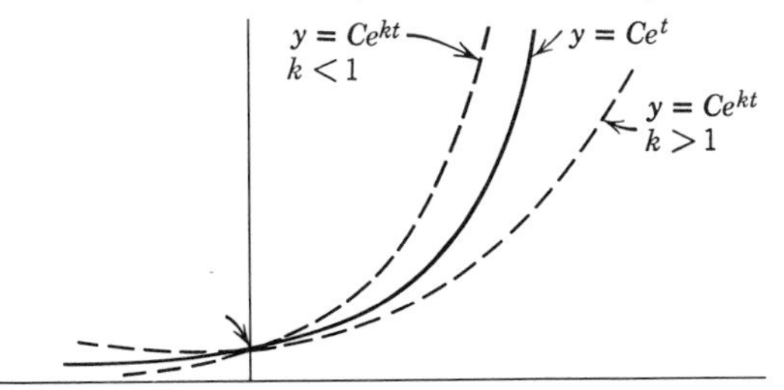

Typical "growth curves" where rate of growth is proportional to number present

important mathematical equation $y = Ce^{-kt}$, where e is the base of the natural logarithms, C and k are constants, and y is the amount of radioactive element present at any time.

The longest half-life of radioactive elements is that of uranium where activity falls to one-half its original value in 5×10^9 years. On the other hand, some radioactive elements have a half-life much less than a second. In the problem of fall-out from nuclear bombs, the half-lifes of the by-products of the bomb become a very important consideration. Such bombs result in many different types of radioactive debris. Fortunately, most of these have short half-lifes and, hence, the radioactive danger is reduced rapidly. Other fission products, such as strontium have a half-life of longer periods, and it is the presence of these, particularly when they are easily absorbed by animals or humans, that has given most concern.

On the positive side, as one of the most effective research tools in many sciences (including medicine), radioactive elements of various specific half-lifes have proved invaluable. Radioactive elements of very short half-life can be injected into organisms, and their course through the various structural parts of the body can be traced through their emission of particles detected by Geiger counters. Another example of the use of radioactive rates of change is in radioactive dating, which will be considered in Chapter 24.

2. RUTHERFORD'S DISCOVERY OF A SMALL NUCLEUS SURROUNDED BY ELECTRONS

Let us review briefly what radioactivity reveals. When atoms of an element "explode," they give off three kinds of particles or rays: alpha, beta, and gamma, with varying penetrating power. The twofold problem is then raised. (*1*) What, exactly, are these rays or particles? (*2*) What is their origin within the structure of the atoms from which they come?

The solution to the first problem was quickly forthcoming. By shooting these rays at a target (either a photographic plate or a screen that fluoresces or scintillates when struck by a ray) and noting the direction in which they are deflected when the paths are subject to electric and magnetic fields, and also the magnitude of such deflection, it was soon determined: (*1*) that the *beta particles* are *electrons,* identical in mass and charge to those already identified by Thomson; (*2*) that the *alpha particles* are particles with *twice* the amount of charge of an electron, but *positive* in sign, and of a mass many thousands of times larger than the electron particles and that they are identical with *atoms of helium* except that they are stripped of electrons (that is, ionized); and (*3*) that the *gamma rays,* undeflected by either magnetic or electric fields, are indeed nothing but X-rays of very high frequency, short wavelength, and of great penetrating power.

It was, therefore, quickly concluded that the natural atom must contain component parts carrying positive and negative charges, since these are emitted by the atoms of radioactive elements. But how are these component parts arranged within the atom? What means are available to investigate a structure on such a submicroscopic scale (about one-millionth the size of a living cell)? What dissection instruments are available? Even the X-rays (which were so quickly used by doctors to "see" what was going on in the structural anatomy of humans) are far too gross to penetrate *within* the atom itself (although they can show the structural arrangement *between* atoms in a crystal).

However, the fact that electrons and alpha particles can penetrate even the most solid materials suggested that they could be used as tools for the diagnosis of what goes on inside the atom. Therefore, scientists turned to such processes to find answers. In our century, similar methods have been rapidly developed until today we have immense and very costly machines, such as the cyclotrons, the bevatrons, and the synchrotrons—all of which shoot alpha, beta, and other particles into materials at high velocity, acting as probes that can reach deeper and deeper into the ultimate constituency of matter. These are all descendants of the first primitive probing tools: the high speed particles emitted by radioactive elements at the beginning of the twentieth century. The tools have become more and more complicated and the theories of explanation more and more mathematical and abstract, but they all stem from the simpler tools of "sealing wax and string" and the simple algebraic equations of the early years when Thomson and Rutherford began their diagnosis of what was "inside the atom."

Essentially, the method is to probe with the high-speed particles that are shot through thin metal foils. Within the metal, the atoms are packed so closely together that they are almost "in contact" with one another (shown by the fact that metals are incompressible and that gases do not flow readily through metal sheets). Therefore, any particle, such as an electron (beta particle) or a helium atom (alpha particle), which travels straight through the metal, must travel through the atoms themselves and not merely through the interstitial spaces.

Before Rutherford conducted his famous experiments with alpha particles, considerable study had been made, using beta particles (electrons) as bullets shot through metals. These electrons went straight through, with very little deflection and with little loss of energy. The early theories proposed by Thomson and others was that the atom must be a kind of jellylike positive mass in which the negative electrons were imbedded, much as raisins are imbedded in a pudding (to use a very common analogy). The electrons could then pass right through the "pudding," rarely meeting another electron. Very few would be deflected or "scattered."

Rutherford, an experimental physicist with a profound instinct for choosing the significant questions, while pondering the results of the beta particle penetration decided that alpha particles would be an even more effective probe into the nature of the atom. These particles with a mass almost 8000 times that of an electron are shot off by radioactive polonium with a velocity of about 2×10^7 meters/sec (about 10,000 miles/sec or an appreciable fraction of the velocity of light—higher than any other available velocities except those reached by the modern cyclotrons). Thus, their energy is tremendous. If the "puddinglike" model was correct, then the particles should pass through the atoms with little deviation. The presence of electrons with their small masses should be negligible (much as a billiard ball on striking a ping pong ball is deviated negligibly from its path). Experimentally, in other ways also alpha particles are good tools. Rutherford and his colleagues had found methods by which *individual* alpha particles striking a target could be detected. Rutherford concluded, therefore, that shooting the alpha particles through metal film would be the most appropriate method of exploring the interior of an atom, and examined the scattering or deflection of the alpha particles as they travel through a gold film about 10^{-5} cm in thickness.

Rutherford, conducting his experiments with his assistants, Geiger and Marsden, had already observed and examined some scattering of alpha particles through very small angles and, in 1910, Geiger suggested that they *look for scattering* through *large angles.* As Rutherford dramatically puts it:

I may tell you in confidence that I did not believe that they would be (scattered through large angles), since we knew that the α particle was a very fast, massive particle, with a great deal of energy, and you could show that if the scattering was due to the accumulative effect of a number of small scatterings, the chances of an alpha-particle being scattered backwards was very small. Then I remember two or three days later Geiger coming to me in great excitement and saying, "We have been able to get some of the alpha-particles coming backward . . ." It was quite the most incredible event that has ever happened to me in my life. It was almost as incredible as if you fired a 15-inch shell at a piece of tissue paper and it came back and hit you.[4]

Rutherford quickly came to the following conclusion, which upon further quantitative analysis and experiment was fully born out: each central positive charge, with its accompanying mass, was concentrated into an extremely small volume. He termed this central body the nucleus of the atom, and estimated it to have a diameter in the neighborhood of 10^{-12} cm, that is, one ten-thousandths of the diameter of the whole atom itself, which had previously been estimated to be about 10^{-8} cm. The conclusion is that the atom is almost entirely "empty space," the small central nucleus being surrounded by electrons at great distances. The "cloud" of electrons of one atom prevent the penetration by another atom through the mutual repulsion between the two electron clouds.

The following analysis, which leads to this conclusion, provides another elegant illustration of how mathematical analysis, physical analogies, and experimental

[4]J. Needham, and W. Pagel, editors, *Background to Modern Science,* copyright Cambridge University Press, Macmillan, New York, 1938, pp. 61–74 (A lecture by Rutherford on atomic structure).

facts may be brought together to expand our knowledge of what lies beyond the reach of our direct perceptions.

Let us first look at the evidence and explanation qualitatively and, then, introduce some rough calculations. Rutherford saw at once that the fact that a small number of alpha particles were actually being scattered or deflected through very large angles implied that the atom could be looked upon neither as a solid impenetrable mass (in which case *all* the alpha particles would be deflected through *large* angles) nor as a puddinglike mass of positive charge with embedded electrons (in which case, *no* alpha particles would be deflected through large angles). The atom then must contain within it a small concentration of mass. When the alpha particle collided with this concentrated mass, it would be sharply deflected and might even bounce directly back toward the source. But this, Rutherford concluded, happened *only occasionally* because the "target" mass or the nucleus was very small compared to the atom itself.

Let us use some approximate figures, similar to those found by Rutherford in his early experiments in which he shot alpha particles through gold leaf. He found that approximately one out of 100,000 (or 10^5) particles were scattered at angles larger than 90°. Now, if the gold foil is 10^{-5} cm thick and if an atom of gold has a diameter of an order of about 10^{-8} cm, then there will be somewhere in the neighborhood of 1000 (10^3) layers of atoms in the gold foil.[5] It follows that if the alpha particles were striking only a *single* layer of atoms, a much smaller number would be deflected over 90°, that is, about one out of 10^8 instead of one out of 10^5. If we assume that this probability is the same as the ratio of the cross-sectional area of the nucleus to the cross section of the atom, then the areas are likewise in the ratio of 1 to 10^8 and the *diameters* are in the ratio of 1 to 10^4. In this way the nucleus was estimated to have one ten-thousandth the diameter of the atom.

Such is a rough picture that a quick calculation presents. But it is by no means complete, and Rutherford continued his analysis and brilliant experimentation to discover a great deal more about the internal structure of the atom by studying many "collisions" of the particles with the nuclei under varying conditions. He recognized that *tremendous forces* must be at play in turning back the alpha particle "bullets" traveling at such high velocities. What is the nature of these forces? What kind of collision is taking place? And how can alternative theories be tested experimentally?

The two simplest types of interaction between the alpha particle and nucleus, which might account for the deflection are: (*1*) a collision such as occurs when two billiard balls or marbles collide and bounce apart, the force only taking

[5]For practice, calculate this estimate on the basis of Avogadro's number. Assume the density of gold = mass/volume = 19.3 gm/cc. Remember that 1 gram-atomic weight contains 6×10^{23} atoms. Calculate the mass of 1 atom of gold. Then calculate the volume of an atom and, assuming the atom is a sphere, obtain an estimate for the diameter.

place over a very short space and a very short time during which the objects are in actual "contact" and distorting one another; or *(2)* an electrostatic repulsion between the two bodies due to their similar electric charges. Presumably such a repulsive force would follow the inverse square law of Coulomb and would be spread out for an appreciable distance as the two bodies carrying the charges came close to one another.

These were the alternatives considered first, since they appeared the simplest. If we recall our experience with marbles or billiards, we will recognize what a more careful mathematical analysis bears out, that with collisions between such objects *most* deflections are through fairly *large* angles. In particular, if a light marble (representing the alpha particle) collides with a much heavier one (representing the nucleus) the light marble is very likely to bounce either backward or at a very large angle from its original direction. It is only rarely that the collision is of a "grazing" nature; then only does the lighter marble suffer a small deflection from its course. On this alternative, therefore, *most* of the alpha particles, which are scattered at all, should be scattered through *large* angles. This, the evidence showed, was not the case (see Fig. 18.3).

What about the second alternative, that is, that the deflection is caused by the coulomb force of repulsion? Can this be analyzed as simply? Fortunately for Rutherford, the analysis was already at hand, having been worked out by Newton and his followers. Again we see an example of a mathematical analysis applied to one situation (in this case, the orbits of comets) at hand ready to be applied to an apparently totally different situation. Note the striking similarity of form between the law of gravitation and Coulomb's law.

$$F = G\frac{m_1 m_2}{d^2} \quad \text{(attraction)}$$

G is a constant, m_1 and m_2 are masses, and d is distance between masses.

$$F = K\frac{Q_1 Q_2}{d^2}$$

(repulsion if Q_1 and Q_2 are similar in sign, as in the case of the positive alpha particle and positive nucleus)

K is a constant, Q_1 and Q_2 are charges, and d is distance between the charges.

In our discussions on astronomy, we saw that the law of gravitation leads to circular or elliptical orbits for planets. They do not have sufficient energy (or velocity) to "escape" entirely from the sun's influence. If, however, a body such as a comet comes in from outer space with a high enough velocity (that is, greater than the "escape velocity" at the sun), it will not go into *an elliptical orbit* about the sun but, instead, will describe a *hyperbolic* orbit with the sun as a focus. It will be deflected from its path somewhat toward the sun, depending on how close it comes and upon its velocity, and will depart again into space never to return. Many comets take such orbits (unlike Halley's comet, which is in an elliptical orbit similar to a planet's).

We need not be concerned with the mathematical details, which involve the calculus, but we can picture the result geometrically as shown in the left-hand diagram of Fig. 18.2, where a comet comes in toward the sun on a hyperbolic orbit and then departs. Now, since the mathematical equation for the interaction of the alpha particle and the nucleus is identical in form (an inverse square law) with the interaction of the comet and the sun, except that repulsion between the charges replaces the gravitational *attraction* between the masses, we expect and do find that the orbits for the scattered alpha particles are exactly similar in their hyperbolic form to comets with high velocities but, in this case, with the nucleus located outside the orbit. The left-hand diagram of Fig. 18.2 shows the orbit of a comet, and the right-hand diagram shows the orbit of the alpha particle as it approaches and recedes from the nucleus.

(It may be asked: Why do not the masses of the nucleus and alpha particle also have an effect of attraction—why do we only consider the electrical forces that are acting? The answer is that gravitational forces are *extremely weak* between the small masses of the particles involved, compared to the electrostatic forces. In other words, the constant G is of a much smaller magnitude for the

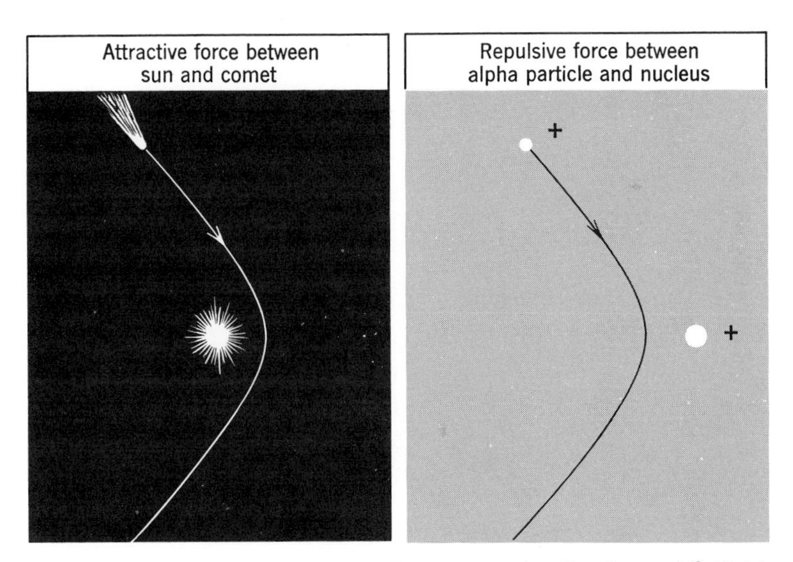

Fig. 18.2. Left, comet's orbit close to the sun, equation $F = Gm_1m_2/d^2$. Right, alpha particle's orbit close to the nucleus, equation $F = KQ_1Q_2/d^2$.

minute mass of the elementary particles compared to the constant K for the charges born by the elementary particles. The constant G is so many billions of times smaller, in fact, that we can neglect gravitational effects between elementary particles altogether. When we examine the forces holding the nucleus together, we shall find that there is still another force many times stronger than even the electrostatic or coulomb force we are considering here.)

The second alternative, therefore, which Rutherford examined in trying to account for the scattering of alpha particles, was the inverse square law hypothesis. He drew these mathematical conclusions: (*1*) that the number of particles reflected backwards at a given angle must be inversely proportional to the square of the velocity of the particle, and (*2*) that the number reflected at a given velocity must be inversely proportional to the fourth power of the sine of ½ θ, where θ is the angle between the original direction of the alpha particle and its final direction. Now, each of these proportionalities could be tested experimentally by the apparatus set up by Rutherford, Geiger, and Marsden. Very close agreement was found between the predictions on the basis of the hypothesis and the experimental tests. Once again we meet a procedure similar to Galileo's. It will be recalled that Galileo could not establish directly the law governing the velocity of falling bodies. He considered various hypotheses. From these, he deduced consequences that could be tested, rejecting all but that hypothesis which led to consequences in agreement with experimental tests. Similarly, Rutherford proceeded to deduce consequences from each of the simplest hypotheses concerning the type of collision occurring between alpha particles and nuclei. For "billiard-balllike" collisions, when the collisions do occur, the deflection or scattering angle should be large for a high proportion of the collisions; for inverse-square interactions, the proportionate number of large-angle deflections should be small. Furthermore, he worked out and tested the precise law of such deflections. Figure 18.3 shows the alternative possibilities.

By a brilliant series of experiments, both the velocities of the alpha particles were varied (this was done by "slowing down" the particles by making them pass through foils of measured thicknesses so as to absorb some of the energy) and the angles of scattering were varied, by means of a specially designed apparatus. The inverse square law was thus carefully established. Furthermore, once this had been done, both the charge on the nucleus and the mass of the nucleus could be established. It was found that the *charge on the nucleus of any element was equal to its atomic number,* that is, the number assigned to that element in the periodic table (Chapter 12). The first step had now been taken to explain this table.

Finally, to get a more precise estimate of the "size of the nucleus" and to give it meaning, the following considerations were used. By "size" or radius of the nucleus, we mean the distance of the closest approach by an alpha particle. The

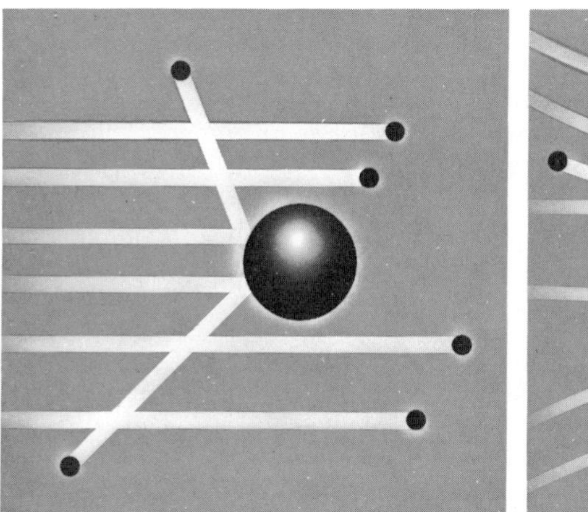

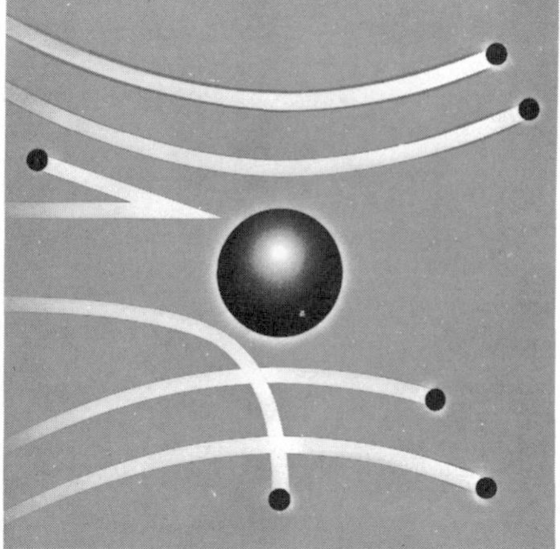

Fig. 18.3. *Left,* contact collisions (When they occur, the angle of scattering is large.) *Right,* "electrostatic" collisions (When they occur, most particles are scattered through small angle.)

approach will be closest when the alpha particle is aimed directly at a nucleus and is reflected directly backward in the same direction from which it came. At the point of closest approach, all its kinetic energy will be transformed into potential energy, the potential energy again being transformed into kinetic energy as the alpha particle bounces back. A simple equation then determines the distance of closest approach as about 10^{-12} cm.[6]

For the reader to appreciate how "empty" the atom becomes under this conception of its structure, an analogy may be useful. We consider our solar system with the central sun and the vast distances between the sun and the most distant planets to contain very little "matter" compared to the volume of space. If the atom could be "blown up" to the same general magnitude as the solar system, relative distances of the electrons from the central nucleus would be comparable to the distances of the planets. (On this scale, however, the central nucleus would be very much smaller than the sun.)

However, such analogies, useful as they may be for pictorial purposes, should not be taken literally. Note, for example, how the *size* of the nucleus is approxi-

[6]The equation determining the distance of closest approach is

$$\tfrac{1}{2}mv^2 = K\frac{Ze^2}{R}$$

kinetic energy (at a distance) = potential energy (at closest approach)

where R is the distance of nearest approach, Z is atomic number, and e is the charge on an electron. The expression on the right-hand side for potential energy is discussed in Chapter 21, Section 4. A noncalculus derivation may be found in Jay Orear's *Fundamental Physics,* Wiley, New York, 1961, p. 87.

mated by the closest approach by an alpha particle, not by a material boundary as in the case of a planet. When we begin to examine nuclear structure, we shall find that other particles, such as neutrons, which carry no charge, can approach even closer. In fact, using different types of particles as bullets to explore the target atoms, the measure of the size of the nucleus is given by the cross section it presents to the "bullet" or particle, and the unit cross section used is called a "barn," which is taken[7] as 10^{-24} cm^2. (Note that if a nucleus had a diameter of 10^{-12} cm its cross-sectional area is $\pi D^2/4$ or roughly 10^{-24} cm.^{2})

3. THE BOHR MODEL OF THE ATOM

With the discovery that the atom consists of a very small but massive nucleus in which all the positive charges are concentrated, surrounded at *relatively* large distances by electrons that contain all the balancing negative charges, the problem now becomes one of explaining the distribution of such charges. Why does not the atom automatically collapse, the positive charges on the nucleus sucking in all the electrons? At the small distance involved, the coulomb force of attraction is a tremendous one, so much greater than the gravitational attraction as to render the latter negligible. The analogy to the solar system was inevitable. Just as the planets remain in their stable orbits with the centripetal attractive force of gravitation exactly equal to that required for their central acceleration toward the sun, so it was thought the electrons might be in orbit at very high velocities about the central nucleus with the coulomb forces of electrical attraction between nucleus and electrons taking the place of the force of gravity between the sun and the planets.

Such a theory looks most elegant and simple. The masses are known, the charges on the nucleus and electrons are known, and the distances can be estimated. Therefore, just as in the case of a circular orbit of a planet about the sun in which, it will be recalled, the gravitational force equals for each planet its mass times its central acceleration, or

$$G\frac{m_s m_p}{d^2} = \frac{m_p v^2}{d}$$

so the equivalent equation for each electron should hold, that is, the coulomb force equals the central acceleration of the electron:

$$K\frac{(Ze)(e)}{d^2} = \frac{mv^2}{d}$$

[7]The "barn" was a code name used during the World War II when the Manhattan project was being pursued. For a neutron, 10^{-24} cm^2 presents a fairly large target. It is as "easy to hit as a barn."

(where Z is the atomic number of the nucleus, and therefore Ze is the charge on the nucleus, and e the charge on the electron, and K the constant for Coulomb's law).

To bring into as sharp a focus as possible the problems that Bohr faced as he tried to develop a theory of the atom, we summarize with the following questions.

Accepting Rutherford's conclusion that an atom had a small central nucleus with electrons at comparatively large distances away, can a "planetary model" of an atom be constructed?

There are two problems associated with using such a model.

First, if the electrons are moving in a circular path, they are accelerating. Why, then, do not the electrons radiate electromagnetic energy, as required by Maxwell's laws (which imply that any charge being accelerated will generate electromagnetic waves) and "fall" rapidly into the central nucleus as they lose energy?

Second, why do atoms of gas emit only certain *discrete* frequencies of radiation instead of a continuous band of frequencies?

Just as Rutherford used alpha-particle scattering as one method to diagnose the internal structure of atoms, so Bohr used another method of diagnosis to extend Rutherford's theory: that of analyzing the spectral lines emitted by heated gases. These are the most effective signals of the internal structure of atoms, once a systematic interpretation becomes understood.

Much work had already been done in spectroscopy. It was already known that every element in the periodic system gives off its own characteristic pattern of spectral lines. The scientist naturally wished to find some systematic way of identifying these patterns (just as the criminologist requires a system for classifying fingerprints). It is an immediate and immense advantage that a definite number representing frequency can be associated with each line. But, even so, the spectral lines for a long time appeared to be a complex jumble with no rhyme or reason. Possibly, it was thought, the lines represented some sort of harmonics. Could it be that just as plucked strings give off *some,* but *not all* frequencies, so do the atoms emit light of some but not all frequencies? With vibrating strings the sound energy emitted has frequencies that lie in a simple numerical series. (The overtones, for example, of low C on the piano, which has a frequency of 128, are in simple numerical ratios 1:2:3:4:5 . . . to this fundamental frequency.) No such simple relationship can be found in spectral lines. For a long time, no formula at all could be discovered.

Finally, in 1885, Balmer, studying the simplest of the spectra, the visible spectrum of hydrogen, did find a numerical relationship. The spectral lines of hydrogen are shown in Fig. 18.4. This series of lines can be fitted into an empirical, numerical formula, called the Balmer series, after its discoverer. In the visible

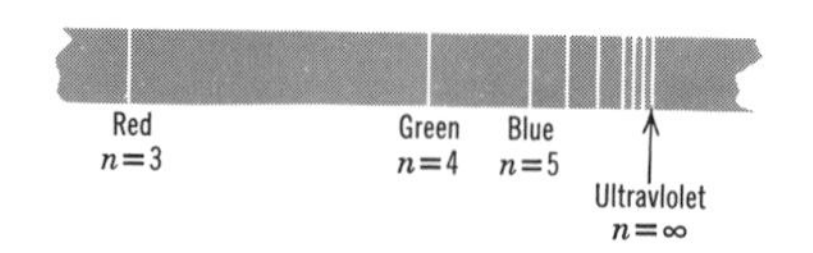

Fig. 18.4. Hydrogen spectrum (Balmer series lines). The lines representing several frequencies in the visible-to-ultraviolet range crowd closer together as n, in the equation $f = R(1/4 - 1/n^2)$, becomes larger. The upper limit is fixed by n = infinity, when $1/n^2$ falls to zero.

range, the frequencies of hydrogen spectral lines are given by

$$f = R\left(\frac{1}{2^2} - \frac{1}{n^2}\right)$$

where R is an experimentally measured constant equal to 3.29×10^{15} sec^{-1}, and n is a *whole* number greater than two.

The series of lines shown in Fig. 18.3 are specified in order as

$$f = R\left(\frac{1}{4} - \frac{1}{9}\right)$$

$$f = R\left(\frac{1}{4} - \frac{1}{16}\right)$$

$$f = R\left(\frac{1}{4} - \frac{1}{25}\right)$$

or, in general,

$$f = R\left(\frac{1}{4} - \frac{1}{n^2}\right)$$

where n = 3, 4, 5 . . . and where R is a constant.

It will be seen that as n increases the term $1/n^2$ becomes smaller and smaller, approaching zero, and the lines become more and more crowded toward one end of the spectrum. Such a series is obviously different from the usual series of musical harmonics, which does not have any limiting values (the overtones of a plucked string do not have any upper limit although the intensity of the higher harmonics may become so faint as not to be easily heard).

Furthermore, in this empirical equation of Balmer, there was no explanation of the peculiar constant R, let alone the other terms of the equation. Once more the physicist was faced with the problem of "explaining" an empirical equation by trying to relate it to basic principles. Even more frustrating was the fact that the very equation seemed to contradict the known laws of mechanics and electromagnetism. Classical laws of physics led to very different results.

In 1913, Bohr took the first fateful and imaginative step toward an explanation. He was well aware of the radical theory of Planck that energy of radiation comes in bundles or quanta. "Might it not be possible," he asked himself, "that the explanation of the discrete frequencies emitted by atoms could be found on the basis of quantum theory?" According to Planck's theory the energy of radiation is discrete, and is always equal to some multiple of the frequency times h. Bohr applied this same idea to the atom. He assumed that the electrons circulating about the atoms would be confined to "stationary" orbits and that, as they "jumped" from orbit to orbit, they would emit or absorb energy only in a single quantum $h \times f$. This would "explain" the existence of *discrete* spectral lines. It was a radical departure from the classical theory explaining the revolution of planets about the sun. In the latter case, *any orbit* is permissible, and energy is lost by a gradual and continuous process (quickly, as in the case of artificial satellites, where the friction caused by space debris—meteorites or thin layers of gas—brings about a relatively rapid loss of energy; very slowly, in the case of planets, where the loss of energy is due primarily to tidal friction operating over millenia causing extremely slow changes in the orbits).

If Bohr's hypothesis of selected "stationary orbits" (later to be more appropriately called "energy levels") is to be taken seriously, three things need to be shown: (*1*) that the form of Balmer's equation may be derived from the hypothesis; (*2*) that the constant R (known as Rydberg's constant) can be explained; and (*3*) that, if possible, previously unknown phenomena can be predicted.

Bohr's theory of the atom accomplished all of these. Its success was quickly acclaimed even though, today, we recognize it as only a first (and, in the end, a somewhat misleading) step toward a satisfactory analysis.

As to the most general problem—why frequencies take on the particular discrete values that they do—Bohr set forth the postulates listed below. (*Note.* He did not *justify* these postulates on the basis of "common sense" or inductive experience any more than Newton justified his postulate of universal gravitation on the basis of inductive experience. The justification lay in the consequences that could be deduced and tested in both cases. We recall that Newton had declared, *"hypotheses non fingo"*—I make no hypotheses—regarding a fundamental explanation of gravity. He felt that the *next* step of explanation should not interfere with the immediate task of finding a general law sufficient to cover all planetary motion. Similarly, Bohr sought for a set of postulates that would be general enough to cover all known phenomena regarding spectral emissions. In Bohr's case, indeed, the postulates appeared artificial, even contrary to the common sense of classical physics. A deeper understanding or explanation had to await the development of quantum or wave mechanics. Meanwhile, his postulates were extremely fruitful in systematizing the known data and suggesting further investigations.)

The postulates of Bohr were as follows.

Postulate 1. The orbits of electrons are circular with the nucleus in the center. The force keeping them in orbit is coulomb force of attraction between the electron and the positively charged nucleus. (This latter charge is equal to Z, the atomic number times e_1, the unit charge.) The force of attraction is therefore

$$F = K\frac{Q_1Q_2}{r^2} \quad \text{or} \quad F = K\frac{(Ze_1)e_2}{r^2}$$

As in classical physics, this force is equal to the force accelerating the electron in a circular path, that is,

$$F = \frac{mv^2}{r}$$

Equating these two forces, we arrive at

$$\frac{mv^2}{r} = K\frac{(Ze_1)e_2}{r^2} = K\frac{Ze^2}{r^2} \qquad (1)$$

(recalling that $e_1 = e_2 = e$, the unit charge of an electron).
Thus far, Bohr has introduced nothing new in his model. The equations are similar in form and meaning to the equations used to explain orbits of planets in which, it will be recalled, the central accelerating force

$$F = \frac{mv^2}{r}$$

was equated to the force of attraction between the sun and the planet.

Postulate 2. Bohr now introduced a radical innovation: that only certain orbits are permissible and that these orbits are determined by the law that the angular momentum times 2π must be equal to a whole number times h. (Remember that the angular momentum of an object traveling in a circle of radius r is equal to the linear momentum times r.) Thus

$$2\pi mvr = nh \qquad (2)$$

where n is a whole number 1, 2, 3, . . . etc.

This, then, is a *quantum* condition. Bohr gave no justification for this postulate; he merely stated it, and showed that it led to fruitful results.

One immediate consequence of the first two postulates is that equations 1 and 2 contain two unknowns: v and r, plus a whole number n. Every other constant is known. We shall eliminate v, and solve the resulting equation for r. (We pass over the actual algebra. Solve the second equation for v, and substitute this value of v in the first equation.)

We arrive at

$$r = \frac{n^2h^2}{4\pi^2KZe^2m}$$

For hydrogen of atom number 1, $Z = 1$. The only constants on the right-hand side are

$h = 6.62 \times 10^{-34}$ joule-sec

$K = 9 \times 10^{9}$ (the coulomb constant arrived at from study of electrostatic attraction between charges)

$e = 1.6 \times 10^{-19}$ coul

$m = 9.1 \times 10^{-31}$ kg

Substituting these values, we find

$$r = (n^2)(5.3 \times 10^{-10} \text{ m})$$

What does this tell us? It signifies that for $n = 1$, when the radius is the smallest and the atom is in its lowest energy or ground state, the radius of the hydrogen electron is 0.53×10^{-10} m or 10^{-8} cm, *which agrees with the values previously obtained from other* estimates of the size. Furthermore, it means that the radius can be 1, 4, 9, 16, etc., times as great as it is in the "ground" state, but that it cannot take on any intermediate values.

Postulate 3. To explain discrete spectral emissions, Bohr introduced his third and most significant postulate, namely, that an electron may "jump" from one orbit to another and in doing so *emits or absorbs energy in quanta.* The energy emitted or absorbed must therefore be whole quanta or $h \times f$ to agree with Planck's theory. Bohr was carrying Planck's hint of discontinuity in nature one step farther: The electron would have to "jump" instantaneously from one orbit to the next, never occupying any intermediate position. The same instant it disappeared from one orbit it would appear in the next.

This postulate was used by Bohr to explain the Balmer series of spectral lines to which reference has been made. The basic idea can be expressed this way: in each orbit the electron has total energy composed of two kinds, its kinetic energy and its potential energy. These energies are completely determined by the first and second postulate, and may be given a simple mathematical expression. The kinetic energy is just $\frac{1}{2}mv^2$; the potential energy depends on the distance between the nucleus and the electron. Therefore, when an electron "jumps" from one orbit to another, it will gain or lose a total amount of energy that is determined by the radii of the orbits. This total energy change must equal hf.

To understand the mathematical expression for potential energy, a slight reorientation from the earlier uses of the term is required. It will be recalled that in handling problems of motion close to the earth's surface, potential energy was given by mgh, where h represented the height above the earth's surface. This implied (*1*) that we arbitrarily selected the surface of the earth as the point of zero potential energy, and (*2*) that g is constant. Now, in the problem we are handling, the position of zero potential energy is best taken as the point where the electron is so far away from the nucleus that no force is acting on it, that is, at infinity. Since it loses potential energy as it comes closer to the nucleus, its potential energy is negative. Furthermore, the distances involved cause a significant change in the force exerted. Therefore, the energy is not proportional to the distance. Analysis[8] shows that the proper expression for the potential energy of the electron at a distance r from the nucleus is

$$\text{potential energy} = -K\frac{Ze^2}{r}$$

The mathematical expression, therefore, for the total energy of the electron in any given orbit becomes

$$\text{total energy} = \tfrac{1}{2}mv^2 - K\frac{Ze^2}{r}$$

Now, once again using the first and second postulates, we can eliminate both v and r by algebra. Doing so, we obtain for the total energy of an electron in the orbit corresponding to any whole number n

$$\text{energy (of } n\text{th orbit)} = \frac{2\pi^2 mK^2Z^2e^4}{n^2h^2}$$

which is a clumsy-looking expression but which contains nothing other than known constants.

Now for the triumph of Bohr's theory. The energy difference between any two orbits should be equal to a quantum of energy hf.

Thus, between an outer orbit where $n = n_0$ and an inner orbit where $n = n_i$

$$hf = \frac{2\pi^2 mK^2Z^2e^4}{(n_0)^2h^2} - \frac{2\pi^2 mK^2Z^2e^4}{(n_i)^2h^2}$$

$$= \frac{2\pi^2 mK^2Z^2e^4}{h^2}\left(\frac{1}{n_i^2} - \frac{1}{n_0^2}\right)$$

whence

$$f = \frac{2\pi^2 mK^2Z^2e^4}{h^3}\left(\frac{1}{n_i^2} - \frac{1}{n_0^2}\right)$$

Lo and behold! The constant R, in the Balmer series, where we have seen

$$f = R\left(\frac{1}{2^2} - \frac{1}{n^2}\right)$$

now finds an explanation. Evaluating $2\pi^2 mK^2e^4/h^3$ (where $Z = 1$), this is exactly the same as R, that is, $3.29 \times 10^{15}\ \text{sec}^{-1}$, within a fraction of one per cent.

Here, then, was a tremendous triumph for the Bohr model. It gave a systematic explanation of spectral lines of hydrogen. More than this, it also could be

[8] See page 200.

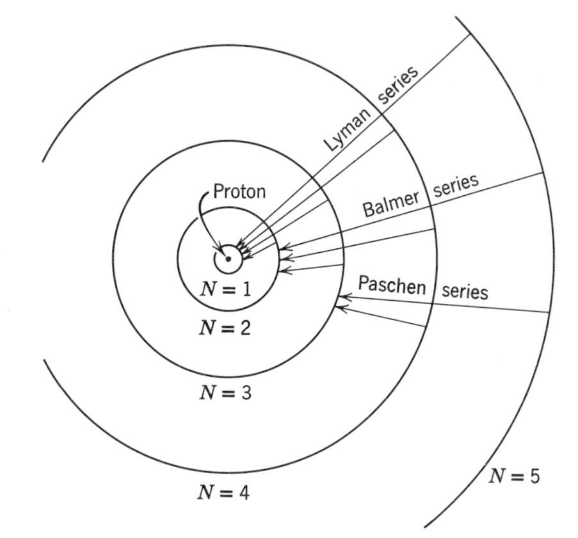

Fig. 18.5. Schematic diagram of electron orbits according to Bohr model.

used for predicting other spectral lines, which had not yet been observed. Note that in the Balmer series $n_i = 2$. The Bohr theory indicated that there should be a series in which $n_i = 1$, in which case the frequencies would be

$$f = R\left(\frac{1}{1} - \frac{1}{n_0^2}\right)$$

This was discovered a year after Bohr announced his model, and was called the Lyman series. (Other series of lines for $n_i = 3$, $n_i = 5$ had already been known.)

Figure 18.5 is a pictorial diagram of how electrons can jump from one outer orbit of higher energy to an inner orbit of lower energy, generating in the process the various spectral series.

Other predictions were forthcoming, which stemmed from Bohr's model. In particular, the spectral lines of a singly ionized helium atom (the helium atom, which normally has two electrons, is singly ionized if one of the two electrons is stripped from it) were calculated by Bohr and found to agree with the evidence.

However, in the case of the heavier elements, although the Bohr model gives a systematic qualitative explanation, the exact calculations of frequencies become extremely difficult. So many electrons are involved that interact with one another, causing such large perturbations in each other's orbits, that ordinary mathematical analysis breaks down. Other methods had to be devised.

4. EXTENSION OF BOHR'S THEORY: QUANTUM NUMBERS

The presence of frequency always implies periodicity of some kind. We have seen that spectral lines suggested to Bohr his model of the atom with electrons traveling about in circular orbits periodically. The radii of these orbits are characterized in each case by a quantum number associated with the energy level of the orbit. However, closer experimental examination of the prominent spectral lines shows that they exhibit a fine structure; that is, a line may be composed of more than one spectral line, less intense but closely spaced. How are these to be accounted for? Is there some other periodicity at work?

Physicists soon developed an answer based on an extension of Bohr's model. First, it was suggested that the orbits might not be circular, but could be elliptical as in the case of the planets. In this case, the principal quantum number is determined by the major axis of the ellipse instead of the radius. This, in itself, would cause no real change. Each magnitude of the allowed major axis would now be "quantized" in place of the radius. However, a very interesting phenomenon now ensues, which introduces another type of periodicity. When an elliptical orbit is very eccentric, the velocity of the electron at perihelion becomes very high (it would be helpful to review the argument for this, recalling the discussion about elliptical planetary motion)—so high, in fact, as to approach that of light. Under these circumstances, by Einstein's theory, the mass of the electron will become larger. This, in turn, will cause the whole ellipse to revolve about its center, a motion that is called precession. This, then, is a *second* periodic motion with *its own independent periodicity.* Again a quantum condition is imposed: the angular momentum times 2π must be a whole number k times h. Thus, a second series of quantum numbers is introduced, called the k quantum numbers.

It did not take long to find still a *third quantum number,* again using the Bohr model. The plane of the elliptical orbits can be revolved in space. A new dimension is introduced. Its effect is to change the *magnetic* orientation of the atom. When electromagnetism was discussed, we saw that a charge moving in a circle in a plane gives rise to a magnetic force at right angles to the plane of the motion. Thus, the electrons, moving in either ellipses or circles in a plane, give rise to a magnetic field. If the plane of the ellipse is now revolved, the magnetic field's direction is changed. Consequently, a third periodicity is introduced, and it is again explicable by a third quantum number, *the magnetic quantum number, m.*

Finally, even a *fourth quantum number* was found. This was explained by assuming that the electron itself can rotate or spin about its own axis clockwise or counterclockwise with respect to the plane of its orbit about the nucleus, introducing still another periodicity. The quantum number associated with the electron's rotation is called the *spin quantum number.* Without going into the reasons, it has been accorded the value either $+\frac{1}{2}$ or $-\frac{1}{2}$.

Let us try to understand the meaning of these four "quantum numbers." We have seen that each is associated in some way with *periodicity*. But any combination of several periodicities can bring about extraordinarily intricate patterns, as anyone familiar with music can testify. In Western music there is the basic periodicity of an octave (numerically determined by the ratio of frequencies 2:1); within the octave there are the harmonic intervals, each associated with numerical ratios (the major fifth, for example, having the frequency ratio of 3 to 2, as measured in vibrations per second). On top of these periodicities are superimposed the periodicities of rhythm. The length of each note may vary within certain rhythmic patterns: "long-short, long-short," etc., with certain well-defined numerical ratios of the time of each note. Similarly, there may be a rhythm in intensity. Now, *if* we could restrict music to the mixture of a limited number of combinations of patterns of rhythmic repetitions of notes, timing, and intensity, we would have a "quantum theory" of music—one describable in terms of an extraordinarily large *but finite number* of possible combinations of these "quantum numbers." Any melodies would have to fall within this number of possibilities. This limitation, however, is excluded from music because there are continuous variations: within notes—not only half tones, etc., but the slightly sharp or flat tones, and, even more important, the slight almost imperceptible variations in time and intensity. Furthermore, the joining of one melody with another by a combination of voices or instruments gives tremendous additional possibility of variety.

This analogy of combination of musical patterns to combination of quantum numbers which form atom properties—the atomic pattern, in turn, forming the vast number of possible molecules (as melodies of many instruments are combined into symphonies)—may be helpful. But, although it is largely a matter of "culture," adaptation, and training that determine those combinations of the periodicities, in music that "go together" or appeal to our ears, it is a fundamental requirement of physical nature that only certain *select* combinations of periodicities represented by quantum numbers are "allowed to exist" within the atom. This is in contrast, for example, to the notes of a violin (which are not "quantized" by nature), in which continuous variations are possible. Quantum mechanics is an attempt to give a systematic account, through mathematical analysis, of all the possible combinations of these periodicities as well as the probability (relative frequency) of each combination.

Just as Pythagoras "discovered" that harmonious combinations with tones are related to the length of plucked strings by simple ratios of whole numbers, without knowing the reason "why" (but later explained by the possible modes of vibration of strings), so did Bohr and his immediate followers discover that there were certain ratios and combinations of the four quantum numbers in terms of which spectral lines for the simplest elements could be systematized. In both cases, the "descriptive," nevertheless numerical, results were later to be better "explained" by a still more profound analysis.

After the development of the four quantum numbers, each defining a kind of periodicity fitting spectral lines, the further problem remained to use these numbers to explain still another general periodicity; that of the periodic table which we have examined in Chapter 12. On the basis of the quantum numbers, how can the periodic table be explained? A most ingenious system was evolved on the principle of the Bohr model, later to be given a more complete mathematical expression. This system actually did reduce the basic problems of chemistry and chemical compounds, with all the associated qualitative properties (valence, affinity, and even such qualities as color, texture, etc.), to those of physics.

The mathematical rules required to accomplish such a systematic explanation, in addition to the previously stated hypotheses of distinct, quantized, energy levels, are these:

(*1*) At any energy level where the principal quantum number is n, the next quantum number k must be one less than n.

(*2*) When k is fixed, the next quantum number m must take on a value not greater than k in absolute value. (For example, if $k = 1$, then m equals either $+1$, 0, or -1.)

(*3*) The fourth quantum number, s, can be either $+\frac{1}{2}$ or $-\frac{1}{2}$.

(*4*) No two electrons can occupy the same quantum state in the same atom. This principle, called the Pauli Exclusion Principle after its discoverer, Wolfgang Pauli, limits the possible configurations of electrons, since each electron must have a unique combination of quantum members. (Perhaps the best "explanation" of the Pauli Exclusion Principle is that if two electrons were in the same state as standing waves inside the atom, the waves would cancel each other.) Were it not for this principle, the number of chemical elements would not be limited and there would not be such a wide variety of distinct properties which can be classified into similar groups as we proceed step by step up the periodic table. (See Chapter 21, Section 1.)

(*5*) The primary determinant of the *chemical* properties of an element is the *outer* electrons. (The inner electrons have their effect mitigated by the offsetting charge of the nucleus.)

5. THE BUILDING-UP PRINCIPLE AND THE PERIODIC TABLE

Such are the general empirical rules that guide us effectively into the *manner* of combining (without "explaining") the *quantum* periodicities into the *chemical* periodicity of the Mendeléeff's table of elements (in which, it will be recalled, as the number of electrons of an atom increase element by element, the chemical properties of the atom change periodically).

A hydrogen atom is pictured now as a single positive charge (nucleus) surrounded by a balancing negatively charged electron. In the lowest energy state,

$n = 1$, $k = 0$, $m = 0$, and $s =$ either $+\frac{1}{2}$ or $-\frac{1}{2}$ (an electron cannot have both). The electron can have either one of two quantum states when it is at the lowest energy level.

A helium atom has two electrons and two positive charges. The lowest energy state is still $n = 1$, $k = 0$, $m = 0$, but now *one* electron must have $s = +\frac{1}{2}$, the other $s = -\frac{1}{2}$. At this point, the basic energy level for $n = 1$ is expressed as a "state of saturation." No other electrons can occupy this state where $n = 1$, since all quantum numbers have been exhausted. Another way of putting it is that the electron "shell" is "full" for $n = 1$.

Look at the next atom lithium, with atomic number 3, that is, with three positive charges and three electrons. Two electrons can "occupy" a shell similar to that occupied by helium electrons, that is, $n = 1$, $k = 0$, $m = 0$, and $s = +\frac{1}{2}$ or $-\frac{1}{2}$. The third electron, however, is excluded from this inner shell for which $n = 1$ because of the fourth postulate above (Pauli's exclusion principle). For this third electron, therefore, $n = 2$, and k may now equal either 0 or 1. Again s may be $+\frac{1}{2}$ or $-\frac{1}{2}$. In summary, therefore, the third lithium electron may have *eight* different "positions" corresponding to $n = 2$ (that is, *all* the combinations for $k = 0$ or $k = 1$: for $k = 0$, $m = 0$, $s = \frac{1}{2}$ or $-\frac{1}{2}$; for $k = 1$, $m = -1$, 0 or 1, $s = +\frac{1}{2}$ or $-\frac{1}{2}$).

As we go up the periodic table of elements, adding electrons one by one, we see that when we have reached a total of ten electrons we have *exhausted all the possibilities.* We say now that both the first and the second "shell" have been "filled" or saturated. It happens that the element with 10 electrons (atomic number 10), with its inner shell of two electrons and outer shell of *eight* electrons, is neon, an inert gas similar in all chemical properties to helium. The chemical "qualities" have thus been repeated; they are periodic, much as an octave in music is periodic.

But, to make the periodicity clearer, let us look at one more element, the next element, sodium, with one more electron (and, therefore, one more positive charge on its nucleus). The new (eleventh) electron cannot "occupy" either the inner shell for $n = 1$ nor the second shell for $n = 2$, since these shells are "saturated." Consequently, it is left off by itself (beginning a new "shell" with a principal energy level $n = 3$).

Because it starts a new "period," it is similar to lithium, which also had a single electron outside the saturated shells in which the "positions" were filled. Sodium and lithium, therefore, are alike (much as in music the note D' following C' an octave above C is similar to D following C itself).

Proceeding down group 1a to the element in period IV, potassium, we see that it too is similar to lithium in its properties (i.e., valence of $+1$, metallic in nature, highly reactive, large specific volume compared to its neighbors in the table, etc.). This is hardly surprising since its electron configuration tells us that

TABLE 1. PERIODIC TABLE OF THE ELEMENTS

KEY TO CHART: Atomic Number → 50; Oxidation States → +2 +4; Symbol → Sn; Atomic Weight → 118.69; Electron Configuration → -18-18-4

Groups → / Periods ↓	1a	2a	3b	4b	5b	6b	7b	8			1b	2b	3a	4a	5a	6a	7a	0	Orbit
I	1 +1 −1 H 1.00797 1																	2 0 He 4.0026 2	K
II	3 +1 Li 6.939 2-1	4 +2 Be 9.0122 2-2											5 +3 B 10.811 2-3	6 +2 +4 −4 C 12.01115 2-4	7 +1 +2 +3 +4 +5 −1 −2 −3 N 14.0067 2-5	8 −2 O 15.9994 2-6	9 −1 F 18.9984 2-7	10 0 Ne 20.183 2-8	K-L
III	11 +1 Na 22.9898 2-8-1	12 +2 Mg 24.312 2-8-2	Transition Elements					Group 8					13 +3 Al 26.9815 2-8-3	14 +2 +4 −4 Si 28.086 2-8-4	15 +3 +5 −3 P 30.9738 2-8-5	16 +4 +6 −2 S 32.064 2-8-6	17 +1 +5 +7 −1 Cl 35.453 2-8-7	18 0 Ar 39.948 2-8-8	K-L-M
IV	19 +1 K 39.102 -8-8-1	20 +2 Ca 40.08 -8-8-2	21 +3 Sc 44.956 -8-9-2	22 +2 +3 +4 Ti 47.90 -8-10-2	23 +2 +3 +4 +5 V 50.942 -8-11-2	24 +2 +3 +6 Cr 51.996 -8-13-1	25 +2 +3 +4 +7 Mn 54.9380 -8-13-2	26 +2 +3 Fe 55.847 -8-14-2	27 +2 +3 Co 58.9332 -8-15-2	28 +2 +3 Ni 58.71 -8-16-2	29 +1 +2 Cu 63.54 -8-18-1	30 +2 Zn 65.37 -8-18-2	31 +3 Ga 69.72 -8-18-3	32 +2 +4 Ge 72.59 -8-18-4	33 +3 +5 −3 As 74.9216 -8-18-5	34 +4 +6 −2 Se 78.96 -8-18-6	35 +1 +5 −1 Br 79.909 -8-18-7	36 0 Kr 83.80 -8-18-8	-L-M-N
V	37 +1 Rb 85.47 -18-8-1	38 +2 Sr 87.62 -18-8-2	39 +3 Y 88.905 -18-9-2	40 +4 Zr 91.22 -18-10-2	41 +3 +5 Nb 92.906 -18-12-1	42 +6 Mo 95.94 -18-13-1	43 +4 +6 +7 Tc (99) -18-13-2	44 +3 Ru 101.07 -18-15-1	45 +3 Rh 102.905 -18-16-1	46 +2 +4 Pd 106.4 -18-18-0	47 +1 Ag 107.870 -18-18-1	48 +2 Cd 112.40 -18-18-2	49 +3 In 114.82 -18-18-3	50 +2 +4 Sn 118.69 -18-18-4	51 +3 +5 −3 Sb 121.75 -18-18-5	52 +4 +6 −2 Te 127.60 -18-18-6	53 +1 +5 +7 −1 I 126.9044 -18-18-7	54 0 Xe 131.30 -18-18-8	-M-N-O
VI	55 +1 Cs 132.905 -18-8-1	56 +2 Ba 137.34 -18-8-2	57* +3 La 138.91 -18-9-2	72 +4 Hf 178.49 -32-10-2	73 +5 Ta 180.948 -32-11-2	74 +6 W 183.85 -32-12-2	75 +4 +6 +7 Re 186.2 -32-13-2	76 +3 +4 Os 190.2 -32-14-2	77 +3 +4 Ir 192.2 -32-15-2	78 +2 +4 Pt 195.09 -32-16-2	79 +1 +3 Au 196.967 -32-18-1	80 +1 +2 Hg 200.59 -32-18-2	81 +1 +3 Tl 204.37 -32-18-3	82 +2 +4 Pb 207.19 -32-18-4	83 +3 +5 Bi 208.980 -32-18-5	84 +2 +4 Po (210) -32-18-6	85 At (210) -32-18-7	86 0 Rn (222) -32-18-8	-N-O-P
VII	87 +1 Fr (223) -18-8-1	88 +2 Ra (226) -18-8-2	89** +3 Ac (227) -18-9-2																-O-P-Q

															Orbit
* Lanthanides	58 +3 +4 Ce 140.12 -19-9-2	59 +3 Pr 140.907 -20-9-2	60 +3 Nd 144.24 -22-8-2	61 +3 Pm (145) -23-8-2	62 +2 +3 Sm 150.35 -24-8-2	63 +2 +3 Eu 151.96 -25-8-2	64 +3 Gd 157.25 -25-9-2	65 +3 Tb 158.924 -26-9-2	66 +3 Dy 162.50 -28-8-2	67 +3 Ho 164.930 -29-8-2	68 +3 Er 167.26 -30-8-2	69 +3 Tm 168.934 -31-8-2	70 +2 +3 Yb 173.04 -32-8-2	71 +3 Lu 174.97 -32-9-2	-N-O-P
** Actinides	90 +4 Th 232.038 -19-9-2	91 +5 +4 Pa (231) -20-9-2	92 +3 +4 +5 +6 U 238.03 -21-9-2	93 +3 +4 +5 +6 Np (237) -22-9-2	94 +3 +4 +5 +6 Pu (242) -23-9-2	95 +3 +4 +5 +6 Am (243) -24-9-2	96 +3 Cm (247) -25-9-2	97 +3 +4 Bk (249) -26-9-2	98 +3 Cf (251) -28-8-2	99 Es (254) -29-8-2	100 Fm (252) -30-8-2	101 Md (256) -31-8-2	102 No (254) -32-8-2	103 Lw	-O-P-Q

Numbers in parentheses are mass numbers of most stable isotope of that element.

(Reproduced from *Handbook of Chemistry and Physics*, 47th edition by permission of the Chemical Rubber Co.)

there is a single electron outside the closed shells. Similarly, the elements in group 2a which have two electrons outside closed shells are closely related to each other but much less so to the elements of group 1a. The groups further to the right behave in a like manner, although one salient difference stands out here. The families toward the center of the table (groups 3a, 4a, 5a, 6a) are not so closely related as those near the edges. For the members of groups 7a and 0, the halogens and inert gases, again we find a group similarity, much as the alkali metals form a closely knit family. It is worth remarking here that the closed shells of the members of group 8 (or 0) lead to the near-absence of chemical reactivity; hence the name *inert gases*. On the other hand, the members of group 4a are not so closely related. This is particularly marked in the case of carbon and silicon which, while similar in some respects, are vastly different in others. Carbon, for example, forms a vast number of compounds with hydrogen (and other elements) while silicon possesses no such tendency. The metallic character also changes as we go across the table, metals appearing on the left, and nonmetals on the right. They are separated by a class of elements called semimetals which behave in some respects like metals and in others like nonmetals. These elements, shown as shaded boxes in Table 1, are not confined to a single group, but run diagonally from upper left center to lower right center.

It remains to discuss two groups, the transition elements and rare earths, which at first glance do not appear to fit into the general scheme. With the aid of our present knowledge of atomic structure it becomes simpler to explain their positions and properties. Reference to Table 1, tells us that the first element in group 3b, scandium, differs from its neighbor calcium by having 9 electrons instead of 8 in the third shell. One would at first expect to find the additional electron in the fourth shell, but this does not occur for two reasons. First, the third shell can hold 18 electrons before it is filled. Secondly, the atom will be in a *lower* energy state if the additional electron occupies the third shell. The question arises as to why the configurations of potassium and calcium aren't 2, 8, 9 and 2, 8, 10 rather than 2, 8, 8, 1 and 2, 8, 8, 2. The answer again is that these atoms are in the lowest possible energy states if one (or two in the case of calcium) electrons occupy the fourth shell. The same explanation holds for the rare earths in which the fourth shell is filled step by step to a total of 32 electrons.

One of the most striking effects of the atomic shell structure is illustrated in Figure 18.6 which shows the ionization potential (energy required to remove an electron from the atom) as a function of atomic number. Atoms with all shells filled (the inert gases) are tightly bound and the ionization potential is large as shown, but for atoms with one electron outside the closed shells, it is small. As one goes to higher atomic number, atoms within the same family are larger and the outer electrons are not so tightly bound.

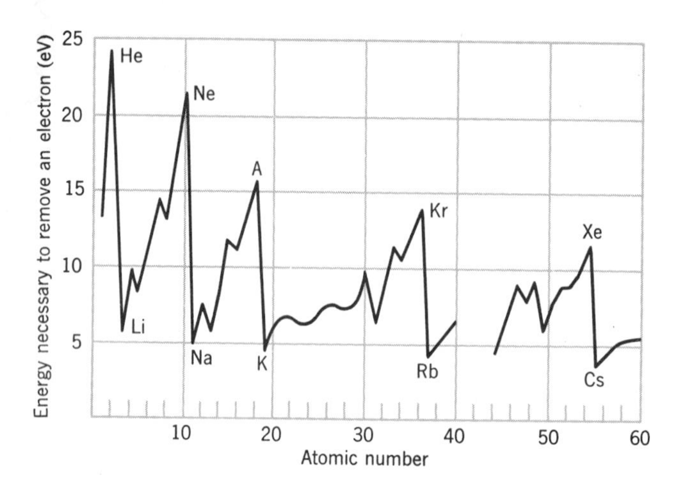

Fig. 18.6. Ionization potentials.

One could go into much greater detail in comparing and contrasting the properties and characteristics of families and members within families, and explaining them in terms of atomic structure. However, the foregoing discussion illustrates the principles involved and we terminate our treatment of the periodic classification at this point.

6. EXPLANATION OF QUANTUM NUMBERS; ELEMENTARY QUANTUM THEORY RELATED TO THE BOHR ATOM

As a first step toward a modern analysis of atomic structure, Bohr's ideas were extraordinarily fertile. He set investigators on the right track, both theoretically and experimentally. But the theory, the model, was short-lived as a basic explanation on three counts. (*1*) It was cumbersome, and could not be used to deal quantitively with elements with more electrons than lithium. The mathematical complications became too formidable. (*2*) Although it seemed to be able to predict the *frequency* of the spectral lines of a few atoms, particularly hydrogen and helium, it could not account for the *intensity* of the lines. (*3*) Above all, the postulates or assumptions upon which the Bohr model were based appeared "strange," *ad hoc,* and not properly tied in with other basic principles of physics.

The first tentative steps for understanding the internal atomic structure had been taken by Bohr and, perhaps, these were the most decisive ones (just as the Pythagorean connection between harmonies and numbers was a more important discovery than the later physical understanding for the reason for this connection). Nevertheless, a more fundamental approach was required, and Bohr took the leadership in insisting upon a deeper look into the problem of explanation.

The first theoretical breakthrough came with de Broglies' suggestion of the wave theory of the electron and its application to the Bohr atom. De Broglies' idea presented an alternative model in terms of which basic periodicity of atomic structure could be understood. Instead of explaining the periodicity by the orbital motion of the electron, de Broglie proposed that the concept of "standing waves" be applied. In this way the existence of discrete, rather than a continuous range of frequencies can be accounted for. Although there is no "natural" limitation to the period of an orbiting body, such as a planet or an electron, there are definite limitations on the possible frequencies of the standing waves of a vibrating body. A plucked string, for example, of a given mass and length, and under the condition of a given tension, can vibrate only in certain frequencies. The frequencies are "quantized," that is, they may take on only those values in which the waves fit the length of the string: there must be nodes (condition of no motion) at both ends. This means, as we have seen in the earlier chapter on waves, that they can only vibrate with certain wavelengths: the length of the string must be equal to a multiple of half wavelengths.

Similarly, de Broglie argued that if the electron has a wavelike nature, the explanation for the quantum states of the electron in the atom can be found in the condition that at any one energy level the wavelength of the electron must "fit" into the circumference of its "orbit." In other words, the circumference $2\pi r$ must be a whole multiple of the wavelength.

The implications of this assumption are quickly drawn from de Broglies' assumption that the wavelength of an electron is given by

$$\lambda = \frac{h}{p}$$

where p is the momentum and h is Planck's constant and the momentum $p = mv$

Therefore, since there must be an integral number of wavelengths in the circumference,

$$n\lambda = 2\pi r$$

or

$$\frac{nh}{mv} = 2\pi r$$

or

$$2\pi rvm = nh$$

This equation states precisely what Bohr's first postulate demanded: that the circumference times linear momentum equals nh. The postulate is thus made to appear more reasonable, and is no longer an artificial hypothesis invented only to bring system into the pattern of spectral lines. The electron as a standing wave cannot collapse into a smaller orbit because a fraction of a standing wave is impossible.

The model of a hoop, shown in Fig. 18.7, yields an oversimplified picture. An electron exists in three dimensions, and is not restricted to moving in a plane. Therefore, a better model might be a vibrating sphere with a complex pattern of standing waves. There would then occur radial vibrations of the spherical surfaces. Each of these radial vibrations could also be accompanied by vibrations in horizontal planes. And, in turn, each of these could be accompanied by vibrations in a meridial plane. In this way, the first three quantum numbers could be represented by a physical model or analogue; the fourth quantum number or spin of the electron doubling all these possibilities. Max Born, one of the physicists responsible for developing quantum mechanics, shows a model of this sort in his *The Restless Universe*.[9] For our purpose we need not examine the details of this three-dimensional model. The important idea to grasp is that the electron has properties that can be accounted for by waves, which can occur in a great many modes of vibration, each with a frequency determined by a quantum number.

Such a conception is an elegant and profound idea. The particles of physics act as "standing waves" subject to certain requirements at the boundaries determined by electrostatic or other forces caused by interaction with other particles, just as the standing waves in a bell, for example, are restricted to certain frequencies by the physical boundaries and elastic forces of the material out of which they are formed. However, the conception carries with it disadvantages

[9] Max Born, *The Restless Universe,* Dover Publications, New York, 1951.

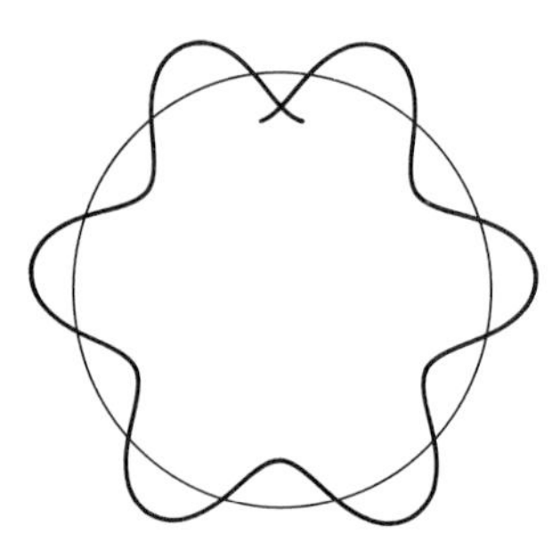

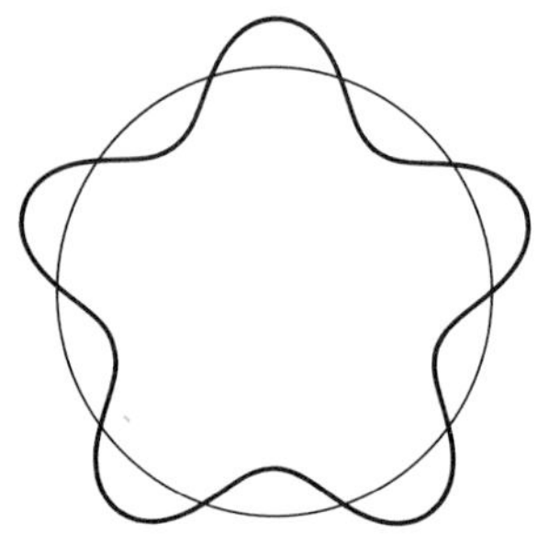

Fig. 18.7. Left, standing wave impossible (the wave does not "fit"). Right, standing wave is possible, $2\pi r = n\lambda$.

when we try to combine it with the older conception of an electron as a particle. Where *is* a wave? It is not a substance having a definite location in space and time such as we ascribe to a billiard ball or a marble. Instead, it is "spread out" over space. We are forced, therefore, to abandon our earlier conception of the electron as a particle in space. Furthermore, even the concept of a wave itself does not apply in the usual sense, although properties are present that have a periodic structure. Certainly these waves are not waves in a material medium such as are sound and water waves, nor are they electromagnetic waves. Practically all that seems to be left of our model is the abstract mathematical structure, which leads to correct experimental observations and suggests the *existence* of electrons and their nuclei. If we try to examine the notion of existence or being, we must return to a definition similar to Plato's "I hold that being is simply power, that is, power to affect others." In modern terms, we say that the fundamental particles are recognized as existing because they interact with each other. The presence of these particles becomes known through the gain and loss of energy in various interactions; the energy gain or loss, and also the frequencies, are subject to objective measurement. The task of the physicist now becomes that of building a logically consistent mathematical structure from which results can be deduced that check with observation. If necessary, all intuitive, mechanical models designed as analogies to the systems of physical objects that are on "our scale," that is, visible to our senses, must be abandoned if they involve inconsistencies. But on our scale we never see objects that act both as particles, with defined positions, and also as waves. The two conceptions, as we consider them on the basis of what we experience, appear to be contradictory. But the physicist must demand of all theories (*1*) that they be consistent, and (*2*) that they lead to conclusions that are experimentally observable. These two "commandments" overrule the natural desire to find an easily understood model. Since, at present, no model has been suggested that fulfills the requirements, quantum mechanics has developed without one and, hence, appears extremely abstract. The Bohr atom, with electrons restricted to well-defined orbits, is no longer adequate to the facts as known.

In these circumstances we require a mathematical expression for a wave, which has certain well defined properties. It must include a reference to discrete energy levels, associated with definite frequencies. These frequencies must, in turn, be related to certain physically measurable properties determined by surrounding conditions, such as force fields that, in turn, are related to positions in space. We shall see that the mathematical function for such a wave equation has no physical meaning in itself. The best we can do is to note that the square of the function does give us a measure of probability of finding a particle such as an electron within a certain volume of space.

In classical physics, the meanings of the wave functions used are clear. For a

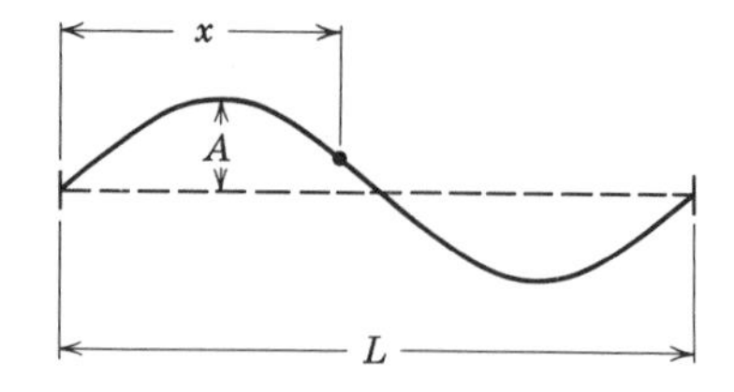

Fig. 18.8. *A* is the amplitude of the wave, *x* is the distance from the origin, and *L* is the wavelength.

standing wave the function describes the measurable amplitude of the wave, showing how it varies from position to position, and with time. Every term that enters into the equation is specified by some measurable quantity. Thus, the mathematical expression maps the physically vibrating object clearly and unambiguously in all its terms.

The classical standing wave function for a vibrating string fixed at both ends can be expressed as

$$y_n = A \sin \frac{\pi n x}{L} \cos \frac{2\pi t}{P} \qquad n = 1, 2, 3 \ldots$$

which gives an expression for the amplitude at any point x, and at any time t for the one mode of vibration. If $t = 0$, for example, we obtain a "snapshot" of the wave at that time. For the fundamental mode of vibration then $y = A \sin \pi n x/L$ and the geometric representation of the wave is an ordinary sine wave, such as we have discussed (Fig. 18.8).

When, however, Schrödinger, the mathematical physicist, first attempted to formulate a corresponding expression[10] for an electron wave (or, as it may be

[10] Some readers may find it interesting (even if it is abstract and not understood) to, at least, see the one-dimensional form of the Schrödinger psi function and a simple application to a standing wave. If an electron were confined in its motion to linear motion in a box, this would be analogous to the standing wave in a string rigidly held at both ends. (This is an artificial model, of course, since the electron actually is confined to a certain volume surrounding the nucleus, not by any physical boundary but by the coulomb forces of attraction between the electron and nucleus. This model is used, however, to show more clearly the analogy between a standing wave on a string, and the standing wave of an electron.) Under these circumstances, the ψ function of the electron is given as

$$\psi = \sqrt{\frac{2}{L}}\, e^{-(2\pi/h)E_n i t} \sin \frac{n\pi x}{L}$$

Note the imaginary quantity i, or $\sqrt{-1}$. If this expression is squared (that is, multiplied by its complex conjugate) and an average taken for $t = 0$, the "snapshot" of the square becomes

$$\psi^2 = \frac{2}{L} \sin^2 \frac{n\pi x}{L}$$

This squared expression then gives the probability of finding the electron at any position x. The probability distribution is, of course, somewhat different for each value of $n = 1, 2, 3 \ldots$ where n is the quantum number.

called, a "matter wave"), he found that the only possibility was a wave expression that has no counterpart in physical reality. There is no way in which all the terms in the psi-function (as this expression is known) can be related to numerically measurable quantities because the function contains the imaginary $\sqrt{-1}$. Now, although classical physics often uses imaginaries, this is done as a convenient mathematical manipulation. The expressions in which the $\sqrt{-1}$ enter can be always translated into real terms susceptible to measurement. In the end, when the mathematics is applied to a physical situation, the imaginaries disappear. But, in the psi-function the $\sqrt{-1}$ enters as an integral part of the expression and cannot be eliminated.

This apparent dilemma, the lack of connection between the mathematical expression of the wave properties of the electron and physical measurements, was overcome finally by noting that *the square of the function* (more precisely the product of the function by its conjugate) is related to the *probability* of finding the electron at a particular position. Probabilities, that is, the relative frequency of occurrences, can be measured if enough events are studied experimentally. Electrons can, for example, be shot through holes in a screen onto a photographic plate. They will form an impression that will reveal a pattern of light and dark, the more intense patches of light indicating that a greater number of electrons reached that position. The *relative intensity* thus becomes a *measure of the probability* of the position of the electron. When experiments of this sort are carried out, the probabilities predicted by the square of the psi function and the measured probabilities are in agreement. The mathematical expression achieves, in this way, a statistical verification in experiment.

On the basis of this wave equation, a marvelous synthesis of known facts emerged, and many predictions could be made that were later verified. In particular, the *intensities* of the spectral lines, about which the Bohr model said nothing, could be explained. The splitting of spectral lines by strong magnetic and electric fields surrounding the radiating atoms also were explained. Additional mathematical developments resulted in alternative methods of handling the wave equation of Schrödinger. Although still complicated, many of the mathematical manipulations became somewhat simplified although even more "abstract," that is, more divorced from a visualizable model. Today, all of these methods and devices constitute that large branch of physics called quantum mechanics.

From our point of view, the important thing is that the results achieved represent a significant departure from the approach of classical Newtonian physics. According to the older physics, if the energy and configuration of any system of particles is known, future configurations can at least theoretically be predicted with precision. In quantum mechanics we can never measure with precision the position of a "particle," having a specific energy. The most we can do is to indicate where it is "likely" to be, giving this likelihood a measure in terms of probability. The predictions, then, are fundamentally statistical in nature. Furthermore, unless quantum mechanics is superseded at some future time, this situation cannot be relieved even theoretically, and there consequently remains a degree of uncertainty attached to all statements regarding individual microscopic events.

This does not mean, however, that classical physics is no longer valid over any range. It is still just as effective a tool for handling macroscopic events where the number of particles is so tremendous that the statistical probabilities become, in effect, certainties. Nor does it imply that there are two independent branches of physics, one for the microscopic and another for macroscopic events. Quantum mechanics rules the roost as the most adequate general theory of physics from the electron to the motion of the planets. The more complicated equations of quantum mechanics turn into the simpler ones of classical physics whenever the quantum of action h is so small relative to other quantities that it can be disregarded.

We have discussed in this chapter the structure of the atom itself. The question of whether the elements of the atom, in turn, have internal structure is taken up in Chapter 21 on nuclear physics, where many of the concepts of quantum mechanics are applied.

7. QUESTIONS

1. If a small object is electrically neutral what is implied, if anything, about the number of protons, neutrons, and electrons of which it is constituted? If a wire is conducting a current does it contain more electrons than the same wire when it is not conducting a current?

2. Outline the experimental evidence and reasoning which led to the rejection of an atomic model in which mass is uniformly distributed throughout the atom. Do the same for a model in which the charges are uniformly distributed.

3. When an atom is "ionized" it carries either an excess or deficiency of negative charges. Is energy required to ionize an atom?

4. Why is the decay time of a radioactive substance measured by its half-life?

5. Why is it that in the Bohr model of the atom, gravitational forces are disregarded?

6. In what way would the classical picture of an electron traveling in an orbit differ from Bohr's model? Even Bohr was not altogether content with his atomic model, which was widely criticized because it contained *ad hoc* hypotheses. What is an *ad hoc* hypothesis? Give other examples from scientific and social theories.

7. What is the relationship between an atomic number and the number of electrons in an atom? Is an atom always neutral?

8. The Bohr atomic model is no longer used in its original form. However, many of the

general ideas have been carried over into modern quantum mechanics. The concept of the electron as a small particle circling about a central nucleus has been abandoned, but the concepts of energy levels, quantization, and frequency have been preserved. Although circular (or elliptical orbits) are no longer considered applicable, the constant π appears to be so closely associated with h, that a special symbol for $h/2\pi$, namely, $\hbar$ has been introduced. Can you suggest a reason for the appearance of π in the absence of such orbits? (*Hint:* Recall the discussion of periodicity and frequency.)

9. Alone, or in combination, elements seem to "seek" the lowest possible energy state of electron configuration. Thus, we "explain" atomic events. Discuss whether this type of "explanation" contains "anthropomorphic" aspects similar to those in Aristotle's explanation of gravity that each object "seeks" its proper place (or that "water seeks its own level" or "nature abhors a vacuum").

10. Why was the Bohr model so successful in predicting *discrete* spectral lines? (A *continuous* spectrum is given off by an incandescent solid, when many more modes of vibration of the molecules are possible.)

11. Explain why, in the Bohr atom the expression for potential energy, namely $-KZe^2/r$ is negative?

12. In terms of energy levels, when do you expect X-rays rather than visible light to be radiated from an atom?

13. Dark spectral lines are seen against the bright continuous spectrum of sunlight. These are called absorption or Fraunhofer lines. Can you think of an explanation of how they arise?

Problems

1. If a certain radioactive element decays in such a way that ⅞ of the original atoms have decayed in 15 minutes, what is its half-life?

2. Assume that a certain radioactive element decays into a single nonradioactive daughter element. Draw the decay curve of the radioactive element and the growth curve of its daughter.

3. Explain how you would calculate the minimum frequency of a photon absorbed by a hydrogen atom in its ground state. Make an *approximate* calculation. Is this photon in the visible range, or above, or below it?

4. Examine the basic units in which the right-hand side of Bohr's expression for the radius of a hydrogen atom $r = n^2h^2/4\pi^2Ke^2m$. (An analysis of this kind is called "dimensional analysis" and often yields insight into how quantities are related.) Be sure to notice that K is a "dimensional" constant and not a pure number. Does the equation for the radius "balance dimensionally?" Bohr could see that if the radius is to be related to e^2, m, and K, it would be necessary to introduce a quantity having the dimensions of energy times time. This suggested introducing h into the equation. Try for yourself various combinations to convince yourself that the radius r cannot be expressed in terms of e^2, m, and K only.

5. Another result of dimensional analysis suggestive and tantalizing to physicists arises out of the combination of equations (1) and (2) page 424. From these equations, the veloc-

ity of the ground state hydrogen electron ($n = 1, Z = 1$) is given by $v = 2\pi Ke^2/h$. If we compare this velocity with the velocity of light, show that the ratio $v/c = 2\pi Ke^2/hc$ to be equal to very nearly $1/137$ and "dimensionless," i.e., a pure number. Why e, h, and c are so combined into a universal constant is a puzzle to physicists, some of whom feel it has significance as to the basic structure of the universe. (Recall the analogously intriguing mathematical equation $e^{i\pi} + 1 = 0$ which combines the basic constants of mathematics into a single equation.)

An additional aspect of the ratio of v/c is that for heavy atoms, say for one with $Z = 90$, this represents a significantly large fraction. As we shall see, for such high velocities, relativity theory requires a correction in mass. Hence this is one reason that Bohr's semiclassical theory requires modification for heavier atoms.

6. If one expresses the Pauli exclusion principle in terms of Schrödinger's "wave function" it takes the following form: When any two electrons of the system are interchanged, the wave function changes sign. What happens to the wave function if two electrons are in the same state i.e., quantum numbers are identical?

7. Using the knowledge acquired from our discussion of the building-up principle and the periodic table, predict the behavior of atomic volume with respect to atomic number and draw an appropriate graph to represent your predictions. Justify on the basis of the modern theory of atomic structure.

8. Without looking ahead to the following chapter try to explain the chemical behavior of the inert gases and the alkali metals in terms of the building-up principle.

Recommended Reading

Theodore A. Ashford, *From Atoms to Stars,* Holt, Rinehart & Winston, New York, 1960, pp. 331–395. An excellent summary of both atomic structure and chemical behavior of the elements. For a detailed review of chemical reactions, this book is highly recommended. A fine group of problems is found at the end of each chapter.

Gerald Holton and Duane Roller, *Foundations of Modern Physical Science,* Addison-Wesley, Reading, Mass., 1958, Chapters 33–35. A good discussion of line emission spectra and their interpretation on the basis of the Bohr atom.

Harvey B. Lemon, *From Galileo to the Nuclear Age,* University of Chicago Press, Chicago, 1946, Chapters 28–34. A good elementary text for nonscientists.

Linus Pauling, *General Chemistry,* W. H. Freeman and Company, San Francisco, 1953. A good standard text of college grade.

J. Needham and W. Pagel, editors, *Background of Modern Science,* Macmillan, New York, 1938, pp. 61–74. Excerpts from Rutherford's famous lecture on the development of the theory of atomic structure.

George Gamow, The Exclusion Principle, Scientific American, July 1959. Page 74 is a lucid discussion by one of the country's outstanding physicists and popularizers of science.

chapter nineteen

Modern chemistry

The modern chemist's extraordinary success in synthesizing new compounds and predicting their properties in advance is based on knowledge of the internal structure of the atom and application of quantum principles. Although chemistry during the 19th century made remarkable progress, it remained largely an empirical and classificatory science, and the mechanism of chemical combination was still shrouded in mystery. True, Dalton's early and crude model had been abandoned, and by the end of the century a conviction had developed that electric interactions must play an important role in chemical combination. But the chemist still depended—almost exclusively—on empirical knowledge. Prediction lagged owing to the lack of a satisfactory model. The chemical bond remained a puzzle, with valence theories offering only a rather fuzzy concept of its true nature. By contrast, today the chemical bond is understood in principle: the problems remaining are due to limitations imposed by mathematical methods used to extend the theory to include the results of experimental observation. Perhaps the most significant discovery to come out of the quantum theory was the discovery that the force responsible for *all* chemical phenomena can be explained in terms of the electromagnetic force between charged particles. Out of this insight the nature of the forces binding atoms into molecules became clear, and the door was opened to exciting new studies such as molecular biology.

1. CHEMICAL BONDING: ELECTROVALENT OR IONIC BONDS

Recall that the Pauli Exclusion Principle implies a "shell" structure to the electron configuration surrounding the nucleus of the atom. For any given energy level, or shell, each electron must have a unique combination of quantum numbers or states. In the case of the helium atom at the lowest energy level, the shell is "full" when it contains two electrons, because the only available quantum states are spin $+\frac{1}{2}$ and spin $-\frac{1}{2}$. For heavier elements the outer shells tend to fill until they have an "octet" of electrons. With eight electrons in the outer shell, the available quantum states are exhausted. The cloud of electrons is tightly bound, the element chemically stable and almost unreactive. The inert gases have this electron configuration.

We have to remember that were it not for the Pauli Exclusion Principle, chemical properties would be far different from what we find. For example, in the case of lithium with three electrons, the third electron would drop into the lowest energy level and the three electrons would be even more closely bound than the two electrons of helium. Similarly with other elements, if electrons could be added without limit to the lowest energy level, the periodic character and the striking variety of chemical elements would be lost. One element would be only slightly different from the next.

Let us examine again the Periodic Table, page 432. Consider as examples the elements in groups 1a, 2a, 6a, and 7a. These elements could achieve a full-shell configuration by the loss or gain of one or two electrons. For example, the alkali metals such as sodium have an electron outside a closed shell, while the halogens (e.g., chlorine) need an electron to reach the octet configuration. Atoms such as chlorine and oxygen will readily attach any free electrons which are present, and electrons which fill the outermost shell actually are bound to the atom just as the other electrons. The binding energy of the extra electron is called the *electron affinity* of the atoms. On the other hand, the metals near the left hand edge of the periodic table have very low ionization potentials, and thus can lose an electron relatively easily in order to attain the closed-shell configuration.

A closely allied concept is that of *electronegativity:* Essentially it is the relative power of an atom *within* a molecule to attract electrons to itself. Electronegativity is more than a qualitative idea and can be expressed quantitatively in electron volts or other units of energy. In fact a table of electronegativity which includes most elements has been available for many years. In general, the elements toward the upper right-hand corner of the periodic table are the most electro-

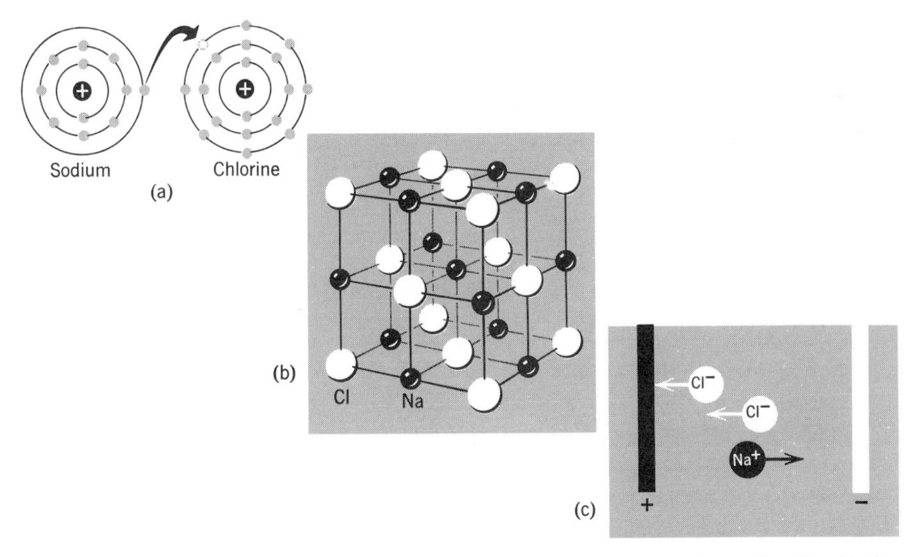

Fig. 19.1. (a) Electrovalent bond. The outer electron of the sodium atom is donated to the chlorine atom, completing the outer shells of both. (b) Crystalline structure of NaCl. (c) Electrolytic solution.

negative. Flourine has the largest electronegativity, followed by oxygen, nitrogen, and chlorine. Electronegativity and electron affinity, though not the same, are closely related.

The atoms characterized by loss or gain of electrons, and thus bearing a net electric charge, are called *ions* (from the Greek word for "wanderer"), and are symbolized by Na^+, Ca^{++}, Cl^-, O^{--}, etc., where + means "has one electron to donate", −− means "has vacancies for two electrons." Bonds characterized by donation from the metal of electrons to the nonmetal, as in NaCl, are said to be *electrovalent* or ionic bonds. Compounds of this type usually exist in *crystalline form* with the ions held rigidly in place by electrostatic forces. However, when dissolved in a suitable solvent such as water, the ions are no longer fixed in position and can *wander.* This behavior is illustrated in Fig. 19.1 which shows a sodium atom donating an electron to a chlorine atom. We thus have the fundamental explanation of Faraday's laws of electrolysis in which it is assumed that each ion carries one or more electronic charges.

2. CHEMICAL BONDING: THE COVALENT BOND

Not all compounds are held together by ionic bonds. In fact, most are characterized by *covalent bonds* in which electrons are *shared* by two atoms rather than transferred from one atom to the other. Nonmetals and semimetals, i.e., elements in the center and toward the right hand side of the periodic table, tend to form covalent bonds, in contrast to the metals which tend to form ionic bonds. In general, the formation of ionic bonds requires large differences in electronegativity among the constituent elements, whereas covalent bonds do not.

The covalent bond was first proposed in 1916 by the American chemist G. N. Lewis who originated the "dot" notation for molecular structure. As an example, we write the dot formula for a hydrogen molecule as

$$H:H$$

which indicates that the hydrogen atoms complete their outer shells by sharing their two electrons. Because Lewis' notation is somewhat cumbersome, it is no longer used. Instead we replace the double dot : by a dash —. The chlorine molecule whose atoms seek the octet configuration is thus denoted by

$$Cl—Cl$$

while the structural formula for the methane molecule (CH_4) is written

```
    H
    |
H—C—H
    |
    H
```

indicating that carbon attains the octet configuration by bonding to four hydrogen atoms. The fact that carbon can share four of its electrons endows it with very peculiar bonding properties which permits the vast number and complexity of organic compounds which we will discuss later.

Although Lewis' work was an important milestone in the development of bond theory, it did not lead directly to understanding the mechanism. In fact, all attempts prior to the introduction of quantum mechanics were fruitless because the Bohr theory of atomic structure was inadequate. The first success—the modern theory of covalent bonding really dates from this achievement—was the theory of the hydrogen molecule proposed in 1927 by Heitler and London. This was a true quantum theory in which the Pauli Exclusion Principle and the spin orientation of the electrons played central roles. In this theory, the two hydrogen atoms were assumed to share their two electrons, and the binding energy i.e., the energy of the electrons in the field of the two protons *less* the binding energy for two free hydrogen atoms, was computed for two cases: electron spins parallel and electron spins opposed. The results showed that only if the spins are *opposed* ("paired") are the atoms bound together. If the spins are *parallel,* they repel each other. This was an extremely important result, which can be generalized as follows: In all covalent bonds the spins of the electrons must be oppositely oriented. It should be emphasized that this effect is not due to any interaction between the spins themselves, but to the underlying Pauli Exclusion Principle.

It will be recalled from the preceding chapter that the square of the wave (or *psi*) function representing the electrons in a system was interpreted as the probability per unit volume (probability density) of finding an electron at a given point. Heitler and London showed that for the case of molecular binding, the probability density appears as in Fig. 19.2a; the nonbonding (repulsive) case is shown in Fig. 19.2b. Note that in (a) there is a large probability of finding the electrons on the internuclear axis. In (b), on the other hand, the electron probability densities tend to be concentrated in spherical "blobs" about each nucleus. The repulsive nature of the intermolecular force is somewhat analogous to the force acting between two rubber balls when they collide. When the "blobs" are deformed by the collision, they try to return to their initial shape by repelling each other. The two cases are really direct consequences of the Pauli Exclusion Principle which permits only case (a) when spins are opposed and case (b) when they are parallel.

It should not be inferred from the foregoing that molecular bonds are always either ionic or covalent. It is quite common, in fact, for a bond to be covalent part of the time and ionic the rest of the time. The relative amounts of time in each are such that the total *bond energy* achieves its maximum possible value. As one would suspect, electronegativity plays an important role in this respect: The greater the disparity in electronegativities of the constituents, the more dominant is the ionic character. In bond mixtures the one which is dominant is used to classify the bond type. (There is also a third type, the metallic bond, but it will not be considered here.)

It frequently happens that two atoms can share four or even six electrons under the proper conditions. A good example of a *double bond* is carbon monoxide for which we write the formula

$$C{=}O$$

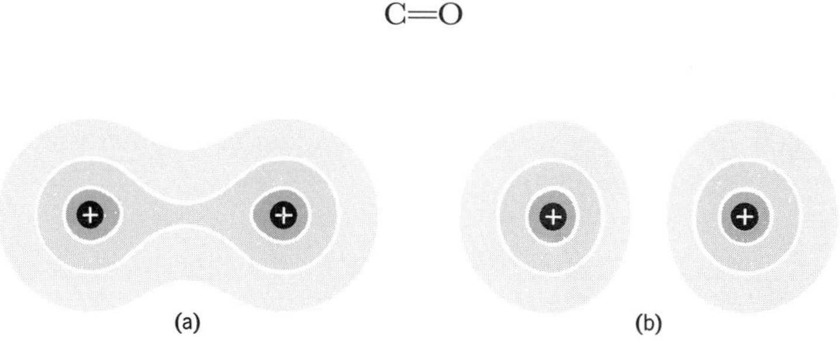

Fig. 19.2. (a) Electron probability density in the hydrogen molecule. (b) Electron probability density about two hydrogen atoms for the case of parallel spin.

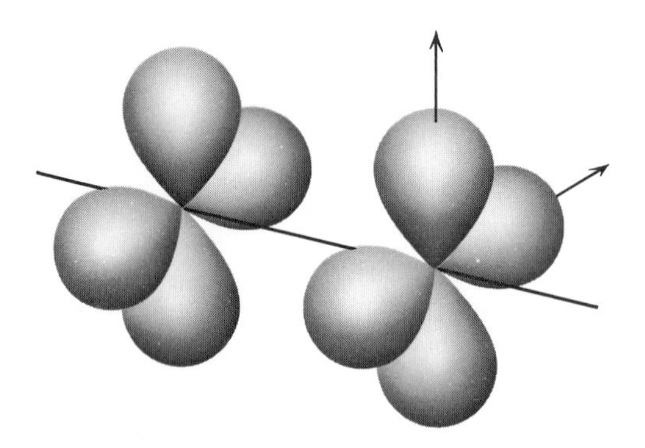

Fig. 19.3. The "Pi" bonds. Opposing loops interlock.

Actually this molecule oscillates or *resonates* among the double bond type, the *single bond* type

$$^{+}C—O^{-}$$

and the *triple bond* type

$$^{-}C{\equiv}O^{+}$$

with the double bond dominating. It is possible, by use of diagrams of the computed electron probability density in different directions relative to the internucleus axis, to give pictorial representations of multiple bonds. The single bond (called a "sigma" bond) appears as in Fig. 19.2a, that is, two nearly spherical clouds which overlap midway between the nuclei. On the other hand, the second and third types (called "pi" bonds) are represented by the overlapping of electron probability density clouds whose axes are perpendicular to the internuclear axis. In each bond shown it should be understood that electron spins are paired. With the aid of quantum mechanics it is possible to show that the mutually perpendicular clouds shown in Fig. 19.3 together with the spherical clouds of Fig. 19.2a are the only ones possible. Hence one does not ordinarily find chemical bonds of order greater than three. Interestingly, the total bond energy is greater than the sum of two simple sigma bonds because of the great strength and stability of the pi bonds.

We are now in a position to discuss the bonding of organic compounds and the nature of chemical reactions.

3. ORGANIC CHEMISTRY

The field of organic chemistry embraces a larger number of compounds than any other branch. Not only are they more numerous, but they are also more complex. The so-called macromolecules such as DNA contain millions of atoms.

Their great complexity is due mainly to the bonding peculiarities of carbon and, to a lesser extent, hydrogen. At one time it was thought that organic compounds could be produced only in living tissue, but in 1828 the German chemist Friedrich Wöhler disproved this idea by synthesizing urea from ammonium acetate and potassium cyanate.

Before discussing the types of organic compounds it is useful to consider the *shapes* of molecules, particularly the shape of the simplest hydrocarbon, methane, which contains one carbon and four hydrogen atoms. It has been known for many years that it has the regular tetrahedral shape shown in Fig. 19.4. This shape was "guessed at" sixty years ago on the basis of indirect evidence and has been confirmed in more recent times by X-ray diffraction experiments. Although the diagram in Fig. 19.4 is the accurate way of portraying the molecule, it does not lend itself to two-dimensional representation. Hence, we simplify it to the notation adopted previously

$$\begin{array}{ccccc} & & H & & \\ & & | & & \\ H & — & C & — & H \\ & & | & & \\ & & H & & \end{array}$$

The possibility of joining together two of these molecules (with the simultaneous loss of an H atom from each CH_4) in the form (ethane)

$$\begin{array}{ccccccc} & & H & & H & & \\ & & | & & | & & \\ H & — & C & — & C & — & H \\ & & | & & | & & \\ & & H & & H & & \end{array}$$

has probably already occurred to the reader. In fact such *chains* which constitute

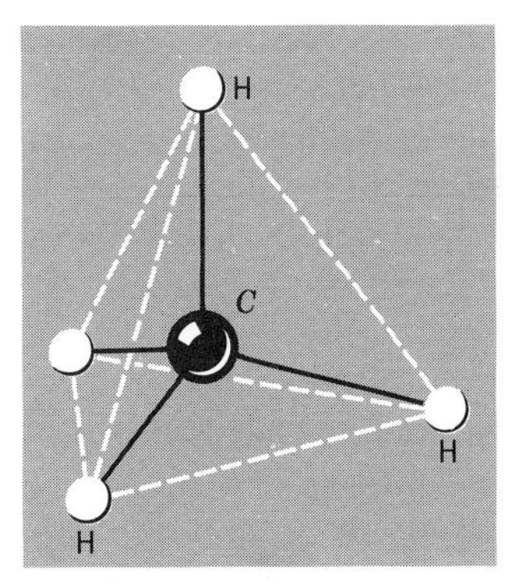

Fig. 19.4. Structure of the methane molecule.

the paraffin family can be very long; for example,

$$\begin{array}{ccccccccccccccccccc}
 & & H & & H & & H & & H & & H & & H & & H & & H & & \\
 & & \mid & & \mid & & \mid & & \mid & & \mid & & \mid & & \mid & & \mid & & \\
H & — & C & — & C & — & C & — & C & — & C & — & C & — & C & — & C & — & H \\
 & & \mid & & \mid & & \mid & & \mid & & \mid & & \mid & & \mid & & \mid & & \\
 & & H & & H & & H & & H & & H & & H & & H & & H & &
\end{array}$$

represents a molecule of the liquid hydrocarbon octane. Incidentally, as the length of the chain increases, the boiling point rises, presumably because the intermolecular forces are stronger for the larger molecules. Hydrocarbon molecules in which the carbon atoms participate only in single bonds hold the largest possible number of hydrogen atoms or other atoms, e.g., methyl chloride

$$\begin{array}{ccccc}
 & & H & & \\
 & & \mid & & \\
H & — & C & — & Cl \\
 & & \mid & & \\
 & & H & &
\end{array}$$

For this reason these compounds are said to be *saturated.*

The carbon atoms can also be doubly and triply bonded to each other. For example,

$$\begin{array}{ccccc}
H & & & & H \\
 & \diagdown & & \diagup & \\
 & & C{=}C & & \\
 & \diagup & & \diagdown & \\
H & & & & H
\end{array}$$

represents ethylene, and

$$H—C{\equiv}C—H,$$

represents acetylene. Obviously both are *unsaturated* hydrocarbons. Because of the nature of the double and triple bonds, illustrated in Fig. 19.3, both of these molecules are resistant to twisting about the carbon-carbon axis. Ethylene, incidentally, is only the first member of a chain family which is quite similar to the paraffins; the next member, propylene, is denoted by

$$\begin{array}{ccccccccc}
H & & & & H & & H & & \\
 & \diagdown & & & \mid & & \mid & & \\
 & & C & = & C & — & C & — & H \\
 & \diagup & & & & & \mid & & \\
H & & & & & & H & &
\end{array}$$

The compounds discussed above, whether saturated or unsaturated, belong to a class called *aliphatic* hydrocarbons. Another general class, called the *aromatic* hydrocarbons, is characterized by a ring structure in which six carbon atoms are linked in an hexagonal pattern

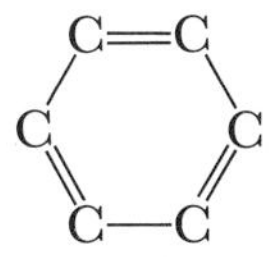

```
   C═C
  /   \
 C     C
  \\  //
   C—C
```

The simplest of these is benzene, each molecule of which contains a single ring

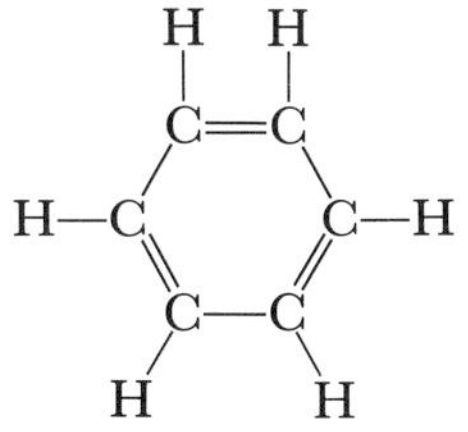

```
      H  H
      |  |
      C═C
     /   \
 H—C      C—H
    \\   //
      C—C
     /   \
    H     H
```

These structures may be linked together to form bigger compounds just as in the paraffin family. The next heavier aromatic compound is naphthalene.

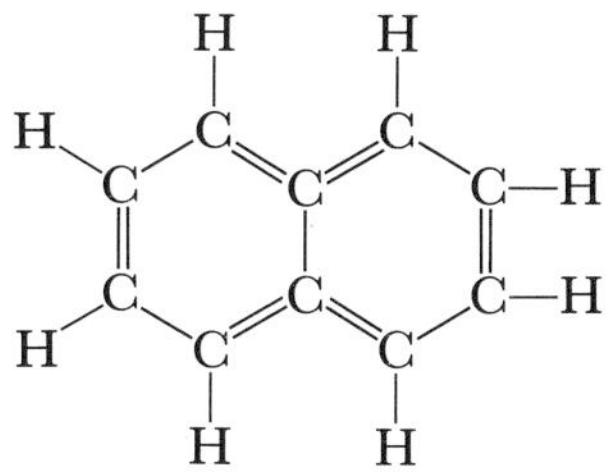

```
       H      H
       |      |
 H     C      C
  \  /   \\ //  \
   C       C      C—H
   ||      |      ||
   C       C      C—H
  /  \\   / \\   /
 H     C      C
       |      |
       H      H
```

This ring structure was first proposed in 1858 by Friedrich Kekulé, one of the architects of structural chemistry. Like many creative concepts, it was borne by fantasy, in this case by mental pictures of snake-like chains which "swallowed their tails," rather than by any conscious endeavor. See p. 172.

Although the basic ring structure of the aromatic compounds has been known for many years, the explanation of their comparative inertness awaited the development of quantum chemistry. We cannot, of course, go into the details of the theory but merely present some plausability arguments. Consider the basic ring structure

```
   C═C
  /   \
 C     C
  \\  //
   C—C
```

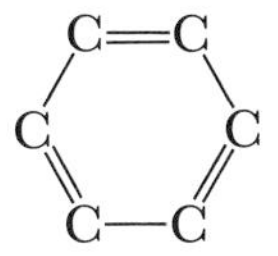

in which alternate carbon atoms are doubly bonded. Since the ring is hexagonally symmetric, there is no reason why the structure cannot be

```
    C—C
   //   \\
  C       C
   \     /
    C==C
```

In fact, both are equally probable and the resultant structure is a superposition of them. Now suppose we try to calculate the bond energy for the ring. We find that it is a *maximum if the ring spends an equal amount of time in each pattern.* This condition, which is a consequence of the quantum mechanical behavior of the molecule, is called *resonance.* It is responsible for the great stability (inertness) of the aromatic compounds.

It may already have been noticed that the chain and ring structures can branch in different ways when they grow large enough. For example, butane ordinarily takes the form

```
    H  H  H  H
    |  |  |  |
 H—C—C—C—C—H
    |  |  |  |
    H  H  H  H
```

but there is no reason why it cannot branch in the following manner

```
     H       H       H
     |       |       |
  H—C———C———C—H
     |       |       |
     H  H—C—H  H
             |
             H
```

In fact it can do so and we call the second form *isobutane* as opposed to the *normal* variety which exists in a simple chain. We call these two patterns *isomers* of butane. Isomers [from the Greek *iso* (equal) + *mer* (part)] are extremely abundant in organic chemistry and can occur with respect to the aromatic rings as well. All of these examples are structural isomers, each of which is distinguished by the way in which the chain structure branches. A more subtle type is composed of the *stereo*-isomers in which the branching is the same for all members, but they are distinguished by rotational or reflection considerations. For example *cis* and *trans* butene have the representations:

```
H         H                       H
 \       /                        |
  C==C                     H      H—C—H
 /       \                  \    /
H—C—H H—C—H                 C==C
  |       |                /    \
  H       H            H—C—H     H
                         |
                         H
   (cis)                    (trans)
```

which are distinguished by rotation through 180° about the C═C axis (remember that double bonds resist twisting). The isomers which are distinguished by their different behaviors under reflection of the coordinate axes are more difficult to portray; in fact it is impossible to do so in two dimensions. We simply mention as an example that some complex molecules have helical shapes; the senses are those of either left-handed or right-handed screws. Hence they are distinguished in their behavior under reflection.

So far we have confined the discussion to compounds of carbon and hydrogen. Actually, atoms of other elements as well as other radical groups can be substituted for hydrogen under appropriate conditions. Perhaps the best known of these *derivatives* are the *alcohols* which are characterized by substitution of an OH (hydroxyl) radical for a hydrogen atom. The simplest of these, methyl alcohol, is represented by

```
    H
    |
H—C—OH
    |
    H
```

while the next heavier (ethyl alcohol) is a derivative of ethane

```
    H  H
    |  |
H—C—C—OH
    |  |
    H  H
```

The hydroxyl radical can also be substituted for hydrogen in the benzene ring structures, for which phenol

```
        OH
        |
H       C       H
 \    /   \\   /
   C         C
H  ||        |  H
 \ C         C /
     \     //
        C
        |
        H
```

is the simplest. Nitrogen, oxygen, and sulphur are particularly important as substituents for hydrogen. For example, the important class of *amino acids* is exemplified by

$$\begin{array}{ccccccc} H & & & H & & & O \\ & \diagdown & & | & & /\!/ & \\ & & N & - C - & C & & \\ & \diagup & & | & & \diagdown & \\ H & & & H & & & OH \end{array}$$

Basically, this is a methyl molecule (center group) with opposing hydrogen atoms replaced by the *amine* radical $\begin{array}{ccc} H & & \\ & \diagdown & \\ & & N \\ & \diagup & \\ H & & \end{array}$ and the carboxyl radical $\begin{array}{ccc} & & O \\ & /\!/ & \\ C & & \\ & \diagdown & \\ & & OH \end{array}$

Amino acids are very important in biological processes because of their relationship to proteins. For example, the latter must be broken down into amino acids by body chemical action in order to be digested. It should be obvious by now that the chemistry of hydrocarbon derivatives is very extensive and complex, embracing an incredibly large number of compounds. Having outlined the general way in which these molecules are formed, we refer the interested reader to one of the references listed at the end of the chapter for more details.

A final topic in organic chemistry that we wish to introduce concerns giant molecules or *polymers.* Such structures are based on relatively simple hydrocarbon molecules such as ethylene. Two of these *monomers* can be joined together to yield the *dimer* butylene

$$\begin{array}{ccccccccc} H & & & H & & H & & H & \\ & \diagdown & & | & & | & & | & \\ & & C = & C & - & C & - & C & - H \\ & \diagup & & & & | & & | & \\ H & & & & & H & & H & \end{array}$$

A third can then be added to the dimer to yield a *trimer,* and so on. When the chains become indefinitely long, they are called *polymers,* the prefix "poly" simply standing for "many." One of the well-known polymers is called polyethylene and has a wide range of industrial, commercial, and home uses. For example, most of the plastic "bathtub" toys which are no doubt familiar to the reader from his childhood days are made of this material.

There are many other polymers of comparable or even greater importance in our highly technological civilization, but we shall mention only one before closing. It occurs naturally in the form of a viscous liquid which is commonly called "natural rubber." Its basis is the unsaturated monomer *isoprene*

```
              H
              |
       H   H—C—H        H
        \     |        /
         C=C—C=C
        /         \
       H          H   H
```

molecules of which are linked to each other in a long chain to yield *polyisoprene.*

```
  H       H  H  H      H  H  H      H
  |       |  |  |      |  |  |      |
—C—C=C—C—C—C=C—C—C—C=C—
  |   |      |  |   |      |  |   |
  H   C      H  H   C      H  H   C
     /|\           /|\           /|\
    H H H         H H H         H H H
```

The *vulcanization* process that leads to the formation of rubber crosslinks these chains by means of sulphur atoms

```
           H            H            H
          H|H          H|H          H|H
           \|/          \|/          \|/
   H  H  H  C  H  H  H  C  H  H  H  C
   |  |  |  |  |  |  |  |  |  |  |  |
 =C—C—C—C—C—C—C—C=C—C—C—C—
      |  |  |  |  |  |        |  |  |
      H  H  |  H  H  H        H  H  |
            |                       |
A ----------+-----------------------+---------- A′
            S                       S
            |          H            |
            |         H|H           |
            |          \|/          |
      H  H  |  H  H  H  C     H  H  |
      |  |  |  |  |  |  |     |  |  |
 =C—C—C—C—C—C—C—C=C—C—C—C—
   |  |  |  |  |  |  |     |  |  |  |
   H  H  H  C  H  H  H     H  H  H  C
           /|\                     /|\
          H H H                   H H H
```

It should be remembered that we are really dealing with three-dimensional arrays, and that the apparent planar character shown is illusory. The familiar mechanical properties of rubber, an elastic, highly deformable solid with considerable strength under tension, compression, and shear are consequences of the crosslinking of long chains of polyisoprene. For example, twisting of the section of polymer represented by the above array about some axis such as AA' will slightly deform the bonds. In the deformed state the total bond energy will be a bit smaller than normal and the polymer will try to return to the unstressed condition in which the bond energy is largest. A similar consideration holds for stretching and compression.

4. CHEMICAL KINETICS

We found in Chapter 12 that the law of conservation of mass was useful in quantitative studies of chemical reactions. Another conservation law which must be satisfied in every chemical process is the law of conservation of energy, i.e., the amount of energy present after the reaction occurs is the same as before. Such energy can take the familiar form of heat energy or it can manifest itself as the bond energy which was discussed in Section 2. For example, the bond energy of an oxygen molecule is 5.13 electron volts. (The conversion of this value to kilocalories per mole is left as a supplementary exercise for the student.)

If the *product* molecules are more tightly bound than the molecules of the *reactants*, energy is released by the reaction, which is then said to be *exothermic*. On the other hand, if the molecules of the reactants are the more tightly bound, energy must be added to the system to cause the reaction, and it is considered *endothermic*. The mere fact that a given reaction is exothermic usually is not sufficient to cause it to start spontaneously. Frequently energy must be added (e.g., a spark in a mixture of hydrogen and oxygen) to initiate the process. This is called the activation energy.

Ever since the general acceptance of the kinetic theory of gases, it has been realized that chemical reaction rates must depend upon the rate of collisions between atoms or molecules of the reactants. For example, in the reaction

$$H_2 + Cl_2 \longrightarrow 2HCl$$

the reaction rate depends upon the rate of collisions or the *collision frequency* between molecules of hydrogen and molecules of chlorine. In the case of reactions such as

$$2H_2 + O_2 \longrightarrow 2H_2O$$

which apparently involve collisions among three molecules simultaneously, the situation is more complicated. Two of them, for example, may remain associated for some time after their initial impact but, before they dissociate, they may collide with the third molecule, allowing the reaction to occur. Even though the molecules did not come together simultaneously, we still consider it to be a three-body collision.

While the collision mechanism had been proposed in the mid-nineteenth century, there was one segment of experimental evidence which could not be explained. In many reactions when the temperature was raised *slightly* the reaction rate increased enormously. As an example, suppose that the rate increases tenfold when the temperature is raised from 300 to 330°K. From the results of Chapter 11 we learn that the mean velocity of the molecules increases by approximately 3 per cent when the temperature is raised 10 per cent. Since the mean collision frequency is proportional to the mean molecular velocity, we would expect the reaction rate to rise by about 3 per cent instead of the observed 1000 per cent. The Swedish chemist Arrhenius proposed that although a given reaction may be exothermic, there must be some minimum amount of molecular kinetic energy present to cause it to occur. This activation energy is best explained by the analogy presented in Fig. 19.5. The gravitational potential energy actually converted into kinetic energy is represented by the vertical distance h. However, the ball must have an initial energy mga in order to carry it over the hill of height a. The energy mga is the activation energy for the system. It is not involved in the exothermo or endothermic nature of the reaction itself but only in its initiation. The concept of activation energy proved to be of enormous importance to chemical kinetics.

We have mentioned that the reaction rate is proportional to the collision frequency. Why should this be so? Consider a chemical system consisting of the two gases, H_2 and Cl_2. A reaction occurs when a hydrogen molecule collides with a chlorine molecule under the proper conditions. On the basis of this statement it is intuitively obvious that the rate at which reaction occurs must be proportional to the rate at which collisions occur. (More rigorous support could be given to this argument but is not necessary for our purposes.) At a given temperature, the collision frequency, and thus the reaction rate, should be proportional to the *concentrations* of the reactants, i.e., to the number of molecules per unit volume present. We can express the reaction rate R (i.e., the rate of formation of reaction products) for two-body or *bimolecular* ractions as

$$R = k_{12}n_1n_2$$

where n_1 and n_2 are the concentrations of the reactants and k_{12} is the associated rate constant. In the cgs system n_1 and n_2 would be expressed in cm^{-3}, k_{12} in $cm^3\ sec^{-1}$, and R in $cm^{-3}\ sec^{-1}$.

In trimolecular and higher-order reactions such as

$$xX + yY + zZ \longrightarrow aA + bB + cC$$

where the capital letters refer to the reactants or products and the lower case

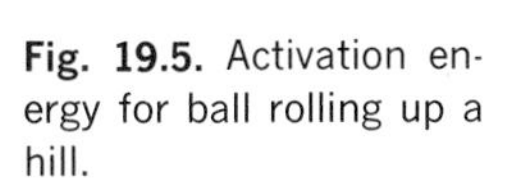
Fig. 19.5. Activation energy for ball rolling up a hill.

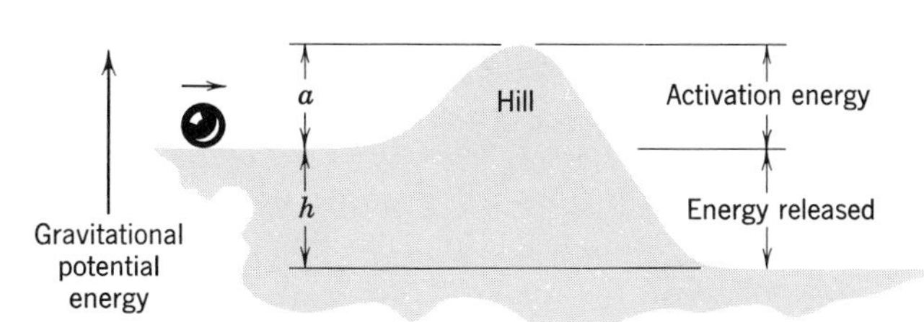

letters to the numbers of molecules involved in *one* reaction, we find that the reaction rate is expressed by

$$R = kn_X{}^x n_Y{}^y n_Z{}^z$$

(The superscripts in this case are exponents of the concentrations.) This expression can be particularized to any reaction simply by specifying the reaction rate coefficient k and the exponents x, y, z.

Under the proper physical conditions (e.g., temperature and concentrations) many chemical reactions are *reversible*. For example, the ammonia produced in the reaction

$$3H_2 + N_2 \longrightarrow 2NH_3$$

can be decomposed by the reverse reaction

$$2NH_3 \longrightarrow 3H_2 + N_2$$

Each one has a characteristic reaction rate. In a given system maintained at the same pressure and total mass, the reaction will go in the forward direction ($\longrightarrow$) until the rate becomes equal to the rate in the reverse ($\longleftarrow$) direction. At this point we have a condition of *dynamic* equilibrium, i.e., no *net* change, although reactions are continually occurring. As an example, let us consider equilibrium in the reversible reaction

$$A + B \rightleftarrows C + D$$

for which the forward rate is

$$R_{\rightarrow} = k_{AB} n_A n_B$$

and the reverse rate is

$$R_{\leftarrow} = k_{CD} n_C n_B$$

At equilibrium $R_{\rightarrow} = R_{\leftarrow}$ and

$$k_{AB} n_A n_B = k_{CD} n_C n_D$$

or, rearranging,

$$n_C n_D / n_A n_B = k_{AB}/k_{CD} = K$$

where K is the *equilibrium constant* for the reaction at a given temperature. Because of the temperature dependence of k_{AB} and k_{CD}, K will also be a function of temperature.

One of the most important questions in chemical kinetics concerns the behavior of a system in equilibrium when properties such as concentration, temperature, and pressure are changed. First consider the effect of increasing the concentration of species A in the above reaction. According to equation 1 the equilibrium constant K will remain constant if the concentrations of C and D increase and the concentrations of species A and B decrease. Hence we find that the increase of reactants A or B shifts the equilibrium point to the right, i.e., the concentration of products C and D is increased. On the other hand, if n_C or n_D is increased, the equilibrium point is shifted to the left. Suppose that the temperature of the system is now raised. If the system is endothermic, this will have the effect of increasing the rate constant k_{AB} for the forward reaction more than that (k_{CD}) for the reverse reaction. The result will be an increase in the equilibrium constant K and an equilibrium shift toward the right. By contrast, an increase of temperature in an exothermic reaction will cause the equilibrium point to shift toward the left.

These are examples of a principle discovered by the French chemist Le Chatelier in 1888. It is more general than we have indicated in the examples above and can be briefly stated: When a chemical system in equilibrium is disturbed, the equilibrium point is shifted in such a way that the system tends to return itself to its original condition. It can be shown that the principle is a consequence of the second law of thermodynamics.

5. ELECTROCHEMISTRY

Everyone is familiar with the existence of commonly occurring solutions such as table salt in water, iodine in alcohol, etc. It is important to clearly distinguish such solutions from other equally familiar liquids which are called colloids.

A colloid is a suspension of finely divided solid material in a fluid (gas or liquid). Typical examples are milk which consists of solid particles suspended in a liquid (water), and smoke which consists of solid particles suspended in a gas (air). Obviously colloids are highly inhomogeneous mixtures. A solution, on the other hand, is a homogeneous mixture of two or more chemical compounds. Usually we think of a solid *solute* dissolved in a liquid *solvent* (e.g., table salt in water), but we may have solids dissolved in solids (e.g., some types of metallic alloys), liquids dissolved in liquids, or gases dissolved in liquids. (In general, the compound present in least concentration is called the solute, while that present in largest concentration is the solvent.)

One of the most striking characteristics of any solution is its *polar* or *nonpolar* nature. A polar compound is one in which the molecules are small electric dipoles ("rods" with opposite electric charges at the ends). Because of asymmetry about some axis, they have their internal positive and negative charges displaced slightly. Thus they may be considered to have oppositely charged ends. A typical example, water, has the configuration shown in Fig. 19.6; note the asymmetry about the axis AA. If the molecule should possess symmetry about all of its main axes, its internal charge is equally distributed, and it does not behave like a

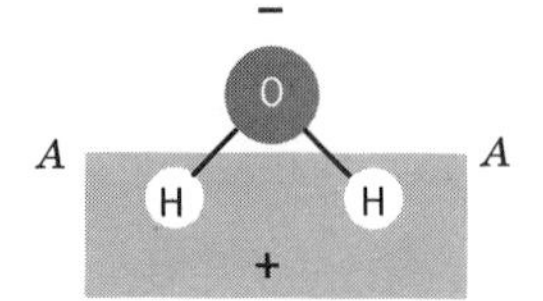

Fig. 19.6. Schematic representation of the water molecule.

dipole. As a general rule, polar solvents dissolve polar solutes, and nonpolar solvents dissolve nonpolar solutes. There are important exceptions, however. For example, water will dissolve sugars, which are nonpolar compounds. There is a simple explanation for this particular case: The water molecules form a very loosely bound chemical compound with the sugar molecules (the *hydration* process).

The explanation for the characteristic that solute and solvent are of the same type for high solubility is also quite simple. Because of strong intermolecular forces which act between polar molecules, they will form a conglomerate which excludes the weakly interacting nonpolar molecules. If the solute and solvent are both strongly polar, their molecules can intermingle freely because of the near-equality of intermolecular forces. Here we have a case of high solubility. However, if the polarity of one is relatively large and the other small (e.g., alcohol in water), we have a condition of low solubility. The effects of the polar character are illustrated in Fig. 19.7. In example (a) the intermolecular forces are strong enough to allow the Na^+ and Cl^- ions to move freely relative to each other. No (+ −) ion pair is identifiable as a molecule. This type of solution, which is said to be *electrolytic,* is of very great importance and has already been discussed in Chapter 13.

Electrolytes are conveniently classified according to whether they are *strong* or *weak,* strong electrolytes being much better conductors of electric currents than weak ones. However, for weak electrolytes the *conductance per mole of solute* increases

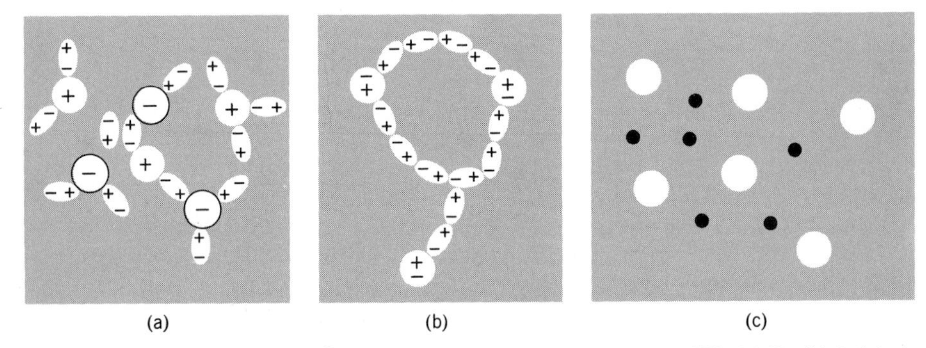

Fig. 19.7. (a) Solute is ionic ⊕⊖(Na^+, Cl^-); solvent is strongly polar (+−) (H_2O). (b) Solute is weakly polar (+−) (CH_3OH); solvent is strongly polar (+−) (H_2O). (c) Solute is nonpolar ● (CCl_4); solvent is nonpolar ○ (C_6H_6).

as its concentration decreases. For a given amount of solute, dilute solutions are much better conductors than concentrated ones. Strong electrolytic solutions also show increasing conductance per mole of solute as the concentration decreases, but the increase is not so fast as for weak electrolytes.

In 1887 Arrhenius developed a theory which satisfactorily accounted for the behavior of weak electrolytes. According to this theory, an electrolytic solute dissociates into positive and negative ions when in solution; this effect is caused by the electric polarization forces of the solvent, water. Furthermore, the dissociation process is a dynamic one. At equilibrium, a balance is struck between dissociation and recombination of the ions. Only a fraction of the solute is dissociated at a given time. Although this fraction indeed approaches unity for very great dilution of strong electrolytes, weak electrolytes are still only slightly dissociated at very great dilution. The assertion that ions are *always* present in an electrolytic solution was quite revolutionary and encountered strong opposition for some time. Faraday, the father of electrolytic chemistry, had assumed that ions existed only during the flow of electric currents.

Although the Arrhenius theory did explain the chemistry of weak electrolytes, it failed to do so for strong electrolytes. Increasing evidence for the complete dissociation of strong electrolytes was adduced which could not be reconciled with the behavior shown when the concentration changed. Debye and Hückel accounted for the peculiar behavior of strong electrolytes by postulating that an ion of one charge sign is surrounded, on the average, by a cloud of ions of the opposite charge. The latter act to shield the former from other ions of like charge. Furthermore, if a current is passing through the electrolyte, the ions of either sign (+ or −) are impeded in their transit by the effect of the surrounding charge cloud which is attracted to the opposite electrode. The effect of this "Debye shielding" is shown in Fig. 19.8.

The theory of Debye and Hückel certainly accounts for the behavior in solution of ionic compounds such as NaCl, but it does not explain that some compounds characterized by strong covalent bonds (such as HCl) are strongly dissociated in aqueous solution. Actually, the polar nature of the water molecule

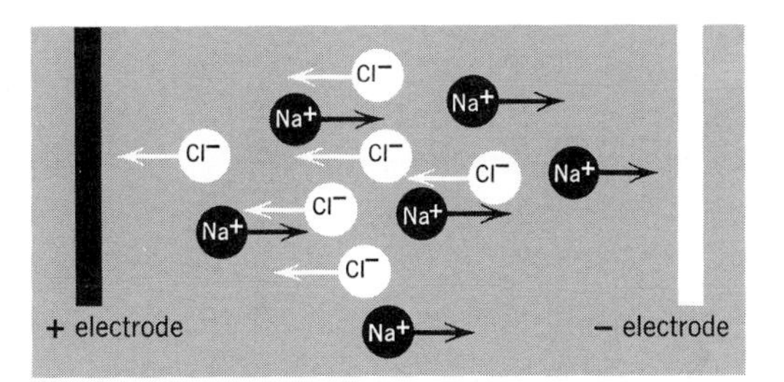

Fig. 19.8. Electrolysis of a strong electrolyte. Note that the Na^+ ions are, on the average, surrounded by a cloud of Cl^- ions, and vice versa. We do not need to include the effect of the polar force of the water molecule in this diagram.

suffices: The dipole forces exerted by the water molecule on the HCl molecules actually break the covalent bonds. As a result, HCl is strongly dissociated into H^+ and Cl^-.

Two of the most important classes of compounds are the *acids* and the *bases.* We introduce them here because they are best described in terms of their electrolytic properties. Briefly, an acid is any substance which can *donate* hydrogen ions (H^+), while a base is characterized by its ability to *accept* hydrogen ions in a chemical change. For example, when HNO_3 is dissolved in water, the reaction

$$HNO_3 + H_2O \longrightarrow H_3O^+ + NO_3^-$$

follows. Because water accepts an H^+ ion in the formation of H_3O^+ (hydronium ion), it is a base. On the other hand, HNO_3 donates an H^+ ion, so it is an acid (nitric acid). Similarly, a solution of NaOH (sodium hydroxide) in water yields the reaction

$$NaOH + H_2O \longrightarrow NaH_2O^+ + OH^-$$

Here, H_2O is the acid and NaOH is the base (Na^+ plays the role of an H^+ ion in this case). The terms "strong" and "weak" when applied to acids and bases have exactly the same meaning as for any other electrolyte. Strong acids or bases are highly dissociated while weak ones are only slightly so.

6. CHEMISTRY OF THE INERT GASES

Until very recently the inert gases were considered to be just that—completely inactive chemically. Then, in 1962, compounds of the gas xenon (atomic number 54) were produced in the laboratory for the first time. Since that time a number of compounds of xenon as well as a few of krypton have been created. The former have, for the most part, proved to be comparatively stable, but not the latter—formed at high temperature, they rapidly decompose into their constituent elements when cooled to room temperature.

The first compound of xenon was produced as a by-product in the study of platinum hexafluoride which, when reacted with oxygen, combined in a one-to-one ratio, that is, one molecule of oxygen joined by ionic bonding to one molecule of platinum hexafluoride

$$O_2^+(PtF_6)^-$$

The existence of an ionic bond in this molecule indicates that PtF_6 must have an extremely large electron affinity, at least 12.2 eV which is the ionization potential of O_2. Since this is larger than the ionization potential of xenon, it seems reasonable that the oxygen atom could be replaced by xenon to yield

$$Xe^+(PtF_6)^-$$

Indeed this proved to be the case, although the actual structure of the molecule is more complex than this simple formula indicates. Subsequent experimental investigations resulted in the production of three fluorides of xenon, XeF_2, XeF_4, and XeF_6, xenon trioxide, XeO_3, and two (unstable) fluorides of krypton, KrF_2, and KrF_4. Fluorine and oxygen, because of their very large electronegativities, are probably the only elements which will bond to xenon. Attempts have been made to substitute chlorine, but without success. Radon, the heaviest inert gas, may be expected to form compounds with chlorine because of its lower ionization potential compared with xenon. Unfortunately, radon's high degree of radioactivity makes it very difficult to employ in experiments of this type, but fluorides of radon have been created in the laboratory.

Work has also progressed on the nature of the bonding mechanism in the inert gases. Although the arguments in the foregoing paragraphs seem to imply ionic bonds, it now appears that they are covalent. For example, in the compound XeF_6 the xenon atom would share its six outermost electrons with six fluorine atoms. According to this scheme, the outer electron clouds of the xenon atom are configured somewhat similarly to those in Fig. 19.3 except that there are six lobes instead of four. Each of these interlocks with a cloud in each fluorine atom. The actual details involved are much more complex, but this picture suffices for a general qualitative understanding. Experimental and theoretical studies of inert gas chemistry are continuing, and probably additional compounds will be produced. However, it is doubtful that they will include the lighter elements, helium, neon, and argon.

7. SOLID STATE PHYSICS; LASERS

Based upon quantum principles, modern chemistry has developed a theoretical explanation of how and why atoms combine. The key idea is that of quantized energy levels. Two or more atoms will combine if thereby a lower energy level is possible than for these atoms dissociated. The so-called covalent bonding of hydrogen, for example, comes about when two hydrogen molecules are close enough together. As we have seen, the pair of electrons then form a single wave configuration patterned about the central protons. The resultant energy level is smaller than the total of the energy levels of two hydrogen separated (dissociated) atoms.

The "molecule" is, in other words, no longer thought of as two hydrogen atoms existing side by side. Extension of this basic idea has shed considerable light on the structure of solids. A whole new field of inquiry called "solid state physics" has flowered as a result, producing in over the last two decades dramatic applications such as transistors and lasers.

In solids, atoms commonly (though not always) join together in regular patterns to form crystals. These are not merely arrays of molecules held together by intermolecular forces. Rather, the individual molecule loses its identity, and each atom is bound to the others in a solid state. In other words, the whole crystal can be regarded as a single giant molecule.

One type of such crystalline structure is sodium chloride, where each ion of Na^+ and of Cl^- is situated at alternative corners of a cubic lattice. The ions are bound together by ionic or electrostatic forces. (See Fig. 17.5.) Such crystals are nonconducting—the electrons stick close to the ions and are not free to move about through the solid. In other cases, such as metals, the atoms of the solids are not bound together by electrostatic forces or ionic bonding, but instead are bound together by covalent bonding, somewhat similar to hydrogen bonding; thus, electrons are not merely shared by adjacent atoms, but rather by the whole group of atoms constituting the metal. There is a whole "sea" of electrons not specifically tied to particular atoms but more or less free to travel over the whole latticework of positive ions. Hence metals are good conductors. Electrons can be drawn out from one point of a metal and pumped into another point. A current then "flows" readily through the metal, losing only a small portion of the kinetic energy which is transferred to the vibrational energy of the metal ions, and results in heat loss of resistance.

Between these two cases, i.e., crystals like NaCl and the conducting solids, there are a number of solids called semi-conductors, in which the electrons are *neither* completely free to wander as in metals *nor* held tightly in fixed sites by the ions as in NaCl crystals. There are two types of semiconductors, both formed by introducing an impurity into a nonconducting solid. In the one case (negative or n-type), a large number of atoms with almost empty outer shells are in the matrix. Electrons in this case can easily "jump" from one atom to the next, carrying negative charges from one point of the crystal to another. In the other type of semiconductor (p-type) there are atoms with outer shells almost filled with electrons. If an electron does move from an outer shell, a "hole" is left which is equivalent to a positive charge. In this case it is convenient to think of the "holes" or positive charges as being in motion.

By combining n-type and p-type semiconductors in sandwiches made of thin slices of "doped" materials, transistors are built. They have a natural "bias," i.e., electrons pass more easily from the n-type to the adjacent p-type semiconductors than vice versa. Under the influence of a small potential difference a large current can be induced. In this way transistors can be used as valves and amplifiers in a manner similar to the diode and triode tubes previously discussed. The great advantages of transistors are that they can be miniaturized (it is difficult to engineer a small triode tube, whereas a sandwich of thin wafers of n-type and p-type semiconductors can be assembled) and that they take no driving power, as do triode tubes. (It will be recalled that for a vacuum tube to work as a value or an amplifier, electrons had to be "boiled off" by heating the cathode. The n-type conductor plays a role similar to the cathode and supplies electrons without heating.)

An equally promising development generated by solid state physicists is the laser, the first of which was designed in 1960. In Chapter 16 the laser was mentioned as the source of intense and coherent light which would not "spread" as does ordinary light. The explanation can now be given in terms of quantum "jumps" of electrons from one energy level to another.

Whenever an electron goes from a higher to a lower energy level, a photon of energy hf is emitted. There are usually many possible "excited" states of the atom above "ground level." Sometimes when an atom is in a highly excited unstable state, an electron will spontaneously drop to an intermediately lower energy level and remain at that level for a time. While staying at this *metastable* energy level, however, it is "ready to jump" to the ground state and this jump may be stimulated by a photon of the right frequency. In fact if a photon which matches the energy-level frequency of the metastable state of the atom strikes it, another photon of exactly the same frequency will be emitted.

The situation can be summarized in this way. An atom can be raised (or "optically pumped") into a highly excited state by absorbing a photon. It then falls into a metastable state where it will remain for a time, although eventually it will, by a random process, fall to the ground state. This fall can be triggered by a photon of the right frequency before the random process would cause it to fall. If enough atoms can be raised to the metastable state and their number predominates, stimulated emissions in turn may well predominate over the absorption of photons taking place in the whole collection of atoms.

The ruby laser (see Fig. 16.12 page 373) is one of the devices that takes advantage of this process. A ruby cylinder is ground optically flat on both ends. These ends are then silvered; one end forms a completely opaque mirror, the other end, being half-silvered, allows a beam of photons to pass through when the number striking it is large enough. The ruby is placed next to a "pumping lamp" which supplies an intense source of light, thus providing the energy to raise the atoms from ground state to the highest possible excited state. These atoms then fall into the metastable state ready to be stimulated into emission. Photons emitted in directions not parallel to the axis of the cylinder are wasted. But those emitted along the axis of the cylinder are reflected back and forth between the mirrors, causing a cascade of photons, all of exactly the same frequency and all in step. The number of photons increases vastly in about a microsecond and finally a short pulse of coherent light bursts forth through the half-silvered face of the ruby.

It should be emphasized that there is no *net* energy gain in the laser. The device merely concentrates the energy emitted into a small time interval and into a beam of very small cross section, a beam, furthermore, which does not spread the way ordinary light does. The net result is a beam which can reach an intensity of more than 5×10^7 watts per square centimeter. Since this occurs in about one microsecond, the total amount of energy is small. It is true that a laser beam can readily melt a hole in steel, because the power is so concentrated, but a laser beam could not, as has been sometimes loosely suggested, melt down, for example, an enemy space missile.

8. QUESTIONS

1. Would you expect the following reaction for recombination of hydrogen atoms

$$H + H \longrightarrow H_2$$

to have an activation energy? Justify your answer.

2. Le Chatelier's principle applies to any system in dynamic equilibrium, not just to chemical systems. On the basis of the principle, can you explain why the vapor pressure of water increases with increasing temperature?

3. Using the principle of Le Chatelier can you explain why the freezing point of water is depressed when salt is added to it?

4. In the chapter on the atmosphere we shall find that electrons and ions are removed in the upper ionosphere by the two-step process

$$O^+ + N_2 \longrightarrow NO^+ + N; \; NO^+ + e \longrightarrow N + O$$

of which the second is the faster. Can you explain why the first reaction controls the rate of the entire process?

5. Which of the following compounds would you expect to be electrovalent?

$CaCl_2$	CS_2
SO_2	Na_2CO_3HCl
$CuSO_4$	

6. How would you expect the electron affinity of the halogens to vary in going from the lighter to the heavier elements?

7. One very interesting chemical bond is that of the hydrogen molecule ion H_2^+. Can you explain the bond mechanism involved here? (Remember that there is only a single electron involved.)

8. Why are the two structures

```
                                  C
                                  |
                                  C
                                  |
C—C—C—C   and   C—C—C—C
  |  |               |
  C  C               C
  |
  C
```

equivalent to each other while the two isomers
are distinct?

```
   H           H                  H      H
                                         |
    \         /                   \    H—C—H
     C ═══ C          and           C ═══ C
    /         \                   /        \
H—C—H     H—C—H               H—C—H        H
  |         |                   |
  H         H                   H
```

Dichlorobenzene is a derivative of benzene which contains two chlorine atoms. Draw the possible configurations (isomers) of this compound.

Problems

1. The binding energy of the two hydrogen atoms in a hydrogen molecule is 104.2 kcal mole^{-1}, while that for the oxygen atoms in an O_2 molecule is 118.3 kcal mole^{-1}. If the total binding energy (energy required to separate H_2O into H + H + O) is 221.2 kcal mole^{-1}, what is the energy in kcal mole^{-1} released by the combustion of hydrogen?

2. Convert your answer to question (1) to electron volts/molecule.

3. Consider the reaction

$$O_2 + O^- \longrightarrow O_3^-$$

How is the rate of formation of the ozone ion O_3^- affected by (a) doubling the concentration of molecular oxygen? (b) Doubling the concentration of oxygen ions? (c) Raising the temperature 30° K?

4. At room temperature the equilibrium constant for the reaction

$$2\ NO_2 \rightleftharpoons N_2O_4$$

is K = 6.3. What are the relative concentrations of NO_2 and N_2O_4 present in the mixture at a given concentration of NO_2? What are the units of K?

5. Draw structural diagrams for the following compounds: SO_3, NH_3, SiF_4, CO_2.
6. Draw a structural diagram of carbon tetrachloride.
7. Explain in detail why the double bond is resistant to twisting about the interatomic axis.

Recommended Reading

Isaac Asimov, *The World of Carbon,* Abelard-Schuman, New York, 1958, is a good popular account of basic organic chemistry.

Kenneth Conrow and Richard McDonald, *Deductive Organic Chemistry,* Addison Wesley, Reading, Mass., 1966, is a good beginning textbook on general organic chemistry; it contains an excellent discussion of the covalent bond.

Henry M. Leicester, *The Historical Background of Chemistry,* Wiley, 1956. Chapters 18, 19 and 21 give an account of the development of organic and physical chemistry in the nineteenth century.

Life Science Library, Giant Molecules, Time, Inc., New York, 1966 is a popular account of the structure, preparation and uses of polymers.

Herman F. Mark, The Nature of Polymeric Materials, Scientific American, Sept. 1967, page 148.

J. R. Partington, A Short History of Chemistry, Harper, New York, Harper Torchbook, New York, 1960, chapters 10, 12, 13 and 14 are concerned with the development of organic and physical chemistry, and with the theory of valence.

Linus Pauling, *The Nature of the Chemical Bond,* 3rd Ed., Cornell University Press, Ithaca, New York, 1960 is the best work in the field by a great scientist. Although most of the book requires considerable scientific sophistication, the first few chapters can be read with profit.

Henry Selig, John G. Malm, and Howard H. Claasen, "The Chemistry of the Noble Gases," *Scientific American,* May 1964, page 66.

chapter twenty

The theory of relativity

Relativity theory is more than a branch of modern physics; it constitutes a new way of looking at the physical sciences—a new attitude that distinguishes the modern scientist from predecessors of the past three centuries. Einstein's analysis of the familiar ideas of space, time, motion, and matter has had many consequences, not only for science but for almost every domain of modern thought.

The difficulty in understanding relativity theory (at least, the special theory) is not because it is abstract (as in the case of quantum mechanics) but because it forces us to re-examine and ultimately to abandon some of our most strongly held notions in terms of which we tend to organize both our day-to-day affairs and our basic ideas about the universe. Re-examination of a traditional point of view is always a struggle for scientists and nonscientists alike. In Einstein's words:

> In the attempt to achieve a conceptual formulation of the confusingly immense body of observational data, the scientist makes use of a whole arsenal of concepts which he imbibed practically with his mother's milk, and seldom if ever is he aware of the eternally problematic character of his concepts.[1]

[1] Foreword by Albert Einstein in *Concepts of Space* by Max Jammer, Harvard University Press, Harper Torchbooks edition, Harper, New York, 1960, pp. xi, xv.

Just as it took time and effort for the scientist and the philosopher in the sixteenth century to get used to the new notion that the earth was not the fixed center of the universe, so it has taken time in our own century to overthrow the Newtonian concepts. Einstein continues:

It required a severe struggle to arrive at the concept of independent and absolute space, indispensable for the development of (Newtonian theory). It has required no less strenuous exertions subsequently to overcome this concept—a process which is probably by no means as yet completed.[2]

1. NEWTON'S SYSTEM OF ABSOLUTE SPACE AND ABSOLUTE TIME; MAXWELL'S SYSTEM OF EQUATIONS

In the center of Newton's system of physics is the concept of uniform motion. In an earlier chapter this concept was defined as the motion of a body in a straight line at a constant velocity. However, such a definition requires some kind of a reference system. An object moving in a straight path relative to the earth is describing a very complicated curve with respect to the sun. In addition, "constant velocity" presumes a constant change in location or place during equal intervals of time. Both space and time intervals must be measured with respect to some fixed reference.

In the Aristotelian-Ptolemaic system, the problem of motion was handled by providing an absolute reference point—the center of the earth, which was also the center of the universe. Two types of motion were possible: the natural motion "up or down," that is, toward or away from the center of the earth; and the motion of the heavenly bodies in uniform rotation about the center of the earth. Motion up or down was caused not by any gravitational attraction but by the elements (earth, air, fire, and water), each seeking its appropriate level relative to the earth's center. Thus, space was not isotropic (that is, it was not considered to have the same properties in every direction). Only on the surface of a sphere with points equidistant from the center was direction considered to be indifferent. Furthermore, motion on such a surface could only be defined relative to another spherical containing surface. Beyond some outermost sphere, space had no reality, no meaning, since there existed no further sphere against which locations could be specified.

Against this background of thinking, Copernicus introduced his new formulation of motion.

All attempts to reconcile the obvious contradiction between the notion of absence of place (or space) for the last sphere and the assumption that it moves (and therefore changes its place,

[2] *Ibid.*, p. xv.

according to Aristotle's definition of motion) were doomed to failure. It remained a major problem in scholastic philosophy until Copernicus finally came to the conclusion that the two ideas were irreconcilable, and that at least one of them would have to be rejected. Either the definition of "place" had to be revised, or the dogma of the motion of the outermost celestial sphere had to be repudiated. As we know Copernicus preferred the second alternative.[3]

For Copernicus, the reference system now became the "fixed stars" and the sun at rest in the center of the universe. But when Newton developed his system, he found a need for a wider frame of reference, which he established by means of an abstract absolute space in terms of which *any* motion could be given meaning: whether of the sun moving among the stars or rotating on its own axis; of the planets revolving about the sun; or of the stars themselves (which move relative to one another). He therefore set forth this well-known description:

Absolute space in its own nature and without regard to anything external, always remains similar and immovable.

This absolute space was conceived of as an empty container which, itself, contains all objects of the universe, extending infinitely in all directions and similar in all directions. It is a three-dimensional Euclidean space in which objects are most effectively located by a Cartesian coordinate system, and move according to the three laws of motion.

In addition to absolute space, Newton also had to assume absolute time, which he described in this way:

Absolute, true and mathematical time, of itself and by its own nature flows uniformly on, without regard to anything external.

Under this definition, the problem of measuring time becomes that of constructing a good clock. To the extent that the clock is efficient, it will be in agreement with absolute time and with all other clocks regardless of location in the universe. Time and space are thus considered completely independent of one another.

Having assumed the existence of absolute space, the problem of identifying it becomes important. At once a very curious distinction is found to arise in Newtonian theory: there is no way in which we may determine whether a body is at rest or is moving uniformly in absolute space; on the other hand, there does seem to be a way of detecting rotation or acceleration. It is important to understand this difference.

We ordinarily detect uniform motion as relative motion, but select the earth as a common frame of reference in terms of which every motion is described. A person traveling by plane from New York to London does not think of London

[3] *Ibid.*, p. 70. This book presents an excellent survey of the history of the concept of space from ancient times to the present. Particularly illuminating is the careful analysis of the shift from the Aristotelian to the Newtonian theory of space.

as coming toward him. And, such a journey is described as being one of about 3000 miles, although if the journey were being measured from the sun as a reference base it would be hundreds of times farther than this. If Newton's theory of absolute space is valid, how can we use the earth as a common reference? The answer is found in the observation that the same laws of motion hold as well on a train, a plane, or any uniformly moving body, as on the earth itself. Hence, the matters of convenience and convention determine which frame of reference is chosen. An object dropped by a passenger on a train will accelerate downward in precisely the same way, relative to the train, that an object dropped on earth accelerates. The same force acting on a mass will cause the same change in velocity that it does on the earth.

There is no *"privileged" frame* of reference with respect to uniform motion. This principle is often referred to as the Newtonian principle of relativity, and all systems at rest or moving uniformly relative to one another are called inertial systems. The choice of the earth as our usual frame of reference is the natural one, since most locations on the earth are fixed relative to each other and we are able to relate positions by simple procedures.

Strictly speaking, even within Newtonian physics, the earth is not an inertial system because it rotates and revolves. The equation $F = ma$ does not hold *strictly* true if the earth is used as a reference, even at small velocities, because of the earth's rotation. Foucault's pendulum does not continue to move in the same plane relative to the earth. The "Coriolis forces," (which will be discussed in Chapter 22) that cause cyclones to move in counterclockwise circles in the Northern Hemisphere and clockwise in the Southern Hemisphere, result, in Newtonian physics, from the rotation of the earth. The inertial guidance systems, involving gyroscopes, used by missiles and submarines are sufficiently sensitive to the earth's rotation as to sense the direction of the axis at all times. The basic inertial system has been taken classically to be the system of fixed stars, and is sometimes called the Galilean frame of reference.

Although an object moving in a straight line with uniform velocity in one inertial system will also move in a straight line with uniform velocity in any other inertial system, the *amount* of velocity will, of course, differ in each system, and it is necessary to be able to describe the velocity in either system. This is done through use of *"transformation equations."* The Newtonian transformation equations are very simple in form and appeal to our intuitive sense. Important among them is the principle of addition of velocities. If a man is walking forward in a train at 2 miles an hour, while the train is traveling in the same direction at 40 miles an hour with respect to the earth, the man's velocity relative to the ground is calculated at 42 miles an hour. (If a man is walking at an angle to the train's motion, then the velocities are added by vector methods. We shall return to examples of vector addition when the Michelson-Morley experiment is discussed in the next section.)

These laws of addition of velocities are based upon the assumption that both intervals of length and intervals of time, as specified in one frame of reference, will remain the same in any other frame of reference. A pendulum seen from a train window, for example, will keep the same time as an identical pendulum on the train. The physicists put this principle in these terms: *length and time intervals are invariant* under Newtonian transformations. Eventually, we shall find that although the principle that the *basic laws* of physics are the same in *all* inertial frames of reference remains valid, the invariance of *time intervals* and *length intervals* has been abandoned in modern physics.

While there is no mechanical way of distinguishing uniform motion from rest relative to absolute space, *rotation* is detectable within the Newtonian system. The classical experiment with Foucault's pendulum demonstrates the rotation of the earth with respect to the fixed stars. If a pendulum is started swinging in a plane at the North Pole, it will continue to swing in the same plane *relative to the fixed stars,* but the plane of its motion relative to the earth will turn through 360° during a full day, indicating, according to Newtonian theory, that the earth has completed a rotation in space during this period.[4]

Other phenomena also give evidence of the rotation of the earth: a satellite launched eastward requires less energy to be placed in orbit than one launched westward; and the Coriolis forces, ascribable to the earth's rotation, cause a deflection of motion of air masses as they rise or descend. Even if we had a perpetual cloud cover, so that the stars were not visible, we could detect evidence of rotary motion.[5] Thus, while there appears to be no "privileged" system of reference with respect to uniform motion, it does make a difference what system of reference is chosen when rotation or acceleration is being considered.

Most physicists in the eighteenth and nineteenth centuries did not trouble themselves greatly about defining exactly what was meant by "absolute space" or how it was to be found. It was rather loosely held that the system of "fixed" stars provided the absolute reference frame. This, obviously, was not a satisfactory system, since the stars are not fixed, but move relative to one another at high velocities. The very foundations of physics remained somewhat in question since no meaning in clear, operational terms could be given to such terms as rotation. Some physicists, such as Carl Newman, Lord Kelvin, and Ernst Mach, did address them-

[4]If such a pendulum is set into motion at any latitude other than the equator, a similar phenomenon occurs, although the change of angle relative to the earth is related to time in a somewhat more complicated fashion.

[5]Strictly speaking, as Ernst Mach pointed out, we are still in a dilemma: the only thing we can say on the basis of observation alone is that there is relative rotary motion between the fixed stars and the earth. We choose to regard the "aggregate" of fixed stars at rest because it makes the equations of physics simpler; we could take the alternative position and declare that these distant masses are rotating about the earth causing the various phenomena of deflection such as occur in Coriolis forces, Foucault's pendulum, etc.

selves to this problem. Among other suggestions, it was proposed that the name *The Body Alpha* be given to a fundamental system of reference:

> . . . W. Thomson (Kelvin) and P. G. Tait in their *Treatise* on *Natural Philosophy* suggested as a basis for specifying the Body Alpha that the centre of gravity of all matter in the universe might be considered to be *absolutely at rest,* and that the plane in which the angular momentum of the universe round its centre of gravity is the greatest, might be regarded as fixed in *direction* in space. Other writers proposed that the Body Alpha should be based on the system of the fixed stars, or the aggregate of all the bodies in existence.[6]

None of these proposed solutions is completely satisfactory. Although physicists continued to believe in the absolute space of Newton, a method of determining just how it was to be specified was not agreed upon. Meanwhile, however, Maxwell had developed his electromagnetic theory of light, and with it arose a new possibility of finding an absolute frame of reference. Maxwell and his early followers believed that the waves of light had to be carried by some material medium, which was called the ether. Later, the attempt to endow ether with the material properties analogous to properties of matter became so complicated that it was abandoned. An explanatory interpretation of the equations was given up, just as had been similar attempts to "explain" Newton's equation of gravity. Newton's followers abandoned any real attempt to provide a material mechanism for the transmission of gravitational forces, being content to express the operation of gravity through an equation. Likewise, Maxwell's followers soon gave up trying to give *material* status to the ether. Hertz, who conclusively established the existence of electromagnetic waves, when pressed for an explanation of the theory of light, stated bluntly that the whole theory was embodied in Maxwell's equations. Hertz's assertion almost echoes Newton's earlier admonition to concentrate on the equation of gravity without worrying about a more basic explanation.

There are, however, very important differences between the equation of gravity and the equations of Maxwell. We have already emphasized one of these differences, that is, that the equation of gravity refers only to *discontinuous masses* separated by an "empty space" and makes no reference to an intervening medium, although Maxwell's equations are *field* equations endowing the intervening space, whatever its nature, with properties that vary *continuously* from point to point. In addition, Maxwell's equations *appear* to carry a certain *asymmetry* with respect to relative motion. For example, if a charged particle is at rest in a given frame of reference, an electric field but no magnetic field is produced, although if the same charged particle is observed from a reference

[6]Sir Edmund Whittaker, *A History of the Theories of Aether and Electricity,* Thomas Nelson & Sons, 1951. Harper Torchbook edition, 1960, Vol. II, p. 28.

frame in relative motion it appears as a current and a magnetic field is produced.[7] In another most significant instance, where light is given off from a source it spreads out from that source in a spherical wave relative to the central source. Maxwell's equations seem to imply that an observer in motion, approaching or receding from the spreading sphere of light, would find the spherical waves distorted. In other words, electromagnetic phenomena, unlike those of mechanics, appear to require an absolute frame of reference fixed in space, irrespective of whether a *material* ether exists. In principle, at least, one should be able to determine uniform motion relative to the space in which light is being propagated. The velocity of light should be greater as one moves in a direction opposite to the direction of propagation than when one moves in the same direction.

Here, at last (toward the end of the nineteenth century), physicists hoped to find an absolute space, a reference frame in terms of which all motions could be clearly defined. The search to detect this ether, soon doomed by negative results, was begun by a whole series of experiments in electrodynamics, the most important of which is the Michelson-Morley experiment.

2. THE MICHELSON-MORLEY EXPERIMENT: INVESTIGATION OF POSSIBLE CHANGES IN VELOCITY OF LIGHT WHEN MOVING TOWARD OR AWAY FROM SOURCE

Maxwell's equations, as we have emphasized, appear to imply that there exists a fundamental asymmetry in electromagnetic phenomena. The equations will lead to one conclusion if they are interpreted in terms of one frame of space-time reference, another if they are interpreted in terms of a frame of reference moving relative to the first. With this background scientists hoped to find a "privileged" frame of reference that would afford a means of identifying Newton's absolute space. According to Maxwell's equations, the velocity of light in this absolute space should be a certain constant c, representing a ratio between electrostatic and electromagnetic constants. If, then, an observer moves in the opposite direction to the motion of the light, the velocity of light, as he observes

[7] It is significant that Albert Einstein's famous paper of 1905, in which he announced his formulation of the special theory of relativity, was entitled "On the Electrodynamics of Moving Bodies," and began: "It is known that Maxwell's electrodynamics—as usually understood at the present time—when applied to moving bodies, leads to asymmetries which do not appear to be inherent in the phenomena." (Einstein, by the phrase "which do not appear to be inherent in the phenomena," means that the theoretical asymmetries of Maxwell's equations had not been observed experimentally.) *The Principle of Relativity,* a collection of original memoirs on the special and general theory of relativity, translated by W. Perrett and G. B. Jeffery, Dodd Mead, New York, 1923, page 37.

it, should be increased. If he moves in the same direction, the light should pass him at a lower velocity.

These conclusions are unavoidable if both Maxwell's equations and the Newtonian principle of the addition of velocities are interpreted according to classical theories of physics. Just as two automobiles approaching each other, going 40 miles an hour relative to the earth, have a relative velocity of 80 miles an hour relative to each other, so, it was felt, light produced by a source at a velocity c, passing an observer traveling toward the source with velocity v should find the velocity of light, relative to him, to be $c + v$.

The velocity of light is so tremendous (3×10^8 m/sec) compared with ordinary velocities (even the velocity of orbiting astronauts today is only about 10^4 m/sec or 30,000 times slower) that an experimental measurement of the change in the velocity of light as one approached or receded from the direction of its source would require extremely precise instruments. Such an instrument did become available. An American physicist, Michelson, and his colleague Morley, devised an apparatus in 1885 to measure any possible changes in the speed of light caused by the motion of the earth about the sun. Continuous improvements were made upon the apparatus until by, 1900, such precision was obtained that the results could be regarded as definitive.

The principle of Michelson's experiment can best be approached by first considering a commonly used analogy. Suppose that a man wishes to row across and back over a stream in which the water is traveling 4 ft/sec. Assume that relative to the water the rowboat travels at 5 ft/sec. Now, compare the time it will take the boat to travel the distance across the stream and back with the time it will take to travel the same distance down the stream and back. Suppose the distance across the stream is 180 ft. Using the diagram in Fig. 20.1 and using Newton's principle of the addition of velocities, we can readily calculate the comparative times it will take, depending on which way the rowboat travels.

To go from the point A to point B the rower must direct his boat upstream. Since he is rowing at 5 ft/sec, he must row in a diagonal direction in such a manner that in 1 sec he will travel relative to the water 5 ft, while the water is sweeping him downstream 4 ft. Therefore, to maintain his progression at right angles to the bank, he rows in such a direction as to move across the stream at the rate of 3 ft/sec. The time it will take him to go across the stream and back will thus be 120 sec.

On the other hand, when he rows downstream and back over the same distance the time he will take is not 120 seconds. Going downstream, he will travel at 9 ft/sec, and will cover 180 ft in 20 sec; rowing upstream, he will travel only 1 ft/sec, and will take 180 sec. This total journey down and back will thus take 200 sec.

This implies that if two boats start off at the same time from the same point

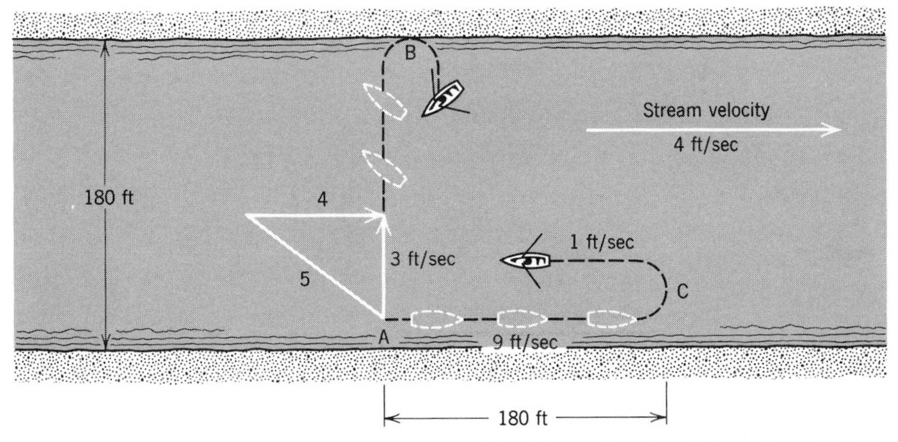

Fig. 20.1. It takes 120 sec to go 180 ft from A to B and back; it takes 200 sec to go 180 ft. from A to C and back.

and travel the same distance but in different directions returning once again to their starting point, they will not arrive at the starting point at the same time. Michelson arranged an apparatus in which two rays of light take the place of the rowboats. The two rays of light start out together, cover equal distances with their paths at right angles, and are again brought together. The test is to see whether they arrive at the same time. If the earth is in motion in space, and if Newton's principle of the addition of velocities applies, then the two rays of light cannot traverse the equal distances in the same time interval.

The apparatus used is shown in Fig. 20.2.

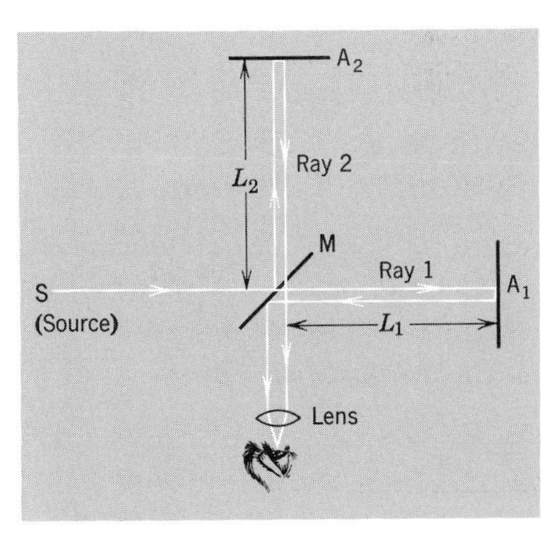

Fig. 20.2. The Michelson-Morley apparatus. The beam of light from the source is divided into two parts by the half-silvered mirror. The two light beams are then made to travel paths L_1 and L_2 at right angles before coming together again at the lens. L_1 and L_2 are made as equal as possible. If there is any changing inequality of path or any changing inequality of time taken by the two light rays, it will appear to the eye in a changing interference pattern of light and dark bands seen by the eye.

In the apparatus a beam of light comes from the source S, and strikes the half-silvered mirror M (by half-silvered is meant a glass very lightly coated with silver so that approximately one half the light proceeds through the glass and half of it is reflected). The light striking the glass M is now split into two separate rays as shown, traveling at right angles to one another over the paths L_1 and L_2, which are so adjusted as to be as nearly equal as possible. The two rays strike the plane mirrors A_1 and A_2, are reflected back to the half-silvered mirror M and, as shown, the two rays are finally brought together, passing through the lens. When these two rays of light are brought together they will produce a pattern of interference fringes such as was discussed in Chapter 16 on light, that is, a circular pattern of bright and dark bands where the rays of light interfere with one another constructively and destructively. The exact appearance of these interference fringes will depend on the difference of time taken by the two rays of light traveling over the respective paths. Any change in the relative length of L_1 and L_2 or in the relative time taken by the two rays would then show up as a visible change in the interference pattern. When the two rays meet exactly "in step," the circular pattern has a bright central spot; if now a change occurs so that they are "out of step" by a half wavelength, the bright central spot becomes a dark one.

The whole apparatus was mounted by Michelson on a heavy concrete slab to give it great inertial stability. This concrete slab in turn was floated in mercury so that it could be rotated smoothly through 360°. The purpose of this was to be able to interchange or vary the positions of the paths L_1 and L_2 relative to the earth. If the earth is moving in space and if light always travels at the same velocity in space, the time taken for light to travel over the paths L_1 and L_2 should vary, depending upon which direction the earth is moving. The arrangement is somewhat similar to the arrangement with the rowboats discussed earlier. Just as one boat going back and forth across the stream moving relative to the bank takes a different time to travel a specified distance than does a boat traveling the same distance downstream and back, so a light ray traveling at right angles to the earth's motion in space was expected to take a different time than one traveling parallel to the earth's motion.

Because of the significance of the Michelson-Morley experiment, some readers may wish to examine the mathematical analysis, which involves nothing more than algebra and an application of Newton's principle of the addition of velocities to Michelson's apparatus.

As the earth revolves about the sun, the apparatus is assumed to be transported with a velocity v relative to the absolute space (or ether) in which light rays are moving with a velocity c. When the whole apparatus is rotated on its axis through 360°, there should exist some angle such that the apparatus is moving in the direction of one of the arms. In that position the other arm will be moving at right angles to absolute space. Assume the apparatus is moving parallel to the arm L_1, through a distance d, during the time t with a velocity v relative to space (of the ether). Then the two light rays will traverse two paths, as shown in Fig. 20.3.

Examine, now, the time taken for the two rays of light to go from the central mirror M, respectively, to the mirrors A_1 and A_2 and back. In accordance with Newton's principle of addition of velocities, Ray 1 will travel from M to the mirror A_1 to the right overtaking the apparatus with the *relative velocity* $c - v$, and to the left with the relative velocity $c + v$. The time t_1 taken by this ray becomes

$$t_1 = \frac{L_1}{c - v} + \frac{L_1}{c + v} = \frac{2L_1 c}{c^2 - v^2}$$

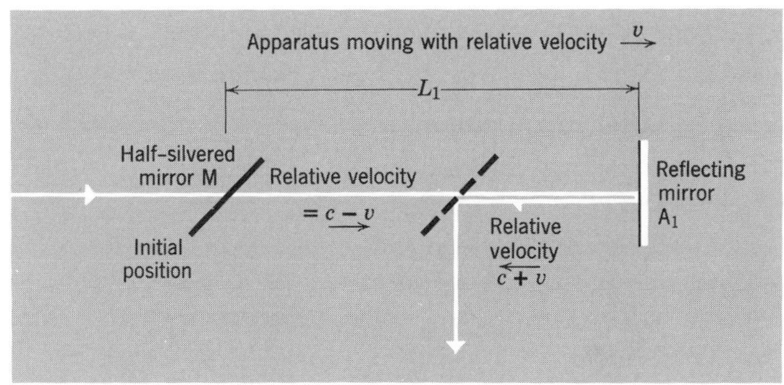

Path of Ray 1, parallel to motion, from M to A_1 and back to M

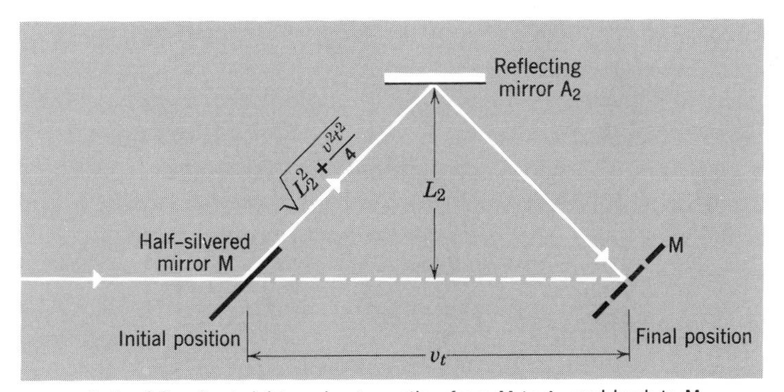

Path of Ray 2, at right angles to motion from M to A_2 and back to M

Fig. 20.3. Diagram of relative velocities and times, as expected by classical principles, for Michelson-Morley experiment. Light signal traveling path 1, parallel to motion, is expected to take a different time, over same distance from light signal traveling path 2, perpendicular to motion.

which, for later purposes, can be expressed as

$$t_1 = \frac{2L_1}{c}\left(\frac{1}{1 - \frac{v^2}{c^2}}\right)$$

Meanwhile, the apparatus has also moved at right angles to the path L_2, through a distance $d = vt$, also shown in the figure. Therefore, the time for the light to go from M to A_2 and back is the time required for the light to travel over the diagonal MA_2, that is,

$$\frac{\sqrt{L_2{}^2 + \frac{v^2 t_2{}^2}{4}}}{c}$$

plus the equal time taken to travel over the diagonal A_2M. The total time for Ray 2 is

$$t_2 = \frac{2\sqrt{L_2{}^2 + \frac{v^2 t_2{}^2}{4}}}{c} = \frac{\sqrt{4L_2{}^2 + v^2 t_2{}^2}}{c}$$

whence

$$t_2{}^2 = \frac{4L_2{}^2 + v^2 t_2{}^2}{c^2}$$

$$c^2 t_2{}^2 = 4L_2{}^2 + v^2 t_2{}^2$$

$$t_2{}^2 = \frac{4L_2{}^2}{c^2 - v^2} \quad \text{or} \quad t_2 = \frac{2L_2}{c\sqrt{1 - \frac{v^2}{c^2}}}$$

The difference in time taken by the two rays of light becomes therefore

$$t_1 - t_2 = \frac{2L_1}{c}\left(\frac{1}{1 - \frac{v^2}{c^2}}\right) - \frac{2L_2}{c}\left(\frac{1}{\sqrt{1 - \frac{v^2}{c^2}}}\right)$$

or if $L_1 = L_2 = L$

$$t_1 - t_2 = \frac{2L}{c}\left[\frac{1}{1 - \frac{v^2}{c^2}} - \frac{1}{\sqrt{1 - \frac{v^2}{c^2}}}\right]$$

So, we see that the time taken for the two rays of light should be different on the basis of classical physics. Furthermore, if the arms L_1 and L_2 are interchanged by rotating the apparatus, the difference in time should be changed, with an interchange of the above expressions L_1 and L_2. Such a difference, caused by rotating the apparatus, should definitely cause a change in the interference pattern. If an approximation using the binomial expansion is used, and such very small terms as (v^4/c^4) are dropped,[8] the expression for the difference in times is, more simply, given by

$$t_1 - t_2 = \left(\frac{2L}{c}\right)\frac{v^2}{2c^2} = \frac{Lv^2}{c^3}$$

The earth travels about the sun at the rate of 3×10^4 m/sec. The apparatus used by Michelson was so delicate and precise that the difference in the time for the two rays of light should have been easily detectable. Even if, by some chance, the earth were at rest relative to space at one point in its orbit (in which case the time difference would not occur), months later it would have to be in motion. Michelson and Morley conducted their experiment over periods longer than six months, constantly rotating their apparatus. Not once did they discover the expected shift in the interference pattern! To Michelson this was a grave disappointment. The experiment was repeated many times after careful modification to insure greater precision. Even a velocity as low as one-thirtieth the velocity of the earth relative to the sun should have been detectable if the earth were traveling through an ether. This best hope of finding absolute space, like so many other attempts by other means, had ended in failure.

3. EINSTEIN'S SPECIAL THEORY OF RELATIVITY

Let us look clearly at the sharp dilemma faced by physicists in the early 1900's. Maxwell's equations, as perfected by Hertz, had stood up well under a multitude of tests and were regarded as valid; Newton's laws of motion had been so well tested over a period of two centuries that they could hardly be doubted. Yet these two apparently valid theories in combination led to predictions that were contradicted not only by the Michelson-Morley experiment but by several other tests as well. What could be wrong?

Two possible escapes were suggested. One was the theory that perhaps the ether of space was somehow dragged along by masses moving through it. For reasons we need not examine, it was quickly shown that this was an inadequate explanation. Another theory was proposed by Lorentz and Fitzgerald: that by some unknown mechanism all objects moving in space or in the ether must contract in length by a factor equal to $\sqrt{1 - v^2/c^2}$. It was shown that such a contraction would explain the negative results of Michelson-Morley's experiment as well as the other negative experimental results. Such a hypothesis, however, seemed to physicists to be much too artificial, and was therefore not widely

[8] See Appendix III.

accepted. (Among other objections it seemed very artificial to suppose materials such as rubber would contract precisely in the same proportions as steel.)

Albert Einstein stepped into this situation with a completely fresh approach, and proposed a solution whose implications have changed the face of physics. He had training not only as a physicist; he had also pondered deeply the philosophical foundations of his subject. In particular, he had studied the work of Ernst Mach, who had developed a penetrating critical analysis of the methods and assumptions of Newtonian mechanics. Einstein had the boldness to consider the dilemma posed by the Michelson-Morley experiment, stripped down to its essentials, without allowing himself any unconscious assumptions. Einstein attended to the contradiction within the following three general statements:

A. The velocity of light is experimentally constant, regardless of any relative movement of the source or the observer.

B. The laws of mechanics are the same for all frames of reference, regardless of any relative uniform motion.

C. Newton's statement of the principle of the addition of velocities remains valid.

Since a contradiction existed (that is, statements *A*, *B*, and *C*, above, could not all be valid at the same time), two possibilities presented themselves: either one or more of the three statements must be reinterpreted by some modification, or one of them must be abandoned. Those who had suggested that the ether was carried along with the moving matter were trying to reinterpret the experimental finding of *A* (the constant velocity of light). Lorentz and Fitzgerald, by their hypothesis of contraction were attempting to reinterpret statement *B*, but doing so in a way that was felt to be unsatisfactory. Einstein decided to abandon the third principle, that of the addition of velocities, and to examine the consequences of holding to statements *A* and *B*. Furthermore, Einstein decided that every mathematical term he used in carrying out this analysis would be subject to the requirement that it could be operationally defined, that is, defined in terms of a specified measuring procedure. Only in this way, he felt, could the unconscious metaphysical assumptions regarding the meaning of each term be avoided.

Now, if we are to abandon the orthodox principle of the addition of velocities when we talk about the same motion as observed from two frames of reference in relative motion, it is necessary to develop some other principle (that is, a transformation equation) to take its place. Only in this way can we obtain a physics that can deal with all events taking place on separated bodies in relative uniform motion. Furthermore, any new principle that is developed should not yield, at low velocities, very different results from the former principle, which had proved so effective in classical physics. Einstein showed that a new principle developed by him relating velocities in separated frames of reference is almost identical with Newton's, when the velocities considered are much smaller than the velocity of light.

In our previous example we said that a person walking forward on a train with a velocity of 2 miles per hour when the train travels 40 miles per hour relative to the ground is moving with a velocity of 42 miles per hour relative to the ground. This may be expressed as $v' = u + v$ where v' is the velocity of the man relative to the ground, u the velocity of the man relative to the train, and v is the velocity of the train relative to the ground. Einstein's analysis ended with the conclusion that this equation must be modified to read

$$v' = \frac{u + v}{1 + \frac{uv}{c^2}}$$

But note that in uv/c^2, uv in miles per second is about 2.2×10^{-2} miles per second (40 miles per hour times 2 miles per hour). Thus

$$\frac{uv}{c^2} = \frac{(2.2 \times 10^{-2})}{(1.8 \times 10^5)^2}$$

or approximately 1.2×10^{-12}. The revised equation for the addition of velocities, at ordinary velocities is so infinitesimally small as to be completely negligible! The finest measuring instruments today could not detect the difference between the new and old equation when applied to these small velocities, even if the change were 1000 times as large. On the other hand, when velocities are so great as to approach the velocity of light, then a very distinct difference appears between Newton's and Einstein's formulation.

In addition, the new equation

$$v' = \frac{u + v}{1 + \frac{uv}{c^2}}$$

leads to a consistent explanation of the results of the Michelson-Morley experiment. Suppose, for example, that an artificial satellite is traveling toward the sun with a uniform velocity of u, and we inquire as to the velocity of a ray of light propagated from the sun with a velocity c. According to Newton's equation, the velocity of the ray of light relative to satellite should be $v + c$. On the other hand, according to the Einstein equation, it will be

$$v' = \frac{v + c}{1 + \frac{vc}{c^2}} = \frac{v + c}{\frac{c + v}{c}} = c$$

In other words, the ray of light will pass the satellite at the same speed regardless of how slowly or how fast the satellite is traveling toward the sun. Even if the satellite's velocity toward the sun were that of light itself (that is, $v = c$), the same results hold. Or, again, if the velocity were negative relative to the sun, that is, if the satellite is traveling away from the sun, the same results would remain valid. In this case,

$$v' = \frac{-v + c}{1 + \left(\frac{-vc}{c^2}\right)} = \frac{-v + c}{\frac{c - v}{c}} = c$$

Thus, it makes no difference how the satellite moves or at what velocity; light will always pass it at the same velocity.

But how did Einstein arrive at this new equation? Is it an artificially assumed or *ad hoc* hypothesis just to explain the Michelson-Morley results? If so, it is no better than the Lorentz-Fitzgerald contraction hypothesis. Actually, Einstein was led to the new transformation equation by straight logical steps from the two postulates: A, the speed of light is constant in empty space for *all* reference frames, and B, the same laws of physics hold for all reference frames regardless of any relative uniform motion. However, in arriving at the new consequences, Einstein was forced to abandon certain unconscious assumptions regarding space and time, which lay at the basis of classical physics. It was necessary to give a careful redefinition of what is meant both by an interval of time and an interval of distance. (If the classical assumptions are maintained, the classical transformation equation for velocities follows, and the contradiction to postulate A and B reappears.)

Let us review the situation carefully, paying attention only to the logic involved. Three statements, A, B, and C, lead to contradiction. Therefore, one or more of them must be abandoned or modified. Attempts to modify A and B had been found unsuccessful. Consequently, Einstein decided C must be abandoned. But C, concerning the principle of the addition of velocities, was itself a logical consequence of two other "self-evident" assumptions, which had never been questioned before, that is, that lengths and time intervals may be ascribed fixed mathematical and physical meaning, regardless of the motion of the measured lengths or clocks specifying time. Einstein had the perspicacity to see that if the principle of the addition of velocities is to be abandoned, this requires a careful re-examination of what is meant by length and time. All metaphysical unconscious prejudices are to be eliminated for the sake of logical consistency. The results were surprising, and were not at first easily accepted by either scientist or nonscientist, any more than were the conclusions of Copernicus accepted readily in the sixteenth century.

It is our common presumption, which was carried over into classical physics, that there is a definite meaning to the conception of "true" time and "true" length, even if our measurements by clocks and measuring rods can only approximate them. Two events in the universe may, common sense tells us, occur at the same time. This is a proposition that appears to be unambiguous. If we have the proper instruments (clocks), we can establish simultaneity in such a way that it would make no difference whether the two events occurred close to us or separated from us. When we speak of an observed event occurring, for example, 1000 years ago, on a distant star (such as a change into a supernova), we must make the proper allowance for the time for light to reach us. Nevertheless, if such an event occurred 1000 years ago in *our own* measurement of time, common sense tells us, as it did the classical physicist, that it would have occurred 1000 years ago on the basis of *anyone's* measurement of time, even if the person were located on another planet, provided only that the clocks used were in agreement with ours, and that the same appropriate allowance was made by the second observer for the time it would take for light to travel to him.

But, in his famous paper of 1905, Einstein shows that the very notion of absolute simultaneity must be abandoned, and that there can be no way of specifying time intervals, which are the same for all observers regardless of their relative motion. And, since there is no possible method of establishing a "privileged" frame of reference, the concept of time intervals must be always defined operationally in terms of a local specific frame of reference. A pendulum on earth will beat out seconds as we watch it, but will beat out periods slower than a second for someone watching the same pendulum from Mars. Furthermore, there will be absolutely no way of determining which time period is "true" for the universe as a whole. All we can do is to insure that we understand how to translate measurements in one system into those of any other.

A good example is that proposed by Einstein in explaining his principles to the public.[9] Suppose we have a train traveling along a straight track as illustrated in Fig. 20.4. At a certain instant of time, the two ends of the trains A' and B' are coincident with the two points A and B on the track. Furthermore, the midpoint between A and B is located at P, coinciding with the point P' located on the train between A' and B'.

Suppose that one bolt of lightning strikes the track at A and another at B. Assume that to the observer on the track, sitting at midpoint P, the flashes of lightning from A and B will reach this point simultaneously and, since he has carefully measured the distances to A and B and found them equal, he will declare that, to him, lightning flashes at A and B are simultaneous. And what about the man on the train? The point P' now will have moved toward the

[9]Einstein's own essay written specifically for the public, entitled *Relativity: The Special and General Theory,* may be found in condensed form in Masterworks of Science, edited by J. W. Knedler, Doubleday and Company, New York, 1947.

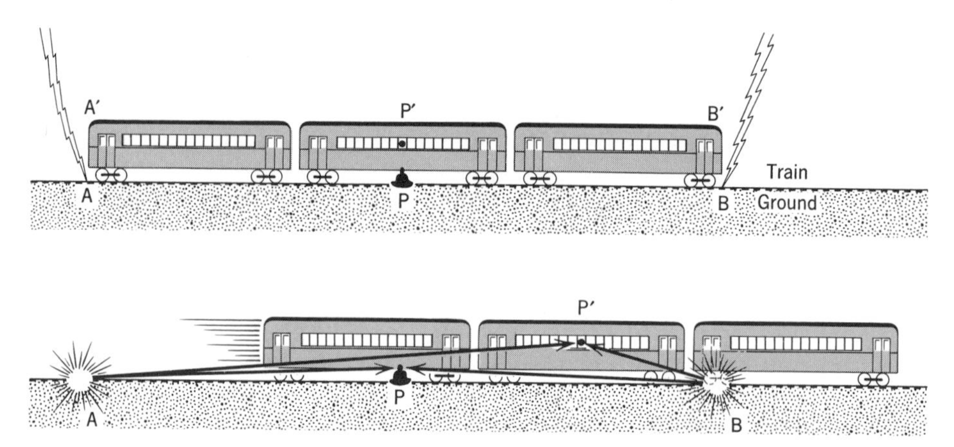

Fig. 20.4. The simultaneity for two observers may differ.

source B. Therefore, although the two flashes reach P on the track at the same time, the flash from B will reach the point P' on the train *before* the flash from A reaches P'. The flashes of lightning cannot be simultaneous with respect to the frame of reference on the train if they are simultaneous with respect to the frame of reference on the ground. In general, events simultaneous in one frame of reference are not simultaneous in another frame which is in motion relative to it. In addition, it should be noted that if the argument about the lightning flashes is reversed, that is, if the light flashes are found to be simultaneous on the train (both reaching the midoint P' at the same time), then the light from A (and its coincident point A') will pass P before it reaches P' and, thus, in advance of the time that the light from B reaches P. In this case, the flashes of lightning are simultaneous relative to the point P' on the train, but not simultaneous relative to the point P on the ground. We must be careful, in other words, not to assume any privileged frame of reference (such as the ground) in which the light flashes are "really" simultaneous. Thus, events simultaneous in one frame of reference are not necessarily simultaneous in another. Furthermore, since we cannot regard the track as fixed in space (it certainly is in motion with respect to the sun, for example) there is no reason to give preference to the judgment of simultaneity as observed from a frame of reference fixed to the railroad track over that as observed from the train. (In fact, these same two events looked at from the sun, for example, might even be seen to take place in reverse order, if the train is moving to the left relative to a frame of reference fixed on the sun.)

This example and similar ones show that *there is no possible way of specifying unique time measurements that will be the same for all systems in the universe.* The time of an event relative to one frame of reference will not be the same for all systems. The time of an event relative to one frame of reference will be different relative to another one that is in motion. If the motion is uniform between the two frames of reference, either one of them may be taken as at rest; hence, either system of recording time is equally valid.

With such considerations in mind, Einstein first considered the problem of how time measured in one system is related to the time measured in another. He began by examining a method of establishing what is meant by time coincidence at two points mutually at rest but separated from one another:

> In order to measure time, we have supposed a clock, u, present somewhere at rest relative to k. But we cannot fix the time, by means of this clock, of an event whose distance from the clock is not negligible; for there are no instantaneous signals that we can use in order to compare the time of the event with that of the clock. In order to complete the definition of time, we may employ the principle of the constancy of the velocity of light in a vacuum. Let us suppose that we place similar clocks at points of the system k, at rest relative to it, and regulated on the following scheme. A ray of light is sent out from one of the clocks u_m, at the instant when it indicates the time t_m, and travels through a vacum a distance r to the clock u_n; at the instant when this ray meets the clock u_n the latter is set to indicate the time $t_n = t_m + r/c$. The principle of the constancy of the velocity of light then states that this adjustment of the clocks will not lead to contradictions.[10]

The notable aspect of Einstein's definition of time in terms of physical measurements is that it requires reference to space and to the velocity of light. The two clocks, separated by a distance, cannot be synchronized in any other way. (Notice that we cannot synchronize the clocks by bringing them together and then separating them without tacitly assuming that moving the clocks will not change the time that they keep.)

Einstein states clearly the reason that light signals rather than any other means are used to establish time relationships between separated events:

> In order to give physical significance to the concept of time, processes of some kind are required which enable relations to be established between different places. It is immaterial what kind of processes one chooses for such a definition of time. It is advantageous, however, for the theory to choose only those processes concerning which we know something certain. This holds for the propagation of light in vacuo in a higher degree than for any other purpose which could be considered, thanks to the investigation of Maxwell and H. A. Lorentz.[11]

Having established a clear, physical, and operational—rather than a "fictitious" or mathematical—definition of time for events at rest in a frame of reference, but separated in space, Einstein determines how time as measured in one system of coordinates will be measured in another that is in relative motion. Similarly, he

[10] Albert Einstein, *The Meaning of Relativity,* fourth edition, Princeton University Press, Princeton, 1953, p. 28.
[11] Ibid., pp. 28–29.

analyzes length, forsaking also the assumption that the distance between two points as measured in a moving co-ordinate system is invariant. His analysis is simple and straightforward, requiring nothing more than algebra. (The mathematical steps are given in Appendix IV. Only the conclusions are summarized here.) As between two inertial systems (that is, systems in which there is no relative acceleration or rotation), moving with a velocity v relative to one another, the following relationships, known as the Lorentz transformations, hold true.

1. Time intervals in a moving system, as observed from a stationary system, are slowed down by an amount specified in the equation:

$$t' = \frac{t}{\sqrt{1 - \frac{v^2}{c^2}}}$$

showing how time intervals t' in the moving system are related to the time intervals t in the "stationary" system.

This implies that any process in nature, such as, for example, the oscillation of a pendulum or the beating of the heart, will be slower by a very small amount (completely negligible with ordinary velocities, even in the case of the seemingly "high" velocities of jet planes or artificial satellites) when it is in motion relative to a frame of reference in which a "standard" clock is maintained. The period of vibration within atoms traveling at very high speeds, on the other hand, is distinctly longer than when the same atoms are at rest or moving slowly.

2. Lengths are similarly contracted by an amount, in the direction of their relative motion, specified by:

$$L' = L\sqrt{1 - \frac{v^2}{c^2}}$$

where L' is the length of a moving object as observed from a stationary system, while L is the rest length.

Again the contraction is negligible for ordinary velocities. With high velocities, however, such as may be given to atomic particles in cyclotrons or other accelerating instruments, the contraction becomes significant.

3. The principle of the addition of velocities becomes modified to the following: if an object travels with a velocity u relative to a frame of reference k, and if the frame of reference k is traveling with a velocity v relative to a frame of reference k', then the velocity of the object relative to the frame of reference k' is given by v' equal to

$$v' = \frac{u + v}{1 + \frac{uv}{c^2}}$$

As a corollary to this principle it follows that no material object can have a velocity greater than the velocity of light. It should also be mentioned that special relativity is "built into" Maxwell's equations, although this is not immediately evident. We should expect this to be the case since "light travels at the speed of light" and any successful theory of light must therefore be a relativistic one.

4. FURTHER CONSEQUENCES OF THE SPECIAL THEORY OF RELATIVITY; SPACE-TIME; MASS AND ENERGY

Many significant consequences flow directly from the special theory of relativity —consequences that have affected the whole philosophy of modern man as well as the physical sciences. Interestingly, one of the earliest results growing out of the formulation of the special theory of relativity was the emphasis that it placed upon its own limitations, thus stimulating an immediate effort to remove those limitations and to generalize the conclusions. The term *special theory* refers to the fact that in this theory, while all the laws of physics, including both mechanics and electrodynamics, can be stated in such a way that they remain the same for all frames of reference that are at rest or moving *uniformly* with respect to each other, no such claim is made with respect to frames of reference that are moving with relative *accelerated* motions. Why should, Einstein asked, the laws of nature depend at all upon what coordinate system an observer chooses in which to describe those laws? Does not the assumed uniformity of nature require that the uniformity can be formulated without reference to where the observer is sitting or how he is moving relative to someone else? Once Einstein had accomplished the proper formulation of physical laws independent of the state of uniform motion, he was impelled to try to do the same for accelerated motion.

This effort occupied him from 1905 until 1915 when he finally announced the general theory of relativity, which will be discussed in Section 5 of this chapter.

Meanwhile, we shall review some of the other results of the special theory. Perhaps the broadest of these, as it affects both philosophy and physics proper, is the new light cast upon the nature of basic scientific concepts—how they are to be defined and how distinguished from mathematical concepts. Actually, we have used this distinction throughout this book, and we have already discussed it at several points. It was Einstein's theory that inaugurated the point of view generally accepted today—a viewpoint that places greater emphasis upon clear operational definitions, and avoids intuitive preconceived notions. For example, *length* and *time,* as now used by physicists, are no longer regarded as mere application of mathematical concepts given to us as *a priori* self-evident intuitions or by Kantian categories or Platonic forms, nor are they regarded, on the other hand, as patterns of sensations impinging on the blank tablet of the mind, as suggested by the early British empirical philosophers. Instead, length and time are seen in a twofold relationship, logically embedded in a mathematical system, but also defined physically in terms of measuring processes specified by rods and clocks. The epistemological problem (that is, how we attain valid knowledge) is certainly not solved, but it is carefully attended to by the modern physicist as well as by the philosopher, and has a different setting than it had before the advent of Einstein's theory. Furthermore, the restatement of the problem in the physical sciences has profoundly affected the social sciences as well.

Another and highly imaginative consequence arising out of the special theory is the recognition that space and time cannot properly be regarded as independent. Already we have noted that in the specification of the time of events separated in space, both distance between events and the velocity of light must be considered; and that, similarly, in the specification of the distance between two events as viewed from different inertial systems in relative motion, time must enter into the equations through the factor of velocity. Minkowski, a few years after Einstein's paper of 1905, formulated an effective mathematical method of blending together space and time into a single coordinate system. In an address delivered before an assembly of German natural scientists of Cologne in 1908, Minkowski began:

> The views of space and time which I wish to lay before you have sprung from the soil of experimental physics, and therein lies their strength. They are radical. Henceforth space by itself, and time by itself, are doomed to fade away into mere shadows, and only a kind of union of the two will preserve an independent reality.[12]

Minkowski asked himself the question: since length is not an invariant quantity, nor is time an invariant quantity as between different inertial systems, can we find some invariant that will replace these as demanded by relativity theory?

It is assumed in classical physics that any two points located in space have an invariant distance between them, regardless of time. Thus, presumably, invariant distance was specified by the Pythagorean theorem:

$$s^2 = x^2 + y^2 + z^2$$

where x, y, and z are the components of the distance in the x, y, and z directions.

This equation for the distance between two points is now replaced by an invariant "separation" between two events:

$$s^2 = c^2t^2 - (x^2 + y^2 + z^2)$$

In the case of the former equation, the quantity s^2 must vary from one inertial system to another, since, as we have seen, the lengths of objects are "con-

[12]Reference 7, p. 75.

tracted" when they are in motion. But, for the latter equation, the quantity s^2 remains the same in all inertial systems.

The fact that time enters into physical laws as a fourth and necessary coordinate is all that is meant by the rather loose phrase that "time is the fourth dimension." Actually, time enters in conjunction with the factor c (the constant velocity of light), and its square is of opposite sign to the squares of the other dimensions. However, with this understanding, Minkowski studied the consequences that would follow from a four-dimensional geometry with coordinates x, y, z, and ict (where $i = \sqrt{-1}$). All the vectors, which previously had been expressed in terms of *three* components, now become *four*-component vectors, including a time component, and the laws of physics assume a greater generality than they had before.

Finally, we mention the best known and most dramatic consequence of the special theory of relativity: the identification of mass and energy, and the famous equation $e = mc^2$, which governs the transformation of mass into energy and vice versa. Through the application of this equation many previously unexplained puzzles of the universe, such as the apparently inexhaustible source of the heat of the sun and stars, transmutation of radioactive elements, and other nuclear processes of a wide variety, became understood; the development of atomic energy and all it implies for mankind was undertaken under the guidance of this equation.[13]

If the equation $e = mc^2$ is not to be regarded in a sense as a magical incantation, we should try to understand precisely what it means and how it originated. Although scientists are sometimes pleased at how the public reacts to their more dramatic discoveries, theoretical or experimental, they would doubtless be happier if the public also were more generally informed about the manner in which the ideas have been reached and the arguments and evidence that support them.

Before undertaking the analysis, let us look qualitatively at the way in which Einstein arrived, in 1905, at his stunning prediction of the equivalence of mass and energy on the basis of inexorable logic guided by bold intuition. Remember that Einstein's only postulates were (*1*) that all physical laws should be expressible in an invariant form independent of uniform motion, and (*2*) that the velocity of light is constant. Among the laws that he felt had to be retained

[13]In 1939, Einstein received a letter from L. Szilard, his friend, and an eminent physicist, who had learned that the Germans were manufacturing "heavy water" and suspected and feared that they were working on the development of an atomic bomb. Einstein was prevailed upon to write a letter to President Roosevelt, calling this development to his attention. He wrote, in part: "Some recent words by E. Fermi and L. Szilard which have been communicated to me in manuscript lead me to expect that the element Uranium may be turned into a new and important source of energy in the immediate future . . . A single bomb . . . exploded in a port . . . might well destroy the whole port . . ."

were the law of the conservation of energy and the law of conservation of momentum.

Consider, now, the problem of expressing the energy of a system in two frames of reference moving relative to one another. Kinetic energy is expressed in classical physics as $\frac{1}{2}mv^2$, and so we have used it up to this point. This involves velocity; but velocity in *relativity theory* is no longer translated from one reference system to another by the simple Newtonian equation of velocities. In classical physics, if one rolls a ball along the deck of a ship, the kinetic energy of the ball as computed from its velocity relative to the ship, plus the kinetic energy of the ball and the ship, computed from their velocity relative to the earth, comes out exactly the same as the alternative computation, that is, the kinetic energy of the ball as computed from its velocity relative to the earth, plus the kinetic energy of the ship without the ball. The problem is simple and straightforward and presents no difficulties. But with the new Einstein equation relating velocities in two different frames, there must be a new formulation of kinetic energy equations, differing from the classical because of the different statement of relative velocities.

Similarly, according to classical theory, a volume of radiant energy (for example, a sphere enclosing a certain amount of light energy) can be computed from Maxwell's equations. This volume of energy was assumed to be the same as measured by all observers regardless of their relative velocities.

Einstein was impelled by his new transformation equations relating time, distance, and velocities as measured by two separated and moving systems of reference, to review both the transformation equations relating to kinetic energy and radiant energy. Therefore, in a second paper in 1905, significantly entitled, "Does the Inertia of a Body Depend Upon Its Energy Content?"[14] Einstein proposes the following problem. Suppose a body emits a certain amount of light energy L, and that we compare the total energy before and after the emission of this light energy, both from a frame of reference stationary with respect to the source and from a frame of reference that is moving with a velocity v relative to the source. (Note that, viewed from a moving frame of reference, the body emitting the radiation will have a relative velocity and will thus have an associated kinetic energy. This kinetic energy may or may not remain constant during light emission, since all that can be said on the basis of relativity postulates is that the *total* energy change before and after the emission of light is zero. In other words, the possibility remains that any gain or loss in kinetic energy is offset by a gain or loss in radiant energy. Each of these may vary even if their sum remains constant.)

The analysis now proceeds as follows. Suppose there is a stationary body in

[14]Reference 7, pp. 69 ff.

one system of reference (x, y, and z), which emits a plane wave of light carrying the energy $\frac{1}{2}L$ along the direction of the positive x axis and, at the same time, another plane wave of light, carrying the energy $\frac{1}{2}L$ in the opposite direction along the negative x axis. If we call the total energy of the body before the emission of light E_0, and the remaining energy after the emission of light E_1, we have

$$E_0 = E_1 + \tfrac{1}{2}L + \tfrac{1}{2}L$$

How will the equation of energy appear from a system moving along the x axis with a velocity v relative to the emitting source? First, let us analyze the amount of light energy as viewed from the two reference frames. The energy density of light in a volume of space must differ. This follows from the fact that lengths contract under relative motion. (A sphere containing a specified volume of energy becomes an ellipsoid when viewed from a moving reference frame, the axis in the direction of motion being shortened according to the transformation equation for length.) We need not go into the detailed analysis of the radiant energy transformation equation but it was found by Einstein to be

$$L' = L\left(\frac{1 \pm \frac{v}{c}}{\sqrt{1 - \frac{v^2}{c^2}}}\right)$$

where L is the light energy in the stationary frame, and L' is the light energy in the moving frame of reference.

Thus, the equation of energy corresponding to

$$E_0 = E_1 + \tfrac{1}{2}L + \tfrac{1}{2}L$$

becomes, in the moving system,

$$H_0 = H_1 + \frac{1}{2}L\left(\frac{1 - \frac{v}{c}}{\sqrt{1 - \frac{v^2}{c^2}}}\right) + \frac{1}{2}L\left(\frac{1 + \frac{v}{c}}{\sqrt{1 - \frac{v^2}{c^2}}}\right)$$

or

$$H_0 = H_1 + L\left(\frac{1}{\sqrt{1 - \frac{v^2}{c^2}}}\right)$$

where H_0 and H_1 are energies, before and after emission, for the moving frame of reference.

Furthermore, we assume that the only effect the radiation of energy can have is on the kinetic energy causing a reduction of kinetic energy K_0 before the emission to K_1 after the emission of light. Therefore

$$(H_0 - E_0) - (H_1 - E_1) = K_0 - K_1$$

But

$$(H_0 - E_0) - (H_1 - E_1) = L\left\{\frac{1}{\sqrt{1 - \frac{v^2}{c^2}}} - 1\right\}$$

The change of kinetic energy, therefore, is given by the expression

$$K_0 - K_1 = L\left\{\frac{1}{\sqrt{1 - \frac{v^2}{c^2}}} - 1\right\}$$

Now, we can simplify the right-hand side of the equation considerably by using the binomial theorem, and neglecting terms such as v^4/c^4, which are negligibly small compared even to v^2/c^2.[15] We use the fact that

$$\frac{1}{1 - \frac{v^2}{c^2}} = \left(1 - \frac{v^2}{c^2}\right)^{-1/2} = 1 + \frac{1}{2}\frac{v^2}{c^2} + \frac{3}{8}\frac{v^4}{c^4} \cdots$$

to conclude that

$$K_0 - K_1 = \frac{1}{2}\frac{Lv^2}{c^2}$$

But, now, note that $K_0 - K_1$ is the change in kinetic energy, that is, $\frac{1}{2}mv^2$ under classical physics. Therefore

$$\frac{1}{2}mv^2 = \frac{1}{2}L\frac{v^2}{c^2} \qquad \text{or} \qquad m = \frac{L}{c^2}$$

From this, Einstein concluded:

If a body gives off the energy L in the form of radiation, its mass diminishes by L/c^2. The fact that the energy withdrawn from the body becomes energy of radiation evidently makes no difference, so that we are led to the more general conclusion that:

The mass of a body is a measure of its energy-content; if the energy changes by L, the mass changes in the same sense by $L/9 \times 10^{20}$, the energy being measured in ergs, and the mass in grams.

[15] The binomial theorem, developed by Newton, is an extremely important tool, constantly used by the modern physicist. See Appendix III for a discussion of how approximation formulas, including the binomial theorem, are used.

It is not impossible that with bodies whose energy-content is variable to a high degree (e.g., with radium salts) the theory may be successfully put to the test.

If the theory corresponds to the facts, radiation conveys inertia between the emitting and absorbing bodies.[16]

Einstein refers constantly in this paper to the *inertia* rather than *mass.* Thus, there culminates a long development, starting with Newton's definition of mass as "quantity of matter," which carried with it so much of our intuitive (and Aristotelian) sense of substance. Inertia, of itself, can hardly be identified as an innate property of objects. It can only be defined in terms of reaction with other bodies. We measure inertia by the effect that one body has upon some other body external to it. And, now, the unity of physics is once more advanced with Einstein's formulation. Light energy possesses inertia just as much as do "ponderable" bodies, which are attracted to other "ponderable, massive bodies." It therefore possesses "mass." Furthermore, just as "energy" possesses mass, we can think of mass in terms of its energy equivalence, and the two terms become a different method of describing the same quantity in the universe. It did not take long for this prediction of Einstein to be fully confirmed; today, the equivalence of mass and energy and the equation of their transformation has become one of the great generalities of modern physical science.

So important is the equation $e = mc^2$ that we might find it instructive to see how it can be arrived at from another approach, which does not involve radiant energy and which is more straightforward and more generally used.

The law of conservation of momentum is another basic law of physics that is to be preserved. Classically, momentum was usually defined as the product of mass times velocity. Here, again, if (*1*) we are to replace the classical statement of the principle of addition of velocities, and (*2*) if the quantity called momentum is to be conserved, (that is, if it is to be an invariant property regardless of any relative motion of the coordinate systems), we are thereby forced to modify the notion of an invariant mass. How is this to be accomplished?

Rather than answer with a strictly rigorous development of how we are forced to change the concept of momentum (which is a vector quantity having components in all three spatial directions), we summarize the gist of the analysis as follows.[17]

Suppose we have two masses m_1 and m_2 constrained to move only in the x direction. They move with velocities u_1 and u_2 relative to a "stationary" frame of reference before collision, and with the velocity u_1' and u_2' after collision. Under classical physics, the momentum P, defined as $(m_1u_1 + m_2u_2)$, is conserved, and

[16]Reference 7, p. 71.

[17]For those who may wish to follow the more rigorous analysis, see *Introduction to Modern Physics,* Richtmyer, Kennard, and Lauritsen, McGraw-Hill, New York, 1955, pp. 64 ff.

is equal to $(m_1u_1' + m_2u_2')$. The total momentum P is an invariant. Furthermore, for any other frame of reference moving uniformly relative to the original frame of reference, *if we use the Newtonian principle of addition of velocities,* the total momentum is likewise conserved for this frame of reference. If this is true, then substituting a new transformation equation for the Newtonian principle of addition of velocities will require either that the law of conservation of momentum be abandoned, or that momentum must somehow be redefined in such a way that it is a function not only of the velocity of the material particle but of the velocity of light as well. The theory of relativity does precisely this, through a redefinition of momentum and mass.

Note that momentum in classical physics is a three-dimensional vector. We should therefore in relativity theory, look for a four-dimensional vector expression for momentum, since space and time measurements in this theory cannot be separated if velocities are high and distances are large. Let us look for appropriate new expression for such a four-dimensional generalization of momentum. Suppose that the mass of a particle starting at $x = 0$, at $t = 0$ travels at a constant velocity v in the direction x with respect to an observer at rest in the x, y, z, and t reference system. Momentum, then, is defined by reference to this system of coordinates as $mv = mx/t$. But, for an observer traveling with the mass, his time coordinate will be different from that of the first observer and must have the value called his "proper" time of t', related to t by

$$t = t'\sqrt{1 - \frac{v^2}{c^2}}$$

If we then redefine momentum, replacing mx/t by m_0x'/t', the new equation for momentum becomes

$$P = \frac{m_0x'}{t'\sqrt{1 - \frac{v^2}{c^2}}} = \frac{m_0v}{\sqrt{1 - \frac{v^2}{c^2}}}$$

To hold to the classical conception of momentum as mass times velocity, then, requires a modified definition of mass, that is,

$$m = \frac{m_0}{\sqrt{1 - \frac{v^2}{c^2}}}$$

Obviously, such a definition is negligibly different from the older one except at very high velocities, since v^2/c^2 is ordinarily such a small fraction.

Passing over the mathematical details, it may easily be shown that with such a redefinition of momentum applied to problems of collision, etc., momentum is conserved for all inertial frames of reference (that is, frames of reference moving with uniform relative velocities), even at high velocities, using the postulate of relativity!

This means, for example, in the collision between two masses m_1 and m_2 traveling with velocities v_1 and v_2 relative to a frame of reference, the sum of the momenta is conserved, that is, that

$$\frac{m_1 v_1}{\sqrt{1 - \frac{v^2}{c^2}}} + \frac{m_2 v_2}{\sqrt{1 - \frac{v^2}{c^2}}}$$

remains constant regardless of their interaction and regardless of what inertial frame of reference is used to specify the velocities.

Furthermore if, following Einstein, we define the total energy of a mass as $E = mc^2$, we may then write for energy

$$E = \frac{m_0 c^2}{\sqrt{1 - \frac{v^2}{c^2}}}$$

As before, for small velocities compared with the velocity of light, the expression

$$\frac{1}{\sqrt{1 - \frac{v^2}{c^2}}} = \left(1 - \frac{v^2}{c^2}\right)^{-1/2}$$

may be expanded by the binomial theorem to the series

$$1 + \frac{1}{2}\frac{v^2}{c^2} + \frac{3}{8}\frac{v^4}{c^4} \cdots$$

with continuously higher powers of v/c. Since v/c is a small fraction, v^4/c^4 and higher terms may be neglected, and the expression for energy then becomes

$$E = m_0 c^2 \left(1 + \frac{1}{2}\frac{v^2}{c^2}\right) = m_0 c^2 + \frac{1}{2} m_0 v^2$$

This was interpreted by Einstein to mean that the energy associated with any particle is composed of two types: its permanent internal "rest" energy $m_0 c^2$ and its kinetic energy $\frac{1}{2} m_0 v^2$. The latter term is of the same form as the classical kinetic energy. In the nuclear reactions, which are studied in the next chapter, this transformation of the rest mass into energy always follows the equation proposed by Einstein.

5. THE GENERAL THEORY OF RELATIVITY

With the widespread acceptance of the special theory of relativity, physicists led by Einstein were naturally impelled toward an even broader generality. If laws of physics can be so stated that they are the same regardless of the relative *uniform* motion of the system of reference, why not attempt to formulate them so that they are independent of *accelerated* motions? The splendid idea occurred to Einstein that perhaps the key was in the equivalence of inertial mass and gravitational mass. It had long been known that the mass of a body as measured by gravitational attraction (that is, its weight) is equivalent to the mass of the same body as determined by its inertia (that is, its resistance to acceleration). Because of the way Newton formulated his law of gravitation

$$F = G\frac{m_1 m_2}{R^2}$$

and perhaps also because he had given up speculating about an explanation for his equation, the remarkable equivalence of inertial mass and gravitational mass had not been given much theoretical attention (although a careful series of experiments by Eotvos had been carried out and the equivalence had been established with great precision).

In 1911, Einstein published his first tentative conclusions in a paper "On the Influence of Gravitation on the Propagation of Light." Here, he not only developed a portion of the general theory of relativity but proposed an experimental test. If mass and energy are fundamentally of the same nature, radiant energy such as light must possess inertia. Further, if inertial mass and gravitational mass are equivalent, then, Einstein concluded, light energy should be acted on by gravity; it should be accelerated (that is, bent out of its path, if it passed close to a large body such as the sun). In his paper, Einstein, on the basis of Newtonian concepts and Euclidean geometry, calculated that the angular deflection of a ray of light grazing close to the sun would be given by the equation $\theta = 2GM/c^2R$, measured in radians (it will be recalled that 2π rad = 360°), where G is the constant of gravitation, M the mass of sun, c the velocity of light, and R the radius of the sun. This first calculation leads to a deflection of 0.87 sec of arc (as we shall see, this calculation was later modified).

But Einstein had thus far only begun to examine the consequences of the equivalence of gravitational mass and inertial mass. He had not yet formulated his theory in terms of which laws of physics are invariant among coordinate systems that may be in relative *accelerated* motion.

In his autobiographical notes Einstein reviews his thinking during this period:

> The fact of the equality of inert and heavy mass thus leads quite naturally to the recognition that the basic demand of the special theory (invariance under Lorentz transformation)

is too narrow, i.e., that an invariance of the laws must be postulated also relative to non-linear transformations of the coordinates in the four-dimensional continuum. This happened in 1908. Why were another seven years required for the construction of the general theory of relativity? The main reason lies in the fact that it is not so easy to free oneself from the idea that coordinates must have an immediate metrical meaning.[18]

The idea was gradually taking shape in Einstein's mind that ordinary Euclidean geometry with its fixed metric must be abandoned as a universal manner of defining events in space and time. Under special relativity, which used a Euclidean geometry of four dimensions, space remained isotropic, that is, it had the same properties in all directions, regardless of the presence of masses. The transformation equations between events in different inertial systems were linear and easily expressible in a Euclidean metric. However, to embrace accelerated systems of reference as well as those in uniform motion into a more general theory, Einstein found that he had to use a non-Euclidean geometry in which the coordinates defining space and time are "curved" by the presence of masses. Under this new geometry, space no longer has the same properties in all directions.

Unfortunately, the detailed development of general relativity theory requires a much more complicated mathematical analysis than does special relativity, and we shall not attempt to present it. The main ideas, however, can be outlined along the lines suggested by Einstein. First, he concentrated on the meaning of the equivalence of gravitational and inertial masses, showing by illustration the impossibility of distinguishing by any physical method between a uniform gravitational field and accelerated motion. As one example, he asks us to suppose than an elevator is suspended above the earth in space. Now, if the elevator were allowed to fall freely, then all objects within it would move downward (relative to the earth) with the same acceleration, and no gravitational effect could be detected. (This would be effectively the condition of "weightlessness," made so familiar to us today by the reports of orbiting astronauts.) If a co-ordinate system is established under these conditions, it becomes *an inertial system.* All of Newtonian dynamical laws would hold; an object set in motion would continue to move with a uniform velocity in a straight line (until it struck the elevator wall) relative to the elevator. Thus, by choosing the proper axes of reference, in this case the elevator, we have removed or "transformed away" the gravitational field. Or, if we pursue the opposite line of thought, if we imagine an elevator (or a rocket ship) somewhere far out in space where the earth's gravitational effect is negligible, and give it an acceleration relative to the distant stars, we can create within the elevator what appears to be a uniform gravitational field. (In some suggested plans for large space ships, in order to reintroduce a gravitational field, which may be

[18]*Albert Einstein, Philosopher-Scientist,* edited by Paul A. Schilpp, Library of Living Philosophers, Inc., Open Court Publishing Co., LaSalle, Ill. 1949, p. 67.

necessary for efficient human operations, it is proposed that the whole space ship be put into rotation, which would create a radially outward field of force.)

From such examples it becomes clear that a uniform gravitational field can always be shown to be equivalent to changing the axes of reference. In classical physics, some "forces" accompanying acceleration, such as "centrifugal forces," were regarded as being due to motion relative to space; other forces were regarded as gravitational forces between masses. Einstein suggested that all such apparent forces are due to the properties of space-time, which properties are influenced by the presence of mass. In short, the Euclidean space-time metric, thought of as an isotropic and empty container, is replaced by a non-Euclidean space-time, in which the metric varies from location to location within the universe depending on the presence or absence of matter.

Working upon this basis, Einstein published his paper in 1916, called *"The Foundation of the General Theory of Relativity."*[19] in which he developed the necessary analytic tools and in which he returned again to the problem of the deflection of a light ray passing close to the sun. He now showed that under the new formulation a ray of light will not only have the deflection predicted on the basis of the special relativity but, due to the non-Euclideanism of space-time close to the sun, there is an additional deflection. The calculated value on this basis comes out as a first approximation to be $\theta = 4GN/c^2R = 1.75$ sec of arc (which was twice as great as the earlier calculation). In several observations during total eclipses of the sun, deflections have been observed. These tend to confirm Einstein's prediction, although they cannot be regarded as definitive because of the large errors of observation that are entailed. In the 1919 observations, the deflections calculated by one expedition were between 1.31 and 1.91 sec of arc; in another, between 1.86 and 2.10 sec of arc. Observations since then give values somewhat higher than Einstein's prediction but are close enough to lend credence to his theory.

Another consequence of the general theory of relativity is that just as "space" is distorted by the presence of mass, so is time. A clock will slow up a very small amount in the presence of a gravitational field. The amount is so slight as to defy ordinary means of detection. However, in 1960 a carefully designed "atomic clock," that is, an instrument that keeps extraordinarily precise frequency measurements based on atomic phenomena, was used in a 70-ft tower built at Harvard University to test whether one clock, closer by 70 ft to the center of the earth than a second clock, would keep slower time. The test agreed with Einstein's prediction.

The third piece of evidence, which is commonly cited in support of Einstein's general theory, concerns the orbit of Mercury. According to Newtonian physics, a planet will travel in a plane in an elliptical orbit. This

[19] *Annalen der Physik,* Vol. 49, 1916.

elliptical orbit will have a fixed configuration in space (relative to the sun) if undisturbed by gravitational forces of other planets. The presence of the other planets, however, causes a "perturbation" of the orbit. In the case of Mercury, which is close to the sun, small in mass, with a relatively high eccentricity of orbit (that is, the orbit departs from a circle, or is more "flattened" than in the case of other planets), the perturbation consists in causing its perihelion (point of closest approach to the sun) to advance a small amount with each revolution. This advance of the perihelion has been observed with great care over many years, and Newtonian theory can almost, *but not quite,* account for it on the basis of the gravitational effect of the other planets. The discrepancy amounted to 43 sec of arc per century. According to Einstein's general theory, the perihelion should advance by

$$\theta = 24\pi^3 \frac{a^2}{T^2c^2(1 - e^2)}$$

(where a is the semimajor axis, e the eccentricity, and T the period of revolution) in each revolution. This equation implies a secular advance of 42 seconds of arc per century, even if no other planets were present. The close agreement between 42 seconds of arc and the "unaccounted for" 43 seconds on the basis of Newtonian theory is striking if nothing else.[20]

However, Einstein's *general* relativity theory, unlike his *special* theory, has had much less solid evidence for its support. The special theory has gained universal acceptance because of the very widespread evidence from innumerable, different, and independent laboratory observations. The evidence for the general theory is narrow and, as yet, slight. As was pointed out in the first chapters of this book, an hypothesis is never *proved* to be true because its consequences are in agreement with observation or evidence. Other hypotheses or theories may equally well lead to the same consequences. It is only when the hypothesis or the theory leads to a large and widespread group of consequences that cannot be "explained" on the basis of any other theory that the hypothesis or theory in question becomes more generally accepted as a law or fundamental principle of the science. Einstein's general theory of relativity has not today reached this status, although his special theory has.[21]

[20]In 1967 an alternative explanation of the anomaly of Mercury's perihelion advance was presented by R. H. Dicke. He calculated that the sun's departure from sphericity could be used to explain the degree of precession of Mercury's orbit. Should Dicke's theory be confirmed the general theory of relativity might have to be modified. Nevertheless the principle of equivalence of inertial and gravitational mass would remain valid and the general theory would still retain its great significance for modern physics.

[21]The essay of E. A. Milne, "Gravitation without General Relativity" in *Albert Einstein, Philosopher-Scientist,* edited by Paul A. Schilpp, Library of Living Philosophers, Inc., Evanston, Ill. 1949, Open Court Publishing Co., La Salle, Ill., and subsequent papers by him may be cited as alternative approaches to the problem of gravitational force fields.

In spite of all these reservations about the final adequacy of Einstein's general theory, its influence has been most significant for both physics and philosophy.

First, it has thrown new light on the nature of geometry itself. It might be said that we have two distinct types of geometry today: the mathematician's systematic logical structure, which is a construction of pure symbolic form having no direct reference to physical reality, and "physical" geometry whose nature is to be determined by empirical tests. To a mathematician, Euclidean and the several types of non-Euclidean geometry are all equally valid. To a physicist, the type of geometry that is ultimately chosen will depend upon the measurements and observations to be carried out. From the approach proposed by Einstein, for example, a satellite in orbit about the earth runs its course not because of an external force caused by gravity acting at a distance between the mass of the earth and the mass of the satellite, but rather because of space-time "curvature." Just as a great circle path is the shortest distance between two points on the surface of the earth determined by the radius and shape of the earth, so the orbit of the satellite is a "minimum" path (called by physicists a *geodesic*), which it travels automatically. Furthermore, if such motion relative to the earth contracts the dimension of the satellite in the direction of its motion, it will be tracing out a circle whose circumference is no longer π times the diameter of the circle, but somewhat less. Thus, we have a physical circle, which has a circumference no longer specified by $2\pi R$ as in Euclidean geometry. The departure may be exceedingly small, so small as to make measurement difficult, but *any* departure is surely of great significance.

Second, the problem of the kind of space and time that dominates the universe gives rise once more to many of the ancient problems of cosmology. Is the space in which we live infinite? If it is found that over very large distances one type of non-Euclidean geometry is valid, the universe would be finite, not in the old Aristotelian sense that it would be bounded by an outer sphere, but in the sense that no matter how one proceeded outward in one direction he would return eventually to his starting point. We shall discuss this and related questions in Chapter 24 on cosmology.

Finally, general relativity has taught scientists to be more critical of the common sense notions that tend to dominate their approach to investigations. Werner Heisenberg, one of the developers of quantum theory and the originator of the uncertainty principle, has emphasized this point. Speaking of the general theory, he says:

> It was the first time that the scientists learned how cautious they had to be in applying the concepts of daily life to the refined experience of modern experimental science . . . This warning later proved extremely useful in the development of modern physics, and it would certainly have been still more difficult to understand quantum theory had not the success

of the theory of relativity warned the physicists against the uncritical use of concepts taken from daily life or from classical physics.[22]

Beyond the realm of science proper, such adventures in critical thought as Einstein's theory have also had their influence because of the emphasis on the necessity of a continuous re-examination by each age of the "common-sense prejudices" it has inherited.

6. QUESTIONS

1. What major contradictions and unsolved problems within classical physics faced the scientist at the turn of the century?

2. What would happen to Einstein's equations of special relativity if the speed of light were infinitely high?

3. Explain why it took so long to discover that Newton's law of the addition of velocities was only approximately true.

4. Consider carefully how you would define operationally an "inertial reference system." In Newton's theory, "absolute space" constituted the basic inertial system. Relative to this reference the velocity of light would vary according to the principle of addition of velocities. If the velocity of light is constant, how can you define two systems (which may be moving with *constant* velocity relative to each other) as being in the same inertial system?

5. What is meant by a "transformation equation?" Using the equation of a circle as an example, how do you "transform" the equation for a circle whose center is at the origin in one set of coordinates to an equation whose center is at the origin of a new set of coordinates, such that $x' = x + h, y' = y + k$. What properties of the circle remain invariant under this coordinate transformation?

6. State as cogently as you can the contradiction between classical theory and experiment that led Einstein to his special theory. Then notice that to escape the contradiction, Lorentz had previously suggested that distances were contracted when an object moved in absolute space. This was an alternative explanation of the negative results of the Michelson-Morley experiment. Why, in time, was Einstein's explanation regarded as superior?

7. Distinguish carefully between the general theory of relativity and the special theory of relativity. Why is the latter generally accepted today by all physicists, although the former is still regarded as subject to modification and reinterpretation.

8. Sometimes those aspects of the universe that are accepted without question lead to the most profound problems. Why was the equivalence of inertial mass and gravitational mass never regarded as puzzling until Einstein considered the problem? Why was Newton's definition of mass as "quantity of matter" circular? State two alternative equations that might be used by classical physics to define mass (consider both gravitation and inertia).

[22] Werner Heisenberg, *Physics and Philosophy,* Harper, New York, 1962, p. 127.

9. A moving picture is a series of discontinuous static pictures. We can "stop" the picture and identify the elementary constituents. Consider the possibility that space and time are ultimately discontinuous. Does this have special meaning? Mathematical meaning? If you wish to examine this perennial problem, look up Zeno's paradoxes and various proposed solutions.

10. Gauss, in one of his letters (referred to by Max Jammer in *Concepts of Space,* Harper, New York, 1960, p. 125), indicated that he believed that the question as to the number of dimensions of space is similar to the problem of whether it is Euclidean or non-Euclidean. In both cases, the decision is to be made on the basis of some external criterion, foreign to pure mathematics. Explain this point of view of Gauss, and suggest the type of criteria that might decide what kind of space we live in.

11. Contrast Gauss's point of view with that expressed by Kant and others who argued that space must be three-dimensional, otherwise gravitational forces would not vary inversely as the square root of the distance.

12. Some elementary particles such as mesons have very short half-lives. They come into our atmosphere with extremely high velocities. On the basis of relativity theory, why would you expect them to have greater half-lives when they are traveling at high velocities? How would such evidence support Einstein's theory that moving clocks slow down?

13. Suppose that gravitational and inertial masses differed by 10 per cent. Can you suggest an experiment which might be performed to establish this difference?

14. Although Einstein's general theory of relativity may ultimately prove unsatisfactory, the development of this theory has had certain philosophical implications, particularly regarding the nature of mathematics. Discuss one of them.

Problems

1. How fast would an object have to travel for its mass to be doubled? Compare this velocity with the tangential speed of an earth satellite.

2. Show that $1/\sqrt{1 - (v^2/c^2)}$ is approximately $1 + \frac{1}{2}(v^2/c^2)$. (*Hint:* Use binomial theorem.) Then show that since for a particle at rest $E = mc^2$, and for a particle in motion $E = mc^2/\sqrt{1 - (v^2/c^2)}$ that the kinetic energy of a particle moving with velocity v approaches the classical kinetic energy equation.

$$\text{k.e.} = \tfrac{1}{2}mv^2$$

3. Plot a classical graph of kinetic energy against velocity and a relativistic graph of kinetic energy against velocity. (On the abscissa, let the values of velocity become greater than c.)

4. What is the mass of a proton which has a velocity of 2×10^8 m/sec?

5. Express the rest mass of a proton in terms of energy.

6. Show by how much the half-life of a meson will be increased if it has a rest mass of 30 MeV and a half-life of 2×10^{-8} seconds.

7. If approximately two small calories per minute of radiant energy reach the earth from the sun, calculate the approximate loss of mass of the sun per minute. What percentage is this of the total mass of the sun?

Recommended Reading

G. Gamow, *Mr. Tompkins in Wonderland,* Cambridge University Press, New York, 1957. An entertaining account of modern physics in story form.

Lincoln Barnett, *The Universe and Dr. Einstein,* Harper, New York, 1948. A good, although elementary, discussion of the theory of relativity. The main ideas are simply discussed, and a minimum of mathematics is involved.

Albert Einstein, *Relativity: The Special and General Theory* (Masterworks of Science), Doubleday, Garden City, N.Y., 1947. Einstein wrote this account of relativity specifically for the layman.

Albert Einstein and Leopold Infeld, *The Evolution of Physics,* Simon & Schuster, New York, 1942. An account of relativity theory in the context of the broad development of physics. Nonmathematical.

Bertrand Russell, *The A B C of Relativity,* a Mentor Book published by the New American Library of World Literature, New York, 1960. A more philosophical approach to the theory of relativity.

Philipp Frank, *Relativity—A Richer Truth,* The Beacon Press, Boston, 1950. An analysis of the broader implications and applications of relativity theory for ethics and social sciences, and an examination of the problem of the relativity of ethics.

Clement V. Durrell, *Readable Relativity,* Harper Torchbooks, Harper, New York, 1960. A serious, careful statement of relativity theory, which makes liberal use of mathematics although calculus is avoided. For a mastery of the ideas in this book, it is recommended that all problems be worked out.

Max Jammer, *Concepts of Space,* Harper Torchbooks, Harper, New York, 1960. A historical account of the developments leading to the modern concept of space.

Peter G. Bergmann, *Introduction to the Theory of Relativity,* Prentice-Hall, Englewood Cliffs, N.J., 1942. A straightforward presentation of relativity theory, involving the use of higher mathematics. For those acquainted with calculus, the early chapters on special relativity can be read with profit.

chapter twenty-one

Nuclear structure

A much-quoted aphorism regarding flux and change has come down to us from Heraclitus. He said, "A man cannot step twice into the same river." To this a logical and humorous extension has been made that "A man cannot step *once* into a river." Although we agree that "man" and "river" change from one instant to the next, both recognition and predictability require some substratum of "reality," whether it be "substance," "thingness," "matter," or "mass." It is this substratum that changes fast or slowly, according to some sort of rules. Even when phenomena change in a way that seems completely baffling to the intelligence, as in dreams, we are not satisfied until some understanding of the principles of change is attempted. To reach greater comprehensiveness and more accurate predictability than can be achieved by "common sense," science has been forced to develop abstract concepts that are only indirectly related to sensed phenomena, and to develop very strict rules of "transformation" in terms of which changes may be precisely defined.

It was the hope in the nineteenth century that at last the ultimate substance of which the universe was constituted could be found in the chemical atoms. The atom, as such, was thought of as an indivisible undifferentiated permanent unit of matter. In earlier chapters we have traced the twentieth-century theory that the atom has a structure and that it is composed of two parts: a positively charged nucleus and a surrounding cloud of electrons. As investigations proceeded from Rutherford's and Bohr's theories, it soon became evident that even the nucleus has a structure that is to be explained. A study of this structure, including the analysis of the particles composing it, the forces binding it together, and the transformations to which it is susceptible, constitutes the very broad field of scientific investigation called nuclear physics. Since the early 1930s, much effort and attention has been devoted to unlocking the secrets hidden in the nucleus. Tremendous successes have been achieved but, for each success, new mysteries have appeared. So broad is the new field of investigation that we can attend to only a few of the more prominent concepts, making use of the ideas about quanta and relativity, which we have already developed.

1. ISOTOPES; DISCOVERY OF THE NEUTRON; NUCLEAR EQUATIONS

Rutherford's scattering experiments provided convincing evidence that the atom is not a simple structure, but is composed of a very small, very massive nucleus surrounded by a cloud of electrons. But the problem remained: What is the "stuff" out of which the nucleus is composed? Is it the primordial "substance" of the universe, and what are its properties? To answer these questions, intensive efforts were made to collect as much information as possible regarding all radioactive processes (which were recognized as involving the nucleus itself rather than the whole chemical atom) and the properties of the various particles shot off during the radioactivity. All the tools of chemistry and physics were used in attempts to unlock the secrets hidden within the nucleus.

One hundred years before Rutherford's work, William Prout suggested a theory that all chemical atoms were made up of multiples of hydrogen atoms, somehow bound together. He had been led to this hypothesis by the fact that most elements have atomic weights nearly equal to an integral number of times the atomic weight of hydrogen. But the theory was finally discarded because very careful measurements showed that some elements have *fractional atomic weights* based upon the atomic weight hydrogen as unity. With a closer study of the radioactive elements, however, Prout's theory of 1815 was again revived when it was shown by Soddy and others in 1913 that certain elements, such as radioactive ionium and radioactive thorium which are chemically identical, have different atomic weights (in this case, 230 and 232). Such elements, with different atomic weights but identical chemical properties, were given the name "isotopes" by Soddy. If such isotopes existed among radioactive substances, it was natural to assume their existence among other chemical elements. Chlorine, for example, which is found in nature always to have an atomic weight of 35.46, might be composed of two isotopes chemically identical but of different atomic weights, which produced the "average" value as determined by the chemists. Methods were soon devised to show this to be true, and that there is one isotope of chlorine with an atomic weight of 35, another with an atomic weight of 37, the former being more abundant in the mixture. Both isotopes of chlorine have 17 electrons, and thus have identical properties. The difference in atomic weight (or, more precisely, in atomic mass) can only be accounted for by a difference in the composition of the nucleus.

Meanwhile, Rutherford was continuing his investigation of the scattering of alpha particles. In 1919, he tried shooting alpha particles into nitrogen gas and studied the effects. Presumably, the alpha particle with an atomic mass of 4, upon striking the heavier nitrogen nucleus, would lose its energy and momentum. The nitrogen atom would be given a smaller velocity and would not travel far. However, Rutherford found, to his surprise, that although the alpha particles (whose presence could be detected when they struck a fluorescent screen) theorectically should not be able to penetrate through nitrogen more than about 7 cm, these particles actually produced some kind of bright spots on a screen that was 28 cm away from the source. Had a new particle been produced when the nitrogen atoms were bombarded by alpha particles? Further investigations were carried out with a "cloud chamber," a device by which the passage of individual particles through moist air can be traced. As the high speed particles rush past and through the atoms of air, they tear off electrons, ionizing the atoms. At once, droplets of fog condense about these ionized atoms and, properly illuminated, the path of the particle is made clearly visible. A study of the cloud chamber photographs (see Fig. 21.1) convincingly revealed that a new particle was formed. When an alpha particle (leaving a wide track in the cloud chamber) collides with a nucleus of nitrogen, a very thin, long track starts off from the point of collision. Rutherford was able to calculate from the characteristics of this track that the new particle formed had a mass of one quarter of the mass of an alpha particle and a positive charge equal in magnitude to that of an electron. This particle was named the *proton*. It was identical in mass with the nucleus of a hydrogen atom. Rutherford's experiment was actually the first example of artificial transmutation. The alpha particle (helium nucleus with atomic mass 4), colliding at high velocity with the nitrogen nucleus (atomic mass 14), produced an oxygen isotope (atomic mass 17) plus a proton (hydrogen nucleus with atomic mass 1).

A theory about the nucleus was now beginning to take shape, reviving Prout's hypothesis of a century before. In 1920, Rutherford suggested that there were both protons and electrons in the nucleus, with some of the protons so closely combined with electrons as to constitute neutral particles to which the name "neutron" was given. (At first it was considered probable that electrons must exist

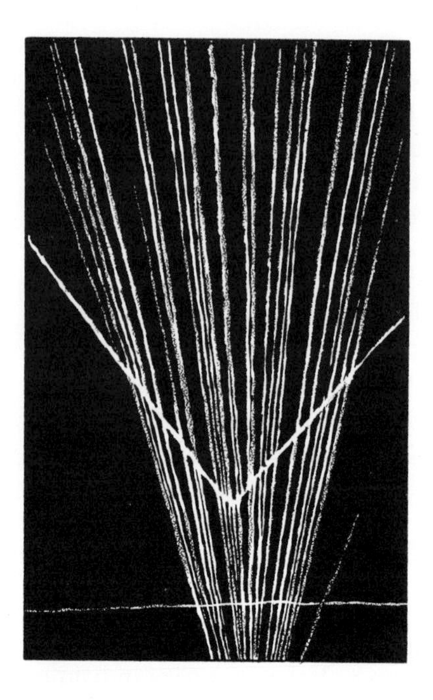

Fig. 21.1. Typical illuminated tracks left by alpha particles passing through a cloud chamber. Collisions and deflections are easily detected and photographed. (Photograph courtesy of P. M. S. Blackett, *Proceedings of the Royal Society,* **107A,** 349, (1925).

within the nucleus, since the emitted beta particles, which had been identified as electrons, had such a high velocity as to preclude their arising from the outer atomic shell of electrons. Later, as we shall see, it was shown that *free* electrons cannot exist inside the nucleus. When the electron is emitted the neutron becomes a proton.) On this theory, the nucleus of every *element* contained a certain number of neutrons and protons. The total number of protons determines the atomic number of the element (designated by Z) and, thus, the number of positive charges in the nucleus. This same number Z is also, of course, the number of electrons outside the nucleus in the neutral atom. Z thus determines the chemical properties of the element. The protons and neutrons were postulated by Rutherford to be identical in mass. Hence, the atomic mass of an element (designated by A) was accounted for by the total number of nucleons, the name given to protons and neutrons.

As has been the case with many particles, the existence of the neutron was suggested by theoretical considerations long before experimental confirmation was reached. A neutron is a very difficult particle to detect, since it carries no charge. It is not subject to magnetic or electric deflection because it has no charge and, hence, it is very penetrating. Furthermore it produces none of the ionizing effects that lead to particle detection in cloud chambers. In all these properties it resembles X-ray radiations. Its presence can be revealed only when it strikes a nucleus. It took 10 years after its existence was predicted before direct evidence for the neutron was actually produced. The apparatus designed by Chadwick in 1932, which confirmed the existence of the neutron, is diagrammed in Fig. 21.2.

Alpha particles are shot out of the radioactive source S and impinge upon a beryllium sheet B. It had been previously discovered by Curie and Joliot that this led to some kind of a reaction within the beryllium causing a very penetrating type of radiation, at first considered to be X-rays. The unidentified radiation impinging upon a paraffin sheet (a hydrocarbon rich in hydrogen atoms) in turn caused protons to be emitted in copious quantity and at high energy, whose velocity could be measured by an ionization chamber. Chadwick showed that for a balance of momentum and energy equations, the radiation from the beryllium could only be accounted for by assuming it consisted of neutral particles having the same mass as a proton. A long series of similar experiments confirmed the existence of the "neutron" predicted by Rutherford.

The basic structure of the nucleus and the principles governing nuclear reactions now became clear, and a systematic way for writing the equations describing such reactions was developed. With each element two numbers are associated: its atomic number Z, written as a subscript, and its atomic

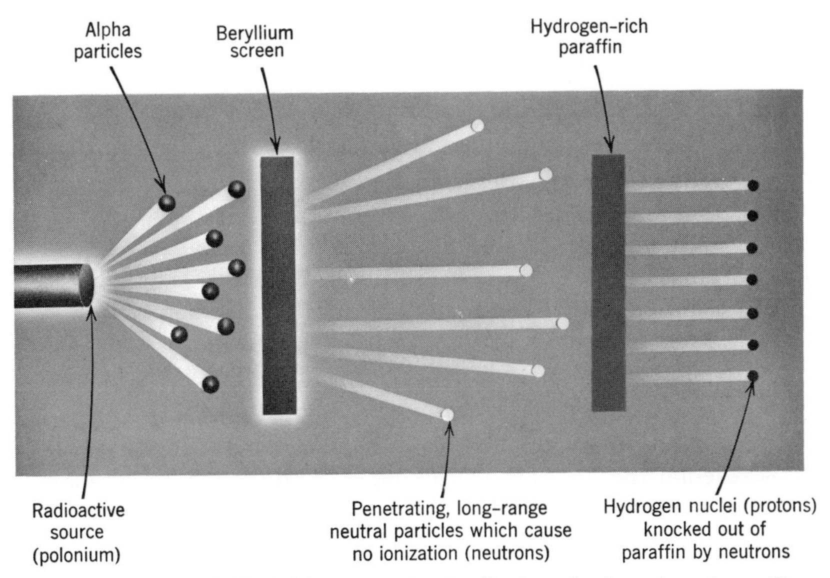

Fig. 21.2. Diagram of Chadwick's apparatus for the investigation of neutrons. The neutral particles cannot be directly detected (they carry no charge), but their presence is revealed when they collide with hydrogen atoms, causing an emission of positively charged protons.

weight A, written as a superscript. (Thus, $_8O^{16}$ signified oxygen with atomic number 8, atomic weight 16.) As an example, the reaction producing neutrons by bombarding beryllium with alpha particles is written as $_4Be^9 + _2He^4 \longrightarrow _6C^{12} + _0n^1$.

The term $_0n^1$ stands for a neutron with atomic number zero (it has no positive charge, and cannot form the nucleus of an atom).

It will be noted that the total number of nuclear particles or nucleons (13 in all) entering into the above reaction is conserved. This is true of all nuclear reactions.

Another example is the nuclear reaction studied by Rutherford, which we have mentioned as the first example of artificial transmutation. This is symbolized as

$$_7N^{14} + _2He^4 \longrightarrow _8O^{17} + _1H^1$$

In this it will be noted that an isotope of ordinary oxygen is produced as a by-product, having an atomic mass of 17 instead of 16 as in the case of normal oxygen. (Since the atomic number Z, that is, the number of positive charges or protons is the same for $_8O^{17}$ as for $_8O^{16}$, the electron cloud surrounding both atoms is identical and the chemical properties are therefore also identical.)

Natural radioactivity usually results in the emission of alpha particles and gamma rays (high energy X-rays that have zero rest mass). Thus radium disintegrates as

$$_{88}Ra^{226} \longrightarrow _{86}Rn^{222} + _2He^4 + \gamma$$

where $_{86}Rn^{222}$ is an inert gas similar to helium except that it is also radioactive and γ represents the gamma or X-ray.

On the other hand, some natural radioactive processes involve the emission of beta particles (high energy electrons emitted from the nucleus). When this occurs, the atomic number Z is *raised,* since the emission of a negative charge leaves an additional positive charge on the nucleus, although the atomic mass is unchanged (the mass of the electron being considered negligible compared to the mass of nucleons). Thus, one isotope of uranium, $_{92}U^{239}$, disintegrates according to the following scheme:

$$_{92}U^{239} \longrightarrow _{93}Np^{239} + e^-$$

In this reaction the product neptunium is of the same atomic mass as the uranium but is higher in atomic number and is therefore chemically distinct. (We shall find later that the reaction written above must be modified somewhat to account for another interesting particle, the neutrino, which is also emitted along with the electron. This additional particle, however, carries no charge and has zero rest mass. It therefore affects neither the atomic number nor the atomic masses of the nuclei involved.)

In the atomic bomb and other fission reactions, many different nuclear processes may take place, since a neutron that strikes an atom of uranium starting off the reaction may cause the uranium to "fission" in many different ways. Typical of fissioning reactions, however, is the following one, involving uranium 238:

$$_0n^1 + {}_{92}U^{238} \longrightarrow {}_{92}U^{239} \longrightarrow {}_{56}Ba^{145} + {}_{36}Kr^{92} + {}_0n^1 + {}_0n^1$$

where the intermediate uranium isotope $_{92}U^{239}$ is first formed and then breaks down into two large fragments, barium and krypton atoms, plus two neutrons, each of which is available as a "bullet" to disintegrate other adjoining atoms of uranium. We shall discuss the fissioning process in the next two sections.

2. ENERGIES AND FORCES WITHIN THE NUCLEUS

Up to now the picture we have drawn of nuclear processes has been qualitative, describing merely the changes that occur as one atom transforms into another with a gain or loss or rearrangement of protons and neutrons within the nucleus. All kinds of questions remain. What are the properties of these particles, the neutrons and protons? What is their size? What is the nature of the interaction between them as they collide with one another or with other nuclei? Above all, what forces bind them together?

Let us first consider the problem of the forces binding protons and neutrons together into the small central nucleus. It will be recalled that Rutherford, from his scattering experiments, estimated the size of the nucleus to be about 10^{-15} m. Many other methods of determining nuclear sizes have been used, and they yield comparable figures.[1]

If two protons are brought together as closely as this, the electrostatic repulsion between them should be tremendous, even for charges of such small magnitude. The use of Coulomb's law of repulsion, that is,

$$F = K\frac{Q_1Q_2}{d^2}$$

yields a force of 230 nt between each pair of protons. This force is extremely large for such small particles. If nothing else were acting to bind them together, we would expect all material things instantly to explode because of the repulsive

[1]The sizes of nuclei vary with the element. An approximate empirical equation has been developed relating the radius of the nucleus, R, to its atomic mass, A, as follows:

$$R = 1.3 \times 10^{-15}\sqrt[3]{A}\ \text{meters}$$

which also indicates that the nuclei have approximately the same "density." Why?

forces within the nuclei. We cannot account for their being bound together by gravitational forces, for the gravitational attraction between such small masses is 10^{35} times smaller than coulomb forces. Comparatively gravitational forces are so small as to be entirely negligible.

An entirely new force must therefore be at play, called the *binding force* between nucleons. This force, however, does not obey the same type of law as the inverse square laws of gravitational, electrostatic, or electromagnetic forces. It is extremely powerful over very short ranges, but becomes negligible at a distance greater than about 3×10^{-15} m, at which point the coulomb forces of repulsion become more powerful.

In order to deal quantitatively with the forces, and the energies resulting from them, it is convenient to use the units of atomic physics: the atomic mass unit (amu) and the electron-volt (eV). Up to now we have considered the neutron and proton to have an equivalent, unchanging mass. This concept must now be modified, and greater precision must be introduced, using Einstein's special theory of relativity and its consequent equation for mass energy equivalence, $e = mc^2$.

The *atomic mass unit* is based upon the most common isotope of carbon, which is arbitrarily given the mass of 12.0000 atomic mass units (being very close to 12 times the atomic mass of hydrogen).[2] On this basis, a proton has a mass of 1.00759 amu, although a neutron has a mass of 1.00898 amu. Furthermore, 1 amu has an equivalent mass equal to 1.66×10^{-27} kg. These values are established with this precision by a number of different experimental methods. One of the methods used in the case of charged particles is the mass spectrograph, which is an apparatus similar to J. J. Thomson's apparatus in which the charged masses are made to follow curved paths produced by magnetic or electric deflections. The principle of the mass spectrometer is diagrammed in Fig. 21.3.

Now, the mass of the neutron and of the proton, as given above, applies to them as individual particles and *not* when they are combined into a nucleus. The *mass of the nucleus differs from the sum of the masses of the constituent nucleons.* For example, an alpha particle (nucleus of helium atom, designated by ${}_2He^4$) has two protons and two neutrons. If we calculate the sum of the masses of two protons and two neutrons, using the figures given above, we arrive at 4.03314 amu. But experimental determination shows that the alpha particle actually has a mass of only 4.00278 amu. There is, therefore, a *mass defect* of 0.03036 amu. What accounts for this? The answer is found in the Einstein mass-energy equivalence. The protons and neutrons, in combining, lose a certain mass, which has been

[2] The International Unions for Chemistry and Physics endorsed the proposal to base the atomic weight scale on carbon 12 in 1960 and 1961. Before this, oxygen was given the exact weight 16. Under the new standard the atomic mass of oxygen becomes 15.9949 in amu.

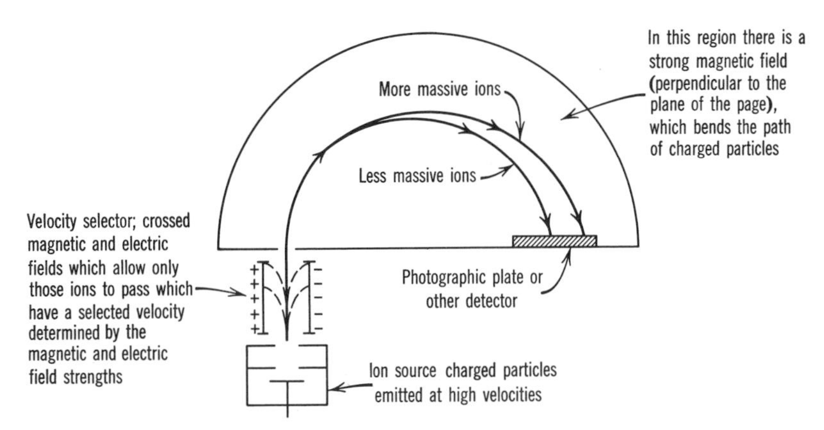

Fig. 21.3. Principle of the mass spectrometer. The charged particles are shot out from the ion source by an electrostatic field. The path that they take depends on (1) mass, (2) electric field strength, and (3) magnetic field strength. (2) and (3) may be varied, and the mass determined.

converted into energy. Let us calculate this energy in the case cited. The mass defect of 0.03036 amu converted into kilograms is $0.03036 \times 1.66 \times 10^{-27}$ or approximately 5×10^{-29} kg. Using the equation, e (in joules) $= m$ (in kg) c^2 (in m/sec)^2, we have $e = (5 \times 10^{-29})(3 \times 10^8)^2 = 45 \times 10^{-13}$ joules less energy in the helium nucleus than in the separate protons and neutrons. When it is considered that there are 6×10^{23} atoms in 4 g of helium, the figure represents a considerable amount of energy. The formation of 4 g of helium out of protons and neutrons would yield 2.7×10^{12} joules of energy (equivalent to about 25 million kilowatt-hours, enough energy to light a small city for several weeks).

Note that, in the combination of nucleons to form the helium nucleus, energy is given up. Thus, in order to split up the nucleus into its constituent parts, energy is required. Hence the term binding energy. To calculate the binding energy for any atom, its mass is first determined and expressed in atomic mass units. This figure is subtracted from the sum of the masses of the nucleons composing the atom (the masses of electrons may be disregarded, since they are comparatively negligible). The mass defect in atomic mass units is then expressed in energy terms.

The most common term in which energy is expressed in nuclear physics, is an electron-volt. It is the amount of energy acquired by an electron falling through a potential difference of 1 volt. In joules this becomes the charge on an electron (1.6×10^{-19} coul) times 1 V (which, it will be remembered, is equal to 1 joule/coul) or 1.6×10^{-19} joule. Since 1 kg of mass is equivalent to $(3 \times 10^8)^2$ joules, the atomic mass unit comes out equal to 931 million electron volts (expressed as 931 MeV).

If we look at the other elements in the periodic table we find in each case that they all have a certain mass defect, which is equivalent to a binding energy. However, the amount of this binding energy differs from element to element. This difference is not only due to the number of nucleons associated with the nucleus of each chemical atom. There is also a difference in binding energy per nucleon. If we plot the binding energy per nucleon against the atomic number we find that the values lie on a curve, which rises sharply for a while as the atomic number increases, reaches a peak and, then, descends toward the heavier elements (Fig. 21.4).

This curve shows the reason why the two types of atom bombs, with which the public is familiar, are possible. If two atoms of the heavy isotopes of hydrogen, $_1H^2 + {_1H^2}$, are combined to form helium $_2He^4$ the process of *fusion* must release energy, since the binding energy per nucleon of helium is greater than that of the hydrogen isotopes. On the other end of the curve, if an atom of uranium is fissioned into two atoms toward the central portion of the scale of atomic numbers, again the binding energy per nucleon becomes greater, and the difference in energy is released. Furthermore, it can be seen from the shape of the curve that there is a proportionately greater release of energy from the fusion than the fission process.

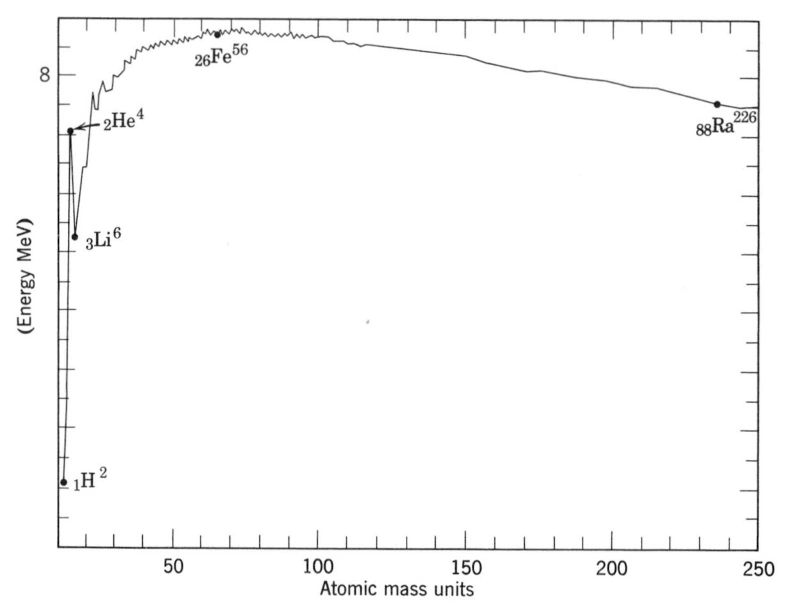

Fig. 21.4. Blinding energy per nucleon.

3. FISSION AND FUSION; ATOMIC BOMBS AND NUCLEAR REACTORS

Although Fig. 21.4 shows whether energy will be released, and how much when one or more atoms are combined or split apart, it tells us nothing about whether such reactions will take place. It is known that the fusion reaction involving the combination of light elements requires tremendous temperatures, such as are found in the interior of stars or the interior of atomic bombs. On the other hand, many of the very heavy atoms, such as radium, thorium, and uranium atoms, tend to undergo a type of natural, nevertheless lopsided, fission into alpha particles and some other lighter element in the process of radioactivity. But radioactivity proceeds at its own pace, and cannot be speeded up or retarded by any known means. Therefore, to obtain energy in copious quantity, artificial transmutation must be induced by shooting "bullets," usually neutrons, at high velocities to collide with the nuclei of the heavier elements, such as uranium, causing them to split into two fragments with the release of energy. This can be done readily on a small scale in the laboratory, but the energy released is small, since the number of collisions obtained involves only a fractional number of the bullets used.

This whole situation was radically changed when it was discovered in 1939 by Otto Hahn and F. Strassmann in Germany that not only could an isotope of uranium, ${}_{92}U^{235}$, be split up into two large fragments by being bombarded with neutrons with a large release of energy, but also that in *the splitting process a side product was the emission of two or more neutrons.* The possibility was at once apparent that a chain reaction might be started within uranium 235, each neutron colliding with another uranium 235 atom producing a new fission, and also generating additional neutrons available to strike other uranium 235 atoms, thus maintaining or extending the fissioning process (Fig. 21.5).

The discovery of the fissioning of U^{235} in the laboratory led scientists also to conclude that an atomic bomb might be possible. The United States government under Franklin D. Roosevelt made funds available to explore such a possibility. Enrico Fermi, an eminent physicist who had fled with his family from Italy was placed in charge of a laboratory project under the direction of Arthur Compton to demonstrate the feasibility of a chain reaction using uranium. On December 2, 1942, working in great secrecy, a group of scientists under Fermi were successful in producing the first chain reaction, after having amassed a large pile of uranium interspersed with graphite and cadmium rods, in a squash court at the University of Chicago.

Eugene Wigner gives an eyewitness account:

When our group assembled on that second of December, the pile was large enough so that it would have been chain-reacting without the control rods. We stood on the balcony of the

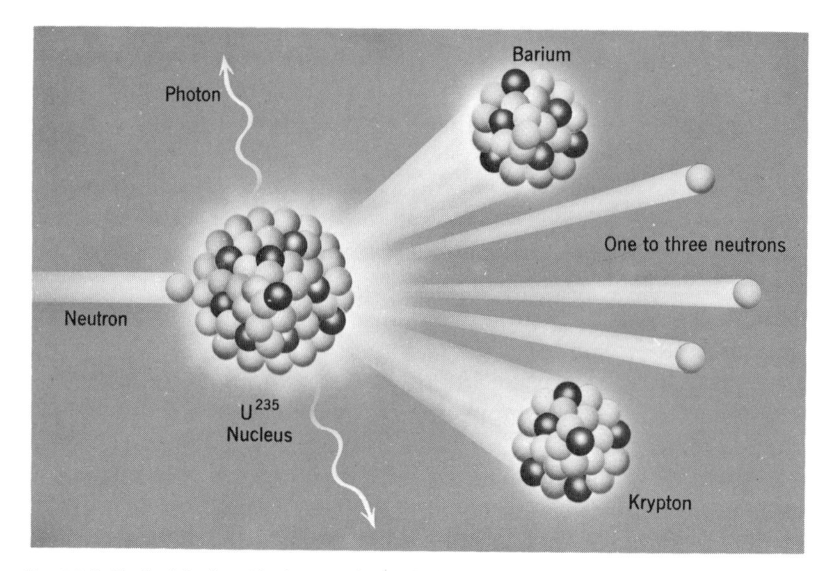

Fig. 21.5. Typical fission. Fission products, including neutrons, have smaller mass than original U^{235} (energy release is about 200 MeV). The released neutrons are available to produce additional fissions, unless they are absorbed.

court and Fermi directed the withdrawal of the rods in steps of about a foot each. After each step, the clicking of the counters started to speed up, but soon leveled off and reached a steady value. This steady clicking rate became higher and higher as the control rods were pulled out further and further. However, this was still "background," for the counting rate was still leveling off. Finally, at the last stage, the clicking (thus the flow of neutrons inside the pile) continued to increase and did not seem to approach a steady state. Left alone, in another few minutes the neutron count would have doubled, then doubled again during the same interval, and so on.

This meant that the self-sustaining chain reaction was established; the pile was "critical." When it was certain that this stage had been reached, Fermi had the control rods reinserted and the clicking died down. . .

I produced a bottle of imported Chianti from a brown paper bag on the balcony floor. Italian wine was appropriate because our leader, Fermi, was of Italian birth. He uncorked the bottle and we toasted the success of the experiment. As we drank the wine from paper cups, we sent up silent prayers that what we had done was the right thing to do.[3]

Thus came about a new chapter in the myth of Prometheus. Up to now, man's energy had always derived directly or indirectly from the sun. Now fire had been "stolen from the gods" within matter itself, and a new degree of independence

[3] *Twentieth Birthday of the Atomic Age* by Eugene P. Wigner in The New York Times Magazine, December 2, 1962, copyright by the New York Times, reprinted by permission.

from the sun was won. Although the immediate motive was the production of a weapon to destroy, man had found a vast new source of energy available for productive enterprises, if he should choose to live in peace.

Let us examine more closely the details of the fissioning process as it occurs in the chain reactions. The general condition for a sustained reaction is, of course, the presence of atoms that will fission and, in the process, will give off neutrons in such abundance and of such energy levels as to maintain further fissions either at a continuing and steady rate (as in nuclear reactors) or at a fast and explosive rate (as in atomic bombs). Not only must each fission result in the production of more than one neutron, but too high a proportion of the neutrons must not be "wasted." Inevitably, many of the neutrons will either escape from the fuel pile altogether or will have their energies dissipated by absorption in nonfission processes. Just the right balance of neutrons used to neutrons wasted must be maintained.

The first requirement is that the fissionable material must be present in a sufficiently large volume; otherwise, too large a fraction of the neutrons generated will escape through the surface of the fuel before they have collided with a neighboring nucleus to cause fission. (It should be recalled that although the atoms may be packed together tightly, as in a solid sphere of metal, the nuclei are separated by relatively large distances. Since neutrons carry no charge they can pass right through atoms with ease, unless they happen to strike the small nucleus of the atom which occupies less than one-ten-thousandths of the volume of the atom itself.) Again we encounter the significance of the ratio of surface area to volume, which plays such an important role in nature. An analogy is useful. There is a minimum or critical size below which warm-blooded animals cannot exist because in small animals the *surface area* through which heat escapes is large relative to the *volume* of the animal, and hence to the volume of food intake that provides requisite energy for heat and locomotion. (The ravenous shrew being probably close to this limit must eat to support its own life almost continuously, and will starve to death if it goes without food for more than a few hours.) Likewise there is a critical size below which fissionable material will not support a chain reaction. In one case, heat escapes faster through the surface than replacement energy can be supplied by food; in the other case, a greater number of neutrons escape through the surface than are being generated by fissioning processes.

Other conditions must also be met that dictate the type of fuel that can be used. When struck by neutrons of sufficiently high energy (that is, high velocity), almost any nucleus may be caused to fission, but the product neutrons, which result, may not have the right energy to produce additional fissions. What is required is a material that will be fissioned by neutrons through a large range of energies, preferably in the low energy or "thermal" range (that is, where the average velocity is about the same as the velocity of molecules due to temperatures of about 20°C). Only three isotopes, uranium^{233}, uranium^{235}, and plutonium^{239}, happen to have this property and, of these, only U^{235} occurs in nature. Natural uranium consists almost entirely of the isotope U^{238} (99.3%) mixed with an extremely small portion of U^{235} (0.7%). Although U^{238} can be fissioned and, in the process, does give off more than one neutron, it cannot sustain a chain reaction, no matter how large a volume is brought together. The reason is twofold: (*1*) to cause U^{238} to fission, a very fast neutron is required; and (*2*) many of the neutrons that are fast enough originally to cause fission are either captured by neighboring U^{238} atoms causing a transmutation to Pu^{239} instead of causing fission, or are slowed down by a series of collisions. Once they are "slowed down" to thermal velocities, the reaction is no longer propagated. Too high a proportion of product neutrons are thus "wasted" for a sustained chain reaction in either U^{238} or natural uranium, unless special provision is made to prevent the undesirable absorption.

It is quite different with U^{235}. In this case, both fast and slow neutrons can cause fission; there is no lower limit or threshold to the process. If, then, a sufficient quanitity of pure U^{235} is brought together into a "critical" mass, a chain reaction will inevitably ensue because the probability of a given neutron causing fission in a neighboring atom is large compared to the probability of its "escape." Below the critical volume, no chain reaction is possible; above this volume, it is statistically certain. Such is the principle used in the construction of the atomic bomb. Subcritical masses of U^{235} are kept apart until the bomb is to be detonated. When the explosion is desired these subcritical masses are brought together with extreme suddenness, either by a mechanical means or by a subsidiary chemical explosion of TNT or gunpowder. This sudden assembly of the subcritical masses into a critical one must be accomplished with great rapidity for an efficient bomb (in less than one ten-thousandth of a second), otherwise as the critical size is reached, the explosion will start prematurely and will blow apart the fuel long before complete fission is obtained. An atomic explosion of this kind requires almost pure U^{235} (or plutonium obtained artificially from reactors). Because this isotope of uranium constitutes such a small fraction of natural uranium (less than 1%) and because there is no *chemical* method for separating it from U^{238}, the amassing of a sufficient quantity of U^{235} for a bomb requires elaborate and costly capital equipment. Up to now, few countries have had both the willingness and resources to undertake such separation. But both the separation of U^{235} from natural uranium and the mechanical construction of atomic bombs are relatively simple (though costly) processes, and cannot be kept secret. These facts should be clearly understood by everyone today. The production of atomic bombs will soon be within the reach of many nations willing so to use their energies and resources.

The problem of building a nuclear reactor for *industrial* energy is considerably different from that of an atomic bomb. Here, the chain reaction process must be carefully controlled so that the process of fissioning will continue at a slow and steady pace, yielding the energy required. Usually, either natural uranium or uranium that has been "enriched" somewhat by increasing the proportion of U^{235} is used as a fuel.[4]

Natural uranium may be used in a reactor if care is taken to keep too many neutrons from being absorbed by U^{238} without fission. This can be accomplished by separating the slugs of uranium by layers of graphite, which act as a "moderator" to slow down the neutrons below that critical velocity at which they are captured by U^{238} without fission. At these slow or thermal velocities there is a high probability that they will collide with a U^{235} nucleus and cause fission. A sufficient number of control rods are inserted into the pile (Fig. 21.6). These are of materials that absorb neutrons; by inserting or withdrawing them, the number of neutrons available for producing fissions is delicately controlled.

[4]Either U^{233}, artificially produced from thorium, or plutonium, artificially produced from uranium ^{238}, may also be used. These two fuels can be produced in what are termed reactor "breeders" which, in operation, yield both energy and additional fuel. Actually, the first breeder reactors were built in order to produce plutonium, which can be used instead of U^{235} in an atomic bomb.

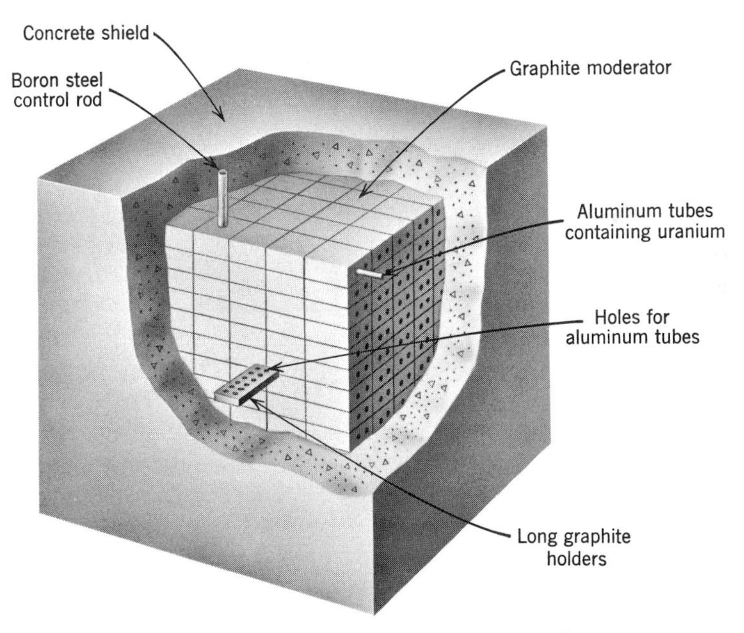

Fig. 21.6. Diagram of a uranium-graphite pile.

A steady operation is then established by arranging that the number of neutrons generated in any interval of time is always sufficient to sustain fission reactions leading to the same number of neutrons being generated in each ensuing interval of time.

The neutrons are, in other words, carefully "budgeted" by the construction of the reactors and by the controls in such a way that the loss of neutrons, in other than fission reactions, is small enough so that exactly one neutron remains *on the average* available from each fission to produce one additional fission. Although reactors differ considerably in design, an illustration (Fig. 21.7) of this "budgeting process" for a typical reactor may make this process clear. Let us start at a given moment with 100 fissions. On the average, this will produce 250 neutrons. About one third of these, or about 85, will be captured by U^{238} atoms, even with the presence of a graphite moderator whose purpose is to reduce such capture as much as possible. An additional 20, on the average, are lost on striking U^{235}

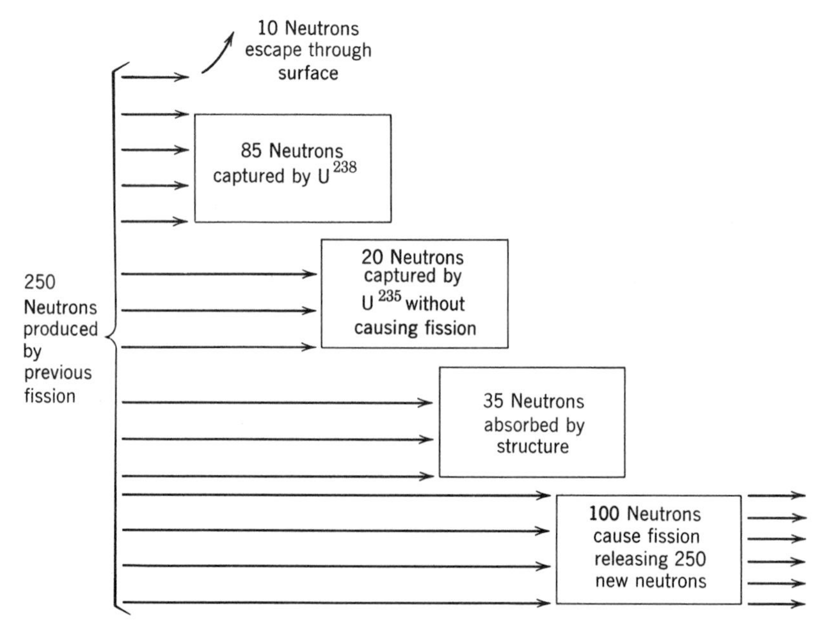

Fig. 21.7. Typical distribution of how neutrons are used in a chain reaction. The reactor is said to have "gone critical" in operation when the number of neutrons produced in each "generation" remains constant. Below the critical point, the reaction dies down; above the critical point, the reaction becomes explosive. Fortunately, in addition to the control rods, commercial reactors have a "built-in" control: When the number of neutrons produced begins to increase, temperature rises, and the fuel atoms are pushed further apart, reducing the probability of nuclear fission and bringing about once more a stable, equilibrium condition.

and being captured without causing fission (instead of fissioning the U^{235}, they merely transform U^{235} into U^{236}). An additional 35 are absorbed by the various structural parts of the reactor, causing transmutations of the elements out of which the reactor is built. And, finally, even with a large volume (small ratio of surface area to volume), about 10 neutrons will escape through the surface. Thus, from the original 250 neutrons produced, 150 are "lost" and cause no additional fissions, but there remain 100 neutrons which again produce 100 fissions and, consequently, a new "generation" of 250 neutrons. The process continues at a steady rate. These figures are, of course, average, but since the number of atoms is so tremendous (recall Avogadro's number, specifying that the number of atoms in one gram atomic mass is 6×10^{23}) the figures are extremely stable on a statistical basis. In time, the uranium fuel must be renewed and wastes disposed of, but the reactor can continue in stable operation for long periods by slight adjustmests of the control rods, which vary the number of neutrons absorbed to a slight but very significant degree. Figure 21.7 summarizes the process that we have described.

Such are the general principles of operation of most nuclear reactors. The energy generated (about 200 MeV per fission) is mostly transformed into the kinetic or thermal energy of the fission fragments, which can be utilized by a conventional power plant to generate electricity. At present, in the United States, the capital costs have been so lowered that nuclear reactors are beginning to compete with plants using fossil fuels. As improvements are made and the capital costs are further reduced, the atomic plants will replace others as they have already done is other countries, such as in England where coal is not abundantly or cheaply available. (One very interesting type of commercial reactor now being developed is the so-called "breeder reactor," which uses uranium as a fuel. In these reactors, in addition to the energy generated, atoms of plutonium are produced, and may be recovered and used as fuel for other reactors, since plutonium itself is a fissionable material. Such reactors may be said, therefore, to produce more fuel than they use. The process, however, cannot be repeated, since the plutonium on being fissioned does not produce further fissionable material.)

Finally, before leaving the subject of atomic energy, brief mention should be made of the *fusion* reaction and its possible use as a source of energy. The fusion of two nuclei of heavy hydrogen into a helium nuclei also causes a loss of mass and hence the production of energy. This process is brought about not by the impact of neutrons but by bringing the two hydrogen nuclei sufficiently close so that the close-range nuclear attractive forces become larger than the repulsive coulomb forces. This can only be done if the two nuclei approach one another at very high velocities with sufficient energy to overcome the coulomb forces. The only large scale method to obtain these velocities is to heat a mass of heavy hydrogen to temperatures in the range of the temperature of the sun. At this temperature the kinetic energy of the hydrogen nuclei is high enough to cause fusion or a thermonuclear reaction.

As is well known, such reactions have been artificially obtained in the so-called hydrogen bomb. The heavy hydrogen is heated to the necessary temperature by the explosion of a small atomic bomb, which triggers the fusion process. Although a great deal of effort has been put into an attempt to develop other methods of bringing about a controlled thermonuclear reaction, none has been successful except on a laboratory scale. The difficulties facing such attempts are tremendous. Obviously, no container can withstand such high temperatures. Therefore, the attempt is made to "heat" a flow of ionized hydrogen gas, which is "contained" within a small volume, by means of very powerful electric and magnetic fields. Momentary success has been reported, but on such a small scale and for such a small duration of time that the practical development of fusion reaction processes for the production of energy is still far off. Because of the vast availability of hydrogen, the energy problems of mankind would be solved once and for all if success ever becomes more than momentary and small.

4. FURTHER INVESTIGATIONS INTO THE NATURE OF NUCLEAR FORCES

Well in advance of the project brought so successfully to completion by Fermi and his colleagues, a tremendous amount of theoretical work had to be done. The amassing of the uranium pile of a critical size with the precise positioning of the uranium within graphite, and the placement and control of the cadmium rods could never have been accomplished if thousands of scientists, the world over, had not been carefully investigating the ultimate nature of matter, using abstract theories to account for their findings. Typically, few of these investigations were carried out with any "practical" end in view. The motive was primarily "creative curiosity," often with unexpressed overtones similar to those more openly revealed by an earlier generation of scientists: seeking to come closer to truth or to find the "harmony of nature or of God." The modern scientist might express different words but the same sentiments as Newton did:

> I do not know what I may appear to the world; but to myself I seem to have been only a boy playing on the seashore, and diverting myself in now and then finding a smoother pebble or a prettier shell than ordinary, whilst the great ocean of truth lay all undiscovered before me.

Before the forces and energies locked inside the nucleus could be unleashed, many facts and relationships pertaining to the inside of the nucleus had to be discovered. A great deal more remains to be found out and explained, so that we are now turning our attention to one of the frontiers of science marked by incomplete speculation, alternative theories, and models vying with each other to construe new experimental facts as they are being accumulated.

At the turn of this century, man had knowledge of three basic forces: gravitational, electrical, and magnetic. All forces, whether they were acting "at a distance" or as "field forces," could always be interpreted as one of these three. Also it was believed that all elastic forces, collision forces, contact forces, and chemical forces could be interpreted in the same terms. It now developed, however, that a fourth force was at work in the universe, far mightier than any of the others and far more difficult to measure and to interpret. With electrostatic, magnetic, and gravitational forces, measurements on a laboratory or macroscopic scale can be made and the laws governing these forces established. Even on a microscopic scale, by indirect measurements such as the scattering experiments of Rutherford, it could be established that electrostatic or coulomb forces obey the same laws at extremely small distances and between extremely small particles as they do on a large scale. But entirely new methods are required in the study of nuclear binding forces; the form of the laws governing such forces has not yet been established on a firm basis.

Before pursuing the analysis in greater detail, we summarize certain conclusions regarding the nuclear forces that bind the neutrons and protons together into the nucleus.

(*1*) The forces between neutron and neutron, neutron and proton, proton and proton are all of the same order of magnitude and are extremely powerful, being about 10^{38} times greater than gravitational forces and about 10^2 times greater than the electrostatic forces of repulsion between proton and proton.

(*2*) The forces cannot be measured. Their magnitude can only be calculated from changes in energy level. It is as though we could not measure the force of gravitation acting on a mass by the process of weighing, but could only calculate it indirectly from the change in potential energy as the mass is moved farther and closer to the earth. Because forces are not directly measured and because energy measurements are possible only when some nuclear event occurs in which one or more particles interact, the physicist today talks about "strong" or "weak" *interactions* instead of forces. Energy changes, being the only measurable quantities, have largely supplanted the classical concept of force.

(*3*) Forces revealing themselves through the changes in energy levels occur only when particles interact at *extremely* short range, and *not* according to any simple equation, such as an inverse square law. Below the distance of about 3×10^{-15} m, the attractive force between nucleons is very powerful; at distances greater than that, it falls off rapidly to zero. Electrostatic forces then become dominant and the nuclear forces become negligible.

(*4*) The functional relationship between the force of attraction and the distance between the particles is not a simple one, and is not, as in the case of the other forces we have studied, a simple case of a central force depending on distance alone, but involves instead other characteristics of the particles which are interacting, such as their "spin" and "magnetic moment." Furthermore, to a degree at least, the forces acting are "saturated," which means that in a collection of several neutrons or protons, the attractive force between one nucleon and the others is not merely given by the addition of all the forces acting on that nucleon. (These phenomena contrast with the case of gravitational forces. For example, we can determine the orbit of the moon by first calculating the gravitational force between the earth and the moon and then simply adding the lesser attraction of the sun or even the other planets. Each force acts independently and additively. This does not occur in the case of nuclear attractions, and makes the problem of analysis more complicated.)

(*5*) Finally, all nuclear particles obey the statistical laws of quantum mechanics instead of the classical laws which apply to larger masses. An example may make this clear. An artificial satellite, traveling in an ellipse about the earth, oscillates in its orbit between perihelion and aphelion. While it is true that if it did go beyond a certain critical distance from the earth, the attractive force of the sun would become larger than the attractive force of the earth, it cannot reach this critical distance without additional energy, and its orbit remains stable (disregarding friction caused by the "debris" of space—the small particles of dust and molecules that exist even above the "atmosphere" of the earth). Using language analagous to that used in nuclear physics, it remains in its orbit because no energy is available to overcome the "binding energy" between earth and satellite. It is very different in the case of a nuclear particle that is inside a nucleus, "oscillating" many millions of times each second, presenting itself, so to speak, to the outer limit of the nucleus again and again. The mere existence of a more stable energy configuration lying beyond a certain distance causes the nuclear particle to "leak out" of the nucleus after a sufficiently long time to the more stable energy configuration, *even if* in classical theoretical terms *there is no extra energy available to enable it to penetrate the barrier.* (It is as if a stone in a well at the top of a mountain could tunnel through the side of the well and then roll down the side of the mountain seeking the more stable ground level of energy at the bottom. It is indeed after this analogy that the escape of a nucleon from a "potential well" to the outside is called a "tunneling effect." It can be explained only in terms of quantum mechanics and statistical probability, not in terms of classical mechanics.)

A clarification of these five summary statements requires an analysis that proceeds boldly by analogy with classical physics, drastically modified, however, by the modern conceptions and new experimental data. Scientists cannot proceed entirely *de novo;* they must build upon familiar conceptions and analogies, modifying them, and expanding or limiting them as new data is gathered and according to the logical requirements of consistency and clarity.

To understand the concepts of nuclear physics, let us first review the connection between force and potential energy. We are by now familiar with the idea of an inverse square law of gravitation, in which the force of attraction varies inversely with the square of the distance between the centers of two spherical masses. Since the potential energy of a mass is the product of the force required to move the mass from one position to another, times the distance through which it is moved, it is clear that the *potential energy per unit mass* (which is called the *potential*) of a small body is the greater the farther it is from a large attracting mass such as the earth. Recall that the gravitational force between two spherical bodies separated by a distance d between their centers of gravity is given by

$$F = G\frac{M_1M_2}{d^2}$$

In this case, because an inverse square law holds, the potential (that is, the potential energy per unit mass) of a body exterior to the surface of the earth is represented by $V = GM/R - GM/x$ where M is the mass of the earth, R is its radius, and x is the distance of the unit mass from the center.

The significance of the use of this expression, replacing the use of force to indicate the interaction between masses, is twofold: (*1*) it represents a scalar difference between two quantities, and has no directional significance; and (*2*) it can be measured by the change in kinetic energy without knowledge of the form of the force equation. (A mass sliding down a frictionless inclined plane from the same vertical height will always gather the same kinetic energy, regardless of the path it pursues or the *direction* of the net force acting on the body.)

Recall the classical expression for the potential of a unit mass lifted to a height h above the surface of the earth. In this case, the potential appears as a simple linear expression $V = gh$, since the work done to lift a mass m to the height h is $W = mgh$. Closer examination reveals that this depends upon two assumptions: (*1*) that the distance h which is equal to $x - R$ is small compared to R, and (*2*) that the ground level is taken as zero potential. The accurate expression is

$$V = \frac{GM}{R} - \frac{GM}{x} = GM\left(\frac{1}{R} - \frac{1}{x}\right) = \frac{GM}{Rx}(x - R).$$

If x is close to R, this expression becomes

$$\frac{GM}{R^2}(x - R).$$

But it will be remembered that $GM = gR^2$ (since $F = GmM/R^2 = gm$). Therefore, $V = mg(x - R)$ or, more simply, mgh, where $V = 0$ when $x = R$. Furthermore, note that for positions *under* the ground level, the potential is *negative,* since R is then greater than x.

The nuclear physicist uses the terms "potential hill" and "potential barrier" quite freely. We can now see clearly its meaning by analogy to gravitational phenomena. Consider a well on top of a mountain with an object at its bottom as shown in Fig. 21.8.

Although the mass at A has a positive potential energy because it is above zero ground level, it cannot escape from the well without being given sufficient additional energy to lift it up to the point B. However, if it could be given additional energy in the form of kinetic energy, the *net* gain in kinetic energy on arrival at point C would be just equal to the potential energy difference between A and the ground level at C.

We can generalize this analysis by removing the condition that x, the distance from the center of the earth, is approximately equal to R. Let us now assume that a hole extends to the center of the earth and that x represents any distance from the center of the earth out to infinity. No longer is the force equal to a constant g times the mass, nor does the potential vary in a linear fashion with distance above or below ground level. A study of this example should bring further insight into how forces and potentials are used in nuclear physics. It will be noted that both the equation specifying attractive force and that specifying the potential undergo a change at the ground level point, although the change is such that the variation of the force and the variation in the potential is not discontinuous. The appropriate equations and graphs for forces and potentials

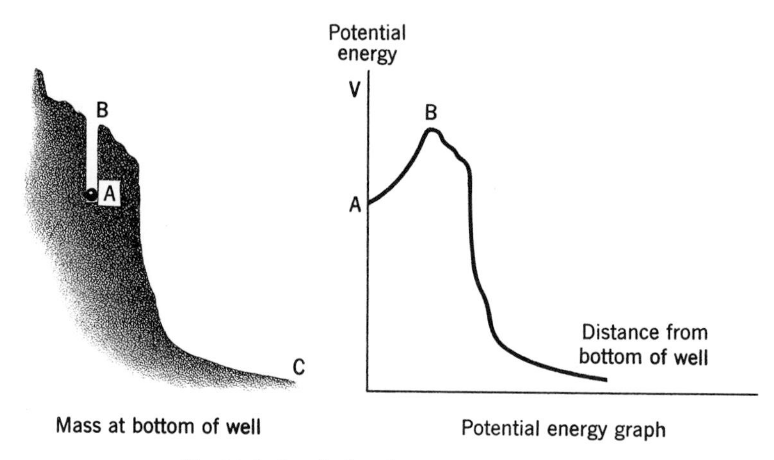

Fig. 21.8. Gravitational potential energy well.

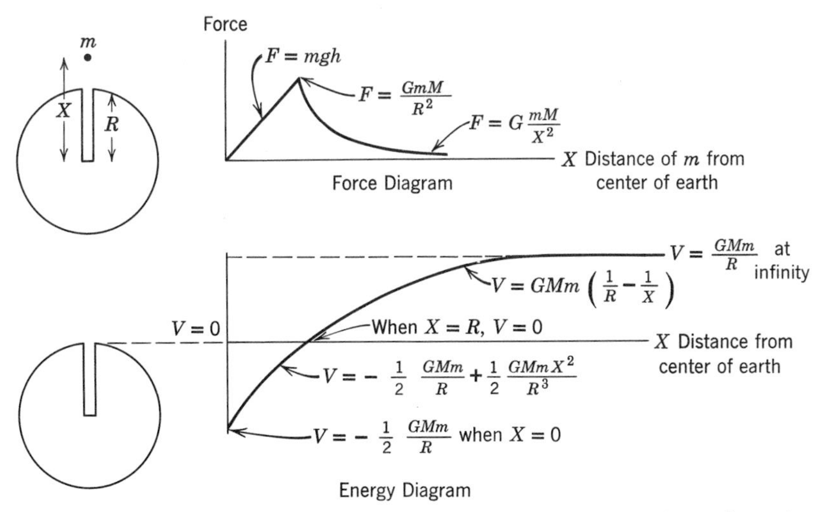

Fig. 21.9. Force and potential energy diagram. The upper diagram shows the variation in force on mass m from the point at the center of the earth.

as they vary with the distance from the center of the earth are shown in Fig. 21.9.

Attention should be given to the fact that the point of zero potential is chosen arbitrarily, since all that is of physical significance is not the absolute value of the potential, but the *difference* in potential. For many problems in both gravitational effects and electrostatic effects, physicists find it more convenient to choose the point of zero potential at infinity. If this is done, then the potential at the surface of the earth is $-GM/R$ instead of zero, meaning that it will take $(GM/R)\,m$ units of energy to remove the mass m to infinity. (On the other hand, if the surface of the earth is regarded as having zero potential, then at infinity the potential is GM/R. The zero point is determined by convenience, since the *difference* in potential energy is the only quantity of significance.)

Up to now we have been considering forces of attraction and the potentials that they give rise to. The same kind of analysis may be applied to repulsive forces. The potential energy between two similarly charged spheres, however, rises as the distance between them decreases. In the case of repulsion it is usual to take the zero point at infinity. The highest point of potential is then at a point just outside the surface of a sphere and descends continuously to zero according to the relationship $V = KQq/x$ where Q is now the charge on the nucleus, and q the charge on a particle, and where K, the coulomb constant, replaces G and where x is once again the distance from the center of the sphere.

A diagram of how nuclear potentials exert their effects is shown in Fig. 21.10. We consider first the case of an alpha particle trapped inside the "potential well"

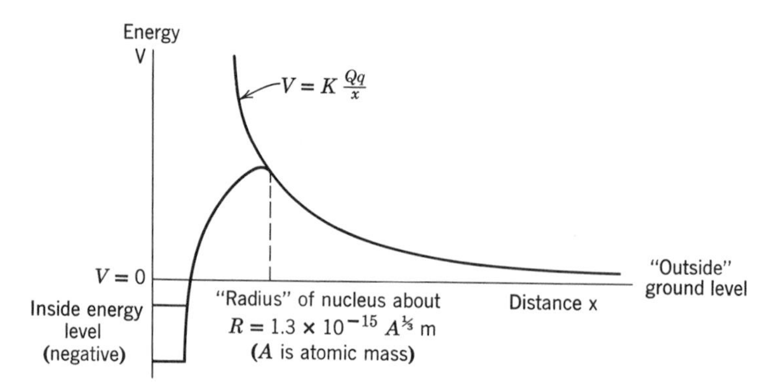

Fig. 21.10. Potential energy diagram about the nucleus.

formed by the nucleus of a fairly heavy atom, for instance of Z equal to about 80. We know, by Rutherford's scattering experiments, that the coulomb law of repulsion remains valid from infinity up to very short distances outside the nucleus (around 5×10^{-14} m). As an alpha particle approaches the nucleus, therefore, the potential energy rises steeply according to the equation KQq/x where $Qq = 2Zee$. At a still closer distance, the strong nuclear forces of attraction take over. The exact form of the potential energy curve is not known, but its depth and general dimensions have been calculated from energy changes.

Once the alpha particle is inside the nucleus, its energy in Fig. 21.10 is less than the energy of an alpha particle outside the nucleus. It is therefore in what is termed a "bound" state. The "binding energy" is shown in the figure, being the algebraic difference between the energy that the alpha particle has within the nucleus and its outside ground level energy. The nucleus shown is stable, and the alpha particle cannot be withdrawn from the nucleus unless it is given sufficient additional energy both to overcome the binding energy and the potential barrier.

Let us look now at the diagram of a *radioactive* nucleus (Fig. 21.11). In this case, the energy of the alpha particle inside the nucleus is above the outside zero ground level. The binding energy is negative. If the alpha particle could penetrate the potential barrier it would have the amount of kinetic energy outside the nucleus as shown. Thus, the nucleus is now in a metastable state corresponding to the stone in a well on a mountain. Neither the stone nor the alpha particle can escape according to the laws of classical physics, because the energy is not available for it to climb the potential barrier even though the energy it has is far higher than zero ground level. However, according to the modern concepts, given enough time the alpha particle does escape by a process of tunneling through the barrier. This has its explanation in terms of quantum mechanics.

Speaking rather loosely, we can conceive of the alpha particle as undergoing periodic oscillations within the nucleus. The particle "strikes" the potential barrier many millions of times each second, usually being reflected back. There is, however, a small but finite probability that it will penetrate through the barrier. This probability can be calculated by applying the Schrödinger wave equation to the particle, and is found to depend on the kinetic energy of the particle and on the "radius" of the potential well. If the energy is large and the radius is small the alpha particle has a high probability of penetration, and the nucleus will disintegrate rapidly, as in the case of radioactive materials with short half-lives. In other cases the energy is less and the radius somewhat larger and the probability of penetration is much smaller. In this case the nucleus has a

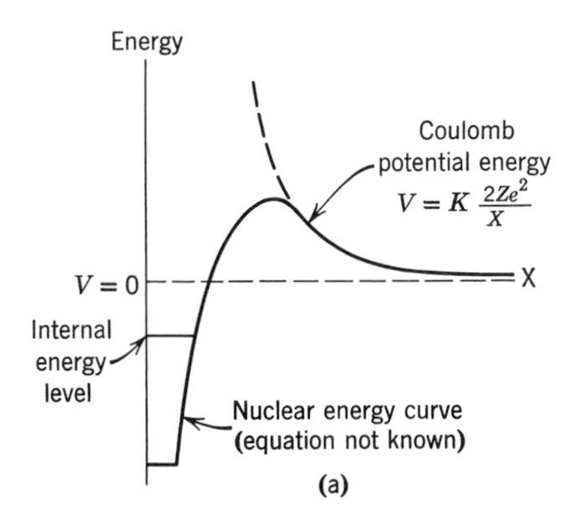

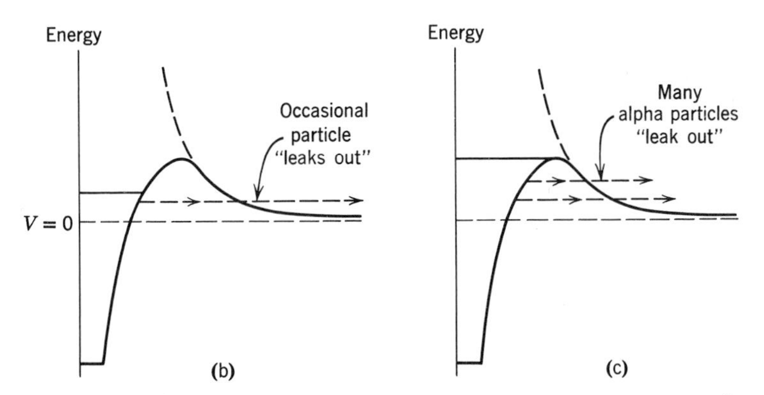

Fig. 21.11. (a) Stable nucleus (internal energy below external ground level); (b) long-life radioactive nucleus (internal energy level slightly above 0); and (c) short-life nucleus. (Note. The charge Q on the nucleus is Ze; the charge q on the alpha particle is $2e$. Therefore, Qq is $(Ze)\,(2e) = 2Ze^2$.)

much longer average lifetime, as in the case of radioactive substances with very long half-lives.

The picture of the nucleus that begins to emerge is a complicated and abstract one. It cannot be given full justice except in terms of higher mathematics. Nevertheless, there are several points that can be brought out. First, the nucleons must be considered in their wavelike as well as their particlelike attributes, and all the implications of quantum mechanics must apply, including the determinations of wavelengths, energy levels, and the probability of transition from one energy level to another. These considerations lead to a picture that departs very significantly from one in which the nucleons are regarded as little balls trapped in a containing well and rattling around in it. Because of their wave nature, the nucleons can only exist at certain discrete energy levels. To each nucleon there is associated a specific wavelength given by the de Broglie equation that we have already considered, that is,

$$\lambda = \frac{h}{mv} = \frac{h}{\sqrt{2mE}}$$

where h is Planck's constant, m the mass, and E the energy of the particle. If, then, the nuclear particles are to exist at all within the potential well they must "fit," that is, they must exist as standing waves such that the radius of the nucleus is equal to one half a wavelength. Since the wavelengths are dependent upon the energy of the nucleon, only certain discrete energy levels are possible (just as a piano string can only vibrate at certain frequencies dependent upon the length of the string). It is significant that on this basis alone (there are additional reasons also) it may be concluded that *an electron cannot exist as a free particle within the nucleus*. Its mass and energy are such that its wavelength cannot "fit" into a radius as small as required. For example, if the radius is taken as being in the neighborhood of 10^{-14} m, then, using de Broglie's equation

$$\lambda = 10^{-14} = \frac{h}{\sqrt{2mE}}$$

the energy would have to be more than 10^{-9} joules, or nearly 10,000 MeV. But beta particles emerge from the nucleus at energies thousands of times less than this. (Even the binding energy per nucleon is only around 5 MeV.) Thus, according to our model, the only elementary particles within the nucleus are the neutron and proton.

In a complicated heavy nucleus, the nucleons occupy many discrete energy levels. As in the case of the electrons in the atomic model, which was discussed in an earlier chapter, there are restrictions as to the number of particles that can occupy the same energy level. (Pauli's exclusion principle). When all the levels are occupied at the lowest possible energy configurations, the nucleus is in its

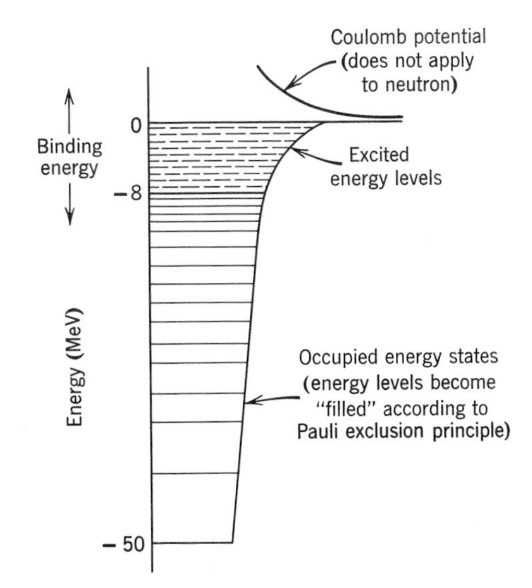

Fig. 21.12. Potential well and binding energy of nucleus. If a nucleon is raised above 0 binding energy, it will, sooner or later, be expelled from the nucleus.

ground state. If energy now enters the nucleus from an external source, a nucleon may be lifted above the internal ground state and the nucleus now goes into an excited state (similar to the excited state of an atom in which an electron is raised above its lowest energy level). Just as an atom returns to a lower energy level in time with the emission of radiant energy, the nucleus makes a similar transition by emitting energy. If the external energy impinging on the nucleus is higher than the binding energy, a nucleon will be given sufficient energy so that it is raised not only above the internal ground state of the atom, but above the external ground state. Under these circumstances, a nucleon or several of them will leave the nucleus entirely. The nucleus has thus undergone a transmutation. Figure 21.12 illustrates the situation.

5. A CLOSER EXAMINATION OF THE FISSION PROCESS; THE COMPOUND NUCLEUS

We can use the concepts developed in the previous section to get a clearer picture of the mechanism that occurs when a nucleus fissions. In our previous discussion of this process we looked at it from the outside, examining only the situation before fission and after fission. Before fission we have the nucleus of the uranium 235 atom with an external neutron coming toward the nucleus. After fission we have the nucleus split into two large fragments (fission products such as barium and krypton) plus a number of smaller fragments (neutrons) and a large amount of energy due to the conversion of mass into energy. But how does this process take place? What happens between the moment of collision of the neutron with the nucleus and the fission that results?

The theory, originally proposed by Bohr, is that the external neutron coming into the nucleus is absorbed by the nucleus forming a compound nucleus whose total energy is now far above what we have termed the external ground level energy. The compound nucleus is thus in an excited and unstable state, and can last in this state only for a certain length of time before it either ejects the excess energy by emitting nucleons or splits into two fragments each of which has a lower and more stable energy configuration.

The process may be thought to take place as follows. The incident neutron colliding with the nucleus gives up its energy, which is rapidly distributed among all the other nucleons. One of two things may now occur. A nuclear particle may, through random fluctuations, acquire enough energy to emerge from the compound nucleus, or a series of oscillations may be set up in the nucleus as a whole in such a way that a deformation occurs which separates two large portions of the nucleus beyond the effective range of the short-range nuclear attractive forces. At this point, the coulomb repulsive force becomes dominant and tears them apart with great energy. Again we can study the process by means of a potential energy diagram as given in Fig. 21.13.

In this diagram at the point C the two fission fragments are apart and the potential energy is at zero ground level. At the point B the fission fragments are just touching. To the left of B the attractive nuclear forces are dominant over the repulsive coulomb forces. The ground state of the two fission fragments in the nucleus is shown at A. If, now, an incident neutron is absorbed having more than the critical energy $(B - A)$, the energy of the two fission fragments will be lifted above the point B and fission will take place. Such a process could theoretically occur in any nucleus above a mass of 90, but such a large critical energy is not available. However, for nuclei of mass about that of U^{235}, the critical energy, is only about 6 MeV. For mass numbers above 260, no critical energy at all is required for fission and such elements would be unstable and undergo

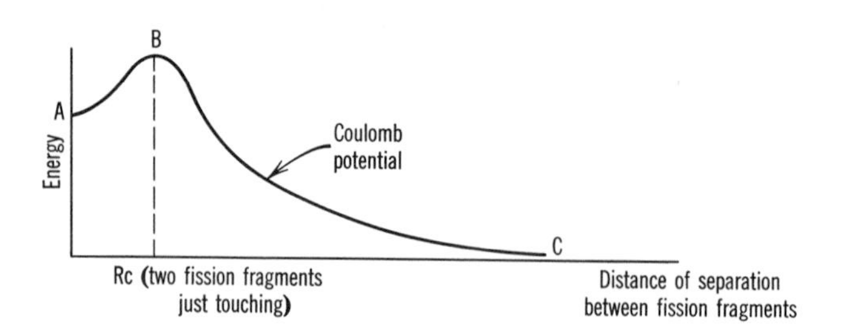

Fig. 21.13. Potential energy, due to combined coulomb and binding forces acting on nuclear particles. Beyond the distance Rc (critical radius), coulomb forces of repulsion dominate. Below this distance, nuclear binding forces dominate.

spontaneous fission in about 10^{-21} sec. (Therefore, the number of new elements remaining to be manufactured in the laboratory beyond the present periodic table is distinctly limited.)

Before examining the exciting aspect of nuclear physics that concerns a wide variety of newly discovered "elementary particles," including the particles of "antimatter" as they are sometimes termed, we shall review some of the general conclusions already reached. In particular, we shall take a closer look at the meaning of the spatial and temporal dimensions assigned to nuclear phenomena.

The term "*radius* of the nucleus" has been rather freely used, and a measured value has been assigned to it of the order of 10^{-15} m (now often given the name one Fermi unit). What does this term "radius" really mean? In our daily macroscopic experience, when we assert that an object is spherical and has a given radius, we imply that there is a well-defined transition point between the "inside" and the "outside," which we call the surface. This transition is often established clearly by a specific operation, involving, for example, the contact forces between parts of a measuring instrument such as a caliper. In other situations it may be difficult to specify dimensions precisely. For example, what are the dimensions of a cloud? Or what is meant by the radius of the earth, including its attached atmosphere? In the latter case it is most difficult to establish, except in an arbitrary manner, where the "atmosphere" leaves off and "empty space" begins.

In the case of the nucleus, the radius is marked by the point of transition between the dominance of the short-range nuclear forces and the dominance of the coulomb forces. This transition point is determined in two independent ways: (*1*) by studying the scattering of "bullets" such as alpha particles, protons, and neutrons directed at the nucleus; and (*2*) by the measurement of energy changes, in particular the energy changes revealed by particles that fly out of the nucleus with measurable velocities. (In the second approach the assumption is made that a particle existing within the nucleus must have an associated wavelength—determined by its mass and energy—which fits the nuclear radius.) These two distinct types of determinations have led to an empirical equation expressing the radius of the nucleus with considerable accuracy, $r = 1.3 \times A^{1/3} \times 10^{-15}$ m, in which A is the atomic mass of the nucleus. It becomes evident that regardless of the element and the constituent structure, all nuclei have approximately the same density (since density is the mass divided by volume and the volume is proportional to the cube of the radius or the cube of $A^{1/3}$). The experimental findings regarding nuclear dimensions are, in fact, such that a much more precise meaning can be given to the term "nuclear radius" than to the term "atomic radius." The atom's size is determined by the "cloud" of electrons surrounding it. It is much more difficult to specify where the cloud begins and ends while atoms interact in various combination.

With regard to temporal dimensions, that is, the duration of events within the nucleus, the approach is based upon quantum considerations. Assuming a given radius and a given mass, a nucleon is assigned a velocity corresponding to its kinetic energy. The neutron, for example, may be regarded as traveling back and forth across the "diameter" of the nucleus within a certain time. Calculation shows that this duration is about 10^{-21} sec. Thus, any event which takes place, for instance, at 10^{-9} sec (one billionth of a second) is a tremendously long period of time so far as nuclear events are concerned. For example, in a radioactive element, a particular configuration that leads to the emission of an alpha particle may be highly improbable as measured by frequency of occurrence, yet because of the long period of time involved on a nuclear time scale, this emission may take place in a very "short" period of time (that is, a few seconds) as measured by our normal laboratory scale of time. The vast duration on a nuclear scale has allowed many billions of configurations to be "tried out" in a random fashion so that the highly improbable configuration inevitably arises. (If 10^{18} or a billion billion hands of bridge were played in the world each second, a hand of 13 spades would occur millions of times each day.)

Finally, we emphasize again that *nuclear events obey quantum rules* and that measurements represent *averages* and not the classical determinations that apply to each individual object and every individual occurrence. It should be remembered that energy transitions are *not,* as in the case of classical theory, continuous, but must always fit configurations that are "allowed." As we speak of the average number of persons of a family being 2¼, while we recognize that no single family can be composed of 2¼ persons, so when we speak of the average energy released per nuclear fission as being 200 MeV, we recognize that this is an *average* and do not mean there is a *continuous* range of possible energy transitions. Throughout nuclear physics the laws and generalizations are founded on probability and statistical considerations.

6. SUBNUCLEAR STRUCTURE: THE NOTION OF ELEMENTARY PARTICLES

The ancient suggestion of Democritus that permanence and change can best be explained by postulating small, indivisible particles of matter, without internal structure, whose "coming together" and "separation" lie at the bottom of all explanations of events, has proved extremely fertile to the sciences in the last century. Triumph after triumph of explanation and prediction have followed one another. The assumption of molecules explained the various gas laws; the assumptions of atoms explained chemical changes; the assumption of the nucleus explained the "spongy" nature of solid materials; the assumption of protons, neutrons, and electrons—together with relativity theory—explained radioactivity, fission, fusion, and a wide variety of laboratory experiments. However, each time a new "elementary" particle has been postulated, it has been later found that it had an internal and complex structure and consequently "subparticles" had then to be postulated. Dean Swift, two centuries ago, in another context, raised the same puzzle that plagues physicists today:

> So, naturalists observe,
> A Flea hath smaller Fleas that on him prey.
> And these have smaller Fleas to bite 'em,
> And so proceed *ad infinitum.*

What is the meaning of the term ultimate or elementary particle, without internal structure? Does this mean without spatial extension, or a kind of "limit" in the mathematical, or perhaps Newtonian sense, such as a "point mass?" If the molecule as once conceived in the kinetic theory was later found to have an internal structure, as were in turn the atom and the nucleus, does there not remain a possibility that the neutron and proton themselves have an internal structure? What about the possibility of an infinite regress, such as suggested by Swift, to smaller and smaller masses and magnitudes not experimentally observed today?

These questions cannot be answered by scientists on the basis of present information. If anything, the scientist has learned that he must use extreme caution in extending (or extrapolating) results beyond the range to which they have been verified. Again and again, the laws that so appealed to common sense, insight, and aesthetic sense have been proved so wanting in the history of science that the scientist hesitates to declare that the generalizations of today will not be superseded or modified tomorrow. Particles that are more elementary than the neutron, proton, electron, and neutrino are not known, but one day they may be revealed.

Therefore, let us examine what is meant by an elementary particle "without internal structure." First, consider briefly why the molecule (or atom) was regarded as elementary in the nineteenth century. According to the kinetic theory of gases, which embraced and explained a wide variety of phenomena, collisions between molecules (or atoms) were regarded as "elastic," that is, no energy during collision was lost to heating of internal parts (because no internal kinetic motion was possible). This conclusion was well substantiated by experimental evidence. Note the wide difference between the collision phenomena involving molecules and those involving macroscopic objects. In the latter case, no collision is ever completely elastic. Energy is *always* lost to some degree, going into the kinetic energy or heat of the particles making up the object. The very fact that a steel ball bearing, dropped from a height onto a massive steel plate, fails to rise to its original height (even after allowance is made for the very small motion of the steel plate) but loses some of its energy to heat, tells us a great deal about the internal structure of the ball. The ball must be composed of particles which can be set into vibration.

When, at the turn of the century, the realm of higher energies was opened up by spectroscopic studies and, at about the same time, the discovery of radioactivity was made, the collisions between high velocity particles and atoms were again investigated, and it was found that by no means all collisions were elastic but that a considerable proportion of them absorbed energy. Such absorption of energy could be accounted for only on the postulate of an internal structure. It turned out later that this deeper level of internal structure had not been made evident before because of the quantum nature of energy. Since energy cannot be absorbed except in quanta, collisions were found to be elastic even though an internal structure was present, because less than a minimal quantum change was not possible.

7. FUNDAMENTAL INTERACTIONS AND ELEMENTARY PARTICLES

Nature is now thought to include four major types of forces by which material particles interact with one another. In descending order of magnitude they are: the "strong" nuclear forces, binding proton to proton and to neutron; the electromagnetic forces; the "weak" forces, exemplified in beta decay; and gravitational forces. Today these forces are called "interactions." The physicist striving for unifying principles seeks a single general model in terms of which he can explain how these interactions take place. For the first two types of forces, a model has now been constructed. The mechanism of these forces can be thought of as a compromise between "action at a distance" and "contact forces." It is found that when one particle acts on another, the action is mediated by an intermediate particle which is emitted by one and absorbed by the other. The resultant interactions are sometimes referred to as "exchange forces." The success of this model is attested to by the discovery of new particles predicted in advance on the basis of the model. In the case of gravitation, however, the force is so extremely weak compared with the others that the mediating particle which is predicted, the "graviton," which presumably would "transmit" gravitational interactions, has not yet been observed experimentally. A somewhat similar situation exists with the "weak" interaction. Intermediate quanta have been suggested, but no evidence for them has yet been found.

We begin with electromagnetic radiation. In the atomic domain, electromagnetic radiation interacts with electrons as if it were concentrated in "quanta" or "photons." If a photon has enough energy, it will, when absorbed by an atom, eject an electron (photoelectric effect). Since photons travel at the speed of light, any theory of electrons and their interaction with radiation must be relativistic. Various attempts to construct a theory embracing relativity were made many years ago, but one proposed by the English physicist, Dirac, was the most successful. Among other consequences, it correctly predicted the "spin" of the electron.

Dirac's mathematical investigations led him inexorably to a surprising paradox. The solution of his relativistic equations required that a "free" electron could exist in either of two energy states, positive or negative. What meaning can be found for a negative energy state? Certainly none within the framework of classical physics. A negative kinetic energy would imply an imaginary velocity! Since $\frac{1}{2}mv^2$ would have to be a negative quantity $(-E)$, it followed that $v = i(\sqrt{E}/2m)$ where $i = \sqrt{-1}$. This was only the beginning of Dirac's bizarre speculation. His theory also required that "positive energy" electrons would tend to fall into negative energy states, emitting energetic photons (gamma rays) during the transition. If such transitions were permitted, all electrons would soon turn into radiation and the world which we know could not exist. At first the negative energy states were simply discarded as "spurious" solutions, but such an arbitrary approach is not really a satisfactory resolution of the problem. Dirac, however, suggested an ingenious interpretation which led him to predict the existence of a new particle. He assumed at the outset that all electrons including those in negative energy states obey the Pauli exclusion principle. Normally the negative energy states, like the electron states in full atomic shells, are occupied. Hence, a positive energy state electron can not "fall" into this vast collection of negative energy states *unless* one is vacant. If there should be such a vacancy, it should be observable. It is expected to have a *negative-negative* (i.e., positive) energy, *negative-negative* (i.e., positive) electric charge, and the rest-mass of an electron. What a strange, but ingenious and logically consistent theory—which predicted correctly!

Although the electrons occupying negative energy states are unobservable, they can interact with electromagnetic radiation. An incident photon can eject an electron from a negative energy state in a manner somewhat analogous to the photoionization effect. The process is illustrated in Fig. 21.14 which also shows the photoelectric analogue. The photon "kicks" the electron from a state of negative energy ($E < -m_0c^2$) (bound state in the case of an atom) across the "forbidden band" to the observable positive energy region (ionized state for an atom), leaving a vacancy or "hole" behind. The situation is complicated by the fact that an atom usually has excited states. Here we assume that it has none, in order to draw the analogy more sharply. In each case the electron appears as an observable negatively charged free particle at the top of the diagram. For the atom, a positively charged ion is left behind, with the vacancy or "hole" being

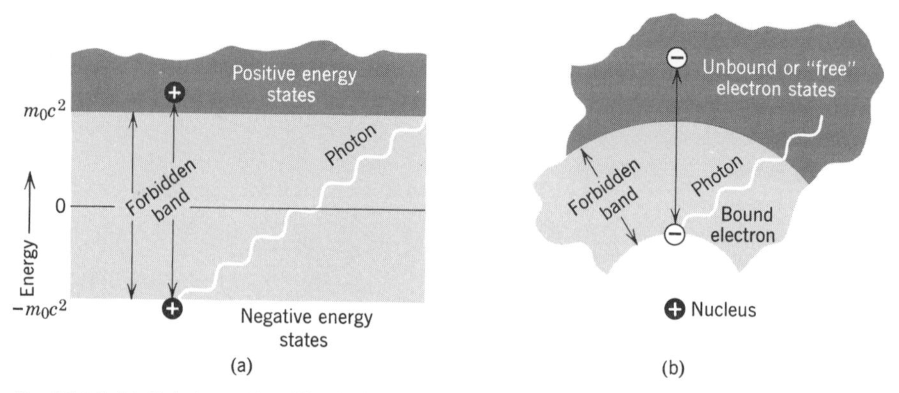

Fig. 21.14. (a) Pair formation. The incident photon raises an electron from a state of negative energy across the "forbidden" band to an observable positive energy state. The "hole" left behind is observed as a positively charged anti-electron or positron. (b) Photoionization. The incident photon raises an electron from a bound state to an unbound or "free" state.

securely bound to the ion. By contrast, the vacant negative energy state appears as a *free* particle with positive energy, positive charge, and the same rest-mass as the electron. In this respect the latter process differs from photoionization. The new particles which are usually called positrons are said to be the "anti" particles with respect to electrons. Perhaps anti-electrons would be a better name. Note the complete symmetry about the zero energy axis in Fig. 21.14a. If we let all electrons become positrons and all positrons become electrons, the system is unchanged. Since the electron has acquired enough energy to carry it all the way across the forbidden band, the incident photon must have rest energy $\geq 2m_0c^2$ or at least 1.02 MeV, just as a photon must have an energy equal to the ionization potential in order to cause photoionization. This process, called pair formation, was first observed in 1933 in a cloud chamber experiment. It was one of the most striking verifications of the special theory of relativity: Mass had been "created" out of energy.

Let us now go a step further and consider the interaction between two electrons. If they are at rest, the force exerted by one on the other is the Coulomb force discussed in Chapter 13. However, electrons which are in motion generate electric currents, and thus magnetic fields, which interact with each other. In fact, one can show that they interact through electromagnetic waves which "mediate" the force. As the energy of the particles is increased, quantum effects become significant. We therefore speak of photons as "interaction quanta" which transmit an electromagnetic impulse between the two electrons. This is somewhat analogous to two athletes who run down a field throwing a heavy medicine ball back and forth. Each time one man throws the ball to the other, he causes a deflection in the paths of both men. Thus the two may be said to "scatter" each other due to the "collision" caused by throwing and catching the medicine ball. The collision process is shown schematically in Fig. 21.15. In the case of the electromagnetic interaction the photon, which corresponds to the ball, is quite elusive; it cannot even be observed. Because it is unobservable, it is called a *virtual photon.* Its unobservability can be explained by plausability arguments using the Heisenberg Uncertainty Principle. A statement equivalent to $\Delta x \, \Delta p \geq h/2\pi$ is that the product of the collision time Δt and uncertainty ΔE in the photon energy is of the order of $h/2\pi$. If one succeeds in detecting the transfer of the photon in the very short duration available for measurement, he finds a large uncertainty in energy. But, because the energy transfer is a definite quantity, the observer does not know whether he has "seen" the collision photon or some other photon. On the other hand, if the measurement is carried out at some instant over a long time span in order to observe the energy precisely, one can not know when the photon is actually transferred. The photon must therefore be virtual.

Following the prediction of their existence in the mid-1930's, a family of strongly interacting particles called the π mesons [from the Greek *mesos* (middle)] were discovered shortly after World War II. They were originally suggested as quanta which mediate the nuclear, or strong, interaction between nucleons in much the same way that photons are thought to mediate the electromagnetic interaction. Since then, a large menagerie of additional strongly interacting "elementary" particles including "antiprotons" and "antineutrons" has been discovered. Some are similar in their essential properties to the π-mesons (e.g., spinless or integral spin), while others (hyperons) [from the Greek *hyper* (over)] behave like nucleons (e.g., half-integer spin). However, they all have a common property, instability. They all decay through what is termed the "weak interaction," exemplified by the emission of a beta particle, and have lifetimes of the order of 10^{-9} seconds. Although 10^{-9} seconds seems to us to be a very short time, it is very long on the subnuclear scale. The lifetime of a system that breaks up through the strong interaction is typically about 10^{-21} seconds. Since the

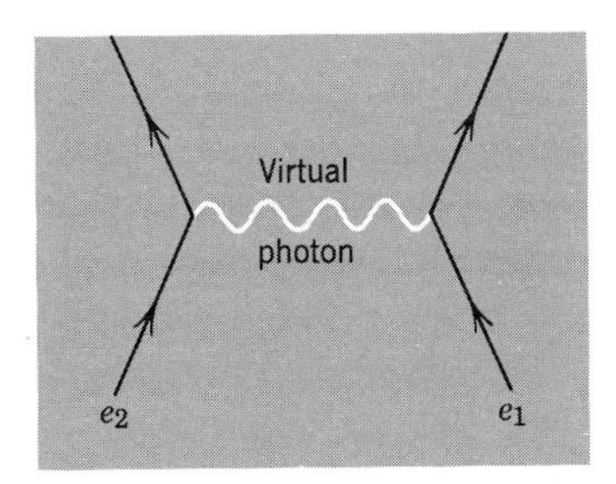

Fig. 21.15. Scattering of two electrons e_1 and e_2 by the exchange of a "virtual" photon.

electromagnetic interaction is about 100 times weaker, corresponding lifetimes are much longer ($\sim 10^{-16}$ to 10^{-20} seconds). Some of the particles which have been discovered are not sufficiently stable (i.e., *metastable*) to decay through the weak interaction. Instead, they immediately break up (i.e., in a time of order 10^{-21} seconds) into the same or a new set of strongly interacting particles. The metastable particles are called "resonances." These resonances are apparently excited states whose spin angular momenta are greater than the lowest possible values for the particles. For example, the π-meson has a resonance with a rest energy of 750 MeV and a spin of $h/2\pi$ while its lowest or *ground state* is spinless and has a rest energy of 137 MeV. Similar resonances exist for the nucleons and hyperons (collectively called baryons) [from the Greek *barys* (heavy)] with the difference that the spins of the resonant states must be half-odd-integer multiples of $h/2\pi(3/2, 5/2, 7/2$, etc.). Actually one can consider all the baryons as excited states of a single particle, the nucleon. Each particle (excited state) is then specified by a set of quantum numbers which we shall describe later. The principal characteristics of the known baryons and mesons are shown in Fig. 21.16 which is analogous to a chart of the energy levels of the hydrogen atom. Baryons are apparently conserved in every interaction if we assign the particles a baryon number of $+1$ and the antiparticles a number of -1. The tracks of some particles in a hydrogen bubble chamber are shown in Fig. 21.17.

If sufficient energy is available, new particles can be created in large numbers during collision processes. In fact, this is the way in which the host of unstable strongly interacting particles was discovered; being unstable, they are not found under normal conditions. The known assembly of particles appears extremely chaotic. Naturally, the physicist would like to establish a pattern which relates each particle to all the others. This has been one of the chief avenues of attack in recent years and has met with considerable success. Returning to Fig. 21.16, we see that the rest-energy difference between neutron and proton is quite small, about 0.78 MeV. Since these two particles are quite similar except for their electromagnetic interactions, we are tempted to attribute this difference to interactions of the electromagnetic field with a fundamental particle, the nucleon. Furthermore, since the possible states of this particle are two in number, perhaps they can be treated like the two possible spin states ($S = \pm\frac{1}{2}h/2\pi$) of the electron. In an atom the two spin states of the electron normally have the same energy. However, if we turn on a reasonably strong magnetic field, the energy of one spin state becomes slightly different from that of the other.

A somewhat similar effect is believed to occur with the nucleon, except that the general electromagnetic interaction, rather than a static magnetic field, is believed to cause the energy (mass) separation of the proton and neutron states. By analogy with ordinary mechanical spin we label these states *isobaric spin states* [from the Greek *iso* (same) + *baro* (weight)] and treat them in much the same way as the mechanical spin states. This leads to a new quantum number,

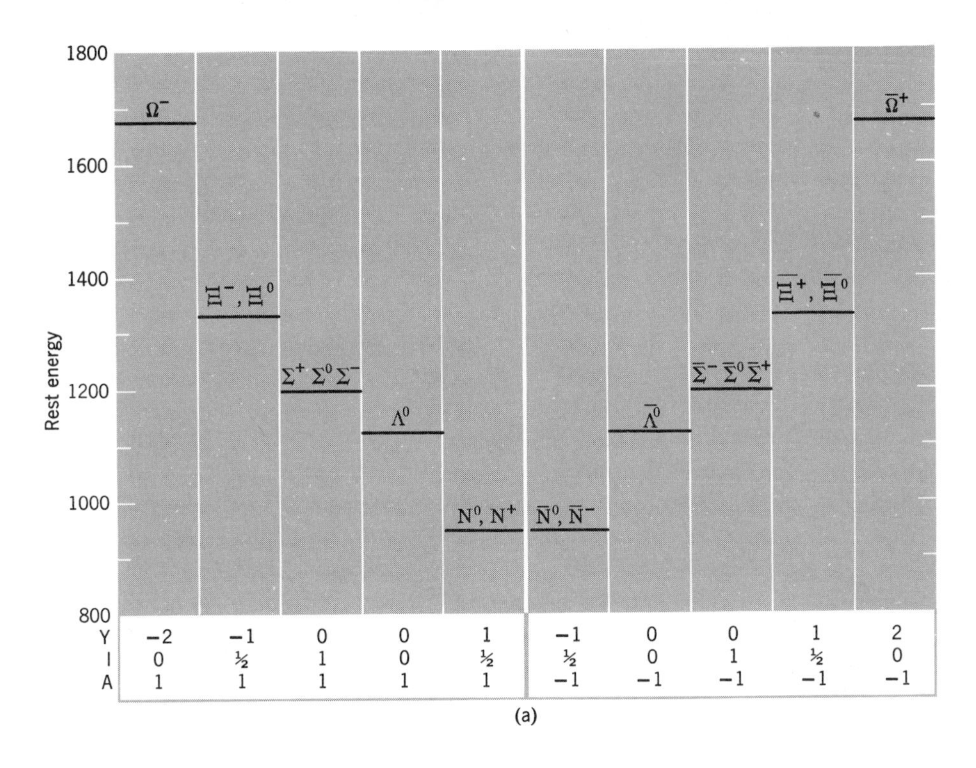

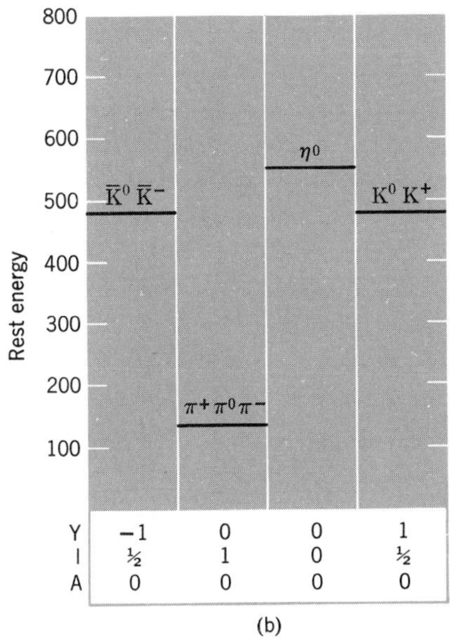

Fig. 21.16. The strongly interacting particles, baryons (a) and mesons (b), showing the various multiplets. Only the lowest energy states of the particles are presented, and the splitting within a multiplet is omitted for ease of illustration. A is the baryon (mass) number.

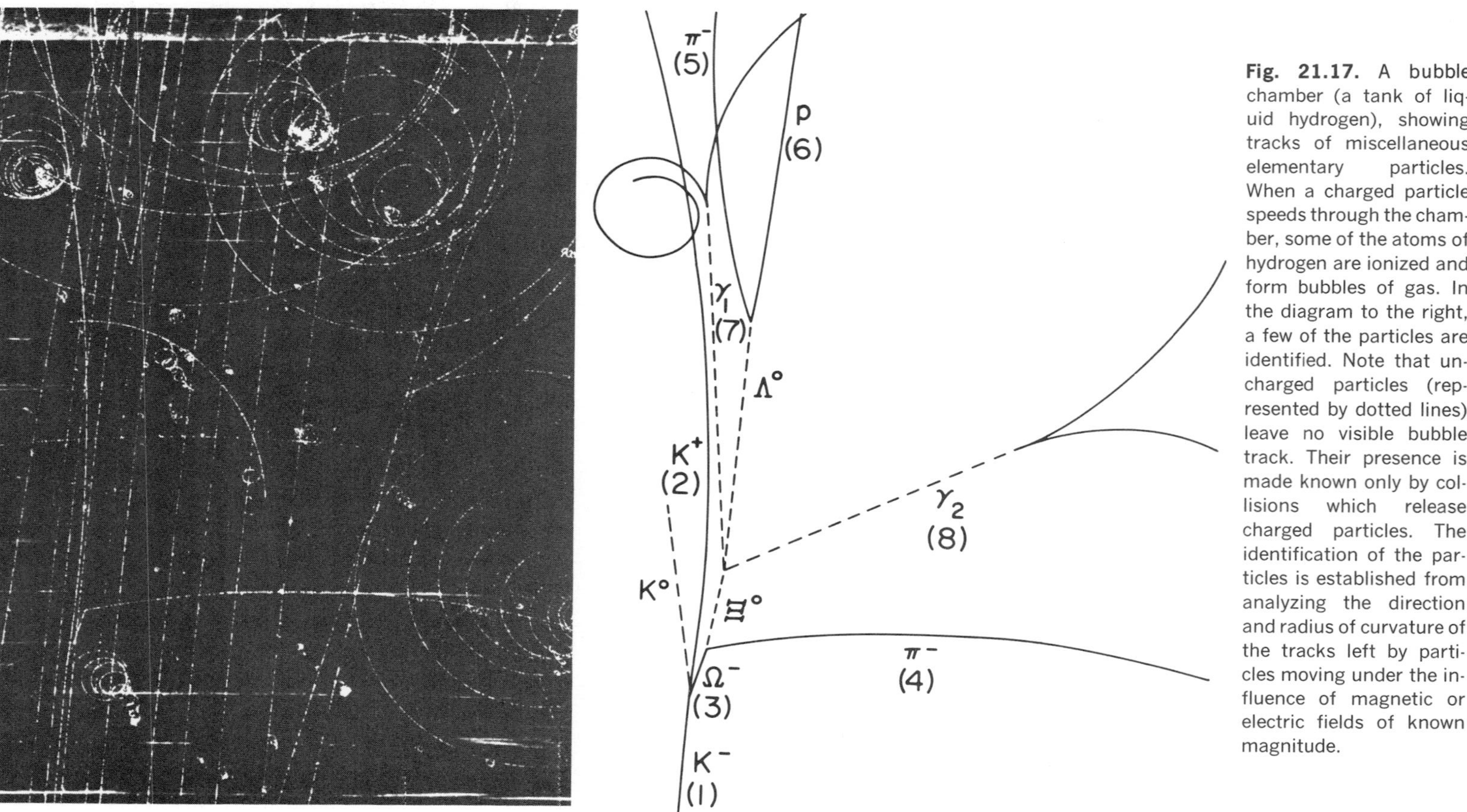

Fig. 21.17. A bubble chamber (a tank of liquid hydrogen), showing tracks of miscellaneous elementary particles. When a charged particle speeds through the chamber, some of the atoms of hydrogen are ionized and form bubbles of gas. In the diagram to the right, a few of the particles are identified. Note that uncharged particles (represented by dotted lines) leave no visible bubble track. Their presence is made known only by collisions which release charged particles. The identification of the particles is established from analyzing the direction and radius of curvature of the tracks left by particles moving under the influence of magnetic or electric fields of known magnitude.

the isobaric spin $I(I = \pm\frac{1}{2})$. It is also possible to classify the charge states of the π and K mesons according to isobaric spin. For example, the mass difference between π^+ or π^- and the neutral pion π^0 is also assumed to be an electromagnetic effect. When the hyperons were discovered, it was found that the mass differences among certain members of the family were also of the same order as the nucleon mass difference. Hence the isobaric spin concept was applied within each set of particles with similar properties (called a multiplet).

However, the mass differences between members of different multiplets was much larger, about 100 MeV rather than a mere 1 to 10 MeV. The implication is that classification according to isobaric spin breaks down when going from one multiplet to another, and recourse must be made to a more complicated classification scheme if some sort of order is to be established. In the early 1960's an American physicist, M. Gell-Mann, and an Israeli Army colonel-turned-physicist, Y. Ne'eman, discovered a way to represent the symmetry of all the known baryons and mesons. In order to do this they had to introduce another quantum number, usually called hypercharge,[5] which is apparently conserved in all strong interactions. The theory passed another test for success: it made a prediction. A particle corresponding to one of the possible hyperon states had not been discovered. The Gell-Mann and Ne'eman theory, sometimes called the "eightfold way," because of eight fundamental states implied by it, predicted the properties of this particle including its mass. The experimenters, knowing what to look for, soon made the discovery. As a useful by-product, the resonances mentioned previously were embraced within the general pattern. In more recent years, attempts have been made to refine the theory in order, for example, to include mechanical spin; however, the attempts have not been satisfactory and the "eightfold way" remains the best approach to the classification of the strongly interacting particles. It is interesting to note that the theory also predicts the existence of particles named "quarks" with extremely unusual properties. Among these are fractional electric charge ($\frac{2}{3}e$ or $-\frac{1}{3}e$) and fractional baryon number ($\frac{1}{3}$). Several attempts to observe quarks have been made recently, but they were not successful, thus leaving quarks as unidentified.

In the preceding discussion of the fundamental particles we mentioned the "weak" interaction in connection with the decay of unstable strongly interacting particles. The term "weak" is used because this force, while not nearly so weak as gravitation, is much weaker than either the strong or the electromagnetic forces. Actually the history of the weak interaction greatly antedates the host of strange particles discovered in recent years.

[5]Hypercharge Y is related to another quantity, "strangeness" S, first suggested some years earlier by Gell-Mann as a necessary quantum number in strong interactions. The relation is $S = Y - A$ where A is the baryon number (see Fig. 21.16).

Of particular interest is the manner in which the study of the weak interactions in beta emission led to the prediction and later discovery of the neutrino, one of the most elusive and dramatic elementary particles. The arguments and steps leading to this discovery illustrate both how conservative is the physicist in his investigation, unwilling lightly to abandon well-established laws, and yet able to entertain in the face of experimental evidence, the possibility that such laws are subject to question.

The problem arose in connection with the nuclear reactions called beta decay, in which an electron is emitted from a "parent nucleus" which then turns into a different "product nucleus" with a higher positive charge and consequently a higher atomic number Z. In these reactions a paradoxical situation was discovered: The energy of the emitted electron is not equal to the difference between the energy of the parent nucleus and the energy of the product nucleus. The most careful measurements showed that, on the average, the emitted electron carried with it only about one-third the energy called for. The *possibility was seriously entertained that at last an exception had been found to the law of conservation of energy.* However, Pauli suggested in 1927 an alternative solution—that a very small particle of negligible mass and without charge was also being emitted, carrying with it the "lost" energy. A careful analysis of this postulate was carried out in 1934 by Fermi, who gave the particle the name "neutrino" using the Italian diminutive to indicate its smallness. It was not until about two decades later that the neutrino made its elusive presence directly known after an elaborate trap had been set up to detect it. A network of particle counters placed beneath a nuclear reactor showed unmistakable evidence of its presence. (The difficulty of finding such a small neutral particle can be appreciated by the fact that a neutrino can easily penetrate the earth with no deviation.) This "neutrino" was endowed with a second function—it conserved angular momentum. The spin of the decaying nucleus was observed always to change by an integer times $h/2\pi$, but the spin of the electron was $h/4\pi$. Somewhere angular momentum of $h/4\pi$ was lost. However, if the spin of the neutrino is also $h/4\pi$, the dilemma is resolved, since the two particles can share the angular momentum lost by the nucleus. Subsequent investigations of beta reactions have revealed that the neutrino probably has zero rest-mass and that it can undergo inverse reactions with nuclei such as

$$\bar{\nu} + p \longrightarrow n + e^{+}$$

Beta interactions are quite slow, being typically of the order of seconds. The weak decay of unstable particles other than the neutron is much more rapid. For example, the π-meson decays in about 10^{-8} seconds into a neutrino and an electron or into a neutrino and another particle called a muon, (sometimes erroneously called the μ meson).

$$\pi^+ \longrightarrow \bar{\mu}^+ + \nu_\mu \text{ or } \bar{e}^+ + \nu_e$$
$$\pi^- \longrightarrow \mu^- + \bar{\nu}_\mu \text{ or } e^- + \bar{\nu}_e$$

where $\bar{\nu}$ denotes an antineutrino. For many years it was believed that ν_μ and ν_e were identical but only a few years ago they were found to be somehow different. The muon itself is an unstable particle decaying in about 10^{-6} seconds into an electron (or positron), a neutrino (ν_μ), and an antineutrino ($\bar{\nu}_e$). The muons, electrons, and neutrinos together with their antiparticles are collectively known as "leptons" (acronym for light particles). Actually, the muon is not really light; it is in fact almost as heavy as the π-meson. These particles are, despite their large mass differences, to be grouped together in a single family because they all participate in the weak, but not the strong, interactions. All existing evidence seems to support the hypothesis of *lepton conservation;* hence we assign *lepton numbers* of $+1$ to electrons, negative muons, and neutrinos, and assign -1 to their antiparticles. Their rest energies are shown graphically in Fig. 21.18.

We have already mentioned that all the strongly interacting particles except the proton are unstable against the weak interaction. Although the neutron and the charged pions decay into leptons, most of the baryons and mesons decay into other strongly interacting particles, as for example:

$$K^+ \longrightarrow \pi^+ + \pi^+ + \pi^-$$
$$\Sigma^+ \longrightarrow p + \pi^0 \text{ or } n + \pi^+$$
$$\Lambda^0 \longrightarrow p + \pi^- \text{ or } n + \pi^0$$

in times of the order of 10^{-9} seconds. Certain peculiarities in K-meson decay led to an important suggestion by T. D. Lee and C. N. Yang in 1957. In order to explain these peculiarities (which we shall not pursue further here, because they

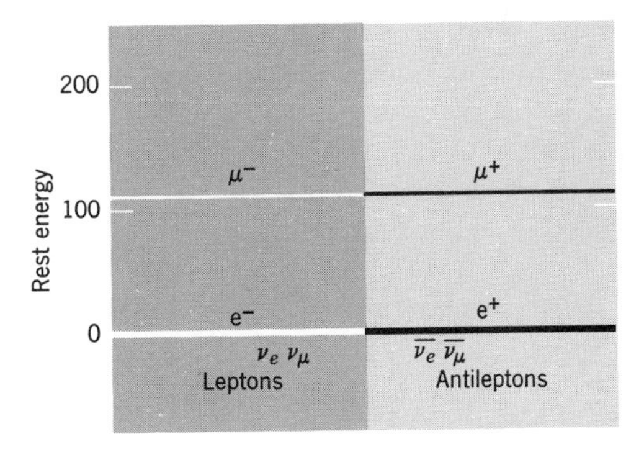

Fig. 21.18. The leptons; ν_e and ν_μ denote the neutrinos associated with electrons, e, and muons, μ, respectively.

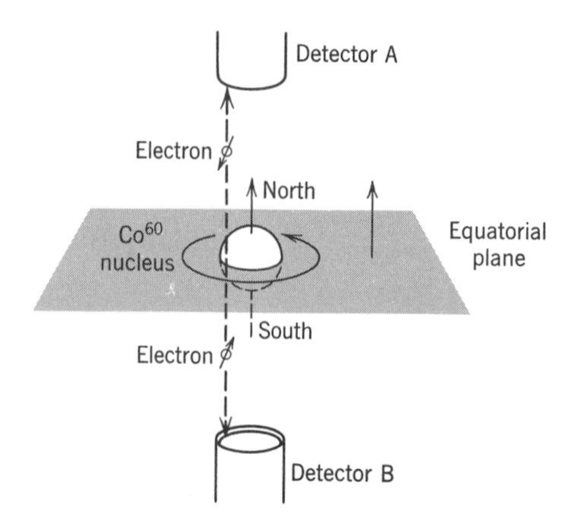

Fig. 21.19. Beta-decay of a Co^{60} nucleus in a magnetic field. The nucleus can emit an electron toward detector A or toward detector B. If the process is symmetric about the nuclear equatorial plane, the counting rate for a large assembly of Co^{60} nuclei should be the same for both detectors. This is not observed.

are no longer of great interest), Lee and Yang proposed that a weakly interacting system does not look the same in a left-handed world as it does in a right-handed one. The behavior of a system with respect to inversion[6] of the coordinate system leads us to the very important concept of "parity." If the system is unchanged by this operation, it is said to possess spatial *inversion* symmetry: It looks essentially the same in both left-handed and right-handed coordinate systems. This was a basic tenet of quantum theory for many years, simply because it seemed so reasonable. Now let us see what is actually observed in nuclear beta-decay, a weak process which is relatively simple to observe.

Suppose we investigate a beta-emitting material which has a non-zero nuclear spin such as Co^{60}. Because of the spin its nuclei act like small magnets and can thus be aligned in strong magnetic fields as indicated in Fig. 21.19. We mount a suitable detector above the "north" poles of the nuclei as shown and measure the decay rate. In order to perform our space inversion we move the detector to a position below the "south" magnetic pole, that is, we "reflect" it to the lower position. The sense of rotation of the nucleus is not affected by the space inversion because the motion of the "equator" of the nucleus is not affected: it "rotates" in the same direction. The decay rate is measured again. If the system is indeed unchanged by the parity operation (coordinate axis inversion), the two measured decay rates should be identical. However, and this was a most surprising discovery, the decay rates are quite different. Subsequent investigations of other weak decays have shown that they too exhibit the same property. (Electromagnetic and strong interactions do not exhibit this property—they

[6]In the inversion process all coordinate axes are reversed. A screw which rises when turned clockwise in one system (the left-handed one) would descend when similarly rotated in the other (right-handed) system.

"conserve parity.") If space inversion by itself changes the system, perhaps there is another type of operation which will, when carried out in combination with the parity operation, leave the system unchanged. In particular let us imagine a magic mirror that converts the Co^{60} to anti Co^{60}. Although such a mirror cannot be constructed, there are ways of observing the consequences of the particle-antiparticle transformation. Apparently weakly interacting systems remain unchanged if we *simultaneously* perform a space inversion and a particle-antiparticle "reflection." Thus, some degree of reflection symmetry is apparently preserved after all. There is a third type of inversion which must be mentioned in our discussion. This is "time reversal" in which we reverse all clocks. There is a very fundamental theorem which is thought to govern certain systems which occur in nature. It requires that a system must be unchanged if we simultaneously carry out the operations of space inversion, time reversal, and particle-antiparticle interchange. This "Pauli-Lüders theorem" (named after its discoverers) has for some years been thought to be a very basic restriction on all physical systems. However, some very recent discoveries in the area of weak decays indicate that either the theorem or time reversal invariance may have to be abandoned. If either eventuality turns out to be true, some further revolutionary shifts in our basic ideas concerning the fundamental interactions will have to be made. [As of this writing, it is still too early to tell.]

8. LAWS OF CONSERVATION IN THE MICROSCOPIC DOMAIN

In this chapter we have been examining events and changes that occur at the microscopic level. It is important to keep in mind the extent to which the "laws of physics," which operate on the macroscopic scale (that is, the scale upon which senses operate in our daily lives and in the laboratory), also operate on the microscopic scale. The general conclusion we reach on the basis of evidence available is that, although there are many new forces and energy relationships involved in reactions taking place on the atomic and nuclear level, there remain many familiar principles that extend down to this level with little change. The more important of these principles can be considered as a set of conservation laws, that is, generalizations about properties and relationships that remain invariant throughout the changes taking place in the world of nature. As we have indicated, these conservation laws are based upon experimental evidence. We cannot conclude that because they hold true within one range of dimensions they will also hold true in a different range.

Actually, most of the conservation laws that appear valid in the macroscopic realm remain valid as far as we know, in the microscopic. The great exception has been that pair of laws—the law of conservation of mass and the law of conservation of energy—formulated by prerelativity physics. We have seen, however, that by their reformulation as the law of conservation of mass-energy (where mass and energy are regarded as equivalent) the two conservation laws are now preserved in a more general single law of conservation. But there is another exception that is alluded to above: the law of conservation of parity which, in 1957, was found to be violated. This violation is significant because it posts the warning that what is taken as an established law or general principle at one level may not hold true universally.

The more important conservation laws at the level of nuclear interactions may be summarized as follows.

1. The law of conservation of mass-energy. No exception has yet been found at any level. For a short time it was considered possible that the law was not valid in the case of beta decay. With identification of the neutrino, however, this conservation law was preserved.

2. The law of conservation of momentum (linear and angular). This law holds true just as firmly at the atomic and nuclear level as at the macroscopic level. Its importance can hardly be exaggerated. As an illustration of its fertility in aiding analysis, one example may be cited, which we have mentioned only briefly. When a positron and electron collide they annihilate one another, their mass turning into radiant energy. Not only must this reaction conserve mass energy; momentum must also be conserved, and photons in appropriate directions must be emitted, with just the right momentum so that the momentum after the reaction is equal to the momentum before.

3. Conservation of charges. As we have seen, all charges, whether in the macroscopic or microscopic domain, are multiples of the charge on an electron. In all reactions that have been studied, the total charge remains constant. Even if an electron is annihilated by a positron, the total charge remains invariant, since the algebraic sum of the positive charge $+e$ on the positron and the negative charge $-e$ on the electron is zero, and therefore the sum remains zero after the reaction. Similarly, if an electron is emitted from a nucleus, a neutron within the nucleus is changed to a proton and the total charge remains constant. Many of the other elementary particles carry charges positive or negative, but in all reactions there is never any net gain or loss of charge.

4. Conservation of particles. The elementary particles can be classified in a number of ways. In one classification they are divided into groups: (*1*) the *heavy particles,* which include neutrons, protons, and hyperons, and their corresponding antiparticles (all of which are sometimes included under the name baryon); (*2*) *lighter particles* including electrons, positrons, neutrinos and antineutrinos, μ mesons and anti-μ mesons (all of which are sometimes classified as leptons); and (*3*) π mesons, K mesons and, most importantly, photons (the last being quanta of radiant energy). There is a conservation law, which applies to the first two groups, but not to the third. The total number of heavy particles (minus the number of antiparticles) is conserved in all reactions. Similarly, the total number of leptons minus the number of antileptons is conserved. As far as is known, for example, a proton can never be changed or decay into a less massive particle. In the case of group (*3*), however, conservation does not apply. Photons, for example, may disappear in a reaction.

5. Conservation of parity (mirror symmetry). The concept of symmetry is important throughout physics, and has many applications. In the macroscopic domain, that which may be described as mirror symmetry applies to all mechanical motions. For example, a system of planets rotating about the sun in an identical manner to our solar system, but in opposite directions (as they would appear in a huge mirror), would obey the same laws of mechanics. It had always been assumed that nature shows no preference for either clockwise or anticlockwise rotations (as viewed from a point toward which the rotating body was moving). Physicists had assumed that this would also be true of the microscopic world. For example, all the particles emitted from a nuclear reaction with a certain spin should not have a spin with any built-in direction preference, so that the reaction could be described equally well in terms of a mirror reflection. In the previous section we pointed out that in the microscopic domain, conservation of purity is indeed violated in the case of certain particle interactions. This, of course, posts the warning that as our knowledge increases, other conservation laws may possibly be brought into question also.

We have now completed the examination of theories regarding what occurs in nature *on the microscopic scale* and can proceed in the opposite direction. The frontiers of science are being pushed outward toward an understanding of the very large as well as the very small, and the huge distant galaxies present as many puzzles as the little neutrino. The scientist today can well echo the poet's intuition with his own emphasis—there are indeed "more things in heaven and earth than have been dreamed of" in our philosophy.

9. QUESTIONS

1. It has been hypothesized by the quantum theorists that the gravitational force is mediated by quanta called *gravitons.* From your knowledge of the algebraic form of the gravitational force (inverse square law) give an estimate of the mass of gravitons. *Hint:* Compare with the Coulomb force.)

2. Some physicists have suggested that another part of the universe may contain an "antimatter world." Compare such phenomena as spectral lines of antihydrogen, chemical bonds, living organisms, etc., in this world with their analogues in our world. Can you associate this with a conservation law?

3. When an electron and a positron annihilate, at least two photons must be created. Why cannot the annihilation result in the formation of only one photon? (*Hint:* Two conservation laws are involved.)

4. A single energetic photon creates an electron-positron pair. The positron combines with an electron and two energetic photons are created. These in turn create electron-positron pairs of lower kinetic energy than the first pair and so on. Can you associate this series of processes with a physical law that we studied earlier in the book?

5. What is the mass difference between an antiproton and an antineutron? Justify your answer.

6. A strong interaction "resonance" does not have a well-defined rest energy. What is the reason for this? If the lifetime is 10^{-23} seconds, what is the uncertainty in the rest energy?

7. The neutron decays by the weak interaction with a lifetime of the order of 10^3 seconds, whereas a typical lifetime of a hyperon against weak decay is only 10^{-9} seconds. Can you think of reasons for this difference?

8. How can a consideration of the coulomb forces between similarly charged particles (protons) be used to explain the fact that heavy elements contain a greater proportion of neutrons than do the lighter ones? How is this related to the fact that all elements of atomic number greater than 92 are unstable?

9. Can two elements with different chemical properties have the same atomic mass? Can they have the same atomic number? Explain. How has the "operational definition" of an element changed from that proposed by Lavoisier and followed by chemists for almost a century? In what sense is the new definition superior to the old?

10. An examination of the curve of binding energies shows that neither iron nor calcium would make a good nuclear fuel. Explain why not.

Problems

1. Complete the following nuclear reactions (in each case only one kind of particle is missing):

$$_{92}\mathrm{U}^{235} + {_0n^1} \longrightarrow {_{56}\mathrm{Ba}^{141}} + {_{36}\mathrm{Kr}^{92}} + ?$$

$$_{93}\mathrm{Np}^{239} \longrightarrow {_{94}\mathrm{Pu}^{239}} + ?$$

$$_{88}\mathrm{Ra}^{226} \longrightarrow {_{86}\mathrm{Rn}^{222}} + ?$$

2. The atomic mass unit is taken as exactly 16.0000 for oxygen (8 neutrons and 8 protons). On the basis of their relative mass, other elements are assigned atomic mass units. One of the typical fission reactions that occurs when uranium^{235} is struck by a neutron is then symbolized as follows:

$$_{92}\mathrm{U}^{235} + {_0n^1} \longrightarrow {_{36}\mathrm{Kr}^{92}} + {_{56}\mathrm{Ba}^{141}} + 3{_0n^1} + Q$$

Q stands for the energy release. Using the atomic mass units given below, calculate (*1*) the mass loss in atomic mass units for the fission of 1 atom of U^{235}; and (*2*) assume that you start with 235 kg of U^{235}, what is the mass loss through the fission of 235 kg of uranium and how much energy in joules or Calories does this represent?

$$_{92}\mathrm{U}^{235} = 235.116 \text{ amu}$$

$$_0n^1 = 1.009 \text{ amu}$$

$$_{36}\mathrm{Kr}^{92} = 91.934 \text{ amu}$$

$$_{56}\mathrm{Ba}^{141} = 140.951 \text{ amu}$$

What is the energy release per kilogram? Compare this to the energy release per kilogram of TNT explosive (about 3500 Calories).

3. Compare the energy release per kilogram of uranium as found in the last problem with the energy release per kilogram of "heavy" hydrogen in a "fusion" reaction symbolized as follows:

$$_1H^2 + {}_1H^2 \longrightarrow {}_2He^4 + Q$$

The atomic mass of $_1H^2$ (deuterium) is 2.0147 amu. The atomic mass of $_2He^4$ (helium) is 4.0039 amu.

4. The half-life of radioactive carbon is 5600 years. It is present in very small quantities in the atmosphere formed by the bombardment by cosmic rays on C^{12} in carbon dioxide already present. A measure of the proportion of C^{14} present in a sample of carbon is achieved by noting the counts per minute per gram of carbon as detected by a Geiger counter. In wooden artifacts that absorb no additional C^{14} by absorption of this isotope from the atmosphere, the proportion of C^{14} present gradually decreases. Suppose that carbon from an ancient artifact shows one-eighth of the number of counts per minute per gram that the carbon from new wood shows. What is the age of the artifact? (For those interested in geology, an elementary account of radioactive dating is given in Patrick M. Hurley's *How Old is the Earth* (Anchor Book), Doubleday, Garden City, N.Y., 1959.)

Recommended Reading

Max Born, *The Restless Universe,* Dover Publications, New York, 1951. A delightfully written, elementary but thorough presentation of many aspects of modern physics, by one of the scientists who helped to develop quantum mechanics.

Morris H. Shamos and George Murphy, editors, *Recent Advances in Science,* New York University Press, New York, 1956. The chapter on "Nuclear Structure" by H. A. Bethe and the one on "Elementary Particles" by V. F. Weisskopf are fine nonmathematical treatments of the two subjects.

Edward U. Condon, "Physics," a chapter in *What Is Science?* edited by James R. Newman, Simon & Schuster, New York, 1955. A short, nonmathematical review of modern developments in physics.

Turning Points in Physics, lectures given at Oxford University, introduction by A. C. Crombie, Harper, New York, 1959 (Harper Torchbooks, 1961). Topics in modern physics with attention to the general significance of the ideas.

Jay Orear, *Fundamental Physics,* Wiley, New York, 1961. This book has already been referred to. Its illustrations and examples are particularly helpful.

Eric Rogers, *Physics for the Inquiring Mind,* Princeton University Press, Princeton, 1960. Another excellent book written for the nonscientist, which has been referred to previously.

Geoffrey F. Chew, Murray Gell-Mann, and Arthur H. Rosenfeld, "Strongly Interacting Particles," *Scientific American,* February 1964, p. 74, is an excellent account of the theory of strong interactions by two of the leading architects.

Victor F. Weisskopf, "The Three Spectroscopies," *Scientific American,* May 1968, p. 15, is a more recent article which draws analogies among atomic, nuclear, and elementary particle structure.

chapter twenty-two

The environment of man: the atmosphere

About the earth there is a very thin shell composed of the earth's crust, the oceans, and the atmosphere. It is within this thin shell, constituting a mere one-half of one percent of the earth's volume, that physical and chemical conditions are maintained in a marvelous and intricate system insuring both the stability and the constant changes necessary for the development of the myriads of living forms the earth has produced. Oxygen, carbon, and hydrogen, elements which are almost unique in their potentiality for producing great numbers of compounds and entering into an endless variety of reactions, abound over the whole surface of the earth and are forever being mixed and stirred by meteorological conditions. Energy flows copiously from the sun towards the earth to provide the motive power for the mixing process and the heat required for sustaining the chemical reactions. The earth's diurnal rotation together with its annual revolution about the sun in the plane of the ecliptic produces wide fluctuations in the energy absorbed, insuring the wide variety of conditions out of which millions of species of living forms have developed.

However, in spite of the constant mixing and the changes that take place, this environment for living forms is also extraordinarily stable. For billions of years the temperature, the composition of the oceans, and the atmosphere have been maintained within very narrow limits. (One of the early objections against Darwin's theory was that evolution would take much longer than the life of the sun. Until nuclear reactions were discovered, it had been calculated that the sun could not have poured out its vast energy for more than a few million years, whereas evolutionary processes required billions of years.)

In the study of meteorology, we shall find beautifully illustrated most of the laws of physics and chemistry that we have studied in previous chapters. Here again we shall see how separate strands of information are woven together to extend knowledge of our immediate environmest in all its intricate complexities. We look at the development of a comparatively new science, meteorology, and finally, we will discover how great and serious is the possibility that man with all his technology and knowledge is moving towards irreversibly destroying the environment in which he and other forms of life have their abode, and finally we will study some of the phenomena which occur in the rarefied atmosphere at great heights.

1. SOLAR AND TERRESTRIAL RADIATION

For many purposes the sun approximates a "black body" (i.e. an object with a large quantity of internally trapped radiation, relatively little of which is escaping), radiating at a temperature of 6000°K. The energy poured out by the sun in all directions in space amounts to about 4×10^{27} calories per minute. Of course, only a small portion (about one part in two billion) reaches the earth. It has been calculated, and measurements sustain the calculation, that about 2 calories per sq cm per minute reach the earth's upper atmosphere. This figure is referred to as the "solar constant" and probably fluctuates no more than one or two percent, even when there are solar flares. The earth and its atmosphere absorb, reflect, and radiate in such a way that the temperature of most of the earth's surface remains on the average close to 300°K.

The fraction of the incident solar electromagnetic energy reflected by the earth and its atmosphere is called the reflectivity or *albedo.* The reflectivity at any given position on the earth depends upon the nature of the surface (snow covered, tropical rain forest, desert, etc.) and degree of cloud cover. Snow, as one may suspect, has the largest albedo of any surface covering, and tropical rain forest has the least. In addition to high reflectivity, cloud cover has another very important atmospheric effect; it traps radiation from the earth. An atmosphere with a large water content is transparent to solar radiation of relatively short wavelength (visible and near infrared). A large fraction of this radiation is absorbed by the earth which radiates as a somewhat imperfect black body (grey body) of temperature 300°K. Now, the lower the temperature of a black body, the longer is the wavelength at which it radiates most of its energy. Hence the solar energy absorbed by the earth is reradiated well into the infrared range where it is absorbed very strongly by water, either in vapor form or in clouds. Of course, clouds reflect a very large fraction of the incident solar radiation, but the fairly large fraction of the long wave energy from the earth is very effectively trapped. It is because of such energy entrapment that cloudy nights are warmer than clear ones; all other factors being equal. For the same reason, regions of low *humidity* (water vapor content) tend to have larger diurnal temperature ranges than regions of high water vapor content. Such radiation trapping is by analogy called the *greenhouse effect.* The glass walls of a greenhouse transmit the short wavelength solar rays. However, the long wavelength rays are effectively trapped by its interior surfaces, which are good absorbers of infrared radiation. The principle is illustrated in Fig. 22.1.

The yearly cyclical change in the amount of solar radiation incident on any given zone of the earth (i.e., the region between two specified parallels of latitude) gives rise to *seasonal changes* of temperature. Such changes in the amount of incident radiation occur for two reasons: (a) changes in the duration of daylight, and (b) changes in the incident intensity. For example, the longer period during which the sun is above the horizon on summer days means that heat energy is absorbed by the earth for a longer time than in winter. The second effect is illustrated in Fig. 22.2. Since in summer the angle of incidence of the rays is smaller than in winter they are more concentrated with the result that more radiation falls on a unit area. Hence we have greater total heating during the

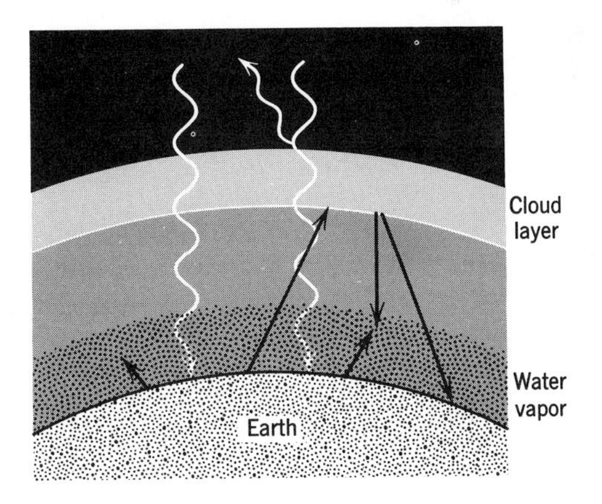

Fig. 22.1. The greenhouse effect. Solar radiation (wavy arrows) easily penetrates the clouds and water vapor, warming the earth's surface. Infrared radiation (heavy arrows) from the earth is absorbed by the water, however. Some solar radiation is back-scattered by the clouds.

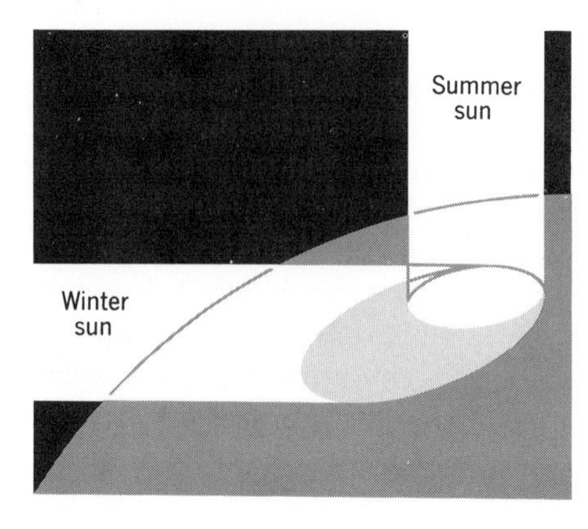

Fig. 22.2. Effect of the solar zenith angle on the intensity of the radiation which is incident on the earth's surface.

summer months, with the greatest daily average occurring at the summer solstice (about June 21st in the northern hemisphere). On the basis of these simple arguments one may expect the maximum yearly temperature to occur at this time. This is not the case—the warmest period of the year usually occurs from one to one and one-half months after the solstice. The reason for this effect, sometimes called the seasonal lag, should be immediately evident to the reader. It lies in the large heat capacity of the earth which leads to a rising daily average surface temperature even after the average heating rate has started to decrease. In addition to the seasonal lag there is a daily lag of temperature which also occurs because of the heat capacity of the earth. Usually the maximum daily temperature occurs between 2 and 3 p.m.

2. ATMOSPHERIC CIRCULATION

Within the atmosphere transfer of energy is accomplished almost entirely by large scale mass motion, usually called the general circulation, which is strongly influenced by the earth's rotation, as we shall see. However, to simplify the presentation, we begin the discussion without considering the effect of planetary rotation.

The expansion of a gas as it is heated was discussed in some detail in Chapter 10. To summarize the conclusions reached there, if the gas pressure is held constant, the density is inversely proportional to the temperature. Hence, the density of a parcel of air heated by absorption of solar radiation decreases with the result that the surrounding denser region will exert an upward force on the parcel (buoyancy) causing it to rise. The rising mass is thus displaced by an inward flowing current of air which in turn is heated, causing it to ascend. As the parcel rises, it cools due to expansion and starts to flow outward. Eventually, it descends to the earth's surface where it joins the current moving toward the rising air stream. This is the familiar air current pattern over a hot stove with which the reader is probably already familiar. It is illustrated in Fig. 22.3 which also shows the ideal pattern for a non-rotating earth.

As the air parcel rises, it leaves a reduced pressure in the region below it. Of course, the air current flowing into the warm region tends to raise the pressure back to normal, but never completely does so. A pressure slightly lower than normal is necessary to cause the air current to flow and the proper balance is struck when the volume of inward moving air is just equal to that which is rising. On the other hand the region of sinking air is at a slightly higher pressure than normal. Because of these observed pressure differences we call the two regions low and high pressure areas, respectively, or simply *lows* and *highs*. Typical pressure ranges above or below normal are of the order of $1/100$ of an atmosphere although, in the "eye" of a hurricane, an intense low, the pressure may be over $1/10$ of an atmosphere below normal. Atmospheric pressure is usually given on weather maps in millibars (1 mb $= 10^3$ dynes cm^{-2}); a pressure of one standard atmosphere is about 1013.4 mb.

The ideal circulation patterns shown in Fig. 22.3 are profoundly influenced by the earth's rotation which causes winds in the northern hemisphere to be deflected toward the right (clockwise); those in the southern hemisphere have the opposite deflection. The *Coriolis effect*, as it is called, is due to the inertia of an air parcel as it moves over the earth's surface. Relative to an observer in an inertial reference frame the parcel tends to move in a straight line subject only to the effect of genuine forces such as pressure differences. However, relative to an observer on the earth, the air appears to describe a curved path because of

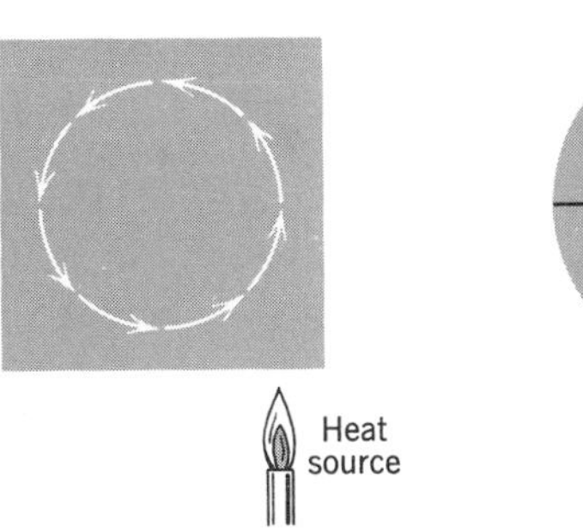

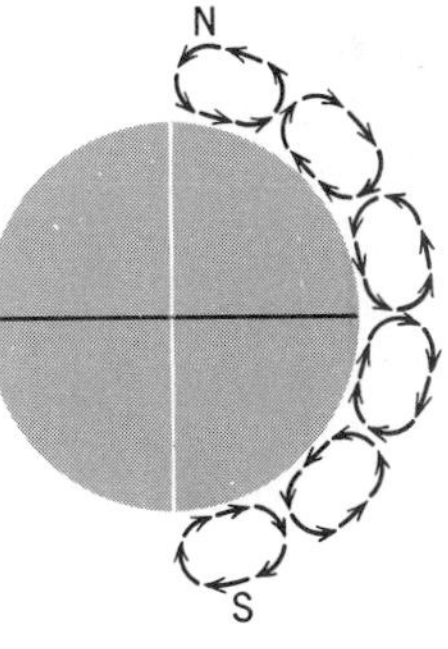

Fig. 22.3. (a) Circulation pattern in a gas. (b) Circulation pattern on a nonrotating earth.

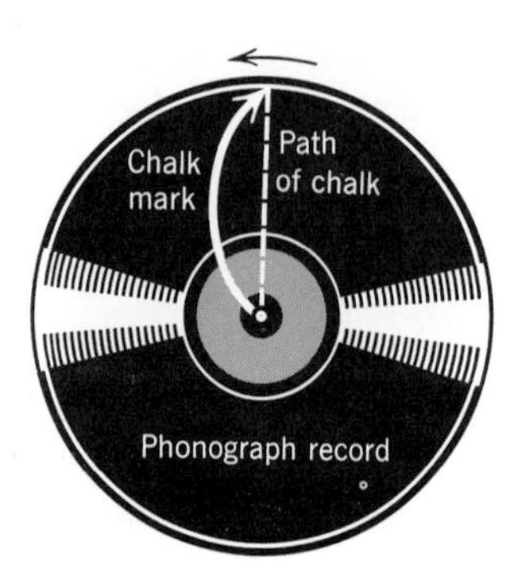

Fig. 22.4. Experiment for demonstrating the Coriolis Effect.

the earth's rotation under it. To such an observer the curvature appears to be the result of an additional force, called the Coriolis force, acting on the parcel. The principle is easily demonstrated with the simple experiment illustrated in Fig. 22.4. Place a phonograph record on the turntable and slowly but steadily rotate it by hand counterclockwise. In the meantime move a piece of chalk from the center (i.e., the "north pole") uniformly outward in a straight line relative to yourself. This traces out the curved path on the record. The curvature of the path is such that, to a person who is riding on the record, the chalk appears to be deflected in clockwise direction. The Coriolis force is thus a fictitious force of the same class as centrifugal force which appears *only* because the observer is in an accelerated reference frame. In general, the Coriolis force is strong enough that the wind direction at an altitude of about 1000 meters above the surface (where friction is not important) is approximately along the direction of lines of equal pressure (isobars) of a weather map. If there were no Coriolis force the flow would be directed into the low and out of the high.

Bearing the foregoing arguments in mind, it is easy to modify the circulation pattern of Fig. 22.3 to fit a rotating earth. The result, together with the names of the wind systems and regions of high and low pressure, is shown in Fig. 22.5 for the equinoxes. Some of the names are of historical significance. For example, the northeast trade winds marked a favorite route for sailing ships before the days of steam. The horse latitudes which separate two wind systems are a high pressure region in which ships carrying horses from Spain to America were frequently becalmed for long periods. When the food supply was exhausted the horses starved and were thrown overboard, sometimes littering the sea over large expanses. The true circulation pattern is not nearly so simple as that shown because of the complications of surface features which have a marked effect on the heating rate of the air. This aspect will be discussed in greater detail in the section on air masses.

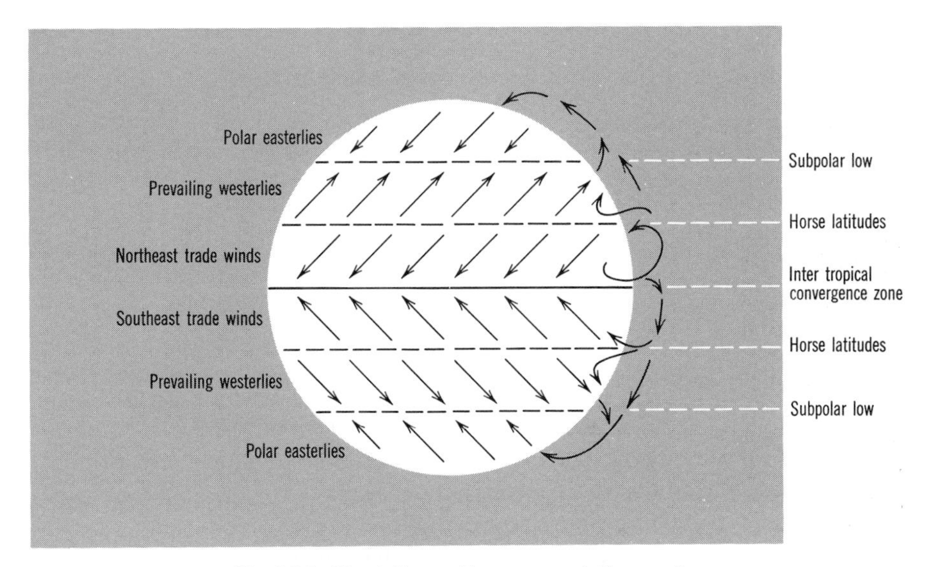

Fig. 22.5. Circulation pattern on a rotating earth.

3. WATER IN THE ATMOSPHERE

The presence in the atmosphere of water vapor and condensed water droplets has a pronounced effect on weather and climate. It was mentioned in Section 1 that water in the atmosphere controls its heating and cooling rates. The condensation and freezing of water vapor is responsible for the formation of rain drops, snow flakes, and hail stones. The descent of water in these forms is called *precipitation.*

The amount of water vapor present in the air is called the humidity. Air can hold only a certain amount of water vapor and when this value is reached it is said to be *saturated.* In general, warm air can hold more water vapor than can cold air. It is quite common to express humidity as the ratio of actual water vapor content to the maximum possible value at the given temperature. This ratio is called the *relative humidity* and is expressed in per cent; as a consequence of the temperature dependence of the saturation point, relative humidity increases as the air cools. When the *dew point temperature* is reached, the vapor begins to condense. Fog is formed when the air close to the earth's surface is cooled below the dew point, and condensed water appears as suspended drops. In addition to relative humidity, meteorologists frequently employ *specific humidity* or mixing ratio which is defined as the number of grams of water contained in a kilogram of air. If one wants to know the thermodynamic behavior at point B of a parcel of air which started at point A, one must know the initial specific humidity. Obviously specific and relative humidities are closely related.

Clouds form when air is cooled below its saturation point. The very small droplets formed during the condensation process are easily supported as a suspension and do not precipitate unless local conditions favor growth. There are a number of ways in which clouds are produced. For example, moist air heated by contact with the earth's surface will rise and cool. When the temperature reaches the dew point, clouds form. Such clouds are characterized by considerable vertical extent and, to the airplane pilot, signify "bumpy" air which is caused by turbulent motions in the rising air currents. A warm moist air mass in horizontal motion may be forced upward by a mountain range which stands in its path. Again, it cools as it rises with condensation beginning when saturation is reached. A particularly important mode of cloud formation which we shall discuss in more detail in a subsequent section is the lifting of warm air by a heavier mass of cold air which pushes under it like a wedge. Can you think of some additional mechanisms for cloud formation?

Clouds are usually classified according to their altitudes and to their vertical or horizontal development. *Cumulo-form* clouds which are formed by rising air currents are of marked vertical extent while *strato-form* (layered) clouds are in the form of horizontal sheets. High altitude clouds or *cirro-form* (those whose bases are about 6,000 meters above the earth's surface) are of three types: cirrus, cirrocumulus, and cirrostratus. Cirrus clouds are very thin and wispy, and are composed entirely of ice crystals. Cirrocumulus are thin, patchy clouds with wave-like patterns while cirrostratus are thin sheets which frequently form halos or luminous circles around the moon and sun. Both types are also composed of ice crystals, which accounts for halo formation. These optical phenomena are somewhat similar to rainbows in that they are dependent upon refraction, in this case by ice crystals. Alto or middle clouds which average about 3000 meters above the earth are of strato or cumulo-form types. Altostratus appears somewhat like cirrostratus but does not form halos because of the absence of ice crystals. Altocumulus appear as patches of puffy white or grey clouds. There are three main kinds of low altitude clouds: stratus which is a quite uniform sheet (it is a fog if in contact with the ground); nimbostratus which is darker than ordinary stratus and is a source of rain; and stratocumulus which are irregular masses of clouds spread out in a rolling layer. Finally, we mention the clouds of great vertical development, cumulonimbus. These are the familiar thunderheads which may extend vertically over 20–25 kilometers although 10 kilometers is probably more typical. Their great height is a consequence of strong vertical air currents. Photographs of clouds which are representative of these types are shown in Fig. 22.6.

The air acquires its water vapor as a result of evaporation, mainly by solar radiation, from oceans, lakes, and rivers. Of course, the atmosphere loses water

Fig. 22.6. Cloud types: (a) cirrus, (b) cirrocumulus, (c) cumulus, (d) altocumulus, (e) stratocumulus, (f) stratus. (Courtesy, Weather Bureau, U.S. Department of Commerce, Washington, D.C.)

(c)

(d)

(e)

(f)

as well as gaining it, and over a long period of time the two effects balance. The loss occurs by precipitation in the form of rain, hail, and snow. The evaporation, transport, and precipitation chain constitutes the water cycle of the atmosphere. According to current theories, rain drops are formed by coalescence, particularly with ice crystals acting as nuclei for the process. Under such conditions the ice crystals tend to grow at the expense of water droplets because of evaporation from the former and condensation on the latter.

Eventually the ice crystals become heavy enough to fall or *precipitate* from the cloud. The apparent necessity for the presence of ice crystals led to the suggestion some years ago that precipitation may be induced in clouds at temperatures below freezing but which contain only water droplets. The agents used are dry ice (solid carbon dioxide) and silver iodide crystals. The former freezes out ice crystals while the latter act as nuclei and apparently cause small droplets to freeze. The mechanism involved is still being researched. These agents are also small enough to remain in suspension. Such *seeding* will not, of course, replace naturally occurring coalescence because of practical limits on the quantity of dry ice and AgI which can be suspended. It is difficult to assess the effects of seeding because of the difficulty in ascertaining that precipitation would not have occurred had the clouds not been seeded. Sometimes precipitation occurs when ice crystals are known to be absent and some other mechanism must be at work. For example, rain in tropical areas is believed to result from coalescence by collision in air which is very turbulent. Turbulent air may also be instrumental in producing heavy precipitation when ice crystals are present as in the upper parts of thunderheads.

Hail stones form when rain drops suddenly encounter strong rising air currents which carry them to very high altitudes where they freeze. They then fall, encounter another rising air current, repeating the process and growing by acretion. Finally they become too heavy to be kept aloft and fall to the ground.

4. THE ATMOSPHERE AND OCEANS AS A HEAT ENGINE

When one analyzes the thermal and mechanical processes which occur in the atmosphere, the resemblance to a heat engine (see Chapter 10) becomes apparent. In a steam engine, water is the "working fluid" by which energy is transferred from one point to another in the system. In the atmosphere both water and air serve in this capacity. Actually our system is incomplete unless the oceans are also included. They act not only as fluid reservoirs but as reservoirs for thermal and mechanical energy as well. Although ocean temperatures are usually rather low and motions (e.g., currents and upswelling) rather slow, the large masses involved means that the energies are tremendous. Most of the heat absorbed by the ocean-atmosphere system is absorbed by the oceans. Part of it drives the ocean currents and part is transferred to the atmosphere by evaporative processes. *The water vapor content of the atmosphere is its principal reservoir of energy,* and evaporation accounts for the effect of ocean currents on local weather.

A good example of the influence of currents on weather is offered by the Gulf Stream which forms in warm water at low latitudes and flows northeastward into the eastern North Atlantic. Prevailing winds carry the air warmed by the current over the land masses of northwest Europe accounting for the relatively mild climates of those high latitude regions. The growth of humidity of the air as it moves past the Gulf Stream is the cause of the dense fogs which are common in the British Isles and the North Sea.

We have already discussed the supply of energy by the sun but we shall repeat and amplify some of our remarks about its absorption by the atmosphere-ocean system. On the average most of the absorption occurs near midday at low latitudes although there is considerable variability depending upon local conditions such as the nature of the surface, topography, water content of the atmosphere, etc. Hence, the heating of the earth's surface is quite uneven. The tendency to approach thermal equilibrium (recall the second law of thermodynamics) results in advective processes which cause the wind systems and ocean currents. Such fluid motion carries kinetic energy which is proportional to the air or water mass and the square of the wind speed. Hence we have converted heat into mechanical energy. Evaporation of water from the oceans, lakes, etc., also adds energy to the air, but it is a kind of chemical potential energy rather than kinetic energy. When the water vapor later condenses, it returns the energy to the air, causing local heating. This is the principal mechanism for driving the cyclonic storms. Although we shall discuss them in more detail later, we mention here that a cyclonic storm involves the rotational motion of a large mass of air about a low pressure center. The kinetic energies involved are quite large and are supplied entirely by condensation of water vapor.

Man has tapped the weather system for energy for many centuries. Windmills date back well over one thousand years while hydraulic dams and water wheels are of more recent origin. Of course, winds do not drive the latter, the energy being supplied by the deposition of rain and snow at relatively high altitudes. Various schemes have also been suggested to harness the oceanic tides and currents, but so far they do not appear practicable.

5. HIGH AND LOW PRESSURE AREAS, AIR MASSES, AND FRONTS

High pressure areas form wherever air cools and sinks because of its increased density; the horse latitudes and the polar high are the most prominent, but others may also occur. As the air sinks and spreads out in a region of high pressure, it gradually acquires a rotating motion due to the Coriolis effect (clockwise in the northern hemisphere, counterclockwise in the southern). In general the wind direction is very close to the direction of the isobars with a slight outward drift. A high is also called an *anticyclone.* Low pressure areas, or *cyclones* on the other hand, occur where there are rising air currents. The wind direction in the cyclone is opposite to that in the anticyclone, i.e., it is counterclockwise in the northern, clockwise in the southern hemisphere, and like the anticyclonic case the air motion is nearly along the isobars, but with a slight inward drift. The most prominent low pressure areas are the doldrums and the sub-polar low. Of course, the highs and lows which really occur do not conform to the simple model shown in Fig. 22.7. Unevenness of heating rate due to differences in surface character cause the formation of many relatively small cyclones and anticyclones as well as considerable variability in the major highs and lows.

Very large cells of air which are more-or-less thermally uniform in a horizontal direction are called air masses. By "thermally uniform" we mean that their temperature and water vapor content vary only slightly over very large distances. The region of origin is one of high pressure. The initial thermal character of an air mass is determined by the nature of the surface over which it forms. For example, one which forms over water can be expected to have a much larger water vapor content than one which forms over land, and one which is a product of the polar regions will be much colder than one created in the tropics. The standard designations for the four types of air masses are continental polar (CP), continental tropical (CT), maritime polar (MP), and maritime tropical (MT). Air masses are characterized not only by the area of origin but also by the surface over which they move and by their "age" or the time lapse since leaving the area of origin. If colder than the surface, they tend to be warmed from the bottom upward, producing rising air currents. Such air is said to be unstable. That is clouds of vertical development are likely to form and pre-

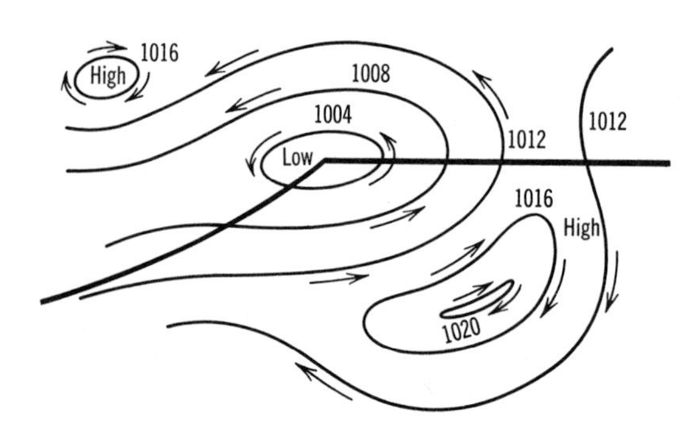

Fig. 22.7. Two adjacent high pressure cells showing the low pressure region and front (heavy line). Isobars (lines of equal pressure) are in millibars.

cipitation may result if sufficient water vapor is available. On the other hand, if the surface is colder than the air mass moving over it, the latter will be cooled from the bottom up. Such air which resists the formation of vertical currents is said to be stable. The "older" the air mass, the more it will differ from its initial state. In this classification scheme the two types of air masses are called cold (k) and warm (w), respectively. A typical designation for an air mass is (MPk). The boundary between two neighboring air masses is called a frontal surface. The intersection with the earth's surface is called a *front.* It is quite thin because air at atmospheric pressure mixes very slowly over distances of meteorological scale.

In order to associate the concept of air masses with the weather, we shall consider two types, CP and MP. A continental polar air mass forms at very high latitudes and is extremely cold. When it pushes southward over Canada in winter it brings very cold and dry weather to that country and to the northern United States. However, if the air already present was warm and moist, the cold air will wedge under it. As a result of this action the moist air is cooled enough to cause precipitation usually in the form of snow. Such a *cold front* will be discussed later. A typical maritime polar air mass forms over the north Pacific and moves southeastward over the western United States. It is usually stable air and brings the winter rain and snow as a result of cooling in passing over the mountains.

A front is said to be strong or weak depending upon the contrast between the air masses. If the contrast is small enough, the front in effect vanishes. Frontal systems are particularly important with regard to storms; the temperature change across the boundary represents the main energy source for the creation of storms. Although they are sometimes *stationary,* fronts are usually characterized by motion. If warm air is displacing cold air, it is a warm front; if the opposite is true it is a cold front.

There is on the front a point at which the pressure is at a minimum. It is the neighboring area which is usually called the associated low pressure system. The character of the isobars, illustrated in Fig. 22.7, shows that they form closed loops about the point of lowest pressure. Since the wind direction is nearly tangent to the isobars, the pattern is circular. This is the reason that it is given the name *cyclonic.* Because of the large thermal contrasts in the vicinity of the cyclone, a considerable amount of energy may be released by condensation of water vapor. It is this effect which accounts for the occurrence of the cyclonic storms discussed in the following section. Typically the cyclonic center propagates along the front with a wave-like motion. In order that the cyclone may develop, it is necessary that the wave motion be *unstable;* that is, the wave must grow. This is the case for wavelengths of 600 to 3000 km. The physics of wave generation is a rather complicated problem in fluid dynamics and we shall not pursue it further here.

There are two basic types of fronts, warm and cold. In the former, advancing warm air rides up over a wedge of cold air which it displaces (see Fig. 22.8). The warm air is cooled as a result of lifting, and condensation results. The type of clouds formed depends upon the stability of the two air masses. For example, if the cold air is stable and the warm air unstable, one finds strato-form clouds in the former and cumulo-form clouds in the latter. By contrast, the advancing air behind a cold front slides under the warm air which it displaces (see Fig. 22.9). As a consequence, the latter is forced upward causing condensation of water vapor. Again, the types of clouds which form along the front depend upon the stability characteristics of the two air masses.[1]

Figure 22.10 shows the development of a cyclonic storm center which occurs along a frontal surface. The wave-like bend (a) which separates warm and cold fronts is quite small but as the front advances (b) the bend grows in size, principally because the speed of the cold front is greater than that of the warm front. Eventually, the former overtakes the latter (c) producing an *occluded front* whose cross section is shown in Fig. 22.11. The air around the occlusion, shown in (c), is eventually deprived of its moisture because the wind system does not feed fresh moisture-laden air into it. When the energy source represented by the moisture of the incoming air is exhausted, the storm decreases in intensity and eventually dies.

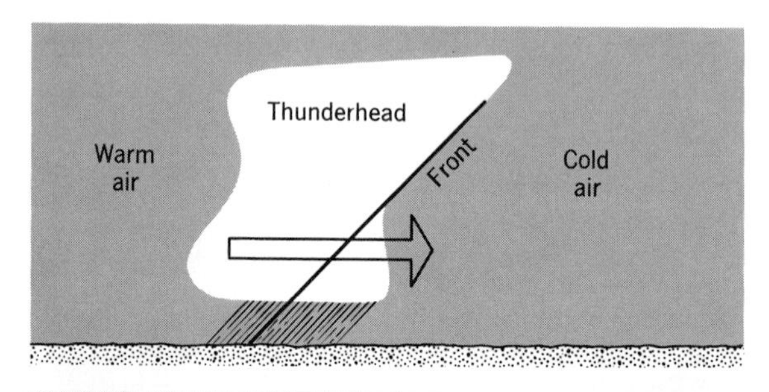

Fig. 22.8. Vertical section of a warm front. The vertical extent is quite exaggerated.

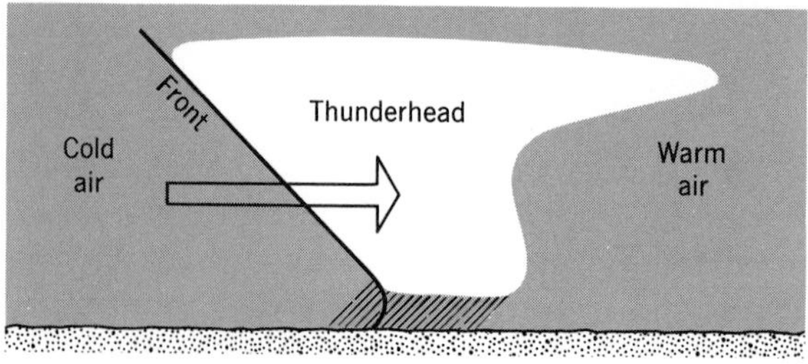

Fig. 22.9. Vertical section through a cold front.

[1] The thunderheads shown in Figs. 22.8, 9, and 11 form only if the air is unstable.

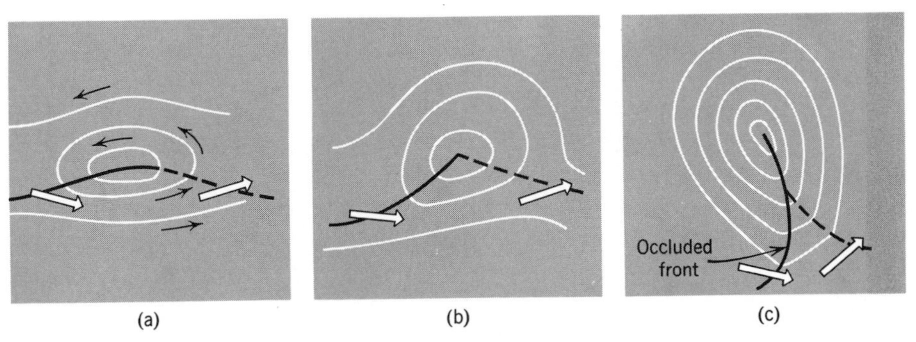

Fig. 22.10. Steps in the development of a cyclonic storm. denotes cold front, warm front, stationary front.

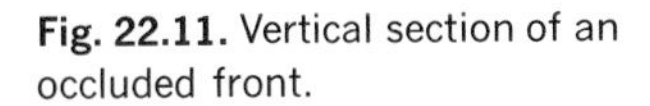

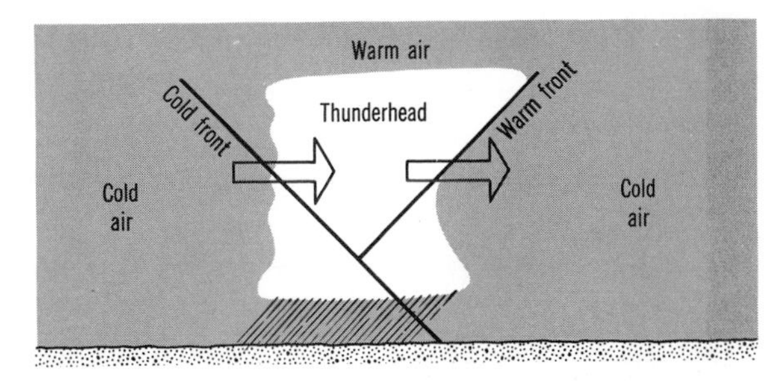

Fig. 22.11. Vertical section of an occluded front.

The motion of frontal systems is particularly important to the weather forecaster because of the strong influence on local weather. Empirical methods of various types have been in use by forecasters for many years with varying degrees of success. Most recently they have made use of large computing facilities to solve (in some approximation) the hydrodynamic and thermodynamic equations which govern atmospheric motion. A portion of a typical weather map is shown in Fig. 22.12.

6. STORMS

In the preceding section we mentioned the cyclone as an important feature of a frontal system. The energy which drives it is supplied by condensation of water vapor in the vicinity of the cyclone center. The heat released by the condensation process is absorbed by the air which becomes buoyant and rises. As the pressure in the center of the system falls, the wind velocity increases, thus feeding moist air at a greater rate. As a consequence of the higher rate of energy supply, the wind increases in velocity until it reaches a limit set by surface friction and the humidity of the air. We now have a full-blown cyclonic storm. Eventually

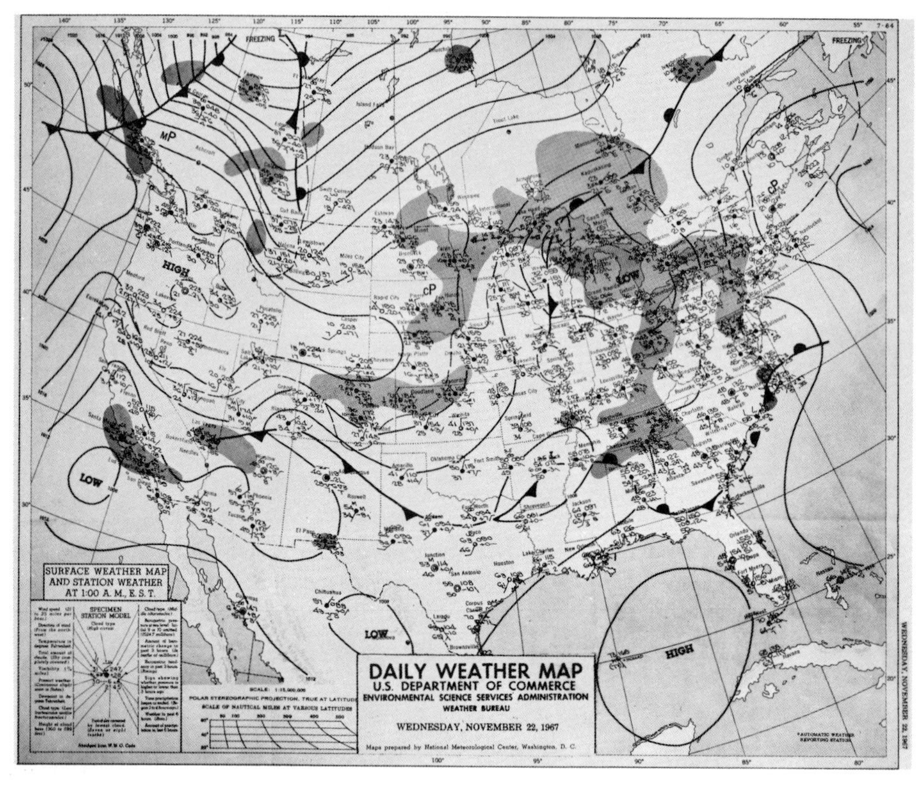

Fig. 22.12. Weather map of United States, November 22, 1967. (Courtesy, Weather Bureau, U.S. Department of Commerce, Washington, D.C.) ▲▼▲ denotes stationary front.

the supply of moist air is exhausted and the storm dies. The development is illustrated in Fig. 22.10.

The hurricane or tropical storm is a particularly violent cyclonic storm which forms as a low pressure instability in the inter-tropical convergence zone where the northeast and southeast trade winds meet. The large amount of moist air available in tropical ocean areas means that an unusually large quantity of heat energy can be transformed into the kinetic energy of storm motion. As a result, wind speeds are frequently in excess of 150 km per hour. The rotating mass of air moves initially in a west northwest to northwest direction (in the northern hemisphere) but gradually veers to the northward and eventually eastward at midlatitude. It frequently happens that hurricane paths are not as regular as we have indicated. Their tracks are sometimes so capricious that they strike the same area twice. The track and wind pattern of a typical Caribbean hurricane

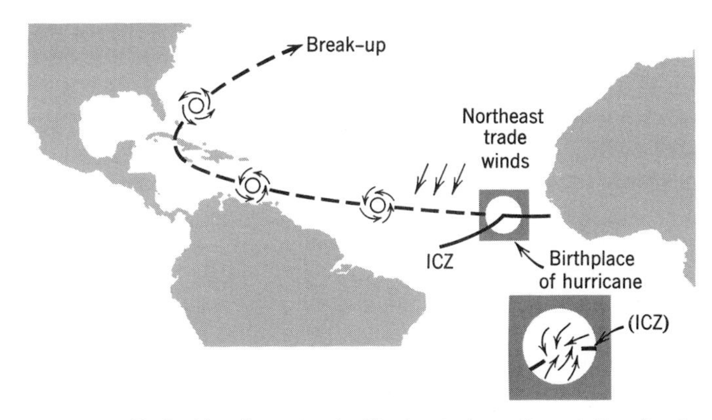

Fig. 22.13. Typical hurricane track. The inset shows the rotational pattern which develops in an instability in the intertropical convergence zone (ICZ).

is shown in Fig. 22.13. Note that the hurricane dies over the north Atlantic where the supply of warm moist air is greatly diminished. If the track should be over a large land mass, the storm will die for the same reason. Hurricanes are also found in other parts of the world including the southern hemisphere. In the Pacific they are called typhoons.

Thunderstorms usually form where there are violent vertical movements of air, that is, in unstable air. Sometimes the clouds build up to heights of 20 km although 10 km is probably more typical. The vertical movement may be due to the action of a cold front, to heating of the air by the ground surface, or to temperature differences between land and ocean. Eventually, if the currents become strong enough, the cooling and turbulence suffice to produce heavy precipitation. Friction between rain drops and the air results in electric charge separation within the cloud. When the electric fields become greater than the breakdown point of moist air, the familiar lightning strokes occur. Thunderclaps are produced by the rapid heating of the air in the vicinity of the discharge. A thunderhead (cumulonimbus) will continue to produce heavy rain or hail as well as lightning until the downcoming precipitation cools the air enough to break up the strong vertical currents.

7. AIR POLLUTION

Air pollution is one of the major problems confronting this nation today and no general account of atmospheric science would be complete without discussing it. It can conveniently be divided into two classes, chemical pollution and radioactive pollution. Although the deadlier of the two, the latter is much the easier to control, at least at the present time: simply avoid releasing the contaminants. Such a solution is not practicable for chemical pollution simply because the atmosphere is the only convenient and available disposal medium for the discharge from autos, chimneys, and industrial activities. The emission rate of the most serious contaminants is susceptible to a considerable degree of control, however, and most efforts to reduce air pollution have been in this direction.

Historically, chemical air pollution has been classed according to the region where it was first observed. The word "smog" was coined in London many years ago to describe the vile combination of smoke, soot, fog, and corrosive substances typical of that city during certain parts of the year. The soot and smoke has its genesis in the millions of soft coal-burning fireplaces that are in general use for residential heating. The large output of sulphur dioxide from these fires reacts with the atmospheric water content of fog for which London is famous to produce relatively large and probably injurious concentrations of sulphuric acid. Although these facts have been known for many years, London's building codes still permit the old fashioned hearth in new construction in some sections of the city. Efforts to control this type of air pollution in the U.S. have been fairly successful, although some problem with respect to sulphur dioxide still remains. This is a consequence of the presence of large quantities of sulphur in most commerical fuels.

A new type of contamination began to appear about twenty years ago, at first in the Los Angeles area in California, and more recently in other population centers. It is characterized by low atmospheric visibility, by crop damage and very noticeable eye irritation, and by the smell associated with atmospheric oxidant (ozone in this case). Early efforts to reduce it were concentrated in the obvious area—industrial emissions. In this respect Los Angeles soon became the most rigidly controlled locale in the world. Nevertheless, the problem of the "Los Angeles smog" as it is usually called continued to get worse. In the early 1950's scientists at Stanford Research Institute in California together with colleagues at other institutions found that the principal contributor to the southern California air pollution was automobile exhaust. As the number of passenger cars in the Los Angeles region steadily increased smog incidence became more frequent and more severe.

The study of the production of the pollutants responsible for low visibility, eye irritation, and crop damage is a complicated problem in photochemistry and we can do no more than touch on it here. Briefly, the nitrogen oxides (e.g., NO_2) present in auto exhaust are dissociated by the sunlight

$$NO_2 + hf \longrightarrow NO + O$$

The free oxygen atom then reacts with unburned hydrocarbons, emitted along

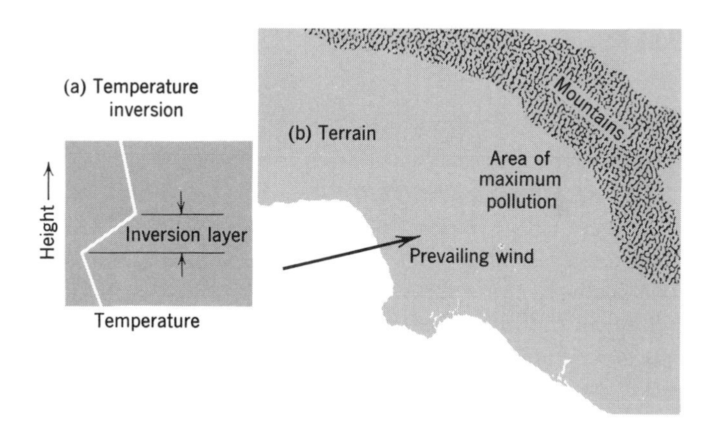

Fig. 22.14. Trapping of air within the Los Angeles basin.

with the nitrogen oxides. The hydrocarbons must be of the unsaturated variety if they are to react chemically with the atomic oxygen. It will be recalled from Chapter 18 that unsaturated molecules contain less than the maximum number of hydrogen atoms which they are capable of bonding. As a result they are highly reactive, i.e., ethylene (C_2H_4), as opposed to the more inert ethane (C_2H_6), a saturated molecule. The products of the latter reactions are ozone and large agglomerates of hydrocarbons. The latter is apparently responsible for low visibility, eye irritation, and crop damage; the former is often associated with crop damage.

So far, little has been said of the influence of weather and climate on the occurrence of Los Angeles type air pollution. It so happens that there are certain natural factors, in addition to sunlight, which make the Los Angeles basin particularly susceptible to smog attacks. In summer and fall the air stagnates in the basin when the prevailing winds are weak. In addition, a high pressure system (anticyclone) moves into the coast from seaward. The air in the anticyclone is heated by compression as it descends. It stops its descent when it meets the stagnant surface air and then flows outwards horizontally. The result is the formation of the temperature *inversion* shown in Fig. 22.14a. The surface air is, of course, quite stable (resistant to upward movement), preventing dispersal of the contaminants by upward convection through the *inversion layer*. The low wind velocity coupled with the temperature inversion leads to a trapping of the air within the basin. As a consequence, the concentration of pollutants builds up until weather conditions change. The general effect of the local topography in the vicinity of Los Angeles is shown in Figure 22.14b.

Few population centers have the peculiar characteristics which favor growth of air contamination as in southern California. Nevertheless some of them do have a serious pollution problem which promises to grow worse in the years ahead, *even though control methods may be greatly improved.* The solution is left as an exercise for the readers for it is their generation which will ultimately have to provide it.

8. THE UPPER ATMOSPHERE

Our emphasis thus far has been on phenomena associated with the lower atmosphere, namely, weather, climate, and air pollution. By lower atmosphere we mean the region below about 17 km altitude. The air above this height does not have much obvious relation to our daily lives. Yet there is much evidence for an intimate coupling between the lower and upper atmosphere, although we do not know its details nor all of its consequences.

We begin our discussion with a brief account of the thermal properties of the upper atmosphere. Measurements of the air temperature at various altitudes have revealed a profile with roughly the shape shown in Fig. 22.15. By "temperature" we refer, of course, to the gas kinetic temperature or mean energy of molecular motion at these altitudes. The outstanding features of this profile are the maximum and minima which occur at about 17, 50, and 85 km and which serve as convenient division points as illutrated in the diagram. The principal mechanism for heating the upper atmosphere at all levels is the absorption of ultraviolet radiation from the sun, mainly by molecular oxygen at higher altitudes and by ozone at lower ones. Photodissociation of O_2 into oxygen atoms

$$O_2 + hf \longrightarrow O + O$$

probably accounts for most of the energy absorbed. Ultimately through recombination of the atomic oxygen, e.g.,

$$O + O + O_2 \longrightarrow O_2 + O_2$$

most of it takes the form of heat at thermospheric altitudes. The purpose of the third body (O_2 in this case) is to carry off enough energy to allow the atoms to recombine. It is mainly such kinetic energy which eventually becomes heat energy. Very little of the radiation capable of directly dissociating oxygen molecules penetrates to the mesosphere and thus cannot deposit energy in it. There is one wavelength band which, by means of a complicated indirect process called predissociation is able to weakly dissociate O_2 in the mesosphere. The oxygen atoms formed in this manner immediately react with O_2 to form ozone

$$O + O_2 + O_2 \longrightarrow O_3 + O_2$$

Although the concentration of ozone never becomes very large (less than one part in 10^5), enough of it is present in the stratosphere and mesosphere to

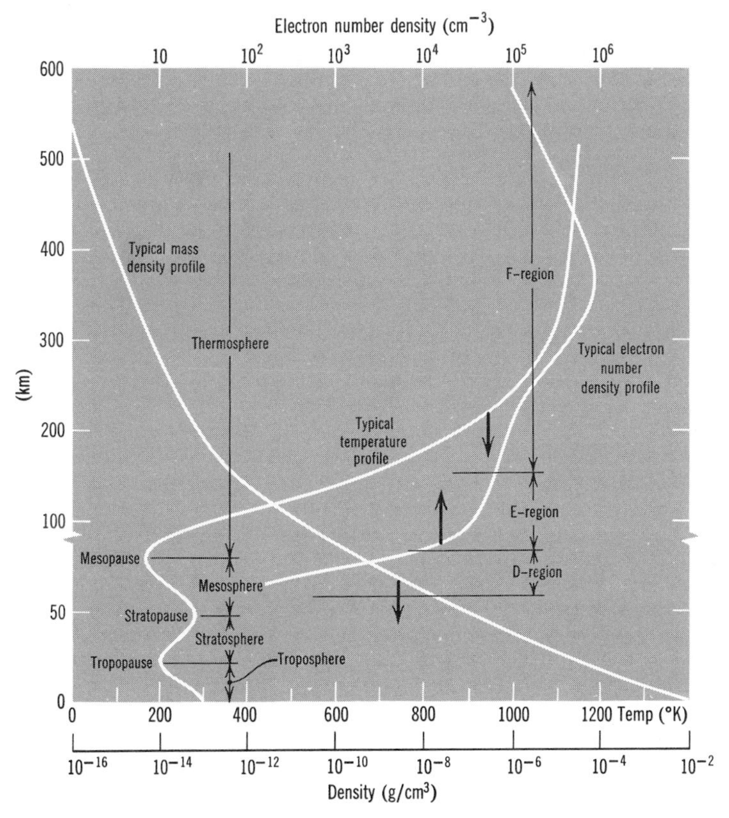

Fig. 22.15. The upper atmosphere. Typical profiles of temperature, mass density, and electron number density.

strongly absorb the *near* ultraviolet radiation from the sun and produce relatively large amounts of heat energy between 30 and 60 km. Note that absorption by O_2 does not directly cause significant heating. However, it results in the formation of ozone which acts as the energy-absorbing medium. Ozone also radiates energy, but it does so in the infrared rather than the ultraviolet region. The infrared radiation emitted by the ozone can be reabsorbed before it escapes from the ozone layer. In fact radiation emitted by a given O_3 molecule may be absorbed and reemitted many times before it leaves the layer. This *radiative transport* process is strongly reminiscent of gaseous diffusion and is treated in much the same way. The atmosphere as a whole is transparent to the infrared emission, and, once it escapes the ozone layer, it passes on into space. Since heat transfer by conduction and convection is quite slow in the mesophere and stratosphere, the ozone layer is quite effectively insulated from the regions above and below it. Hence, a relatively high temperature can be maintained at about 50 km (the stratopause) due to the trapping of radiation by the ozone. Ozone is also important because it absorbs most of the near-ultraviolet radiation from the sun, thus preventing it from reaching the surface of the earth.

Besides ozone, there are other minor constituents that are quite important in certain respects. Among these are nitric oxide which is formed by free nitrogen atoms reacting with O_2

$$N + O_2 \longrightarrow NO + O,$$

As we shall see later, NO plays an important role in the formation of the lower ionosphere. Also important are sodium and the hydroxyl radical; they contribute to the air glow which will be discussed in a later section. The chemistry and distribution of the minor constituents present in the upper air is indeed a fascinating topic but its further pursuit would take us too far afield.

The atmosphere, like any fluid in a state close to rest is in *hydrostatic equilibrium,* that is, the upward force exerted by the surrounding atmosphere on a column of air is equal to the weight of the air in the column. Since air is a gas, it is compressible; hence, with ascending altitude the weight of air in a column of unit length and cross section becomes less and less. Suppose we have a column of unit cross section. The upward force exerted on the bottom is just the air pressure (force acting on a surface of unit area). The pressure difference between points a and b in Fig. 22.16 is equal to the average air density ρ in the column multiplied by the gravitational acceleration g and the height of the column h

$$P_a - P_b = g\rho h$$

With the aid of the ideal gas law PV = nkT (see Chapter 11) and the definition of density, we can rewrite the above (hydrostatic or barometric) equation as

$$P_a - P_b = \frac{mg}{kT} Ph$$

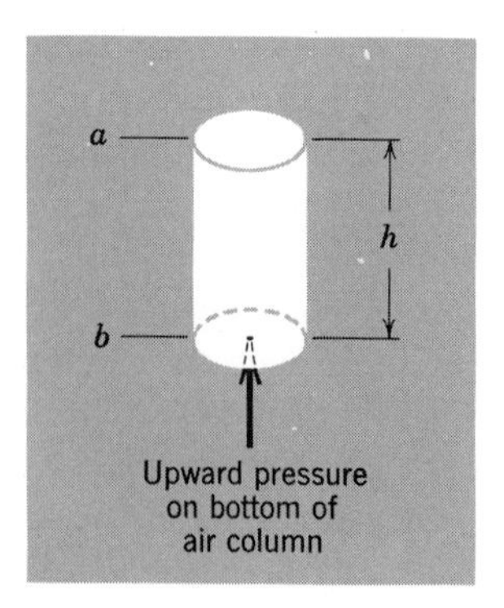

Fig. 22.16. Hydrostatic equilibrium in an air column.

where m is the mass of one molecule of the gas and P is the average pressure in the column. In order to proceed further, one must resort to use of the calculus. However, it is possible to gain more insight into the significance of the hydrostatic equation by defining the *scale height* $H = kT/mg$; then

$$\frac{P_a - P_b}{P} = \frac{h}{H}$$

Use of the calculus would show that the pressure changes by a factor of about 2.72 in one scale height. (For the reader familiar with the calculus the factor 2.72+ is the base of the natural logarithms.) The scale height which is a function of molecular weight, absolute temperature, and gravitational acceleration, is probably the most important parameter in the physics of the upper atmosphere because of its close relationship to thermodynamic properties.

At altitudes below about 110 km the atmospheric components apparently tend to mix in proportion to their natural abundances. For some constituents, of course, this tendency never gets very far because of chemical processes which interfere with it. At altitudes above 110 km molecular motion dominates and we have a condition of free molecular diffusion. Under confined laboratory conditions, molecular diffusion leads to mixing of gases; but with the unbounded vertical extent of the upper atmosphere it leads to separation. This is a result of the effect of the gravitational field of the earth and is called gravitational separation. One can compare the effect to cream separation in an undisturbed bucket of fresh milk. Again, the definition of scale height provides convenient insight into the situation: each component has its own scale height which is larger for the lighter gases than for the heavier (recall the appearance of the molecular mass in the equation for H). Hence the *partial pressures* of the lighter gases decrease more slowly with altitudes than do those of the heavier gases. As a consequence, molecular nitrogen is most abundant at 120 km, atomic oxygen at 300 km, and helium at 800 km.

Numerous measurements of upper air properties have been made in recent years with the aid of rockets and satellites.[2] However, we shall mention here only two methods for measuring the density; a typical density profile obtained by these and other techniques is shown in Fig. 22.15. Both methods are based on the measurement of atmospheric "drag" (i.e., resistance to motion) on projectiles, namely spheres ejected from rockets, and satellites. Because of the large number of the latter placed in orbit since 1957, the coverage of the earth in the altitude range 150 to 800 km has been excellent. Satellite orbits are too unreliable to permit use of the method below 150 km and air drag is too small above approximately 800 km. At altitudes of about 75 to 130 km, the so-called falling sphere method has proved quite successful. Briefly, the technique consists of releasing a sphere

[2] See Chapter 23 for a discussion of these vehicles.

from a rocket at about 70 to 80 km and observing its subsequent ascent and descent. Accelerometers inside the sphere measure the instantaneous acceleration, and a small transmitter radios the information to ground. Air drag and thus density is then computed from the recorded data. Many other techniques are now available for measuring properties of the upper air and have led to very important advances in our knowledge of that region. However, lack of space forbids their discussion.

9. THE IONOSPHERE

At the mesospheric altitudes (50–90 km) and above, solar ultraviolet radiation and X-rays produce significant ionization of atmospheric gases. The mechanism is a form of the photoelectric effect studied in Chapter 17 except that the electrons are ejected from the individual gas molecules rather than from a metallic surface. Two such photo processes are represented by the reactions

$$\mathrm{O} + hf \longrightarrow \mathrm{O}^{+} + \mathrm{e}$$
$$\mathrm{N_2} + hf \longrightarrow \mathrm{N_2}^{+} + \mathrm{e}$$

The positive ions and electrons (negative charges) eventually neutralize each other at a rate about equal to the rate of ionization. Thus during daylight hours at least, the concentration or number density of charged particles approaches an equilibrium condition very closely. At the lower altitudes the proportion of charged particles is as small as one part in 10^{12} but at an altitude of 500 km the proportion may reach one part in 10^{2}. Such an ionized gas which is in thermal equilibrium (i.e., has a well-defined temperature) and is electrically neutral (on a large scale) is called a *plasma*. The lower ionosphere is classed as a "weak" plasma because collisions of the charged particles with neutral ones is more important than collisions of charged particles among themselves. On the other hand the upper ionosphere is a "strong" plasma because collisions among the charged particles is the more significant. We shall discuss plasmas at greater length in the following chapter.

From early investigations of the ionosphere by observing radio waves reflected from it, most scientists believed it to be characterized by discrete layers in each of which the ion number density reached a maximum. Since the early 1950's, rockets with instrumentation for measuring the electron and ion number densities have been flown through the ionosphere; they have definitely proved that, with one exception, there are no well-defined layers (peaks) there. A more accurate descriptive term is "regions" because of the absence of well-defined layers. A typical profile of electron number density is shown in Fig. 22.15.

The existence of the ionosphere was postulated over a century ago but it was not directly observed until about 1925. Realizing the possibility of layered structure, the late Sir Edward Appleton named the region which he discovered the "E-layer," E—representing the electric field. It was later found that a lower lying D-region and an F-layer at higher altitudes also existed. The E-region (90 to 160 km) is formed by photoionization of the molecular nitrogen and atomic oxygen present there by UV radiation and X-rays. The positive molecular ions recombine with electrons by the quite rapid dissociative process which is typified by

$$O_2^+ + e \longrightarrow O + O$$

However, the charge on the oxygen atoms must be transferred to a molecular ion in order that recombination may occur rapidly. Hence O^+ is removed by the sequential reactions

$$O^+ + N_2 \longrightarrow NO^+ + O$$
$$NO^+ + e \longrightarrow N + O$$

The ionized region between about 65 and 90 km is called the D-region. Several types of radiation are responsible for its formation, cosmic rays from outside the solar system ("galactic" cosmic rays) at the lower altitudes, ionization of nitric oxide (a minor constituent) by Lyman α radiation (1216Å) and, when the sun is active, X-rays above about 70 km. Since solar radiation and cosmic rays are discussed in some detail in the following chapter, we defer further consideration of them to that point. Free electrons are removed by the dissociative recombination processes described above and by *attachment* to oxygen atoms and molecules. As we mentioned in Chapter 19, these constituents have a strong affinity for electrons. Finally, the *negative ions* such as O_2^- are neutralized by reactions of the type

$$O_2^- + O_2^+ \longrightarrow O_2 + O_2$$

The F-layer (200 to 500 km altitudes) is the only region to demonstrate a well-defined peak, usually at 350 to 400 km. At these altitudes the dominant atmospheric component is atomic oxygen which is neutralized in the two-step process already described (in the paragraph on the E-layer). Because of the low atmospheric density there, electrons and ions can diffuse quite freely. Of course they must do so together because any charge separation would create electric fields strong enough to bring them together again. For this reason, the transport of the charged particles is usually called *ambipolar diffusion.* Although we cannot go into details here, it is worth remarking that it is the competition between recombination and ambipolar diffusion which accounts for the F-region peak shown in Fig. 22.15. It should be noted that the diffusive motion of electrons and ions is influenced by the earth's magnetic field. Because charged particles tend to move in spirals along magnetic lines of force, the geomagnetic field exerts pronounced control on the motion of the ions and electrons in the F-layer. This effect accounts for many peculiarities in the structure of the upper ionosphere.

As we shall see in the following chapter, the sun occasionally produces momentary bursts of X-rays and energetic protons when it is active. The X-rays are absorbed mainly in the D-region where they enhance the degree of ionization by as much as a factor of 10^3. The resultant *sudden ionospheric disturbance* which occurs over the entire sunlit hemisphere can cause severe disruption of radio communications for times up to several hours, depending upon its severity. The same solar eruptions (flares) which produce X-rays sometimes accelerate protons to quite high energies (10 MeV to over 1 GeV). Because of deflection by the geomagnetic field, they are able to enter the atmosphere only in the polar regions. That is, they are funneled into the polar "caps" where they cause intense ionization at D-layer altitudes (polar cap events). Again, the result is blackout of radio communications but only on paths through the high latitudes. If the solar flare which caused the event was particularly severe, this condition may persist for several days.

The ionosphere plays a very important role in radio communications. In fact, "short wave" communications (wavelengths of about 10 to 200 meters) would be impossible were it not for the ionosphere. We can explain this as follows. Because radio waves travel more rapidly in the ionosphere than in the neutral atmosphere, or even in free space,[3] they are *refracted* back to earth. In general, one finds that the larger the electron number density, the greater is the refraction. The situation is illustrated in Fig. 22.17 which shows the refraction back to earth of a high frequency electromagnetic wave. Note that the wave must enter the ionosphere at an angle greater than some critical value if it is to be bent back toward the earth. This accounts for the "skip zone" in which the radio signals cannot be received. It also explains Marconi's demonstration in 1901 that radio waves can be transmitted over the horizon. If the wavelength is too short (less than about 8 meters), refraction is too small at every angle and the wave is not returned to earth.

In contrast to "short waves," the "very long waves" which are used, for example, by the Navy to communicate with the Polaris submarines, are reflected by the ionosphere. These wavelengths are typically of the order 20 km which is greater than the effective "thickness" of the lower D-layer. Hence, the VLW are

[3]This would seem to be a violation of the theory of relativity. However, this theory deals with the propagation of *signals,* not the waves themselves. One can show that the speed of signals carried by the wave is never greater than the velocity of light in free space. Hence there is no contradiction.

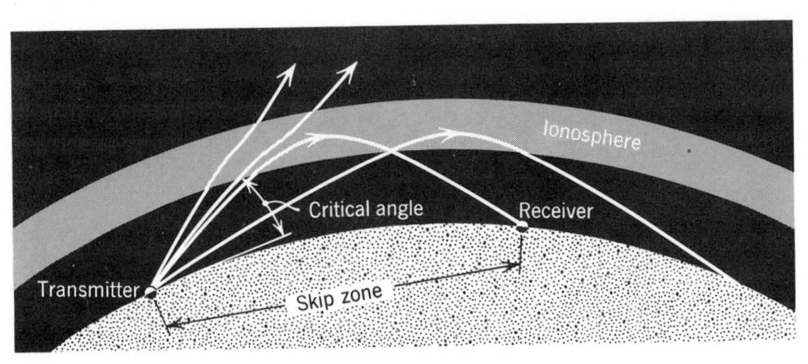

Fig. 22.17. Refraction of radio waves by the ionosphere. The earth's curvature is exaggerated for illustrative purposes.

reflected in much the same way that light is reflected by a metallic mirror. Since the earth's surface is also a good reflector, a very long wave can be efficiently trapped between the earth and the ionosphere, allowing it to propagate over very long paths.

In addition to refraction, radio waves also experience absorption in the ionosphere. When an electromagnetic wave passes an electron, it causes it to oscillate. The wave loses no energy in this case because the electron reradiates the energy, albeit slightly out of step with the incident wave (it is this "phase" shift which leads to refraction). However, if the electron should strike an air particle, its "phase" relationship with the incident wave is lost and the energy cannot be recovered by the wave. Obviously, the greater the electron number density and number density of air molecules, the greater the radio wave absorption. This is the cause of radio blackout during sudden ionospheric disturbances and polar cap events.

Since the ionosphere contains free ions and electrons, it is able to conduct electric currents if there is a difference of potential present. As a result of atmospheric tides electric fields exist, causing ionospheric currents to flow. Unlike the ocean tides, atmospheric tides are produced mainly by solar heating which produces expansion of the atmosphere on the sunlit side of the earth. Pressure differences which result from tidal oscillations cause winds which sweep the ionized air along with the neutral particles. When the stream of charged particles moves across the earth's magnetic lines of force, an electric field perpendicular to the field lines and to the wind direction is created. This causes the path of the charged particles to curve. In fact it eventually curves back on itself, completing the electric circuit. This current pattern is strong enough to produce detectable periodic changes in the geomagnetic field intensity. It was the occurrence of the regular magnetic fluctuations that led Balfour Stewart to suggest the existence of the ionosphere nearly a hundred years ago.

10. OPTICAL EFFECTS—AURORA, AIRGLOW AND NOCTILUCENT CLOUDS

In the northern hemisphere at least, the aurora has been known and recorded in local mythology far back into antiquity. It was, of course, associated with the supernatural until modern times. We still do not know the precise mechanism responsible, but we are sophisticated enough to be sure that it can be entirely accounted for by the known laws of physics. Some typical auroral forms are shown in Fig. 22.18.

Fig. 22.18. The Aurora: (a) corona, (b) drapery. (Courtesy, Dr. V. P. Hessler, University of Alaska, College, Alaska.)

Any reader who has experienced visual aurora will agree that it is indeed an impressive sight. Although it is sometimes seen during daylight hours, it is usually a nocturnal effect, most of the luminous emission occurring before local midnight. The aurora can take many forms but is usually characterized by arc-like appearances. Frequently the arcs exhibit ray structure (or *striations*) aligned with the earth's magnetic field. The aurora is also characterized by rapid motion of the glowing regions across the sky. If the altitude of maximum luminescence moves rapidly upward, we speak of "flaming aurora." Although there is no well-

(*b*)

defined upper boundary, the lower edges of the aurora usually lie between 80 and 150 km.

As we have said, the aurora is a high latitude effect, in the southern as well as in the northern hemisphere. However, it does not extend over the polar regions. For nearly 100 years the aurora has been observed by scientifically trained observers and the region of most frequent occurrence is now well delineated. Each auroral zone is oval in shape, centered approximately at the respective magnetic pole with the center parallel at about 65° north or south (magnetic latitude) and is about 12° wide. Figure 22.19 shows the location of that part of the northern auroral zone which lies over North America.

The aurora are produced principally by energetic electrons bombarding the high atmosphere. We know that the electrons are accelerated in the earth's magnetic field but the precise site and nature of the acceleration mechanism is

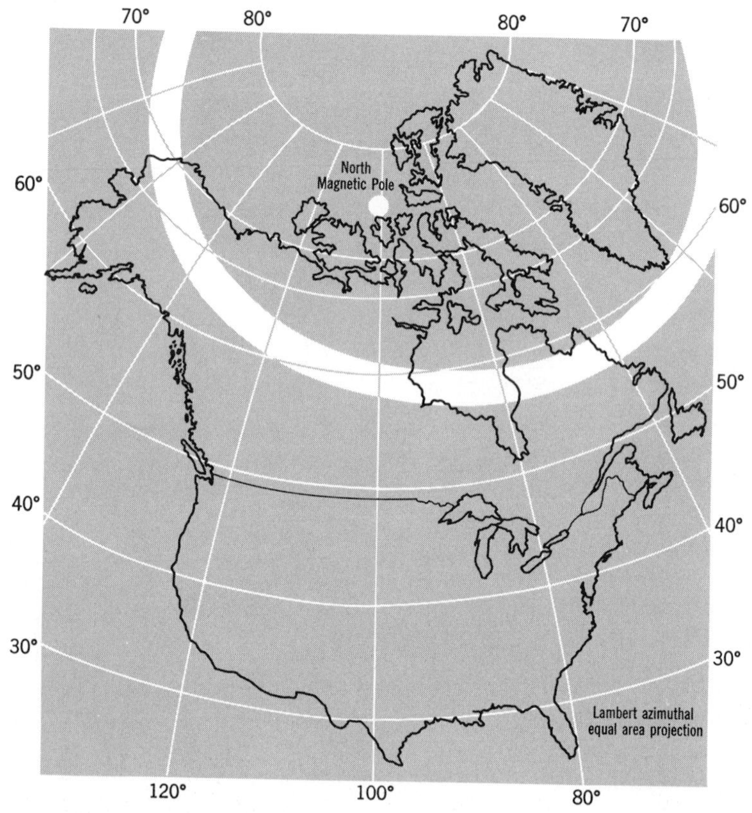

Fig. 22.19. The Western half of the North Auroral Zone. (Shown as a dark belt over Alaska and northern Canada.)

unknown. Conflicting evidence places it in the ionosphere and in the magnetosphere (see Chapter 23); perhaps it is associated with both. During periods of intense solar activity, auroral activity and emission also increase. Many measurements over the years have established a very close correlation between geomagnetic disturbances (see Chapter 23) and intense auroral activity. In fact during the most severe magnetic storms, the auroral zone may undergo a sizable temporary shift to lower altitudes, and it is even occasionally observed as far south as Washington, D.C. Evidently the magnetic fluctuations are closely related to the electron accelerating mechanism.

Spectrographic methods have been used for many years to analyze the auroral light, leading to precise identification of the emitting species which include atomic and molecular oxygen, molecular nitrogen, N_2^+ and O_2^+. Sometimes the Balmer α line of atomic hydrogen, normally occurring at 6566A, is also a prominent feature. However, it is displaced to a shorter wavelength due to the Doppler effect caused by the rapid movement of hydrogen atoms. The obvious conclusion is that the emission is from hydrogen atoms which started as protons, captured electrons, and are still moving rapidly to lower altitudes.

Since the aurora leads to increased ionization in the upper atmosphere, one would expect to encounter associated radio effects. Indeed on high frequency radio communication paths crossing the auroral zones there is usually severe disruption during strong auroral activity. The propagation path may even be changed to non-great circle due to refraction or scattering of the wave.

Even at low latitudes, one finds considerable emission from the night sky. For many years it was thought that all of the light incident on the surface of the earth during a moonless night was starlight. Then it was conclusively proved that starlight cannot possibly account for the observed intensity. It is now known that the luminescence is produced in the upper atmosphere as a result of chemical reactions.

The principal emitters are atomic oxygen and sodium. The former is evidently excited as a result of the recombination process

$$O + O + O \longrightarrow O_2 + O$$

The excited atom which acts as a "third body" in the reaction emits mainly in the green (5577Å) but also has two prominent red lines at 6300 and 6364Å. The excitation mechanism for the sodium is not as well understood but the emission in the well known "D-lines" at 5590 and 5896Å is very strong. This is the same yellow light that one sees emitted by table salt (sodium chloride) when it is placed in a flame. Other species also contribute to the glow (e.g., OH, O_2, NO) and the excitation processes are, for the most part, reasonably well understood.

The last optical effect which we shall discuss is the occurrence of noctilucent clouds (NLC). These clouds, some of which are shown in Fig. 22.20, are seen

Fig. 22.20. Noctilucent clouds. (Courtesy, Dr. B. T. Fogle, National Center for Atmospheric Research, Boulder, Colorado.)

only during twilight at high latitudes. Nevertheless they are undoubtedly present through the day and night; because the intensity of the sunlight scattered by them is low, they can only be seen against a dark background. Rather extensive observations during the past few years have revealed a remarkable correlation between the zones of maximum NLC occurrence and the auroral zone. Whether or not there is a causal connection between the two effects is a tantalizing question.

The altitude range in which NLC are found has been observed to lie very close to the mesopause. The clouds themselves are now thought to be some form of water which condenses or freezes at this altitude. How the water manages to reach a height of 90 km has not yet been satisfactorily answered. The fact that NLC usually occur during summer months where the high latitude mesopause is colder is probably associated with an upward transport process although we cannot be sure. It is suspected that meteoric dust particles serve as condensation centers or nuclei. However, experimental verification is difficult and the few results that have been obtained are very controversial.

11. QUESTIONS

1. The radiation from the sun corresponds to a black body temperature of 6000°K while that from the earth corresponds to a temperature of 300°K. Why couldn't the reverse be true, i.e., solar radiation –300°K, earth radiation –6000°K?

2. When the infrared radiation emitted by the earth is absorbed by water molecules, it is converted into some form of molecular energy. What are the various mechanisms for absorption? Why is electronic excitation not likely to be significant?

3. If the period of rotation of the earth (i.e., length of a day) were doubled, approximately how long would you expect the "seasonal lag" to be?

4. Discuss the effect of the mean molecular weight of the atmosphere on its circulation. What would happen to the rate of circulation if the atmosphere were composed entirely of helium? of argon?

5. How would the wind systems illustrated in Fig. 22.5 be affected if the speed of rotation of the earth were doubled, if the sense of rotation were reversed?

6. Discuss some of the ways in which seas and land masses cause the circulation patterns to depart from the simple one shown in Fig. 22.5.

7. Using the concepts developed in Chapter 11, explain why water boils at a lower temperature in Denver than in San Francisco. What happens to the dew (condensation point) when one goes from lower to higher altitudes?

8. During the bombing of Tokyo and Hamburg during World War II, the large scale fires created a "fire storm" which produced violent winds and precipitation. Discuss the reasons for the occurrence of the latter effects.

9. Discuss the mechanism for fog formation in the vicinity of the Gulf Stream.

10. Figure 22.10 shows steps in the formation of an occluded front. Discuss the possible fate of such a front, using diagrams as necessary.

11. Give reasons why a hurricane usually does not form at about the time of the summer solstice (*Hint.* The intertropical convergence zone is very close to the equator at this time).

12. At a given distance from the "eye" some sectors of hurricanes are more dangerous than others to mariners. Using Fig. 22.13, show which sectors are more dangerous and which are less so.

13. Select a metropolitan area in the U.S. other than Los Angeles (e.g., Chicago, New York, San Francisco Bay) and on the basis of population factors, local climatology, and terrain, diagnose the susceptibility of the area to air pollution.

14. Write some possible chemical reaction equations for the destruction of O_3; of NO.

15. Compute the scale height at an altitude in the upper atmosphere where the dominant constituent is atomic oxygen and the temperature is 1000°K. Is the scale height in the thermosphere likely to be higher or lower at night than during the day? Why?

16. Relatively large quantities of helium are found in the high atmosphere. What is its probable source?

17. The region of the upper air where the mean distance between molecular collisions

exceeds the scale height is called the exosphere. On the basis of this definition and the principles of the kinetic theory of gases discuss some of the probable characteristics of this region.

18. Explain qualitatively why ambipolar diffusion and electron-ion loss by recombination cause a peak in the F-region.

19. Why are auroral rays or striations aligned with the earth's magnetic field lines?

20. Why would you expect water vapor to experience great difficulty in rising to the mesosphere? Why is the mesopause a "trapping" region?

Recommended Reading

(The Lower Atmosphere)

Louis J. Battan, *The Nature of Violent Storms,* Doubleday, Garden City, N.Y., 1961 discusses the origins and characteristics of storms; a paperbound volume for the layman.

J. Haagen-Smit, "The Control of Air Pollution," *Scientific American,* January 1964, p. 24 is a general account of this topic by one of the leading investigators.

Clarence E. Koeppe and George DeLong, *Weather and Climate,* McGraw-Hill, New York, 1958 is an elementary meteorology text on the freshman level; intended mainly for non-science majors.

Paul E. Lehr, R. Will Burnett, and Herbert S. Zim, *Weather,* Golden Press, New York, 1957 is an elementary profusely illustrated paperbound book on fundamental meteorological concepts.

T. Morris Longstreth, *Understanding the Weather,* Macmillan, New York, 1961 is a very good elementary discussion of weather, clouds, rain, fronts, wind, and particularly storms.

Sir D. Graham Sutton, *The Challenge of the Atmosphere,* Harper, New York, 1961 is an excellent treatment of elementary meteorology by the head of the Meteorological Office, United Kingdom.

Albert Miller, *Meteorology,* Merrill Books, Inc., Columbus, Ohio, 1966, is a short but very readable text on the elementary level.

(The Upper Atmosphere)

Sun-Ichi Akasofu, "The Aurora," *Scientific American,* December 1965, p. 54 is a comprehensive discussion of our current knowledge of the aurora, its appearance, occurrence and causes, written by one of the world's outstanding investigators.

R. L. F. Boyd, *Space Research by Rocket and Satellite,* Harper, New York, 1960, chapter 5 is a now-dated but still useful account of upper atmospheric and ionospheric structure.

S. T. Butler, "Atmospheric Tides," *Scientific American,* December 1962, p. 48.

R. E. Newell, "The Circulation of the Upper Atmosphere," *Scientific American,* March 1964. Both are excellent articles on upper air phenomena.

R. K. Soberman, "Noctilucent Clouds," *Scientific American,* June 1963, p. 30.

chapter twenty-three

The environment of man: space beyond the atmosphere

Technological applications of modern physics have spawned two major engineering developments. One is in the nature of a partial "declaration of independence" from the sun, nuclear energy rapidly eroding man's dependence on fossil fuels, which captured the sun's energy during eons of geologic history. The second, an even more fantastic development, is the possibility of journeys and explorations far beyond Earth, man's natural and generally accepted environment. In this endeavor, scientific curiosity, challenge to imagination and adventure, and competitive national pride have been combined. Whether the magnitude of the effort and the prodigious assignment of engineering and scientific skills to this endeavor are warranted will be judged by later generations. Certainly, the results are giving a new understanding of the universe in which we live. The meagre beginnings, only a little more than a decade since inception, have already brought —by a direct inspection through satellites—new understandings and new discoveries about the lower atmosphere and about the "busy" spatial environment above the atmosphere, filled as it is with "cosmic" rays, magnetic fields, and bits of materials and particles of great variety. And when, at last, telescopes are landed on the moon, man will undoubtedly discover much more about the mysteries of the universe which until now have been perceived only through "eyes" clouded by the atmosphere.

In the preceding chapter, we examined man's immediate environment, including the upper atmosphere to an altitude of about 1000 km. Beyond lies what is roughly referred to as "space." This study includes the properties of the outer portion of the geomagnetic field, called the magnetosphere, as well as the nature of interplanetary space, the physics of the sun, and cosmic rays. In this chapter, we will concentrate largely on the newest discoveries, brought about by combining older telescopic observations with new methods and new observations from satellite and space probes.

1. SPACE PROBES AND VEHICLES

No one knows exactly when the first rocket was launched, but it is a matter of historical record that they were used in China as early as the thirteenth century. Until recently their principal application was as signaling devices and as fireworks for celebrations. The former use is familiar to nearly everyone in the phrase "by the rockets red glare" from our national anthem. During World War II they were developed into very effective weapons by the Germans, and to a lesser degree by the British and the Americans. The more recent development of long range ballistic missiles, one of which is shown in Fig. 23.1, is common knowledge and needs no further elaboration.

The rocket is designed to expel high velocity gas through a nozzle, which produces thrust in the direction opposite to its motion. The basic principle behind the rocket, then, must be Newton's third law of motion: *To every action there is always opposed an equal reaction.* One can also look upon this from the standpoint of the law of conservation of momentum; the gas expelled from the rocket steadily carries away this mechanical quantity. Since the total change in the momentum of the gas-rocket system must be zero, the rocket itself must gain an amount of momentum in the forward direction equal to that acquired by the exhaust in the backward direction. It is thus easy to see why, relative to an observer on the ground, the rocket can be accelerated to speeds well in excess of the exhaust gas velocity; only the motion of the two parts of the system relative *to each other* is important. The presence or absence of an observer in a given inertial frame is irrelevant to the argument.

The motion of a rocket is complicated by the fact that the mass of the vehicle is steadily decreasing. For this reason one must have recourse to the calculus

Fig. 23.1. The Minuteman Missile. (Photo by The Boeing Company, Seattle, Washington.)

which is really beyond the scope of this book. Hence, the derivation will not be given. Instead we merely write down an equation for the final velocity of the rocket V_f in terms of the exhaust jet velocity v and the initial and final masses, M_1 and M_2,

$$\frac{M_1}{M_2} = 10^{kV_f/v}$$

where $k = 0.435$.

Obviously, in order to reach high velocities, a very large fraction of the initial mass of the rocket must be in the form of fuel. This is why such large launching rockets are necessary to place relatively small satellites in orbit. Here we have an excellent illustration of the mechanical effects of time-varying mass on the motion of body. In this case the restricted form of Newton's second law, that the force acting on a body is equal to the product of its mass and acceleration, is no longer true. Instead one must substitute the more general statement that the force is equal to the time rate of change of momentum (see Section 2 of Chapter 7.)

The rocket engine itself is nothing more than an *expanding nozzle* which imparts a high velocity to the gases issuing from a combustion chamber. The increase in velocity occurs because the temperature of the gas decreases as a result of its expansion. A simplified sketch of such an arrangement is shown in Fig. 23.2. The gases form in the combustion chamber at very high temperature (T_2). After they pass the throat of the nozzle, they expand in volume with a simultaneous drop in temperature (to T_1). The thermal energy thus lost by the exhaust jet must go somewhere by virture of the first law of thermodynamics. In fact it is converted to kinetic energy of the gas stream. If maximum efficiency were attained, the jet velocity v would be given by

$$\tfrac{1}{2}\,\rho v^2 = \tfrac{3}{2}\,kN(T_2 - T_1)$$

where ρ is the mass density of the gas, N is its number density (number of molecules per unit volume) and k is Boltzmann's constant. Obviously the higher that we can raise the initial temperature, the greater the jet velocity. Rocket engineers have been seeking fuels which yield high initial temperatures for a number of years.

The first artificial earth satellite ("Sputnik I") was placed in orbit by a large Russian rocket on October 10, 1957. Since then an enormous number have been sent aloft. Of these a few have been recovered, a substantial number have been destroyed by reentering the atmosphere, and some are still yielding useful information. The remainder together with spent rocket frames that went into orbit along with their satellites constitute a veritable celestial junkyard. Fortunately, because of the large amount of space available and because of the precision with which the orbits are known, there is little likelihood of collision with vehicles which will be launched in the future.

Satellite circular orbits were discussed at some length in Section 6 of Chapter

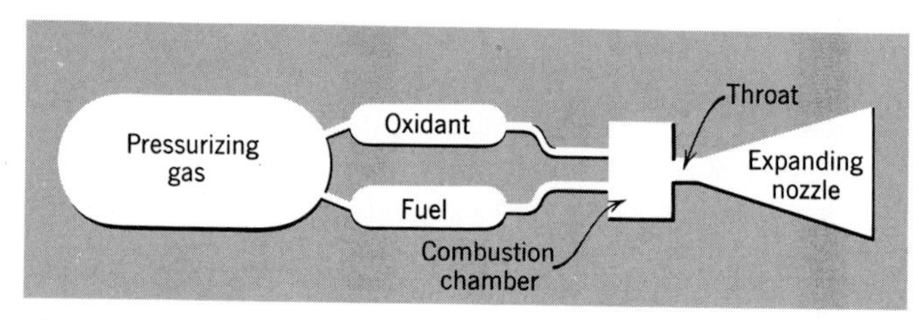

Fig. 23.2. Schematic diagram of a simple rocket engine (liquid fueled).

7 and there is no need to review the principles here. It is worth remarking, however, that satellite orbits are seldom of circular configuration. Usually they are approximate ellipses, the perturbations caused by the gravitational attraction of the moon and sun, and nonuniformities in the internal structure of the earth resulting in departures from perfect elliptical form. Artificial satellites of widely varying sizes, shapes, and uses have been placed in orbit by the United States and the Soviet Union. Although the manned flights have been the most spectacular, the unmanned instrumented satellites have been the most useful from a scientific point of view. The information so gathered includes data on upper air densities, the intensity of the Earth's magnetic field at great distances, energetic charged particles trapped in the geomagnetic field, the spectrum of X-rays emitted by the sun, etc. Some of these will be discussed in the following sections.

If the vehicle is given sufficient velocity by its launching rocket, it can escape the Earth's gravitational field and go on into "outer space"; we shall refer to these as "spacecraft." One can easily find the velocity necessary to escape the Earth's gravitational field by use of the law of conservation of energy. In this case the kinetic energy of a body of mass m and velocity v_e relative to the surface of the earth

$$E_k = \tfrac{1}{2}\, m v_e^2$$

is exactly equal to the magnitude of the potential energy

$$E_p = \frac{GmM_E}{R_E}$$

Since M_E and R_E, the mass and radius of the earth, respectively, are well known, one can solve the equations for v_e

$$v_e = \frac{(2GM_E)^{1/2}}{R_E}$$

Using known values for the constants, it is left to the reader to verify that this velocity is about 11 km/sec or 7 miles/sec, and to compare this escape velocity from the earth with the escape velocity from the moon.

Although the spacecraft may be on an escape trajectory, it is still strongly influenced by the earth until it has moved far into space. Of course, it is still in the gravitational field of the sun and, unless raised to a "solar escape velocity" will go into orbit about that body. Should it move into the vicinity of the moon or one of the planets, it will be strongly affected by them. However, it cannot be captured unless it is decelerated by a "retro-rocket." Let us briefly consider a spacecraft launched on a trajectory about the moon. As it travels outward from the earth it steadily loses kinetic energy and gains potential energy until the moon's attraction becomes stronger than the earth's; at this point it begins to accelerate. If it does not hit the moon it will swing in behind it, and if properly

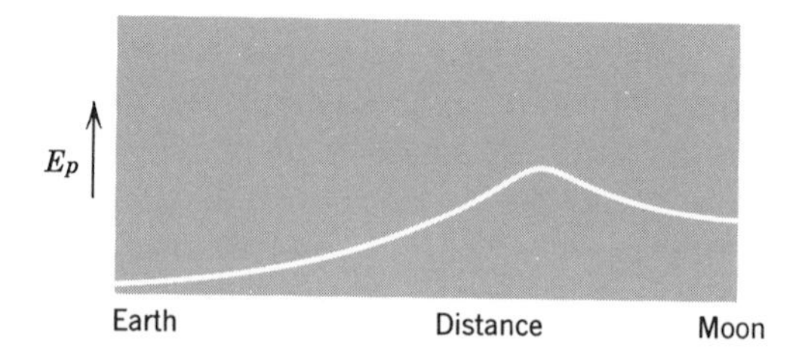

Fig. 23.3. Gravitational potential energy E_p as a function of distance from the earth and from the moon.

maneuvered, will travel back toward the earth. The energetics are illustrated in Fig. 23.3 in which the motion of the satellite is compared to a ball rolling up a hill. The top of the hill is the point at which the attraction of earth and moon are equal.

Numerous manned satellites have been launched since 1961; in fact manned voyages to the moon have already occurred, and the day is not far distant when man will go to Mars. Although the technological achievement is a dazzling one, the utility of such space travel has been questioned by many informed people.

2. THE SUN

The reader has probably been aware of the general effect of the sun on our planet for so much of his life that it seems almost trite to say anything of this nature. Instead we shall discuss in a very brief manner its general characteristics before launching into a more detailed discussion of its structural features and its radiations.

The sun is a large mass (about 3.3×10^5 earth masses) of very high temperature ionized gas or plasma. The reader will recall from the preceding chapter that the term "plasma" refers to an ionized gas which is in thermal equilibrium and is, insofar as its large-scale properties are concerned, electrically neutral. The gas is composed mainly of hydrogen and helium with a small admixture of heavier elements. Deep within the solar interior the temperature and density are high enough to "burn" the hydrogen into helium by nuclear processes. Additional reactions are believed to result in the building of elements by similar reactions. However, further discussion of possible reactions of this type is deferred to the next chapter. The surface of the sun is much cooler than the interior, corresponding quite closely to a black body temperature of about 6000°K over a wide range of wavelengths. At wavelengths in the far untraviolet region (less than 2000 Å), however, the spectrum consists mainly of single lines characteristic of atoms and ions present in the surface layer. At still shorter wavelengths (X-ray region) the spectrum is of both the line and continuous types, with the continua dominating in the long wavelength portions of the X-ray spectrum.

In addition to the background continuum characteristic of a black body, the visible solar spectrum exhibits a series of dark lines, called the Fraunhofer lines after their nineteenth century discoverer, as well as some bright lines. The dark lines are caused by absorption of radiation by atomic species in the surface layers. This *absorption spectrum* is characteristic of elements such as calcium, carbon, hydrogen, sodium, and a host of others. On the other hand, the bright lines constitute emission spectra of various elements, particularly calcium and hydrogen. The hydrogen emission line at 6560 Å and the calcium emission lines are frequently employed as indicators of solar activity. At short wavelengths in the ultraviolet the Fraunhofer lines become less and less distinct; at wavelengths less than 1850 Å they are not detectable.

The amount of solar energy incident on the earth per unit time is represented by a quantity called the *solar constant* which is usually stated in watts per square centimeter, specifically 0.139 watt cm^{-2}. The fraction of solar radiation intercepted by the earth is quite small because of the large orbital radius of the earth and the small planetary diameter. In fact it is only about one part in 10^9.

In discussing radiation emission by the sun or more accurately, the solar surface and atmosphere, it is convenient to consider the emission characteristic of each *level.* These levels are the *photosphere,* the *chromosphere,* and the *corona.* The photosphere which is the source of the continuous (i.e., black body) spectrum as well as a weaker line spectrum is the lowest lying which can be observed directly. It is about 300 km thick and exhibits a granular appearance which is a result of temperature variations from one point to the next over the sphere. These variations are due to the constraining effect of local magnetic fields. When charged particles try to move across the lines of force, they are curved back toward their original direction, thus providing a magnetic barrier against thermal conduction. This effect is illustrated in Fig. 23.4 In addition to the granular appearance, the photosphere is also characterized by the appearance of *sunspots* and *faculae.* Sunspots are circular regions of considerably lower temperature (about 1500°K lower)

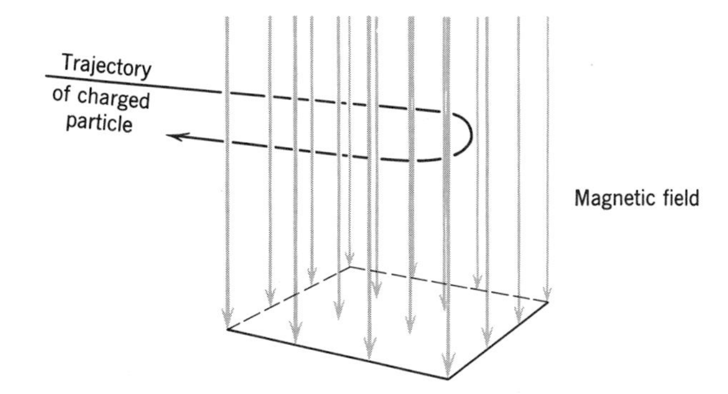

Fig. 23.4. The trajectory of a moving charged particle in a magnetic field.

than the surrounding disc; thus they appear dark when seen against the surrounding 6000°K surface. Faculae on the other hand are regions of relatively high temperature, thus accounting for their bright appearance. They characteristically appear as irregular radial streaks surrounding sunspots near the solar limbs. A photograph of the sun showing photospheric structure is shown in Fig. 23.5.

The chromosphere which is usually observed only during eclipses is seen as a reddish ring about 10^4 km thick. The emission is characterized by a line rather than a continuous spectrum; with the lines characteristic of H, He, and Ca accounting for nearly all the emission. The principal features of the chromosphere are *limb prominences* which are large loops of glowing gas seen on the solar limb,

Fig. 23.5. Photograph of the sun in the light of Hα. Note the prominences on the limbs of the disc. (Courtesy Dr. A. K. Pierce, Kitt Peak National Observatory, Tucson, Arizona.)

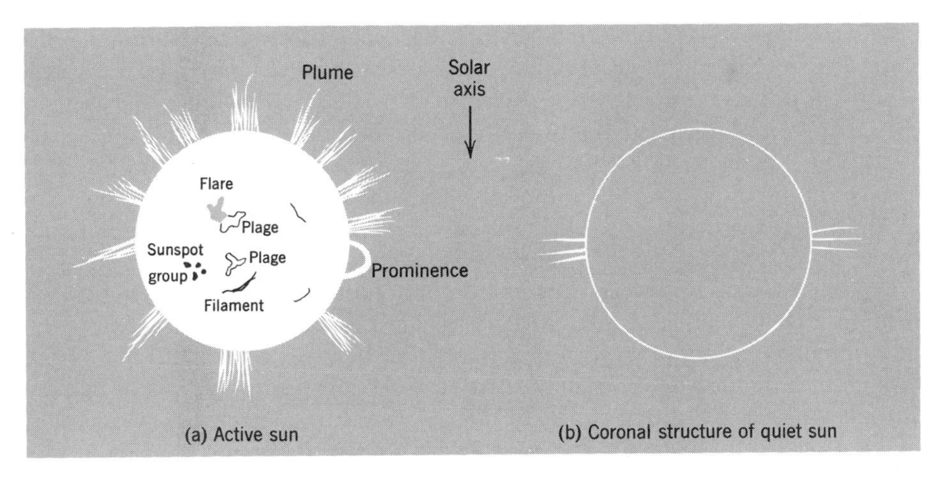

Fig. 23.6. Solar surface and coronal features.

spicules or irregular "spikes" of glowing gas also seen on the limb, and *plages* or bright regions which appear in the vicinity of sunspots. Plages are usually the seat of solar *flares* which will be discussed later in this section.

The corona is the true atmosphere of the sun, extending out to millions of kilometers from the solar disc. Depending upon how one chooses to define the corona, it can be described as extending out to at least the orbit of the earth. The spectrum is that associated with electronic transitions in highly ionized species such as that of iron from which thirteen electrons have been stripped. Such a high degree of ionization requires a high temperature; the corona is indeed typified by a gas kinetic temperature of about 10^6 °K. By "gas kinetic" we refer to the energy of random motion of the particles in the corona. A body such as a satellite which moves through this region would not necessarily be raised to a high temperature because of the very low coronal density. X-ray emission also occurs in the corona although the production processes are probably the result of bombardment by accelerated rather than thermal electrons. The corona can be seen only when the main disc of the sun is obscured such as during an eclipse or with the aid of a special instrument called a coronagraph. When the corona is observed in this manner, it appears as an irregular halo of streamers and plumes which are more numerous when the sun is active. The *solar wind,* a steady outward stream of protons and electrons is also thought to be of coronal origin. The sketch of the sun in Fig. 23.6 shows the coronal structure as well as some of the surface features discussed previously.

We have already referred to the solar cycle in the foregoing paragraphs but have not discussed its detailed characteristics. Briefly, it is characterized by the periodic increase and decrease of the number of sunspots observed on the solar disc.[1] The duration of the cycle is approximately eleven years from peak to peak of sunspot number although this period may vary somewhat from one cycle to the next. During this cycle the intensity of the total solar radiation is quite constant. The ultraviolet flux does vary, however, and it is this factor which relates changes in the upper atmosphere to solar activity. The mechanisms involved are quite well understood. Not so well understood is the connection between climatic changes and the solar cycle. That there is indeed a connection is evidenced by the close correlation between sunspot number and the thickness of growth rings in trees. The influence of solar activity on the lower atmosphere is one of the most important areas of meteorology remaining to be unlocked. It is interesting to note that the famous English economist of the last century, William Stanley Jevons, attributed business cycles to sunspot variations. Again the relationship of solar activity to weather, and thus agriculture, was assumed to be the key. Modern economic forces in advanced societies are not determined by crop yields, however, and Jevons' theory is no longer taken seriously.

The origin of the solar cycle is also poorly understood although several partially successful theories have been advanced. The most recent and most satisfactory one is based on the differential rotation of the sun; that is, the period of rotation is shorter at the equator than at the poles. Since the sun's surface is a highly conductive plasma, a magnetic line of force once embedded in it tends to become "frozen in." Suppose the line of force does start to move through the conducting fluid. Figure 23.7 shows how the relative motion induces an electric current in accordance with Faraday's induction law. The magnetic field of the induced current reinforces the original magnetic field on the side in the direction of motion and opposes it on the other.[2] From our earlier discussion of the interaction of magnetic fields, it is evident that the effect is one of resistance to the motion of the line of force through the "blob" of conducting fluid, or equivalently, of the "blob" past the field lines. As a consequence the field lines and plasma must move together. This is what we mean by "freezing in." Due to the more rapid rotation of the equatorial zone, the magnetic lines of force which are embedded in the solar surface become wound around the sun. In addition, bundles of these lines become twisted like strands in a rope. This analogy is in fact an excellent one. Try winding the midpoint of some untwisted strands about a post, simultaneously holding the midpoint tightly; the strands will quickly become twisted as in ordinary rope. It happens that magnetic fields distorted in this manner tend to form local "kinks." When these break through

[1] The associated quantitative measure is the *Wolf sunspot number* $R = K(10\,g + f)$ in which K is a constant of order unity, g is the number of sunspot groups, and f is the total number of spots. Since spots occurring in groups are more closely associated with solar "activity" then individual sunspots, the former must be given much greater weight.

[2] cf Fig. 13.8.

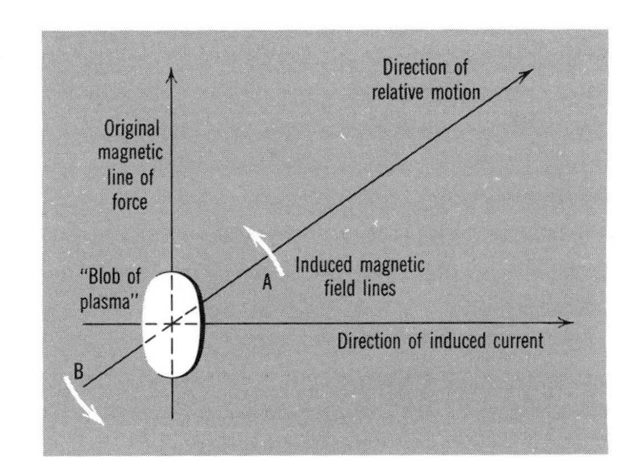

Fig. 23.7. The mechanism which leads to freezing of magnetic lines of force into a perfectly conducting plasma. The original magnetic field is reinforced on side A and opposed on side B. This distortion tends to "push" the field line back to its initial position.

to the surface of the sun, they form "bottles" which screen the ionized gas inside from the rest of the solar surface by the same mechanism illustrated in Fig. 23.4. The gas inside the bottle is cooler because some of its kinetic energy of random motion has been transferred to the potential energy of the magnetic field. The complete theory of sunspots and the solar cycle is more involved and more comprehensive than we have indicated, but our discussion is adequate for a general idea of the mechanism.

The most outstanding characteristic of strong solar activity apart from the complex structure of the corona shown in Fig. 23.6 is the rather frequent occurrence of surface eruptions or chromospheric flares. These phenomena usually occur near complex sunspot groups, particularly in the vicinity of plages. Only rarely are they seen in white light, but even the very small ones are visible in the "Balmer α" or "Hα" line of atomic hydrogen (Fig. 23.5). Historically, a flare was defined as a local brightening of this spectral line. However, we now know that at altitudes well above the chromosphere (i.e., in the corona) copious quantities of X-rays are frequently produced. When these impinge upon the Earth's upper atmosphere, they produce the sudden ionospheric disturbances discussed in the preceding chapter. The larger flares may also result in the production of very energetic protons and alpha particles (usually called solar cosmic rays). When these particles eventually encounter the Earth's magnetic field, they are "funneled" into the polar regions where they cause the "polar cap absorption" events discussed earlier.

The flare mechanism is poorly understood at best. Current thinking is that magnetic fields near sunspots rapidly collapse due to instabilities. Electrons, protons and alpha particles are simultaneously accelerated, the heavy particles escaping into interplanetary space. The electrons, however, are stopped by collision processes before they escape or are guided by magnetic field lines back into dense regions of the corona and chromosphere. Electron bombardment of hydrogen atoms in the chromosphere produces the enhancement in Hα intensity, while X-rays are probably formed by electron bombardment of ions in the corona. Beyond these rather vague ideas, we presently have no adequate theory of flare occurrence.

It has been known for many years that the tails of comets point away from the sun. At first this was attributed to pressure exerted by solar radiation falling on the material in the comet. More recently it was realized that this mechanism is inadequate. As an alternative, it was suggested that the deflection is a result of the impingement of high speed protons and electrons streaming outward from the sun. These particles must, of course, have sufficient energy to escape the gravitational field of the sun. With the aid of spacecraft this *solar wind*, as it has come to be called, has been observed to have speeds of the order of 350 to 750 km sec^{-1} and number densities of 1 to 10 protons and electrons per cm^3. The gas kinetic temperature (i.e., the "spread" in particle velocities—see Chapter 11) is quite high, of the order of 1,000,000°K. We shall have more to say about the solar wind in a subsequent section.

3. THE MAGNETOSPHERE

In Chapter 13 we discussed at some length the discovery by William Gilbert that the earth is a giant magnet. Gilbert actually thought of it as a bar magnet; as we shall see, the field departs quite drastically from the bar magnet ("dipole") shape at great distances from the earth. Undoubtedly the source of the field lies in the flow of strong electric currents in the core of the earth. According to current theory, the earth's core, which is composed largely of strongly magnetic iron, nickel and cobalt, is kept molten by heat from the radioactive decay of heavy elements such as radium, thorium, and uranium. The rotation of the earth produces sufficient turbulence in this metallic liquid to maintain the flow of electric currents which give rise to the observed field.

Unlike the earth, some of our celestial neighbors do not have detectable magnetic fields. The moon, Mars, and Venus have been probed on a number of occasions by spacecraft launched by our National Aeronautics and Space Administration as well as by space activities in the USSR. All the evidence is indicative of no permanent magnetic fields. Jupiter, on the other hand, apparently has a very intense field, although direct measurements of its intensity are not yet possible. The absence of a magnetic field means, for example, that the solar wind impinges directly on the planetary atmosphere[3] and that aurora cannot

[3]Of course, the moon has no atmosphere and the solar wind bombards the lunar surface except when it is screened in the magnetic tail of the earth.

occur. Undoubtedly there are other atmospheric phenomena whose occurrence is not even suspected at the present time. The effect of magnetic fields on the planetary environments promises to be a very significant area for future research.

Near the earth the field has the shape shown in Fig. 23.8a; but far out in space, it is severely distorted by the pressure of the solar wind. The solar wind, being a highly conductive plasma, carries magnetic lines of force along with it. When these lines of force impinge upon the side of the *magnetospheric cavity* or

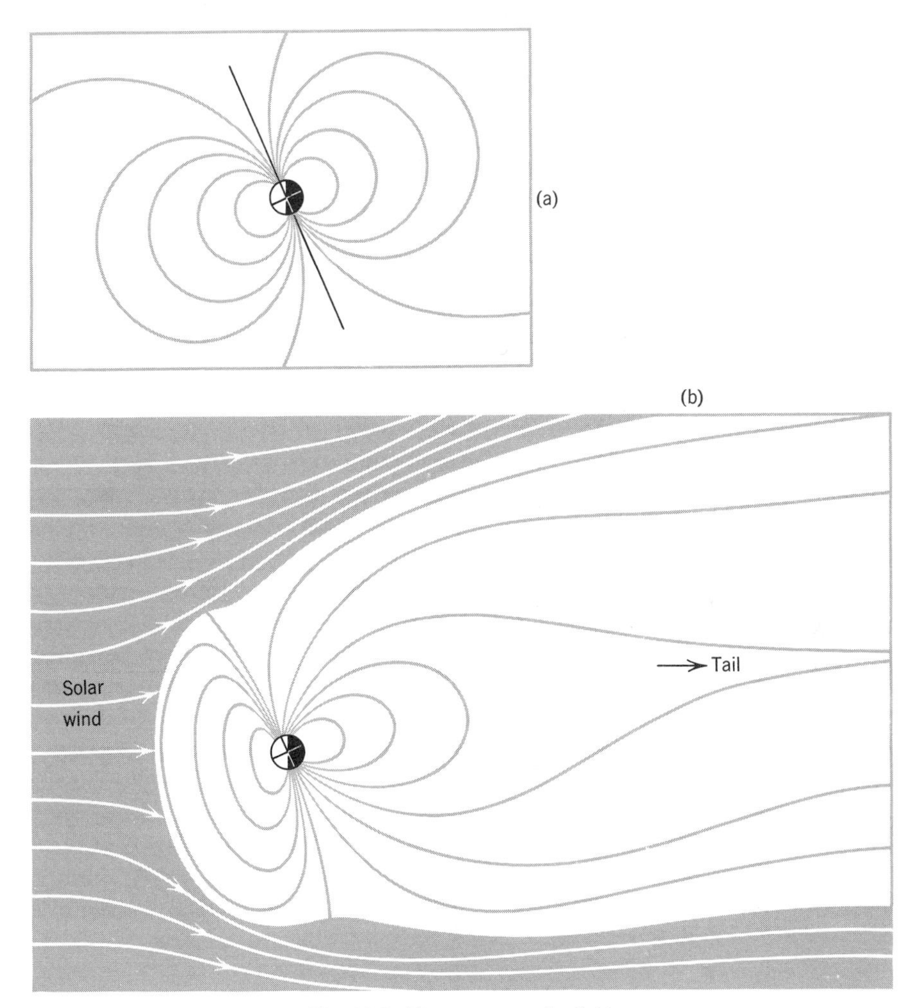

Fig. 23.8. The geomagnetic field.

magnetosphere toward the sun, they compress it. Simultaneously, the "downwind" side of the cavity is elongated in the shape of a teardrop or tadpole as shown in Fig. 23.8b. We do not yet know the exact length of the teardrop, but satellite observations indicate that it is at least 100 earth radii.

The solar wind is stronger during periods of marked solar activity than during quiescent periods, thus causing greater distortion of the earth's magnetic field near the maximum of the solar cycle. At such times, large variations in solar wind intensity are frequent. These are associated with the magnetic storms to be discussed later in this section.

When the first American satellite was successfully launched in early 1958, it was found by Dr. James Van Allen and his associates that charged particle counters intended for cosmic ray observations were "blacked out" by very intense radiation during part of the orbit. Furthermore, these sections of the orbit were always in the same locations relative to the earth (low latitudes, 1000 to 8000 km above the earth). At first they thought the counters were for some reason not working properly. However, one of Van Allen's graduate students concluded that charged particles were trapped in the earth's magnetic field and that they were intense enough to jam the instruments. Such an effect had in fact been predicted by the great Norwegian geophysicist Carl Störmer as far back as 1904. Trapping theory was developed in some detail by the Swedish physicist, Hannes Alfvén in more recent years. Further investigation showed that these particles were predominantly protons with typical energies in the range near 10 MeV.

A second "belt" of charged particles was found later in 1958 at still greater altitudes, typically 15,000 km. Two years later it was shown to be quite distinct from the lower one and to consist mainly of energetic electrons. A cross section of the trapped particle belts is shown in Fig. 23.9.

We still have no completely satisfactory theory of the origin of the charged particles. It was proposed some years ago that the protons in the inner belt have their origin in cosmic ray bombardment of the atmosphere. According to this hypothesis, energetic cosmic rays (mainly protons) eject neutrons from nitrogen nuclei. The neutrons which escape the atmosphere have a fairly large probability for beta-decay (see Chapter 21) while they are passing through the magnetosphere. The protons produced by the decay are then trapped owing to their electric charge. While there are some objections to this theory, it is the best one offered so far to account for the population of the inner belt. The theory of the outer belt on the other hand is in a less satisfactory state. Undoubtedly the electrons are accelerated in the magnetosphere, but the details of the mechanism are unknown. They are probably associated in some manner with the acceleration of auroral electrons, but again the relationship is a mystery. This is not due to lack of ideas but rather to the inadequacy of data and to the enormous theoretical complexity involved.

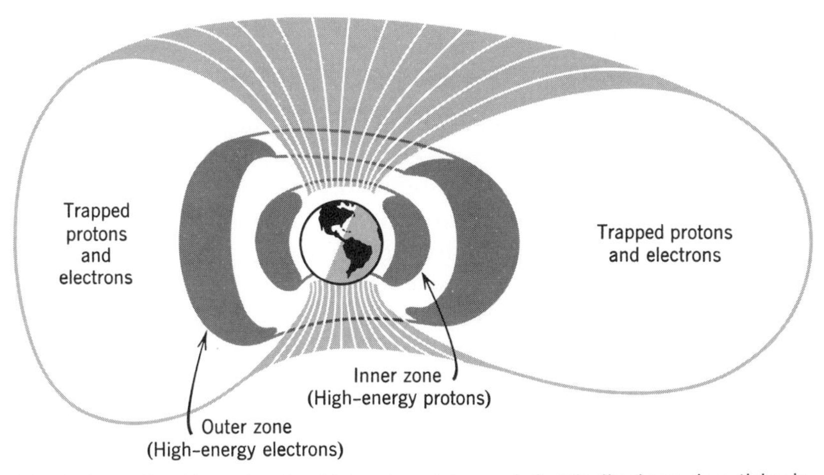

Fig. 23.9. Section through a toroidal region of trapped electrically charged particles in Earth's magnetosphere. (From Glasstone's *Source Book on the Space Sciences*, Copyright 1965, D. Van Nostrand Co., Inc., Princeton, New Jersey.)

The nature of the trapping mechanism is shown in Fig. 23.10. The charged particles tend to move in helices or spirals about the lines of force due to the character of the interaction between charged particles and a magnetic field. As they move toward the poles toward which the field lines converge, the helices become smaller and flatter. Eventually the particles reach a point at which the helix is perfectly flat (vanishing "pitch" angle) and they are reflected. They become smaller because the particle orbits must always enclose the same number of lines of force (or magnetic flux). According to Faraday's induction law, work would have to be done on the particle to cause its orbit (which is really a small

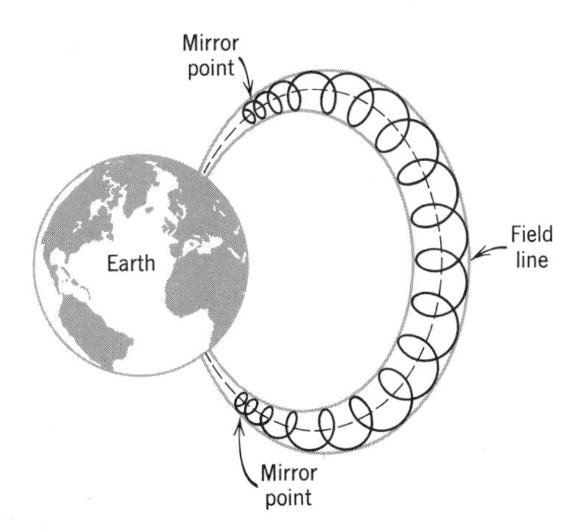

Fig. 23.10. Spiral motion and mirror points of charged particle in Earth's magnetic field. (From Glasstone's *Source Book on the Space Sciences*, Copyright 1965, D. Van Nostrand Co., Inc., Princeton, New Jersey.)

electric circuit) to enclose more flux. Since there is no agency to do this work, the flux through the orbit is constant. One can arrive at the reason for the flattening of the helices by noting that as the spiraling charged particles penetrate into a more and more intense magnetic field, their kinetic energy of longitudinal motion is transformed into potential energy of position in the magetic field. Eventually they lose all of their longitudinal kinetic energy. Then the particles cannot penetrate any further and are reflected. Of course, the associated flattening of the helices occurs gradually, not all at once. They then move back down the field line toward the opposite pole until they are again reflected. Superposed on this motion between reflection points is a drift motion, the protons drifting from east to west, and the electrons west to east. Hence, if a large quantity of charged particles is injected in one geographic location, it will (rather rapidly) form a shell or belt about the earth. This can be, and in fact has been done, by high altitude nuclear explosions. The Starfish explosion above Johnson Island in the Pacific Ocean on July 9, 1962, produced just such a shell.

When the Starfish test was first proposed, it caused an international furor. Astronomers were worried that electromagnetic radiation from the trapped electrons would seriously interfere with the observation of radio emissions from celestial bodies such as the sun, Jupiter, and stars. Some scientists, however, contended that the artificially produced trapped particle belt would decay rapidly, causing little injury to the efforts of the radio astronomers. The actual effect lay somewhere between the predictions of the pessimists and optimists. It did degrade radio astronomical observations for a time, but the shell decayed rapidly enough so that the effect was not serious. On the other hand, we did gain considerable knowledge of the particle injection and removal mechanisms as well as additional knowledge of the earth's ionosphere. On balance, it proved to be a very useful geophysical experiment from a scientific point of view.

We mentioned earlier that Jupiter has a very strong magnetic field. This fact was determined from observations of radio "noise" emitted by that planet. The radiation characteristics indicated that it must have been produced by trapped electrons. The observations have been so precise that we know very accurately the field strength at the planet's surface, and the inclination of the magnetic to the rotational axis.

Very frequently after a severe solar eruption there is an abrupt increase in the intensity of the solar wind. When this plasma arrives at the boundary of the magnetosphere, quite severe disturbances of the geomagnetic field and the ionosphere occur on a worldwide scale. We call such disturbances *magnetic storms*. Sometimes they are associated with no visible solar eruptions at all, although we believe that they are still of solar origin. At the present time we have no satisfactory theory of the relation of such storms to activity on the sun.

We shall defer the discussion of the behavior of the plasma cloud associated with magnetic storms to the following section and concentrate instead on the storm itself. By way of introduction it is necessary to discuss the propagation of *hydromagnetic waves* through a magneto-plasma such as the magnetosphere. Consider a local disturbance in the magnetic field which results in the "kinking" of a line of force. The line of force acts like a string under tension and tries to straighten itself. Since it is frozen into the surrounding plasma it has an associated mass density just like a real mechanical string and will vibrate. As we saw in Chapter 15, such a disturbance will travel along the string as a wave, in this case a hydromagnetic one. The imaginative student may even conceive of a hydromagnetic violin playing hydromagnetic music. Hydromagnetic waves are of great importance in many aspects of plasma physics including magnetic storms to which we now return. When the plasma cloud or the abrupt increase in solar wind intensity impinges on the boundary of the magnetosphere, severe local magnetic disturbances occur, traveling through the magnetospheric cavity as hydromagnetic waves. When they arrive at the earth, they are manifested by an abrupt increase in the magnetic field intensity called a sudden commencement. A typical rise time for sudden commencements is five minutes. It is worth remarking that not all magnetic storms begin with sudden commencements.

After the sudden commencement has spent itself, the storm enters the *main phase* which is characterized by a reduction in the magnetic field intensity. According to the storm theory developed many years ago by Sydney Chapman and Jules Bartels, an electric current (*ring current*) is formed in the magnetosphere in such a way as to decrease the magnetic field intensity. This hypothesis is still in vogue today, with the ring current occurring as result of the drift of protons and electrons in the trapped particle belts. The electric fields which enhance the drift are induced by the flow of solar plasma around the magnetospheric cavity. Eventually the intensity of the solar wind returns to normal; as it does, the ring current becomes weaker, allowing the geomagnetic field to recover. Hence we term this the *recovery phase* of the storm. The physics of magnetic storms is reasonably well understood today, although there are some details which are not yet clear. The most important factor in our development of magnetic storm theory was the artificial earth satellite which provided the observations on which much of the theory is based.

4. INTERPLANETARY SPACE

Contrary to popular belief, the space between the sun and the planets is not an empty void. True, it is a region of very low density, but it is also the home of the solar wind and of very energetic corpuscular radiation emitted by the sun. We have already noted that the solar wind carries its own magnetic lines of force which are frozen into it. Because of turbulence (disordered fluid motion) in the wind the lines of force become tangled, presenting a barrier to the penetration of charged particles, regardless of their direction of incidence. It is only reasonable to expect that the intensity of the magnetic fields as well as the degree of turbulence will increase with increasing solar activity. This is indeed the case. Charged particles from the sun or from outside the solar system are impeded in their transit by the mechanism shown in Fig. 23.4.

According to one proposal, a large magnetic "bottle" expands outward into space from an active region on the solar surface. This magnetic loop, illustrated in Fig. 23.11, traps the high energy protons much as the earth's magnetic field traps charged particles. Eventually the loop spreads outward far enough to envelop the earth. When this occurs, the particles are able to enter the atmosphere, causing the polar cap absorption events discussed in the preceding chapter. Note the curvature of the "center line" of the loop. This occurs because the rotation of the sun causes the inner portion of the loop to lead the outer part. It resembles the motion of the stream from a revolving garden hose and is in fact called the "garden hose effect." Are solar protons guided to the earth by such magnetic "bottles" or is there some other mechanism at work? So far we do not have enough information to be sure, but for a number of reasons the latter reason seems the more likely.

Before closing this section we must emphasize that high energy protons from the sun constitute a very significant hazard to space travelers, particularly during periods of high solar activity. Observations of charged particles from several

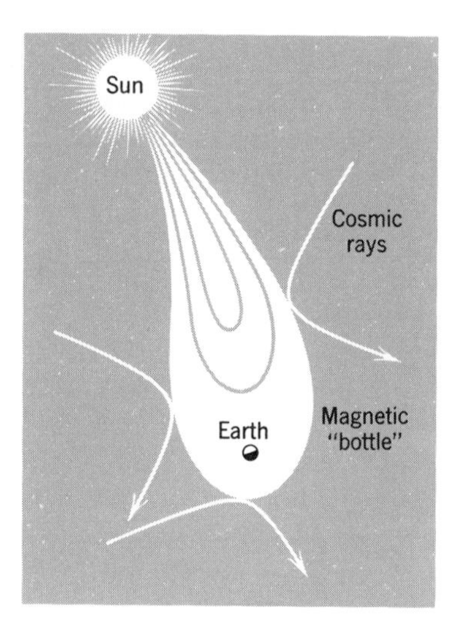

Fig. 23.11.

large eruptions showed that, had astronauts been in interplanetary space with very light shielding they would have received about 5000 radiation units. For comparison, 450 units is considered lethal.

5. GALACTIC COSMIC RAYS

It has been known ever since the first cloud chamber experiments were carried out that the earth is bombarded by extra terrestrial particles of very high energy. The American physicist Robert A. Millikan who developed the famous oil drop experiment discussed in Chapter 14 as well as the German physicist Viktor Hess were two of the leading investigators in these early years. In fact it was Millikan who named the particles "cosmic rays." The nature of the bombarding particles was difficult to establish because usually only fragments from nuclear reactions which occurred high in the atmosphere were observed. By performing the experiments on mountain tops and later in balloons, much of the effect due to the presence of the atmosphere was eliminated and the particles were correctly identified as principally protons with small components of heavier particles (e.g., alpha particles) and electrons. Nevertheless, the experiments at lower altitudes were still important because until about 15 to 20 years ago cosmic rays were the only source of high energy nuclear particles. Many of the most significant discoveries, such as the muon and the π-meson (see Chapter 21) were in fact made in atmospheric "showers" resulting from cosmic ray bombardment.

Although we now know many of the characteristics of cosmic rays, we still do not know the nature of their origin. They were first suggested to be of solar origin, but this was shown to be untenable because it was impossible to account for the very large particle energies (as high as 10^{20} electron volts). For comparison the largest energies produced in an artificial accelerator are of the order of 7×10^{10} electron volts. The sun does indeed occasionally produce high energy protons (usually called solar cosmic rays) during flare events, but their maximum energies are relatively very small. If they are not emitted by the sun, they must come from outside the solar system. The logical alternative choice is somewhere else in the galaxy, either in stars or interstellar space. Two acceleration mechanisms have been suggested. In that proposed by the late Nobel laureate Enrico Fermi the particles are accelerated by repeated collisions with interstellar magnetic fields contained in clouds of plasma which reflect them in the manner illustrated in Fig. 22.10. Depending upon the velocity relative to the plasma cloud, the cosmic rays can gain or lose energy. Because the probability for collision with a plasma cloud is random, it is to be expected that the gain or loss of energy will be statistical in nature. It is possible to show that the probability for energy gain is the greater. Hence, after bouncing between plasma clouds long enough, the particles can acquire the very high observed energies.

An alternative suggestion for the galactic origin of cosmic rays is associated with *supernovae* or exploding stars (see the following chapter for a description). During the explosion process, a blast wave forms about the stellar material, accelerating the protons and the heavier components of cosmic rays to very high energies. Of the two, the supernovae theory seems to be more satisfactory at the present time. Several years ago it was suggested that cosmic rays may be of extragalactic origin, coming from sources of very strong radio emission far out in space. As yet we have insufficient information to evaluate this hypothesis. We shall have more to say about cosmic radio sources in Chapter 24.[4]

Assuming that comic rays are indeed charged particles entering the solar system *isotropically* (i.e., equally from all directions) from outer space, one would expect them to be affected by interplanetary magnetic fields and by the earth's magnetic field. According to electromagnetic theory, the intensity of galactic cosmic radiation should be smaller near sunspot maximum (when interplanetary magnetic fields are more intense) than near sunspot minimum. This is indeed the case. Furthermore, the intensity should decrease at the onset time of a magnetic storm. Such an effect was discovered many years ago and is called the *Forbush decrease* after its American discoverer, Scott Forbush.

Finally the geomagnetic field itself produces an observable effect on cosmic ray intensity. Depending upon the electric charge of the particles, one would expect a longitudinal *asymmetry* in the incidence of cosmic rays, more positively charged particles entering from the west of the local meridian than from the east, and conversely for negatively charged particles. More particles do enter the atmosphere from the west, indicating that they predominantly bear a positive charge.

6. QUESTIONS

1. Some years ago, a popular news medium stated "if the Russians had successfully launched a rocket into space, it was only reasonable to assume that they could recover one from space" by atmospheric reentry. What is the fallacy behind this line of reasoning?

2. Discuss the difficulties encountered in executing the rendezvous of two satellites.

3. Discuss the mechanism by which the kinetic energy of random motion in a plasma can be stored as potential energy in a local magnetic field.

4. It can be shown that the energy density of a gas is equal to its pressure while the energy density of a magnetic field is equal to $B^2/8\pi$ where B is the magnetic flux density. With the aid of these relationships and the law of conservation of energy, discuss the nature of magnetic field "pressure."

[4]As this book goes to press, it appears likely that Thomas Gold's recent proposal that pulsars (see p. 649) are the source of galactic cosmic rays will meet the tests of observation.

5. As one goes out radially from the earth, the earth's magnetic field intensity becomes weaker. Describe qualitatively how this behavior of the geomagnetic field would cause trapped particles to drift to the east or west. Which way is the net current flow?

6. Find a most favorable year with respect to probable radiation dosage for astronauts to attempt a trip to the moon.

7. Using the result of exercises No. 11 and the fact that in this convention, magnetic lines of force are directed from the north pole toward the south pole, show that the trapped particle "ring current" tends to decrease the geomagnetic field intensity.

8. Explain in detail with qualitative arguments why solar cosmic rays tend to funnel into the polar caps. How does this tendency depend upon particle energy and mass?

Problems

1. If the ratio of initial to terminal mass of a rocket is 100, what is the final velocity relative to an observer on the earth if the exhaust jet velocity is 1 Km sec^{-1}?

2. Compute the rocket gas jet velocity if the temperatures before and after expansion are 3000 and 300°K, respectively, and the mean molecular weight of the gases is 10 g/mole.

3. Calculate the orbital radius of a synchronous satellite (one with the same period of resolution as the period of the earth's rotation).

4. Calculate the velocity required to escape the earth's gravitational field.

5. Find the approximate numerical ratio of the intensity of radiation emitted per unit area in a sunspot to that emitted by a unit area on the bright part of the solar disc.

6. How much further will a proton of momentum 1 GeV/c penetrate into a uniform magnetic field than will a proton of momentum 1 MeV/c?

Recommended Reading

R. L. F. Boyd, *Space Research by Rocket and Satellite,* Harper, New York, 1960 contains an excellent discussion of rockets in Chapter 3 and of space experiments in Chapter 6.

Eric Burgess, *Frontier to Space,* Collier Books, New York, 1955 is a rather old but still useful paper bound account of space science.

William C. Livingston, "Magnetic Fields on the Quiet Sun," *Scientific American,* November 1966, p. 54, gives an excellent but simple discussion of the theory of the solar cycle discussed in this text.

Brian J. O'Brien, "Radiation Belts," *Scientific American,* May 1963, p. 84 is a fairly up to date account of the nature of the trapped particle belts.

E. N. Parker, "The Solar Wind," *Scientific American,* April 1964, p. 66. The author is one of the leading theorists in the field.

Sam C. Phillips, Apollo 8: "A Most Fantastic Voyage," *National Geographic,* May 1969, p. 593.

C. S. Warwick, "Solar particles in interplanetary space," *Sky and Telescope,* September 1962, p. 133.

chapter twenty-four

Cosmology and cosmogony

The majesty and mystery of the universe has had a more compelling power over man's attention than the microscopic aspects that we have been studying in earlier chapters. The century-long and patient labors of science to decipher the macroscopic universe have always been difficult both because of the vast immensity and of the inaccessibility of this far-stretching "ocean" of space. Although we can probe into the depth of the atom with our proton and electron bullets, we can extend our probes into space only a puny distance, even with the use of satellites and radar.

In the 20th century, through the development of modern instruments and under the guidance of such physical theories as those involving relativity, quantum mechanics, and nuclear reactions, our understanding has taken a great leap, far beyond the solar system. Until recently the Copernican theory, even with its full development by Kepler, Galileo, Newton, and their successors, was still a *local* theory. Traumatic as the shock must have been to the medieval mind to discover that the sun and not the earth was the center of the solar system, the primary focus of attention remained riveted on the tiny heliocentric system, presumed to be governed by laws similar to those that govern all machines. Kant's and Laplace's speculations on the origin of the universe primarily concentrated upon the origin of the solar system. Little was known about the stars, except their positions, and these were studied and mapped primarily for the purpose of understanding the motions of the planets.

In the twentieth century there has been a tremendous extension of astronomical knowledge, with effects as revolutionary as those of the age of Copernicus. The sun is now regarded as a run-of-the-mill star, not even at the center of our own galaxy, the Milky Way. And the Milky Way, a collection of millions of stars, is now known to be but one small speck, an island among millions of other galaxies scattered widely in space. Spatial distances are not thousands but *millions* of times greater than had been considered possible 100 years ago; temporal durations have been extended again and again until we now estimate the *minimum* age of the universe to be about 7 billion years. "Empty" space, no longer a mere container in which stars and planets move, has taken on properties of a physical nature; indeed, space is now regarded as abounding everywhere with radiant energy, dust, atomic particles, cosmic rays, electrical, and magnetic fields. The "celestial fires" of the stars no longer burn mysteriously: their source of energy is understood; their birth, growth, and death in an evolutionary scheme have been surmised.

Such is the tremendous burst of new knowledge about the universe. Is it surprising that there has been a resurgent interest in cosmological and cosmogonic theories? Several are the rival theories that have been proposed. Although each theory contains speculative aspects and each entails difficulties, scientific cosmology while still in infancy has, at last, taken its place among the rational sciences. Undoubtedly, the ultimate questions about the origin and structure of the universe will never receive complete answers. But our knowledge and understanding is expanding and, at the very least, we are disabusing ourselves of many earlier misconceptions—even of some that had become firmly embedded in our cultural traditions.

In this chapter we shall consider the broad problems of the structure and the origin of the universe. To some degree, each generation is inevitably trapped in its own cultural tradition, influenced by old ideas and metaphysical prejudices. Therefore, let us examine ancient origins out of which present knowledge has grown. Scientific theories of cosmologies, no less than myths, involve creative imagination and interpretation in bringing the chaotic appearance of phenomena into meaningful unity. As the ancient myths have gradually evolved into more scientific and rational theories, the purely dramatic and anthropomorphic elements have only slowly been purged by the dominant critical requirement of science: that any theory must be both logical in its structure and based upon positive knowledge founded upon observation. Even within these boundaries, modern cosmologists find alternative choices and unconscious psychological impulses; and metaphysical assumptions are ever present in affecting the choices that are made. No theory springs complete in its glory into the mind of man, but evolves only after long struggle of trial and error out of old ideas. To guard against too naive an enchantment with present "facts" and "interpretations" it is always well to review the history of the struggle of past generations with regard to important ideas. Cosmic perspective based on present knowledge must be tempered by human perspective regarding the development of thought.

1. TYPICAL ANCIENT CREATION MYTH: THE BABYLONIAN LEGEND

As an example of ancient cosmological theory we can study the Babylonian legend, chosen both because of its deep impression upon our tradition and because it typifies so many myths and theories in a combination of shrewd, systematic observation, plus dramatic and anthropomorphic appeal (see Fig. 24.1).

In the Babylonian myth, the world slowly evolved out of a primeval deep. Dark chaos, personified in the mother goddess Tiomat, intermingled with the father ocean, Arfu. Out of the union germinated the gods of light who multiplied rapidly. Tiomat, dismayed at seeing her progeny expanding and threatening to take over her dominions, decided to defend her territories. Marduk, bravest of the

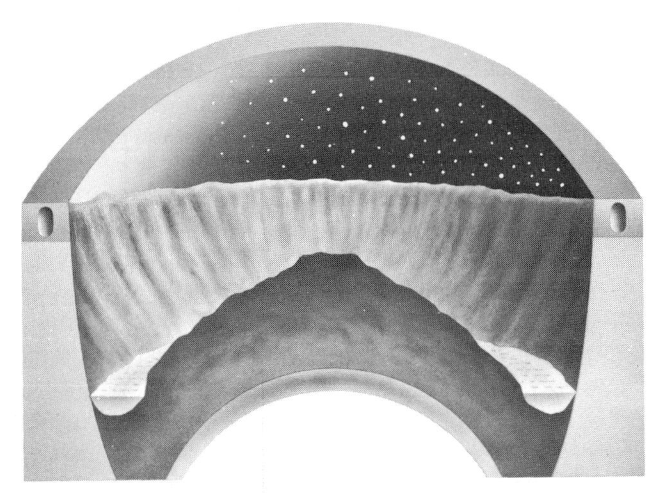

Fig. 24.1. Chaldean picture of the world (drawn by Faucher-Gudin, from Svante Arrhenius's, *The Life of the Universe,* Harper, New York, 1909, p. 21). The continent in the center rises up to Mt. Ararat out of the surrounding water. The heaven above, created by Marduk, is vaulted, shining by the sun in daytime, appearing dark and studded with stars at night. During the night, the sun traveled through a gallery (whose openings can be seen to left and right) in the northern portion of the sky and, therefore, could not be seen until it issued from the eastern opening in the morning.

progeny and son of the god of wisdom, Ea, gave her battle. Tiomat temporarily took the form of a dragon but to no avail, for Marduk overcame her body, chaos, and clove her asunder, forming heaven from one half and the earth from the other. Marduk then arranged the stars and sun and moon in an orderly pattern, and *provided the laws under which each would run its course,* establishing the rules of order and destiny to govern the universe. Plants, animals, and finally men were then created. In this myth, divine law, generation, creation, and orderly laws are all dramatically set forth. Undoubtedly the myth sprang from observations regarding the generation of life, the seasons, and the rhythmic patterns of organic life—all against a background of the changing positions of the stars and the sun, moon and planets. Many are the elements in this myth which appear in later Hebraic cosmogony, as Milton K. Munitz has pointed out in his comment:

> Here by the way is to be found the original etymological meaning of the Hebrew verb bara (to create) in Genesis; it points to the meaning "to cut" or "to carve."[1]

The naive animism of the Babylonian myth is of course modified by the later Hebrews and God is endowed with a more majestic and abstract nature, but the story has, if not identical, at least parallel elements.

In close association with the theological myth, which left its imprint on our traditional theology, was the concomitant development of astronomy, which had such a profound effect on other branches of our cultural heritage through the Greek development of science. Historians tend to emphasize the extraordinary flowering of Greek art and science as the "Greek miracle," but certainly the development of their mathematics and astronomy was most heavily dependent on a very long tradition of earlier Babylonian and Egyptian science. Extraordinary Babylonian tablets have been found dating back at least 2000 years before Ptolemy, which reveal a wealth of systematic observations and interpretations upon which later Greek astronomers developed their theories. The Zodiac was a Babylonian invention. The stars had been named, numbered, and mapped, and long tables have been found regarding the motions of the moon, the sun, and Venus. Eclipses of both the sun and the moon could be foretold with considerable accuracy. While this ancient astronomy was written from an astrological point of view and was closely related to the mythical cosmology, it did provide both a broad basis for further observation and a systematic interpretation of motions without which Greek science could hardly have advanced as it did.[2]

[1] Milton K. Munitz, *Space, Time and Creation,* Crowell-Collier, New York, 1961, p. 19. This book is highly recommended to those who are interested in the broader philosophical problems underlying modern cosmology.

[2] George Sarton, an outstanding historian of science, has written a delightful essay on the relationship of ancient Babylonian and Egyptian science to that of Greece in his "East and West in the History of Science" found in *The Life of Science,* Indiana University Press, Bloomington, Indiana, 1960.

It is out of such early speculations, colored it is true by moral and anthropomorphic prejudices and yet already related to systematic observation, that there gradually developed the science of cosmology in which physical observations and mathematical relationships predominated. Historians of science distinguish four periods in this development. The first period, extending over thousands of years, is marked by the building of creation myths. In the second came the development of the great Aristotelian-Ptolemaic system so closely tied to theology. In the third period the Copernican and Newtonian revolution came about. Finally, in this century, the fourth great advance in cosmological theory has developed from new knowledge of the distant nebulae, new concepts of stellar evolution, and better understanding of energy and radiation phenomena. In some ways, the new horizons are a return to a more majestic conception of the universe than the simpler, narrower mechanical one that had dominated thought over the last four centuries. Perhaps once again a modern Dante can give expression to the newly discovered mystery and drama being played on a cosmic scale about us.

2. ON LARGE NUMBERS, DISTANCES, AND DURATIONS

The egocentric predicament in which man finds himself cannot be escaped. At best it can only be understood. Science should in its way contribute to our rational perspectives as religion should contribute to our moral perspectives. But the scientific, spatial, and temporal perspectives are difficult to make meaningful because of the limited scale upon which humans live. For example, if some cosmic giant could hold the earth in his hand, it would appear to him almost as smooth and spherically perfect as a bowling ball, and the mountain ridges, the deep canyons, and the flattening at the poles, all of which we emphasize, would hardly be perceptible. Man cannot help but be "the measure of all things" unless he proceeds imaginatively and indirectly toward a wider understanding.

Therefore, before embarking upon a discussion of modern observations and methods of measurement applied to the universe, let us review once again the meaning of the very large numbers and dimensions with which we shall be confronted. It is difficult to make familiar the orders of magnitude involved, because they are so far beyond our own scale; a few illustrations, however, may be helpful.

Pick up a stone that fits nicely into your hand. The number of such stones that would be required to constitute a volume as large as the earth would be about 10^{24}. Such a number would be roughly equal to the number of atomic particles constituting the stone itself, and would also approximate recent estimates of the minimum number of stars in the universe. This number—so hard to grasp in understanding—is more than a million times a billion times a billion, which is a number larger than all grains of sand on all the beaches of the world, a number larger than all drops of water in all the oceans of the world.

Consider next a few distances. In this age of satellite and space travel, the dimensions of our own large but still very local solar system take on familiar meaning. An artificial satellite or a space ship can circumnavigate the earth in one hour and a half. At this same speed, such a space ship could traverse the diameter of the earth's orbit about the sun in less than two years. It could reach the most distant major planet Neptune within the lifetime of a man. But the very provincial and local nature of our solar system becomes clearer perhaps if we realize that it would take almost a *million* years for a space ship traveling at this same speed to reach *the nearest star.* Our solar system is a tiny speck in the Milky Way. A journey to the nearest star compared with a journey across our solar system would be similar to comparing a transatlantic journey to a jaunt of one half a block to the corner drugstore.

But this is only a beginning. Consider distances *beyond* our own galaxy. Even if we could, by some scientific magic, move with the speed of light from star to star within our own galaxy, we would even then be caught in a provincial localism. The Milky Way itself is a diminutive island in the vast ocean of space. To go out from our own local collection of stars to the nearest galaxy requires another leap in magnitude as great as going from the solar system to the nearest star. Even at the speed of light, to reach the "next door" galaxies would take a space ship 1000 times longer than the whole span of man's career on earth. Vast dimensions such as these are just as difficult to visualize as are the microscopic dimensions of atomic particles, which we have studied in earlier chapters.

Finally, the conception of time-spans is also hard to comprehend in relation to the human scale. On the human scale, we are familiar with the unit of one year because of the seasons, and the unit of one second because of our heartbeat and the period of our steps. There are over 30 million seconds in a year. All written history embraces perhaps 10,000 years. The length of written history has thus a magnitude of around 10^{11} heartbeats (or seconds). But, in the universe, cosmological time is now measured in units *millions* of times longer than the span of written history (that is, in billions of years). In summary: on the one hand, cosmological time is measured in units about 10^{17} times larger than the beat of a heart; and, on the other hand, as we have seen in earlier chapters, the beat of a heart is itself 10^{21} times longer than the natural unit of time in nuclear events (recalling that it takes but 10^{-21} seconds for a neutron to "cross the diameter of a nucleus"). The span from the shortest to the longest duration is thus a ratio of about 10^{39}, about the same ratio as the span from the smallest dimension (atomic nucleus) to the greatest observable distance (as observed in telescopes). These are large numbers worth pondering.

3. ELEMENTS OF MODERN COSMOLOGY

Measurement of Distances. How are such figures for the vast dimensions of our universe arrived at? It will be recalled that no estimate of the distance of the nearest stars could be made for 300 years after Copernicus' great work on astronomy. The only available method of measurement was by triangulation, using the diameter of the earth's orbit as a base line and seeking to observe the parallax angles of nearby stars. Even with this base line (186 million miles), the nearest star is so distant that its parallax angle is only about one second of arc (that is, one thirty-six hundredths of a degree). This entails extraordinary precision of measurement. Our measuring instruments have become more and more precise, but even with the most modern instruments, including photographic apparatus, an angular displacement of a star of less than 0.05 of a second of arc is difficult to detect, let alone measure with accuracy. Such a method of measurement is therefore only possible for distances less than about 100 light-years (a distance which embraces a very small portion of the Milky Way, which we now know extends a distance more than 100,000 light-years). At first, it would seem that man could never reach out to measure any distance greater than this, because no line longer than the diameter of the earth's orbit is available as a base for triangulation.

There is, however, an ingenious method, which has been contrived in the last two decades based upon a careful study of the types of stars as revealed by the

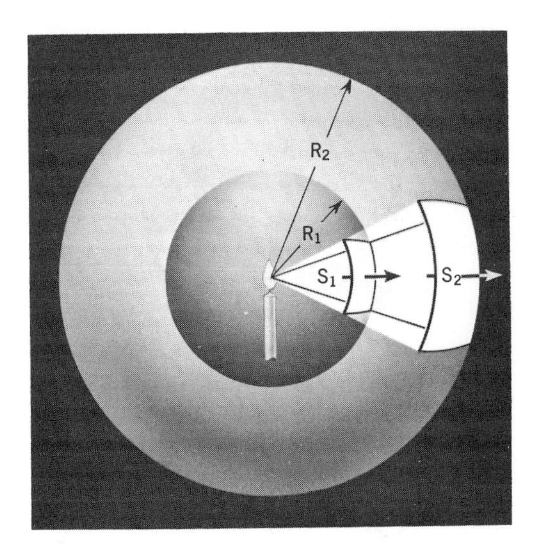

Fig. 24.2. The inverse square law of radiation. If the number of photons crossing spherical surface S_1 of radius R_1 and the number crossing spherical surface S_2 of radius R_2 are the same, the number reaching surface S_2 per unit area of surface will be one fourth of those reaching S_1 per unit area of surface, if $R_2 = 2R_1$.

spectroscope. This new method of measurement provides another good example of how conclusions that cannot be reached directly by obvious traditional methods can be sometimes arrived at by using a happy combination of mathematical analysis and patient observation. Basically, the measurement depends upon combining an observation regarding variable stars with the fundamental law of radiation—which states that when light is radiated from a spherical source in all directions the intensity of the light (that is, the amount of energy per unit time falling on a specified cross-sectional area) varies inversely as the square of the distance from the source. Experimentally this law has always been found valid, and it is also a theoretical consequence of classical electromagnetic and quantum theory. The diagram in Fig. 24.2 illustrates the principle of the inverse square law.

If the number of photons crossing the sphere of radius R_1 and the number crossing the spherical surface of radius R_2 are the same, the number reaching spherical surface R_2 per area of surface will be ¼ of those reaching surface R_1, if $R_2 = 2R_1$, since $S_2/S_1 = R_2^2/R_1^2$.

It is apparent that the inverse square law could be used to determine the distance of light sources *if we could measure accurately two quantities,* that is, the intensity of light reaching the earth from a given source and the intensity of light given off by the source, by using the equation:

$$I_e = \frac{I_s}{D^2}$$

where I_e is the intensity of light reaching the earth, I_s is the intensity of light given off by the star, and D is the distance between the star and the earth.

Now the intensity of light from a star *as it reaches the earth* can be measured by the brightness of the star as it appears to our eyes (or, more precisely, by its effect on photographic or photoelectric equipment). Our first problem is to specify clearly a measure of brightness. There is an ancient scale of brightness, which has come down to us from Ptolemy and which astronomers still use with quantitative refinements. The brightest stars, including about 15 stars visible from positions in the Northern Hemisphere, are called *first magnitude stars.* The scale then falls off in steps of just visually distinguishable differences until the sixth magnitude stars, which are those at the lower limit of visibility to the naked eye.

But the apparent magnitude of a star does not bear a simple mathematical relationship to the intensity of light reaching the earth. The light from a star of second magnitude is not half as intense as the light from a star of first magnitude. Instead, the scale of magnitudes has a relationship of intensity similar to the relationship between musical intervals and the physical frequency of vibrations. It will be recalled that an interval of an octave represents an interval in which frequency is doubled. Since there are 12 intervals in the musical scale, the ratio of frequencies between two successive notes is $2^{1/12}$. Thus, if the note A is taken as

440 cycles per sec, the next note is *A*♯ at 440 × $2^{1/12}$ = 486 cycles per second. A similar mathematical relationship has been established for the scale of magnitudes of stars. A star of the fifth magnitude has 100 times less intensity than one of the first magnitude. The interval of one magnitude is thus taken to correspond to a ratio of brightness of $100^{1/5}$ (equal to about 2.5). A star of the first magnitude is thus 2.5 times brighter than one of the second magnitude.

We now have a precisely designated physical scale of *apparent magnitudes* or brightness of the light reaching the earth, which can be extended in both directions and which will allow for fractional magnitudes. The brightest star, Sirius, and some of the planets (the moon and the sun) all have *negative* magnitudes, while stars invisible to the eye but visible through the telescope have apparent magnitudes ranging up to the twenty-third magnitude. As an illustration, a few of the brighter stars are listed in Table 1, with their apparent magnitudes. Distances are also given in light-years. The last column gives what is called the *absolute magnitude* of stars (next to be explained).

Obviously, *apparent* magnitude, determined by the amount of light *reaching us* from a distant star, is no measure of *intrinsic brightness* or of the energy the star gives off in an interval of time. Stars vary in distance, in size, and in temperature. We require, therefore, a standard of *absolute magnitude* related to the energy that they give off, per unit of time. This standard is chosen arbitrarily (just as in the case of music the note *A* is determined by international agreement as being 440 vibrations per second). Astronomers designate as the absolute magnitude of a star, the magnitude it would have if it were placed at a distance of 10 parsecs or 3.26 light years from us. The most convenient and direct measurement of distances used by astronomers is the parsec. One parsec is the distance of a star whose

TABLE 1. THE TEN BRIGHTEST STARS VISIBLE FROM THE UNITED STATES

Star	Constellation	Distance *D* (light-years)	Parallax angle (in seconds)	Apparent magnitude (*m*)	Absolute magnitude (*M*)
Sun	—	1.5×10^{-7}	—	−26.7	+4.87
Sirius	Canis Major	9	0.38	−1.52	+1.36
Vega	Lyra	27	0.12	+0.12	+0.5
Capella	Auriga	45	0.07	+0.21	−0.5
Arcturus	Bootes	36	0.09	+0.24	0.0
Rigel	Orion	650	0.005	+0.34	−6.2
Procyon	Canis Minor	11	0.29	+0.53	+2.8
Altair	Aquila	16	0.20	+0.89	+2.4
Betelgeuse	Orion	650	0.005	+0.92	−5.6
Aldebaran	Taurus	68	0.05	+1.06	−0.5

parallax is one second. The simple equation of distance then is given by $D = 1''/p$, where p is the parallax in seconds. One parsec is equal to about 3.3 light-years. The absolute magnitude of a star is its apparent magnitude as viewed from 10 parsecs.

What is accomplished by these definitions and relationships? It is this: we now have three quantities, *apparent magnitude m, absolute magnitude M,* and *distance D,* which can be related by a mathematical equation. The equation (which is derived in Appendix for those who are familiar with logarithms) takes the following form:

$$5 \log \left(\frac{D}{10}\right) = m - M, \text{ where } D \text{ is the distance in parsecs.}$$

Typical relationships between m, M, and D are shown in Table 2.

Notice that if two of the three quantities M, m, or D are known, the third can be found. For hundreds of the nearest stars distances and apparent magnitudes may be measured. For these stars, then, we can calculate the *absolute magnitudes.* If now we can only discover, from a careful and intensive study of such stars, an *independent* way of measuring absolute magnitude, then we can use the combination of absolute and apparent magnitude to determine the distances of stars whose parallaxes we cannot measure.

The first necessary scientific breakthrough was achieved in 1917 when Henrietta Leavitt at Harvard and Harlow Shapley at the Mount Wilson Observatory discovered that there is a kind of pulsating or variable star, called the Cepheid

TABLE 2. RELATIONSHIP BETWEEN APPARENT MAGNITUDE, ABSOLUTE MAGNITUDE, AND DISTANCE

$m - M$ (apparent magnitude minus absolute magnitude)	Distance	
	Parsecs	Light-years
−5	1	3.26
−2	4	13
−1	6	20
0	10	33
+1	16	52
+2	25	81
+3	40	130
+4	63	205
+5	100	326
+6	170	570
+10	1000	3,260
+15	10,000	32,600
+20	100,000	326,000

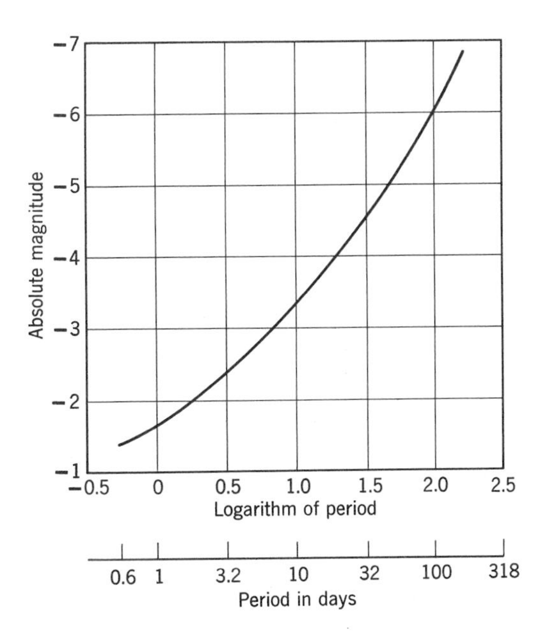

Fig. 24.3. Relationship between absolute magnitude and period of Cepheid variables. From this curve the absolute magnitude of a Cepheid variable can be determined after its period is found by observation. The apparent magnitude can be measured independently. Knowing both the apparent magnitude and the absolute magnitude, the distance can be determined from Table 2 or from the equation, $5 \log D/10 = m - M$.

variable, named for the first such star studied in the constellation Cepheus, in which the *period* of the star's pulsating light *is related to the absolute magnitude* of the star. Stars having a longer period have a greater absolute magnitude. A chart of the variation between absolute magnitude and period is shown on Fig. 24.3.

This discovery of an independent way of determining absolute magnitudes gave an extraordinary new tool for measuring vast distances in space. Wherever variable Cepheid stars could be found, their absolute magnitudes could now be determined and, since their apparent magnitude could also be measured by photographic plates mounted in telescopes, it became possible to determine distances. For example, let us apply the new yardstick to a typical Cepheid variable in the constellation Auriga, which has an observed mean apparent magnitude of 10, and which has an observed period of 10 days. From Fig. 24.3, we see that this corresponds to an absolute magnitude of -3.4. We now can either use the equation for distance, that is, $5 \log D/10 = m - M$, or read off the distance from Table 2. We conclude that this star is at a distance of about 5×10^3 parsecs or about 16,000 light-years.

With such a measuring tool at hand, not only can the size of our galaxy, the Milky Way, be determined as approximately 100,000 light-years in its longer dimension and about 20,000 light-years in thickness, but even more important, distances to other galaxies far outside the Milky Way can be determined whenever the modern telescopes can detect Cepheid variables. By 1924, such variable stars were found in a great many nearby galaxies and their distances determined.

By further methods—including studies of novae, supernovae, and clusters of galaxies—other standards for estimating absolute magnitudes have become available in the last two decades. We shall not go into the details of these additional methods, but shall merely mention that they have been sufficiently refined as to give reasonably reliable estimates of distances over a *billion light-years*. (The order of magnitude, at least, can be fairly well depended on. It is true that with such indirect methods the estimated distances may be in error as much as 100% or more. For example, in 1952, Baade introduced a refinement that doubled all previously estimated distances of galaxies. Nevertheless, we can say that the most distantly observed galaxies are *billions of light-years distant,* even if we cannot reliably tell, for example, whether a particular galaxy is 2 billion rather than 3 billion light-years distant. We know, in other words, the order of magnitude, that is, the power of 10, even if estimates may be in error by almost 100%.)

Measurement of Age. In some editions of the Bible (King James version), a chronology is appended in which the date of 4004 B.C. is stated as the date of the creation of the world. To scientists this represents a singularly poor mixture of theology and mathematics. Have scientists themselves a more convincing answer as to the age of the earth? They have, if certain principles upon which all scientific knowledge is founded is granted to them. Although the age of the earth cannot be stated with complete precision, all modern estimates indicate that it cannot be younger than 3 billion years, and is more probably at least 4 billion years old. Likewise, on the basis of modern studies, there is good reason to believe that many stars have an age of at least 6 or 7 billion years. Perhaps the universe of stars is older than this, if we accept one of the theories of cosmology (the *steady state* theory), but on almost any scientific theory of cosmology we can accept 6 billion years as a *minimum*. How are these figures arrived at? Again we find how powerful the methods of science and analysis can be when patiently applied to unraveling problems that seem at first impossible of solution.

There are several ways in which the age of the earth can be estimated, and they provide important checks on each other, so that in combination they yield results in which we place considerable confidence. Probably the most dependable of the methods uses the rate of radioactive decay of uranium to determine the age of rocks in which the uranium is present. (A similar kind of dating has also been used, with appropriate modification, to determine the age of wooden and other organic relics from ancient cultures. In this case, the radioactive isotope of carbon, C^{14}, is used as the indicator of age. In geology, similar methods are used to date the age of fossils.)

The use of the decay rate of uranium 238 depends on the general equation that applies to all radioactive decay (see Chapters 18 and 20 for a more complete discussion):

$$n = N_0 e^{-\lambda t}$$

where N_0 is the original number of atoms of uranium present, n is the number present at time t, e is a constant, that is, the base of natural logarithms and equal to 2.73, and λ is a decay constant which depends on the element. The equation can be rewritten

$$\frac{n}{N_0} = e^{-\lambda t}$$

Since the decay constant λ can be determined in the laboratory with considerable accuracy, and since this constant is completely unaffected by changes in pressure or temperature, if we have a way of determining the fraction n/N_0, that is, the ratio of the number of radioactive atoms present after an elapse of time t to the number originally present, we then can calculate the elapse of time needed to reach this ratio. Figure 21.3 shows a graph of how this fraction varies with time in the case of uranium 238. It will be noted that the fraction reaches the value one-half in 4.5 billion years; hence, we designate the "half-life" of uranium as 4.5 billion years.

It turns out that this ratio (number of atoms now present to number of atoms originally present) can be determined in samples of uranium-bearing rock. Since uranium 238 goes through a series of relatively rapid radioactive transformations, eventually ending up as a stable isotope of lead (lead 206), it is possible to determine the fractional number of U^{238} atoms which remain today, and hence to determine the time that has elapsed since the original uranium was trapped in the rock formation. Suppose, for example, that in a given sample the ratio of lead to uranium is 1 to 3 by weight. Using Avogadro's number, this yields a ratio of the number of lead atoms now present to the number of uranium atoms now present as roughly 1 to 2.8. But every atom of lead is descended from an original atom of uranium. Therefore, the fraction of the present number of uranium atoms to the original number is 2.8 to 3.8 (that is, $n/N_0 = 2.8/3.8 = 0.74$). Looking, now, at the graph of Fig. 24.4, we see that this fraction corresponds to an age of about 1½ billion years.

Using this general method, and also checking it by other means including the use of radioactive decay series involving U^{235}, which we need not here enter

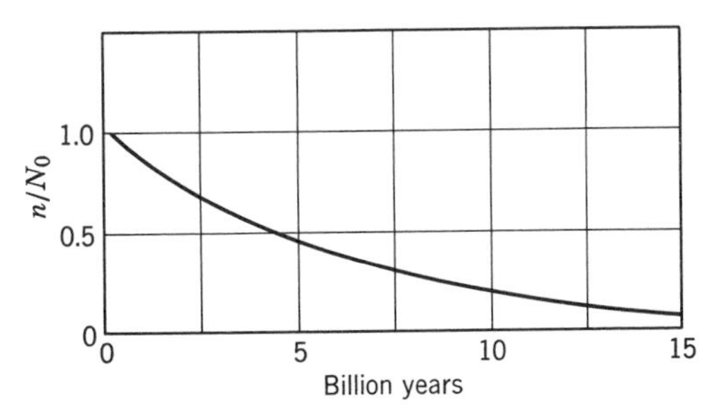

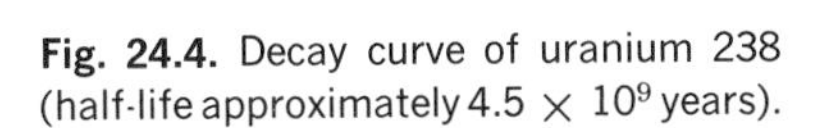
Fig. 24.4. Decay curve of uranium 238 (half-life approximately 4.5×10^9 years).

into, the age of certain sedimentary rocks in South Africa has been determined at about 3 billion years. But these rocks are by no means the oldest on the earth and we can estimate that the earth itself was probably formed a billion years earlier, placing its age at around 4 billion years.

It is interesting that this estimate checks rather closely with other entirely different methods of determination. For example, it is known that the moon is slowly drawing away from the earth, gaining an angular momentum from the earth which is being slowed by tides. Calculations first made by Sir George Darwin, son of the biologist Charles Darwin, show that this constant tidal action would account for the moon's present distance from the earth if it had been formed close to the earth (either out of the earth itself or as another already separate body when the earth was formed) at a time approximately 4 billion years ago.

Finally, concerning the minimum age at which most of the stars were formed, estimates vary between 8 and 14 billion years. The higher estimates depend on the validity of current theories of stellar evolution to be discussed briefly below; the lower estimates are based upon an analysis of the motion of distant galaxies. Since both estimates are arrived at independently, there is considerable assurance that the correct order of magnitude has been determined.

Energies of Stars. A long-standing puzzle of the nineteenth century was the problem of accounting for the energy of the sun and the stars. It was obvious that ordinary chemical reactions such as the burning of hydrogen or carbon could not be an explanation. The sun would burn itself up in a matter of thousands of years. Even an alternative suggestion—that the heat was generated by friction as the matter of the sun, for example, became more condensed in its contraction under the immense gravitational forces—did not seem adequate. On this basis, Kelvin calculated at the end of the nineteenth century that the sun's age must be about 40 million years. Evidence from geology as well as astronomy indicates such a time period is much too small.

A much more satisfying answer was finally found with the discovery of atomic energy. The celestial fires are fueled by the conversion of their mass into energy. Let us apply the equation $e = mc^2$ to determine the rate at which the sun is losing its mass. First, we can determine the total energy given off by the sun by measuring the heat reaching a specified area on the earth during a period of time. Making the necessary correction for the heat lost as the sun's rays pass through the atmosphere, we arrive at a figure for the heat which reaches a small known segment of the surface of a sphere whose radius is 93 million miles from the sun. From this the *total* energy passing outward from the sun into space during an interval of time can be easily calculated. This amount is, of course, vast—approximately 4×10^{26} joules/sec. Such an amount of energy can only come from the conversion of matter into energy in accordance with the mass-energy conversion formula. The loss of mass by the sun in each second (about 5 million tons) seems a staggering quantity, but even this amount is negligible relative to the total mass of the sun. So large is the mass of the sun that such a rate of loss could continue for billions of years with little change.

Although shortly after the discovery of radioactivity and Einstein's formulation of the mass-energy relationship, both occurring around the turn of the century, astronomers became convinced that the source of energy of stars did come from some kind of nuclear reaction; just what sort of reaction remained puzzling. Various types of information concerning the temperature and pressure conditions within stars as well as the types of atomic nuclei present were necessary to arrive at the solution. It is interesting that these studies in astrophysics, pursued for their own sake, eventually led physicists to recognize the possibility of building a hydrogen bomb. More recently the studies have been important in their contribution to the attempt to construct controlled fusion reactors, which may one day solve the problem of man's energy shortage. The study of distant stars, a pure science pursued for its own sake, has already led to many practical consequences and may in the future lead to still more extraordinary consequences in the conduct of day-to-day affairs of man.

Three salient features should be kept in mind if we are to understand the nuclear reactions within stars: (*1*) the chemical composition of the nuclei in the stars; (*2*) the temperature conditions; and (*3*) the physical condition of the matter composing the stars.

By means of spectroscopes, the types of nuclei existing in stars and their abundance can be determined. We have seen how atoms radiate specific wavelengths of light and heat depending on the neutron-proton composition of the nuclei. When the spectra of stars are examined, the presence of certain combinations of lines indicates the wavelength of the radiated energy, and the intensity of those lines indicates the abundance of the elements present. Although stars vary considerably in composition, in general hydrogen with its simple nucleus is by far the most abundant, accounting for over 90% of the mass of stars. The next most abundant element is helium, which accounts for the balance except for a small 1% which is made up of all the other elements. The abundance of elements heavier than helium falls off very rapidly with increasing atomic weight. Helium, an inert stable gas that is so rare on earth, was actually first identified as an element by spectroscopic studies of the sun *before* it was found on earth. The reasons why hydrogen and helium have such an overwhelmingly dominant place in the composition of the universe as a whole, although the heavier elements are so abundant on the earth, we shall leave for a later discussion.

The *temperature* of the sun and stars can be determined with considerable precision. The principle used depends on an observation that is familiar to all of us. If a piece of metal is heated, it first glows with a dull red color, but as the temperature increases the color changes to yellow and finally to white. The reason is that the relative *intensity* of various wavelengths (color) changes with temperature. Planck, who originated the quantum theory, made a careful quantitative analysis of the relationship between temperature and intensity of radiation at different wavelengths in continuous spectra, and out of this analysis astronomers have developed *a color index* to temperature. The color index expresses a measurement of the difference in intensity between two wavelengths, namely, 4.25×10^{-7} m and 5.28×10^{-7} m in the continuous spectrum. This color index, I, is related to the temperature by the equation $T = 7200/I + 0.64$ (where T is the absolute temperature), and applies nicely over the range of temperature between 3500° and 15,000° absolute. For higher temperatures, other measures are used which need not be discussed. The temperature of the sun at its surface has been determined as being about 6,000° absolute or 11,000° Fahrenheit. The interior temperature of the sun is not known with accuracy but it is certainly very much higher, probably *at least* 13 million degrees absolute as calculated from another equation to be discussed farther on.

As for the *physical conditions* that exist within the sun, some fairly reliable conclusions can be drawn. The mass of the sun can be determined from the period of the earth's revolution and its radius of revolution. The diameter of the sun can be determined from a knowledge of its distance and from observing its angular diameter as seen from the earth. Knowing the mass and volume, its density is readily determined. At the sun's high temperatures, all of its matter must be in a gaseous state. But the gas is of a different form from that with which we are familiar because of its tremendous density and extraordinarily high temperature (a pint of the "gaseous" material of some of the densest stars would weigh many tons). The gaseous state that exists at these temperatures is one in which the particles are not bound together into structured atoms, but form a kind of gas of nuclei and free electrons (which has been given the name "plasma"). (See Chapter 23.)

There must exist for long periods of time a kind of equilibrium within the sun and most stars. Gravity is extremely high because of the masses involved. The gaseous material does not completely collapse into a still smaller space because of outward forces which balance the inward force of gravity. These outward forces arise from the pressure of the gas and the pressure of radiation which keeps the mass distended as a great ball of incandescent gas. To the extent that radiation pressure can be neglected, the general gas law applies. Under conditions of high temperature the radiation pressure assumes greater significance because it varies with the fourth power of the temperature.

Estimates of the central temperature of stars are arrived at from observations regarding their mass, surface temperature, luminosity, chemical composition, and opacity of material. The opacity is a measure of the obstruction to the outward flow of radiant energy offered by the material composing the star and has been found by the methods of radiation transfer. The mass of stars can be determined by a relationship known as the mass-luminosity law, an empirical relationship originally studied by Sir Arthur Eddington in 1924 on the basis of an analysis of the luminosity of binary stars whose mass could be calculated by Kepler's laws. (A theoretical equation has since been established for this law.)

Once the temperatures and pressures of stars have been estimated, the problem of energy production and internal equilibrium can be analyzed. From Chapter 21 on nuclear physics, we learned of the potential or energy barrier surrounding a nucleus which ordinarily prevents other protons from entering the nucleus in a fusion process. At any but extremely high velocities, the coulomb forces of repulsion prevent the penetration of the nucleus by external protons. If, however, velocities are high enough, fusion reaction will take place. The velocities depend upon temperature and pressure. At room temperature, an air molecule travels at an average velocity of between one third and one quarter of a mile a second, but at a temperature of 15 million degrees absolute and at the pressures existing in the interior of the stars, the velocities approach 60 miles per second, and nuclear reactions will occur.

Hans Bethe, who with Oppenheimer and Teller first suggested the feasibility of constructing a hydrogen bomb, developed a theory for two possible reactions that can account for the energy production of the sun and the stars. Both of these use hydrogen as a fuel, converting it into helium. The first process is called the proton-proton reaction and occurs in the following steps (where e^+ represents a positron and ν stands for the neutrino, γ for a gamma photon).

$$
\begin{aligned}
{}_1\mathrm{H}^1 + {}_1\mathrm{H}^1 &\longrightarrow {}_1\mathrm{H}^2 + e^+ + \nu \\
{}_1\mathrm{H}^2 + {}_1\mathrm{H}^1 &\longrightarrow {}_2\mathrm{He}^3 + \gamma \\
{}_2\mathrm{He}^3 + {}_2\mathrm{He}^3 &\longrightarrow {}_2\mathrm{He}^4 + {}_1\mathrm{H}^1 + {}_1\mathrm{H}^1 + \gamma
\end{aligned}
$$

In each of these reactions, mass is converted into energy, sustaining the interior temperature of the star.

The second reaction proposed by Bethe also probably occurs, but requires somewhat higher temperatures. It is known as a carbon cycle reaction. According to this suggestion, the details of which we shall not enter, a beryllium isotope of atomic mass 8 is formed out of two helium atoms. Although this is an unstable isotope, it can exist in sufficient abundance so that an occasional carbon atom of mass 12 is then formed by a combination of a beryllium atom (mass 8) plus a helium atom (mass 4). Once carbon atoms have been formed, a whole series of reactions is possible with carbon entering as an intermediary. Such reactions have been reproduced and studied on a small scale in the laboratory using high velocity helium nuclei as bullets. In fact, many possible reactions have been continuously analyzed in the laboratory to gain a knowledge of what occurs in the stars, just as the analysis of phenomena in the stars has suggested further studies in the laboratory. As far back as 1920, Sir Arthur Eddington, whose work on the evolution of stars has been so influential, is said to have remarked after Lord Rutherford's discovery of the transmutation of nuclei, "What is possible in the Cavendish Laboratory *may* not be too difficult in the sun."

Now both these processes, the proton-proton reaction and the carbon cycle reaction, may occur at once in different proportions depending upon the temperature conditions as well as the abundance of hydrogen. The higher the temperature, the more important becomes the carbon cycle reaction.

The analysis of energy production that occurs within stars leads naturally into a consideration of stellar evolution. Stars have been carefully classified according to their luminosity, temperature, and mass. The largest group of such stars of which our sun is a representative member is called the "main sequence." Such stars range considerably in temperature, mass, and radius, but their energy is produced by the proton-proton and carbon cycle reactions. Outside of this group of stars, there fall several other classifications, the most important of which are the red giants, the novae and supernovae, supergiants and the white dwarfs. See Fig. 24.5.

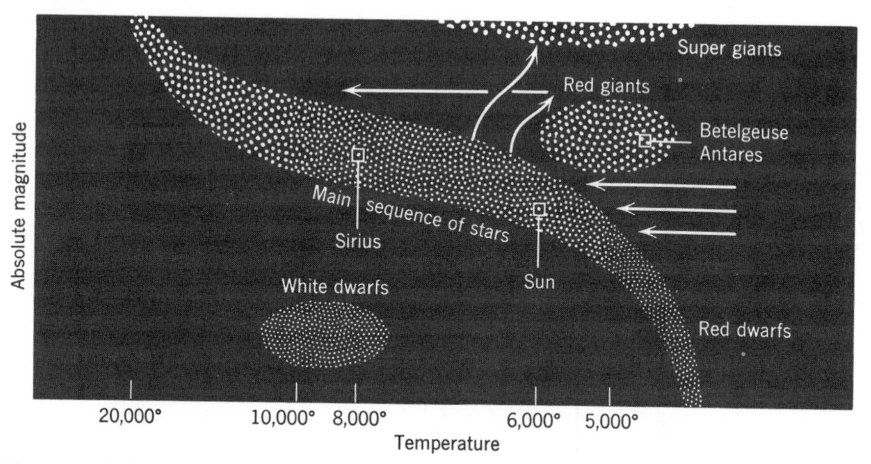

Fig. 24.5. A schematic diagram (called the Hertzsprung Russell or H-R diagram) of the absolute magnitudes of stars plotted against their temperature (measured by spectral types). The arrows show schematically the general direction of stellar evolution. Dust collects to form main sequence stars. When these stars have used up their fuel they move upward to the giant stage. Ultimately they either explode as nova or gradually sink into a white dwarf state. The particular evolutionary course a star takes is thought to depend upon its mass.

To account for the variations in characteristics, a theory of stellar evolution has been fairly well established in its more important points. We cannot very well observe the evolution taking place because of the tremendous time spans involved. Astronomers are somewhat in the same position as biologists who have to construct a theory of evolution on the basis of a wide variety of observed species, except that astronomers do not have the advantage of as precisely dated fossil remains. However, a mechanism of stellar evolution is beginning to be understood in terms of energy production by nuclear reactions, just as the mechanism of biological evolution is understood today in terms of genetic mutations.

What is the fate of a main sequence star such as the sun? Obviously, equilibrium between the outward gas and radiation pressure and the inward pressure of gravitation cannot continue forever, since the fuel supply (hydrogen) will eventually be used up. (It is estimated, however, that the sun's equilibrium can be maintained without substantial change for a period of *at least* 10 billion more years.) Now these nuclear reactions are taking place only at the central core of the star, the central hydrogen supply being continuously refreshed by convection currents which carry hydrogen from the outside toward the center. This process is significantly dependent on the rotation of the star, since this is the most effective means of keeping the material stirred up. (Just as a candle kept burning by being supplied constantly with fresh oxygen by the convection currents of air would soon sputter and then go out in a gravitationless environment, a star that is not rotating would have a much reduced life.) Sooner or later, however, the supply of hydrogen begins to be exhausted, and a central core of the helium "ash" grows larger and larger. At this point, the equilibrium of the star begins to be upset. The core, no longer kept distended by internal radiant energy from nuclear reactions, begins to shrink under the force of gravity. The temperature of the core then rises, due to the tremendous frictional energy released, and this in turn causes an increase in outward radiation pressure, distending the outer envelope gas to an enormous size. This outer shell of gas now grows cooler because of expansion, and glows with a red color instead of its previous whiteness. Thus, a red giant is born.

At the same time the temperature in the central helium core continues to rise until it reaches hundreds of millions of degrees. At these temperatures a new series of nuclear reactions can set in, for now in turn the helium nuclei have reached velocities high enough to overcome the potential barriers between them. Thus, many of the lighter elements are born. Finally, when the helium itself is used up, another shrinkage occurs and again temperatures are raised to more than a billion degrees, at which point all the elements up to iron are born in a variety of nuclear reactions.

What happens beyond the point when the remaining helium begins to become exhausted is more problematical, and the further evolution probably depends primarily upon the original mass of the star. If it is less than approximately 1.4 times the mass of the sun, the red giant will most probably become a nova undergoing a series of explosions as the central core shrinks more and more, throwing off a portion of its mass every time the temperature reaches a critical point. Finally, when all the fuel for nuclear reactions is used up, the star contracts further and further into a very dense "degenerate" gas under such high pressure that atomic nuclei as such do not exist. The star now becomes a "white dwarf" lit only by the energy of contraction. When this energy is gone and no further contraction is possible, it becomes a dark cinder in space.

If, on the other hand, the mass of the giant is above 1.4 times that of the sun, it has been suggested that now the central pressure may rise to a level of 10^{22} pounds per square inch and that, at this pressure, a mighty helium bomb is formed and all the heavier elements up to and beyond uranium are created by a large variety of transmutations and exploded outward into space. According to the British astronomer Hoyle, this accounts for the existence of all the heavier elements scattered widely as debris from ancient supernovae explosions. (As we shall see, however, a more acceptable theory of the widespread existence of the heavier elements has been suggested by another physicist, George Gamow.)

Such supernovae explosions are fairly rare. It is estimated that they occur about once every 400 years in any one galaxy of stars. Many of these have been noted in galaxies external to our Milky Way, but the last two observed in this galaxy were the supernova of 1572—which so excited Tycho Brahe that it provided the motivation for his devotion to astronomy—and the supernova of 1604, studied by Kepler. Previous to this, a supernova was recorded in 1054 by Japanese and Chinese documents although, curiously, no mention is made in any records of the Western Hemisphere.

In a supernova the luminosity of a star is increased many millions of times. If one of these were placed at about the distance of the nearest star, it would be so bright that it would shine spectacularly even in broad daylight, since it would be about 14 times as bright as the moon.

The Expanding Universe. One of the most significant discoveries of astronomy in this century has been that of Hubble who found in 1929 that the spectral lines of galaxies shift toward the red end of the spectrum and that the amount of this shift is proportional to the distance of the galaxies as measured by their luminosity. The most plausible explanation of this shift, now almost universally accepted, is that all the galaxies are receding into space at very high velocities and that their velocities of recession are nearly proportional to their distances. Out of these observations has come the concept of the "expanding universe." Any cosmological theory of structure or origin must take this conception into account.

To understand the relationship of the spectral shift and the velocity of recession, it is convenient to refer to a similar relationship that has long been known

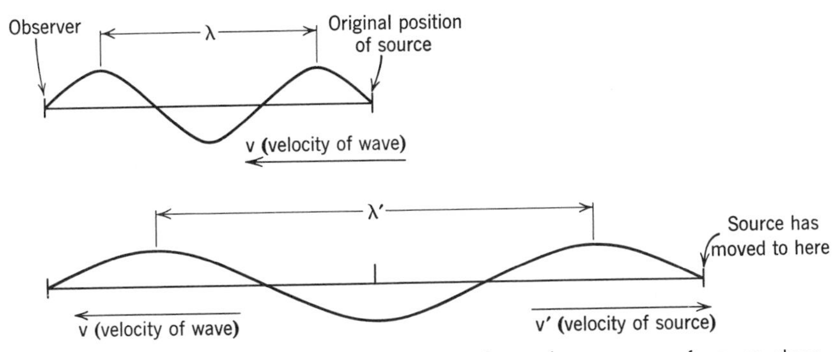

Fig. 24.6. Doppler's principle. When the source of sound moves away from an observer, the wavelength is lengthened, as shown from λ to λ'.

to exist with respect to sound, a relationship called the "Doppler effect." Most of us have noted this effect when we listen to the change in pitch that occurs when a sounding body, such as a locomotive whistle or an automobile horn, approaches and then recedes from us. The pitch is higher than normal as the sounding object approaches, and drops to lower than normal as it recedes.

The explanation can be illustrated by Fig. 24.6. The source of sound gives off waves at a constant frequency. The number of waves reaching the observer in a given period of time is a measure of this frequency. If the source of sound is stationary, the wavelength is related to the velocity by the equation

$$\lambda = \frac{v}{f}$$

where f is the frequency and v the velocity of the wave.

But if the source of sound is moving away from the observer with a constant velocity v', since it is emitting the same number of waves per second and since during a second of time it travels over a certain distance, the waves emitted and eventually reaching the listener must be spread out over this longer distance. The wavelength must consequently be lengthened. (This also implies that the frequency of pitch also must be lowered if the velocity of the wave is constant.) From this analysis, it follows that the ratio of wavelengths is

$$\frac{\lambda'}{\lambda} = \frac{v + v'}{v} = 1 + \frac{v'}{v}$$

$$\frac{\lambda'}{\lambda} - 1 = \frac{v'}{v}$$

or

$$\frac{\lambda' - \lambda}{\lambda} = \frac{v'}{v}$$

Expressed in percentages, this implies that a percentage increase in wavelength is equal to the percentage ratio of the velocity of the motion of the source to the velocity of the wave. This equation must apply to all wave propagation. For example, a shift in a given spectral line from its laboratory measurement, for instance 4.210×10^{-10} m, to what it is when measured in the light from a galaxy, for instance 4.214×10^{-10} m, represents a shift of 0.1%. Such a shift, therefore, implies a velocity of the galaxy of about 186 miles per second (that is, 0.1% of the velocity of light).

Hubble and another astronomer, Milton Humanson, carried out a long study of galaxies from 1928 to 1936, with the 100-inch Mount Wilson telescope. Since 1951, the studies of more distant galaxies have been continued with the 200-inch telescope at Mount Palomar. The relationship found between the receding velocities and distances is nearly linear for the closer galaxies. For the most distant galaxies (where the observations are much more difficult and where the calculation of distances entails more assumptions), a slight upward departure from the linear relationship appears to take place. For example, galaxies that are more than a billion light-years away appear to be receding at about 6000 miles a second faster than in direct proportion to their distance. If this becomes definitely confirmed and the interpretation of an "expanding universe" is correct, it would mean that the universe was once expanding faster than it is now. We shall return to this problem in our discussion of cosmologies.

4. MODERN THEORIES OF COSMOLOGY AND COSMOGONY

Before reviewing briefly some of the contemporary theories about the structure and origin of the universe, we make some general comments.

First, we detect again how closely interwoven are "observed facts" and theoretical interpretation in science. Consider, for example, how indirectly distances and motions are measured in interstellar space. All we can actually observe are a few lines and a few smudges on photographs taken at intervals over a period of hours or days. Even the ancient determination of the distance and the motion of the moon, sun, and planets depended upon a wealth of theoretical assumptions and interpretations. (For example, in applying triangulation methods it must be assumed that light travels in a straight line, and that the base line is known with sufficient accuracy, which in turn requires a knowledge of the earth's curvature, etc.) With the development of modern instruments and with a greater dependence on a whole series of relations prescribed by equations —sometimes theoretical, sometimes empirical—the number of assumptions made has become much greater and hence the vulnerability of any measurement is also increased. A dramatic warning of how cautiously scientific conclusions must be taken, particularly in the field of astronomy, is illustrated in a recent revision of all estimates of galactic distances, which were almost tripled as a result of careful studies by Walter Baade and others. It will be recalled that Shapley first developed what appeared to be a reliable estimate of the absolute luminosity of distant stars and variables on the basis of the relationship between the period of Cepheid variables and their absolute luminosity. In 1953, however, Walter Baade and his associates found that there are two distinct types of Cepheid variables, each with its own period-luminosity equation. All earlier estimates of distances were thus proven to be mistaken, since these were based on an average period-luminosity equation which was applied to all Cepheid variables without discrimination. The consequent revision not only changed the estimate of the distances of galaxies significantly, but indirectly changed other estimates such as galactic velocity-distance relationship, and hence the "minimum" age of an "expanding" universe. For example, Hubble, the discoverer of the linear relationship between the red-shift and luminosities, had remained skeptical about whether this relationship proved that the universe was expanding. Such a conclusion, under Shapley's older scale of distances, had implied that if the universe were expanding, the age of galaxies could be no more than about 2 billion years, although the evidence from radioactive rocks indicated the age of the earth to be much older. Under Baade's revision of distances, the recession velocities now implied that the universe was about three times older. The theory of an "expanding universe" again became more plausible. This example is cited to emphasize the caution with which estimates (particularly of astronomical quantities) should be regarded. Hubble is appropriately quoted by Jagjit Singh in this connection: "Our knowledge fades" and we are obliged "to search among the ghostly errors of observation for landmarks that are scarcely more substantial."[3]

Professional scientists are themselves aware of the many factors and assumptions that enter into the various quantitative estimates that underly their theories. Unfortunately, when these theories are set before the public, the tentativeness of their conclusions, based upon long chains of reasoning intermingled with observations, is easily overlooked; and alas the nonscientist, impressed or enchanted by the conclusions, is not in a position to use the same critical judgment and caution that is second nature to the scientist. A bizarre scientific theory (often marked by imaginative ingenuity) may capture the public's attention and be-

[3] Quoted by Jagjit Singh in his scholarly but popularly written summary of cosmological theories: *Great Ideas and Theories of Modern Cosmology,* Dover, New York, 1961, p. 101. All serious nonexperts of cosmology are urged to read this book. It provides a critical, authoritative and easily readable account of the competing modern theories of cosmology. This book should be read as a supplement to those excellent books by Hoyle, Gamow, and Bondi, each expounding a particular theory and a particular point of view for the general reader.

lief without any understanding of the tenuous nature of the reasoning upon which the theory is based.

The second general observation has almost an opposite import to the first, that is, that although caution and tentativeness must be the handmaidens of scientific inquiry, the siren slogan *ignorabimus* (We *shall* be ignorant) must be repudiated. Auguste Comte, in an early edition of his *Positive Philosophy,* cited the chemical constitution of the sun as an example of what can never be discovered by any scientific method. Today we recognize that his dictum regarding what is "beyond knowledge" has been contradicted by the development of spectroscopic analysis. We probably know more clearly the over-all chemical composition of the sun than that of the earth. That which seems completely inaccessible to observation, reason, or scientific inquiry in one generation may, in another, become most familiar. There can be no set limits or bounds to the reach of science. There may be, indeed there have been, many pauses in the development of one branch of science in which man's knowledge seems to have reached a natural limit. But, sooner or later, new combinations of facts and new theories inevitably develop, and the search takes on new life.

So, in 1917, when Einstein began publishing his work on general relativity, an upsurge of interest in cosmology took place and a new look was given the universe whose form seemed so precisely outlined in the mechanics of Newtonian physics. Einstein's general theory raised the possibility of an alternative way of looking at the universe. With the introduction of a space-time metric, which might well be non-Euclidean or "curved" because of the existence of matter, the problem again arose as to what kind of a large-scale universe we are living in. One soon arrives at several alternative formal possibilities, a choice among which can only be determined by accumulating much more data than are now available from observation, particularly with respect to the density and distribution of matter. If the universe has a non-Euclidean space-time structure, its curvature, which is determined by the density of matter, may be either positive or negative. If it is negative, the universe remains infinite but shaped as a four-dimensional hyperbolic ("saddle-shaped") surface extending without limit in all directions. If the curvature is positive, it will be bent back upon itself and be finite in one or more directions. Einstein, in his earlier papers, was inclined to the view that the universe had a cylindrical structure. The Dutch astronomer, de Sitter, soon afterward propounded the theory that the universe was spherical.

It is always a little difficult to explain to the nonmathematician what is meant by curved space-time. Repeating portions of earlier sections of this book, the conception essentially comes to this: space and time are no longer empty containers without internal structure but are fused together, and this space-time continuum takes on physical properties which affect the paths of moving bodies and light rays. A light ray, for example, now follows what is termed a "geodesic," that is, the shortest path between two points. If the universe is non-Euclidean, such a path may well be other than a "straight line" continuing forever into infinity. It may turn back upon itself if the curvature is positive and the universe is spherical, much as a person traveling the surface of the earth over a geodesic path will eventually return to his original position (the shortest path he can pursue is a great circle). No parallel lines are now possible because all geodesics or great circles intersect.

On the other hand, if the curvature is negative and the space-time continuum is hyperbolic, an infinite number of "parallel" lines is possible (that is, we can draw an infinite number of lines through a point outside a given line without any of them intersecting the given line.)

To make this conception clearer we use the two-dimensional analogy (Fig. 24.7), as is commonly done to illustrate these alternatives.

It should be emphasized that in Fig. 24.7 we are looking at a two-dimensional surface embedded in a three-dimensional space so that the curvature can thus be given a visual interpretation. But we cannot so represent the curvature of space itself, much less space-time, since we cannot "step out" into a fourth dimension and "see" the curvature. Instead, we need an intrinsic mathematical expression for this curvature and, if possible, a method of measuring it.

This problem can perhaps best be understood if we use a simple approach suggested by H. P. Robertson in an essay on "Geometry as a Branch of Physics."[4]

First, Robertson asks how we would determine the curvature of the earth on the basis of surface measurements alone (that is, without going outside the surface of the earth as we do when we use, for example, light rays in space to reveal the earth's curvature). If a circle is drawn on the surface of the earth it is found that, for large radii, the correct formula for the perimeter and area of such circles is given not by $L = 2\pi r$ and $A = \pi r^2$ but, rather, in a manner dependent upon r, the radius of the earth, or upon its curvature K which is defined as $K = 1/r^2$.

The new equations, which are derived from spherical geometry, are

$$L = 2\pi r\left(1 - \frac{Kr^2}{6} + \cdots\right)$$

$$A = \pi r^2\left(1 - \frac{Kr^2}{12} + \cdots\right)$$

[4] This essay, which involves very few and elementary mathematical equations, contains a simple and clear discussion of the curvature of space. It is one of the essays contained in *Albert Einstein: Philosopher-Scientist,* edited by Paul A. Schilpp, Library of Living Philosophers, Open Court Publishing Co., La Salle, Ill., 1949. Reference, however, should also be made to the philosophical criticism of this essay contained in Milton K. Munitz's *Space, Time and Creation,* Free Press, 1957, reprinted as Collier Books, Crowell-Collier, New York, 1961, pp. 114–117.

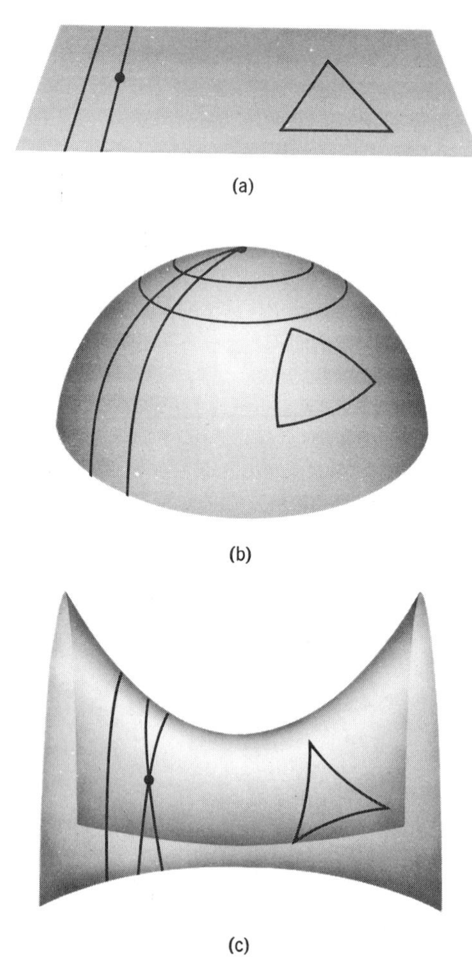

Fig. 24.7. Two-dimensional analogy, picturing three types of curvature. (A) Plane surface: (1) curvature is zero; (2) only one "geodesic" (shortest distance between two points, that is, a "straight" line in case of a plane) can be drawn parallel (that is, not intersecting) to a given straight line through a point not on line; (3) sum of angles of triangle = 180°; and (4) surface extends to "infinity." (B) Spherical surface: (1) curvature is positive; (2) all geodesics intersect (the shortest distances are arcs of great circles), therefore there are no "parallels"; (3) sum of angles of a triangle is greater than 180°; and (4) surface closes upon itself and is finite. (C) Hyperbolic surface: (1) curvature is negative; (2) many parallels can be drawn through a point not on a given geodesic without intersecting it; (3) sum of angles of a triangle is less than 180°; and (4) surface extends to infinity.

Similarly, for a three-dimensional space which is curved, mathematicians arrive at the following equations for the surface and volume of a sphere:

$$S = 4\pi r^2\left(1 - \frac{Kr^2}{3} + \cdots\right)$$

$$V = \frac{4}{3}\pi r^3\left(1 - \frac{Kr^2}{5} + \cdots\right)$$

It is now proposed that we look at nebulae and count their number at various distances from the earth. Robertson points out: "If all the nebulae were of the same intrinsic brightness, then their apparent brightness as observed from the

Earth should be an indication of their distance." Robertson, in illustration (but recognizing the many modifications required for actual application to astronomical data), goes on to say that:

Since the practical procedure in determining d the distance of a star is equivalent to assuming that all this light (i.e., the light from all equally distant nebulae) lies on the surface of a Euclidean sphere of radius d, it follows immediately that the relationship between the "distance" d used in practice and the distance r dealt with in the geometry is given by the equation

$$4\pi d^2 = 4\pi r^2\left(1 - \frac{Kr^2}{3} + \cdots\right)$$

whence by approximation, (by solving for r in place of d and neglecting higher terms)

$$r = d\left(1 + \frac{Kd^2}{6} + \cdots\right)$$

This finally yields, as Robertson points out,

$$V = \frac{4}{3}\pi d^3\left(1 + \frac{3}{10}Kd^2 + \cdots\right)$$

and now on plotting N against inferred "distance" d and comparing this empirical plot with the formula, it should be possible operationally to determine the curvature K.[5]

When Robertson wrote his essay, the empirical plot seemed to indicate that the number of nebulae increased faster than the third power of d and, therefore, he concluded that the curvature of space is positive. Presumably, this would mean that a light signal would ultimately return to us from the opposite direction after having traversed the limits of the universe.

However, the discussion presented above is an extreme oversimplification, and is given only to illustrate the general approach. Many other considerations have to enter into any calculation of curvature. All that we can conclude on the basis of the most recent observations, and this only with *great tentativeness,* is that, if anything, the universe probably has a negative curvature and extends infinitely in all directions. (See Table 3.)

It was along these general lines that Einstein developed his cosmological theory. His general relativity equations appeared to work in the neighborhood of the sun and explained, as discussed in Chapter 19, certain observations that could not be explained by classical physics. Einstein attempted to extend the same field equations of relativity to the universe as a whole. In doing so, however, he found that he could not develop a solution that guaranteed stability to the universe. He had to introduce a certain cosmological constant, which seemed to many to be quite arbitrary and which was later abandoned by most theorists.

[5] *Albert Einstein, Philosopher-Scientist,* edited by Paul A. Schilpp, Library of Living Philosophers Inc., Evanston, Ill., 1949, Open Court Publishing Co., La Salle, Ill.

TABLE 3. ALTERNATIVE POSSIBILITIES FOR SPACE

	Curvature	Surface analogy	Sum of angles of triangle	Parallel line through a point outside a line	Increase of number of nebulae with distance
Euclidean	$K = 0$	plane	$= 180°$	one only	proportional to d^3
Non-Euclidean (positive)	$K > 1$	spherical	more than 180°	none	more than proportional to d^3
Non-Euclidean (negative)	$K < 1$	hyperbolic	less than 180°	more than one	less than proportional to d^3

Einstein's analysis inspired several other variations, among them theories of an exploding or oscillating universe, in particular those developed by Friedman and Abbé Lemaitre. Lemaitre postulated that the whole universe had at one time been contracted into a small sphere of enormous density. Because of the density, a tremendous explosion took place expelling the conglomeration of matter in all directions. According to this theory, Hubble's finding of the recession of the galaxies is quite understandable. The most distant nebulae, consisting of condensing collections of matter, are those which were originally expelled at the highest velocities and hence have traveled the farthest since the original explosion.

This theory, sometimes referred to humorously as the "big-bang" theory of creation has been modified and extended by George Gamow, who brought to bear not only the mathematics of relativity but a study of how the various atoms of the universe could have been formed. Gamow's theory can be summarized as follows.[6] From the data on the red-shift gathered by Hubble and his co-workers, we can calculate that the present expansion of the universe must have begun about seven billion years ago. (Even this estimate, which is a revision of earlier estimates, may still be too short. Allan Sandage, who has recently been conducting an exhaustive study of distant galaxies, now suggests that the universe may well be over 13 billion years old.) At this "creation date" of the universe, Gamow calculates that all matter must have been squeezed into a radius not much larger than about eight times the sun's radius. The original

[6] The treatment here is not intended as a substitute for reading the excellent text itself: George Gamow, *The Creation of the Universe,* The New American Library, New York, 1957.

state of this condensation of matter was a "hot nuclear gas," according to Gamow, of unbelievable density and at temperatures of many billions of degrees, with neutrons, protons, and electrons all mixed together into the enormous primeval atom. This mighty cauldron began to expand explosively and, as it did so, the temperature dropped rapidly. But, in those few early seconds, the temperatures were high enough and the neutron flux great enough to produce all the atoms up to the heaviest. The whole process of atom building is calculated as having taken less than one hour's time. Upon this theory, Gamow, Alpher, and Herman calculated what the relative abundance of all the atoms should be, and this agrees with the abundance of atoms now observed to exist in the universe from spectroscopic studies.

Such is one of the evolutionary theories of the origin of the universe. Another quite different theory has been proposed by H. Bondi and modified by Hoyle.[7] Under their theories the universe is conceived of as being in a "steady state" instead of in a state of evolution. The starting point is the methodological assumption of a "cosmological principle," which is essentially the assumption that the universe must remain the same both in space and time. Although galaxies grow and die and constantly recede from one another, they are continuously being replaced by new galaxies. The universe as observed from any location and at any time remains essentially unchanged except for momentary local changes. The theory asserts that as the galaxies recede from one another the space between them becomes emptier and emptier until it reaches a critical point. When this point is reached a hydrogen atom automatically comes into existence to fill the void. When enough new hydrogen is created, it collects and a new galaxy is formed after billions of years. The mean density of matter in the universe is calculated by Hoyle as remaining at approximately 5×10^{-27} g/cc in accordance with his equations, which requires that new hydrogen must be created at the rate of about 10^{32} tons/sec in the whole observable universe. However, the observable universe is so vast that this implies the appearance of a single hydrogen atom in a volume as large as the Empire State building approximately every century.

Part of the evidence for this alternative theory as against the big-bang theory of Gamow was found by Hoyle in his interpretation of the curve of the abundance of elements. Hoyle emphasized the fact that in the sun and the stars the heavier elements are not present in the same proportion as on the earth. If they were all created at the same time, he argued, their abundance should be uniform throughout the universe. Hoyle proposed as part of his theory that the sun once had a companion star, and was thus part of a binary system. This larger companion star eventually exploded as a supernova. Before its final explosion,

[7] H. Bondi, *Cosmology,* Cambridge Press, 1952, and F. Hoyle, *The Nature of the Universe,* Oxford Press, 1949.

all the heavier elements were found at the high temperatures that then existed in the supernova's interior. The explosion scattered this material far and wide but left a gaseous debris out of which the earth and planets were formed.

There is still another rather intriguing type of evolutionary theory of cosmology, which has several variants—each based upon the existence of certain important ratios, which dominate modern physics. We give a few examples of these ratios, which have impressed Eddington, Dirac, Jordan, and others. The ratio of the coulomb force between two particles (protons or electrons) to the force of gravity between them is e^2/GM_eM_p, which comes out to be about 2×10^{39}. The ratio of the cosmic time (based on Hubble's recession law) to the unit of nuclear time is, as we saw earlier, about this same order of magnitude. On the other hand, other important ratios such as the total estimated mass of the universe to the mass of elementary particles is of the order of magnitude 10^{78}, or the square of the first type of ratio. Eddington, Dirac, and more recently, P. Jordan have suggested that the appearance of these ratios in the universe can hardly be due to pure chance, but are somehow built into the very structure of our universe and call for an explanation. Furthermore, the suggestion is made, particularly by Jordan, that some of the "constants" themselves may be slowly changing. For example, it is suggested that the "constant" of gravitation (as the constant appears in relativity theory) $K = 8\pi G/c^2$, where G is the Newtonian constant of gravity and c is the velocity of light, may itself be a variable quantity and may be slowly decreasing at the rate of 10^{-8} or 10^{-9} parts per year. Jordan has applied this hypothesis to the explanation of geophysical changes. According to his analysis, there has been a very slow expansion of the earth, which accounts for the massive geological changes in the contour of the earth's surface. Jordan accounts for this expansion on the basis of a weakening of gravitational forces. He goes so far as to assert:

> I think our present knowledge of the earth, interpreted in the ways indicated makes the correctness of Dirac's hypothesis an established fact, including also the puzzling circumstances that during the formation of the Earth, K (i.e. $8\pi G/c^2$) must have been considerably greater than now.[8]

We call attention in the above quotation to the phrase "makes the correctness of Dirac's hypothesis *an established fact.*" Scientists are delightfully human and, like Pygmalion, they can become just as enamored of their own creations as the rest of us. Upon finding the slightest local evidence to support a theory to which they are already committed, they often cast caution aside. However, it is perhaps only by such commitment that satisfactory theories are eventually

[8] P. Jordan, "Geophysical Consequences of Dirac's Hypothesis," *Review of Modern Physics,* October, 1962, p. 600.

arrived at. If Kepler had begun to doubt his theory of the "harmony of the spheres," he never would have had the persistent motivation to find the three laws now named in his honor.

Even bolder—in fact, almost outlandish and bizarre—appear a few other scientific speculations about the origin of the universe. For example, there is the theory propounded by Goldhaber, as a modification of Gamow's theory, that out of the primeval exploding atom there was born not only a universe of matter as we know it, but a twin universe, an "anticosmos," composed of antiprotons, antineutrons, and antielectrons, all obeying the same but symmetrically reversed laws that are followed by the particles of matter in the universe in which we live. According to this theory the primeval atom of Gamow contained both nucleons and antinucleons, and when it exploded, seven billion years ago, the twin universes were shot apart, one composed of matter and the other of antimatter. Thus is preserved the over-all principle of particle-antiparticle symmetry and a reason is now given for the apparent lack of symmetry between the mass of the proton and that of the electron. We mention this theory as an example of the heights to which the imaginative and speculative flight of science may reach. Such theories should not be condemned, out of hand, for many such speculations in the past have eventually proved their mettle. They must, however, always undergo the laborious, logical, and observational tests that all scientific theories must undergo before they can be taken with any great seriousness.

Finally, let us turn from these broad cosmic considerations to the somewhat more limited although closely related problem of origins. Out of galaxies, stars and solar systems are born. What is the mechanism of their formation? This problem also gives rise to a wide variety of theories, but it is more limited than the problem of cosmic origin and may be solved in the near future. Certainly both the problem and the nature of the evidence required for its solution have now been stated with precision, and considerable progress has already been made.

It is generally assumed that out of a vast cloud of "dust" particles (mostly hydrogen atoms), galaxies begin to condense (Fig. 24.8). Out of these are generated stars and solar systems. The mechanism of condensation is influenced by a number of factors, each having an individual role in the process, which may be given a different emphasis depending on the theory advanced: (*1*) the play of gravitational forces exerted by each particle upon every other particle; (*2*) the laws of conservation of momentum and of energy, the first of which affects the type of linear and rotary motions and consequent turbulent eddies which arise, the second of which leads to temperature changes and ultimately to the types of nuclear reactions which arise; (*3*) the interaction of radiant

Fig. 24.8. "Whirlpool" galaxy, a spiral nebula in the constellation Canes Venatice.

energy and matter, giving rise to a wide variety of phenomena including radiation pressure; and (*4*) the magnetic and electrical forces created by the high velocities of ions and charged particles moving in space.

Let us begin by assuming a vast cloud of cosmic dust, whatever its origin, extending over unimaginable distances. Inexorably, through the operation of the laws of physics, we know that this cloud will begin to condense because of gravitational forces between the particles. (If there is any radiant energy present in the universe at this state, it will only accelerate the process of condensation.) As the condensation proceeds, eddies are formed producing centers of condensation, each with its own angular momentum (according to Whipple, Weizsäcker, and Kuiper), or the condensation is influenced by magnetodynamic forces (according to Alfven).

Out of the large agglomerations of condensing material, collapsing at differential velocities toward their respective centers, are born the nuclei of stars in singlets or pairs or triplets, each condensation impelled by gravitational forces and following the familiar laws of momentum. The condensation and "collapse" toward centers become rapid, but always subject to the laws of linear and angular conservation of momentum and the electromagnetic forces generated by the motion of the charged particles. Streamers of gaseous particles are left behind as turbulent eddies. While the collapse and condensation of the major portion of the mass continues at a more and more accelerated pace, vast frictional energy is released, and the temperature of the central mass rises to a higher and higher degree until the nuclear reactions at last begin and a star is born. Equilibrium now becomes established. The inward-acting gravitational pressure becomes balanced by the outward pressure of the energy radiating from the internal nuclear reactions. The residue streamers, meanwhile, are themselves condensing into spherical masses, picking up the remaining debris of the streamers left by the central body or bodies. They sweep the surrounding spaces clean of almost all the remaining "dust," and nothing is left in the end but a central body (or bodies) and a group of smaller spheres.

According to this general explanation of the origin of the solar system (which is a variant of the older nebular hypothesis of Kant, now developed, however, with considerable mathematical attention to the problem of the angular momentum of its parts), the materials composing the planets originated in the same cloud of gas out of which the sun is formed. We have already mentioned Hoyle's alternative theory, that is, that the materials composing the earth are formed out of the residue left by an exploding supernova which has now disappeared. Hoyle's theory does, however, include a condensation process out of which the planets are formed much the same as that outlined above.

There is another alternative and very different theory of the origin of the solar system. According to this theory, advanced by Otto Schmidt and Boris Levin, the planets were formed not out of the same material of either the sun or of the companion supernova, but out of external matter swept up as a trailing eddying mass by the sun as it pursued its course through intergalactic dust. The greatest advantage to this theory is that it is in closer accord with recent studies of Harold Urey on the chemistry of the earth, which indicates that the earth never was a hot molten mass, but could, at most, have been subject to temperatures no more than a few hundred degrees centigrade. Urey bases this conclusion on the relative abundance on the earth's surface of volatile elements such as boron, which would have boiled away at high temperatures. To offset this significant advantage of Schmidt's theory, however, there are other serious difficulties. In particular, it has been shown that in the process of picking up an accumulation of dust, the sun would have to be traveling very slowly, in which case it would pick up such a large mass of material as to turn it into a red-giant star. To get around this difficulty, Schmidt proposed that while the sun was picking up its accumulation of eddying gases, another star was in proximity and prevented too great an accumulation by the sun. The theory, therefore, becomes less plausible because of the extraordinary improbability of such a combination of appropriate circumstances. Again, we must end our quest for understanding even of the local problem of the origin of our solar system with no single convincing theory. A firmer decision awaits more evidence, which may very well be forthcoming in the next decade as interplanetary probes are carried forth with greater frequency and to greater depths.[9]

We shall pursue the perennially intriguing problems of cosmology and cosmogony no further. We have stated that as a rational science this branch of human inquiry is still in its infancy. Both the nature of the problems discussed and the fact that the science is so young contribute to its fascination. It may be somewhat frustrating to review so many competing theories, each one with its difficulties and advantages, but doing so makes us more aware of how science does develop. It is not a mere accumulation of observations. Interpretation is necessary at every step of the way if the observations are to be given a structured unity. The very statement of theory almost inevitably suggests the direction in which to look for further evidence for its verification or disproof. Scientists would hardly know what kind of instruments to pack in the space-probing rockets that are being sent aloft unless they had rather specific questions to which they wished to find answers. Sometimes the results may be so completely new and surprising as to assume the nature of "accidental" discoveries. And so they sometimes are, in a sense; but, almost always, these accidental discoveries are found during the process of an investigation guided rather clearly by some precisely stated question.

No mention has been made in this chapter about the possibility of life elsewhere in the universe, because so little can be said that goes beyond the speculations contained in the many articles in newspapers and periodicals. The conditions surrounding organic life, as we know it, are so extremely limited as to temperature and chemical environment that it appears quite improbable that any but the most simple form of vegetable life could exist on any other planet of our solar system. On the other hand, as many astronomers have pointed out on a strictly statistical basis, there is a tremendously high probability that among the billions upon billions of stars in the observable universe, many other solar systems exist with planets having a history, structure, and chemical composition closely approximating that of our own earth. If, as the biologists tell us, organic chains or molecules can evolve out of the proper combination of chemical and physical conditions, it is probable that living forms have evolved on thousands or possibly millions of other planets. Up to now, the nearest stars are so distant as to make a direct observation of any outside planetary system impossible. There has been detected evidence of dark unseen companions of other stars, very much

[9] For a more detailed review of the theories of the origin of the solar system, reference is again made to Jagjit Singh's *Great Ideas and Theories of Modern Cosmology,* previously cited in footnote 3. The present author is indebted to this critical review of cosmological theories for many suggestions.

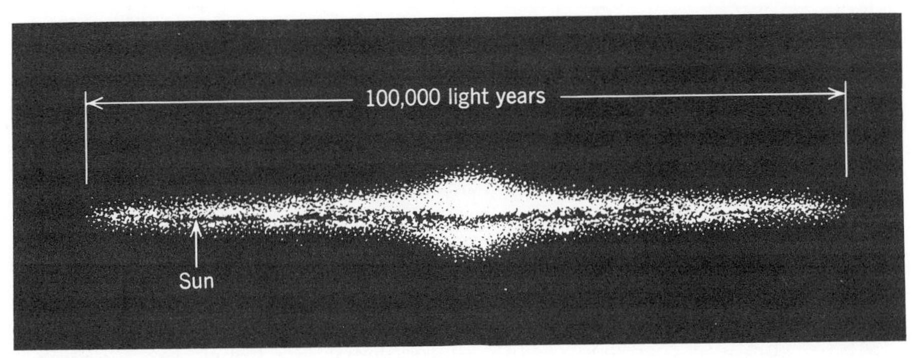

Fig. 24.9. Our galaxy with its 10^{11} stars, with the sun rotating about the center at a speed of about 130 miles per second. The "Milky Way" is the group of stars we see as we look along the diameter of the galaxy.

smaller than the mass of the sun, as inferred from the motions of certain stars. Their existence lends further plausibility to the argument from probability that other planetary systems with living populations do exist.

5. NEW FRONTIERS OF ASTRONOMY: RADIO, X-RAYS, "COSMIC," AND NEUTRINO WAVES; QUASARS AND PULSARS

Astronomy and cosmology are in a ferment today because so many unexpected and still inexplicable discoveries have been made during the last two decades. Almost every month a new "signal" is reported as coming in from outer space. The theories accounting for the energy, frequency, and waveform of these electromagnetic waves only recently detected have their brief day and are discarded. New discoveries excite and baffle the astronomer and physicist.

The main reason for these new and unexpected discoveries is that the "window" into outer space has been opened wide. The atmosphere is opaque to most of the electromagnetic spectrum. The range of frequency that the optical telescope can handle is small, but today with instruments mounted on satellites outside the earth's shielding atmosphere, we can "view" these other radiations pouring from star and galaxy. On the other end of the electromagnetic spectrum, where the optical telescope is "blind," radio telescopes have been developed which sense a wider but lower range of frequencies, fully as important, as those of visible light. The longer radio waves are not as easily scattered by clouds, dust, and moisture, nor are they "drowned out" by the "noise" of sunlight.

The first radio telescope was developed by Grote Reber, an American engineer, in 1937, following upon the first discovery of radio-frequency radiations from stars by another American, Karl Jansky, in 1931, who was investigating "noise" at the time. But the real developments have taken place in the 1950's and 1960's.

When these telescopes were turned skyward, day and night, and impervious to clouds and weather, what a store of surprises were uncovered for astronomers! Not only do the "visible" stars emit radiation, but there are found many sources, invisible to the telescope, from which radiation seems to outpour large quantities of energy. The interpretation of these sources of energy and the quantity of energy remain great puzzles to the astrophysicist.

Most dramatic of these sources of radiated energy are the *quasars,* the clumsy shortened name for "quasi-stellar-objects," (Fig. 24.10).

The first hint of the extraordinary nature of these astronomical objects began with an attempt to find a *visual* identification of a source of radio emission. Astronomers at Cambridge, England, were able, after great effort, to identify the source as a 16th magnitude "star" catalogued as 3 C-48 as a radio emitter. Soon

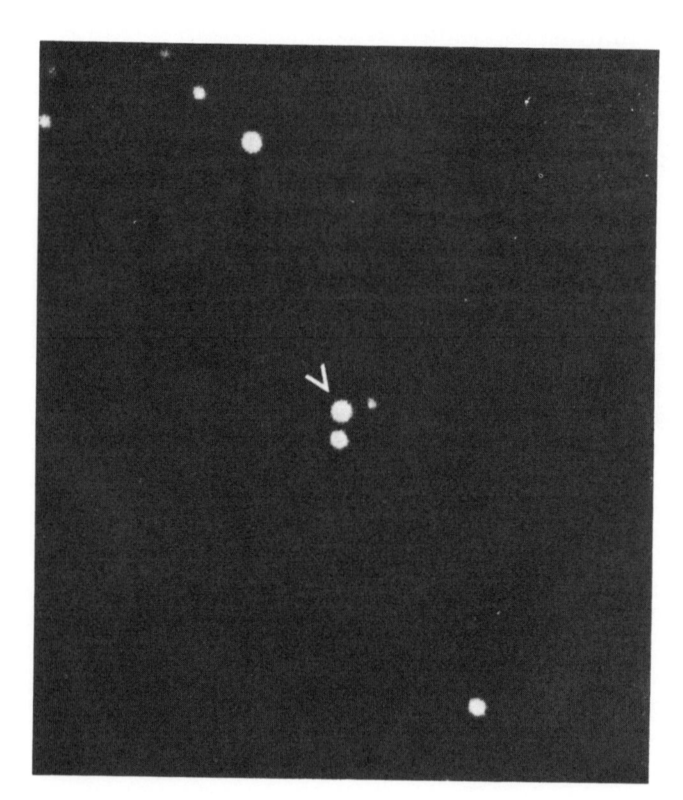

Fig. 24.10. Quasi-stellar radio source 3C.

afterwards in 1963, M. Schmidt at the Palomar observatory began to observe a similar source of radio emission, 3 C-273. When the spectrum was examined Schmidt made a startling discovery. The "red shift" of the spectrum was an amazing 16 percent. If Hubble's law and the Doppler principle held, this would place the object at a distance of some 6 billion light years!

No ordinary star could have a luminosity sufficient to be seen at this distance. (Until 1962, most astronomers had maintained that these radio emitters must be within our own galaxy and no more than a hundred or so light years away.) Are these objects distant galaxies then? Hardly, and for two reasons. In the first place, a galaxy, with its share of spreadout stars has a discernible angular diameter and does not appear as a star-like point. In the second place these quasistellar objects vary in luminosity in the shorter wavelengths as much as 20 percent in one or two years. There is a periodic variation, in other words, which could not reach over a greater diameter in this period more than a few light years. The conclusion reached by most astronomers is that these objects are some kinds of distant "stars" of tremendous mass—millions of times greater than the mass of the most massive stars heretofore known. Also they must be in the process of undergoing some extraordinary transformation in which energy is being emitted at a rate millions of times greater than it is emitted even by any known supernova.

The release of such energy over a short space of time is baffling. It is far too great for an explanation by nuclear reactions. Possibly a kind of catastrophic gravitational collapse may account for the energy, but there are difficulties here too. ("Gravitational collapse" is a term used to explain a catastrophically rapid concentration of mass about a center. Potential energy is changed into kinetic energy of the particles rapidly accelerating towards the central point and the kinetic energy in turn changes into thermal energy until a temperature is reached at which electromagnetic radiation occurs.) The condition for such a "gravitational collapse" is the presence of a sufficiently high concentration of matter. Possibly such an explanation for the quasar's emission of energy is adequate. But it is doubtful because the calculated acceleration required is too high to account for the rate of emission of energy.

Whatever ultimate theory will be found satisfactory to account for this extraordinary emission of energy is bound to affect modern cosmology. (Already Fred Hoyle, author of the "continuous creation" theory of matter, has suggested the possibility that instead of matter appearing as in his original suggestion, in small quantities, perhaps it appears in the form of huge and concentrated lumps which by some process transform into a quasar.)

Another mystery revealed in 1966 are the "pulsars." From some source in space come pulsed radio emissions of great and unaccounted for regularity. It was suggested by some when they were first detected that they might be signals from some intelligent life on a planet, but that idea was soon discarded. Instead a tentative explanation has been building up on the basis of Thomas Gold's theory proposed in 1968, namely, that the pulses of energy arise out of the radiation from a rapidly rotating "neutron star."

A neutron star, unobserved by any emitted signal until 1967, was hypothesized as the last stage of a dying star which has used up all its fusion and fission energy. It could be, it was thought, the final remnants of a supernova explosion. Having used up all its nuclear energy, radiation pressure from nuclear reactions would fall to zero, and the star would suffer a gravitational collapse resulting eventually in an extraordinary density of a sphere, possibly no more than 10 km in diameter and consisting only of neutrons. Such a sphere of matter, so compressed (thousands of tons per cubic inch) would no longer resemble ordinary matter at all. But out of such a neutron star, calculations show that the remaining magnetic fields in the star's interior could still generate radio waves of energy. If the star were rotating and emitting energy from only parts of its surface, the radio waves would be emitted in pulses much as a beacon light from a lighthouse emits a periodic flash in any one direction. Furthermore, with the emission of such energy, the rotations should slow down and the pulses should be gradually reduced in frequency. Such was Gold's theory, which in 1969 appears to have received some confirmation.[11] There was a supernova explosion in 1045 A.D. Its remnants are found in what is called the crab nebula in the constellation Taurus. From this point radio pulses have emanated, and surprisingly even light flashes have been recently identified (1969). Significantly the frequency of these pulses has been found to be decreasing slowly in agreement with the theory that they are emitted by a rotating neutron star which is losing angular momentum. Careful reexamination of some of the other 27 known pulsars reveals that there is a minute but perceptible slowing down of frequency, and the theory that the final stage of a star is a neutron star seems well on its way to confirmation.

In 1965 Penzias and Wilson of the Bell Telephone Laboratories discovered an isotropic cosmic background radiation in the millimeter region of the electromagnetic spectrum. Preliminary measurements suggested that this radiation was characteristic of a blackbody at 3°K. Such properties are just what one would expect if the radiation were a remnant of a cosmic expansion which began about ten billion years ago. If such an interpretation were valid, it would probably be the death-knell of the steady-state theory. However, only recently, Friedman and his associates at the U.S. Naval Research Laboratory made some observations at submillimeter wavelengths and found that the radiation intensity is not characteristic of a blackbody at all. Unfortunately, such measurements are extremely difficult to make and there is still uncertainty about their reliability. As of this writing the steady-state theory still survives.

As these and other new phenomena are detected, many aspects of cosmology will undoubtedly be cleared up. New telescopes are needed, both optical and radio.

[11] See footnote 4, page 610.

The telescope at Mt. Palomar is the only one of its size. Furthermore, it can scan the stars only in the northern celestial hemisphere. X-ray telescopes will be required, either mounted on satellites or on the moon. And finally there is a possibility, not yet achieved, of designing a neutrino telescope whose purpose would be again to extend the range of signals received from space. Neutrinos do pour in from outer space, but from what directions and with what flux is not well known.

6. RETROSPECT AND PROSPECT

The major conceptions of twentieth century physics have been presented in the last several chapters. We have traced their genesis and development out of older ideas. We have examined their internal, logical structure as well as their relationship to the evidence of experiment and observation. It is fitting that we now stand aside once again to consider them in a broader perspective.

If, as seems likely on the basis of modern evidence and speculation, the physical universe is indeed expanding, our scientific knowledge about the universe is expanding at an even more rapid rate. Myriads of imaginative connections between phenomena are made today, which were undreamed of a few generations ago. And the concepts that provide these connections, in terms of which scientific experience is organized and by means of which technology is guided and advanced, go far beyond the reach of what we can directly sense. Consider but two examples of how the human mind ranges beyond the limits of direct observation: on the one hand, from a few smudges and lines on a photographic plate mounted in a 200-in. telescope, we can define the conditions existing within a distant galaxy, interpreting the signals emitted billions of years ago but just now reaching us over the vast spread of space; on the other hand, by an exquisitely elaborate theoretical and instrumental trap, we detect the presence of the elusive neutrino whose flitting and ghostly passage through the densest material takes but a microsecond. And, always, as we extend and elaborate our ideas on such scaffoldings of abstract conceptions and delicate instrumentation, probing the universe to deeper and deeper levels, we discover that nature ever becomes more vivid, varied, and mysterious. The more the scientists of our century expand their knowledge, the more aware they become of their own ignorance.

Not only does nature, thus questioned, continuously present novelties, but every scientific theory, no matter how successful, remains forever provisional. This is a built-in limitation of science. It arises out of the nature of proof and of implication in a manner first discussed in the earliest chapters of this book. This point is important enough to bear still further repetition. As we explained earlier, if one set of propositions A implies another set B, the "truth" of B—

judged by agreement with observation—does not imply the truth of A. We have now seen that the scientific theories (which constitute sets of propositions we designate by A) cannot be tested directly because of their abstract nature; only their consequences B can be so tested against observation. It follows from this analysis that no scientific theory is ever proved. Other alternative theories might well be constructed; there is an infinite variety of possibilities awaiting the creative intuition of some scientist. The best at any moment that can be said regarding a particular theory is that we are satisfied with it as being better than any other theory that has as yet been thought up.

It follows also from this analysis that our basic theories, no longer being tied to any *a priori* principles or any direct observations, may be developed with greater freedom. They may consequently depart very far from intuition, common sense, or traditional first principles. We have seen several examples of such theories. One final and striking example of how far a theory may depart from classical principles may be cited. It is one that also illustrates the practical application of noncommutative algebras discussed in Chapter 3. It will be recalled that the postulates of classical algebra include the postulate that $A \cdot B = B \cdot A$ or $A \cdot B - B \cdot A = 0$. In quantum mechanics, however, a fundamental equation appears that relates two quantities (called operators) referring to distance x and momentum p, which takes the form $p \cdot x - x \cdot p = ih/2\pi$. The modern physicist, not being tied down by the necessity of starting from classical postulates, has a freedom to postulate whatever he wishes, provided only that his logic is sound and that the consequences deduced agree with observation.

Can we not conclude from considerations such as these that one of the main attitudes that we can learn from modern science is one of freedom and tolerance —freedom to examine and to entertain many alternative theories, regardless of the dictates of tradition? In this sense, twentieth century science has opened to us a universe in which our imaginations are given a new freedom.

No matter how provisional scientific knowledge is regarded, or how far it is from approaching reality in any final fashion, it still remains one of the most powerful weapons known to man in his quest to understand himself and nature. Confidence toward science is steadily growing, founded upon the progressive generality that marks the march of science as well as upon the successful applications in technology, both indicating that the complex symbolic structures developed by scientists do reflect in *some* fashion, even though ambiguously and with distortions, the complex order and structure of nature. More and more inclusive generalizations are continually being created. Older theories based on a narrower set of observations rarely are completely discarded. Instead, they are embraced under new, broader, and abstract syntheses.

Language, maps, the arts, equations—all forms of symbolic structure and patterns—guide men in understanding and action. Man is unique in this symbol-creating, symbol-following activity. It is out of this practice that he binds together the past and the present with visions of the future. But the practice of handling symbols effectively is relatively young. It was about two million years ago that man first walked erect upon the face of the earth; it was a brief ten thousand years ago that he first learned to handle the symbols of art and writing effectively; but it was just a wink of time—four centuries ago—that science in the modern sense had its birth.

It is not surprising that we are still novices in handling and understanding the variety of symbolic structures and in finding the proper role of science. But we realize that each of us adjusts and responds as much to symbols, their meaning and suggestiveness, as to our raw biological and physical environment, which we sense directly. Poetry, mathematics, art, music, religion, and science constitute an extraordinary variety of symbolic expressions. None of them exhausts "reality," each has its genre, its scope, its limitations. But they all contribute to the unity of knowledge—the passage from "what I know" to "what we know."

Whether we are conscious of it, scientific knowledge penetrates deeply today into our whole conception of nature. And the way man thinks about nature affects the way he thinks about himself. In this situation, the new scientific symbolism and the older symbolism of the arts, language, and religion should complement each other, support each other, affect each other. It is true that art and science move in opposite directions. The scientist craves for generalities, no matter how abstract, finding interest—as a scientist—in the particular only as it represents an exciting confirmation or contradiction of theory. Otherwise, particular events tend to be regarded as trivial. In contrast, the artist, seeking to keep in close contact with immediate human experience, focuses upon particular patterns of sensation, seeking to evoke or suggest by his imagery, his arrangement of words, sounds, or colors, some universal mood or wordless understanding. He is glad to sacrifice precise meaning for an intense aesthetic intuition of the world. It is folly for a scientist to demand explicit and precise meaning in a poem such as he demands in his equations. It is equal folly for a poet to recoil from abstractions as colorless, unreal, and divorced from human experience. Surely both aspects of reality—one apprehended best by intellectual cognition; the other by direct, aesthetic intuition—are required. The sciences and the arts each have their own role in drawing man toward his common humanity in which he becomes more sensitively and compassionately aware of the identity of what he is, what he knows, what he feels, and what he does.

7. QUESTIONS

1. Detection of parallax by telescopic and photographic means at a certain observatory is considered to be precise to 1/20 of one second of arc. The parallax angle of a certain star is found to be 0.3 seconds of arc. What is the distance of the star? What are the limits of

accuracy? (Today, about 5000 stars are near enough for direct measure of distance by parallax methods. This constitutes what percentage of all the stars in the Milky Way, estimated as 100 billion or 10^{11}?)

2. Explain three different methods of measuring distances of celestial objects. What are the assumptions upon which each method is based? Which method is least certain?

3. An old paradox in astronomy was first proposed by Olber. The argument ran: (*1*) the universe is infinite; (*2*) the stars are distributed more or less uniformly throughout the vast reaches of space; and (*3*) although light from a star varies inversely as the square of the distance, since the number of stars in all directions is infinite, the sky should be equally *and brilliantly* lit day and night. This follows from the fact that although the intensity of light from any *one* star varies inversely as the square of the distance, the number of stars increases as the square of the distance and one effect cancels the other. This paradox is known as Olber's paradox. What are some of the ways of escaping from it?

4. The wavelength of a line observed in the spectrum of a star is 5001 angstrom units (one angstrom unit is 10^{-10} meters). In the laboratory the wavelength of the same spectral line is found to be 5000 angstrom units. In which direction is the star moving, and with what velocity?

5. The "magnitude" of a star is an ordinal scale roughly measuring our sensation of the relative brightness of stars. Our sensation of brightness, however, is not related in a linear fashion to the light energy received from a star. A difference of five magnitudes actually corresponds to a difference of 100 times as much energy reaching our eyes each second, and thus a first magnitude star is 100 times brighter than a sixth magnitude star (or approximately 2.55 times (that is, $\sqrt[5]{100}$) as bright as a second magnitude star). The moon has a magnitude of -12.5. How does it compare (approximately) in brightness to the brightest star Sirius, which has a magnitude of -1.5? How does the sun (magnitude -27) compare with Sirius in brightness?

6. The density of a star may seem to be extraordinarily high (6000 times as heavy as lead). Compare such a density, however, to the density of a proton, calculated from estimates of its mass and radius.

7. Hubble's law states that the velocity of recession of stars and galaxies increases in direct proportion to their distance. If Hubble's constant of proportionality in the equation $v = kd$ where v is velocity in miles per second and d is distance in millions of light years is taken as approximately $k = 14$, what would be the farthest distance into space that we could ever "see," in principle, with even the most powerful optical or radio telescope?

8. The spectroscopic "red shift" indicates that stars and galaxies are receding in every direction. Does this imply that the solar system is at the center of the universe? Give an alternative explanation.

9. Two forces are probably at work, causing dust particles in space to condense. One of these is gravitational attraction between the particles; the other is radiation pressure of light from already existing stars. Explain the operation of radiation pressure on the particles.

10. What reasons can you give for the assumption that stars are approximately spherical?

11. What is the "cosmological principle" upon which some astronomers, Hermann Bondi in particular, have based their cosmological theories? What is the justification for such a

principle? (See Milton K. Munitz, *Space, Time, and Creation,* Collier Books, New York, 1961, for a critical and philosophical analysis of the principle.)

12. In the realm of the microscopic we have seen the operation of laws that are novel (that is, seem to differ from those that work on our own scale of perception and experience). Are there any similar unfamiliar principles that may be at work on a vast macroscopic scale?

13. A planet hundreds of times the size of Jupiter, if it were in orbit about the nearest star other than the sun, would give off so little reflected light that it could not be seen by the most powerful telescope. Yet, the studies of the astronomer Peter van de Kamp have yielded indication of unseen companions of certain stars whose masses are about one-hundredth of the mass of the sun. Can you surmise how this conclusion is reached? (*Hint:* Consider the motion of binary stars.)

Recommended Readings

The New Astronomy, Scientific American Book, Simon & Schuster, New York, 1955. A series of essays on topics in astronomy by contemporary astronomers.

Hermann Bondi, *Cosmology,* second edition, Cambridge University Press, New York, 1960.

Hermann Bondi *et al; Rival Theories of Cosmology,* Oxford University Press, New York, 1960.

Fred Hoyle, *The Nature of the Universe,* Mentor Paperback, New York, 1950. These three books lay the foundation for the theory of continuous creation in an expanding universe. Hoyle's book, in particular, is popularly written, and forms an exciting introduction to cosmology. A more careful exposition of Hoyle's ideas can be found in his *Frontiers of Astronomy,* Mentor Paperback, New York, 1957.

George Gamow, *The Creation of the Universe,* revised edition, Viking Press, New York, 1961. A delightfully written exposition setting forth the "big-bang" theory of creation, in opposition to Hoyle's theory.

W. Kruse, and W. Dreckvoss, *The Stars,* University of Michigan Press, Ann Arbor, 1957. Provides a good summary of the way in which astronomers find out about the distance, size, temperature, and structure of the stars.

Thorton Page *et al; Stars and Galaxies,* Prentice-Hall, Englewood Cliffs, N.J., 1962. A series of essays on the frontiers of astronomy, including a section on radio-astronomy.

Milton K. Munitz, *Space, Time, and Creation,* Collier Books, New York, 1961. A summary of cosmological theory with special attention to philosophical implications.

Jagjit Singh, *Great Ideas and Theories of Modern Cosmology,* Dover Publications, New York, 1961. A fine summary of the material upon which modern cosmologies are based.

Shklovskii and Sagan, *Intelligent Life in the Universe,* Holden Day, San Francisco, 1966. A careful review of the possibilities of life in the universe, written in collaboration by a Russian and an American astronomer. A wealth of data is analyzed and well presented for the layman interested in this question. As a bonus to the reader, there are many hints as to both similarities and differences in attitudes between Russian and American scientists.

Robert Jastrow, *Red Giants and White Dwarfs,* Harper and Row, New York, 1967. The central ideas of modern astronomy presented in a lively style by the director of the Goddard Institute for Space Studies of NASA.

appendix one

A. Mathematical constants:

Values (precise to at least 0.1%)	Approximate values (Use unless otherwise specified)
$\pi = 3.1415 \cdots$	$\pi = 22/7$
$\pi^2 = 9.896 \cdots$	$\pi^2 = 10$
$\sqrt{2} = 1.4142 \cdots$	$\sqrt{2} = 1.4$
$\sqrt{3} = 1.732 \cdots$	$\sqrt{3} = 1.73$
$\epsilon = 2.718 \cdots$ (base of natural logarithms)	$\epsilon = 2.72$
1 radian = 57.296°	1 radian = 57°

B. Conversion units:

Values (precise to at least 0.1%)	Approximate values
1 meter = 39.37 in.	1 meter = 40 in. = 3.3 ft
1 inch = 2.54 cm	1 inch = 2.5 cm
1 mile = 1.609 km	1 mile = 1.6 km
1 centimeter = 0.3937 in.	1 cm = 0.4 in.
1 kilogram = 2.205 mass of 1 lb	1 kg = 2.2 lb
1 kilogram weighs 9.81 newtons	1 kg weighs 10 newtons
1 newton = 0.225 lb	1 newton = 0.22 lb
1 horsepower = 746 watts = 550 ft-lb/sec	1 hp = 746 watts
1 parsec = 1.92×10^{13} miles = 3.261 light-years	1 psc = 2×10^{13} mi = 3.3 light-years
1 joule = 10^7 ergs	1 J = 10^7 ergs
1 angstrom unit = 10^{-10} m	1 Å = 10^{-10} m = 10^{-4} microns
1 liter = 10^3 cm^3 = 0.2642 gallon	1 liter = 10^3 cm^3 = ¼ gallon

C. Physical constants:

Symbol	Values (precise to at least 0.1%)	Approximate values
c	Velocity of light (in vacuum) $(2.99776 \pm 0.00004) \times 10^8$ m/sec	$c = 3 \times 10^8$ m/sec
g	Acceleration due to gravity: 9.80665 m/sec^2 (sea level 0° latitude)	$g = 10$ m/sec^2
G	Gravitational constant $(6.670 \pm 0.005) \times 10^{-11}$ nt-m/kg^2	$G = 6.7 \times 10^{-11}$
N_0	Avogadro's number: $(6.0228 \pm 0.0011) \times 10^{23}$ per mole	6×10^{23} per mole
h	Planck constant: $(6.624 \pm 0.002) \times 10^{-34}$ joule-sec	6.6×10^{-34} joule-sec

V_0	Volume of ideal gas (0°C, P_0) (22.414 ± 0.0006) liters per mole	22.4 liters per mole
P_0	Standard atmosphere (1.013246 ± 0.000004) × 10^5 newtons per m^2	10^5 newtons/m^2
M_e	Rest mass of an electron: (9.10721 ± 0.00025) × 10^{-31} kg	9 × 10^{-31} kg
e	Charge on an electron: (1.601864 ± 0.000024) × 10^{-19} coulombs	1.6 × 10^{-19} coulombs
amu	1 atomic mass unit 1.660 × 10^{-27} kg	1.66 × 10^{-27} kg
	Mass of proton 1.00758 amu	
	Mass of neutron 1.00896 amu	
	Mass of neutral hydrogen (1.0081284 ± .000003) amu	1 amu
R	Universal gas constant 8.314 joules per degree per mole	8.3 joules per degree per mole
k	Boltzmann's constant 1.380 × 10^{-23} joules per degree	1.4 × 10^{-23} joules per degree
J	Mechanical equivalent of heat	
	4.1855 ± 0.0004 joules per calorie	4.2 joules per calorie
	4185.5 joules per K calorie	4.2 × 10^3 joules per kilocalorie
	Average diameter of earth 7918 miles or 12,543 km	8000 miles or 12,500 km
	Average diameter of moon 3476 km	2100 miles or 3.5 km
	Synodic month (new moon to new moon) 29.53 days	29.5 days
	Sidereal month 27.322 days	27.3 days
	Average distance of moon	240,000 miles or 380,000 km
	Average distance to sun (one astronomical unit)	93 × 10^6 miles or 150 × 10^6 km
	Mass of earth 5.98 × 10^{24} kg	6 × 10^{24} kg
	Mass of sun $3\frac{1}{3}$ × 10^5 times mass of earth	
	Mass of moon $\frac{1}{81}$ times mass of earth	

appendix two a. THE SPRING CONSTELLATIONS

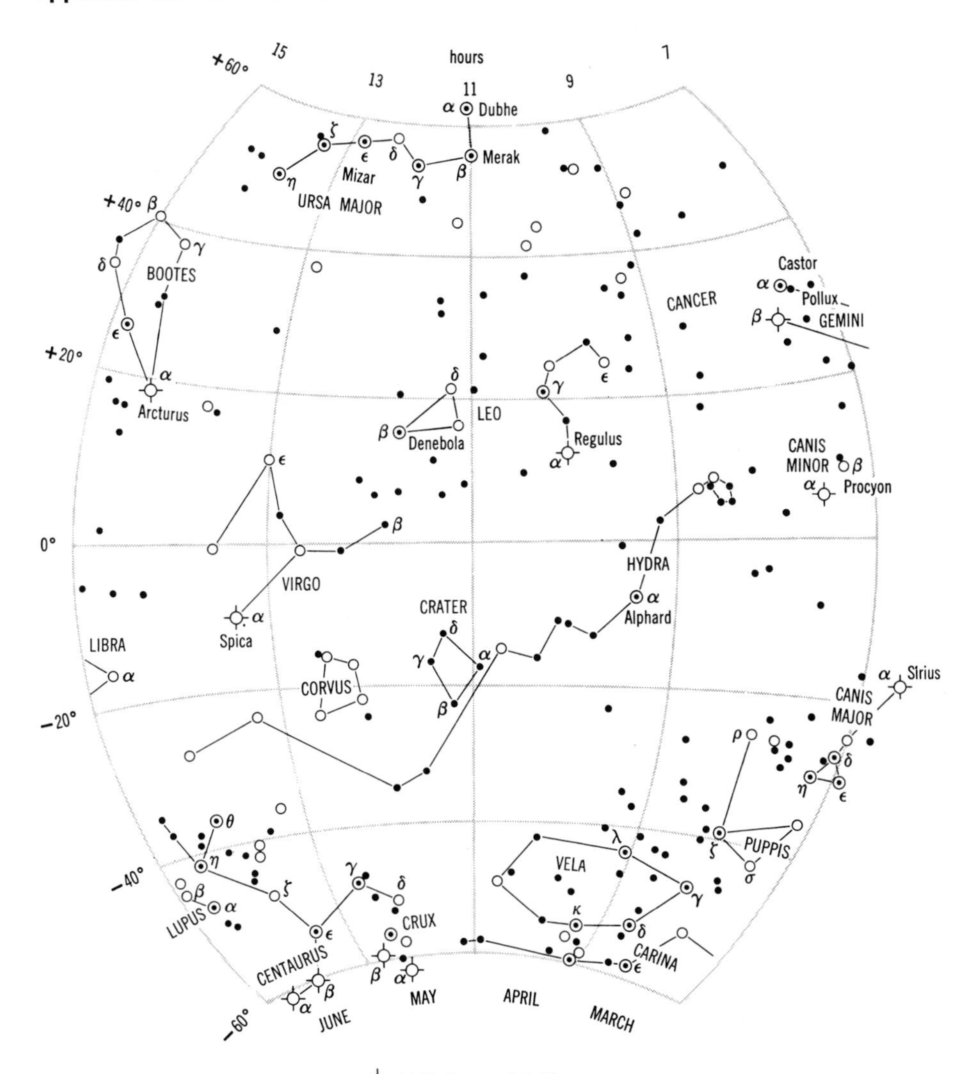

- brighter than magnitude 1.5
- between magnitudes 1.5 and 2.5
- between magnitudes 2.5 and 3.5
- fainter than magnitude 3.5

appendix two b. THE SUMMER CONSTELLATIONS

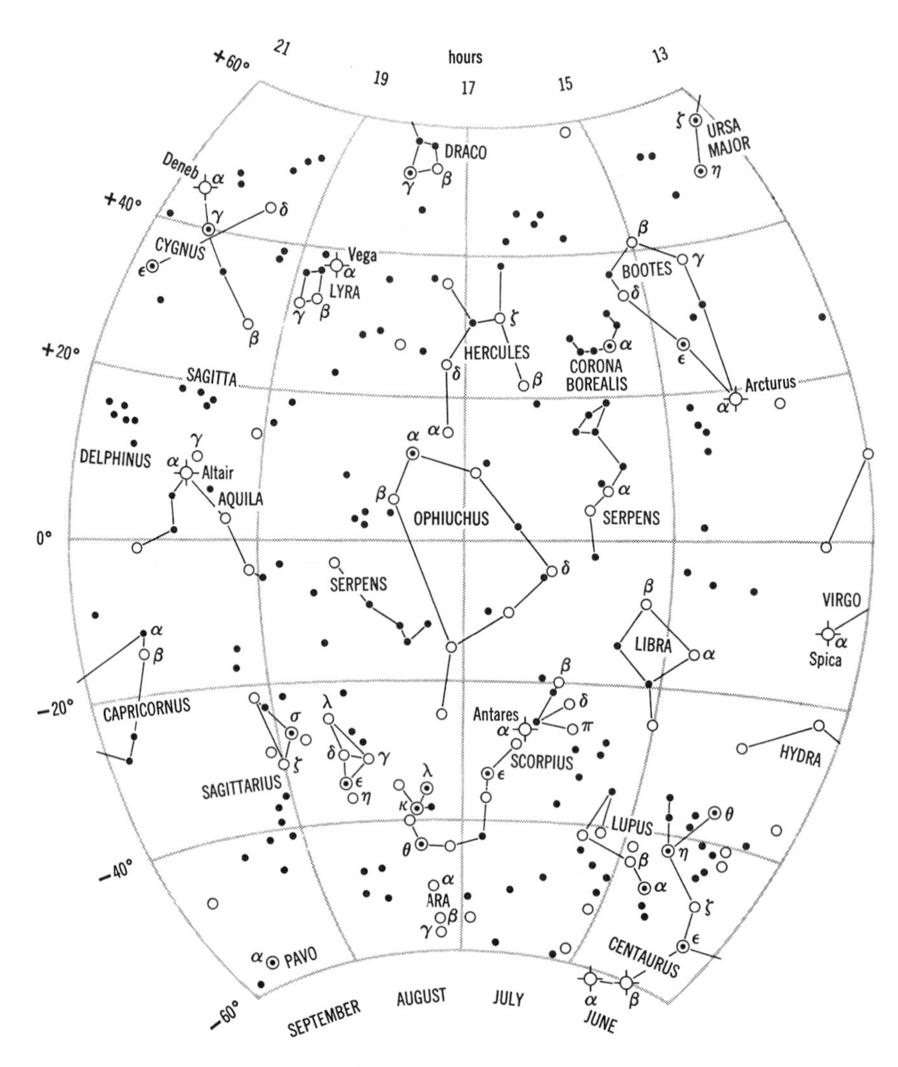

- brighter than magnitude 1.5
- between magnitudes 1.5 and 2.5
- between magnitudes 2.5 and 3.5
- fainter than magnitude 3.5

appendix two c. THE AUTUMN CONSTELLATIONS

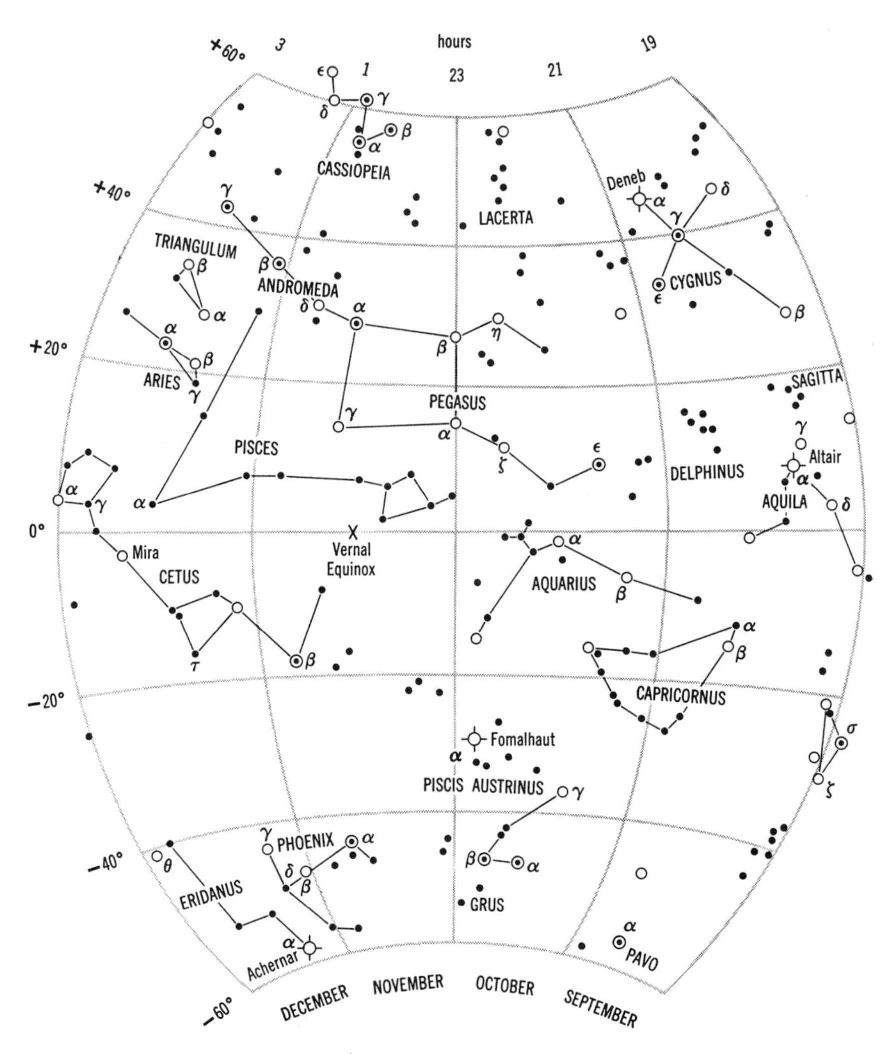

appendix two d. THE WINTER CONSTELLATIONS

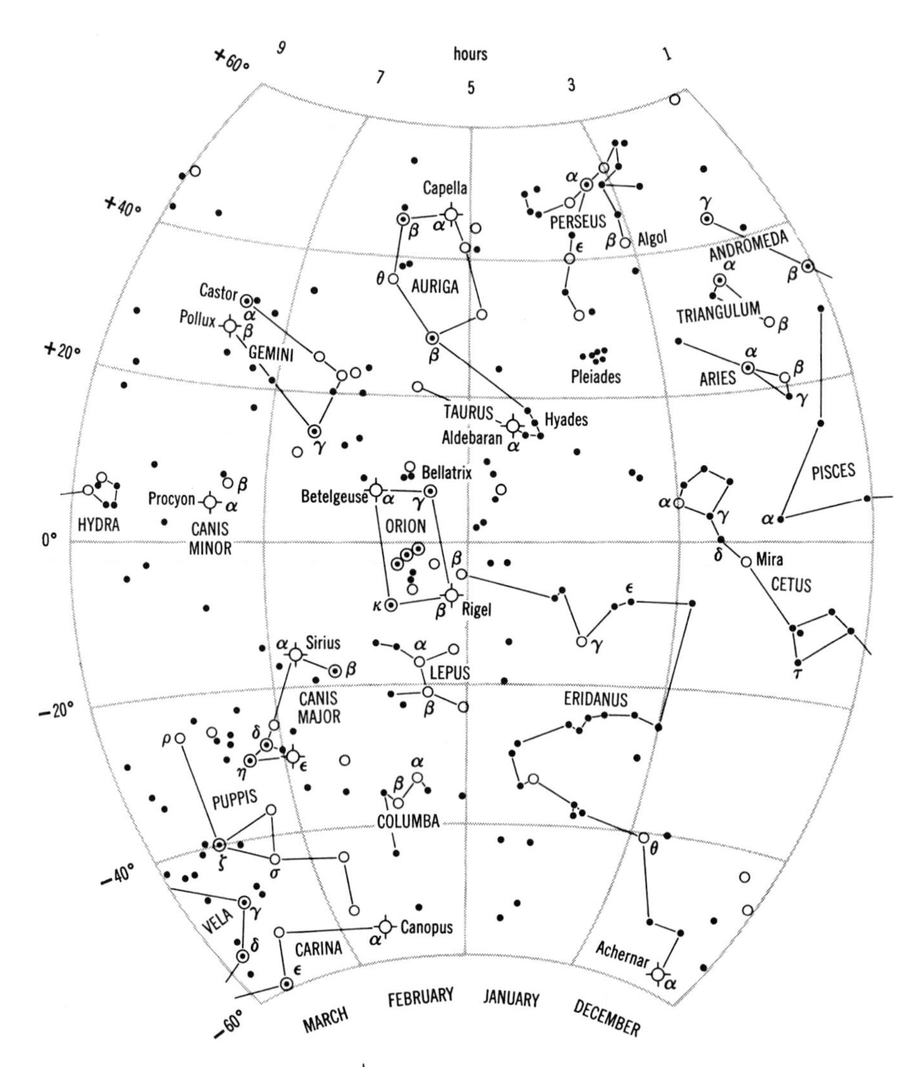

- brighter than magnitude 1.5
- between magnitudes 1.5 and 2.5
- between magnitudes 2.5 and 3.5
- fainter than magnitude 3.5

appendix three. THE BINOMIAL THEOREM

By simple multiplication it can be shown that

$$(a + b)^2 = a^2 + 2ab + b^2$$
$$(a + b)^3 = a^3 + 3a^2b + 3ab^2 + b^3$$
$$(a + b)^4 = a^4 + 4a^3b + 6a^2b^2 + 4ab^3 + b^4$$
$$(a + b)^5 = a^5 + 5a^4b + 10a^3b^2 + 10a^2b^3 + 5ab^4 + b^5$$

Apparently some patterns govern these binomial expansions. Newton found and proved that the general formula for the expansion of a binomial raised to any power n can be expressed as

$$(a + b)^n = a^n + \frac{n}{1} a^{n-1} \cdot b + \frac{n(n-1)}{1 \cdot 2} a^{n-2} \cdot b^2 + \frac{n(n-1)(n-2)}{1 \cdot 2 \cdot 3} a^{n-3} \cdot b^3 + \cdots$$

This formula is proved in algebra for n equal to any whole positive number by the method of mathematical induction. By use of the calculus, the theorem may be proved for any value of n, including negative numbers and fractions. Where n is a fraction, the series is an infinite one. Such an infinite series, however, can be very useful if each succeeding term becomes rapidly smaller and smaller. Since, in the application of mathematics to science, there is no significance to the pursuit of an accuracy beyond the precision to which we can measure, it is often convenient to disregard certain quantities as negligible. (We do this all the time. π is an infinite series, the decimals continuing in an unending series of smaller and smaller values:

$$\pi = 3 + \frac{1}{10} + \frac{4}{100} + \frac{1}{1000} + \cdots$$

We only use that value for π which is justified by the precision of our measurements.)

Similarly, if we wish to use the expression arising in relativity theory,

$$k = \frac{1}{\sqrt{1 - v^2/c^2}} = \left(1 - \frac{v^2}{c^2}\right)^{-1/2}$$

we use the binomial expansion formula,

$$1^{-1/2} + \left(-\frac{1}{2}\right)(1)^{-3/2}\left(-\frac{v^2}{c^2}\right) + \frac{\left(-\frac{1}{2}\right)\left(-\frac{3}{2}\right)(1)^{-5/8}}{1.2}\left(-\frac{v^2}{c^2}\right)^2 + \cdots$$

or

$$1 + \frac{1}{2}\frac{v^2}{c^2} + \frac{3}{8}\frac{v^4}{c^4} + \cdots$$

If, as is most often the case, v/c is less than 1/10, then the term v^4/c^4 becomes less than 1/10,000 and can be neglected. (When the fraction is greater than, for example 1/10, we can then include one or more additional terms but neglect still later terms in the series.)

appendix four. LORENTZ TRANSFORMATION EQUATIONS DERIVED FROM RELATIVITY POSTULATES

One of the simplest ways of deriving the Lorentz transformation equations of special relativity, without using the calculus, is to consider how two observers with a relative constant velocity view the same spherical wave front of light. Under the classical theories of physics, if the wave front of light spreads out in a sphere as viewed from an observer at rest relative to the source, the same wave front would appear distorted from its spherical shape when viewed from an observer in relative motion, because the velocity of light would be different to such an observer. Under the postulate of relativity, however, the velocity of light is constant for both observers. Hence, both must, *with respect to their own frame of reference,* observe a spherical wavelength expanding outward from its source with the same speed c.

We assume two observers moving with a constant velocity v relative to each other. Furthermore, we shall assume that at a certain instant of time their two frames of reference coincide; that is, $x = x' = 0$, $y = y' = 0$, $z = z' = 0$, when $t = t' = 0$, where the unprimed coordinates refer to one observer, the primed coordinates to the other. At the instant of coincidence, a flash of light is set off at the origin. For simplicity of analysis, we further assume that the relative velocity v is in the x direction.

The expanding sphere of light may now be written as follows for each observer:

$$x^2 + y^2 + z^2 = c^2t^2 \tag{1}$$

$$x'^2 + y'^2 + z'^2 = c^2t'^2 \tag{2}$$

since the radial distance from the point of origin must be given by ct and ct', respectively. Note that c alone is identical in both equations. (In classical physics it would be assumed that $t = t'$, and that x and x' would be connected by the equation $x = x' + vt$ if the velocity were in the x direction. This would necessarily imply a different c in the two coordinate systems.)

We assume that the two sets of axes remain parallel while the relative motion at velocity v takes place in the x direction. Since the set of axes is parallel we can assume that y will continue to equal y', z will continue to equal z', and we need inquire only as to the relationship between x and x', t and t'. Because of our restriction that the relative velocity is in the x direction only, the equations 1 and 2 (which can be rewritten as $x'^2 + y'^2 + z'^2 - c^2t'^2 = 0 = x^2 + y^2 + z^2 - c^2t^2$) will simplify to

$$x'^2 - c^2t'^2 = x^2 - c^2t^2 \tag{3}$$

which must remain an identity for all values of x, t, x', and t'.

We seek the simplest equation between x and x', t and t', which will satisfy our requirements. The equation should not contain squared or cubed terms in x and t for, in this case, the space and time coordinates would not be uniquely defined and linear for both observers. The simplest equation is

$$x = k(x' + vt') \tag{4}$$

where k is some constant to be evaluated. The inverse equation, required for symmetry, is

$$x' = k(x - vt) \tag{5}$$

with the same k, since neither frame of reference is "privileged," and v from one frame of reference is the same as the negative v from the other frame of reference.

Our first step is to use equations 4 and 5 to express the primed terms x' and t' as functions of the unprimed terms. Simple algebra yields

$$x' = kx - kvt \tag{6}$$

$$t' = \frac{1}{kv}\{x(1 - k^2) + k^2vt\} \tag{7}$$

We now substitute these values in equation 3, and try to find an expression for k in terms of v, c, and t. Making these substitutions, we get

$$(kx - kvt)^2 - \frac{c^2}{k^2v^2}\{x(1 - k^2) + k^2vt\}^2 = x^2 - c^2t^2 \tag{8}$$

Now, actually, the carrying out of such an expansion in full (that is, squaring the terms on the left) would yield a very cumbersome expression. Fortunately, we need not carry out the full operation if we recall two conditions: (1) the left- and right-hand side of the equation represent an identity (that is, the equation is true for all values of x and t); (2) in a quadratic identity, the coefficients of all like powers of the variables on both sides of the equation must be equal. For example, if the equation

$$ax^2 + bx + cy^2 + d = 3x^2 + 2x + 5y^2 + 6$$

is an identity for *all* values of x and y, then $a = 3$, $b = 2$, $c = 5$, and $d = 6$.

Using these principles, and *attending only* to the terms of equation 8 containing squared powers of t, we get, by equating the coefficients of t^2,

$$k^2v^2 - c^2k^2 = -c^2$$

or

$$k^2 = \frac{c^2}{c^2 - v^2} = \frac{1}{1 - v^2/c^2}$$

whence, finally,
$$k = \frac{1}{\sqrt{1 - v^2/c^2}}$$

This result now leads directly to the Lorentz transformation equations as a consequence of the special relativity postulates. Making the substitutions in equations 4 and 5, we get

$$x = \frac{x' + vt'}{\sqrt{1 - v^2/c^2}} \qquad t = \frac{t' + vx'/c^2}{\sqrt{1 - v^2/c^2}}$$

$$y = y'$$

$$z = z'$$

(Note how the time and space coordinates remain mingled.)

Using these equations, it can be readily shown that a length (in the direction of motion) of a moving object is reduced by this amount:

$$L = L_0\sqrt{1 - v^2/c^2}$$

where L_0 is the length measured by an observer at rest relative to the object, and L is the length measured by an observer in motion relative to the object.

This relationship is readily proved as follows. Let

$$L_0 = x_2 - x_1, L = x'_2 - x'_1$$

Then

$$x_2 - x_1 = \frac{x_2' + vt'}{\sqrt{1 - v^2/c^2}} - \frac{x_1' + vt'}{\sqrt{1 - v^2/c^2}} = \frac{x_2' - x_1'}{\sqrt{1 - v^2/c^2}}$$

or

$$L_0 = \frac{L}{\sqrt{1 - v^2/c^2}}$$

whence

$$L = L_0\sqrt{1 - v^2/c^2}$$

In a similar manner it may be shown that a time interval, as measured by a clock in motion, will be found longer by a clock at rest, as follows:

$$t = \frac{t_0}{\sqrt{1 - v^2/c^2}}$$

and, finally, if an object travels with a velocity u, relative to one frame of reference A, which is, in turn, traveling with the velocity v relative to a second frame of reference B, then the velocity relative to frame of reference B is

$$V = \frac{u + v}{1 + \dfrac{uv}{c^2}}$$

(This equation is simply derived by letting $V = (x_2 - x_1)/(t_2 - t_1)$ and $u = (x_2' - x_1')/(t_2' - t_1')$, and using the Lorentz transformation equation.)

appendix five. THE RELATIONSHIP BETWEEN APPARENT MAGNITUDE, ABSOLUTE MAGNITUDE, AND DISTANCE OF STARS

The apparent magnitude of stars is a measure of their visual brightness, as seen from the earth. In ancient star catalogues, stars were grouped into six categories of brightness, the most brilliant stars being assigned a magnitude of 1, and those just visible to the naked eye being assigned a magnitude of 6. To give a more precise designation of magnitude yet one in substantial agreement with the older classification, an index has been developed on the basis of the measurable amount of energy flux reaching the earth from a star. On this basis, the ratio between two successive magnitudes approximates $100^{-1/5}$, the negative sign indicating that the higher the magnitude the lower the energy flux that is being received. Accordingly, an equation defining the apparent magnitude of stars in terms of the relative intensity of the light received (the energy received per unit of time and per unit of cross-sectional area) has been established:

$$\frac{I_e \text{ (intensity of light from a star of magnitude } m)}{I_0 \text{ (intensity of light from a star assigned zero magnitude)}} = 100^{-m/5} \quad (1)$$

A few very bright stars, such as Vega, have an apparent magnitude close to zero. Some heavenly bodies have negative magnitudes (Sirius, and the planets Venus and Jupiter, as well as the sun and the moon). The moon has a negative magnitude of -12.5, which implies, from the above equation, that the light energy flux reaching us from the moon is $100^{-(-12.5/5)} + 100^{+2.5}$ or one hundred thousand times the light flux reaching us from a star such as Vega with a magnitude of zero. On the other hand, certain spiral nebulas have magnitudes of about 20. The light-energy flux received from such nebulas is $100^{-20/5} = 10^{-8}$, or one hundred millionth of the light energy received from a star of zero magnitude.

A proper comparison of the actual energy flux being given off from different stars naturally requires that the comparison be made as if the stars were at the same distance. To accomplish this, an index of *absolute magnitudes* has been developed. If the sun were 10 parsecs distant from the earth (about 33 light-years), it would have an apparent magnitude of zero. We therefore assign an absolute magnitude of zero to a star which would have zero apparent magnitude at a distance of 10 parsecs. Hence, we can now write the defining equation for absolute magnitude, M, as

$$\frac{I_M \text{ (intensity of a star at a distance of 10 parsecs)}}{I_0 \text{ (intensity of a star of zero apparent magnitude)}} = 100^{-M/5} \quad (2)$$

Between equations 1 and 2, we can eliminate I_0 to get

$$\frac{I_M}{I_e} = \frac{100^{-M/5}}{100^{-m/5}} = 100^{(m-M)/5} \tag{3}$$

We now use the fact, mentioned in the text, that the relative intensity of the light reaching the earth from a star at distance D to the light which would reach us from a distance of 10 parsecs is in inverse proportion to the square of the distances. Therefore

$$\frac{I_M}{I_e} = \frac{D^2}{10^2} = \left(\frac{D}{10}\right)^2$$

Combining equation 3 and 4, we arrive at

$$\left(\frac{D}{10}\right)^2 = 100^{(m-M)/5}$$

Taking the common logarithm of both sides, we finally get

$$2 \log \left(\frac{D}{10}\right) = \left(\frac{m-M}{5}\right) \log 100$$

or

$$5 \log \left(\frac{D}{10}\right) = m - M$$

Answers to selected questions and problems

(Note: g is taken as 10 m/sec^2 throughout.)

CHAPTER 2

Questions

1. $k = 0.0175$.
3. The points must be on the arc of a great circle. Explain why.
4. Use earth's diameter and sun's distance.
5. Approximately only for small angles.

Problems

1. 1000 miles/hr; 700 miles/hr.
2. $\sin 30° = ½$; $\cos 30° = \frac{\sqrt{3}}{2}$, etc.
3. Use $\frac{\text{arc length}}{2\pi R} = \frac{\theta}{360°}$ and definition of radian.
5. If distance between pupils is 6 cm, show that angle for A is 1.2° and for B is 0.6°.
6. Over 3 miles.
9. Check your observation against a star chart.

CHAPTER 3

Questions

5. Volume is increased 8 times. Why? Test it with a toy balloon.
8. Consider changes in surface area relative to volume. Illustrate.
9. Pi (π) is not a repeating decimal, hence not a fractional number.

Problems

1. 111; 18.6; 0.09.
2. $x = \frac{7}{4}, y = 3$.

4. 6.7×10^{25}.
5. Base estimate on surmise of average depth of ocean and surface area of ocean.
6. 2.3×10^{13} miles; 3.8×10^{16} m.
10. Almost one ft.

CHAPTER 4

Questions

1. About 7.5°. Explain. What is exact figure?
3. Consider sizes of shadows.
11. Planets do not twinkle (why not?), are relatively bright, and move in zodiacal belt.

Problems

2. Declination on Dec. 21 of 23½° S, right ascension 18 hr.
3. 38½°.
5. 0.8 second of arc. How many degrees?
6. $d = \frac{1}{n}$ where d is in parsecs.
7. 3.26 light years = 1 parsec.

CHAPTER 5

Questions

2. Synodic period = 29½ days. Sidereal period = 27⅓ days.
5. In Tycho's system, sun revolved about earth and planets about the sun.

Problems

1. About 25,000 miles.
2. $k = \frac{1}{(9.3 \times 10^7)^3}$ in $\frac{\text{years}^2}{\text{miles}^3}$.
3. Take sun's diameter as 8.64×10^5 miles. The variation is 2% from about 0.53 degree to 0.54 degree.
5. An *average* of about ¼ degree per day.
7. A little more than 5 astronomical units.

CHAPTER 6

Problems

1. 88 ft/sec or 26.5 m/sec.
2. 11.4 miles/sec^2.
3. 30 miles/hr; 40 miles/hr.
4. (c) 15 m/sec.
5. 7.1 m/sec; 2.5 m.
6. 80 ft/sec; 80 ft/sec; about 273 ft.

7. Fit straight line graph and find slope.
8. For second part, use basic equation $s = (½)(10)\left(10 - \frac{s}{300}\right)^2$.
9. 5.45×10^3 m/sec^2.

CHAPTER 7

Problems

1. 2800 newtons; 616 lb.
2. 3.3×10^3 newtons.
5. g on moon is about 5 ft/sec^2.
7. 5 m/sec^2.
8. 14,000 newtons; 6,000 newtons.
9. $25g$; over 5000 lb.
11. 23 newtons.

CHAPTER 9

Problems

4. 6.3 m/sec; 150 newtons.

CHAPTER 10

Questions

3. Consider compressibility.
4. Remember to include air pressure at *both* points.
13. Why does a candle require gravitation to continue burning?

Problems

2. About 33 ft.
3. 3550 lb/in.^{2}
5. 77 Calories (kilocalories).
6. 3×10^{-4} Calories (kilocalories).
7. 90 Calories (kilocalories). (Assuming 60 miles/hr is approximately 27 m/sec.)
8. 4.1 cm.
9. 1.2 ton/in.^{2} (round heel).
11. Yes (air temperature taken as 20°C).
13. 3.6×10^6 joules; 8.6×10^2 Calories (kilocalories).

CHAPTER 11

Problems

1. Joules per mole-degree K.
3. 6.7×10^{22} atoms in a gram of water.
4. 1.7×10^3 times.

7. 1.14 cm.
9. 1.8×10^3 m/sec.
10. 3.37×10^{-27} kg.
13. 10^{-16}; 2.7×10^9; 2.7×10^3.

CHAPTER 12

Problems

3. 18.7 liters; 20.5 liters; 5×10^{23}.
5. Na_3PO_4, trisodium phosphate; Na_2HPO_4, disodium phosphate; CS_2, carbon disulphide; $Ca(OH)_2$, calcium hydroxide.
7. 126g; 65.5g; 71g; 40.5g.
9. ⅔.

CHAPTER 13

Problems

1. 2.88×10^5 joules; 0.08 kWh; 69 cal.
2. 2.5×10^{19}.
4. 1.6×10^{-10} joules.
5. $F_g = 2.4 \times 10^{-44} F_e$.
6. 4.2 amp.
8. 3×10^{17} coulombs.
9. 0.001 joule; 45 m/sec.
10. 1 g.

CHAPTER 14

Problems

1. 8.9×10^{14} m/sec^2.
2. 1.33×10^7 m/sec; 8×10^{-17} joules; 500 eV.
4. 9.65×10^{-31} kg.
5. $V_p = .022 \times V_e$.
6. 2.4×10^{-3} m.

CHAPTER 15

Problems

1. 20 m/sec.
3. 5; 90; ⅔.
6. 440 cycles per second.
8. 264 m/sec.
10. About 10 octaves
11. 440; 660.

13. 0.1 W/m^2.
14. 2.7×10^{-8} m.
15. 8×10^{-5} °C.

CHAPTER 16

Problems

1. 186,000 miles/sec; 1 mile.
2. 4.3×10^3; 7.5×10^3.
4. 3 ft.
5. 2.25×10^8 m/sec.
6. $V_t = 10^{-10} S_t$.
9. 1.5×10^4.
10. 2.9×10^{10} m.
11. 1.25×10^{-6} m.

CHAPTER 17

Problems

1. 2×10^{-9}.
2. 3.
3. 1.7×10^6 m/sec.
4. Doubled.
5. 1.2×10^{-12} m.
6. 1.25×10^{-12} m.
8. No; 3.3×10^5 m/sec.
9. 1.1×10^{-29} kg − m/sec.
10. 4×10^{-37} kg.
11. 2×10^{-8} kg − m/sec; 3.3×10^{-10} kg − m/sec^2.

CHAPTER 18

Problems

1. 5 min.
3. 2.4×10^{15}.

CHAPTER 19

Problems

1. 58.0 kcal $mole^{-1}$.
2. 2.52 eV.
3. (a) doubled; (b) doubled; (c) will *probably* increase.
4. $\frac{[NO_2]}{[N_2O_4]} = \frac{K}{[NO_2]}$.

5. SO_3

$$\begin{array}{l} O \\ \quad \diagdown \\ \quad\quad S{=}O \\ \quad \diagup \\ O \end{array}$$

NH_3

$$\begin{array}{l} H \\ \quad \diagdown \\ \quad\quad N{-}H \\ \quad \diagup \\ H \end{array}$$

SiF_4

$$\begin{array}{c} F \\ | \\ F{-}Si{-}F \\ | \\ F \end{array}$$

CO_2 $O{=}C{=}O$

CHAPTER 20

Problems

1. 0.87 *c*.
4. 2.24×10^{-27} kg.
5. 1.5×10^{-10} joules.
7. About 2.5×10^{11} kg/min.

CHAPTER 21

Problems

1. $3{}_0n^1$; e^-; ${}_2He^4$.
2. 0.213 amu; 1.9×10^{16} joules; 8×10^{13} joules/kg.
3. $4/1$.
4. 16,800 years.

CHAPTER 23

Problems

2. 2.6×10^3 m/sec.
4. 1.11×10^4 km/sec.
6. 10^3.

CHAPTER 24

Questions

1. 11 light years.
4. 0.6×10^5 m/sec recession.
5. Sun is 10^{10} times brighter.
6. Density of a proton is over a billion times greater.

Name index*

* n indicates footnote reference; italicized page numbers indicate quoted material.

Subject index